Model-Based Testing for Embedded Systems

Computational Analysis, Synthesis, and Design of Dynamic Systems Series

Series Editor

Pieter J. Mosterman

MathWorks
Natick, Massachusetts

McGill University
Montréal, Québec

Model-Based Testing for Embedded Systems

EDITED BY **Justyna Zander,**
Ina Schieferdecker, and Pieter J. Mosterman

CRC Press
Taylor & Francis Group
Boca Raton London New York

CRC Press is an imprint of the
Taylor & Francis Group, an **informa** business

CRC Press
Taylor & Francis Group
6000 Broken Sound Parkway NW, Suite 300
Boca Raton, FL 33487-2742

© 2012 by Taylor & Francis Group, LLC
CRC Press is an imprint of Taylor & Francis Group, an Informa business

First issued in paperback 2017

No claim to original U.S. Government works
Version Date: 20110804

ISBN 13: 978-1-138-07645-7 (pbk)
ISBN 13: 978-1-4398-1845-9 (hbk)

Library of Congress Cataloging-in-Publication Data

Model-based testing for embedded systems / edited by Justyna Zander, Ina Schieferdecker, and Pieter J. Mosterman.
    p. cm. -- (Computational analysis, synthesis, and design of dynamic systems ; 13)
    Includes bibliographical references and index.
    ISBN 978-1-4398-1845-9 (hardcover : alk. paper)
    1. Embedded computer systems--Testing. I. Zander, Justyna. II. Schieferdecker, Ina. III. Mosterman, Pieter J. IV. Title. V. Series.

TK7895.E42M636 2011
004.1--dc22   2011010401

Visit the Taylor & Francis Web site at
http://www.taylorandfrancis.com

and the CRC Press Web site at
http://www.crcpress.com

Contents

Part III   Integration and Multilevel Testing

Part IV   Specific Approaches

Part V   Testing in Industry

Part VI   Testing at the Lower Levels of Development

Preface

The ever-growing pervasion of software-intensive systems into technical, business, and social areas not only consistently increases the number of requirements on system functionality and features but also puts forward ever-stricter demands on system quality and reliability. In order to successfully develop such software systems and to remain competitive on top of that, early and continuous consideration and assurance of system quality and reliability are becoming vitally important.

To achieve effective quality assurance, model-based testing has become an essential ingredient that covers a broad spectrum of concepts, including, for example, automatic test generation, test execution, test evaluation, test control, and test management. Model-based testing results in tests that can already be utilized in the early design stages and that contribute to high test coverage, thus providing great value by reducing cost and risk. These observations are a testimony to both the effectiveness and the efficiency of testing that can be derived from model-based approaches with opportunities for better integration of system and test development.

Model-based test activities comprise different methods that are best applied complementing one another in order to scale with respect to the size and conceptual complexity of industry systems. This book presents model-based testing from a number of different perspectives that combine various aspects of embedded systems, embedded software, their models, and their quality assurance. As system integration has become critical to dealing with the complexity of modern systems (and, indeed, systems of systems), with software as the universal integration glue, model-based testing has come to present a persuasive value proposition in system development. This holds, in particular, in the case of heterogeneity such as components and subsystems that are partially developed in software and partially in hardware or that are developed by different vendors with off-the-shelf components.

This book provides a collection of internationally renowned work on current technological achievements that assure the high-quality development of embedded systems. Each chapter contributes to the currently most advanced methods of model-based testing, not in the least because the respective authors excel in their expertise in system verification and validation. Their contributions deliver supreme improvements to current practice both in a qualitative as well as in a quantitative sense, by automation of the various test activities, exploitation of combined model-based testing aspects, integration into model-based design process, and focus on overall usability. We are thrilled and honored by the participation of this select group of experts. They made it a pleasure to compile and edit all of the material, and we sincerely hope that the reader will find the endeavor of intellectual excellence as enjoyable, gratifying, and valuable as we have.

In closing, we would like to express our genuine appreciation and gratitude for all the time and effort that each author has put into his or her chapter. We gladly recognize that the high quality of this book is solely thanks to their common effort, collaboration, and communication. In addition, we would like to acknowledge the volunteer services of those who joined the technical review committee and to extend our genuine appreciation for their involvement. Clearly, none of this would have been possible had it not been for the continuous support of Nora Konopka and her wonderful team at Taylor & Francis. Many thanks to all of you! Finally, we would like to gratefully acknowledge support by the Alfried Krupp von Bohlen und Halbach Stiftung.

Justyna Zander
Ina Schieferdecker
Pieter J. Mosterman

Editors

Justyna Zander is a postdoctoral research scientist at Harvard University (Harvard Humanitarian Initiative) in Cambridge, Massachusetts, (since 2009) and project manager at the Fraunhofer Institute for open communication systems in Berlin, Germany (since 2004).

She holds a PhD (2008) and an MSc (2005), both in the fields of computer science and electrical engineering from Technical University Berlin in Germany, a BSc (2004) in computer science, and a BSc in environmental protection and management from Gdansk University of Technology in Poland (2003).

She graduated from the Singularity University, Mountain View, California, as one of 40 participants selected from 1200 applications in 2009. For her scientific efforts, Dr. Zander received grants and scholarships from institutions such as the Polish Prime Ministry (1999–2000), the Polish Ministry of Education and Sport (2001–2004), which is awarded to 0.04% students in Poland, the German Academic Exchange Service (2002), the European Union (2003–2004), the Hertie Foundation (2004–2005), IFIP TC6 (2005), IEEE (2006), Siemens (2007), Metodos y Tecnologia (2008), Singularity University (2009), and Fraunhofer Gesellschaft (2009–2010). Her doctoral thesis on model-based testing was supported by the German National Academic Foundation with a grant awarded to 0.31% students in Germany (2005–2008).

Ina Schieferdecker studied mathematical computer science at Humboldt-University Berlin and earned her PhD in 1994 at Technical University Berlin on performance-extended specifications and analysis of quality-of-service characteristics. Since 1997, she has headed the Competence Center for Testing, Interoperability and Performance (TIP) at the Fraunhofer Institute on Open Communication Systems (FOKUS), Berlin, and now heads the Competence Center Modelling and Testing for System and Service Solutions (MOTION).

She has been a professor on engineering and testing of telecommunication systems at Technical university Berlin since 2003.

Professor Schieferdecker has worked since 1994 in the area of design, analysis, testing, and evaluation of communication systems using specification-based techniques such as unified modeling language, message sequence charts, and testing and test control notation (TTCN-3). Professor Schieferdecker has written many scientific publications in the area of system development and testing. She is involved as editorial board member with the *International Journal on Software Tools for Technology Transfer*. She is a cofounder of the Testing Technologies IST GmbH, Berlin, and a member of the German Testing Board. In 2004, she received the Alfried Krupp von Bohlen und Halbach Award for Young Professors, and she became a member of the German Academy of Technical Sciences in 2009. Her work on this book was partially supported by the Alfried Krupp von Bohlen und Halbach Stiftung.

Pieter J. Mosterman is a senior research scientist at MathWorks in Natick, Massachusetts, where he works on core Simulink® simulation and code generation technologies, and he is an adjunct professor at the School of Computer Science of McGill University. Previously, he was a research associate at the German Aerospace Center (DLR) in Oberpfaffenhofen. He has a PhD in electrical and computer engineering from Vanderbilt University in Nashville, Tennessee, and an MSc in electrical engineering from the University of Twente, the Netherlands. His primary research interests are in Computer Automated Multiparadigm Modeling (CAMPaM) with principal applications in design automation, training systems, and fault detection, isolation, and reconfiguration. He designed the Electronics Laboratory Simulator, nominated for the Computerworld Smithsonian Award by Microsoft Corporation in 1994. In 2003, he was awarded the IMechE Donald Julius Groen Prize for a paper on HyBrSim, a hybrid bond graph modeling and simulation environment. Professor Mosterman received the Society for Modeling and Simulation International (SCS) Distinguished Service Award in 2009 for his services as editor-in-chief of SIMULATION: Transactions of SCS. He is or has been an associate editor of the *International Journal of Critical Computer Based Systems*, the *Journal of Defense Modeling and Simulation*, the *International Journal of Control and Automation, Applied Intelligence*, and *IEEE Transactions on Control Systems Technology*.

MATLAB Statement

MATLAB® is a registered trademark of The MathWorks, Inc. For product information, please contact:

The MathWorks, Inc.
3 Apple Hill Drive
Natick, MA 01760-2098 USA
Tel: 508 647 7000
Fax: 508-647-7001
E-mail: info@mathworks.com
Web: www.mathworks.com

Contributors

Fredrik Abbors
Department of Information Technologies
Åbo Akademi University
Turku, Finland

Veli-Matti Aho
Process Excellence
Nokia Siemens Networks
Tampere, Finland

Lee Barford
Measurement Research Laboratory
Agilent Technologies
Reno, Nevada
and
Department of Computer
    Science and Engineering
University of Nevada
Reno, Nevada

Thomas Bauer
Fraunhofer Institute for Experimental
    Software Engineering (IESE)
Kaiserslautern, Germany

Fevzi Belli
Department of Electrical Engineering
    and Information Technology
University of Paderborn
Paderborn, Germany

Fabrice Bouquet
Computer Science Department
University of Franche-Comté/INRIA
    CASSIS Project
Besançon, France

Manfred Broy
Institute for Computer Science
Technische Universität München
Garching, Germany

Mirko Conrad
The MathWorks, Inc.
Natick, Massachusetts

Frédéric Dadeau
Computer Science Department
University of Franche-Comté/INRIA
    CASSIS Project
Besançon, France

Thao Dang
VERIMAG CNRS (French National
    Center for Scientific Research)
Gieres, France

Gregor Engels
Software Quality Lab—s-lab
University of Paderborn
Paderborn, Germany

Robert Eschbach
Fraunhofer Institute for Experimental
    Software Engineering (IESE)
Kaiserslautern, Germany

Luca Ferro
TIMA Laboratory
University of Grenoble, CNRS
Grenoble, France

Michael Gall
Siemens Industry, Inc.
Building Technologies Division
Florham Park, New Jersey

Jens Grabowski
Institute for Computer Science
University of Goettingen
Goldschmidtstraße 7
Goettingen, Germany

Jürgen Großmann
Fraunhofer Institute FOKUS
Kaiserin-Augusta-Allee 31
Berlin, Germany

Stefan Gruner
Department of Computer Science
University of Pretoria
Pretoria, Republic of South Africa

Axel Hollmann
Department of Applied Data Technology
Institute of Electrical and Computer
    Engineering
University of Paderborn
Paderborn, Germany

Antti Huima
President and CEO
Conformiq-Automated Test Design
Saratoga, California

Antti Jääskeläinen
Department of Software Systems
Tampere University of Technology
Tampere, Finland

Jacques Julliand
Computer Science Department
University of Franche-Comté
Besançon, France

Marko Kääramees
Department of Computer Science
Tallinn University of Technology
Tallinn, Estonia

Stefan Kaiser
Fraunhofer Institute FOKUS
Berlin, Germany

Mika Katara
Department of Software Systems
Tampere University of Technology
Tampere, Finland

Jani Koivulainen
Conformiq Customer Success
Conformiq
Espoo, Finland

Bogdan Kowalczyk
Delphi Technical Center Kraków
ul. Podgórki Tynieckie 2
Kraków, Poland

Andres Kull
ELVIOR
Tallinn, Estonia

Bruno Legeard
Research and Development
Smartesting/University of Franche-Comté
Besançon, France

Lan Lin
Department of Electrical Engineering
    and Computer Science
University of Tennessee
Knoxville, Tennessee

Philip Makedonski
Institute for Computer Science
University of Goettingen
Goldschmidtstraße 7
Goettingen, Germany

Maili Markvardt
Department of Computer Science
Tallinn University of Technology
Tallinn, Estonia

Abel Marrero Pérez
Daimler Center for Automotive IT
    Innovations
Berlin Institute of Technology
Berlin, Germany

Pierre-Alain Masson
Computer Science Department
University of Franche-Comté
Besançon, France

Stephen P. Masticola
System Test Department
Siemens Fire Safety
Florham Park, New Jersey

Pieter J. Mosterman
MathWorks, Inc.
Natick, Massachusetts
and
McGill University
School of Computer Science
Montreal, Quebec, Canada

Sebastian Oster
Real-Time Systems Lab
Technische Universität Darmstadt
Darmstadt, Germany

Sascha Padberg
Department of Applied Data Technology
Institute of Electrical and Computer
    Engineering
University of Paderborn
Paderborn, Germany

Mirosław Panek
Delphi Technical Center Kraków
ul. Podgórki Tynieckie 2
Kraków, Poland

Virginia Papailiopoulou
INRIA
Rocquencourt, France

Ioannis Parissis
Grenoble INP—Laboratoire de Conception
    et d'Intégration des Systémes
University of Grenoble
Valence, France

Fabien Peureux
Computer Science Department
University of Franche-Comté
Besançon, France

Laurence Pierre
TIMA Laboratory
University of Grenoble, CNRS
Grenoble, France

Jesse H. Poore
Ericsson-Harlan D. Mills Chair in Software
    Engineering
Department of Electrical
    Engineering and Computer Science
University of Tennessee
Knoxville, Tennessee

Alexander Pretschner
Karlsruhe Institute of Technology
Karlsruhe, Germany

Kullo Raiend
ELVIOR
Tallinn, Estonia

Ina Schieferdecker
Fraunhofer Institute FOKUS
Kaiserin-Augusta-Allee 31
Berlin, Germany

Holger Schlingloff
Fraunhofer Institute FIRST
Kekulestraße
Berlin, Germany

Andy Schürr
Real-Time Systems Lab
Technische Universität Darmstadt
Darmstadt, Germany

Besnik Seljimi
Faculty of Contemporary Sciences and
    Technologies
South East European University
Tetovo, Macedonia

Paweł Skruch
Delphi Technical Center Kraków
ul. Podgórki Tynieckie 2
Kraków, Poland

Jaroslav Svacina
Fraunhofer Institute FIRST
Kekulestraße
Berlin, Germany

Tommi Takala
Department of Software Systems
Tampere University of Technology
Tampere, Finland

Risto Teittinen
Process Excellence
Nokia Siemens Networks
Espoo, Finland

Régis Tissot
Computer Science Department
University of Franche-Comté
Besançon, France

Dragos Truscan
Department of Information Technologies
Åbo Akademi University
Turku, Finland

Mark Utting
Department of Computer Science
University of Waikato
Hamilton, New Zealand

Jüri Vain
Department of Computer Science/Institute
  of Cybernetics
Tallinn University of Technology
Tallinn, Estonia

Bruce Watson
Department of Computer Science
University of Pretoria
Pretoria, Republic of South Africa

Stephan Weißleder
Fraunhofer Institute FIRST
Kekulestraße 7
Berlin, Germany

Hans-Werner Wiesbrock
IT Power Consultants
Kolonnenstraße 26
Berlin, Germany

Andreas Wübbeke
Software Quality Lab—s-lab
University of Paderborn
Paderborn, Germany

Justyna Zander
Harvard University
Cambridge, Massachusetts
and
Fraunhofer Institute FOKUS
Kaiserin-Augusta-Allee 31
Berlin, Germany

Technical Review Committee

Lee Barford

Fevzi Belli

Fabrice Bouquet

Mirko Conrad

Frédéric Dadeau

Thao Dang

Thomas Deiss

Vladimir Entin

Alain-Georges Vouffo Feudjio

Gordon Fraser

Ambar Gadkari

Michael Gall

Jeremy Gardiner

Juergen Grossmann

Stefan Gruner

Axel Hollmann

Mika Katara

Bogdan Kowalczyk

Yves Ledru

Pascale LeGall

Jenny Li

Levi Lucio

José Carlos Maldonado

Eda Marchetti

Steve Masticola

Swarup Mohalik

Pieter J. Mosterman

Sebastian Oster

Jan Peleska

Abel Marrero Pérez

Jesse H. Poore

Stacy Prowell

Holger Rendel

Axel Rennoch

Markus Roggenbach

Bernhard Rumpe

Ina Schieferdecker

Holger Schlingloff

Diana Serbanescu

Pawel Skruch

Paul Strooper

Mark Utting

Stefan van Baelen

Carsten Wegener

Stephan Weißleder

Martin Wirsing

Karsten Wolf

Justyna Zander

Book Introduction

Justyna Zander, Ina Schieferdecker, and Pieter J. Mosterman

The purpose of this handbook is to provide a broad overview of the current state of model-based testing (MBT) for embedded systems, including the potential breakthroughs, the challenges, and the achievements observed from numerous perspectives. To attain this objective, the book offers a compilation of 22 high-quality contributions from world-renowned industrial and academic authors. The chapters are grouped into six parts.

- The first part comprises the contributions that focus on key test concepts for embedded systems. In particular, a taxonomy of MBT approaches is presented, an assessment of the merit and value of system models and test models is provided, and a selected test framework architecture is proposed.

- In the second part, different test automation algorithms are discussed for various types of embedded system representations.

- The third part contains contributions on the topic of integration and multilevel testing. Criteria for the derivation of integration entry tests are discussed, an approach for reusing test cases across different development levels is provided, and an *X-in-the-Loop* testing method and notation are proposed.

- The fourth part is composed of contributions that tackle selected challenges of MBT, such as testing software product lines, conformance validation for hybrid systems and nondeterministic systems, and understanding safety-critical components in the passive test context.

- The fifth part highlights testing in industry including application areas such as telecommunication networks, smartphones, and automotive systems.

- Finally, the sixth part presents solutions for lower-level tests and comprises an approach to validation of automatically generated code, contributions on testing analog components, and verification of SystemC models.

To scope the material in this handbook, an *embedded system* is considered to be a system that is designed to perform a dedicated function, typically with hard real-time constraints, limited resources and dimensions, and low-cost and low-power requirements. It is a combination of computer software and hardware, possibly including additional mechanical, optical, and other parts that are used in the specific role of actuators and sensors (Ganssle and Barr 2003). *Embedded software* is the software that is part of an embedded system. Embedded systems have become increasingly sophisticated and their software content has grown rapidly in the past decade. Applications now consist of hundreds of thousands or even millions of lines of code. Moreover, the requirements that must be fulfilled while developing embedded software are complex in comparison to standard software. In addition, embedded systems are often produced in large volumes, and the software is difficult to update once the product is deployed. Embedded systems interact with the physical environment, which often requires models that embody both continuous-time and discrete-event behavior. In terms of software development, it is not just the increased product complexity that derives from all those characteristics, but it combines with shortened development cycles and higher customer expectations of quality to underscore the utmost importance of software testing (Schäuffele and Zurawka 2006).

MBT relates to a process of test generation from various kinds of models by application of a number of sophisticated methods. MBT is usually the automation of black-box testing (Utting and Legeard 2006). Several authors (Utting, Pretschner, and Legeard 2006; Kamga, Herrmann, and Joshi 2007) define MBT as testing in which test cases are derived in their entirety or in part from a model that describes some aspects of the system under test (SUT) based on selected criteria. In addition, authors highlight the need for having dedicated test models to make the most out of MBT (Baker et al. 2007; Schulz, Honkola, and Huima 2007).

MBT clearly inherits the complexity of the related domain models. It allows tests to be linked directly to the SUT requirements, makes readability, understandability, and maintainability of tests easier. It helps to ensure a repeatable and scientific basis for testing and has the potential for known coverage of the behaviors of the SUT (Utting 2005). Finally, it is a way to reduce the effort and cost for testing (Pretschner et al. 2005).

This book provides an extensive survey and overview of the benefits of MBT in the field of embedded systems. The selected contributions present successful test approaches where different algorithms, methodologies, tools, and techniques result in important cost reduction while assuring the proper quality of embedded systems.

Organization

This book is organized in the six following parts: (I) Introduction, (II) Automatic Test Generation, (III) Integration and Multilevel Testing, (IV) Specific Approaches, (V) Testing in Industry, and (VI) Testing at the Lower Levels of Development. An overview of each of the parts, along with a brief introduction of the contents of the individual chapters, is presented next. The following figure depicts the organization of the book.

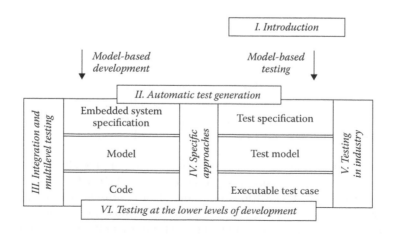

Part I. Introduction

The chapter "A Taxonomy of Model-Based Testing for Embedded Systems from Multiple Industry Domains" provides a comprehensive overview of MBT techniques using different dimensions and categorization methods. Various kinds of test generation, test evaluation, and test execution methods are described, using examples that are presented throughout this book and in the related literature.

In the chapter "Behavioral System Models versus Models of Testing Strategies in Functional Test Generation," the distinction between diverse types of models is discussed extensively. In particular, models that describe *intended system behavior* and models that describe *testing strategies* are considered from both practical as well as theoretical viewpoints. It shows the difficulty of converting the system model into a test model by applying the mental and explicit system model perspectives. Then, the notion of polynomial-time limit on test case generation is included in the reasoning about the creation of tests based on finite-state machines.

The chapter "Test Framework Architectures for Model-Based Embedded System Testing" provides reference architectures for building a *test framework*. The test framework is understood as a platform that runs the test scripts and performs other functions such as, for example, logging test results. It is usually a combination of commercial and purpose-built software. Its design and character are determined by the test execution process, common quality goals that control test harnesses, and testability *antipatterns* in the SUT that must be accounted for.

Part II. Automatic Test Generation

The chapter "Automatic Model-Based Test Generation from UML State Machines" presents several approaches for the generation of test suites from UML state machines based on different coverage criteria. The process of abstract path creation and concrete input value generation is extensively discussed using graph traversal algorithms and boundary value analysis. Then, these techniques are related to random testing, evolutionary testing, constraint solving, model checking, and static analysis.

The chapter "Automated Statistical Testing for Embedded Systems" applies statistics to solving problems posed by industrial software development. A method of modeling the population of uses is established to reason according to first principles of statistics. The Model Language and Java Usage Model Builder Library is employed for the analysis. Model validation and revision through estimates of long-run use statistics are introduced based on a medical device example while paying attention to test management and process certification.

In the chapter "How to Design Extended Finite State Machine Models in Java" extended finite-state machine (EFSM) test models that are represented in the Java programming language are applied to an SUT. ModelJUnit is used for generating the test cases by stochastic algorithms. Then, a methodology for building a MBT tool using Java reflection is proposed. Code coverage metrics are exploited to assess the results of the method, and an example referring to the GSM 11.11 protocol for mobile phones is presented.

The chapter "Automatic Testing of LUSTRE/SCADE Programs" addresses the automation of functional test generation using a LUSTRE-like language in the LUTESS V2 tool and refers to the assessment of the created test coverage. The testing methodology includes the definitions of the domain, environment dynamics, scenarios, and an analysis based on safety properties. A program control flow graph for SCADE models allows a family of coverage criteria to assess the effectiveness of the test methods and serves as an additional basis for the test generation algorithm. The proposed approaches are illustrated by a steam-boiler case study.

In the chapter "Test Generation Using Symbolic Animation of Models," symbolic execution (i.e., animation) of B models based on set-theoretical constraint solvers is applied to generate the test cases. One of the proposed methods focuses on creation of tests that reach specific test targets to satisfy structural coverage, whereas the other is based on manually designed behavioral scenarios and aims at satisfying dynamic test selection criteria. A smartcard case study illustrates the complementarity of the two techniques.

Part III. Integration and Multilevel Testing

The chapter "Model-Based Integration Testing with Communication Sequence Graphs" introduces a notation for representing the communication between discrete-behavior software components on a meta-level. The models are directed graphs enriched with semantics for integration-level analysis that do not emphasize internal states of the components, but rather focus on events. In this context, test case generation algorithms for unit and integration testing are provided. Test coverage criteria, including mutation analysis, are defined and a robot-control application serves as an illustration.

In the chapter "A Model-Based View onto Testing: Criteria for the Derivation of Entry Tests for Integration Testing" components and their integration architecture are modeled early on in development to help structure the integration process. Fundamentals for testing complex systems are formalized. This formalization allows exploiting architecture models to establish criteria that help minimize the entry-level testing of components necessary for successful integration. The tests are derived from a simulation of the subsystems and reflect behaviors that usually are verified at integration time. Providing criteria to enable shifting effort from integration testing to component entry tests illustrates the value of the method.

In the chapter "Multilevel Testing for Embedded Systems," the means for a smooth integration of multiple test levels artifacts based on a continuous reuse of test models and test cases are provided. The proposed methodology comprises the creation of an invariant test model core and a test-level specific test adapter model that represents a varying component. Numerous strategies to obtain the adapter model are introduced. The entire approach results in an increased optimization of the design effort across selected functional abstraction levels and allows for the easier traceability of the test constituents. A case study from the automotive domain (i.e., automated light control) illustrates the feasibility of the solution.

The chapter "Model-Based X-in-the-Loop Testing" provides a methodology for technology-independent specification and systematic reuse of testing artifacts for *closed-loop testing* across different development stages. Simulink®-based environmental models are coupled with a generic test specification designed in the notation called *TTCN-3 embedded*. It includes a dedicated means for specifying the stimulation of an SUT and assessing its reaction. The notions of time and sampling, streams, stream ports, and stream variables are paid specific attention as well as the definition of statements to model a control flow structure akin to hybrid automata. In addition, an overall test architecture for the approach is presented. Several examples from the automotive domain illustrate the vertical and horizontal reuse of test artifacts. The test quality is discussed as well.

Part IV. Specific Approaches

The chapter "A Survey of Model-Based Software Product Lines Testing" presents an overview of the testing that is necessary in software product line engineering methods. Such methods aim at improving reusability of software within a range of products sharing a common set of features. First, the requirements and a conception of MBT for software product lines are introduced. Then, the state of the art is provided and the solutions are compared to each other based on selected criteria. Finally, open research objectives are outlined and recommendations for the software industry are provided.

The chapter "Model-Based Testing of Hybrid Systems" describes a formal framework for conformance testing of hybrid automaton models and their adequate test generation algorithms. Methods from computer science and control theory are applied to reason about the quality of a system. An easily computable coverage measure is introduced that refers to testing properties such as safety and reachability based on the equal-distribution degree of a set of states over their state space. The distribution degree can be used to guide the test generation process, while the test creation is based on the rapidly exploring random tree algorithm (Lavalle 1998) that represents a probabilistic motion planning technique in robotics. The results are then explored in the domain of analog and mixed signal circuits.

The chapter "Reactive Testing of Nondeterministic Systems by Test Purpose Directed Tester" provides a model-based construction of an online tester for black-box testing. The notation of nondeterministic EFSM is applied to formalize the test model. The synthesis algorithm allows for selecting a suboptimal test path at run time by finding the shortest path to cover the test purpose. The rules enabling an implementation of online reactive planning are included. Coverage criteria are discussed as well, and the approach is compared with related algorithms. A feeder-box controller of a city lighting system illustrates the feasibility of the solution.

The chapter "Model-Based Passive Testing of Safety-Critical Components" provides a set of passive-testing techniques in a manner that is driven by multiple examples. First, general principles of the approach to passive quality assurance are discussed. Then, complex software systems, network security, and hardware systems are considered as the targeted domains. Next, a step-by-step illustrative example for applying the proposed analysis to a concurrent system designed in the form of a *cellular automaton* is introduced. As passive testing usually takes place *after* the deployment of a unit, the ability of a component to monitor and self-test in operation is discussed. The benefits and limitations of the presented approaches are described as well.

Part V. Testing in Industry

The chapter "Applying Model-Based Testing in the Telecommunication Domain" refers to testing practices at Nokia Siemens Networks at the industrial level and explains the state of MBT in the trenches. The presented methodology uses a behavioral system model designed in UML and SysML for generating the test cases. The applied process, model development, validation, and transformation aspects are extensively described. Technologies such as the MATERA framework (Abbors, Bäcklund, and Truscan 2010), UML to QML transformation, and OCL guideline checking are discussed. Also, test generation, test execution aspects (e.g., load testing, concurrency, and run-time executability), and the traceability of all artifacts are discussed. The case study illustrates testing the functionality of a Mobile Services Switching Center Server, a network element using offline testing.

The chapter "Model-Based GUI Testing of Smartphone Applications: Case S60™ and Linux®" discusses application of MBT along two case studies. The first one considers built-in applications in a smartphone model S60, and the second tackles the problem of a media player application in a variant of mobile Linux. Experiences in modeling and adapter development are provided and potential problems (e.g., expedient pace of product creation) are reported in industrial deployment of the technology for graphical user interface (GUI) testing of smartphone applications. In this context, the TEMA toolset (Jääskeläinen 2009) designed for test modeling, test generation, keyword execution, and test debugging is presented. The benefits and business aspects of the process adaptation are also briefly considered.

The chapter "Model-Based Testing in Embedded Automotive Systems" provides a broad overview of MBT techniques applied in the automotive domain based on experiences from Delphi Technical Center, Kraków (Poland). Key automotive domain concepts specific to

MBT are presented as well as everyday engineering issues related to MBT process deployment in the context of the system-level functional testing. Examples illustrate the applicability of the techniques for industrial-scale mainstream production projects. In addition, the limitations of the approaches are outlined.

Part VI. Testing at the Lower Levels of Development

The chapter "Testing-Based Translation Validation of Generated Code" provides an approach for model-to-code translation that is followed by a validation phase to verify the target code produced during this translation. Systematic model-level testing is supplemented by testing for numerical equivalence between models and generated code. The methodology follows the objectives and requirements of safety standards such as IEC 61508 and ISO 26262 and is illustrated using a Simulink-based code generation tool chain.

The chapter "Model-Based Testing of Analog Embedded Systems Components" addresses the problem of determining whether an analog system meets its specification as given either by a model of correct behavior (i.e., the system model) or of incorrect behavior (i.e., a fault model). The analog model-based test follows a two-phase process. First, a pretesting phase including system selection, fault model selection, excitation design, and simulation of fault models is presented. Next, an actual testing phase comprising measurement, system identification, behavioral simulation, and reasoning about the faults is extensively described. Examples are provided while benefits, limitations, and open questions in applying analog MBT are included.

The chapter "Dynamic Verification of SystemC Transactional Models" presents a solution for verifying logic and temporal properties of communication in transaction-level modeling designs from simulation. To this end, a brief overview on SystemC is provided. Issues related to globally asynchronous/locally synchronous, multiclocked systems, and auxiliary variables are considered in the approach.

Target Audience

The objective of this book is to be accessible to engineers, analysts, and computer scientists involved in the analysis and development of embedded systems, software, and their quality assurance. It is intended for both industry-related professionals and academic experts, in particular those interested in verification, validation, and testing. The most important objectives of this book are to help the reader understand how to use Model-Based Testing and test harness to a maximum extent. Various perspectives serve to:

- Get an overview on MBT and its constituents;

- Understand the MBT concepts, methods, approaches, and tools;

- Know how to choose modeling approaches fitting the customers' needs;

- Be able to select appropriate test generation strategies;

- Learn about successful applications of MBT;

- Get to know best practices of MBT; and

- See prospects of further developments in MBT.

References

Abbors, F., Bäcklund, A., and Truscan, D. (2010). MATERA—An integrated framework for model-based testing. In *Proceedings of the 17th IEEE International Conference and Workshop on Engineering of Computer-Based Systems (ECBS 2010)*, Pages: 321–328. IEEE Computer Society's Conference Publishing Services (CPS).

Baker, P., Ru Dai, Z., Grabowski, J., Haugen, O., Schieferdecker, I., and Williams, C. (2007). *Model-Driven Testing, Using the UML Testing Profile.* ISBN 9783-5407-2562-6, Springer Verlag.

Ganssle, J. and Barr, M. (2003). *Embedded Systems Dictionary*, ISBN-10: 1578201209, ISBN-13: 978-1578201204, 256 pages.

Jääskeläinen, A., Katara, M., Kervinen, A., Maunumaa, M., Pääkkönen, T., Takala, T., and Virtanen, H. (2009). Automatic GUI test generation for smartphone applications—an evaluation. *Proceedings of the Software Engineering in Practice track of the 31st International Conference on Software Engineering (ICSE 2009)*, pp. 112–122. IEEE Computer Society (companion volume)

Kamga, J., Herrmann, J., and Joshi, P. (2007). Deliverable: D-MINT automotive case study—Daimler, Deliverable 1.1, *Deployment of Model-Based Technologies to Industrial Testing*, ITEA2 Project.

Lavalle, S.M. (1998). *Rapidly-Exploring Random Trees: A New Tool for Path Planning.* Computer Science Dept, Iowa State University, Technical Report 98–11. http://citeseer.ist.psu.edu/311812.html.

Pretschner, A., Prenninger, W., Wagner, S., Kühnel, C., Baumgartner, M., Sostawa, B., Zölch, R., and Stauner, T. (2005). One evaluation of model-based testing and its automation. In *Proceedings of the 27th International Conference on Software Engineering*, St. Louis, MO, Pages: 392–401, ISBN: 1-59593-963-2. ACM New York.

Schäuffele, J. and Zurawka, T. (2006). *Automotive Software Engineering*, ISBN: 3528110406. Vieweg.

Schulz, S., Honkola, J., and Huima, A. (2007). Towards model-based testing with architecture models. In *Proceedings of the 14th Annual IEEE International Conference and Workshops on the Engineering of Computer-Based Systems* (ECBS '07). IEEE Computer Society, Washington, DC, Pages: 495–502. DOI=10.1109/ECBS.2007.73 http://dx.doi.org/10.1109/ECBS.2007.73.

Utting, M. (2005). Model-based testing. In *Proceedings of the Workshop on Verified Software: Theory, Tools, and Experiments VSTTE 2005*.

Utting, M. and Legeard, B. (2006). *Practical Model-Based Testing: A Tools Approach.* ISBN-13: 9780123725011. Elsevier Science & Technology Books.

Utting, M., Pretschner, A., and Legeard, B. (2006). *A Taxonomy of Model-Based Testing*, ISSN: 1170-487X.

Part I

Introduction

1

A Taxonomy of Model-Based Testing for Embedded Systems from Multiple Industry Domains

Justyna Zander, Ina Schieferdecker, and Pieter J. Mosterman

CONTENTS

1.1 Introduction

This chapter provides a taxonomy of Model-Based Testing (MBT) based on the approaches that are presented throughout this book as well as in the related literature. The techniques for testing are categorized using a number of dimensions to familiarize the reader with the terminology used throughout the chapters that follow.

In this chapter, after a brief introduction, a general definition of MBT and related work on available MBT surveys is provided. Next, the various test dimensions are presented. Subsequently, an extensive taxonomy is proposed that classifies the MBT process according to the MBT foundation (referred to as MBT basis), definition of various test generation techniques, consideration of test execution methods, and the specification of test evaluation. The taxonomy is an extension of previous work by Zander and Schieferdecker (2009) and it is based on contributions of Utting, Pretschner, and Legeard (2006). A summary concludes the chapter with the purpose of encouraging the reader to further study the contributions of the collected chapters in this book and the specific aspects of MBT that they address in detail.

1.2   Definition of Model-Based Testing

This section provides a brief survey of the selected definitions of MBT available in the literature. Next, certain aspects of MBT are highlighted in the discussion on test dimensions and their categorization is illustrated.

MBT relates to a process of test generation from models of/related to a system under test (SUT) by applying a number of sophisticated methods. The basic idea of MBT is that instead of creating test cases manually, a selected algorithm is generating them automatically from a model. MBT usually comprises the automation of black-box test design (Utting and Legeard 2006), however recently it has been used to automate white-box tests as well. Several authors such as Utting (2005) and Kamga, Hermann, and Joshi (2007) define MBT as testing in which test cases are derived in their entirety or in part from a model that describes some aspects of the SUT based on selected criteria. Utting, Pretschner, and Legeard (2006) elaborate that MBT inherits the complexity of the domain or, more specifically, of the related domain models. Dai (2006) refers to MBT as model-driven testing (MDT) because of the context of the model-driven architecture (MDA) (OMG 2003) in which MBT is proposed.

Advantages of MBT are that it allows tests to be linked directly to the SUT requirements, which renders readability, understandability, and maintainability of tests easier. It helps ensure a repeatable and scientific basis for testing. Furthermore, MBT has been shown to provide good coverage of all the behaviors of the SUT (Utting 2005) and to reduce the effort and cost for testing (Pretschner et al. 2005).

The term *MBT* is widely used today with subtle differences in its meaning. Surveys on different MBT approaches are provided by Broy et al. (2005), Utting, Pretschner, and Legeard (2006), and the D-Mint Project (2008), and Schieferdecker et al. (2011). In the automotive industry, MBT describes all testing activities in the context of Model-Based Design (MBD), as discussed for example, by Conrad, Fey, and Sadeghipour (2004) and Lehmann and Krämer (2008). Rau (2002), Lamberg et al. (2004), and Conrad (2004a, 2004b) define MBT as a test process that encompasses a combination of different test methods that utilize the executable model in MBD as a source of information. As a single testing technique is insufficient to achieve a desired level of test coverage, different test methods are usually combined to complement each other across all the specified test dimensions (e.g., functional and structural testing techniques are frequently applied together). If sufficient test coverage has been achieved on the model level, properly designed test cases can be reused for testing the software created based on or generated from the models within the framework of back-to-back tests as proposed by Wiesbrock, Conrad, and Fey (2002). With this practice, the functional equivalence between the specification, executable model, and code can be verified and validated (Conrad, Fey, and Sadeghipour 2004).

The most generic definition of *MBT* is testing in which the *test specification* is derived in its entirety or in part from both *the system requirements and a model* that describe selected functional and nonfunctional aspects of the SUT.

The test specification can take the form of a model, executable model, script, or computer program code. The resulting test specification is intended to ultimately be *executed* together with the SUT so as to provide the test results. The SUT again can exist in the form of a model, code, or even hardware.

For example, in Conrad (2004b) and Conrad, Fey, and Sadeghipour (2004), no additional test models are created, but the already existing functional system models are utilized for test purposes. In the test approach proposed by Zander-Nowicka (2009), the system models are exploited as well. In addition, however, a *test specification model* (also called *test case specification*, *test model*, or *test design* in the literature (Pretschner 2003b, Zander et al. 2005, and Dai 2006) is created semi-automatically. Concrete test data variants are then automatically derived from this test specification model.

The application of MBT is as proliferate as the interest in building embedded systems. For example, case studies borrowed from such widely varying domains as medicine, automotive, control engineering, telecommunication, entertainment, or aerospace can be found in this book. MBT then appears as part of specific techniques that are proposed for testing a medical device, the GSM 11.11 protocol for mobile phones, a smartphone graphical user interface (GUI), a steam boiler, smartcard, a robot-control application, a kitchen toaster, automated light control, analog- and mixed-signal electrical circuits, a feeder-box controller of a city lighting system, and other complex software systems.

1.2.1   Test dimensions

Tests can be classified depending on the characteristics of the SUT and the test system. In this book, such SUT features comprise, for example, safety-critical properties, deterministic and nondeterministic behavior, load and performance, analog characteristics, network-related, and user-friendliness qualities. Furthermore, systems that exhibit behavior of a discrete, continuous, or hybrid nature are analyzed in this book. The modeling paradigms for capturing a model of the SUT and tests combine different approaches, such as history-based, functional data flow combined with transition-based semantics. As it is next to impossible for one single classification scheme to successfully apply to such a wide range of attributes, selected dimensions have been introduced in previous work to isolate certain aspects. For example, Neukirchen (2004) aims at testing communication systems and categorizes testing in the dimensions of *test goals*, *test scope*, and *test distribution*. Dai (2006) replaces the test distribution by a dimension describing the different *test development phases*, since she is testing both local and distributed systems. Zander-Nowicka (2009) refers to *test goals*, *test abstraction*, *test execution platforms*, *test reactiveness*, and *test scope* in the context of embedded automotive systems.

In the following, the specifics related to *test goal*, *test scope*, and *test abstraction* (see Figure 1.1) are introduced to provide a basis for a common vocabulary, simplicity, and a better understanding of the concepts discussed in the rest of this book.

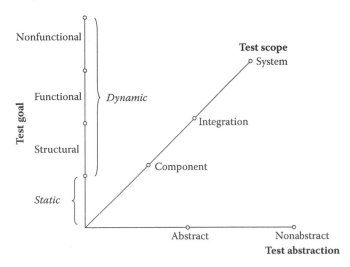

FIGURE 1.1
Selected test dimensions.

1.2.1.1   Test goal

During software development, systems are tested with different purposes (i.e., goals). These goals can be categorized as static testing, also called *review*, and dynamic testing, where the latter is based on test execution and further distinguishes between structural, functional, and nonfunctional testing. After the *review* phase, the test goal is usually to check the functional behavior of the system. Nonfunctional tests appear in later development stages.

- *Static test:* Testing is often defined as the process of finding errors, failures, and faults. Errors in a program can be revealed without execution by just examining its source code (International Software Testing Qualification Board 2006). Similarly, other development artifacts can be reviewed (e.g., requirements, models, or the test specification itself).

- *Structural test:* Structural tests cover the structure of the SUT during test execution (e.g., control or data flow), and so the internal structure of the system (e.g., code or model) must be known. As such, structural tests are also called white-box or glass-box tests (Myers 1979; International Software Testing Qualification Board 2006).

- *Functional test:* Functional testing is concerned with assessing the functional behavior of an SUT against the functional requirements. In contrast to structural tests, functional tests do not require any knowledge about system internals. They are therefore called black-box tests (Beizer 1995). A systematic, planned, executed, and documented procedure is desirable to make them successful. In this category, functional safety tests to determine the safety of a software product are also included.

- *Nonfunctional test*: Similar to functional tests, nonfunctional tests (also called extra-functional tests) are performed against a requirements specification of the system. In contrast to pure functional testing, nonfunctional testing aims at assessing nonfunctional requirements such as reliability, load, and performance. Nonfunctional tests are usually black-box tests. Nevertheless, internal access during test execution is required for retrieving certain information, such as the state of the internal clock.

  For example, during a *robustness test*, the system is tested with invalid input data that are outside the permitted ranges to check whether the system is still safe and operates properly.

1.2.1.2   Test scope

Test scopes describe the granularity of the SUT. Because of the composition of the system, tests at different scopes may reveal different failures (Weyuker 1988; International Software Testing Qualification Board 2006; and D-Mint Project 2008). This leads to the following order in which tests are usually performed:

- *Component:* At the scope of component testing (also referred to as unit testing), the smallest testable component (e.g., a class in an object-oriented implementation or a single electronic control unit [ECU]) is tested in isolation.

- *Integration:* The scope of integration test combines components with each other and tests those as a subsystem, that is, not yet a complete system. It exposes defects in the interfaces and in the interactions between integrated components or subsystems (International Software Testing Qualification Board 2006).

- *System:* In a system test, the complete system, including all subsystems, is tested. Note that a complex embedded system is usually distributed with the single subsystems connected, for example, via buses using different data types and interfaces through which the system can be accessed for testing (Hetzel 1988).

1.2.1.3 Test abstraction

As far as the abstraction level of the test specification is considered, the higher the abstraction, the better test understandability, readability, and reusability are observed. However, the specified test cases must be executable at the same time. Also, the abstraction level should not affect the test execution in a negative way. An interesting and promising approach to address the effect of abstraction on execution behavior is provided by Mosterman et al. (2009, 2011) and Zander et al. (2011) in the context of complex system development. In their approach, the error introduced by a computational approximation of the execution is accepted as an inherent system artifact as early as the abstract development stages. The benefit of this approach is that it allows eliminating the accidental complexity of the code that makes the abstract design executable while enabling high-level analysis and synthesis methods. A critical enabling element is a high-level declarative specification of the execution logic so that its computational approximation becomes explicit. Because it is explicit and declarative, the approximation can then be consistently preserved throughout the design stages. This approach holds for test development as well. Whenever the abstract test suites are executed, they can be refined with the necessary concrete analysis and synthesis mechanisms.

1.3   Taxonomy of Model-Based Testing

In Utting, Pretschner, and Legeard (2006), a broad taxonomy for MBT is presented. Here, three general *classes* are identified: *model*, *test generation*, and *test execution*. Each of the classes is divided into further *categories*. The model class consists of *subject*, *independence*, *characteristics*, and *paradigm categories*. The test generation class consists of *test selection criteria* and *technology categories*. The test execution class contains execution options.

Zander-Nowicka (2009) completes the overall view with test evaluation as an additional class. Test evaluation refers to comparing the actual SUT outputs with the expected SUT behavior based on a test oracle. Such a test oracle enables a decision to be made as to whether the actual SUT outputs are correct. The test evaluation is divided into two categories: *specification* and *technology*.

Furthermore, in this chapter, the test generation class is extended with an additional category called *result of the generation*. Also, the semantics of the class model is different in this taxonomy than in its previous incarnations. Here, a category called MBT basis indicates what specific element of the software engineering process is the basis for MBT process.

An overview of the resulting MBT taxonomy is illustrated in Figure 1.2. All the categories in the presented taxonomy are decomposed into further elements that influence each other within or between categories. The "A/B/C" notation at the leaves indicates mutually exclusive options.

In the following three subsections, the *categories* and *options* in each of the *classes* of the MBT taxonomy are explained in depth. The descriptions of the most important *options* are endowed with examples of their realization.

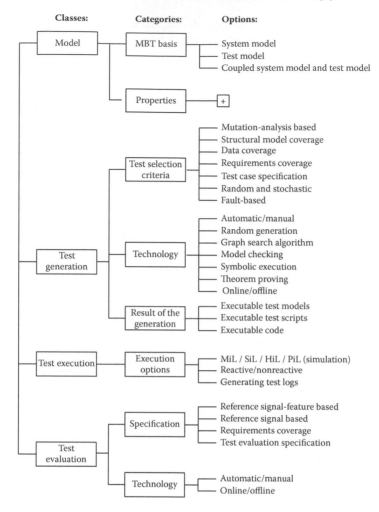

FIGURE 1.2
Overview of the taxonomy for Model-Based Testing.

1.3.1 Model

The models applied in the MBT process can include both system-specific and test-specific development artifacts. Frequently, the software engineering practice for a selected project determines the basis for incorporating the testing into the process and thus, selecting the MBT type. In the following, selected theoretical viewpoints are introduced and join points between them are discussed.

To specify the system and the test development, the methods that are presented in this book employ a broad spectrum of notations such as Finite State Machines (FSM) (e.g., Chapter 2), Unified Modeling Language (UML®) (e.g., state machines, use cases), UML Testing Profile (UTP) (see OMG 2003, 2005), SysML (e.g., Chapter 4), The Model Language (e.g., Chapter 5), Extended FSM, Labeled State Transition System notation, Java (e.g., Chapter 6), LUSTRE, *SCADE®* (e.g., Chapter 7), B-Notation (e.g., Chapter 8), Communication Sequence Graphs (e.g., Chapter 9), Testing and Test Control Notation, version 3 (TTCN-3) (see ETSI 2007), TTCN-3 embedded (e.g., Chapter 12), Transaction Level Models, Property Specification Language, SystemC (e.g., Chapter 22), Simulink® (e.g., Chapter 12, Chapter 19, or Chapter 20), and so on.

Model-Based Testing basis

In the following, selected options referred to as the MBT basis are listed and their meaning is described.

- *System model*: A system model is an abstract representation of certain aspects of the SUT. A typical application of the system model in the MBT process leverages its behavioral description for derivation of tests. Although this concept has been extensively described in previous work (Conrad 2004a; Utting 2005), another instance of using a system model for testing is the approach called architecture-driven testing (ADT) introduced by Din and Engel (2009). It is a technique to derive tests from *architecture viewpoints*. An architecture viewpoint is a simplified representation of the system model with respect to the structure of the system from a specific perspective. The architecture viewpoints not only concentrate on a particular aspect but also allow for the combination of the aspects, relations, and various models of system components, thereby providing a unifying solution. The perspectives considered in ADT include a functional view, logical view, technical view, and topological view. They enable the identification of test procedures and failures on certain levels of detail that would not be recognized otherwise.

- *Test model:* If the test cases are derived directly from an abstract test model and are decoupled from the system model, then such a test model is considered to constitute the MBT basis. In practice, such a method is rarely applied as it requires substantial effort to introduce a completely new test model. Instead, the coupled system and test model approach is used.

- *Coupled system and test model:* UTP plays an essential role for the alignment of system development methods together with testing. It introduces abstraction as a test artifact and counts as a primary standard in this alignment. UTP is utilized as the test modeling language before test code is generated from a test model. Though, this presupposes that an adequate system model already exists and will be leveraged during the entire test process (Dai 2006). As a result, system models and test models are developed in concert in a coupled process. UTP addresses concepts, such as test suites, test cases, test configuration, test component, and test results, and enables the specification of different types of testing, such as functional, interoperability, scalability, and even load testing.

  Another instantiation of such a coupled technique is introduced in the Model-in-the-Loop for Embedded System Test (MiLEST) approach (Zander-Nowicka 2009) where Simulink system models are coupled with additionally generated Simulink-based test models. MiLEST is a test specification framework that includes reusable test patterns, generic graphical validation functions, test data generators, test control algorithms, and an arbitration mechanism all collected in a dedicated library.

  The application of the same modeling language for both system and test design brings about positive effects as it ensures that the method is more transparent and it does not force the engineers to learn a completely new language.

A more extensive illustration of the challenge to select a proper MBT basis is provided in Chapter 2 of this book.

1.3.2 Test generation

The process of test generation starts from the system requirements, taking into account the test objectives. It is defined in a given test context and results in the creation of test cases. A number of approaches exist depending on the test selection criteria, generation technology, and the expected generation results. They are reviewed next.

1.3.2.1 Test selection criteria

Test selection criteria define the facilities that are used to control the generation of tests. They help specify the tests and do not depend on the SUT code. In the following, the most commonly used criteria are investigated. Clearly, different test methods should be combined to complement one another so as to achieve the best test coverage. Hence, there is no best suitable solution for generating the test specification. Subsequently, the test selection criteria are described in detail.

- *Mutation-analysis based:* Mutation analysis consists of introducing a small syntactic change in the source of a model or program in order to produce a mutant (e.g., replacing one operator by another or altering the value of a constant). Then, the mutant behavior is compared to the original. If a difference can be observed, the mutant is marked as killed. Otherwise, it is called equivalent. The original aim of the mutation analysis is the evaluation of a test data applied in the test case. Thus, it can be applied as a foundational technique for test generation. One of the approaches to mutation analysis is described in Chapter 9 of this book.

- *Structural model coverage criteria:* These exploit the structure of the model to select the test cases. They deal with coverage of the control-flow through the model, based on ideas from the flow of control in computer program code.

  Previous work (Pretschner 2003) has shown how test cases can be generated that satisfy the modified condition/decision coverage (MC/DC) coverage criterion. The idea is to first generate a set of test case specifications that enforce certain variable valuations and then generate test cases for them.

  Similarly, safety test builder (STB) (GeenSoft 2010a) or Reactis Tester (Reactive Systems 2010; Sims and DuVarney 2007) generate test sequences covering a set of Stateflow® test objectives (e.g., transitions, states, junctions, actions, MC/DC coverage) and a set of Simulink test objectives (e.g., Boolean flow, look-up tables, conditional subsystems coverage).

- *Data coverage criteria:* The idea is to decompose the data range into equivalence classes and select one representative value from each class. This partitioning is usually complemented by a boundary value analysis (Kosmatov et al. 2004), where the critical limits of the data ranges or boundaries determined by constraints are selected in addition to the representative values.

  An example is the MATLAB® Automated Testing Tool (MATT 2008) that enables black-box testing of Simulink models and code generated from them by Real-Time Workshop® (Real-Time Workshop 2011). MATT furthermore enables the creation of custom test data for model simulations by setting the types of test data for each input. Additionally, accuracy, constant, minimum, and maximum values can be provided to generate the test data matrix.

  Another realization of this criterion is provided by Classification Tree Editor for Embedded Systems (CTE/ES) implementing the Classification Tree Method (Grochtmann and Grimm 1993; Conrad 2004a). The SUT inputs form the classifications in the roots of the tree. From here, the input ranges are divided into classes according to the equivalence partitioning method. The test cases are specified by selecting leaves of the tree in the combination table. A row in the table specifies a test case. CTE/ES provides a way of finding test cases systematically by decomposing the test scenario design process into steps. Visualization of the test scenario is supported by a GUI.

- *Requirements coverage criteria:* These criteria aim at covering all informal SUT requirements. Traceability of the SUT requirements to the system or test model/code aids in the realization of this criterion. It is targeted by almost every test approach (Zander-Nowicka 2009).

- *Test case definition:* When a test engineer defines a test case specification in some formal notation, the test objectives can be used to determine which tests will be generated by an explicit decision and which set of test objectives should be covered. The notation used to express these objectives may be the same as the notation used for the model (Utting, Pretschner, and Legeard 2006). Notations commonly used for test objectives include FSMs, UTP, regular expressions, temporal logic formulas, constraints, and Markov chains (for expressing intended usage patterns).

  A prominent example of applying this criterion is described by Dai (2006), where the test case specifications are derived from UML models and transformed into executable tests in TTCN-3 by using MDA methods (Zander et al. 2005). The work of Pretschner et al. (2004) is also based on applying this criterion (see symbolic execution).

- *Random and stochastic criteria:* These are mostly applicable to environment models because it is the environment that determines the usage patterns of the SUT. A typical approach is to use a Markov chain to specify the expected SUT usage profile. Another example is to use a statistical usage model in addition to the behavioral model of the SUT (Carter, Lin, and Poore 2008). The statistical model acts as the selection criterion and chooses the paths, while the behavioral model is used to generate the oracle for those paths.

  As an example, Markov Test Logic (MaTeLo) (All4Tec 2010) can generate test suites according to several algorithms. Each of them optimizes the test effort according to objectives such as boundary values, functional coverage, and reliability level. Test cases are generated in XML/HTML format for manual execution or in TTCN-3 for automatic execution (Dulz and Fenhua 2003).

  Another instance, Java Usage Model Builder Library (JUMBL) (Software Quality Research Laboratory 2010) (cf. Chapter 5) can generate test cases as a collection of test cases that cover the model with minimum cost, by random sampling with replacement, by interleaving the events of other test cases, or in order by probability. An interactive test case editor supports creating test cases by hand.

- *Fault-based criteria:* These rely on knowledge of typically occurring faults, often captured in the form of a fault model.

1.3.2.2 Test generation technology

One of the most appealing characteristics of MBT is its potential for automation. The automated generation of test cases usually necessitates the existence of some form of test case specifications.

In the proceeding paragraphs, different technologies applied to test generation are discussed.

- *Automatic/Manual technology:* Automatic test generation refers to the situation where, based on given criteria, the test cases are generated automatically from an information source. Manual test generation refers to the situation where the test cases are produced by hand.

- *Random generation:* Random generation of tests is performed by sampling the input space of a system. It is straightforward to implement, but it takes an undefined period of time to reach a certain satisfying level of model coverage as Gutjahr (1999) reports.

- *Graph search algorithms:* Dedicated graph search algorithms include node or arc coverage algorithms such as the Chinese Postman algorithm that covers each arc at least once. For transition-based models, which use explicit graphs containing nodes and arcs, there are many graph coverage criteria that can be used to control test generation. The commonly used are all nodes, all transitions, all transition pairs, and all cycles. The method is exemplified by Lee and Yannakakis (1994), which specifically addresses structural coverage of FSM models.

- *Model checking:* Model checking is a technology for verifying or falsifying properties of a system. A property typically expresses an unwanted situation. The model checker verifies whether this situation is reachable or not. It can yield counterexamples when a property is falsified. If no counterexample is found, then the property is proven and the situation can never be reached. Such a mechanism is implemented in safety checker blockset (GeenSoft 2010b) or in Embedded *Validator* (BTC Embedded Systems AG 2010).

  The general idea of test case generation with model checkers is to first formulate test case specifications as reachability properties, for example, "eventually, a certain state is reached or a certain transition fires." A model checker then yields traces that reach the given state or that eventually make the transition fire. Wieczorek et al. (2009) present an approach to use Model Checking for the generation of Integration Tests from Choreography Models. Other variants use mutations of models or properties to generate test suites.

- *Symbolic execution:* The idea of symbolic execution is to run an executable model not with single input values but with sets of input values instead (Marre and Arnould 2000). These are represented as constraints. With this practice, symbolic traces are generated. By instantiation of these traces with concrete values, the test cases are derived. Symbolic execution is guided by test case specifications. These are given as explicit constraints and symbolic execution may be performed randomly by respecting these constraints.

  Pretschner (2003) presents an approach to test case generation with symbolic execution built on the foundations of constraint logic programming. Pretschner (2003a and 2003b) concludes that test case generation for both functional and structural test case specifications reduces to finding states in the state space of the SUT model. The aim of symbolic execution of a model is then to find a trace that represents a test case that leads to the specified state.

- *Theorem proving:* Usually theorem provers are employed to check the satisfiability of formulas that directly occur in the models. One variant is similar to the use of model checkers where a theorem prover replaces the model checker.

  For example, one of the techniques applied in Simulink® Design Verifier™ (The MathWorks®, Inc.) uses mathematical procedures to search through the possible execution paths of the model so as to find test cases and counterexamples.

- *Online/offline generation technology:* With online test generation, algorithms can react to the actual outputs of the SUT during the test execution. This idea is exploited for implementing reactive tests as well.

  Offline testing generates test cases before they are run. A set of test cases is generated once and can be executed many times. Also, the test generation and test execution can be performed on different machines, at different levels of abstractions, and in different environments. If the test generation process is slower than test execution, then there are obvious advantages to minimizing the number of times tests are generated (preferably only once).

1.3.2.3 Result of the generation

Test generation usually results in a set of test cases that form test suites. The test cases are expected to ultimately become executable to allow for observation of meaningful verdicts from the entire validation process. Therefore, in the following, the produced test cases are described from the execution point of view, and so they can be represented in different forms, such as test scripts, test models, or code. These are described next.

- *Executable test models:* Similarly, the created test models (i.e., test designs) should be executable. The execution engine underlying the test modeling semantics is the indicator of the character of the test design and its properties (cf. the discussion given in, e.g., Chapter 11).

- *Executable test scripts:* The test scripts refer to the physical description of a test case (cf., e.g., Chapter 2). They are represented by a test script language that has to then be translated to the executables (cf. TTCN-3 execution).

- *Executable code:* The code is the lowest-level representation of a test case in terms of the technology that is applied to execute the tests (cf. the discussion given in, e.g., Chapter 6). Ultimately, every other form of a test case is transformed to a code in a selected programming language.

1.3.3 Test execution

In the following, for clarity reasons, the analysis of the test execution is limited to the domain of engineered systems. An example application in the automotive domain is recalled in the next paragraphs. Chapters 11, 12, and 19 provide further background and detail to the material in this subsection.

Execution options

In this chapter, execution options refer to the execution of a test. The test execution is managed by so-called test platforms. The purpose of the test platform is to stimulate the test object (i.e., SUT) with inputs and to observe and analyze the outputs of the SUT.

In the automotive domain, the test platform is typically represented by a car with a test driver. The test driver determines the inputs of the SUT by driving scenarios and observes the reaction of the vehicle. Observations are supported by special diagnosis and measurement hardware/software that records the test data during the test drive and that allows the behavior to be analyzed offline. An appropriate test platform must be chosen depending on the test object, the test purpose, and the necessary test environment. In the proceeding paragraphs, the execution options are elaborated more extensively.

- *Model-in-the-Loop (MiL):* The first integration level, MiL, is based on a behavioral model of the system itself. Testing at the MiL level employs a functional model or implementation model of the SUT that is tested in an open loop (i.e., without a plant model) or closed loop (i.e., with a plant model and so without physical hardware) (Schäuffele and Zurawka 2006; Kamga, Herrman, and Joshi 2007; Lehmann and Krämer 2008). The test purpose is prevailingly functional testing in early development phases in simulation environments such as Simulink.

- *Software-in-the-Loop (SiL):* During SiL, the SUT is software tested in a closed-loop or open-loop configuration. The software components under test are typically implemented in C and are either handwritten or generated by code generators based on implementation models. The test purpose in SiL is mainly functional testing (Kamga, Herrmann, and Joshi 2007). If the software is built for a fixed-point architecture, the required scaling is already part of the software.

- *Processor-in-the-Loop (PiL):* In PiL, embedded controllers are integrated into embedded devices with proprietary hardware (i.e., ECU). Testing on the PiL level is similar to SiL tests, but the embedded software runs on a target board with the target processor or on a target processor emulator. Tests on the PiL level are important because they can reveal faults that are caused by the target compiler or by the processor architecture. It is the last integration level that allows debugging during tests in an inexpensive and manageable manner (Lehmann and Krämer 2008). Therefore, the effort spent by PiL testing is worthwhile in most any case.

- *Hardware-in-the-Loop (HiL):* When testing the embedded system on the HiL level, the software runs on the target ECU. However, the environment around the ECU is still simulated. ECU and environment interact via the digital and analog electrical connectors of the ECU. The objective of testing on the HiL level is to reveal faults in the low-level services of the ECU and in the I/O services (Schäuffele and Zurawka 2006). Additionally, acceptance tests of components delivered by the supplier are executed on the HiL level because the component itself is the integrated ECU (Kamga, Herrmann, and Joshi 2007). HiL testing requires real-time behavior of the environment model to ensure that the communication with the ECU is the same as in the real application.

- *Vehicle:* The ultimate integration level is the vehicle itself. The target ECU operates in the physical vehicle, which can either be a sample or be a vehicle from the production line. However, these tests are expensive, and, therefore, performed only in the late development phases. Moreover, configuration parameters cannot be varied arbitrarily (Lehmann and Krämer 2008), hardware faults are difficult to trigger, and the reaction of the SUT is often difficult to observe because internal signals are no longer accessible (Kamga, Herrmann, and Joshi 2007). For these reasons, the number of in-vehicle tests decreases as MBT increases.

In the following, the execution options from the perspective of test reactiveness are discussed. Reactive testing and the related work on the reactive/nonreactive are reviewed. Some considerations on this subject are covered in more detail in Chapter 15.

- *Reactive/Nonreactive execution:* Reactive tests are tests that apply any signal or data derived from the SUT outputs or test system itself to influence the signals fed into the SUT. As a consequence, the execution of reactive test cases varies depending on the SUT behavior. This contrasts with the nonreactive test execution where the SUT does not influence the test at all.

  Reactive tests can be implemented in, for example, AutomationDesk (dSPACE GmbH 2010a). Such tests react to changes in model variables within one simulation step. Scripts that capture the reactive test behavior execute on the processor of the HiL system in real time and are synchronized with the model execution.

  The Reactive Test Bench (SynaptiCAD 2010) allows for specification of single timing diagram test benches that react to the user's Hardware Description Language (HDL) design files. Markers are placed in the timing diagram so that the SUT activity is recognized. Markers can also be used to call user-written HDL functions and tasks within a diagram.

Dempster and Stuart (2002) conclude that a dynamic test generator and checker are not only more effective in creating reactive test sequences but also more efficient because errors can be detected immediately as they happen.

- *Generating test logs:* The execution phase can produce test logs on each test run that are then used for further test coverage analysis (cf. e.g., Chapter 17). The test logs contain detailed information on test steps, executed methods, covered requirements, etc.

1.3.4 Test evaluation

The test evaluation, also called the test assessment, is the process that relies on the test oracle. It is a mechanism for analyzing the SUT output and deciding about the test result. The actual SUT results are compared with the expected ones and a verdict is assigned. An oracle may be the existing system, test specification, or an individual's expert knowledge.

1.3.4.1 Specification

Specification of the test assessment algorithms may be based on different foundations depending on the applied criteria. It generally forms a model of sorts or a set of ordered reference signals/data assigned to specific scenarios.

- *Reference signal-based specification:* Test evaluation based on reference signals assesses the SUT behavior comparing the SUT outcomes with the previously specified references.

  An example of such an evaluation approach is realized in MTest (dSPACE GmbH 2010b, Conrad 2004b) or SystemTest™ (MathWorks®, 2010). The reference signals can be defined using a signal editor or they can be obtained as a result of a simulation. Similarly, test results of back-to-back tests can be analyzed with the help of MEval (Wiesbrock, Conrad, and Fey 2002).

- *Reference signal-feature-based specification:* Test evaluation based on features of the reference signal* assesses the SUT behavior by classifying the SUT outcomes into features and comparing the outcome with the previously specified reference values for those features.

  Such an approach to test evaluation is supported in the time partitioning test (TPT) (Lehmann 2003, PikeTec 2010). It is based on the scripting language Python extended with some syntactic test evaluation functions. By the availability of those functions, the test assessment can be flexibly designed and allow for dedicated complex algorithms and filters to be applied to the recorded test signals. A library containing complex evaluation functions is available.

  A similar method is proposed in MiLEST (Zander-Nowicka 2009), where the method for describing the SUT behavior is based on the assessment of particular signal features specified in the requirements. For that purpose, an abstract understanding of a *signal* is defined and then both test case generation and test evaluation are based on this

---

*A signal feature (also called signal property by Gips and Wiesbrock (2007) and Schieferdecker and Großmann (2007) is a formal description of certain defined attributes of a signal. It is an identifiable, descriptive property of a signal. It can be used to describe particular shapes of individual signals by providing means to address abstract characteristics (e.g., increase, step response characteristics, step, maximum) of a signal.

concept. Numerous signal features are identified, and for all of these, feature extractors, comparators, and feature generators are defined. The test evaluation may be performed online because of the application of those elements that enable active test control and unlock the potential for reactive test generation algorithms.

The division into reference-based and reference signal-feature-based evaluation becomes particularly important when continuous signals are considered.

- *Requirements coverage criteria:* Similar to the case of test data generation, these criteria aim to cover all the informal SUT requirements, but in this case with respect to the expected SUT behavior (i.e., regarding the test evaluation scenarios) specified during the test evaluation phase. Traceability of the SUT requirements to the test model/code provides valuable support in realizing this criterion.

- *Test evaluation definition:* This criterion refers to the specification of the outputs expected from the SUT in response to the test case execution. Early work of Richardson, O'Malley, and Tittle (1998) already describes several approaches to specification-based test selection and extends them based on the concept of test oracle, faults, and failures. When a test engineer defines test scenarios in a certain formal notation, these scenarios can be used to determine how, when, and which tests will be evaluated.

1.3.4.2   Technology

The technology selected to implement the test evaluation specification enables an automatic or manual process, whereas the execution of the test evaluation occurs online or offline. Those options are elaborated next.

- *Automatic/Manual technology:* The *execution option* can be interpreted either from the perspective of the test evaluation definition or its execution. Regarding the specification of the test evaluation, when the expected SUT outputs are defined by hand, then it is a manual test specification process. In contrast, when they are derived automatically (e.g., from the behavioral model), then the test evaluation based on the test oracle occurs automatically. Typically, the expected reference signals/data are defined manually; however, they may be facilitated by parameterized test patterns application.

  The test assessment itself can be performed manually or automatically. Manual specification of the test evaluation is supported in Simulink® Verification and Validation™ (MathWorks® 2010), where predefined assertion blocks can be assigned to test signals defined in a Signal Builder block in Simulink. This practice supports verification of functional requirements during model simulation where the evaluation itself occurs automatically.

- *Online/Offline execution of the test evaluation:* The online (i.e., "on-the-fly") test evaluation happens already during the SUT execution. Online test evaluation enables the concept of test control and test reactiveness to be extended. Offline means that the test evaluation happens after the SUT execution, and so the verdicts are computed after analyzing the execution test logs.

  Watchdogs defined in Conrad and Hötzer (1998) enable online test evaluation. It is also possible when using TTCN-3. TPT means for online test assessment are limited and are used as watchdogs for extracting any necessary information for making test cases reactive (Lehmann and Krämer 2008). The offline evaluation is more sophisticated in TPT. It offers means for more complex evaluations, including operations such as comparisons with external reference data, limit-value monitoring, signal filters, and analyses of state sequences and time conditions.

1.4 Summary

This introductory chapter has extended the Model-Based Testing (MBT) *taxonomy* of previous work. Specifically, test dimensions have been discussed with pertinent aspects such as test goals, test scope, and test abstraction described in detail. Selected classes from the taxonomy have been illustrated, while all categories and options related to the test generation, test execution, and test evaluation have been discussed in detail with examples included where appropriate.

Such perspectives of MBT as cost, benefits, and limitations have not been addressed here. Instead, the chapters that follow provide a detailed discussion as they are in a better position to capture these aspects seeing how they strongly depend on the applied approaches and challenges that have to be resolved. As stated in Chapter 6, most published case studies illustrate that utilizing MBT reduces the overall cost of system and software development. A typical benefit achieves 20%–30% of cost reduction. This benefit may increase up to 90% as indicated by Clarke (1998) more than a decade ago, though that study only pertained to test generation efficiency in the telecommunication domain.

For additional research and practice in the field of MBT, the reader is referred to the surveys provided by Broy et al. (2005), Utting, Pretschner, and Legeard (2006), Zander and Schieferdecker (2009), Shafique and Labiche (2010), as well as every contribution found in this collection.

References

All4Tec, *Markov Test Logic—MaTeLo*, commercial Model-Based Testing tool, http://www.all4tec.net/ [12/01/10].

Beizer, B. (1995). *Black-Box Testing: Techniques for Functional Testing of Software and Systems*. ISBN-10: 0471120944. John Wiley & Sons, Inc., Hoboken, NJ.

Broy, M., Jonsson, B., Katoen, J. -P., Leucker, M., and Pretschner, A. (Editors) (2005). *Model-Based Testing of Reactive Systems*, Editors: no. 3472. In *LNCS*, Springer-Verlag, Heidelberg, Germany.

BTC Embedded Systems AG, Embedded *Validator*, commercial verification tool, http://www.btc-es.de/ [12/01/10].

Carnegie Mellon University, Department of Electrical and Computer Engineering, *Hybrid System Verification Toolbox for MATLAB—CheckMate*, research tool for system verification, http://www.ece.cmu.edu/~webk/checkmate/ [12/01/10].

Carter, J. M., Lin, L., and Poore, J. H. (2008). Automated Functional Testing of Simulink Control Models. In *Proceedings of the 1st Workshop on Model-based Testing in Practice—MoTip 2008*, Editors: Bauer, T., Eichler, H., Rennoch, A., ISBN: 978-3-8167-7624-6, Fraunhofer IRB Verlag, Berlin, Germany.

Clarke, J. M. (1998). Automated Test Generation from Behavioral Models. In the *Proceedings of the 11th Software Quality Week (QW'98)*, Software Research Inc., San Francisco, CA.

Conrad, M. (2004a). *A Systematic Approach to Testing Automotive Control Software*, Detroit, MI, SAE Technical Paper Series, 2004-21-0039.

Conrad, M. (2004b). *Modell-basierter Test eingebetteter Software im Automobil: Auswahl und Beschreibung von Testszenarien*. PhD thesis. Deutscher Universitätsverlag, Wiesbaden (D). (In German).

Conrad, M., Fey, I., and Sadeghipour, S. (2004). Systematic Model-Based Testing of Embedded Control Software—The MB³T Approach. In *Proceedings of the ICSE 2004 Workshop on Software Engineering for Automotive Systems*, Edinburgh, United Kingdom.

Conrad, M., and Hötzer, D. (1998). Selective Integration of Formal Methods in the Development of Electronic Control Units. In *Proceedings of the ICFEM* 1998, 144-Electronic Edition, Brisbane Australia, ISBN: 0-8186-9198-0.

Dai, Z. R. (2006). *An Approach to Model-Driven Testing with UML 2.0, U2TP and TTCN-3*. PhD thesis, Technical University Berlin, ISBN: 978-3-8167-7237-8. Fraunhofer IRB Verlag.

Dempster, D., and Stuart, M. (2002). *Verification methodology manual, Techniques for Verifying HDL Designs*, ISBN: 0-9538-4822-1. Teamwork International, Great Britain, Biddles Ltd., Guildford and King's Lynn.

Din, G., and Engel, K. D. (2009). An Approach for Test Derivation from System Architecture Models Applied to Embedded Systems, In *Proceedings of the 2nd Workshop on Model-based Testing in Practice (MoTiP 2009)*, In Conjunction with the 5th European Conference on Model-Driven Architecture (ECMDA 2009), Enschede, The Netherlands, Editors: Bauer, T., Eichler, H., Rennoch, A., Wieczorek, S., CTIT Workshop Proceedings Series WP09-08, ISSN 0929-0672.

D-Mint Project (2008). Deployment of model-based technologies to industrial testing. http://d-mint.org/ [12/01/10].

dSPACE GmbH, *AutomationDesk*, commercial tool for testing, http://www.dspace.com/de/gmb/home/products/sw/expsoft/automdesk.cfm [12/01/2010a].

dSPACE GmbH, *MTest*, commercial MBT tool, http://www.dspaceinc.com/ww/en/inc/home/products/sw/expsoft/mtest.cfm [12/01/10b].

Dulz, W., and Fenhua, Z. (2003). MaTeLo—Statistical Usage Testing by Annotated Sequence Diagrams, Markov Chains and TTCN-3. In *Proceedings of the 3rd International Conference on Quality Software*, Page: 336, ISBN: 0-7695-2015-4. IEEE Computer Society Washington, DC.

ETSI (2007). European Standard. 201 873-1 V3.2.1 (2007-02): *The Testing and Test Control Notation Version 3; Part 1: TTCN-3 Core Language*. European Telecommunications Standards Institute, Sophia-Antipolis, France.

GeenSoft (2010a). *Safety Test Builder*, commercial Model-Based Testing tool, http://www.geensoft.com/en/article/safetytestbuilder/ [12/01/10].

GeenSoft (2010b). *Safety Checker Blockset*, commercial Model-Based Testing tool, http://www.geensoft.com/en/article/safetycheckerblockset_app/ [12/01/10].

Gips C., Wiesbrock, H. -W. (2007). Notation und Verfahren zur automatischen Überprüfung von temporalen Signalabhängigkeiten und -merkmalen für modellbasiert entwickelte Software. In *Proceedings of Model Based Engineering of Embedded Systems III*, Editors: Conrad, M., Giese, H., Rumpe, B., Schätz, B.: TU Braunschweig Report TUBS-SSE 2007-01. (In German).

Grochtmann, M., and Grimm, K. (1993). Classification Trees for Partition Testing. In *Software Testing, Verification & Reliability*, 3, 2, 63–82. Wiley, Hoboken, NJ.

Gutjahr, W. J. (1999). Partition Testing vs. Random Testing: The Influence of Uncertainty. In *IEEE Transactions on Software Engineering*, Volume 25, Issue 5, Pages: 661–674, ISSN: 0098–5589. IEEE Press Piscataway, NJ.

Hetzel, W. C. (1988). *The Complete Guide to Software Testing*. Second edition, ISBN: 0-89435-242-3. QED Information Services, Inc., Wellesley, MA.

International Software Testing Qualification Board (2006). *Standard glossary of terms used in Software Testing*. Version 1.2, produced by the Glossary Working Party, Editor: van Veenendaal E., The Netherlands.

IT Power Consultants, *MEval*, commercial tool for testing, http://www.itpower.de/ 30-0-Download-MEval-und-SimEx.html [12/01/10].

Kamga, J., Herrmann, J., and Joshi, P. Deliverable (2007). D-MINT automotive case study—Daimler, Deliverable 1.1, *Deployment of model-based technologies to industrial testing*, ITEA2 Project, Germany.

Kosmatov, N., Legeard, B., Peureux, F., and Utting, M. (2004). Boundary Coverage Criteria for Test Generation from Formal Models. In *Proceedings of the 15th International Symposium on Software Reliability Engineering*. ISSN: 1071–9458, ISBN: 0-7695-2215-7, Pages: 139–150. IEEE Computer Society, Washington, DC.

Lamberg, K., Beine, M., Eschmann, M., Otterbach, R., Conrad, M., and Fey, I. (2004). Model-Based Testing of Embedded Automotive Software Using MTest. In *Proceedings of SAE World Congress*, Detroit, MI.

Lee, T., and Yannakakis, M. (1994). Testing Finite-State Machines: State Identification and Verification. In *IEEE Transactions on Computers*, Volume 43, Issue 3, Pages: 306–320, ISSN: 0018–9340. IEEE Computer Society, Washington, DC.

Lehmann, E. (then Bringmann, E.) (2003). *Time Partition Testing, Systematischer Test des kontinuierlichen Verhaltens von eingebetteten Systemen*, PhD thesis, Technical University Berlin. (In German).

Lehmann, E., and Krämer, A. (2008). Model-Based Testing of Automotive Systems. In *Proceedings of IEEE ICST 08*, Lillehammer, Norway.

Marre, B., and Arnould, A. (2000). Test Sequences Generation from LUSTRE Descriptions: GATEL. In *Proceedings of ASE of the 15th IEEE International Conference on Automated Software Engineering*, Pages: 229–237, ISBN: 0-7695-0710-7, Grenoble, France. IEEE Computer Society, Washington, DC.

MathWorks®, Inc., Real-Time Workshop®, http://www.mathworks.com/help/toolbox/rtw/ [12/01/10].

MathWorks®, Inc., *Simulink® Design Verifier*™, commercial Model-Based Testing tool, MathWorks®, Inc., Natick, MA, http://www.mathworks.com/products/sldesignverifier [12/01/10].

MathWorks®, Inc., *Simulink®*, MathWorks®, Inc., Natick, MA, http://www.mathworks.com/products/simulink/ [12/01/10].

MathWorks®, Inc., *Simulink® Verification and Validation*™, commercial model-based verification and validation tool, MathWorks®, Inc., Natick, MA, http://www.mathworks.com/products/simverification/ [12/01/10].

MathWorks®, Inc., *Stateflow®*, MathWorks®, Inc., Natick, MA, http://www.mathworks.com/products/stateflow/ [12/01/10].

MathWorks®, Inc., *SystemTest*™, commercial tool for testing, MathWorks®, Inc., Natick, MA, http://www.mathworks.com/products/systemtest/ [12/01/2010].

MATLAB Automated Testing Tool—MATT. (2008). The University of Montana, research Model-Based Testing prototype, http://www.sstc-online.org/Proceedings/2008/pdfs/JH1987.pdf [12/01/10].

Mosterman, P. J., Zander, J., Hamon, G., and Denckla, B. (2009). Towards Computational Hybrid System Semantics for Time-Based Block Diagrams. In *Proceedings of the 3rd IFAC Conference on Analysis and Design of Hybrid Systems (ADHS'09)*, Editors: A. Giua, C. Mahulea, M. Silva, and J. Zaytoon, pp. 376–385, Zaragoza, Spain, Plenary paper.

Mosterman, P. J., Zander, J., Hamon, G., and Denckla, B. (2011). A computational model of time for stiff hybrid systems applied to control synthesis, *Control Engineering Practice Journal (CEP)*, 19, Elsevier.

Myers, G. J. (1979). *The Art of Software Testing.* ISBN-10: 0471043281. John Wiley & Sons, Hoboken, NJ.

Neukirchen, H. W. (2004). *Languages, Tools and Patterns for the Specification of Distributed Real-Time Tests*, PhD thesis, Georg-August-Universiät zu Göttingen.

OMG. (2003). MDA Guide V1.0.1. http://www.omg.org/mda/mda_files/MDA_Guide_Version1-0.pdf [12/01/10 TODO].

OMG. (2003). UML 2.0 Superstructure Final Adopted Specification, http://www.omg.org/cgi-bin/doc?ptc/03-08-02.pdf [12/01/10].

OMG. (2005). UML 2.0 Testing Profile. Version 1.0 formal/05-07-07. Object Management Group.

PikeTec, *Time Partitioning Testing—TPT*, commercial Model-Based Testing tool, http://www.piketec.com/products/tpt.php [12/01/2010].

Pretschner, A. (2003). Compositional Generation of MC/DC Integration Test Suites. In *Proceedings TACoS'03*, Pages: 1–11. *Electronic Notes in Theoretical Computer Science* 6.

Pretschner, A. (2003a). Compositional Generation of MC/DC Integration Test Suites. In *Proceedings TACoS'03*, Pages: 1–11. *Electronic Notes in Theoretical Computer Science* 6. http://citeseer.ist.psu.edu/633586.html.

Pretschner, A. (2003b). *Zum modellbasierten funktionalen Test reaktiver Systeme*. PhD thesis. Technical University Munich. (In German).

Pretschner, A., Prenninger, W., Wagner, S., Kühnel, C., Baumgartner, M., Sostawa, B., Zölch, R., and Stauner, T. (2005). One Evaluation of Model-based Testing and Its Automation. In *Proceedings of the 27th International Conference on Software Engineering*, St. Louis, MO, Pages: 392–401, ISBN: 1-59593-963-2. ACM New York.

Pretschner, A., Slotosch, O., Aiglstorfer, E., and Kriebel, S. (2004). Model Based Testing for Real—The Inhouse Card Case Study. In *International Journal on Software Tools for Technology Transfer*. Volume 5, Pages: 140–157. Springer-Verlag, Heidelberg, Germany.

Rau, A. (2002). *Model-Based Development of Embedded Automotive Control Systems*, PhD thesis, University of Tübingen.

Reactive Systems, Inc., *Reactis Tester*, commercial Model-Based Testing tool, http:// www.reactive-systems.com/tester.msp [12/01/10a].

Reactive Systems, Inc., *Reactis Validator*, commercial validation and verification tool, http://www.reactive-systems.com/reactis/doc/user/user009.html, http://www. reactive-systems.com/validator.msp [12/01/10b].

Richardson, D, O'Malley, O., and Tittle, C. (1998). Approaches to Specification-Based Testing. In *Proceedings of ACM SIGSOFT Software Engineering Notes*, Volume 14, Issue 8, Pages: 86–96, ISSN: 0163–5948. ACM, New York.

Schäuffele, J., and Zurawka, T. (2006). *Automotive Software Engineering*, ISBN: 3528110406. Vieweg.

Schieferdecker, I., and Großmann, J. (2007). Testing Embedded Control Systems with TTCN-3. In *Proceedings Software Technologies for Embedded and Ubiquitous Systems SEUS 2007*, Pages: 125–136, *LNCS* 4761, ISSN: 0302–9743, 1611–3349, ISBN: 978-3-540–75663-7 Santorini Island, Greece. Springer-Verlag, Berlin/Heidelberg.

Schieferdecker, I., Großmann, J., and Wendland, M.-F. (2011). Model-Based Testing: Trends. Encyclopedia of Software Engineering DOI: 10.1081/E-ESE-120044686, Taylor & Francis.

Schieferdecker, I., and Hoffmann, A. (2011). Model-Based Testing. Encyclopedia of Software Engineering DOI: 10.1081/E-ESE-120044686, Taylor & Francis.

Schieferdecker, I., Rennoch, A., and Vouffo-Feudjio, A. (2011). Model-Based Testing: Approaches and Notations. Encyclopedia of Software Engineering DOI: 10.1081/ E-ESE-120044686, Taylor & Francis.

Shafique, M., and Labiche, Y. (2010). *A Systematic Review of Model Based Testing Tool Support*, Carleton University, Technical Report, SCE-10-04, http://squall.sce. carleton.ca/pubs/tech_report/TR_SCE-10-04.pdf [03/22/11].

Sims S., and DuVarney D. C. (2007). Experience Report: The Reactis Validation Tool. In *Proceedings of the ICFP '07 Conference*, Volume 42, Issue 9, Pages: 137–140, ISSN: 0362–1340. ACM, New York.

Software Quality Research Laboratory, *Java Usage Model Builder Library—JUMBL*, research Model-Based Testing prototype, http://www.cs.utk.edu/sqrl/esp/jumbl.html [12/01/10].

SynaptiCAD, *Waveformer Lite 9.9 Test-Bench* with Reactive Test Bench, commercial tool for testing, http://www.actel.com/documents/reactive_tb_tutorial.pdf [12/01/10].

Utting, M. (2005). Model-Based Testing. In *Proceedings of the Workshop on Verified Software: Theory, Tools, and Experiments VSTTE 2005*.

Utting, M., and Legeard, B. (2006). *Practical Model-Based Testing: A Tools Approach.* ISBN-13: 9780123725011. Elsevier Science & Technology Books.

Utting, M., Pretschner, A., and Legeard, B. (2006). *A taxonomy of model-based testing*, ISSN: 1170-487X, The University of Waikato, New Zealand.

Weyuker, E. (1988). The Evaluation of Program-Based Software Test Data Adequacy Criteria. In *Communications of the ACM*, Volume 31, Issue 6, Pages: 668–675, ISSN: 0001-0782. ACM, New York, NY.

Wieczorek, S., Kozyura, V., Roth, A., Leuschel, M., Bendisposto, J., Plagge, D., and Schieferdecker, I. (2009). Applying Model Checking to Generate Model-based Integration Tests from Choreography Models. *21st IFIP Int. Conference on Testing of Communicating Systems (TESTCOM)*, Eindhoven, The Netherlands, ISBN 978-3-642-05030-5.

Wiesbrock, H. -W., Conrad, M., and Fey, I. (2002). Pohlheim: Ein neues automatisiertes Auswerteverfahren für Regressions und Back-to-Back-Tests eingebetteter Regelsysteme. In *Softwaretechnik-Trends*, Volume 22, Issue 3, Pages: 22–27. (In German).

Zander, J., Dai, Z. R., Schieferdecker, I., and Din, G. (2005). From U2TP Models to Executable Tests with TTCN-3—An Approach to Model Driven Testing. In *Proceedings of the IFIP 17th Intern. Conf. on Testing Communicating Systems (TestCom 2005)*, ISBN: 3-540–26054-4, Springer-Verlag, Heidelberg, Germany.

Zander, J., Mosterman, P. J., Hamon, G., and Denckla, B. (2011). On the Structure of Time in Computational Semantics of a Variable-Step Solver for Hybrid Behavior Analysis, 18th World Congress of the International Federation of Automatic Control (IFAC), Milano, Italy.

Zander, J., and Schieferdecker, I. (2009). Model-Based Testing of Embedded Systems Exemplified for the Automotive Domain, Chapter in *Behavioral Modeling for Embedded Systems and Technologies: Applications for Design and Implementation*, Editors: Gomes, L., Fernandes, J. M., DOI: 10.4018/978-1-60566-750-8.ch015. Idea Group Inc. (IGI), Hershey, PA, ISBN 1605667501, 9781605667508, pp. 377–412.

Zander-Nowicka, J. (2009). *Model-Based Testing of Embedded Systems in the Automotive Domain*, PhD Thesis, Technical University Berlin, ISBN: 978-3-8167-7974-2. Fraunhofer IRB Verlag, Germany. http://opus.kobv.de/tuberlin/volltexte/2009/2186/pdf/zandernowicka_justyna.pdf.

2

Behavioral System Models versus Models of Testing Strategies in Functional Test Generation

Antti Huima

CONTENTS

An important dichotomy in the concept space of model-based testing is between models that directly describe *intended system behavior* and models that directly describe *testing strategies*. The purpose of this chapter is to shed light on this dicothomy from both practical as well as theoretical viewpoints. In the proceedings, we will call these two types of models "system models" and "tester models," respectively, for brevity's sake.

When it comes to the dividing line between system models and tester models (1) it certainly exists, (2) it is important, and (3) it is not always well understood. One reason why the reader may not have been fully exposed to this distinction between the two types of models is that for *explicit finite-state models* the distinction disappears by sleight of hand of deterministic polynomial-time complexity, something not true at all for more expressive modeling formalisms.

The main argument laid forth in this chapter is that converting a system model into a tester model is computationally hard, and this will be demonstrated both within practical as well as theoretical frameworks. Because system designers and system tester already have some form of a mental model of the system's correct operation in their minds (Pretschner 2005), this then means that constructing tester models must be hard for humans also. This observation correlates well with the day-to-day observations of test project managers: defining, constructing, and expressing testing strategies is challenging regardless of whether the strategies are eventually expressed in the form of models or not.

Even though eventually a question for cognitive psychology, it is generally agreed that, indeed, test designers create or possess mental models of the systems under test. Alexander Pretschner writes (Pretschner 2005):

> Traditionally, engineers form a vague understanding of the system by reading the specification. They build a mental model. Inventing tests on the grounds of these mental models is a creative process...

This line of reasoning leads immediately to the result that creating system models should be "easier" than tester models—and also reveals the other side of the coin, namely that software tools that generate tests from system models are much more difficult to construct than those generating tests from tester models.

To illustrate this, consider Figure 2.1. Testing experts have mental models of the systems they should test. They can convert those mental models into explicit (computer-readable) system models (arrow 1). These system models could be converted into tester models algorithmically (arrow 2); we do not need to postulate here anything about the nature of those algorithms as for this discussion it is enough to assume for them to exist.* Alternatively, test engineers can create tester models directly based on their mental models of the systems (arrow 3).

Now let us assume that engineers could create good tester models "efficiently," that is tester models that cover all the necessary testing conditions (however they are defined) and are correct. This would open the path illustrated in Figure 2.2 to efficiently implement the

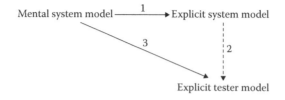

FIGURE 2.1
Conversions between mental and explicit models.

---

*And they of course do—a possible algorithm would be to (1) guess a tester model, (2) limit the test cases that can be generated from the tester model to a finite set, and then (3) verify that the generated test cases pass against the system model. For the purposes of showing existence, this is a completely valid algorithm—regardless of it being a nondeterministic one.

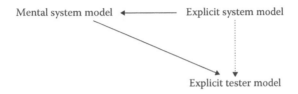

FIGURE 2.2
Illustration of the difficulty of constructing tester models.

arrow number 2 from Figure 2.1. Now clearly this alternative path is not algorithmic in a strict sense as it involves human cognition, but still if this nonalgorithmic path would be efficient, it would provide an efficient way to implement an algorithmically very difficult task, namely the conversion of a computer-readable system model into a computer-readable tester model. This would either lead to (1) showing that the human brain is an infinitely more potent computational device than a Turing machine, thus showing that the Church–Turing thesis* does not apply to human cognition, or (2) the collapse of the entire computational hierarchy. Both results are highly unlikely given our current knowledge.

To paraphrase, as long as it is accepted that for combinatorial problems (such as test case design) the human brain is not infinitely or categorically more efficient and effective than any known mechanical computing devices, it must be that tester model construction is very difficult for humans, and in general much more difficult than the construction of explicit system models. Namely, the latter is a translation problem, whereas the former is a true combinatorial and computational challenge.

In summary, the logically derived result that tester models are more difficult to construct than system models but easier to handle for computers leads to the following predictions, which can be verified empirically:

1. Tester model-based test generation tools should be easier to construct; hence, there should be more of them available.

2. System models should be quicker to create than the corresponding tester models.

For the purposes of this chapter, we need to fix some substantive definitions for the nature of system models and tester models, otherwise there is not much that can be argued. On an informal level, it shall be assumed that tester models can be executed for test generation efficiently, that is, in polynomial time in the length of the produced test inputs and predicted test outputs. The key here is the word *efficiently*; without it, we cannot argue about the relative "hardness" of constructing tester models from system models. This leads to basic complexity-theoretic[†] considerations in Section 2.2.

---

*The Church–Turing thesis in its modern form basically states that everything that can be calculated can be calculated by a Turing machine. However, it is in theory an open question whether the human brain can somehow calculate Turing-uncomputable functions. At least when it comes to combinatorial problems, we tend to believe the negative. In any case, the argument that tester model design is difficult for humans is eventually based, in this chapter, on the assumption that for this kind of problems, the human brain is not *categorically* more efficient than idealized computers.

[†]This chapter is ultimately about the system model versus tester model dichotomy, not about complexity theory. Very basic computational complexity theory is used in this chapter as a vehicle to establish a theoretical foundation that supports and correlates with practical field experience, and some convenient shortcuts are taken. For example, even though test generation is eventually a "function problem," it is mapped to the decision problem complexity classes in this presentation. This does not affect the validity of any of the conclusions presented here.

This polynomial-time limit on test case generation from tester models is a key to understanding this chapter. Even though in practice system model and tester model-driven approaches differ substantively, on an abstract level both these approaches fit the model-based testing pattern "model → tests." It is most likely not possible to argue definitively about the differences between the two approaches based on semantical concepts alone, as those are too much subject to interpretation. The complexity-theoretic assumption that tester models generate the corresponding test suites efficiently (1) matches the current practice and (2) makes sense because tester models encode (by their definition) testing strategies previously construed by humans using their intelligence, and thus it can be expected that the actual test case generation should be relatively fast.

2.1   Introduction

In this section, we will proceed through a series of examples highlighting how the expressivity of a language for modeling systems affects the difficulty of constructing the corresponding tester models. This is not about constructing the models *algorithmically*, but an informal investigation into the growing *mental* challenge of constructing them (by humans). Namely, in the tester model-driven, model-based testing approach, the tester models are actually constructed by human operators, and as the human operators already must have some form of a mental model about the system's requirements and intended behavior, all the operators must go through an analogous process in order to construct the tester models.

In this section, we shall use formal (computer-readable) system models and their algorithmic conversion into tester models as an expository device to illustrate the complexity of the tester model derivation process regardless of whether it is carried out by a computer or a human operator.

2.1.1   Finite-state machines

Consider the system model, in the form of a finite-state machine, shown in Figure 2.3.

Each transition from one state to another is triggered by an input (lowercase letters) and produces an output (uppercase letters). Inputs that do not appear on a transition leaving from a state are not allowed. Assume the testing goal is to test a system specified like this and to verify that every transition works as specified. Assume further that the system under test has an introspective facility so that its internal control state can be verified easily. After reset, the system starts from state 1, which is thus the initial state for this model.

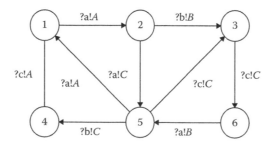

FIGURE 2.3
A finite-state machine.

One approach for producing a tester model for this system is to enumerate test cases explicitly and call that list of test cases a tester model. In the case of a finite-state machine, test cases can be immediately constructed using simple path traversal procedures, such as Chinese Postman tours (Edmonds and Johnson 1973, Aho et al. 1991). One possible way to cover all the transitions in a single test case here is shown in Figure 2.4.

Here, the numbers inside parentheses correspond to verifying the internal control state of the system under test. Note also that inputs and outputs have been reversed as this is a test case—what is an input to the system under test is an output for the tester.

The path traversal procedures are so efficient for explicitly represented finite-state machines (polynomial time) that, however, the system model *itself* can be repurposed as a tester model as illustrated in Figure 2.5.

Two things happen here: inputs and outputs are switched, and the state numbers in the tester model now refer to internal states of the system under test to be verified. Path traversal procedures can now be executed for this tester model in order to produce test cases, which can be either executed online (during their construction) or offline (after they have been constructed). This results in the polynomial-time generation of a polynomial-size test suite.

Obviously, the system model and the tester model are the same model for all practical purposes, and they are both computationally easy to handle. This explains the wide success of finite-state machine approaches for relatively simple test generation problems and also the confusion that sometimes arises about whether the differences between tester models and system models are real or not. For explicitly represented finite-state models, there are no fundamental differences.

2.1.2 Arithmetics

Let us now move from completely finite domains into finite-state control with some numerical variables (extended finite-state machines). For example, consider the extended state machine shown in Figure 2.6. The state machine communicates in terms of rational numbers $\mathbb{Q}$ instead of symbols, can do conditional branches, and has internal variables (x, y, and z) for storing data.

(1)	!a?A	(2)	!b?B	(3)	!c?C	(6)	!a?B	
(5)	!c?C	(3)	!c?C	(6)	!a?B	(5)	!b?C	
(4)	!c?A	(1)	!a?A	(2)	!a?C	(5)	!a?A	(1)

FIGURE 2.4

A tour through the state machine.

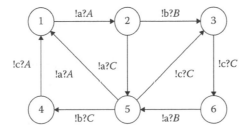

FIGURE 2.5

Repurposing the system model as a tester model.

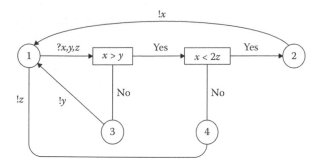

FIGURE 2.6
State machine with internal variables.

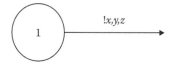

FIGURE 2.7
Prefix of the state machine when inverted.

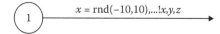

FIGURE 2.8
Randomized input selection strategy.

This model cannot be inverted into a tester model directly in the same manner as we inverted the finite-state machine in the previous section because the inverted model would start with the prefix shown in Figure 2.7.

Notably, the variables x, y, and z would be unbound—the transition cannot be executed.

A simple way to bind the variables would be to use a random testing approach (Duran and Ntafos 1984), for instance to bind every input variable to a random rational number between, say, -10 and 10, as illustrated in Figure 2.8.

The corresponding tester model would then require randomized execution. In this case, the probability for a single loop through the tester model to reach any one of the internal states would be relatively high. However, if the second condition in the original machine would be changed as shown in Figure 2.9, then the same strategy for constructing the tester would result in a tester model that would have very slim chances of reaching the internal state (2) within a reasonable number of attempts. This exemplifies the general robustness challenges of random testing approaches for test generation.

2.1.3   Advanced tester model execution algorithms

Clearly, one could postulate a tester model execution algorithm that would be able to calculate forwards in the execution space of the tester model and resolve the linear equations backwards in order to choose suitable values for x, y, and z so that instead of being initialized to random values, they would be initialized more "intelligently." Essentially, the same *deus ex machina* argument could be thrown in all throughout this chapter. However, it would violate the basic assumption that tester models are (polynomial time) efficient to execute

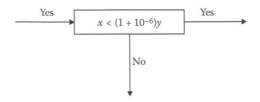

FIGURE 2.9
Changed condition in the state machine illustrates the challenges with random test data selection.

because even though it is known that linear equations over rational numbers can be solved in polynomial time, the same is not true for computationally harder classes of data constraints showing up in system models, as will be discussed below.

Another, slightly more subtle reason for rejecting this general line of reasoning is the following. If one argues that (1) a system model can be always converted straightforwardly into a tester model by flipping inputs and outputs and (2) a suitably clever algorithm will then select the inputs to the system under test to drive testing intelligently, then actually what is claimed is system model-based test generation—the only difference between the system model and the tester model is, so to speak, the lexical choice between question and exclamation marks! This might be even an appealing conclusion to a theorist, but it is very far from the actual, real-world practice on the field, where system models and tester models are substantially different and handled by separate, noninterchangeable toolsets.

At this point, it could be possible to raise the general question whether this entire discussion is somehow unfair, as it will become obvious that a system model-based test generation tool must have exactly that kind of "intelligent" data selection capacity that was just denied from tester model-based tools! This may look like a key question, but actually, it is not and was already answered above.

Namely, a tester model is a model of a testing strategy, a strategy that has been devised by a human. The reason why humans labor in creating testing strategy models is that afterwards the models are straightforward to execute for test generation. The cost of straigthforward execution of the models is the relative labor of constructing them. The "intelligent" capacities of system model-based test generation tools *must exist in order to compensate for the eliminated human labor* in constructing testing strategies. A tester model-based test generation tool with all the technical capacities of a system model-based one is actually equivalent to a system model-based test generation tool, as argued above. Therefore, testing strategy models exist because of the unsuitability, unavailability, undesirability, or cost of system model-based test generation tools in the first place and are characterized by the fact that they can be straightforwardly executed. This analysis corresponds with the practical dicothomy between tester model- and system model-based tools in today's marketplace.

2.1.4 Harder arithmetics

Whereas linear equations over rationals admit practical (Nelder and Mead 1965) and polynomial-time (Bland et al. 1981) solutions, the same is not true for more general classes of arithmetical problems, which may present themselves in system specifications. A well-known example is linear arithmetics over integers; solving a linear equation system restricted to integer solutions is an NP-complete problem (Karp 1972, Papadimitriou 1981). This shows that the "flipping" of a system model into a tester model would not work in the case of integer linear arithmetics because then the postulated "intelligent" tester model execution algorithm would not be efficient. Similarly, higher-order equations restricted to integer solutions (Diophantine equations) are unsolvable (Matiyasevich 1993) and thus do not admit any general testing strategy construction algorithm.

2.1.5   Synchronized finite-state machines

The reachability problem for synchronized finite-state machines is PSPACE-complete (Demri, Laroussinie, and Schnoebelen 2006), resulting in that a system model expressed in terms of synchronized finite-state machines cannot be efficiently transformed into an efficient tester model. The fundamental problem is that a system model composed of n synchronized finite-state machines, each having k states, has internally k^n states, which is exponential in the size of the model's description. Again, a synchronized system model cannot be "flipped" into a tester model by exchanging inputs and outputs because then hypothetical test generation algorithm running on the flipped model could not necessarily find all the reachable output transitions in polynomial time.

When a human operator tries to create testing strategies for distributed and multi-threaded systems, the operator tends to face the same challenge mentally: it is difficult to synchronize and keep track of the states of the individual components in order to create comprehensive testing strategies. In the context of distributed system testing, Ghosh and Mathur (1999) write:

> Test data generation in order to make the test sets adequate with respect to a coverage criteria is a difficult task. Experimental and anecdotal evidence reveals that it is difficult to obtain high coverage for large systems. This is because in a large software system it is often difficult to design test cases that can cover a low-level element.

2.2   Simple Complexity-Theoretic Approach

In this section, we approach the difficulty of tester model creation from a complexity-theoretic viewpoint. The discussion is more informal than rigorous and the four propositions below are given only proof outlines. The reader is directed to any standard reference in complexity theory such as Papadimitriou (Papadimitriou 1993) for a thorough exposition of the theoretical background.

In general, test case generation is intimately related to reachability problems in the sense that being able to solve reachability queries is a prerequisite for generating test cases from system models. This is also what happens in the mind of a person who designs testing strategies mentally. For instance, in order to be able to test deleting a row from a database, a test engineer must first figure out how to add the row to be deleted in the first place. And to do that, the engineer must understand how to log into the system. The goal to test row deletion translates into a reachability problem—how to reach a state where a row has been deleted—and the test engineer proceeds mentally to solve this problem.

This is significant because there is a wide body of literature available about the complexity of reachability problems. As a matter of fact, reachability is one of the prototypical complexity-theoretic problems (Papadimitriou 1993).

In this chapter, we introduce a simple framework where system models and tester models are constructed as Turing machines. This framework is used only to be able to argue about the computational complexity of the system model $\rightarrow$ tester model conversion and does not readily lead to any practical algorithms. A practically oriented reader may consider jumping directly to the discussion in Section 2.2.4.

In order for us to be able to analyze tester construction in detail, we must settle on a concept for test coverage. Below, we shall use only one test coverage, criterion, namely covering output transitions in the system models (as defined below). This is a prototypical model-based black-box testing criterion and suits our purpose well.

2.2.1 Tester model-based coverage criteria

The reader may wonder why we do not consider *tester* model-driven black-box testing criteria, such as covering all the transitions of a tester model. Such coverage criteria are, after all, common in tester model-based test generation tools. This section attempts to answer this question.

Tester model-based coverage criteria are often used as practical means to select a subset of all possible paths through a (finite-state) tester model. In that sense, tester model-based coverage is a useful concept and enables practical tester model-driven test generation. However, in order to understand the relationship between tester model-based coverage criteria and the present discussion, we must trace deeper into the concept of model-driven black-box coverage criteria.

It is commonly accepted that the purpose of tests is to (1) detect faults and (2) prove their absence. It is an often repeated "theorem" that tests can find faults but cannot prove their absence. This is not actually true, however, because tests can for example prove the absence of *systematic* errors. For instance, if one can log into a database at the beginning of a test session, it proves that there is no *systematic* defect in the system, which would *always* prevent users from logging into it. This is the essence of conformance and feature testing, for instance.

Given that the purpose of testing is thus to test for the presence or absence of certain faults, it is clear that tests should be selected so that they actually target "expected" faults. This is the basis for the concept of a *fault model*. Informally, a fault model represents a hypothesis about the possible, probable, or important errors that the system under test may contain. Fault models can never be based on the implementation of the system under test alone because if the system under test works as the sole reference for the operation of itself, it can never contain any faults.

Model-based test coverage criteria are based on the—usually implicit—assumption that the system under test resembles its model. Furthermore, it is assumed that typical and interesting faults in the system under test correlate with, for example, omitting or misimplementing transitions in a state chart model or implementing arithmetic comparisons, which appear in a system model, incorrectly in the system under test. These considerations form the basis for techniques such as boundary-value testing and mutant-based test assessment.

A tester model does not bear the same relationship with the system because it does not model the system but a testing strategy. Therefore, for instance, attempting to cover all transitions of a testing strategy model does not have similar, direct relationship with the expected faults of the system under test as the transitions of a typical system model in the form of state machine. In other words, a testing strategy model already encodes a particular fault model or test prioritization that is no longer based on the correspondence between a system model and a system under test correspondence but on the human operator's interpretation. So now in our context, where we want to argue about the challenge of creating efficient and effective tests, we will focus on system model-based coverage criteria because they bear a direct relationship with the implicit underlying fault model; and as a representative of that set of coverage criteria, we select the simple criterion of covering output transitions as defined below.

2.2.2 Multitape Turing Machines

We will use multitape Turing machines as devices to represent system and tester models. The following definition summarizes the technical essence of a multitape Turing machine. Computationally, multitape Turing machines can be reduced via polynomial-time transformations into single-tape machines with only a negligible (polynomial) loss of execution efficiency (Papadimitriou 1993).

Definition 1 (Turing Machines and Their Runs). A multitape deterministic Turing machine with k tapes is a tuple $\langle Q, \Sigma, q, b, \delta \rangle$, where Q is a finite set of *control states*, Σ is a finite set of symbols, named the *alphabet*, $q \in Q$ is the *initial state*, $b \in \Sigma$ is *blank symbol*, and $\delta : Q \times \Sigma^k \to Q \times (\Sigma \times \{l, n, r\})^k$ is the partial *transition function*.

A configuration of a k-multitape Turing machine is an element of $Q \times (\Sigma^* \times \Sigma \times \Sigma^*)^k$. A configuration is mapped to a next-state configuration by the rule $\langle q, \langle \langle w_1, \sigma_1, u_1 \rangle, \ldots, \langle w_k, \sigma_k, u_k \rangle \rangle \rangle \to \langle q', \langle \langle w'_1, \sigma'_1, u'_1 \rangle, \ldots, \langle w'_k, \sigma'_k, u'_k \rangle \rangle \rangle$ iff it holds that $\delta(q, \langle \sigma_1, \ldots, \sigma_k \rangle) = \langle q', \langle \langle \alpha_1, \kappa_1 \rangle, \ldots, \langle \alpha_k, \kappa_k \rangle \rangle \rangle$ and for every $1 \leq k \leq i$, it holds that

$$\kappa_i = n \Longrightarrow w'_i = w_i \wedge u'_i = u_i \wedge \sigma'_i = \alpha_i \tag{2.1}$$

$$\kappa_i = l \Longrightarrow w'_i \sigma'_1 = w_i \wedge u'_i = \alpha_i u_i \tag{2.2}$$

$$\kappa_i = r \Longrightarrow w'_i = w_i \alpha_i \wedge (\sigma'_i u'_i = u_i \vee (\sigma'_i = b \wedge u_i = \epsilon \wedge u'_i = \epsilon)) \tag{2.3}$$

In the sequel, we assume Σ to be fixed and that it contains a designated separator symbol ⌐, which is only used to separate test case inputs and outputs (see below).

A run of a Turing machine is a sequence of configurations starting from a given initial configuration c and proceeding in a sequence of next-state configurations $c \to c_1 \to c_2 \to \cdots$. If the sequence enters a configuration c_k without a next-state configuration, the computation *halts* and the computation's length is k steps; if the sequence is infinitely long, then the computation is *nonterminating*.

A k-tape Turing machine can be encoded using a given, finite alphabet containing more than one symbol in $\mathcal{O}(|Q||\Sigma|^k (\log Q + k \log |\Sigma|))$ space. □

A tape of a multitape Turing machine can be restricted to be an *input tape*, which means that the machine is not allowed to change its symbols. A tape can be also an *output tape*, denoting that the machine is not ever allowed to move left on that tape.

2.2.3 System models and tester models

A system model is represented by a three-tape Turing machine. One of the tapes is initialized at the beginning of the model's execution with the entire input provided to the system. The machine then reads the input on its own pace. Similarly, another tape is reserved for output and the machine writes, during execution, symbols that correspond to the system outputs. The third tape is for internal bookkeeping and general computations. This model effectively precludes *nondeterministic* system models because the input must be fixed before execution starts. This is intentional as introducing nondeterministic models (ones which could produce multiple different outputs on the same input) would complicate the exposition. However, the issue of nondeterministic models will be revisited in Section 2.6.

Definition 2 (System Model). A *system model* is a three-tape deterministic machine, with one input tape, one output tape, and one tape for internal state. One run of the system model corresponds to initializing the contents of the input tape, clearing the output tape and the internal tape, and then running the machine till it halts. At this point, the contents of the output tape correspond to the system's output. We assume that all system models eventually halt on all inputs.* □

The next definition fixes the model-based test coverage criteria considered within the rest of this section.

Definition 3 (Output Transition). Given a system model, an output transition in the system model is a $\langle q, \sigma_i, \sigma_o, \sigma \rangle$ iff $\delta(q, \sigma_i, \sigma_o, \sigma)$ is defined (the machine can take a step forward) and the transition moves the head on the output tape right (thus committing the machine to a new output symbol). A halting[†] run of a system models *covers* an output transition if a configuration matching an output transition shows up in the run. □

We define two complexity metrics for system models: their run-time complexity (how many steps the machine executes) and their testing complexity (how long an input is necessary to test any of the output transitions). Making a distinction between these two measures helps highlight the fact that even constructing short tests in general is difficult.

Definition 4 (Run-Time and Testing Complexity for System Models). Given a family of system models S, the family has

- *run-time complexity f* iff every system model $s \in S$ terminates in $\mathcal{O}(f(|i|))$ steps on any input i, and

- *testing complexity f* iff every output transition in s is either unreachable or can be covered by a run over an input i such that the length of i is $\mathcal{O}(f(|s|))$. □

Whereas a system model is a three-tape machine, a ester model has only two tapes: one for internal computations and one for outputting the test suite generated by the tester model traversal algorithm, which has been conveniently packed into the tester model itself. This does not cause any loss of generality because the algorithm itself can be considered to have a fixed size, so from a complexity point of view its effect is an additive constant and it thus vanishes in asymptotic considerations.

A tester model outputs a test suite in the form of a string $w_1 \lrcorner u_1 \lrcorner w_2 \lrcorner u_2 \cdots$, where w_i are inputs and u_i are expected outputs. Thus, a test is an input/output pair and a test suite is a finite collection of them. Note that this does not imply in any way that the analysis here is limited to systems that can accept only a single input because the input *string* in the current framework can denote a stream of multiple messages, for instance. The fact that the entire input can be precommitted is an effect of the determinism of system models (see Section 2.6 for further discussion).

Definition 5 (Tester Model). A *tester model* is a two-tape deterministic machine, with one output tape and one tape for internal state. The machine is deterministic and takes no input, so it always carries out the same computation. When the machine halts, the output tape is supposed to contain pairs of input and output words separated by the designated separator symbol $\lrcorner$. A tester model is valid with respect to a system model if it produces a sequence of tests that would pass against the system model. This means that if the model produces the output $w_1 \lrcorner u_1 \lrcorner \cdots \lrcorner w_n \lrcorner u_n \lrcorner$, then for every w_i that is given as an input to the system model in question, the system model produces the output u_i. The outputs are called *test suites*. □

---

*Assuming that system models halt on all inputs makes test construction for system models, which can be tested with bounded inputs a decidable problem. This assumption can be lifted without changing much of the content of the present section. It, however, helps strengthen Proposition 1 by showing that even if system models eventually terminate, test construction in general is still undecidable.

[†]Every run of a system model is assumed to halt.

The run-time complexity of a tester model is measured by how long it takes to produce the output given the output length.

Definition 6 (Run-Time Complexity for Tester Models). Given a family of tester models T, the family has run-time complexity f if any $t \in T$ executes $\mathcal{O}(f(|o|))$ steps before halting and having produced the machine-specific output string (test suite) o.   □

Finally, we consider tester construction strategies, that is, algorithms for converting system models into tester models. These could be also formulated as Turing machines, but for the sake of brevity, we simply present them as general algorithms.

Definition 7 (Tester Construction Strategy). A *tester construction strategy* for a family of system models S is a computable function $\mathfrak{S}$ that deterministically maps system models in S into valid and complete tester models, that is, tester models that generate tests that pass against the corresponding system models, and that cover all the reachable output transitions on those system models.   □

The run-time complexity of a tester construction strategy is measured in how long it takes to produce the tester model given the size of the input system model, as defined below.

Definition 8 (Run-Time Complexity for Tester Construction Strategies). Given a tester construction strategy $\mathfrak{S}$ for a family of system models S, the strategy has run-time complexity f if for any $s \in S$ the strategy computes the corresponding tester model in $\mathcal{O}(f(|s|))$ steps, where $|s|$ is a standard encoding of the system model.   □

This completes our framework. In this framework, tester model-based test generation is divided into two tasks: (1) construct a tester model machine and (2) execute it to produce the tests that appear on the output tape of the tester model machine. Step (1) is dictated by a chosen tester construction strategy and step (2) is then straightforward execution of a deterministic Turing machine.

Complexity-wise, the computational complexity of system models is measured by how fast the machines execute with respect to the length of the input strings. We introduced another complexity measure for system models also, namely their testing complexity—this measures the minimum size of any test suite that covers all the reachable output transitions of a given system model. Tester models are measured by how quickly they output their test suites, which makes sense because they do not receive any inputs in the first place. A tester construction strategy has a run-time complexity measure that measures how quickly the strategy can construct a tester model given an encoding of the system model as an input.

We shall now proceed to argue that *constructing tester models is difficult* from a basic computational complexity point of view.

2.2.4   Tester models are difficult to construct

2.2.4.1   General case

In general, tester models are impossible to construct. This is a consequence of the undecidability of the control state reachability problem for general Turing machines.

Proposition 1. *Tester construction strategies do not exist for all families of system models.*

Proof outline. This is a consequence of the celebrated Halting Theorem. We give an outline of a reduction to the result that the reachability of a given Turing machine's given halting control state is undecidable. Let m be a one-tape deterministic Turing machine with a halting control state q (among others). We translate m to a system model s by (1) adding a "timer" to the model, which causes the model to halt after n test execution steps, where n is a natural number presented in binary encoding on the input tape and (2) adding a control state q' and an entry in the transition function which, upon entering the control state q, writes symbol "1" on the output tape, moves right on the output tape, and then enters the control state q' that has no successors, that is, q' is a halting state. Now, clearly, this newly added output transition $(q \to q')$ can be covered by a tester model if and only if q is reachable in s given large enough n on the input tape. Note that the generated test suite would be either empty or "$b_1 \cdots b_k \lrcorner 1 \lrcorner$" depending on the reachability of q, where $b_1 \cdots b_k$ is a binary encoding of n. □

2.2.4.2 Models with bounded test complexity

We consider next the case of system models whose test complexity is bounded, that is, system models for which there is a limit (as function of the model size) on the length of the input necessary to test any of the output transitions.

Proposition 2. *Tester construction is decidable and* R*-hard* for families of system models with bounded test complexity.*

Proof outline. Let S be a family of system models whose test complexity is given by a f. For any $s \in S$, every reachable output transition can be reached with a test input whose length is thus bounded by $f(|s|)$. Because the system model is assumed to halt on all inputs, it can be simulated on any particular input. Thus, a tester model can be constructed by enumerating all inputs within the above bound on length, simulating them against the system model, and choosing a set that covers all the reachable output transitions. This shows that the problem is computable. To show that it is R-hard, it suffices to observe that in a system model s, the reachability of an output transition can still depend on *any* eventually halting computation, including those with arbitrary complexities with respect to their inputs, and the inputs themselves can be encoded within the description of s itself in polynomial space. □

2.2.4.3 Polynomially testable models

We will now turn our attention to "polynomially testable" models, that is, models that require only polynomially long test inputs (as function of model size) and which can be executed in polynomial time with respect to the length of those inputs.

Definition 9. A family of system models S is polynomially testable if there exist univariate polynomials $\mathscr{P}_1$ and $\mathscr{P}_2$ such that (1) the family has run-time complexity $\mathscr{P}_1$ and (2) the family has testing complexity $\mathscr{P}_2$. □

The next definition closes a loophole where a tester model could actually run a super-polynomial (exponential) algorithm to construct the test suite by forcing the test suite to be of exponential length. Since this is clearly not the intention of the present investigation, we will call tester construction strategies "efficient" if they do not produce tester models that construct exponentially too large test suites.

Definition 10. A tester construction strategy $\mathfrak{S}$ for a family S of system models is efficient if for every $s \in S$, the tester $\mathfrak{S}(s) = t$ produces, when executed, a test suite whose size is $\mathcal{O}(\mathscr{P}(f(|s|)))$ where f is the testing complexity of the family S. □

---

*This means that the problem is as hard as *any* computational problem that is still decidable. R stands for the complexity class of recursive functions.

Given a family of polynomially testable system models, it is possible to fix a polynomial $\mathscr{P}$ such that every system model s has a test suite that tests all its reachable output transitions and can be executed in $\mathscr{P}(|s|)$ steps. This is not a new definition but follows logically from the definitions above and the reader can verify this.

The following proposition now demonstrates that constructing efficient testers for polynomially testable models is NP-complete. This means that it is hard, but some readers might wonder if it actually looks suspiciously simple as NP-complete problems are still relatively easy when compared for example to PSPACE-complete ones, and it was argued in the introduction that for example the reachability problem for synchronized state machines is already a PSPACE-complete problem. But there is no contradiction—the reason why tester construction appears relatively easy here is that the restriction that the system models must execute in polynomial time is a complexity-wise severe one and basically excludes system models for computationally heavy algorithms.

Proposition 3. *Efficient tester construction is NP-complete for families of polynomially testable system models, that is, unless* P = NP, *efficient tester construction strategies for polynomially testable system models cannot in general have polynomial run-time complexity bounds.*

Proof outline. We give an outline of reductions in both directions. First, to show NP-hardness, consider a family of system models, each encoding a specific Boolean satisfaction problem (SAT) instance. If the length of an encoded SAT instance is ℓ, the corresponding system model can have ℓ control states that proceed in succession when the machine starts and write the encoding one symbol at a time on the internal tape. Then, the machine enters in a general portion that reads an assignment of Boolean values to the variables of the SAT instance from the input tape and outputs either "0" or "1" on the output tape depending on whether the given assignment satisfies the SAT instance or not. It is easy to see that the system model can be encoded in a time polynomial in ℓ, that is, $\mathcal{O}(\mathscr{P}(\ell))$ for a fixed polynomial $\mathscr{P}$. If there would exist a polynomial-time tester construction strategy $\mathfrak{S}$ for this family of models, SAT could be solved in polynomial time by (1) constructing the system model s as above in $\mathcal{O}(\mathscr{P}(\ell))$ time, (2) running $\mathfrak{S}(s)$ producing a tester model t in time polynomial in $|s|$ as it must be that $|s| = \mathcal{O}(\mathscr{P}(\ell))$ also, and (3) running t still in polynomial time (because $\mathfrak{S}$ is efficient) and producing a test suite that contains a test case to cover the output transition "1" iff the SAT instance was satisfiable.

To show that the problem is in NP, first note that because the output transitions of the system model can be easily identified, the problem can be presented in a form where every output transition is considered separately—and the number of output transitions in the system model is clearly polynomial in the size of the machine's encoding. Now, for every single output transition, a nondeterministic algorithm can first guess a test input that covers it and then simulate it against the system model in order to verify that it covers the output transition in question. Because the system model is polynomially testable, this is possible. The individual tester models thus constructed for individual output transitions can be then chained together to form a single tester model that covers all the reachable output transitions.   □

2.2.4.4   System models with little internal state

In this section, we consider system models with severely limited internal storage, that is, machines whose internal read/write tape has only bounded capacity. This corresponds to system models that are explicitly represented finite-state machines.

Proposition 4. *Polynomial-time tester construction strategies exist for all families of system models with a fixed bound B on the length of the internal tape.*

Proof outline. Choose any such family. Every system model in the family can have at most $|Q||\Sigma|^B$ internal states and the factor $|\Sigma|^B$ is obviously constant in the family. Because $|Q| = \mathcal{O}(|s|)$ for every system model s, it follows that the complete reachability graph for the system model s can be calculated in $\mathcal{O}(\mathscr{P}(|s|))$ time for a fixed polynomial $\mathscr{P}$. The total length of the test suite required to test the output transitions in the reachability graph is obviously proportional to the size of the graph, showing that the tester can be constructed in polynomial time with respect to $|s|$. $\qquad\square$

2.2.5 Discussion

In general, constructing tester models is an undecidable problem (Proposition 1). This means that in general, it is impossible to create tester models from system models if there is a requirement that the tester models must actually be able to test all reachable parts of the corresponding system models.

If there is a known bound on the length of required tests, tester model construction is decidable (because system models are assumed to halt on all inputs) but of unbounded complexity (Proposition 2). This shows that tester construction is substantially difficult, even when one must not search for arbitrarily large test inputs. The reason is that system models can be arbitrarily complex internally even if they do not consume long input strings.

In the case of system models, which can be tested with polynomial-size test inputs and can be simulated efficiently, tester construction is an NP-complete problem (Proposition 3). This shows that even when there are strict bounds on both a system model's internal complexity as well as the complexity of required test inputs, tester model construction is hard.

In the case of explicitly represented finite-state machines, tester models can be constructed in polynomial time, that is, efficiently (Proposition 4), demonstrating the reason why the dichotomy that is the subject of this chapter does not surface in research that focuses on explicitly represented finite-state machines.

Thus, it has been demonstrated that *constructing tester models is hard*. Assuming that producing system models is a translation problem, it can be concluded that computer-readable tester models are harder to construct than computer-readable system models.

We will now move on to examine the practical approaches to system model- and tester model-driven test generation. One of our goals is to show that the theoretically predicted results can be actually empirically verified in today's practice.

2.3 Practical Approaches

In this section, we survey some of the currently available or historical approaches for model-based testing with emphasis on the system model versus tester model question. Not all available tools or methodologies are included as we have only chosen a few representatives. In addition, general academic work on test generation from finite-state models or other less expressive formalisms is not included.

2.3.1 System-model-driven approaches

At the time of writing this chapter, there are three major system-model-driven test generation tools available on the market.

2.3.1.1   Conformiq Designer

Conformiq Designer is a tool developed by Conformiq that generates executable and human-readable test cases from behavioral system models. The tool is focused on functional black-box testing. The modeling language employed in the tool consists of UML statecharts and Java-compatible action notation. All usual Java constructs are available, including arbitrary data structures and classes as well as model-level multithreading (Huima 2007, Conformiq 2009a).

The tool uses internally constraint solving and symbolic state exploration as the basic means for test generation. Given the computational complexity of this approach, the tool encounters algorithmic scalability issues with complex models. As a partial solution to this problem, Conformiq published a parallelized and distributed variant of the product in 2009 (Conformiq 2009b).

2.3.1.2   Smartesting Test Designer

Smartesting Test Designer generates test cases from behavioral system models. The tool uses UML statecharts, UML class diagrams, and OCL-based action notation as its modeling language. In the Smartesting tool, the user must enter some of the test data in external spreadsheets instead of getting the data automatically from the model (Fredriksson 2009), but the approach is still system model driven (Smartesting 2009).

The tool uses internally constraint solving and symbolic state exploration as the basic means for test generation.

2.3.1.3   Microsoft SpecExplorer

Microsoft SpecExplorer is a tool for generating test cases from system models expressed in two possible languages Spec# (Barnett et al. 2005) and the Abstract State Machine Language (AsmL) (Gurevich, Rossman, and Schulte et al. 2005). Spec# is an extended variant of C#. Developed originally by Microsoft Research, the tool has been used to carry out model-based testing of Windows-related protocols inside Microsoft.

The tool works based on system models but avoids part of the computational complexity of test case generation by a methodology the vendor has named "slicing." In practice, this means reducing the system model's input data domains into finite domains so that a "slice" of the system model's explicitly represented state space can be fully calculated. The approach alleviates the computational complexity but puts more burden on the designer of the system model. In practice, the "slices" represent the users' views regarding the proper testing strategies and are also educated guesses. In that sense, the present SpecExplorer approach should be considered a hybrid between system-model- and tester-model-driven approaches (Veanes et al. 2008, Microsoft Research 2009).

2.3.2   Tester-model-driven approaches

As predicted in the discussion above, there are more usable tester-model-driven tools available on the market than there are usable system-model-driven tools. In this section, we mention just some of the presently available or historical tester-model-driven tools and approaches.

2.3.2.1   UML Testing Profile

The UML Testing Profile 1.0 (UTP) defines a "language for designing, visualizing, specifying, analyzing, constructing and documenting the artifacts of test systems. It is a test modeling language that can be used with all major object and component technologies and applied to testing systems in various application domains. The UML Testing Profile can be used stand alone for the handling of test artifacts or in an integrated manner with UML for a handling of system and test artifacts together" (Object Management Group 2005).

As in itself, UTP is not a tool nor even exactly a methodology but mostly a language. However, methodologies around UTP have been defined later, such as by Baker et al. (2007). For example, UML message diagrams can be used to model testing scenarios and statecharts to model testing behaviors. There exist also tools, for example, for converting UTP models into TTCN-3 code templates.

2.3.2.2   ModelJUnit

ModelJUnit is a model-based test generation tool that generates tests by traversal of extended finite-state machines expressed in Java code. It is clearly a tester-model-driven approach given that the user must to implement, as part of model creation, methods that will "include code to call the methods of your (system under test) SUT, check their return value, and check the status of the SUT" (Utting et al. 2009). The tool itself is basically a path traversal engine for finite-state machines described in Java (Utting and Legeard 2006).

2.3.2.3   TestMaster

The Teradyne Corporation produced a model-based testing tool called TestMaster, but the product has been since discontinued, partially because of company acquisitions, which orphanized the TestMaster product to some extent. The TestMaster concept was based on creating finite-state machine models augmented with handcrafted test inputs as well as manually designed test output validation commands. The tool then basically generated different types of path traversals through the finite-state backbone and collected the input and output commands from those paths into test scripts. Despite of being nowadays discontinued, the product enjoyed some success at least in the telecommunications domain.

2.3.2.4   Conformiq Test Generator

Conformiq Test Generator is a historical tool from Conformiq that is no longer sold. On a high level, it was similar to TestMaster in concept, but it focused on online testing instead of test script generation. Also, the modeling language was UML statecharts with a proprietary action notation instead of the TestMaster's proprietary state machine notation. Conformiq Test Generator was also adopted by a limited number of companies before it was discontinued.

2.3.2.5   MaTeLo

MaTeLo is a tool for designing testing strategies and then generating test cases using a statistical, Markov-chain related approach (Dulz and Zhen 2003). This approach is often called statistical use case modeling and is a popular method for testing, for instance, user interfaces.

2.3.2.6   Time Partition Testing

Time Partition Testing (TPT) is a method as well as a tool from Piketec. It combines test case or test strategy modeling with combinatorial generation of test case variation based on the testing strategy model. In addition to the test inputs, the user also implements the desired output validation criteria (Bringmann and Krämer 2006). TPT has special focus on the automotive industry as well as continuous-signal control systems.

2.3.3   Comparison

The approaches based on a tester model that are presented above are all fundamentally based on path traversals of finite-state machines, even though in the MaTeLo approach the traversals can be statistically weighted and the TPT system adds finite, user-defined combinatorial control for path selection. In all these approaches, the user must define output validation actions manually because no system model exists that could be simulated with the generated test inputs to produce the expected outputs. The path generation procedures are well understood, relatively simple to implement, and of relatively low practical complexity.

The system-model-driven approaches, on the other hand, are all, based on reports from their respective authors, founded on systematic exploration of the system model's state space, and they aim at generating both the test inputs as well as the expected test outputs automatically by this exploration procedure. For the modeling formalisms available in Conformiq Designer, Smartesting Test Designer, and SpecExplorer, the reachability problem and henceforth test generation problem is undecidable. A finite-state machine path traversal cannot be applied in this context because every simulated input to the system model can cause an infinite or at least hugely wide branch in the state space, and it can be difficult to derive *a priori* upper bounds on the lengths of required test inputs (Huima 2007).

The tester-model-driven tools have two distinguishing benefits: (1) people working in testing understand the concept of a (finite-state) scenario model easily and (2) the tools can be relatively robust and efficient because the underlying algorithmic problems are easy. Both these benefits correspond to handicaps of the system-model-driven approach as (1) test engineers can feel alienated toward the idea of modeling the system directly and (2) operating the tools can require extra skills to contain the computational complexity of the underlying procedure.

At the same time, the system-model-driven test generation approach has its own unique benefits, as creating system models is straightforward and less error prone than constructing the corresponding tester models because the mental steps involved in actually designing testing strategies are simply omitted.

2.4   Compositionality

An important practical issue which has not been covered above is that of *compositionality* of models. Consider Figure 2.10, which shows two subsystems A and B that can be composed together to form the composite system C. In the composite system, the two subsystems are connected together via the pair of complementary interfaces b and b'. The external interfaces

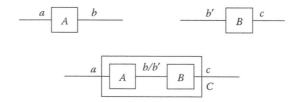

FIGURE 2.10
System composition.

of C are then a and c, that is, those interfaces of A and B that have not been connected together and are thus hidden from direct observation and control.

It is intuitively clear that a system model for A and system model for B should be able to be easily connected together to form the system model for the composite system C, and this has been also observed in practice. This leads to a compositional approach for model-based testing where the same models can be used in principle to generate component, function, system, and end-to-end tests.

When system models are composed together basically all the parts of the component models play a functional role in the composite model. However, in the case of test models, it is clear that those parts of the models that are responsible for verifying correct outputs on interfaces hidden in the composite system are redundant; removing them from the composite model would not change how tests can be generated. This important observation hints toward the fact that testing strategy models are not in practice compositional—a practical issue that will be elaborated next.

The two key problems are that (1) tester models do not predict system outputs fully, but usually implement predicates that check only for certain properties or parts of the system outputs, and that (2) tester models typically admit only a subset of all the possible input sequences.

To consider the issue (1) first, suppose that Figures 2.11 and 2.12 represent some concrete testing patterns for the components A and B, respectively.

Assuming $M_2 = P_1$ and $M_3 = P_4$, it appears that these two patterns can be composed to form the pattern shown in Figure 2.13.

For system models, it is easy to describe how this kind of composition is achieved. Consider first A as an isolated component. A system-model-driven test generator first "guesses" inputs M_1 and M_3 (in practice using state-space exploration or some other method) and can then simulate the system model with the selected inputs in order to obtain the predicted outputs M_2 and M_4. Now, the system model for A is an executable description of how the component operates on behavioral level, so it is straightforward to connect it with

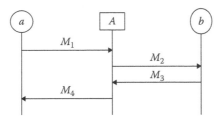

FIGURE 2.11
A testing pattern.

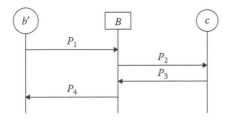

FIGURE 2.12
A testing pattern.

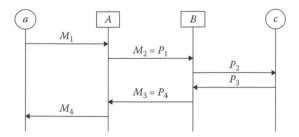

FIGURE 2.13
A composed testing pattern.

the system model for B in the same way as two Java classes, say, can be linked together by cross-referencing to create a compound system without any extra trickery.

Now, the system model generator first "guesses" inputs M_1 and P_3 using exactly the same mechanism (the integrated model containing both A and B is simply another system model). Then, the integrated model is simulated to calculate the expected outputs. First, the chosen value for M_1 is sent to that part of the integrated model which corresponds to the component model A. Then, this component sends out the predicted value for M_2. It is not recorded as an expected output but instead sent as an input to the component model for B; at this point, the output M_2 becomes the input P_1. Then, the simulated output from component B with input P_1 becomes the expected value for the output P_2. The output value M_4 is calculated analogously. It is clear that this is a compositional and modular approach and does not require extra modeling work.

For tester models, the situation is less straightforward. Consider, for the sake of argument, an extended finite-state model where every admissible (fulfilling path conditions) path through the state machine generates a test case. Now, some such paths through a tester model for the component A generate test cases for the scenario above. Instead of being a sequence of input and output messages, it is actually a sequence of (1) input messages and (2) predicates that check the corresponding outputs. Thus, a particular test case corresponding to the component A scenario looks like

$$!m_1, ?\phi_2, !m_3, ?\phi_4, \qquad (2.4)$$

where m_1 is a value for M_1 and m_3 is value for M_3, but ϕ_2 and ϕ_4 are user-defined predicates to check the correctness of the actual outputs corresponding to M_2 and M_4.

One problem now is that ϕ_2 in practice returns *true* for multiple concrete outputs. This is how people in practice create tester models because they want to avoid the labor of checking for details in observed outputs, which actually do not contribute toward testing the present testing purposes. Namely, suppose now that we also have a tester model for the component B, and similarly we can generate test sequences consisting of inputs and validation predicates from that model. Let such a sequence be denoted

$$!p_1, ?\psi_2, !p_3, ?\psi_4 \qquad (2.5)$$

analogously to what was presented above. Is it now true that if $\phi_2(p_1)$ and $\psi_4(m_3)$ evaluate to *true* the sequence

$$!m_1, ?\psi_2, !p_3, ?\phi_4$$

is guaranteed to be a valid test case for the compound system? The answer is no! Consider the first output check ψ_2 in the compound system test case. It checks the actual value of the output P_2 produced by the component B. This value in general depends on the input P_1, which must have the value p_1 from (2.5) in order for ψ_2 to return *true* because the predicate ψ_2 was derived assuming that input in the first place. However, there is no guarantee that the component A will produce output p_2 when given the input m_1 as the only guaranteed assertion is that the output after m_1 fulfills ϕ_2, while ϕ_2 and p_1 do not necessarily have any relationship to each other as they come from two independent and separate models.

This discussion has so far focused on that compositionality breaks for tester models that do not predict system outputs fully. The other problem mentioned previously was that tester models do not usually admit all required input sequences. Continuing the present example, it is plausible that for a particular test sequence (2.5), there are no test sequences (2.4) that could be generated from the tester model for component A such that the output triggered by m_1 would actually match to p_1—a necessary condition for an integrated test case to exist. The reason why this is possible is that a tester model can be very far from admitting all possible input sequences. As a matter of fact, the entire idea of use case modeling which is very near to tester modeling is to focus the models on a set of representative, interesting input sequences. Now when the component models for A and B are created possibly independently, there are no guarantees that the outputs from the model A would match the inputs generated from the model B. But to even get to this point would first require the tester models to generate the full expected outputs in order to avoid the problem with partially specified outputs, which was highlighted first.

It can be hence concluded that the current tester model-based tooling and methodology leads to noncompositionality of tester models. We conjecture that fully compositional tester models are indistinguishable from the corresponding system models because it seems that it is the capability to predict system outputs that makes models compositional in the usual sense of component-based software compositionality.

On the other hand, the compositionality of system models may also be of limited value because testing subsystems (such as A and B) separately based on their models leads to stronger testing than testing an integrated system (such as the composite formed of A and B) based on the integrated model. The reason is that in the integrated system, it can be impossible to trigger, for example, error conditions around the internal, hidden interfaces.

2.5 Scalability

The underlying problem of deriving test cases from a (mental) system model is difficult, so both the system model as well as the tester model approaches must suffer from scalability challenges. This is a natural consequence of the material presented above. The scalability challenges, however, differ. The main scalability challenge for tester-model-driven test generation is human operators' ability to construct good tester models for increasingly complex systems under test. For system-model-driven test generation the major scalability challenge is the algorithmic infeasibility of deriving a comprehensive test suite from increasingly complex system models. This leads to the situation depicted in Figure 2.14. For the pure system model-driven paradigm, the main scalability problem is algorithmic complexity. The complexity grows when the system under test becomes more complex from a testing perspective (leftmost dashed arrow). For the tester-model-driven approach, the main scalability issue is the cognitive difficulty of producing and maintaining good tester models (rightmost dashed arrow). Some solutions, SpecExplorer for instance, provide a hybrid methodology aiming to strike a balance between the two ends of the spectrum (dotted line in the figure).

This leads to two practical, context-dependent questions: (1) should one embrace the system-model-driven or the tester-model-driven approach and (2) how these two methodologies can be predicted to evolve in the future?

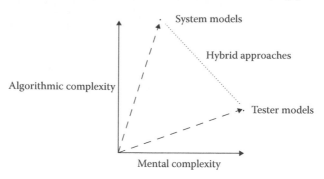

FIGURE 2.14
Scalability issues.

The best answer to (1) is that tools and methodologies that work best should be embraced. The system-model-driven approach is theoretically attractive, compositional, and requires less manual work than the tester-model-driven approach, but it can fail because of inadequate tooling. Given our current affiliation with Conformiq Inc., we can be open about the fact that *none* of the currently available system-model-driven test generation tools are known to scale to complex models without challenges, even though the exact nature of those challenges are tool specific. In contrast, the tester-model-driven approach is easily supported by robust (both free and commercial) tools, but it still leaves the design of testing strategies to the user, and thus provides less upside for productivity improvement. In some contexts, the present processes may enforce for example a use case centric test design methodology, which may make certain types of tester-model-driven tools attractive.

To answer (2), observe that given that the main challenge for system-model-driven test generation is of algorithmic complexity, it can be predicted that this approach will gain in capacity and popularity in the future. It will follow the same trajectory as, for instance, programming languages and hardware circuit design methods have followed in the past. When high-level programming language compilers came, they eventually replaced hand-written object code. Automatic circuit layout replaced manual layout, and also since 1990s digital systems are verified not by hand but by computerized methods based on recent advances in algorithmic problems such as Boolean satisfiability. There is no fundamental reason to believe that the same transition would not take place in due time around the algorithmic challenges of system-model-driven test generation.

2.6   Nondeterministic Models

The complexity-theoretic analysis above excludes nondeterministic system models, that is, system models whose behavior is not completely determined by their inputs.

Most of the offline test generation tools, that is, tools which generate executable test scripts, do not support nondeterministic systems because the test scripts are usually linear. However, SpecExplorer supports nondeterministic system models even though the slicing requirement forces the nondeterminism on the system's side to cause only finite and in practice relatively narrow branching in the state space. SpecExplorer exports the computed testing strategy as a finite-state model, and it can be then executed by a test execution subsystem that supports branching based on the system under test's variable responses.

Conformiq Designer originally supported online testing of nondeterministic systems based on nondeterministic system models, but the support was removed later. Similarly, Conformiq Test Generator, a tester-model-driven tool, was capable of running online tests against a nondeterministic system.

The two main reasons why nondeterministic systems were excluded above are that (1) it is complicated to define what is a "complete" test suite for a nondeterministic system because the actual model-based test coverage can be calculated only during test execution and depends on how the system under test implements its nondeterministic choices; and (2) in practice, it is a recognized principle that a system should be as deterministic as possible in order to admit good testing. This second point is certainly a methodological one and is related more to practice than theory.

That being said, the conclusions presented previously in this chapter can be transferred to the case of nondeterministic systems also: tester models are difficult to construct also for nondeterministic systems (even more so), and generating tester models from system models of nondeterministic systems is computationally difficult (even more so). So the main arguments stand.

2.7 Conclusions

The tester-model-driven test generation approach is well established in the industry and has been adopted in its different forms by a multitude of engineering teams. The system-model-driven solution in its modern form was developed in the early 2000s by several, independent research teams and is currently, at the time of writing this chapter, in its early adoption phase.

Ultimately, the system model versus tester model dichotomy is about the computation platform's capability to deliver an algorithmically complex result (system-model-driven test generation) versus organizations' capability and desire to spend human labor to carry out the same task by hand. Thus, it is a choice between either an industrial, mechanical solution

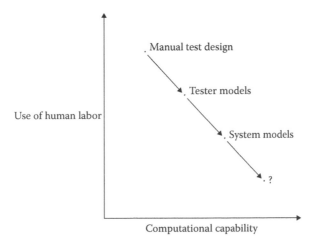

FIGURE 2.15
Industrialization of test design.

or a solution using human labor. To put it succintly, the choice is a matter of industrialization.

Thus, given the history of industrialization in general as well as in the area of software engineering, it can be safely and confidently predicted that as science continues its perpetual onward march, the system-model-driven test generation approach will ultimately become dominant over the tester-model-driven solution. This would be a logical conclusion of the presented facts and illustrated by Figure 2.15.

The timescale of that transition is, however, yet veiled from our eyes.

References

Aho, A. V., Dahbura, A. T., Lee, D., and Uyar, M. Ü. (1991). An optimization technique for protocol conformance test generation based on UIO sequences and rural chinese postman tours. *IEEE transactions on communications*, 39(11):1604–1615.

Baker, P., Dai, Z. R., Grabowski, J., Haugen, Ø., Schieferdecker, I., and Williams, C. (2007). *Model-Driven Testing Using the UML Testing Profile*. Springer.

Barnett, M., Rustan,K., Leino, M., and Schulte, W. (2005). The Spec# programming system: An overview. In *Construction and Analysis of Safe, Secure, and Interoperable Smart Devices*, Lecture Notes in Computer Science, Pages: 49–69.

Bland, R. G., Goldfarb, D., and Todd, M. J. (1981). Ellipsoid method: A survey. *Operations Research*, 29(6):1039–1091.

Bringmann, E., and Krämer, A. (2006). Systematic testing of the continuous behavior of automotive systems. In *International Conference on Software Engineering*, Pages: 13–20. ACM.

Conformiq. (2009a). http://www.conformiq.com/.

Conformiq. (2009b). http://www.conformiq.com/news.php?tag=qtronic-hpc-release.

Demri, S., Laroussinie, F., and Schnoebelen, P. (2006). A parametric analysis of the state-explosion problem in model checking. *Journal of Computer and System Sciences*, 72 (4):547–575.

Dulz, W., and Zhen, F. (2003). MaTeLo—statistical usage testing by annotated sequence diagrams, Markov chains and TTCN-3. In *Third International Conference On Quality Software*, pages 336–342. IEEE.

Duran, J. W., and Ntafos, S. C. (1984). Evaluation of random testing. *IEEE Transactions on Software Engineering*, SE–10(4):438–444.

Edmonds, J., and Johnson, E. L. (1973). Matching, Euler tours and the Chinese postman. *Mathematical Programming*, 5(1):88–124.

Fredriksson, H. (2009). Experiences from using model based testing in general and with Qtronic in particular. In Fritzson, P., Krus, P., and Sandahl, K., editors, *3rd MODPROD Workshop on Model-Based Product Development*. See http://www.modprod.liu.se/workshop_2009.

Ghosh, S., and Mathur, A. P. (1999). Issues in testing distributed component-based systems. In *First International ICSE Workshop on Testing Distributed Component-Based Systems*.

Gurevich, Y., Rossman, B., and Schulte, W. (2005). Semantic essence of AsmL. *Theoretical Computer Science*, 343:370–412.

Huima A. (2007).Implementing Conformiq Qtronic. In Alexandre Petrenko et al., editor, *Testing of Software and Communicating Systems*, number 4581/2007 in LNCS, Pages: 1–12. Springer: Berlin/Heidelberg.

Karp, R. M. (1972). Reducibility among combinatorial problems. In Miller, R. E., and Thatcher, J., editors, *Complexity of Computer Computations*, Pages: 85–103. Plenum.

Matiyasevich, Y. (1993). *Hilbert's 10th Problem*. The MIT Press.

Microsoft Research. (2009). http://research.microsoft.com/en-us/projects/specexplorer/.

Nelder, J. A., and Mead, R. (1965). A simplex method for function minimization. *The Computer Journal*, 7(4):308–313.

Object Management Group. (2005). UML testing profile 1.0. Published standard.

Papadimitriou, C. H. (1981). On the complexity of integer programming. *Journal of the ACM*, 28(4):765–768.

Papadimitriou, C. H. (1993). *Computational Complexity*. Addison Wesley.

Pretschner, A. (2005). Model-based testing in practice. In *Proc. Formal Methods 2005*, number 3582 in LNCS, Pages: 537–541. Springer.

Smartesting. (2009). http://www.smartesting.com/.

Utting, M., and Legeard, B. (2006). *Practical Model-Based Testing: A Tools Approach*. Morgan Kauffman.

Utting, M., Perrone, G., Winchester, J., Thompson, S., Yang, R., and Douangsavanh, P. (2009). http://www.cs.waikato.ac.nz/ marku/mbt/modeljunit/.

Veanes, M., Campbell, C., Grieskamp, W., Schulte, W., Tillmann, N., and Nachmanson L. (2008). Model-based testing of object-oriented reactive systems with Spec Explorer. In Hierons R. M. et al., editors, *Formal Methods and Testing*, number 4949/2007 in LNCS, Pages: 39–76.

3

Test Framework Architectures for Model-Based Embedded System Testing

Stephen P. Masticola and Michael Gall

CONTENTS

3.1 Introduction

Model-based testing (MBT) (Dias Neto et al. 2007) refers to the use of models to generate tests of components or entire systems. There are two distinct kinds of MBT, which we will call here *behavior based* (Utting and Legeard 2006) and *use based*, (Hartmann et al. 2005) depending on whether we are modeling, respectively, the system under test (SUT) itself or the use cases which the SUT is intended to support. Regardless of the modeling technique used, once the tests are generated from the model, they must be executed. If possible, automated execution is preferable to reduce cost and human error. Automation usually requires some sort of a test harness around the SUT. This is especially true for embedded systems.

In designing test harnesses, we are concerned not with the modeling itself, but with executing the tests that are generated from the models. Test engineers will frequently also want to write some tests manually, as well as generate them from models.

Test harnesses for embedded systems are employed in both production testing to identify manufacturing defects and engineering testing to identify design defects. Modern test harnesses for either type of application are almost universally controlled by software. Software control allows the test harness to exercise the SUT thoroughly and repeatably.

It is useful to divide test harness software into two categories: *test scripts*, which specify how the SUT is to be exercised, and *test framework*, which runs the scripts and performs other "housekeeping" functions such as logging test results. Test scripts are almost always written by the test team that must test the SUT. The test framework is usually some combination of commercial and purpose-built software. Figure 3.1 shows the test framework and the types of script creation that must be supported in MBT. To some degree, the test framework is typically customized or custom designed for the SUT.

3.1.1 Purpose and structure of this chapter

This chapter describes a reference architecture for a test framework for embedded systems. The frameworks we describe are model based, in the sense that the systems under test are explicitly modeled in specialized scripting languages. This naturally supports a use-based MBT context.

The remainder of Section 3.1 presents several common quality goals in software test frameworks that control test harnesses for testing embedded and mechatronic systems and testability "antipatterns" in the systems under test that such frameworks must support. The rest of this chapter describes a practical way in which a test system architect can meet

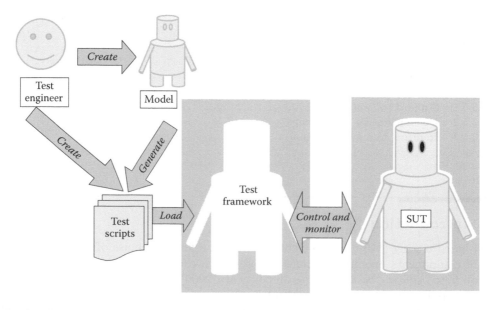

FIGURE 3.1
Test generation and execution process.

these quality goals in a test framework, and some methods to work around the antipatterns in the SUT.

Section 3.2 describes the preliminary activities of the test system architect, such as gathering requirements for the test framework, evaluating existing infrastructure for reuse, selecting a test executive, modeling the SUT, and prototyping the architecture. From this, we show in Section 3.3 how to architect the test harness. In particular, we present this in detail as a specialization of a reference architecture for the test framework. Section 3.4 then presents a brief description of an implementation of the reference architecture more fully described in Masticola and Subramanyan (2009). Finally, Section 3.5 reviews supporting activities, such as iterating to a satisfactory architecture, planning for support software, and creating and presenting documentation and training materials.

Throughout this chapter, we will note "important points" and "hints." The important points are facts of life in the sort of MBT automation frameworks that we describe. The hints are methods we have found useful for solving specific problems.

3.1.2   Quality attributes of a system test framework

The quality of the test framework's architecture has a major impact on whether an automated testing project succeeds or fails. Some typical quality attributes of a test framework architecture include:

- *Ease of script development and maintenance,* or in other words, the cost and time required to write a test suite for the SUT. This is a concern for manually developed scripts, but not for scripts automatically generated from a model. If scripts are manually written, then they must be as easy as possible to write.

- *Ease of specialization to the SUT*, or the cost and time to adapt the test framework to the SUT. This is especially a concern if multiple systems have to be tested by the same test group.

- *Endurance.* This is necessary to support longevity or soak testing to make sure that the performance of the SUT remains acceptable when it is run for long periods of time. The test framework thus must be able to run for such long periods, preferably unattended.

- *Scalability.* If the SUT can scale, then the test framework must support a corresponding scaling of the test harness.

- *Ability to interchange and interoperate simulators and field hardware.* Simply modifying the test framework may not be enough. For best productivity in test development, the test framework should support such interchange without modifying the test scripts. Easy integration with simulation environments is a necessary precondition.

- *Support for diagnosis.* The test framework must support engineers in diagnosing failures from logged data in long-running tests, without sacrificing ease of script development. These failures include both failures in the SUT and failures in the test harness. Diagnosis support from logs is particularly important in scenarios where the tests are run unattended.

- *Timing accuracy.* The requirements for time measurement accuracy for a specific SUT sometimes contain subtleties. For example, it may be necessary to measure reaction time to only 100 ms, but synchronization to 5 ms. Monitoring overhead and jitter should be kept as low as practicable.

- *MBT environment support.* Easy integration to MBT environments, especially to test generation facilities, enhances the value of both the test framework and the MBT tools.

- *Flexible test execution.* Support for graphical user interface (GUI), manual test execution, and automated test execution, may be needed, depending on the SUT.

3.1.3   Testability antipatterns in embedded systems

Embedded systems have certain recurring particular problems. We cannot always avoid these "antipatterns" (Brown et al. 1998). Instead, the best we can do is live with them. We list some testability antipatterns and workarounds for them here.

The SUT does not have a test interface. Designing, programming, and maintaining a test interface in an embedded system cost money and other resources that are often in short supply. When budgets must be tightened, test interfaces are often one of the features that are considered disposable.

In this case, we are forced to either modify the SUT or, somehow, mimic its human users via automation. At the very worst, we can try to use robotics technology to literally push the buttons and watch the screen. Sometimes, though, the situation is better than that, and we can use extensibility mechanisms (such as buses for plug-in modules or USB interfaces) that already exist in the SUT to implement our own test interface.

The SUT has a test interface, but it does not work well enough. Controllability and runtime state sensing of the SUT may not be sufficient to automate the tests you need, or may not be set up in the correct places in the system. Alternatively, the communication format might be incorrect—we have seen systems in which it was possible to subscribe to state-update messages, but where we could never be sure that we had ever obtained a baseline state to update. We know of no elegant solution to this antipattern short of the redesign of the SUT's test interface.

Modifying the SUT for test is not possible. In the absence of a reasonable test interface, it is tempting to try to solder wires onto the SUT, or otherwise modify it so that tests can be automated. There are a variety of reasons why this may not be possible. The technology of the SUT may not permit this, or the SUT may be an expensive or one-of-a-kind system such that the stakeholders may resist. The stakeholders might resist such modifications for fear of permanent damage.

Regardless of the reason, we are forced to test the SUT in a way that does not modify it, at least not permanently. Stakeholders may often be willing to accept temporary modifications to the SUT.

The SUT goes into an unknown state, and the test framework must recover. Here, we are trying to run a large test suite and the SUT becomes unstable. In high-stakes and agile projects, we cannot afford to let the SUT remain idle. We have to force it to reset and resume testing it.

Working around this testability antipattern often requires extra support in the hardware and software of the test harness. Putting this support into place can, with good fortune, be done without running afoul of any other antipattern. In addition, before we initiate recovery, we must ensure that all diagnostic data are safely recorded.

3.2   Preliminary Activities

Before we architect a test framework for the SUT, we must conduct some preliminary information-gathering and creative activities. This is true regardless of whether we are custom creating a test framework, specializing one from a software product line, or, indeed, creating the product line.

3.2.1 Requirements gathering

The first activity is to gather the requirements for testing the SUT (Berenbach et al. 2009). These testing requirements are different from the functional and quality requirements for the SUT itself and often are not well developed before they are needed. However, the testing requirements will always support verification of the SUT requirements.

Some examples of the testing requirements that must be gathered are the following:

- *The SUT's external interfaces.* These interfaces between the SUT and its environment drive some aspects of the low-level technology selection for the test framework.

- *The scenarios that must be tested.* Frequently, project requirements documents for the SUT will list and prioritize operational scenarios. These can be used as a starting point. Be aware, though, that not all scenarios can necessarily be found in the project documents. One such scenario is interchanging components with different firmware versions to test firmware compatibility.

  Important Point: When you capture these scenarios, keep track of the major components of the SUT with which the tester directly interacts and the actions that he or she performs to test the system.

  These will become the "nouns" (i.e., the grammatical subjects and objects) and "verbs" (i.e., actions) of the *domain model* of the SUT, which is a model of the SUT's usage in, and interaction with, the domain for which it is intended. If you are capturing the scenarios as UML sequence diagrams (Fowler 2003), then the tester-visible devices of the SUT and the test harness are represented as the business objects in the sequence diagrams, and the actions are the messages between those business objects. The classes of the business objects are also usually captured, and they correspond to device classes in the test framework, as described in Section 3.3.3. As we will see in Section 3.2.4, these scenarios are crucial in forming the test model of the SUT.

- *Performance requirements of the test harness.* These derive from the performance requirements of the SUT. For example, suppose the SUT is an intelligent traffic system. If it is supposed to scale to support 1000 traffic sensors, then the test harness must be capable of driving 1000 real or simulated traffic sensors at their worst-case data rates. If the SUT is a hard real-time or mechatronic system (Broekman and Notenboom 2003), then the time-measuring accuracy of the test harness will be important. These performance requirements will drive such basic decisions as whether to distribute control among multiple test computers.

- *Test execution speed.* This requirement usually stems from the project parameters. If, for example, you are using the test harness to run smoke tests on the SUT for an overnight build and you must run 1000 test cases, then you will be in trouble if each test case takes any longer than about 30 s to run, including all setup and logging.

- *Quiescent state support.* The SUT may, sometimes, have one or more idle states when it is running but there is no activity. We call such a state a *quiescent* state. It is very useful, at the start and end of a test case, to verify that the system is in a quiescent state. If a quiescent state exists, you will almost certainly want to support it in your test framework.

- *Data collection requirements.* In addition to simple functional testing, the test framework may also be called upon to gather data during a test run. This data may include the response time of the SUT, analog signals to or from the SUT, or other data streams. You will have to determine the data rates and the required precision of measurement. For very large systems in which the test framework must be distributed, clock synchronization of the test framework computers during data collection may become an important issue. You may also have to reduce some of the collected data while a test is running, for example, to determine pass/fail criteria for the test. If so, then supporting this data reduction at run time is important.

This set of requirements is not exhaustive but only a starting point. Plan, at the start of the project, to enumerate the testing scenarios you will have to support. Be as specific as you can about the requirements of the test framework.

The system test plan and system test design for the SUT can provide much of the information on the requirements of the test framework. For example, the system test plan will often have information about the SUT's external interfaces, data collection requirements, performance requirements of the test harness, and at least a few scenarios to be tested. The system test design will provide the remainder of the scenarios. If the test framework is being designed before the system test plan, then the requirements gathered for the test framework can also provide information needed for the system test plan and at least some typical scenarios for the system test design.

Interview the stakeholders for the test harness and people with knowledge you can use. The manual testers who are experienced with the SUT or systems like it are often the best source of information about what it will take to test the SUT. Developers can often tell you how to work around problems in the test interfaces because they have had to do it themselves. Project managers can provide the project parameters that your test harness must support.

3.2.2   Evaluating existing infrastructure

Few projects are built in "green fields" anymore. There will almost always be existing test artifacts that you can reuse. Inventory the existing hardware and software infrastructure that is available for testing your system. Pay special attention to any test automation that has been done in the past. Evaluate ways in which improvement is needed—check with the stakeholders. Find out what the missing pieces are and what can be reused.

You will have to make a decision about whether to use existing infrastructure. Be aware that some stakeholders have "pet" tools and projects, and they may plead (or demand) that you use them. You will have to evaluate on a case-by-case basis whether this is sensible.

3.2.3   Choosing the test automation support stack

We define the test automation support stack as the generic hardware and software necessary to implement a test harness. The test automation support stack consists of three major subsystems:

- The *test executive*, which parses and controls the high-level execution of the test scripts. The test executive may also include script editing and generic logging facilities.

- The *adaptation software*, which adapts the high-level function calls in the test scripts to low-level device driver calls. The software test frameworks that this chapter describes are implemented in the adaptation software.

- The low-level device drivers necessary for the interface between the test system and the hardware interface to the SUT.

*Hint: **When deciding on the support stack, prefer single vendors who provide the entire stack, rather than mixing vendors.***

Vendors who provide the entire stack have a strong motivation to make the entire stack work together. Be prepared, however, to do some architectural prototyping to evaluate the support stack for your specific needs. See Section 3.2.5 for a description of architectural prototyping.

A *test executive* is a system that executes test scripts and logs results. It is a key component of the test framework. Some examples of test executives include Rational System Test, National Instruments TestStand, and Froglogic Squish. Many test executives are patterned around integrated development environments and support features such as breakpoints, single-stepping, and variable inspection in the test scripts. It is the part of a test framework that the test engineer sees the most. Test executives are also often designed to support different markets, for example, GUI testing versus test automation for embedded and mechatronic products. You will have to decide whether to buy one or build one. Each of these choices leads to other decisions, for example, which test executive to buy, which scripting language platform to build your own test executive upon, etc.

If you are going to buy the test executive, then evaluate the commercial off-the-shelf (COTS) alternatives. The choice you make should demonstrate, to your satisfaction, that it supports the requirements you have gathered. In most cases, you will want to experiment with evaluation versions of each COTS test executive to decide whether it will meet the testing requirements.

If you decide that no COTS test executive meets your requirements, you will be forced to build one. In special circumstances, this may be a wise decision. If you make this decision, be aware that building a commercial-quality test executive is a lengthy and expensive process, and you will probably not be able to build one that is as sophisticated as the ones you can buy.

In addition to deciding on your test executive, you may also have to decide on the *scripting language* that test engineers will use to write tests and that the MBT generator will emit. Many test executives allow you to choose from among several scripting languages, such as VBScript, Python, Perl, etc. The preferences and knowledge of the test team play a large part in making this choice. Vendors generally add at least some test-executive-specific functions to the scripting language's standard library. Changes to syntax for purposes of vendor lock-in (commonly known as "vendorscripts") should be avoided.

Adaptation software is the software that glues the SUT to the test harness. Most of the rest of this chapter is about the architecture of the adaptation software.

*Important Point: **The adaptation software should support a strong object-oriented development model.***

This is necessary in order to make the implementation of the test framework (as described here) tractable.

At the lowest level of the adaptation, software is the interface between the test framework and the SUT (or the electronic test harness attached to the SUT). Often, a change of software technology is forced at this interface by availability of test harness components, and different software technologies do not always interoperate well. If the test harness supports it, you will also want to report events in the SUT to the test framework without having to poll the state of the SUT. The test framework must function efficiently through this interface, so you will likely have to do some architectural prototyping at the low-level interface level to make sure that it will. (Refer to Section 3.2.5 for details on architectural prototyping.)

3.2.4   Developing a domain-specific language for modeling and testing the SUT

A *domain-specific language* (DSL) (Kelly and Tolvanen 2008) is a computer language that has been created or adapted to model a specific domain. DSLs are not necessarily programming languages because their users are not necessarily doing programming.

In test automation frameworks, a *test automation DSL* is a DSL that represents the SUT and the test framework in test scripts. A *test automation DSL* is often represented in the test framework as a library of functions that are used to access, control, and check the SUT. The *test automation DSL* is usually specialized for a specific SUT or class of similar SUTs.

In the requirements gathering phase of creating a test framework (Section 3.2.1), you should start to understand what parts of the SUT you will have to control and monitor and what information must be passed. You should always assume that some scripting will have to be done manually to prevent forcing the test team to depend solely on the model-based test generator. Therefore, the *test automation DSL* will have to have a good level of developer friendliness.

The scenarios that you identified in the requirements gathering phase are the key input into developing the *test automation DSL*. They contain the objects of the SUT that the tester interacts with and the kinds of interactions themselves.

Important Point: The test automation DSL should be specified from the point of view of a tester testing the system.

Do not waste time specifying the entire *test automation DSL* in the early stages. It will change as the team learns more about automating the tests for the SUT.

A *test modeling DSL* is used in use-based test modeling (as defined in Section 3.1). It is typically created by customizing a generic use-based test modeling language to the SUT. The design of the *test automation DSL* should be coordinated with the design of the test modeling DSL, as we mentioned in Section 3.2.1. Fortunately, this coordination can be done quite efficiently since the same requirements engineering and customization process can be used to create the test modeling DSL and the test automation DSL. A single DSL can thus be used for both test modeling and test automation, as long as the DSL includes the semantics necessary for both the test modeling and test automation tasks. Using the same DSL for both improves project efficiency.

Often, a modeler will want to represent the same objects and activities in the models that a test developer is representing in their tests. Doing this will make both the models and the test scripts intuitively analogous to the SUT.

3.2.5   Architectural prototyping

Before starting serious construction of the test framework, it is advisable to do some *architectural prototyping* (Bardram, Christensen, and Hansen 2004) to make sure that the known technical risks are addressed. An architectural prototype is a partial implementation of a risky part of the system and is created to ensure that the issue can be solved. Functionality and performance risks—in particular, requirements for scalability and timing accuracy—can often be addressed early by architectural prototyping.

Architectural prototyping helps you work out the known problems, but not the unknown ones. Only a complete and successful implementation of the test framework can totally eliminate any risk that it will fail. The goal of architectural prototyping is not that stringent. Through architectural prototyping, you will eliminate the severe risks that you know about before development starts.

3.3 Suggested Architectural Techniques for Test Frameworks

Once the preliminary activities have been completed, the test framework is architected and implemented. Here, we describe some architectural techniques that are useful in creating test frameworks for embedded systems, especially in model-driven environments.

3.3.1 Software product-line approach

A *software product line* (Clements and Northrup 2002) is a set of related systems that are specialized from reusable, domain-specific *core assets* (see Figure 3.2). Specialization of the core assets to a product may be done by hand or may be assisted by special-purpose tools. The great business advantage of a product-line approach is that it can allow the enterprise to efficiently create related systems or products.

A product line of test frameworks would thus consist of core assets that could be specialized for a particular class of SUTs. If the SUT is configurable, if it is one of several similar systems you must test, or if the SUT is developed using a product-line approach,* then consider a product-line approach for the test framework.

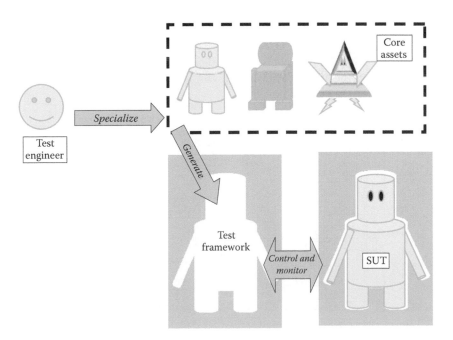

FIGURE 3.2
Software product-line approach for test framework.

---

*If the SUT is part of a product line, then the product line will usually include the test framework as a specializable core asset.

Trying to build a software product line before the first SUT has been placed under test will probably result in a suboptimal design. Generally, at least three instances of any product (including test frameworks) are necessary before one can efficiently separate core assets from specialized assets.

3.3.2 Reference layered architecture

For the following example, we assume that the SUT is a typical embedded system that contains several components that the test engineer must exercise. These include the following:

- A human–machine interface (HMI). We will assume that the HMI is interfaced to the test harness electronics via custom hardware.

- Several sensors of various types. These may be either actual hardware as intended for use in the SUT, or they may be simulators. Actual hardware is interfaced via custom electronics, while simulators are interfaced via a convenient standard bus.

- Several effectors of various types. Again, these may be actual or simulated.

- Adapters of various types for interfacing the SUT to other systems. These may be based on either proprietary technology or standard interface stacks.

- A logging printer, interfaced via RS-232.

Figure 3.3 shows a layered architecture for the test framework software. As we shall see, this layering system can accommodate product-line approaches or one-of-a-kind test framework implementations.

The layers are as follows:

- *The Test Script Layer (TSL)* contains the test scripts that execute on the test executive. These are written in the scripting language of the test executive, which has been extended with the test automation DSL.

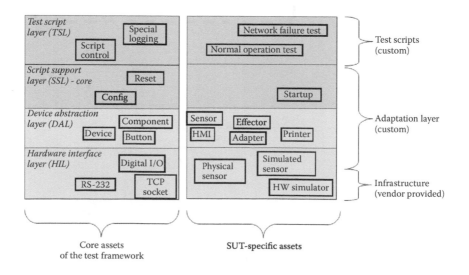

FIGURE 3.3
Layered architecture of the test framework.

Important Point: Test scripts should contain descriptions of tests, not descriptions of the SUT. Represent the SUT in the layers under the TSL.

To maintain this clear boundary between the TSL and other layers, it may prove helpful to implement the TSL in a different language than the adaptation layer software uses, or to otherwise limit access to the adaptation layer software from test scripts. Fortunately, testers seem to prefer scripting languages to object-oriented (OO) languages, and the reverse is true for software developers, so this decision is culturally easy to support.

- *The Script Support Layer (SSL)* contains helper functions for the test scripts. Some examples we have used include system configuration, resetting to a quiescent state, and checking for a quiescent state. Again, it is important to maintain clear layer boundaries. Since the SSL is generally written in the same programming language as the two layers below, it is more difficult to ensure separation of the layers.

Hint: The SSL can be a good place to put test-script-visible functionality that concerns multiple devices or the entire SUT. If the functionality concerns single devices or components of devices, consider putting it in the layer(s) below.

Functionality that concerns multiple devices does not always have to be kept in the SSL, though. For example, we have found that there is a need for Factory design patterns (Gamma et al. 1995) that yield concrete Device instances, even at the lowest layers. This functionality concerns multiple devices, but is not part of the test automation DSL and should not be accessed through the test scripts. In fact, we have put information-hiding techniques into place to prevent test scripts from directly accessing the device factories either accidentally or deliberately.

Hint: You can also use the SSL to hide the layers below from the test scripts of the TSL.

We have, for instance, used the SSL to hide incompatible technology in the lower layers from the test scripts, or to simplify the test automation DSL, by defining a common interface library within the SSL to act as a façade for the lower layers.

- *The Device Abstraction Layer (DAL)* contains abstract classes that represent the devices in the SUT, but with an abstract implementation. Continuing the example from above, the device class representing an HMI would typically have methods for activating touchscreens and buttons, but the interface to the hardware that actually does this work must be left unspecified. The DAL contains both core assets and system-specific abstract device classes.

- *The Hardware Interface Layer (HIL)* contains one or more concrete *implementor class* for each abstract device class. These implementor classes form the "view" of the SUT that is represented more abstractly in the DAL. Again continuing the same example, if the sensors in the SUT are actual field hardware interfaced by analog I/O devices, then a class HardwareSensor in the HIL would implement the class Sensor in the DAL. HardwareSensor would also contain configuration information for the analog I/O devices. If the sensors in the SUT are simulated, then a class SimulatedSensor would implement Sensor and would contain configuration information about the simulator. The decision of which implementor of Sensor to use can be deferred until run time, when the test framework is being initialized, by using Bridge and Factory design patterns (Gamma et al. 1995). This allows dynamic configurability of the test framework. (See Figure 3.8 for an example of the Bridge pattern.)

As software projects age, layers and other information-hiding mechanisms have a tendency to become blurred. If possible, the test system architect should therefore put some mechanism into place to defer this blurring as long as possible.

Hint: Use the directory structure of the source code to help keep the layers separated and to keep products cleanly separated in a product line.

We have had good success in encouraging and enforcing layer separation by including the lower layer directories within the upper layer directories. Sibling directories within each layer directory can be further used to separate core assets from SUT-specific assets and to separate assets for different SUTs from each other, in a product-line approach. Directory permissions can be used to limit access to specific groups of maintainers to further avoid blurring the structure of the framework. Figure 3.4 shows an example of a directory structure that encourages layer separation and product separation.

3.3.3   Class-level test framework reference architecture

Figure 3.5 shows a class-level reference architecture for a test framework. We have successfully implemented this reference architecture to meet the quality attribute requirements outlined in Section 3.1.2 (Masticola and Subramanyan 2009).

From the SSL, the script-visible functionality is kept in a package labeled TestExecInterface. All the functionality in this package is specific to a particular test executive. Much of it is also specific to a particular SUT product type in a software factory.

In the DAL (and hence in the HIL), we have found it helpful to define two abstract classes that represent the SUT: Device and Component. An instance of a Device can be individually referenced by the test script and configured to a particular hardware interface in the SUT.

Components exist only as pieces of devices and cannot be individually configured. Often, a Device serves as a container for Components, and there are cases in which this is the only functionality that the Device has.

Important Point: The test steps in the script address the Devices by their names. Therefore, Devices have names, Components do not.

Since Components do not exist independently of Devices, it is not necessary (from the standpoint of test framework architecture) to give them names. However, it may be convenient to assign them handles or similar identifiers in some specific implementations.

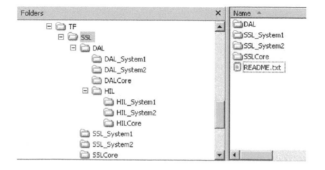

FIGURE 3.4
Directory structure used to encourage layer separation.

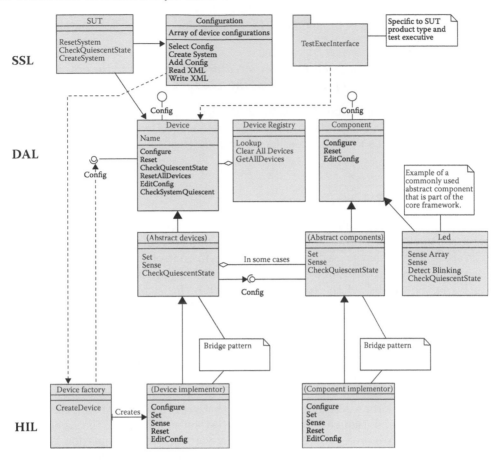

FIGURE 3.5
Reference class architecture for a test framework.

Important Point: If a tester-visible unit of the SUT must be addressed individually by the test script, make it a Device.

Hint: The business classes in the testing scenarios that were captured in the requirements gathering phase of Section 3.2.1 identify the tester-visible devices of the SUT.

The Device instances will, again, mostly correspond to the business objects in the SUT that are identified in the scenario capture of Section 3.2.1. Continuing the example of Section 3.3.2, the scenarios will identify the HMI as a business object in the testing process. The test developer will therefore have to access the HMI in test scripts. As the test system architect, you might also decide that it's more useful to represent and configure the parts of the HMI together as a single device than to treat them individually. Under these circumstances, HMI can usefully be a subclass of Device.

Hint: Introduce Components when there is repeated functionality in a Device, or when the Device has clearly separable concerns.

Again, from the continuing example of Section 3.3.2, a typical HMI may have a video display, touchscreen overlaying the video, "hard" buttons, LEDs, and a buzzer. The HMI class may then be usefully composed of HmiDisplay, HmiTouchscreen, HmiButtons, HmiLEDs, and HmiBuzzer subclasses of Component. HmiButtons and HmiLeds are containers for even smaller Components.

Hint: In a product-line architecture, create generic components in the core architecture if they are repeated across the product line.

Continuing the example from Section 3.3.2, HmiButtons and HmiLeds can be composed of generic Button and Led subclasses of Component, respectively. Buttons and LEDs are found in so many SUTs that it is sensible to include them in the core assets of the test framework.

3.3.4   Methods of device classes

English sentences have a "subject-verb-object" structure and the *test automation DSL* has the same parts of speech. The tester-visible Device objects make up the "subjects" of the test automation DSL that is used in the test scripts to control the test state. The methods of these tester-visible Device objects which are visible from the test scripts correspond, indirectly, to the "verbs" of the test automation DSL. Expected results or test stimulus data are the "object phrases" of the test automation DSL.*

Hint: The actions in the testing scenarios that were captured in the requirements gathering phase of Section 3.2.1 can help identify the "verbs" of the test automation DSL. However, direct translation of the actions into test automation DSL verbs is usually not desirable.

You will have to form a *test automation DSL* that represents everything that the verbs represent. Keep in mind, though, that you want a test automation DSL that is easy to learn and efficient to use. We present here some common verbs for test automation DSLs.

Setting and sensing device state. The test script commands some devices in the SUT, for example, pushing buttons or activating touchscreen items in the HMI, or controlling simulated (or physical) sensor values. For this, Device objects often have Set methods.

Because different types of devices are controlled differently, the Set methods cannot typically be standardized to any useful type signature. Furthermore, not all objects have Set methods. For instance, effectors would typically not since the SUT is controlling them and the test framework is only monitoring the behavior of the SUT in controlling them.

Sense methods are similar to Set methods in that they cannot be standardized to a single-type signature. On the other hand, devices may have more than one Sense method. If the scripting language supports polymorphism, this can help in disambiguating which Sense method to call. Otherwise, the disambiguation must be done by naming the "Sense-like" methods differently. One could envision abstract classes to represent the Actual State and Expected State arguments of Set and Sense methods, but our persuasion is that this is usually overengineering that causes complications in the test automation DSL and scripting language and may force the test engineers to do extra, superfluous work.

For these reasons, the base Device and Component classes of Figure 3.5 do not define Set or Sense methods. Instead, their subclasses in the DAL should define these methods and their type signatures, where necessary.

---

*We have never had a situation in which a Device was part of an "object phrase" in the test automation DSL, but it is possible. An example would be if two Devices had to be checked for a state that is "compatible," in some sense that would be difficult or tedious to express in the scripting language. Since the corresponding "verb" involves an operation with multiple devices, any such "object phrase" properly belongs in the test executive interface package in the SSL.

Hint: Do not include Set and Sense methods in the Device base class. However, do keep the abstract device classes that have Sense and Set methods as similar as you can.

Such regularity improves the ease with which developers can understand the system.

Checking for expected state. In embedded systems, the usual sequence of test steps is to apply a stimulus (via setting one or more device states) and then to verify that the system reaches an expected state in response. If the actual state seen is not an expected one, then the actual and expected states are logged, the cleanup procedure for the test case is executed, and the test case ends with a failure. Many commercial test executives support this control flow in test cases.

This sequence may appear to be simple, but in the context of a test framework, there are some hidden subtleties in getting it to work accurately and efficiently. Some of these subtleties include:

- The response does not usually come immediately. Instead, there are generally system requirements that the expected state be reached within a specified time. Polling for the expected state is also usually inefficient and introduces unnecessary complexity in the test script.

   Hint: Include expected states and a timeout parameter as input parameters in the Sense methods. The Sense methods should have at least two output parameters: (a) the actual state sensed and (b) a Boolean indicating whether that state matched the expected state before the timeout.

- There may not be a single expected state, but multiple (or even infinite) acceptable expected states.

   Hint: Add a "don't care" option to the expected states of the low-level components. Hint: If the expected state of a device cannot be easily expressed as the cross-product of fixed states of its components with don't-cares included, add extra check functions to the SSL or the abstract device to support checking the expected state.

   For example, suppose that two effectors in a single device must represent the sine and cosine of a given angle whose value can vary at run-time. An extra check function would verify that the effectors meet these criteria within a specified time and precision.

   Hint: Add a singleton "SUT" class at the SSL layer to support test-harness-wide "global" operations (see **Figure 3.5***).*

   The SUT class is useful when there are system states that exist independently of the configuration of the system, such as a quiescent state. We will talk more about how to specifically implement this in Section 3.3.5.

- The test frameworks described here are designed around devices. A check for an arbitrary expected state of several devices must therefore be composed of individual checks for the states of one or more individual devices. Note that in some special cases, such as a system-wide quiescent state, we add support for checking the state of the entire system. In testing, we usually want to check that several devices, or the entire system, reach their expected states within a specified time. Most scripting languages are basically sequential, so we list the devices to check in some sequence. If we check devices sequentially, we often cannot be certain which one is going to reach its expected state first. If any device has to wait, all the devices after it in the sequence may have to wait, even if they are already in their expected states. Hardcoding the individual device wait times in the test script may thus result in spurious test failures (or successes) if the devices do not reach their expected states in the order they are checked.

Hint: Sample a "base time" when the stimulus is applied and sense that each device reaches its expected state within a specified interval from the base time.

This mechanism does not always guarantee that slow-responding devices will not cause spurious failures.

- Executing the Sense methods takes some time. Even with the base time approach, it is possible that a slow Sense method call may cause a later Sense method call to start later than its timeout with respect to the base time. The result is that the second Sense method call instantly times out.

Hint: Consider adding a SenseMultiple container class to the SSL.

(Not shown in Figure 3.5.) This is a more robust alternative to the "base time" idea above. Before the stimulus is applied, the test script registers devices, their expected states, and their timeout parameters with the SenseMultiple container. After the stimulus is applied, the SenseMultiple container is called in the test script to check that all its registered devices reach their expected states before their individual timeouts. Such a SenseMultiple container class may require or benefit from some specific support from the devices, for example, callbacks on state change, to avoid polling.

Some SUTs may also have a global "quiescent state" that is expected between test cases. It is useful to add support to the test framework for checking that the system is in its quiescent state.

Hint: Include overridable default CheckQuiescentState methods in the Device and Component base classes to check that these are in a quiescent state. Implement these methods in the DAL or HIL subclasses, as appropriate.

The default methods can simply return true so that they have no effect if they are unimplemented. The use of these CheckQuiescentState methods to reset the system is described in Section 3.3.5.

Accumulating data for later checking and analysis. In some cases, we wish to record data for later analysis, but not analyze it as the test runs. One example of such data would be analog signals from the effectors that are to be subjected to later analysis. Sometimes it is desirable to start and stop recording during a test. Usually the data recording must at least be configured, for example, to specify a file to store the data.

Hint: If a data stream is associated with a particular kind of device, include methods for recording and storing it in the Device class. Otherwise, consider adding a Device class for it as a pseudo-device.

The pseudo-device will do nothing but control the data recording.

Configuring the device. Configuring a device involves declaring that an instance of the abstract device exists, how the test scripts can look the instance up, the concrete implementor class, and how the implementor is configured. These are the four parameters to the Configure method of an abstract device.

Hint: Include abstract Configure methods in the Device and Component base classes.

Variant device configuration parameters to these Configure methods may be serialized or otherwise converted to a representation that can be handled by any device or component. The Configure method must be implemented in the concrete device class in the HIL. A DeviceFactory class in the HIL produces and configures concrete device instances and registers them with the framework.

Resetting the device. As mentioned before, it is very convenient to be able to sense that the SUT is in a quiescent state. It is likewise convenient to be able to put the SUT back into a known state (usually the quiescent state) with a minimum amount of special-purpose scripting when a test is finished. Returning the SUT to its quiescent state can sometimes be done by individually resetting the devices.

Hint: Include overridable default Reset methods in the Device and Component base classes to return the test harness controls to their quiescent values. Override these methods in the DAL or HIL subclasses, as appropriate.

In some common cases, however, resetting the individual devices is insufficient. The SUT must be navigated through several states (for example, through a series of calls to the HMI Set and Sense methods) to reach the quiescent state, in a manner similar to, though usually much simpler than, recovery techniques in fault-tolerant software (Pullum 2001). This is usually the case if the SUT is stateful. The use of these Reset methods to reset the system is described in Section 3.3.5.

Hint: If necessary, include in the SSL a method to return the SUT to a quiescent state, by navigation if possible, but by power cycling if necessary.

If the system fails to reset automatically through navigation, the only way to return it to a quiescent state automatically may be to power cycle it or to activate a hard system reset. The test harness must include any necessary hardware to support hard resets.

3.3.5   Supporting global operations

The test framework must support a small number of global operations that affect the entire SUT. We present some of these here. Specific SUTs may require additional global operations, which often can be supported via the same mechanisms.

Finding configured devices. Since the Device classes address specific devices by name, it is necessary to look up the configured implementors of devices at almost every test step. This argues for using a very fast lookup mechanism with good scalability ($O(\log n)$ in the number of configured devices) and low absolute latency.

Hint: Create a Device Registry class in the DAL.

The Device Registry, as shown in Figure 3.5, is simply a map from device names to their implementors. The Device base class can hide the accesses to the Device Registry from the abstract device classes and implementors and from the test scripts.

Occasionally, we also wish to find large numbers of individual Devices of a specific type without hardcoding the Device names into the test script. This is useful for, for example, scalability testing. If this sort of situation exists, the Device Registry can also support lookup by other criteria, such as implementor type, regular expression matching on the name, etc.

Configuring the system. The Configure method described above configures individual devices. To configure the entire system, we usually read a configuration file (often XML) and create and configure the specific Device implementors.

Hint: Create a Configuration singleton class in the SSL.

The Configuration singleton will parse the test harness configuration file and create and configure all devices in the system. We have also used the Configuration class to retain system configuration information to support configuration editing. Avoid retaining information on the configured Devices in the Configuration class as this is the job of the Device Registry.

Hint: Create a Device Factory singleton class in the HIL.

This class creates Device implementors from their class names, as mentioned above, and thus supports system configuration and configuration editing. The Device Factory is implementation specific and thus belongs in the HIL.

Resetting the system. Some controllable Devices have a natural quiescent state as mentioned above and can be individually supported by a Reset method.

Testing for common system-wide states. The same strategy that is used for Reset can be used for testing a system-wide state, such as the quiescent state.

Hint: Create static ResetAllDevices and CheckQuiescentState methods in the Device class. Expose the system-wide ResetSystem and CheckQuiescentState methods to test scripts via the static SUT class in the SSL.

The system-wide Reset method should simply call the Reset methods of all registered Devices. Similarly, the system-wide CheckQuiescentState method should simply calculate the logical "and" of the CheckQuiescentState methods of all registered Devices. We suggest that these methods may be implemented in the Device Registry and exposed to scripts through a Façade pattern in the Device class (Gamma et al. 1995). This avoids exposing the Device Registry to scripts.

The test harness can provide additional hardware support for resetting the SUT, including "hard reset" or power cycling of the SUT. If the test harness supports this, then ResetAllDevices should likewise support it. Alternatively, ResetAllDevices might optionally support navigation back to the quiescent state without power cycling, if doing so would be reasonably simple to implement and robust to the expected changes in the SUT.

3.3.6   Supporting periodic polling

Polling is usually done to support updates to the state of values in the test harness that must be sensed by the test script. Even though polling is usually inefficient, it will likely be necessary to poll periodically. For example, polling will be necessary at some level if we must sense the state of any component that does not support some kind of asynchronous update message to the test harness.

There are good ways and bad ways to support polling. It is usually a very bad idea to poll in the test script. Where it is necessary, the test framework should support polling with as little overhead as possible. Additionally, the Devices and Components that require polling should not be hard-coded into the test framework.

Hint: Create an abstract Monitor class that represents something that must be polled and a concrete Monitor Engine class with which instances of the class Monitor are registered. Subclass Monitor to implement configuration and polling (this is not shown in Figure 3.5).

The Monitor Engine is initiated and runs in the background, calling the Poll method on each of its registered Monitors periodically. Monitors are created, configured, and registered with the Monitor Engine when the Devices or Components they support are configured.

An alternative, though somewhat less flexible, method to support polling would be to implement callback functions in Devices or Components. Monitors can support polling at different rates, and thus a Device or Component using Monitors can support polling at different rates. A Device or Component with a polling method would have to be polled at a single rate. Using Monitors also helps prevent races within Devices or Components by cleanly separating the data to be used in the primary script thread and the one in which the Monitor Engine runs.

3.3.7 Supporting diagnosis

The SUT can fail tests, and we wish to log sufficient information to understand the reason for the failure, even in a long-running test. It is therefore important to log sufficient information from the SUT to diagnose failures. This is primarily a requirement of the test harness design.

Hint: Create Monitors to periodically log the important parts of the SUT state.

Since Monitors run in the background, this avoids having to clutter the test scripts with explicit logging steps.

Important Point: In addition to the SUT, the test harness can also fail.

Failure can occur because of a mechanical or electronic breakdown or a software error. For this reason, errors must be reported throughout the test framework.

Hint: Create an Error class and add Error In and Error Out parameters to all methods in the test framework (this is not shown in Figure 3.5).

This is the standard methodology used in National Instruments' LabVIEW product (Travis and Kring 2006), and it can also be applied to procedural and OO languages. Any specific information on the error is logged in the Error object.

If the Error In parameter of any method indicates that an error has occurred previously, then the method typically does nothing and Error Out is assigned the value of Error In. This prevents the test framework from compounding the problems that have been detected.

Important Point: Error logging should include as much information as is practical about the context in which the error occurred.

Ideally, the calling context would be similar to that provided by a debugger (i.e., the stack frames of all threads executing when the error occurred, etc).

Hint: If there is a reasonably good source-level debugger that works in the language(s) of the adaptation software, consider using it to help log the error context.

However, in normal testing, do not force the test execution to stop because of an error. Instead, just log the debug information, stop executing the current test case, restore the system to quiescent state, and continue with the next test case. It may alternatively be worthwhile to try to recover from the error rather than aborting the current test case.

It is also worth mentioning that the test framework should be designed defensively. Such a defensive design requires you to know the type, severity, and likelihood of the failures in the test harness that can stop the execution of a test. The countermeasures you choose will depend on the failure modes you expect. For example, if power failure may cause the test harness to fail, then you may have to install an uninterruptible power supply.

Hint: Consider conducting a risk analysis of the test harness, to identify the scenarios for which you must support error recovery. (Bach 1999)

If you are testing a high-cost system, the number of scenarios may be considerable because stopping testing once it has begun can incur the cost of idle time in the SUT.

3.3.8   Distributed control

If the SUT is very large or complex, it is possible that a single test computer will be inadequate in control it. If this is the case, you will have to distribute control of the test framework. If distributed control is not necessary, then it is best to avoid the added expense and complexity.

Important Point: If you think that it may be necessary to distribute your test framework, then conduct experiments in the architectural prototyping phase to determine whether this is necessary indeed.

It is better to learn the truth as early as possible. If distributed control proves to be necessary, some additional important technical issues will have to be resolved.

Remoting system. Often the test automation support stack will determine what remoting system(s) you may use for distributed control. There still may be multiple choices, depending on the level in the stack at which you decide to partition.

Partitioning scheme. There are several ways to divide the load among control computers. You can have the test script run on a single control computer and distribute the Device instances by incorporating remoting information into the Device Registry entries. Alternatively, you can partition the test script into main and remote parts.

Hint: Prefer partitioning at the test script level if possible.

This can avoid creating a bottleneck at the main control computer.

Load balancing. You will have to balance resource utilization of all types of resources (processor, memory, file I/O, network traffic) among the control computers in a distributed framework.

Important Point: Consider all the different types of load that your test framework will generate when balancing.

This includes network and disk traffic because of data collection and logging, as well as the more obvious CPU load.

Clock synchronization and timing. The Network Time Protocol (NTP) (Mills, RFC 1305—Network Time Protocol (Version 3) Specification, Implementation and Analysis 1992) and Simple Network Time Protocol (SNTP) (Mills, RFC 1361—Simple Network Time Protocol (SNTP) 1992) are widely used to synchronize clocks in distributed computer systems. Accuracy varies from 10 ms to 200 μs, depending on conditions. NTP and SNTP perform better when the network latency is not excessive.

Important Point: Dedicate one computer in the test harness as an NTP or SNTP server and set it up to serve the rest of the harness.

The time server should probably not be the main control computer. Instead, make it a lightly loaded (or even dedicated) computer.

Hint: Minimize network delays between the test harness computers.

Designing the local area network (LAN) for the test harness as an isolated subnet with minimum latency will allow more accurate time synchronization.

Sensing and verifying distributed state. You will still have to sense whether the SUT is in a particular state to verify it against expected results. This becomes more complicated when the system is distributed, if there are Byzantine cases in which parts of the system go into and out of the expected state. In a naïve implementation, the framework may report that the SUT was in the expected state when it never completely was. Although the likelihood of such an error is low, it is good practice to eliminate it to the extent possible.

Hint: Instead of just returning a Boolean pass–fail value, consider having the Sense methods for the distributed computers report the time intervals during which the SUT was in the expected state.

This at least ensures that, whenever the main control computer reports that the SUT as a whole was in a given state, it actually was in that state, within the limits of time synchronization.

The downsides of reporting time intervals rather than pass–fail values are increased data communication and processing loads. Slightly longer test step times are also necessary because of the need for the Sense methods to wait for a short interval after reaching their expected state to provide a safe overlap.

3.4 Brief Example

This section presents an example instance of the test framework reference architecture outlined in this chapter. The test framework in this example was designed for testing complex fire-safety systems and is more fully described in Masticola and Subramanyan (2009).

Figure 3.6 shows an example script which uses the test automation DSL of the example test framework. The test executive is National Instruments TestStand (Sumathi and Sureka 2007). The view of the script shows the test steps, but not the test data, which are visible in a separate window in the TestStand user interface, as shown in Figure 3.7.

For this specific test automation system, we chose to relax the recommendation, in Section 3.2.3, against vendorscripts in an engineering tradeoff against the other benefits of the automation stack. We also confirmed, through interviews with other users of TestStand,

FIGURE 3.6

Example script using a test automation DSL for fire-safety system testing.

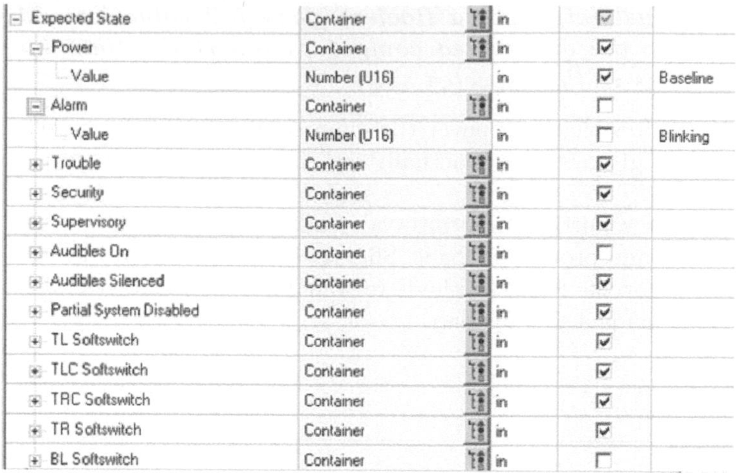

⊟ Expected State	Container		in	☑	
⊟ Power	Container		in	☑	
Value	Number (U16)		in	☑	Baseline
⊟ Alarm	Container		in	☐	
Value	Number (U16)		in	☐	Blinking
⊞ Trouble	Container		in	☑	
⊞ Security	Container		in	☑	
⊞ Supervisory	Container		in	☑	
⊞ Audibles On	Container		in	☐	
⊞ Audibles Silenced	Container		in	☑	
⊞ Partial System Disabled	Container		in	☑	
⊞ TL Softswitch	Container		in	☑	
⊞ TLC Softswitch	Container		in	☑	
⊞ TRC Softswitch	Container		in	☑	
⊞ TR Softswitch	Container		in	☑	
⊞ BL Softswitch	Container		in	☐	

FIGURE 3.7
Variables window in TestStand.

that the vendor typically provided reasonable pathways for customers to upgrade customer software based on their test automation support stack. We decided that the risk of planned obsolescence of the scripting language was sufficiently small, thus TestStand was selected.

Referring to the reference class architecture of Figure 3.5, we can see how some of the classes and methods are filled in. CreateSystem, in the SUT class in the SSL, is implemented using the DeviceFactory class in the HIL. CheckQuiescentState in the SUT class is implemented using the individual CheckQuiescentState methods of the device implementers in the HIL and a collection of configured devices in the SSL. The "nouns" in the script shown in Figure 3.6 correspond to components of the fire-safety system and of the test automation DSL (RPM, PMI, ZIC4A, ZIC8B, etc.). The "nouns" also correspond to subclasses of Device in the DAL. The "verbs" correspond to the actions that the test executive is to take with these components. For example, "Check ZIC4A state" is a "sense" method and "Clear Fault on ZIC4A_5" is a "set" method.

Figure 3.8 shows the concrete realization of a portion of the reference class architecture. Many of the devices in the SUT have repeated components, which are implemented by subclasses of Component. For example, Person–Machine Interface (PMI) has several LEDs and function keys. Classes representing a LED and a function key exist as subclasses of Component.

In the HIL, Device and Component implementor classes map the abstract PMI, and other devices, and their components, to concrete digital I/O devices that control and monitor the corresponding physical devices. For example, a LED can only be sensed, so HardwareLed class maps a particular digital input line to a particular LED on the PMI.

3.5 Supporting Activities in Test Framework Architecture

Once the preliminary test framework is architected and implemented, support software, documentation, and training material must be prepared. In this section, we describe typical test framework support software, provide key points to include in the test framework documentation, and discuss iterating the test framework architecture to a good solution.

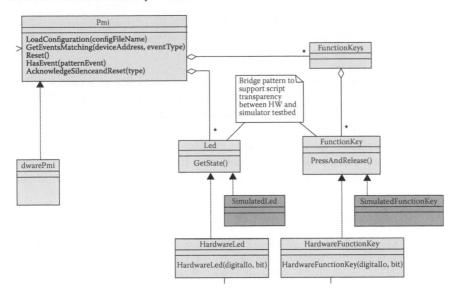

FIGURE 3.8

Concrete realization of a part of the test framework (the device and component classes are not shown). (Reproduced from Masticola, S., and Subramanyan, R., Experience with developing a high-productivity test framework for scalable embedded and mechatronic systems, 2009 ASME/IEEE International Conference on Mechatronic and Embedded Systems and Applications (MESA09), © 2009 IEEE.)

3.5.1 Support software

The test framework you create is likely going to require some support software to perform functions in addition to test execution. Some examples of support software that we have developed on past projects include:

- *Editing configuration files.* Manually editing complicated XML configuration files is both tedious and error-prone. Unless the test framework has been designed to be robust in the presence of mistakes in its configuration files, it makes good economic sense to provide tool support to reduce the likelihood of such errors.

    Hint: Incorporate support for configuration editing into the test framework.

    One way we did this was to require each device implementor to implement EditConfig and GetConfig functions (Masticola and Subramanyan 2009). EditConfig pops up an edit page with the current configuration values for the device and allows this information to be edited. A similar GetConfig method returns a variant containing the current (edited) configuration of the device. Editing the configuration was thus implemented by actually editing the set of configured devices in a Config Editor GUI.

- *Importing configuration files* from the tools that manage the configuration of the SUT. The SUT configuration can usually provide some of the information necessary for test framework configuration. For example, the list of Devices present in the SUT may be derivable from the SUT configuration. Generally, the SUT configuration will contain information not required by the test framework (e.g., the configuration of devices not interfaced to the test framework) and vice versa (e.g., the classes of the device implementors). The import facility also helps to keep the test framework configuration synchronized with the SUT configuration.

- *Test log analysis tools.* For example, you may wish to perform extract-transform-load operations on the test logs to keep a project dashboard up to date, or to assist in performance analysis of the SUT.

- *Product builders.* These are the tools in a software product line (Clements and Northrup 2002) that create components of the software product from "core assets." Here, we consider the test framework as the software product we are specializing. A core asset in this context would probably not be part of the core test framework, but might instead be a template-like component class, device class, or subsystem. If you are implementing the test framework as part of a software product line, then it may be economical to create product builders to automate the specialization of the test framework to a SUT product type.

3.5.2   Documentation and training

Two forms of documentation are necessary: documentation for test automation developers and documentation for the software developers of the test framework. These groups of users of the test framework have different goals. A single document will not serve both purposes well.

A document for test developers will typically include:

- Details on how to execute tests, read the test log, and resolve common problems with the test harness.

- A description of the test automation DSL you have created to automate the testing of the SUT.

- Examples of test scripts for common situations that the test developers may face, including examples of the test automation DSL.

- Information on how to configure the test framework to match the configuration of the test harness and SUT.

- Information on how to use the support software described in Section 3.5.1.

  A document for the software developers of the test framework will typically include:

- The architectural description of the core test framework and an example specialization for a SUT product. These may be described in terms of architectural views (Hofmeister, Nord, and Soni 2000).

- Detailed descriptions of the core components of the test framework.

- The conventions used in maintaining the test framework, such as naming conventions and directory structure.

- An example of how to specialize the core test framework for a particular SUT product. This will include specializing the Device and Component classes, configuration and editing, etc.

Documentation is helpful, but it is seldom sufficient to ensure efficient knowledge transfer. Hands-on training and support are usually necessary to transfer the technology to its users. If your organization has the budget for it, consider organizing a hands-on training for your test framework when it is mature enough for such a training to be beneficial.

3.5.3 Iterating to a good solution

Hint: Be prepared to have to iterate on the test framework architecture.

You will probably have to rework it at least once.

On a recent project (Masticola and Subramanyan 2009), for example, we explored three different technologies for the software test framework via architectural prototyping before we found the choice that we considered the best. Even after that, we did two major reworks of the framework architecture before we reached results that would realize all of the quality attributes listed in Section 3.1.2. The test framework works well as intended, but with greater load or tighter requirements, it may require another revision.

Hint: Implement the test framework in small steps. Learn from your mistakes and previous experience.

Small steps reduce the cost of a rewrite. You can learn from your own experience, from that of your colleagues on the project, and from experts outside of your organization, such as the authors of this book.

The authors welcome comments and suggestions from other practitioners who have architected model-based frameworks for test automation control, especially of embedded and mechatronic systems.

References

Bach, J. Heuristic risk-based testing. *Software Testing and Quality Magazine*, November 1999:99.

Bardram, J., Christensen, H., and Hansen, K. (2004). Architectural Prototyping: An Approach for Grounding Architectural Design and Learning. Oslo, Norway: Fourth Working IEEE/IFIP Conference on Software Architecture (WICSA 2004).

Berenbach, B., Paulish, D., Kazmeier, J., and Rudorfer, A. (2009). *Software & Systems Requirements Engineering in Practice*. McGraw-Hill: New York, NY.

Broekman, B., and Notenboom, E. (2003). *Testing Embedded Software*. Addison-Wesley: London.

Brown, W., Malveau, R., McCormick, H., and Mowbray, T. (1998). *AntiPatterns: Refactoring Software, Architectures, and Projects in Crisis*. Wiley: New York, NY.

Clements, P., and Northrup, L. (2002). *Software Product Lines: Practices and Patterns*. Addison-Wesley: Boston, MA.

Dias Neto, A., Subramanyam, R., Vieira, M., and Travassos, G. (2007). A survey on model-based testing approaches: a systematic review. Atlanta, GA: 1st ACM International Workshop on Empirical Assessment of Software Engineering Languages and Technologies.

Fowler, M. (2003). *UML Distilled: A Brief Introduction to the Standard Object Modeling Language*. Addison-Wesley: Boston, MA.

Gamma, E., Helm, R., Johnson, R., and Vlissides, J. M. (1995). *Design Patterns: Elements of Reusable Object-Oriented Software*. Addison-Wesley: Reading, MA.

Graham, D., Veenendaal, E. V., Evans, I., and Black, R. (2008). *Foundations of Software Testing: ISTQB Certification.* Intl Thomson Business Pr: Belmont, CA.

Hartmann, J., Vieira, M., Foster, H., and Ruder, A. (2005). A UML-based approach to system testing. *Innovations in Systems and Software Engineering*, Volume 1, Number 1, Pages: 12–24.

Hofmeister, C., Nord, R., and Soni, D. (2000). *Applied Software Architecture.* Addison-Wesley: Reading, MA.

Kelly, S., and Tolvanen. J. -P. (2008). *Domain-Specific Modeling: Enabling Full Code Generation.* Wiley-Interscience: Hoboken, NJ.

Masticola, S., and Subramanyan, R. (2009). Experience with Developing a High-Productivity Test Framework for Scalable Embedded and Mechatronic Systems. San Diego, CA: 2009 ASME/IEEE International Conference on Mechatronic and Embedded Systems and Applications (MESA09).

Mills, D. (1992). RFC 1305—Network Time Protocol (Version 3) Specification, Implementation and Analysis. *Internet Engineering Task Force.* http://www.ietf.org/rfc/rfc1305.txt?number=1305.

Mills, D. (1992). RFC 1361—Simple Network Time Protocol (SNTP). *Internet Engineering Task Force.* http://www.ietf.org/rfc/rfc1361.txt?number=1361.

Pullum, L. (2001). *Software Fault Tolerance Techniques and Implementation.* Artech House: Boston, MA.

Sumathi, S., and Sureka, P. (2007). *LabVIEW based Advanced Instrumentation Systems.* Springer: New York, NY.

Travis, J., and Kring, J. (2006). *LabVIEW for Everyone: Graphical Programming Made Easy and Fun (3rd Edition).* Prentice Hall: Upper Saddle River, NJ.

Utting, M., and Legeard, B. (2006). *Practical Model-Based Testing: A Tools Approach.* Morgan Kaufmann: San Francisco, CA.

Part II

Automatic Test Generation

4

Automatic Model-Based Test Generation from UML State Machines

Stephan Weißleder and Holger Schlingloff

CONTENTS

4.1 Introduction

Model-based testing is an efficient testing technique in which a system under test (SUT) is compared to a formal model that is created from the SUT's requirements. Major benefits of model-based testing compared to conventional testing techniques are the automation of test case design, the early validation of requirements, the traceability of requirements from model elements to test cases, the early detection of failures, and an easy maintenance of test suites for regression testing.

This chapter deals with state machines of the Unified Modeling Language (UML) [91] as a basis for automated generation of tests. The UML is a widespread semiformal modeling language for all sorts of computational systems. In particular, UML state machines can be used to model the reactive behavior of embedded systems. We present and compare several approaches for the generation of test suites from UML state machines.

For most computational systems, the set of possible behaviors is infinite. Thus, complete testing of all behaviors in finite time is impossible. Therefore, the fundamental question of every testing methodology is when to stop the testing process. Instead of just testing until the available resources are exhausted, it is better to set certain quality goals for the testing process and to stop testing when these goals have been met. A preferred metrics for the quality of testing is the percentage to which certain aspects of the SUT have been exercised; these aspects could be the requirements, the model elements, the source code, or the object code of the SUT. Thus, test generation algorithms often strive to generate test suites satisfying certain coverage criteria. The definition of a coverage criterion, however, does not necessarily entail an algorithm how to generate tests for this criterion.

For model-based testing, coverage is usually measured in terms of covered model elements. The standard literature provides many different coverage criteria, for example, focusing on data flow, control flow, or transition sequences. Most existing coverage criteria had been originally defined for program code and have now been transferred and applied to models. Thus, these criteria can be used to measure the quality of test suites that are generated from models. Test generation algorithms can be designed and optimized with regard to specific coverage criteria. In this chapter, we present several test generation approaches that strive to satisfy different coverage criteria on UML state machines.

This chapter is structured as follows: in the following, we give an introduction to UML state machines and present the basic ideas of testing from UML state machines. Subsequently, we describe abstract path generation and concrete input value generation as two important aspects in automatic test generation from state machines: the former is shown in Section 4.2 by introducing graph traversal techniques. The latter is shown in Section 4.3 by presenting boundary value analysis techniques. In Section 4.4, we describe the relation of these two aspects to other techniques. We go into random testing, evolutionary testing, constraint solving, model checking, and static analysis.

4.1.1   UML state machines

The UML [91] is a widely used modeling language standardized and maintained by the Object Management Group (OMG). In version 2, it comprises models of 13 different diagrams, which can be grouped into two general categories: Structure diagrams are used to represent information about the (spatial) composition of the system. Behavior diagrams are used to describe the (temporal) aspects of the system's actions and reactions. All UML diagram types are defined in a common meta model, so the same modeling elements may be used in different types of diagrams, and there is no distinct separation between the various diagram types. Among the behavior diagrams, state machine diagrams are the most common way to specify the control flow of reactive systems. Intuitively, a UML state machine can be seen as a hierarchical parallel automaton with an extended alphabet of actions. In order to precisely describe test generation algorithms, we give a formal definition of the notion of UML state machines used in this chapter.

A *labeled transition system* is a tuple $M = (\mathcal{A}, S, T, s_0)$, where $\mathcal{A}$ is a finite nonempty alphabet of labels, S and T are finite sets of states and transitions, respectively, $T \subseteq S \times \mathcal{A} \times S$, and s_0 is the *initial* state. In UML, the initial state is a so-called pseudostate (not belonging to the set of states) and marked by a filled circle. Assume a set E of *events*, a set C of *conditions*, and a set A of *actions*. A *simple state machine* is a labeled transition system where $\mathcal{A} = 2^E \times C \times 2^A$, that is, each label consists of a set e of input events, a condition c, and a set a of output actions. The input events of a transition are called its *triggers*, the condition is the *guard*, and the set of actions is the *effect* of the transition. The transition $(s, (e, c, a), s')$ is depicted as $\boxed{\text{s}} \xrightarrow{e[c]/a} \boxed{\text{s}'}$, where sets are just denoted by their elements, and empty triggers, guards, and effects can be omitted. States s and s' are the *source* and *target* of the transition, respectively.

A (finite) *run* of a transition system is any word $w = (s_0, t_0, s_1, t_1, \ldots, t_{n-1}, s_n)$ such that s_0 is the initial state, and $(s_i, t_i, s_{i+1}) \in T$ for all $i < n$. The *trace* of a run is the sequence $(t_0, t_1, \ldots, t_{n-1})$. For a simple state machine, we assume that there is an evaluation relation $\models \subseteq S \times C$ that is established iff a condition $c \in C$ is satisfied in a state $s \in S$. A word w is a run of the state machine if in addition to s_0 being initial, for all $i < n$ and $t_i = (e_i, c_i, a_i)$ it holds that $s_i \models c_i$. Moreover, it must be true that

1. $e_i = \emptyset$ and $(s_i, t_i, s_{i+1}) \in T$, or

2. $e_i = \{e\}$ and $(s_i, (e_i', c_i, a_i), s_{i+1}) \in T$ for some e_i' containing e, or

3. $e_i = \{e\}$, $(s_i, (e_i', c_i', a_i'), s_{i+1}') \notin T$ for any e_i' containing e, and $s_{i+1} = s_i$

These clauses reflect the semantics of UML state machines, which allows for the following

1. Completion transitions (without trigger).

2. Transitions being enabled if any one of its triggers is satisfied.

3. A trigger being lost if no transition for this trigger exists.

In order to model data dependencies, simple state machines can be extended with a concept of variables. Assume a given set of domains or classes with Boolean relations defined between elements. The domains could be integer or real numbers with values 0, 1, $<$, $\leq$, etc. An *extended state machine* is a simple state machine augmented by a number of variables $(x, y, \ldots)$ on these domains. In an extended state machine, a guard is a Boolean expression involving variables. For example, a guard could be $(x > 0 \land y \leq 3)$. A transition effect in the state machine may involve the update (assignment) of variables. For example, an effect could be $(x := 0; y := 3)$. The UML standard does not define the syntax of assignments and Boolean expressions; it suggests that the Object Constraint Language (OCL) [90] may be used here. For our purposes, we rely on an intuitive understanding of the relevant concepts.

In addition to simple states, UML state machines allow a hierarchical and orthogonal composition of states. Formally, a *UML state machine* consists of a set of *regions*, each of which contains vertices and transitions. A *vertex* can be a state, a pseudostate, or a connection point reference. A *state* can be either simple or composite, where a state is composite if it contains one or more regions. *Pseudostates* can be, for example, initial or fork pseudostates where *connection point references* are used to link certain pseudostates. A *transition* is a connection from a source vertex to a target vertex, and it can contain several triggers, a guard, and an effect. A *trigger* references an event, for example, the reception of a message or the execution of an operation. Similar to extended state machines, a *guard* is a Boolean condition on certain variables, for instance, class attributes. Additionally, UML also has a number of further predicates that may be used in guards. Finally, an *effect* can be, for example, the assignment of a value to an attribute, the triggering of an event, or a postcondition defined in OCL. In Figure 4.1, this syntax is graphically described as part of the UML meta model, a complete description of which can be found in [91].

The UML specification does not give a definite semantics of state machines. However, there is a generally agreed common understanding on the meaning of the above concepts. A state machine describes the behavior of all instances of its context class. The status of each instance is given by the values of all class attributes and the configuration of the state machine, where a *configuration* of the machine is a set of concurrently active vertices. Initially, all those vertices are active and are connected to the outgoing transitions of the initial pseudostates of the state machine's regions. A transition can be traversed if its source vertex is active, one of the triggering events occurs, and the guard evaluates to true. As a

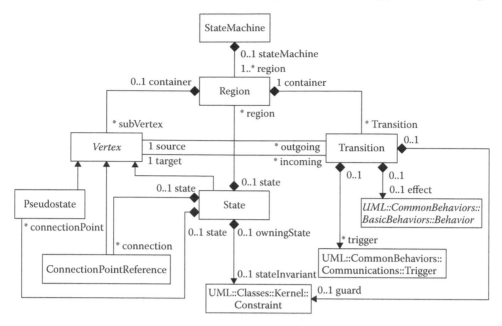

FIGURE 4.1
Part of the meta model for UML state machines.

consequence, the source vertex becomes inactive, the actions in the effect are executed, and the target vertex becomes active. In this way, a sequence of configurations and transitions is obtained, which forms a run of the state machine. Similarly as defined for the labeled transition system, the semantics of a state machine is the set of all these runs.

4.1.2   Example—A kitchen toaster

State machines can be used for the high-level specification of the behavior of embedded systems. As an example, we consider a modern kitchen toaster. It has a turning knob to choose a desired browning level, a side lever to push down the bread and start the toasting process, and a stop button to cancel the toasting process. When the user inserts a slice of bread and pushes down the lever, the controller locks the retainer latch and switches on the heating element. In a basic toaster, the heating time depends directly on the selected browning level. In more advanced products, the intensity of heating can be controlled, and the heating period is adjusted according to the temperature of the toaster from the previous toasting cycle. When the appropriate time has elapsed or the user pushes the stop button, the the heating is switched off and latch is released. Moreover, we require that the toaster has a "defrost" button that, when activated, causes to heat the slice of bread with low temperature (defrosting) for a designated time before beginning the actual toasting process.

In the following, we present several ways of describing the behavior of this kitchen toaster with state machines: we give a basic state machine, a semantically equivalent hierarchical machine, and an extended state machine that makes intensive use of variables.

First, the toaster can be modeled by a simple state machine as shown in Figure 4.2. The alphabets are $I =\{push, stop, time, inc, dec, defrost, time\_d\}$ and $O = \{on, off\}$. The toaster can be started by pushing (*push*) down the latch. As a reaction, the heater is turned on (*on*). The toaster stops toasting (*off*) after a certain time (*time*) or after the stop button (*stop*) has been pressed. Furthermore, the toaster has two heating power levels, one of which

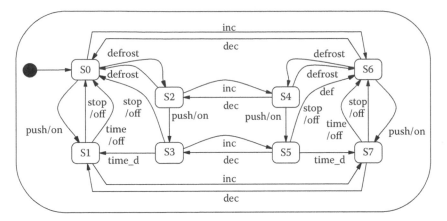

FIGURE 4.2
Simple state machine model of a kitchen toaster.

can be selected by increasing (*inc*) or decreasing (*dec*) the heating temperature. The toaster also has a defrost function (*defrost*) that results in an additional defrosting time (*time_d*) of frozen toast. Note that time is used in our modeling only in a qualitative way, that is, quantitative aspects of timing are not taken into account.

This simple machine consists of two groups of states: $s_0 \ldots s_3$ for regular heating and $s_4 \ldots s_7$ for heating with increased heating power. From the first group of states, the machine accepts an increase of the heating level, which brings it into the appropriate high-power state; vice versa, from this state, it can be brought back by decreasing the heating level. Thus, in this machine only two heating levels are modeled. It is obvious how the model could be extended for three or more such levels. However, with a growing number of levels the diagram would quickly become illegible.

It is clear that this modeling has other deficits as well. Conceptually, the setting of the heating level and defrosting cycle are independent of the operation of latch and stop button. Thus, they should be modeled separately. Moreover, the decision of whether to start a preheating phase before the actual toasting is "local" to the part dealing with the busy operations of the toaster. Furthermore, the toaster is either inactive or active and so active is a superstate that consists of substates defrosting and toasting.

To cope with these issues, UML offers the possibility of orthogonal regions and hierarchical nesting of states. This allows a compact representation of the behavior. Figure 4.3 shows a hierarchical state machine with orthogonal regions. It has the same behavior as the simple state machine in Figure 4.2. The hierarchical state machine consists of the three regions: *side latch*, *set temperature*, and *set defrost*. Each region describes a separate aspect of the toaster: in region *side latch*, the reactions to moving the side latch, pressing the stop button, and waiting for a certain *time* are described. The state *active* contains the substates *defrosting* and *toasting*, as well as a choice pseudostate. The region *set temperature* depicts the two heating levels and how to select them. In region *set defrost*, setting up the defrost functionality is described. The defroster can only be (de)activated if the toaster is not currently in state *active*. Furthermore, the defroster is deactivated after each toasting process.

Both the models in Figures 4.2 and 4.3 are concerned with the control flow only. Additionally, in any computational system the control flow is also influenced by data. In both of the above toaster models, the information about the current toaster setting is encoded in the states of the model. This clutters the information about the control flow and leads to

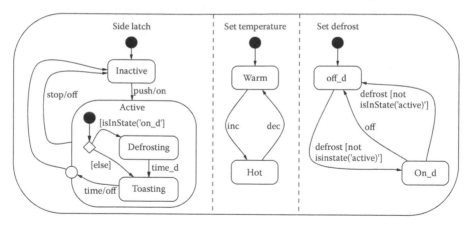

FIGURE 4.3

A hierarchical state machine model.

an excessive set of states. Therefore, it is preferable to use a data variable for this purpose. One option to do so is via extended finite state machines, where for instance, the transitions may refer to variables containing numerical data.

Figure 4.4 shows a more detailed model of a toaster that contains several variables. This model also contains the region *residual heat* to describe the remaining internal temperature of the toaster. Since a hot toaster reaches the optimal toasting temperature faster, the internal temperature is used in the computation of the remaining heating time. The state machine consists of the four regions: *side latch*, *set heating time*, *residual heat*, and *set defrost*. The names of the regions describe their responsibilities. The region *side latch* describes the reaction to press the side latch: If the side latch is pushed (*push*), the heater is turned on (releasing the event *on* and setting $h = true$). As a result, the toaster is in the state *active*. If the defrost button has been pressed ($d = true$), the toaster will be in the state *defrosting* for a certain time (*time_d*). The heating intensity (*h_int*) for the toasting process will be set depending on the set heat (*s_ht*) for the browning level and the residual heat (*r_ht*). The details of regulating the temperature are described in the composite state *toasting*: Depending on the computed value *h_int*, the toaster will raise the temperature (fast) or hold it at the current level. The toaster performs these actions for the time period *time* and then stops toasting. As an effect of stopping, it triggers the event *off* and sets $h = false$. The region *set heating time* allows to set the temperature to one of the levels 0 to 6. In the region *residual heat*, the heating up and the cooling down of the internal toaster temperature are described. The region *set defrost* allows to (de)activate the defrost mode. After completing one toasting cycle, the defrost mode will be deactivated.

4.1.3   Testing from UML state machines

Testing is the process of systematically experimenting with an object in order to detect failures, measure its quality, or create confidence in its correctness. One of the most important quality attributes is *functional correctness*, that is, determining whether the SUT satisfies the specified requirements. To this end, the requirements specification is compared to the SUT. In model-based testing, the requirements are represented in a formal model, and the SUT is compared to this model. A prominent approach for the latter is to derive test cases from the model and to execute them on the SUT. Following this approach, requirements, test cases, and the SUT can be described by a validation triangle as shown in Figure 4.5.

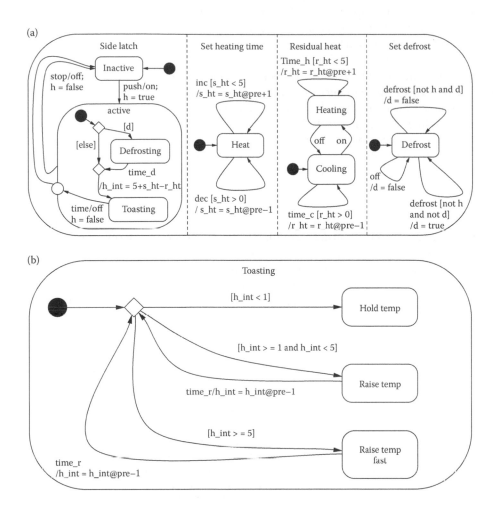

FIGURE 4.4
A UML state machine with variables.

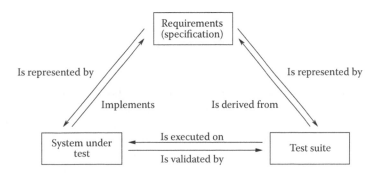

FIGURE 4.5
Validation triangle.

A *test case* is the description of a (single) test; a *test suite* is a set of test cases. Depending on the aspect of an SUT that is to be considered, test cases can have several forms—see Table 4.1. This table is neither a strict classification nor exhaustive. As a consequence, systems can be in more than one category, and test cases can be formulated in many different ways. Embedded systems are usually modeled as deterministic reactive systems, and thus, test cases are sequences of events. The notion of test execution and test oracle has to be defined for each type of SUT. For example, the execution of reactive system tests consists of feeding the input events into the SUT and comparing the corresponding output events to the expected ones.

For our example, the models describe the control of the toaster. They specify (part of) its observable behavior. Therefore, the observable behavior of each run of the state machine can be used as a test case. We can execute such a test case as follows: If the transition is labeled with an input to the SUT (pushing down the lever or pressing a button), we perform the appropriate action, whereas if it is labeled with an output of the SUT (locking or releasing the latch, turning heating on or off), we see whether we can observe the appropriate reaction. As shown in Table 4.2, model-based tests can be performed on various interface levels, depending on the development stage of the SUT.

An important fact about model-based testing is that the same logical test cases can be used on all these stages, which can be achieved by defining for each stage a specific

TABLE 4.1

Different SUT Aspects and Corresponding Test Cases

SUT Characteristics	Test Case
functional	pair (input value, output value)
reactive	sequence of events
nondeterministic	decision tree
parallel	partial order
interactive	test script or program
real time	timed event structure
hybrid	set of real functions

TABLE 4.2

Model-Based Testing Levels

Acronym	Stage	SUT	Testing Interfaces
MiL	Model-in-the-Loop	System Model	Messages and events of the model
SiL	Software-in-the-Loop	Control software (e.g., C or Java code)	Methods, procedures, parameters, and variables of the software
PiL	Processor-in-the-Loop	Binary code on a host machine emulating the behavior of the target	Register values and memory contents of the emulator
HiL	Hardware-in-the-Loop	Binary code on the target architecture	I/O pins of the target microcontroller or board
	System-in-the-Loop	Actual physical system	Physical interfaces, buttons, switches, displays, etc.

test adapter that maps abstract events to the concrete testing interfaces. For example, the user action of pushing the stop button can be mapped to send the event *stop* to the system model, to call of Java AWT ActionListener class method `actionPerformed(stop)`, to write *1* into address `0x0CF3` in a certain emulator running, for example, Java byte code, or to set the voltage at pin `GPIO5` of a certain processor board to high. System-in-the-loop tests are notoriously difficult to implement. In our example, we would have to employ a robot that is able to push buttons and observe the browning of a piece of toast.

4.1.4   Coverage criteria

Complete testing of all possible behaviors of a reactive system is impossible. Therefore, an adequate subset has to be selected, which is used in the testing process. Often, coverage criteria are used to control the test generation process or to measure the quality of a test suite. Coverage of a test suite can be defined with respect to different levels of abstraction of the SUT: requirements coverage, model coverage, or code coverage. If a test suite is derived automatically from one of these levels, coverage criteria can be used to measure the extent to which it is represented in the generated test suite.

In the following, we present coverage criteria as a means to measure the quality of a test suite. Experience has shown that there is a direct correlation between the various coverage notions and the fault detection capability of a test suite. The testing effort (another quality aspect) is measured in terms of the size of the test suite. In practice, one has to find a balance between minimal size and maximal coverage of a test suite.

Model coverage criteria can help to estimate to which extent the generated test suite represents the modeled requirements. Usually, a coverage criterion is defined independent of any specific test model, that is, at the meta-model level. Therefore, it can be applied to any instance of that meta-model. A *model coverage criterion* applied to a certain test model results in a set of *test goals*, which are specific for that test model. A test goal can be any model element (state, transition, event, etc.) or combination of model elements, for example, a sequence describing the potential behavior of model instances. A test case achieves a certain test goal if it contains the respective model element(s). A test suite *satisfies* (or *is complete for*) a coverage criterion if for each test goal of the criterion there is a test case in the suite that contains this test goal. The *coverage* of a test suite with respect to a coverage criterion is the percentage of test goals in the criterion, which are achieved by the test cases of the test suite. In other words, a test suite is complete for a coverage criterion iff its coverage is 100%. Typical coverage criteria for state machine models are as follows:

1. All-States: for each state of the machine, there is a test case that contains this state.

2. All-Transitions: for each transition of the machine, there is a test case that contains this transition.

3. All-Events: the same for each event that is used in any transition.

4. Depth-n: for each run $(s_0, a_1, s_1, a_2, \ldots, a_n, s_n)$ of length at most n from the initial state or configuration, there is a test case containing this run as a subsequence.

5. All-n-Transitions: for each run of length at most n from any state $s \in S$, there is a test case that contains this run as a subsequence (All-2-Transitions is also known as All-Transition-Pairs; All-1-Transitions is the same as All-Transitions, and All-0-Transitions is the same as All-States).

6. All-Paths: all possible transition sequences on the state machine have to be included in the test suite; this coverage criterion is considered infeasible.

In general, satisfying only All-States on the model is considered too weak. The main reason is that only the states are reached but the possible state changes are only partially covered. Accordingly, All-Transitions is regarded a minimal coverage criterion to satisfy. Satisfying the All-Events criterion can also be regarded as an absolute minimal necessity for any systematic black-box testing process. It requires that every input is provided at least once, and every possible output is observed at least once. If there are input events that have never been used, we cannot say that the system has been thoroughly tested. If there are specified output actions that could never be produced during testing, chances are high that the implementation contains a fault. Depth-n and All-n-Transitions can result in test suites with a high probability to detect failures. On the downside, the satisfaction of these criteria also often results in big test suites.

The presented coverage criteria are related. For instance, in a connected state machine, that is, if for any two simple states there is a sequence of transitions connecting them, the satisfaction of All-Transitions implies the satisfaction of All-States. In technical terms, All-Transitions subsumes All-States.

In general, coverage criteria *subsumption* is defined as follows: if any test suite that satisfies coverage criterion A also always satisfies the coverage criterion B, then A is said to *subsume* B. The subsuming coverage criterion is considered *stronger* than the subsumed one. However, this does not mean that a test suite satisfying the coverage criterion A necessarily detects more failures than a test suite satisfying B.

All-Transition-Pairs subsumes All-Transitions. There is no such relation for All-Events and All-Transitions. There may be untriggered transitions that are not executed by a test suite that calls all events; likewise, a transition may be activated by more than one event, and a test suite that covers all transitions does not use all of these events. Likewise, Depth-n is unrelated to All-Events and All-Transitions. For practical purposes, besides the All-Transitions criterion often the Depth-n criterion is used, where n is set to the diameter of the model. The criterion All-n-Transitions is more extensive; for $n \geq 3$, this criterion often results in a very large test suite. Clearly, All-n-Transitions subsumes Depth-n, All-$(n+1)$-Transitions subsumes All-n-Transitions for all n, and All-Paths subsumes all of the previously mentioned coverage criteria except All-Events. Figure 4.6 shows the corresponding subsumption hierarchy. The relation between All-n-Transitions and Depth-n is dotted because it only holds if the n for All-n-Transitions has at least the same value as the n of Depth-n.

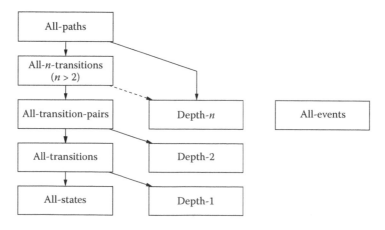

FIGURE 4.6
Subsumption hierarchy of structural coverage criteria.

Beyond simple states, UML state machines can contain orthogonal regions, pseudostates, and composite states. Accordingly, the All-States criterion can be modified to entail the following:

1. All reachable configurations,

2. All pseudostates, or

3. All composite states.

Likewise, other criteria such as the All-Transitions criterion can be modified such that all triggering events of all transitions or all pairs of configurations and outgoing transitions are covered [69]. Since there are potentially exponentially more configurations than simple states, constructing a complete test suite for all reachable configurations is often infeasible.

Conditions in UML state machine transitions are usually formed from atomic conditions with Boolean operators {and, or, not}, so the following control-flow-based coverage criteria focused on transition conditions have been defined [115]:

1. Decision Coverage, which requires that for every transition guard c from any state s, there is one test case where s is reached and c is true, and one test case where s is reached and c is false.

2. Condition Coverage, which requires the same as Decision Coverage for each atomic condition of every guard.

3. Condition/Decision Coverage, which requires that the test suite satisfies both Condition Coverage and Decision Coverage.

4. Modified Condition/Decision Coverage (MC/DC)[32, 31], which additionally requires to show that each atomic condition has an isolated impact on the evaluation of the guard.

5. Multiple Condition Coverage, which requires test cases for all combinations of atomic conditions in each guard.

Multiple Condition Coverage is the strongest control-flow-based coverage criterion. However, if a transition condition is composed of n atomic conditions, a minimal test suite that satisfies Multiple Condition Coverage may require up to 2^n test cases. MC/DC [32] is still considered very strong, and it is part of DO178-B [107] and requires only linear test effort. The subsumption hierarchy of control-flow-based coverage criteria is shown in Figure 4.7.

There are further coverage criteria that are focused on the data flow in a state machine, for example, on the definition and use of variables.

4.1.5 Size of test suites

The existence of a unique, minimal, and complete test suite, for each of the coverage criteria mentioned above, cannot be guaranteed. For the actual execution of a test suite, its size is an important figure. The size of a test suite can be measured in several ways or combinations of the following:

1. The number of all events, that is, the lengths of all test cases.

2. The cardinality, that is, the number of test cases in the test suite.

3. The number of input events.

At first glance, the complexity of the execution of a test suite is determined by the number of all events that occur in it. At a closer look, resetting the SUT after one test in order to run the next test turns out to be a very costly operation. Hence, it may be

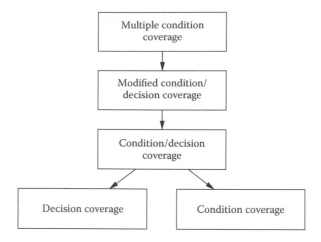

FIGURE 4.7
Subsumption hierarchy of condition-based coverage criteria.

advisable to minimize the number of test cases in the test suite. Likewise, for manual test execution, the performance of a (manual) input action can be much more expensive than the observation of the (automatic) output reactions. Hence, in such a case the number of inputs must be minimized. These observations show that there is no universal notion of minimality for test suites; for each testing environment different complexity metrics may be defined. A good test generation algorithm takes these different parameters into account. Usually, the coverage increases with the size of the test suite; however, this relation is often nonlinear.

4.2   Abstract Test Case Generation

In this section, we present the first challenge of automatic test generation from UML state machines: creating paths on the model level to cover test goals of coverage criteria. State machines are extended graphs, and graph traversal algorithms can be used to find paths in state machines [3, 62, 80, 82, 84, 88]. These paths can be used as *abstract test cases* that are missing the details about input parameters. In Section 4.3, we present approaches to generate the missing input parameters.

Graph traversal has been thoroughly investigated and is widely used for test generation in practice. For instance, Chow [33] creates tests from a finite state machine by deriving a testing tree using a graph search algorithm. Offutt and Abdurazik [92] identify elements in a UML state machine and apply a graph search algorithm to cover them. Other algorithms also include data flow information [23] to search paths. Harman et al. [67] consider reducing the input space for search-based test generation. Gupta et al. [61] find paths and propose a relaxation method to define suitable input parameters for these paths. We apply graph traversal algorithms that additionally compute the input parameter partitions [126, 127].

Graph traversing consists of starting at a certain start node n_{start} in the graph and traversing edges until a certain stopping condition is satisfied. Such stopping conditions are, for example, that all edges have been traversed (see the Chinese postman problem in [98]) or a certain node has been visited (see structural coverage criteria [115]). There are many different approaches to graph traversal. One choice is whether to apply *forward* or *backward* searching. In forward searching, transitions are traversed forward from the start state to the target state until the stopping condition is satisfied, or it is assumed that the criterion cannot be satisfied. This can be done in several ways such as, for instance, breadth-first, depth-first, or weighted breadth-first as in Dijkstra's shortest path algorithm. In backward searching, the stopping condition is to reach the start state. Typical nodes to start this backward search from are, for example, the states of the state machine in order to satisfy the coverage criterion All-States.

Automated test generation algorithms strive to produce test suites that satisfy a certain coverage criterion, which means reaching 100% of the test goals according to the criterion. The choice of the coverage criterion has significant impact on the particular algorithm and the resulting test suite. However, none of the above described coverage criteria uniquely determines the resulting test suite; for each criterion, there may be many different test suites achieving 100% coverage. For certain special cases of models, it is possible to construct test suites that satisfy a certain coverage criterion while consisting of just one test case. The model is strongly connected, if for any two states s and s' there exists a run starting from s and ending in s'. If the model is strongly connected, then for every n there exists a one-element test suite that satisfies All-n-Transitions: from the initial state, for all states s and sequence of length n from s, the designated run traverses this sequence and returns to the initial state.

An Eulerian path is a run that contains each transition exactly once, and a Hamilton path is a run that contains each state exactly once. An Eulerian or Hamiltonian cycle is an Eulerian or Hamiltonian path that ends in the initial state, respectively. Trivially, each test suite containing an Eulerian or Hamiltonian path is complete for All-Transitions or All-States, respectively. There are special algorithms to determine whether such cycles exist in a graph and to construct them if so.

In the following, we present different kinds of search algorithms: Dijkstra's shortest path, depth-first, and breadth-first. The criteria of when to apply which algorithm depend on many aspects. Several test generation tools implement different search algorithms. For instance, the Conformiq Test Designer [38] applies forward breadth-first search, whereas ParTeG [122] applies backward depth-first search.

4.2.1 Shortest paths

Complete coverage for All-States in simple state machines can be achieved with Dijkstra's single-source shortest path algorithm [108]. Dijkstra's algorithm computes for each node the minimal distance to the initial node via a greedy search. For computing shortest paths, it can be extended such that it also determines each node's predecessor on this path. The algorithm is depicted in Figures 4.8 and 4.9: Figure 4.8 shows the algorithm to compute shortest path information for all nodes of the graph. With the algorithm in Figure 4.9, a shortest path is returned for a given node of the graph. The generated test suite consists of all maximal paths that are constructed by the algorithm, that is the shortest paths for all nodes that are not covered by other shortest paths. For our toaster example in Figure 4.2, this algorithm can generate the test cases depicted in Figure 4.10.

The same algorithm can be used for covering All-Transitions by inserting a pseudostate in every transition as described in [124]. Furthermore, the generated path is extended by the outgoing transition of the just inserted pseudostate. In the generated test suite, only those sequences must be included, which are not prefixes (initial parts) of some other path. This set can be constructed in two ways:

1. In decreasing length, where common prefixes are eliminated.

2. In increasing length, where new test cases are only added if their length is maximal.

```
01 void Dijkstra(StateMachine sm, Node source) {
02   for each node n in sm {
03     dist[n] = infinity;              // distance function from source to n
04     previous[n] = undefined;         // Previous nodes determine optimal path
05   }
06   dist[source] = 0:                  // initial distance for source
07   set Q = all nodes in sm;
08   while Q is not empty {
09     u = node in Q with smallest value dist[u];
10     if (dist[u] = infinity)
11       break;                         // all remaining nodes cannot be reached
12     remove u from Q;
13     for each neighbor v of u {
14       alt = dist[u] + dist_between(u, v);
15       if alt < dist[v] {
16         dist[v] = alt;
17         previous[v] = u;
18 } } } }
```

FIGURE 4.8
Computing shortest distance for all nodes in the graph by Dijkstra.

```
01 Sequence shortestPath(Node target) {
02   S = new Sequence();
03   Node u = target;
04   while previous[u] is defined {
05     insert u at the beginning of S;
06     u = previous[u];
07 } }
```

FIGURE 4.9
Shortest path selection by Dijkstra.

```
TC1: (s0, (push, , on), s1, (dec, , ), s7)
TC2: (s0, (defrost, , ), s2, (push, , on), s3, (inc, , ), s5)
TC3: (s0, (inc, , ), s6, (defrost, , ), s4)
```

FIGURE 4.10
Test cases generated by the shortest path algorithm by Dijkstra.

The presented shortest path generation algorithm is just one of several alternatives. In the following, we will introduce further approaches.

4.2.2   Depth-first and breadth-first search

In this section, we describe depth-first and breadth-first graph traversal strategies. We defined several state machines that describe the behavior of a toaster. Here, we use the flat state machine of Figure 4.2 to illustrate the applicability of depth-first and breadth-first. The algorithm to find a path from the initial pseudostate of a state machine to certain state s via depth-first search is shown in Figure 4.11. The returned path is a sequence of transitions. The initial call is $depthFirstSearch(initialNode, s)$.

```
01 Sequence depthFirstSearch(Node n, Node s) {
02   if(n is equal to s) { // found state s?
03     return new Sequence();
04   }
05   for all outgoing transitions t of n { // search forward
06     Node target = t.target; // target state of t
07     Sequence seq = depthFirstSearch(target, s);
08     if(seq is not null) { // state s has been found before
09       seq.addToFront(t); // add the used transitions
10       return seq;
11   } }
12   if(n has no outgoing transitions) // abort depth-search
13     return null;
14 }
```

FIGURE 4.11
Depth-first search algorithm.

```
TC: (s0, (push, , on), s1, (inc, , ), s7, (time, , off), s6, (defrost, , ),
     s4, (dec, , ), s2, (push, , on), s3, (inc, , ), s5, (stop, , off), s6)
```

FIGURE 4.12
Test case generated by ParTeG for All-States.

```
01 Sequence breadthFirstSearch(Node n, Node s) {
02   TreeStructure tree = new TreeStructure();
03   tree.addNode(n);
04   while(true) { // run forever (until sequence is returned with this loop)
05     NodeSet ls = tree.getAllLeaves(); // get all nodes without outgoing transitions
06     for all nodes/leaves l in ls {
07       if(l references s) { // compare to searched state
08         Sequence seq = new Sequence();
09         while (l.incoming is not empty) { // there are incoming transitions
10           seq.addToFront(l.incoming.get(0)); // add incoming transition
11           l = l.incoming.get(0).source; } // l is set to l's predecessor
12         return seq;
13       } // else
14       for all outgoing transitions t of l { // search forward - build tree
15         Node target = t.target; // target state of t
16         new_l = tree.addNode(target); // get tree node that references target
17         tree.addTransitionFromTo(t, l, new_l); // add an edge from node l
18           // to node new_l; this new edge references transition t
19 } } } }
```

FIGURE 4.13
Breadth-first search algorithm.

For the example in Figure 4.2, ParTeG generates exactly one test case to satisfy All-States. Figure 4.12 shows this test case in the presented notation.

Figure 4.13 shows an algorithm for breadth-first search. Internally, it uses a tree structure to keep track of all paths. Just like a state machine, a tree is a directed graph with nodes and edges. Each node has incoming and outgoing edges. The nodes and edges of the tree reference nodes and edges of the state machine, respectively. It is initiated with the call *breadthFirstSearch(initialNode, s)*.

Both algorithms start at the initial pseudostate of the state machine depicted in Figure 4.2. They traverse all outgoing transitions and keep on traversing until *s* has been visited. Here, we present the generated testing tree for breadth-first search in the toaster example. We assume that the goal is to visit state *S5*. The testing tree is shown in Figure 4.14. It contains only edges and nodes; events are not presented here.

Because of loops in transition sequences, the result may be in general an infinite tree. The tree, however, is only built and maintained until the desired condition is satisfied, that is, the identified state is reached. In this example, the right-most path reaches the state *S5*.

A finite representation of this possibly infinite tree is a reachability tree, where each state is visited only once. Figure 4.15 shows such a reachability tree for the toaster example. Again, the figure depicts only edges and nodes, but no event or effect information.

Graph traversal approaches can also be applied to hierarchical state machines such as presented in Figure 4.3. For each hierarchical state machine, there exists an equivalent simple state machine; for instance, the models in Figures 4.3 and 4.2 have exactly the same behavior. Basically, each state in the flat state machine corresponds to a state configuration, that is, a set of concurrently active states, in the parallel state machine.

Extended state machine such as the one presented in Figure 4.4 can contain variables on infinite domains, and transitions can have arithmetic guard conditions and effects of

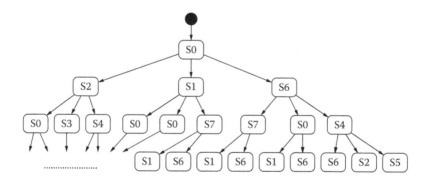

FIGURE 4.14
Testing tree that shows the paths for breadth-first search.

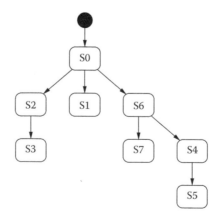

FIGURE 4.15
Reachability tree that shows only the paths to reach all states.

arbitrary complexity. The problem of reaching a certain state or transition in an extended state machine is therefore non trivial and, in the general case, undecidable. Therefore, for such models, the state set is partitioned into equivalence classes, and representatives from the equivalence classes are selected. These methods will be described in the next section.

4.3 Input Value Generation

In this section, we present the second challenge for automatic test generation: selecting concrete input values for testing.

All previously presented test generation techniques are focused on the satisfaction of coverage criteria that are applied to state machines. The corresponding test cases contain only the necessary information to traverse a certain path. Such test cases are called *abstract*—information about input parameters is given only partly as a partition of the possible input value space. Boundary value analysis is a technique that is focused on identifying representatives of partitions that are as close as possible to the partition boundaries. In the following, we present partition testing, as well as static and dynamic boundary value analysis.

4.3.1 Partition testing

Partition testing is a technique that consists of defining input value partitions and selecting representatives of them [64, 128, 89, 24, page 302]. There are several variants of partition testing. For instance, the category partition method [96] is a test generation method that is focused on generating partitions of the test input space. An example for category partitioning is the classification tree method (CTM) [60, 46], which enables testers to manually define partitions and to select representatives. The application of CTM to testing embedded systems is demonstrated in [83]. Basanieri and Bertolino use the category classification approach to derive integration tests using case diagrams, class diagrams, and sequence diagrams [13]. Alekseev et al. [5] show how to reuse classification tree models. The Cost-Weighted Test Strategy (CoWTeSt) [14, 15] is focused on prioritizing test cases to restrict their absolute number. CoWTeSt and the corresponding tool CowSuite have been developed by the PISATEL laboratory [103]. Another means to select test cases by partitioning and prioritization is the risk-driven approach presented by Kolb [79].

For test selection, a category partition table could list the categories as columns and test cases as rows. In each row, the categories that are tested are marked with an X. For the toaster, such a category partition table could look like depicted in Table 4.3. There are two test cases *TC1* and *TC2* that cover all of the defined categories.

Most of the presented partition testing approaches are focused on functional black-box testing that are solely based on system input information. For testing with UML state machines, the structure of the state machine and the traversed paths have to be included in

TABLE 4.3

Category Partition Table

Test Cases	Defrost	No Defrost	High Browning Level	Low Browning Level
TC1		X		X
TC2	X		X	

the computation of reasonable input partitions. Furthermore, the selection of representatives from partitions is an important issue. Boundary value analysis (BVA) consists of selecting representatives close to the boundaries of a partition, that is, values whose distances to representatives from other partitions are below a certain threshold. Consider the example in Figure 4.4. For the guard condition $s\_ht > 0$, 1 is a meaningful boundary value for s_ht to satisfy the condition, and 0 is a meaningful value to violate the condition. The task is to derive these boundary values automatically.

Here, we present two approaches of integrating boundary value analysis and automatic test generation with UML state machines: static and dynamic boundary value analysis [125].

4.3.2   Static boundary value analysis

In *static boundary value analysis*, BVA is included by static changes of the test model. For model-based test generation, this corresponds to transforming the test model. Model transformations for including BVA in test generation from state machines have been presented in [26]. The idea is to, for example, split a guard condition of the test model into several ones. For instance, a guard $[x >= y]$ is split into the three guards $[x = y]$, $[x = y + 1]$, and $[x > y + 1]$. Figure 4.16 presents this transformation applied to a simple state machine. The essence of this transformation is to define guard conditions that represent boundary values of the original guard's variables. As a consequence, the satisfaction of the transformed guards forces the test generator to also select boundary values for the guard variables. This helps to achieve the satisfaction of, for example, All-Transitions [115, page 117] requires the satisfaction of each guard and thus the inclusion of static BVA. There are such approaches for model checkers or constraint solvers that include the transformation or mutation of the test model. As one example, the Conformiq Test Designer [38] implements the approach of static BVA.

The advantages of this approach are the easy implementation and the linear test effort. However, this approach has also several shortfalls regarding the resulting test quality. In [125], we present further details.

4.3.3   Dynamic boundary value analysis

In *dynamic boundary value analysis*, the boundary values are defined dynamically during the test generation process and separately for each abstract test case. Thus, in contrast to static BVA, the generated boundary values of dynamic BVA are specific for each abstract test case. There are several approaches to implement dynamic BVA. In this section, we present a short list of such approaches.

In general, for dynamic boundary value analysis no test model transformations are necessary. For instance, an evolutionary approach can be used to create tests that cover certain parts of the model. In this case, a fitness function that returns good fitness values for parameters that are close to partition boundaries results in test cases with such input parameters that are close to these boundaries. Furthermore, any standard test generation approach can

FIGURE 4.16
Semantic-preserving test model transformation for static BVA.

be combined with a constraint solver that is able to include linear optimization, for example, lp_solve [19] or Choco [112], for generating input parameter values. There are many constraint solvers [58, 53, 11, 117, 48] that could be used for this task. Besides the presented approaches to dynamic BVA, there are industrial approaches to support dynamic BVA for automatic test generation with UML or B/Z [81, 110]. All these approaches to dynamic BVA are based on searching forward.

Another approach of searching backward instead of forward is called *abstract backward analysis*. It is based on the weakest precondition calculus [49, 129, 30] and on searching backward. During the generation of abstract test cases, all guards to enable the abstract test case are collected and transformed into constraints of input parameters. As a result, the generated abstract test case also contains constraints about the enabling input parameters. These constraints define partitions and thus can be used for BVA. This approach has been implemented in the model-based test generation prototype ParTeG [122, 126, 123]. In this implementation, the test generation algorithm starts at certain model elements that are specified by the applied structural coverage criterion and iterates backward to the initial node. As a result, the corresponding structural [115] and boundary-based [81] coverage criteria can be combined.

4.4   Relation to Other Techniques

The previous two sections dealt with the basic issues of generating paths in the state machine and selecting meaningful input data, respectively. In this section, we show several other techniques that may be used to support the two basic issues. In the following, we present random testing in Section 4.4.1, evolutionary testing in Section 4.4.2, constraint solving in Section 4.4.3, model checking in Section 4.4.4, and static analysis in Section 4.4.5.

4.4.1   Random testing

Many test generation approaches put a lot of effort in generating test cases from test models in a "clever" way, for instance, finding a shortest path to the model element to cover. It has been questioned whether this effort is always justified [104]. Any sort of black-box testing abstracts from internal details of the implementation, which are not in the realm of the test generation process. Nevertheless, these internals could cause the SUT to fail.

Statistical approaches to testing such as *random testing* have proven to be successful in many application areas [21, 85, 97, 34, 116, 35, 36]. Therefore, it has been suggested to apply random selection also to model-based test generation. In random testing, model coverage is not the main concern. The model abstracts from the SUT, but it is assumed that faults are randomly distributed across the entire SUT. Thus, random testing has often advantages over any kind of guided test generation. The model is used to create a large number of test cases without spending much effort on the selection of single tests. Therefore, random algorithms quickly produce results, which can help to exhibit design flaws early in the development process, while the model and SUT are still under development.

There are several publications on the comparison of random test generation techniques and guided test generation techniques. Andrews et al. [8] use a case study to show that random tests can perform considerably worse than coverage-guided test suites in terms of fault detection and cost-effectiveness. However, the effort of applying coverage criteria cannot be easily measured, and it is still unclear which approach results in higher costs. Mayer and Schneckenburger [86] present a systematic comparison of adaptive random testing techniques. Just like Gutjahr [63], Weyuker and Jeng [128] also focus their work on the comparison of random testing to partition testing. Major reasons for the success of random testing techniques are that other techniques are immature to a certain extent or that the used requirements specifications are partly faulty. Finally, developers as well as testers make errors (see Beizer [17] for the prejudice *Angelic Testers*). For instance, testers can forget some cases or simply do not know about them.

Random test generation can also be applied to model-based testing with UML state machines. For instance, this approach can be combined with the graph traversal approach of the previous section so as the next transition to traverse is selected randomly. Figure 4.17 shows one possible random test generation algorithm. First, it defines the desired length of the test case (line 03). Then, it selects and traverses one of the current node's outgoing transitions (line 06). This step is repeated until the current node has no outgoing transitions (line 07) or the desired test length has been reached (line 05). The resulting sequence is returned in line 13.

Figure 4.18 shows several randomly generated test cases for our toaster example in Figure 4.2.

4.4.2 Evolutionary testing

Evolutionary test generation consists of adapting an existing test suite until its quality, for example, measured with a fitness function, reaches a certain threshold. The initial test suite can be created using any of the above approaches. Based on this initial test suite, evolutionary testing consists of four steps: measuring the fitness of the test suite, selecting only the fittest test cases, recombining these test cases, and mutating them. In evolutionary testing, the set of test cases is also called population. Figure 4.19 depicts the process of evolutionary test generation. The dotted lines describe the start and the end of the test

```
01 Sequence randomSearch(Node source) {
02    Sequence seq = new Sequence();
03    int length = random();
04    Node currentNode = source;
05    for(int i = 0; i < length; ++i) {
06       transitions = currentNode.getOutgoing();
07       if (transitions.isEmpty()) { break; }
08       traverse = randomly select a representative of transitions;
09       seq.add(traverse);
10       // set current node to target node of traverse
11       currentNode = traverse.getTarget();
12    }
13    return seq;
14 }
```

FIGURE 4.17
Random search algorithm.

```
TC1: (s0, (push, , on), s1)
TC2: (s0, (inc, , ), s6, (dec, , ), s0, (push, , ), s1, (stop, , ), s0)
TC3: (s0, (inc, , ), s6, (push, , ), s7, (dec, , ), s1)
```

FIGURE 4.18
Randomly generated test cases.

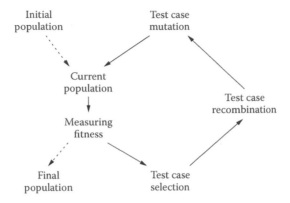

FIGURE 4.19
Evolutionary testing process.

generation process, that is, the initial population and—given that the measured fitness is high enough—the final population.

There are several approaches to steer test generation or execution with *evolutionary approaches* [87, 99, 78, 68, 119]. An initial (e.g., randomly created or arbitrarily defined) set of test input data is refined using mutation and fitness functions to evaluate the quality of the current test suite. For instance, Wegener et al. [120] show application fields of evolutionary testing. A major application area is the area of embedded systems [111]. Wappler and Lammermann apply these algorithms for unit testing in object-oriented programs [118]. Bühler and Wegener present a case study about testing an autonomous parking system with evolutionary methods [25].

Baudry et al. [16] present *bacteriological algorithms* as a variation of mutation testing and as an improvement of genetic algorithms. The variation from the genetic approach consists of the insertion of a new memory function and the suppression of the crossover operator. They use examples in Eiffel and a .NET component to test their approach and show its benefits over the genetic approach for test generation.

4.4.3 Constraint solving

The constraint satisfaction problem is defined as a set of objects that must satisfy a set of constraints. The process of finding these object states is known as *constraint solving*. There are several approaches to constraint solving depending on the size of the application domain. We distinguish large but finite and small domains. For domains over many-valued variables, such as scheduling or timetabling, *Constraint Programming* (CP) [106], *Integer Programming* (IP) [105], or *Satisfiability Modulo Theories* (SMT) [12] with an appropriate theory is used. For extensionally representable domains, using solvers for *Satisfiability* (SAT-Solver) [20] and *Answer Set Programming* (ASP) [10, 57] is state of the art. SAT is often used for hardware verification [50].

There are many tools (*solvers*) to support constraint solving techniques. Examples for constraint programming tools are the Choco Solver [112], MINION [58], and Emma [53]. Integer programming tools are OpenOpt [94] and CVXOPT [45]. An example for SMT solvers is OpenSMT [109]. There are several competitions for solvers [11, 117, 48]. Constraint solving is also used for testing. Gupta et al. [61] use a constraint solver to find input parameter values that enable a generated abstract test case. Aichernig and Salas [4] use constraint solvers and mutation of OCL expressions for model-based test generation. Calame et al. [27] use constraint solving for conformance testing.

4.4.4   Model checking

Model checking determines whether a model (e.g., a state machine) satisfies a certain property (e.g., a temporal logic formula). The model checking algorithm traverses the state space of the model and formula to deduce whether the model meets the property for certain (e.g., the initial or all) states. Typical properties are deadlock- or live-lock-freedom, absence of race conditions, etc.

If a model checker deduces that a given property does not hold, then it returns a path in the model as a counter example. This feature can be used for automatic test generation [7, 56, 55]. For that, each test goal is expressed as a temporal logic formula, which is negated and given to the model checker. For example, if the test goal is to reach "state_6," then the formula expresses "state_6 is unreachable." The model checker deduces that the test model does not meet this formula and returns a counter example. In the example, the counter example is a path witnessing that state_6 is indeed reachable. This path can be used to create a test case. In this way, test cases for all goals of the coverage criterion can be generated such that the resulting test suite satisfies the coverage criterion.

For our toaster example, the hierarchical state machine model depicted in Figure 4.3 can be coded in the input language of the NuSMV model checker as shown in Figure 4.20. The property states that states "toasting" and "on_d" are not reachable simultaneously. NuSMV finds that this is not true and delivers the path (test case) shown in Figure 4.21.

Model checking and test generation have been combined in different ways. Our example above is based on the work described in Hong et al. [72], which discuss the application of model checking for automatic test generation with control-flow-based and data-flow-based coverage criteria. They define state machines as Kripke structures [37] and translate them to inputs of the model checker SMV [73]. The applied coverage criteria are defined and negated as properties in the temporal logic CTL [37]. Callahan et al. [28] apply user-specified temporal formulas to generate test cases with a model checker. Gargantini and Heitmeyer [56] also consider control-flow-based coverage criteria. Abdurazik et al. [1] present an evaluation of specification-based coverage criteria and discuss their strengths and weaknesses when used with a model checker. In contrast, Ammann et al. [7] apply mutation analysis to measure the quality of the generated test suites. Ammann and Black [6] present a set of important questions regarding the feasibility of model checking for test generation. Especially, the satisfaction of more complex coverage criteria such as MC/DC [32, 31] is difficult because their satisfaction often requires pairs of test cases. Okun and Black [93] also present a set of issues about software testing with model checkers. They describe, for example, the higher abstraction level of formal specifications, the derivation of logic constraints, and the visibility of faults in test cases. Engler and Musuvathi [51] compare model checking to static analysis. They present three case studies that show that model checking often results in much more effort than static analysis although static analysis detects more errors than model checking. In [76], a tool is demonstrated that combines model checking and test generation. Further popular model checkers are the SPIN model checker [18], NuSMV [74], and the Java Pathfinder [70].

4.4.5   Static analysis

Static analysis is a technique for collecting information about the system without executing it. For that, a verification tool is executed on integral parts of the system (e.g., source code) to detect faults (e.g., unwanted or forbidden properties of system attributes). There are several approaches and tools to support static analysis that vary in their strength from analyzing only single statements to including the entire source code of a program. Static analysis is known as a formal method. Popular static analysis tools are the PC-Lint tool [59]

```
MODULE main
VAR state_sidelatch : {inactive, active_defrosting, active_toasting};
    state_settemp : {warm, hot};
    state_setdefrost : {off_d, on_d};
    action : {push, stop, inc, dec, defrost, on, off, time, time_d};
ASSIGN
 init(state_sidelatch) := inactive;
 init(state_settemp) := warm;
 init(state_setdefrost) := off_d;
 next(state_sidelatch) := case
  state_sidelatch=inactive & action=push & state_setdefrost=on_d : active_defrosting;
  state_sidelatch=inactive & action=push : active_toasting;
  state_sidelatch=active_defrosting & action=time_d : active_toasting;
  state_sidelatch=active_toasting & action=time : inactive;
  state_sidelatch=active_defrosting & action=stop : inactive;
  state_sidelatch=active_toasting & action=stop : inactive;
  1 : state_sidelatch; esac;
 next(state_settemp) := case
  state_settemp=warm & action=inc : hot;
  state_settemp=hot & action=dec : warm;
  1 : state_settemp; esac;
 next(state_setdefrost) := case
  state_setdefrost=off_d & action=defrost & state_sidelatch=inactive : on_d;
  state_setdefrost=on_d & action=off : off_d;
  state_setdefrost=on_d & action=defrost & state_sidelatch=inactive : off_d;
  1 : state_setdefrost; esac;
 next(action) := case
  state_sidelatch=inactive & action=push : on;
  state_sidelatch=active_toasting & action=time : off;
  state_sidelatch=active_defrosting & action=stop : off;
  state_sidelatch=active_toasting & action=stop : off;
  1 : {push, stop, inc, dec, defrost, on, off, time, time_d}; esac;
 SPEC AG ! (state_sidelatch=active_toasting & state_setdefrost=on_d)
```

FIGURE 4.20
SMV code for the hierarchical state machine toaster model.

for C and C++ or the IntelliJ IDEA tool [77] for Java. There are also approaches to apply static analysis on test models for automatic test generation [22, 95, 44, 100, 101]. Abdurazik and Offutt [2] use static analysis on UML collaboration diagrams to generate test cases. In contrast to state-machine-based approaches that are often focused on describing the behavior of one object, this approach is focused on the interaction of several objects. Static and dynamic analysis are compared in [9]. Ernst [52] argues for focusing on the similarities of both techniques.

4.4.6 Abstract interpretation

Abstract interpretation was initially developed by Patrick Cousot. It is a technique that is focused on approximating the semantics of systems [40, 42] by deducing information without executing the system and without keeping all information of the system. An abstraction of the real system is created by using an *abstraction function*. Concrete values can be represented as abstract domains that describe the boundaries for the concrete values. Several properties of the SUT can be deduced based on this abstraction. For mapping these

```
>NuSMV.exe toaster-hierarch.smv
*** This is NuSMV 2.5.0 zchaff (compiled on Mon May 17 14:43:17 UTC 2010)
-- specification AG !(state_sidelatch = active_toasting & state_setdefrost = on_d)
-- is false as demonstrated by the following execution sequence
Trace Description: CTL Counterexample
Trace Type: Counterexample
-> State: 1.1 <-
  state_sidelatch = inactive
  state_settemp = warm
  state_setdefrost = off_d
  action = push
-> State: 1.2 <-
  state_sidelatch = active_toasting
  action = on
-> State: 1.3 <-
  action = time
-> State: 1.4 <-
  state_sidelatch = inactive
  action = off
-> State: 1.5 <-
  action = defrost
-> State: 1.6 <-
  state_setdefrost = on_d
  action = push
-> State: 1.7 <-
  state_sidelatch = active_defrosting
  action = on
-> State: 1.8 <-
  action = time_d
-> State: 1.9 <-
  state_sidelatch = active_toasting
```

FIGURE 4.21
Result of NuSMV for the above example property.

properties back to the real system, a *concretization function* is used. The abstractions can be defined, for example, using Galois connections, that is, a widening and a narrowing operator [41]. Abstract interpretation is often used for static analysis. Commercial tools are, for example, Polyspace ® [113] for Java and C++ or ASTRE [43]. Abstract interpretation is also used for testing [39, 102].

4.4.7   Slicing

Slicing is a technique to slice parts of a program or a model by removing unnecessary parts and simplify, for example, test generation. The idea is that slices are easier to understand and to generate tests than from the entire program or model [65]. Program slicing was introduced in the Ph thesis of Weiser [121]. De Lucia [47] discusses several slicing methods (dynamic, static, backward, forward, etc.) that are based on statement deletion for program engineering. Fox et al. [54] present backward conditioning as an alternative to conditioned slicing that consists of slicing backward instead of forward. Whereas conditioned slicing provides answers to the question for the reaction of a program to a certain initial configuration and inputs, backward slicing finds answers to the question of what program parts can possibly lead to reaching a certain part or state of the program. Jalote et al. [75] present a framework for program slicing.

Slicing techniques can be used to support partition testing. For instance, Hierons et al. [71] use the conditioned slicing [29] tool ConSIT for partition testing and to test given input partitions. Harman et al. [66] investigate the influence of variable dependence analysis on slicing and present the corresponding prototype VADA. Dai et al. [46] apply partition testing and rely on the user to provide input partitions. Tip et al. [114] present an approach to apply slicing techniques to class hierarchies in C++. In contrast to the previous approaches, this one is focused on slicing structural artifacts instead of behavioral ones.

4.5 Conclusion

Model-based test generation from state-based models is a topic that has already been dealt with for many years. Several books about different modeling languages, application scenarios, and test generation approaches have been published. In this chapter, we presented an introduction to UML state machines as one kind of state-based models and showed several approaches to apply model-based test generation. For the interested reader, we provided several references for further studies.

In summary, we presented an introduction to automatic model-based test generation from UML state machines. For that, we gave a short introduction to UML state machines, presented a running example, and described how to generate tests from UML state machines. Then, we sketched different approaches to derive abstract test cases, that is, paths on graphs such as state machines, and described several approaches to the generation of concrete input parameter values. Finally, we presented the core ideas of related techniques such as constraint solving and model checking and how to apply them to model-based test generation.

References

1. Abdurazik, A., Ammann, P., Ding, W., and Offutt, J. (2000). Evaluation of three specification-based testing criteria. *IEEE International Conference on Engineering of Complex Computer Systems*, 179.

2. Abdurazik, A., and Offutt, J. (2000). Using UML collaboration diagrams for static checking and test generation. In Evans, A., Kent, S., and Selic, B., editors, *UML 2000— The Unified Modeling Language. Advancing the Standard. Third International Conference, York, UK, October 2000, Proceedings*, Volume 1939, Pages: 383–395. Springer.

3. Afzal, W., Torkar, R., and Feldt, R. (2009). A systematic review of search-based testing for non-functional system properties. *Information and Software Technology*, 51(6):957–976.

4. Aichernig, B. K. and Pari Salas, P. A. (2005). Test case generation by OCL mutation and constraint solving. *International Conference on Quality Software*, 64–71.

5. Alekseev, S., Tollkühn, P., Palaga, P., Dai, Z. R., Hoffmann, A., Rennoch, A., and Schieferdecker, I. (2007). Reuse of classification tree models for complex software projects. *Conference on Quality Engineering in Software Technology (CONQUEST)*.

6. Ammann, P. and Black, P. E. (2000). Test generation and recognition with formal methods. citeseer.ist.psu.edu/ammann00test.html.

7. Ammann, P. E., Black, P. E., and Majurski, W. (1998). Using model checking to generate tests from specifications. In *ICFEM'98: Proceedings of the Second IEEE International Conference on Formal Engineering Methods*, Page: 46, IEEE Computer Society, Washington, DC.

8. Andrews, J. H., Briand, L. C., Labiche, Y., and Namin, A. S. (2006). Using mutation analysis for assessing and comparing testing coverage criteria. *IEEE Transactions on Software Engineering*, 32:608–624.

9. Artho, C. and Biere, A. (2005). Combined static and dynamic analysis. In *AIOOL'05: Proceedings of the 1st International Workshop on Abstract Interpretation of Object-Oriented Languages*. Elsevier Science, ENTCS, Paris, France.

10. Baral, C. (2003). *Knowledge Representation, Reasoning and Declarative Problem Solving*. Cambridge University Press.

11. Barrett, C., Deters, M., Oliveras, A., and Stump, A. (2008). Design and results of the 3rd annual satisfiability modulo theories competition (SMT-COMP 2007). *International Journal on Artificial Intelligence Tools*, 17(4):569–606.

12. Barrett, C. W., Sebastiani, R., Seshia, S. A., and Tinelli, C. Satisfiability modulo theories. In Biere et al. [20], 825–885.

13. Basanieri, F., and Bertolino, A. (2000). A practical approach to UML-based derivation of integration tests. *4th International Software Quality Week Europe*.

14. Basanieri, F., Bertolino, A., and Marchetti, E. (2001). CoWTeSt: a cost weighted test strategy. In *In Escom-Scope 2001*, 387–396.

15. Basanieri, F., Bertolino, A., Marchetti, E., Ribolini, A., Lombardi, G., and Nucera, G. (2001). An automated test strategy based on UML diagrams. *Proceeding of the Ericsson Rational User Conference*.

16. Baudry, B., Fleurey, F., Jezequel, J.-M., and Le Traon, Yves. (2002). Automatic test cases optimization using a bacteriological adaptation model: application to .NET components. *Proceedings of ASE'02: Automated Software Engineering, Edinburgh*.

17. Beizer, B. (1990). *Software Testing Techniques*. John Wiley & Sons, Inc., New York, NY.

18. Bell Labs. (1991). SPIN Model Checker. http://www.spinroot.com/.

19. Berkelaar, M., Eikland, K., and Notebaert, P. (2004). lp_solve 5.1. http://lpsolve.sourceforge.net/5.5/.

20. Biere, A., Heule, M., van Maaren, H., and Walsh, T., editors. (2009). *Handbook of Satisfiability*, volume 185 of *Frontiers in Artificial Intelligence and Applications*. IOS Press.

21. Bird, D. L. and Munoz, C. U. (1983). Automatic generation of random self-checking test cases. *IBM Systems Journal*, 22(3):229–245.

22. Bozga, M., Fernandez, J.-C., and Ghirvu, L. (2000). Using static analysis to improve automatic test generation. In *TACAS'00: Proceedings of the 6th International Conference on Tools and Algorithms for Construction and Analysis of Systems*, Pages: 235–250. Springer-Verlag, London, UK.

23. Briand, L. C., Labiche, Y., and Lin, Q. (2005). Improving statechart testing criteria using data flow information. In *ISSRE'05: Proceedings of the 16th IEEE International Symposium on Software Reliability Engineering*, Pages: 95–104. IEEE Computer Society, Washington, DC.

24. Broy, M., Jonsson, B., and Katoen, J. P. (2005). *Model-Based Testing of Reactive Systems: Advanced Lectures (Lecture Notes in Computer Science)*. Springer.

25. Bühler, O. and Wegener, J. (2004). Automatic testing of an autonomous parking system using evolutionary computation.

26. Burton, S. (2001). *Automated Generation of High Integrity Tests from Graphical Specifications*. PhD thesis, University of York.

27. Calame, J. R., Ioustinova, N., van de Pol, J., and Sidorova, N. (2005). Data abstraction and constraint solving for conformance testing. In *APSEC '05: Proceedings of the 12th Asia-Pacific Software Engineering Conference*, Pages: 541–548. IEEE Computer Society, Washington, DC.

28. Callahan, J., Schneider, F., and Easterbrook, S. (Aug 1996). Automated software testing using model-checking. In *Proceedings 1996 SPIN Workshop*. Also WVU Technical Report NASA-IVV-96-022.

29. Canfora, G., Cimitile, A., and Lucia, A. D. (1998). Conditioned program slicing. *Information & Software Technology*, 40(11–12):595–607.

30. Cavalcanti, A., and Naumann, D. A. (2000). A weakest precondition semantics for refinement of object-oriented programs. *IEEE Transactions on Software Engineering*, 26(8):713–728.

31. Chilenski, J. J. (2001). MCDC forms (unique-cause, masking) versus error sensitivity. In *white paper submitted to NASA Langley Research Center under contract NAS1-20341*.

32. Chilenski, J. J., and Miller, S. P. (1994). Applicability of modified condition/decision coverage to software testing. *Software Engineering Journal*, 9: 193–200.

33. Chow, T. S. (1995). Testing software design modeled by finite-state machines. *Conformance testing methodologies and architectures for OSI protocols*, 391–400.

34. Ciupa, I., Leitner, A., Oriol, M., and Meyer, B. (2006). Object distance and its application to adaptive random testing of object-oriented programs. In *RT'06: Proceedings of the 1st International Workshop on Random Testing*, Pages: 55–63. ACM Press New York, NY.

35. Ciupa, I., Leitner, A., Oriol, M., and Meyer, B. (2007). Experimental assessment of random testing for object-oriented software. In *ISSTA'07: Proceedings of the International Symposium on Software Testing and Analysis 2007*, 84–94.

36. Ciupa, I., Pretschner, A., Leitner, A., Oriol, M., and Meyer, B. (2008). On the predictability of random tests for object-oriented software. In *ICST'08: Proceedings of the First International Conference on Software Testing, Verification and Validation*.

37. Clarke, E. M., Grumberg, O., and Peled, D. A. (2000). *Model Checking.* MIT Press.

38. Conformiq. Qtronic. http://www.conformiq.com/.

39. Cousot, P. (2000). Abstract interpretation based program testing. In *In Proc. SSGRR 2000 Computer & eBusiness International Conference*, Compact disk paper 248 and electronic proceedings http://www.ssgrr.it/en/ssgrr2000/proceedings.htm, 2000. Scuola Superiore G. Reiss Romoli.

40. Cousot, P. (2003). Automatic verification by abstract interpretation. In *VMCAI'03: Proceedings of the 4th International Conference on Verification, Model Checking, and Abstract Interpretation*, Pages: 20–24. Springer-Verlag, London, UK.

41. Cousot, P., and Cousot, R. (1992). Comparing the Galois connection and widening/narrowing approaches to abstract interpretation, invited paper. In Bruynooghe, M. and Wirsing, M., editors, *Proceedings of the International Workshop Programming Language Implementation and Logic Programming (PLILP '92)*, Leuven, Belgium, 13–17 August 1992, Lecture Notes in Computer Science 631, Pages: 269–295. Springer-Verlag, Berlin, Germany.

42. Cousot, P. and Cousot, R. (2004). *Basic Concepts of Abstract Interpretation*, Pages: 359–366. Kluwer Academic Publishers.

43. Cousot, P., Cousot, R., Feret, J., Mauborgne, L., Min, A., and Rival, X. (2003). ASTRE Static Analyzer. http://www.astree.ens.fr/.

44. Csallner, C., and Smaragdakis, Y. (2005). Check 'n' crash: combining static checking and testing. In *ICSE'05: Proceedings of the 27th International Conference on Software Engineering*, Pages: 422–431. ACM, New York, NY.

45. Dahl, J., and Vandenberghe, L. (2009). CVXOPT 1.1.1. http://abel.ee.ucla.edu/cvxopt/.

46. Dai, Z. R., Deussen, P. H., Busch, M., Lacmene, L. P., Ngwangwen, T., Herrmann, J., and Schmidt, M. (2005). Automatic test data generation for TTCN-3 using CTE. In *International Conference Software and Systems Engineering and their Applications (ICSSEA)*.

47. de Lucia, A. (2001). Program slicing: methods and applications. In *First IEEE International Workshop on Source Code Analysis and Manipulation*, Pages: 142–149. IEEE Computer Society Press, Los Alamitos, California, USA.

48. Denecker, M., Vennekens, J., Bond, S., Gebser, M., and Truszczyński, M. (2009). The second answer set programming competition. In Erdem, E., Lin, F., and Schaub, T., editors, *Proceedings of the Tenth International Conference on Logic Programming and Nonmonotonic Reasoning (LPNMR'09)*, Volume 5753 of *Lecture Notes in Artificial Intelligence*, Pages: 637–654. Springer-Verlag.

49. Dijkstra, E. W. (1976). *A Discipline of Programming.* Prentice-Hall.

50. Drechsler, R., Eggersglüß, S., Fey, G., and Tille, D. (2009). *Test Pattern Generation using Boolean Proof Engines.* Springer.

51. Engler, D., and Musuvathi, M. (2004). Static analysis versus software model checking for bug finding. citeseer.ist.psu.edu/engler04static.html.

52. Ernst, M. D. Static and dynamic analysis: synergy and duality. In *WODA'03: ICSE Workshop on Dynamic Analysis*, Pages: 24–27. Portland, Oregon. May 9, 2003.

53. Eve Software Utilities. (2009). Emma 1.0. http://www.eveutilities.com/products/emma.

54. Fox, C., Harman, M., Hierons, R., Ph, U., and Danicic, S. (2001). Backward conditioning: a new program specialisation technique and its application to program comprehension. citeseer.ist.psu.edu/fox01backward.html.

55. Fraser, G., and Wotawa, F. (2008). Using model-checkers to generate and analyze property relevant test-cases. *Software Quality Journal*, 16(2):161–183.

56. Gargantini, A., and Heitmeyer, C. (1999). Using model checking to generate tests from requirements specifications. *ACM SIGSOFT Software Engineering Notes*, 24(6): 146–162.

57. Gelfond, M. (2008). Answer sets. In Lifschitz, V., van Hermelen, F., and Porter, B., editors, *Handbook of Knowledge Representation*, Chapter 7. Elsevier.

58. Gent, I., Jefferson, C., Kotthoff, L., Miguel, I., Moore, N., Nightingale, P., Petrie, K., and Rendl, A. (2009). MINION 0.9. http://minion.sourceforge.net/.

59. Gimpel Software. (1985). PC-Lint for C/C++. http://www.gimpel.com/.

60. Grochtmann, M., and Grimm, K. (1993). Classification trees for partition testing. *STVR: Software Testing, Verification and Reliability*, 3(2):63–82.

61. Gupta, N., Mathur, A. P., and Soffa, M. L. (1998). Automated test data generation using an iterative relaxation method. In *SIGSOFT'98/FSE-6: Proceedings of the 6th ACM SIGSOFT International Symposium on Foundations of Software Engineering*, Pages: 231–244. New York, NY.

62. Gupta, N., Mathur, A. P., and Soffa, M. L. (1999). UNA based iterative test data generation and its evaluation. In *ASE'99: Proceedings of the ACM, 14th IEEE International Conference on Automated Software Engineering*, Page: 224. IEEE Computer Society, Washington, DC.

63. Gutjahr, W. J. (1999). Partition testing vs. random testing: the influence of uncertainty. *IEEE Transactions on Software Engineering*, 25(5):661–674.

64. Hamlet, D., and Taylor, R. (1990). Partition testing does not inspire confidence (Program Testing). *IEEE Transactions on Software Engineering*, 16(12):1402–1411.

65. Harman, M., and Danicic, S. (1995). Using program slicing to simplify testing. *Software Testing, Verification & Reliability*, 5(3):143–162.

66. Harman, M., Fox, C., Hierons, R., Hu, L., Danicic, S., and Wegener, J. (2003). VADA: a transformation-based system for variable dependence analysis. *IEEE International Workshop on Source Code Analysis and Manipulation*.

67. Harman, M., Hassoun, Y., Lakhotia, K., McMinn, P., and Wegener, J. (2007). The impact of input domain reduction on search-based test data generation. In *ESEC-FSE'07: Proceedings of the the 6th Joint Meeting of the European Software Engineering Conference and the ACM SIGSOFT symposium on The Foundations of Software Engineering*, Pages: 155–164. ACM, New York, NY.

68. Harman, M. and McMinn, P. (2007). A theoretical & empirical analysis of evolutionary testing and hill climbing for structural test data Generation. In *ISSTA'07: Proceedings of the 2007 International Symposium on Software Testing and Analysis*, Pages: 73–83. ACM, New York, NY.

69. Haschemi, S. (2009). Model transformations to satisfy all-configuration-transitions on statecharts. In *6th Workshop on Model-Based Design, Verification and Validation (MoDeVVa 2009)*.

70. Havelund, K., Visser, W., Lerda, F., Pasareanu, C., Penix, J., Mansouri-Samani, M., O'Malley, O., Giannakopoulou, D., Mehlitz, P., and Dillinger, P. (1999). Java pathFinder. http://javapathfinder.sourceforge.net/.

71. Hierons, R. M., Harman, M., Fox, C., Ouarbya, L., and Daoudi, M. (2002). Conditioned slicing supports partition testing. *Software Testing, Verification and Reliability*.

72. Hong, H., Lee, I., Sokolsky, O., and Cha, S. (2001). Automatic test generation from statecharts using model checking. In *In Proceedings of FATES'01 Workshop on Formal Approaches to Testing of Software, volume NS-01-4 of BRICS Notes Series*.

73. ITC-IRST and Carnegie Mellon University and University of Genoa and University of Trento. (1998). SMV. http://www.cs.cmu.edu/ modelcheck/smv.html.

74. ITC-IRST and Carnegie Mellon University and University of Genoa and University of Trento. (1999). NuSMV. http://nusmv.fbk.eu/.

75. Jalote, P., Vangala, V., Singh, T., and Jain, P. (2006). Program partitioning: a framework for combining static and dynamic analysis. In *WODA'06: Proceedings of the 2006 International Workshop on Dynamic Systems Analysis*, Pages: 11–16. ACM Press New York, NY.

76. Jéron, T., and Morel, P. (1999). Test generation derived from model-checking. In Halbwachs, N. and Peled, D., editors, *CAV'99: Proceedings of the 11th International Conference on Computer Aided Verification*, volume 1633 of *LNCS*, Pages: 108–122. Springer-Verlag, London, UK.

77. JetBrains. (2000). IntelliJ IDEA. http://www.jetbrains.com/.

78. Khor, S., and Grogono, P. (2004). Using a genetic algorithm and formal concept analysis to generate branch coverage test data automatically. In *ASE'04: Proceedings of the 19th IEEE International Conference on Automated Software Engineering*, Pages: 346–349. IEEE Computer Society, Washington, DC.

79. Kolb, R. (2003). A risk-driven approach for efficiently testing software product lines. citeseer.ist.psu.edu/630355.html.

80. Korel, B. (1990). Automated software test data generation. *IEEE Transactions on Software Engineering*, 16(8):870–879.

81. Kosmatov, N., Legeard, B., Peureux, F., and Utting, M. (2004). Boundary coverage criteria for test generation from formal models. In *ISSRE'04: Proceedings of the 15th International Symposium on Software Reliability Engineering*, Pages: 139–150. IEEE Computer Society. Washington, DC.

82. Lakhotia, K., Harman, M., and McMinn, P. (2008). Handling dynamic data structures in search based testing. In *GECCO'08: Proceedings of the 10th Annual Conference on Genetic and Evolutionary Computation*, Pages: 1759–1766. ACM, New York, NY.

83. Lamberg, K., Beine, M., Eschmann, M., Otterbach, R., Conrad, M., and Fey, I. (2005). Model-based testing of embedded automotive software using MTest.

84. Mansour, N., and Salame, M. (2004). Data generation for path testing. *Software Quality Control*, 12(2):121–136.

85. Mayer, J. (2005). On testing image processing applications with statistical methods. *Software Engineering*, 69–78.

86. Mayer, J., and Schneckenburger, C. (2006). An empirical analysis and comparison of random testing techniques. In *ISESE'06: Proceedings of the 2006 ACM/IEEE International Symposium on Empirical Software Engineering*, Pages: 105–114. ACM Press, New York, NY.

87. Mcgraw, G., Michael, C., and Schatz, M. (1997). Generating software test data by evolution. *IEEE Transactions on Software Engineering*, 27:1085–1110.

88. McMinn, P. (2004). Search-based software test data generation: a survey: research articles. *STVR: Software Testing, Verification and Reliability*, 14(2):105–156.

89. Ntafos, S. C. (2001). On comparisons of random, partition, and proportional partition testing. *IEEE Transactions on Software Engineering*, 27(10):949–960.

90. Object Management Group. (2005). Object Constraint Language (OCL), version 2.0. http://www.uml.org.

91. Object Management Group. (2009). Unified Modeling Language (UML), version 2.2. http://www.uml.org.

92. Offutt, J., and Abdurazik, A. (1999). Generating tests from UML specifications. In France, R. and Rumpe, B., editors, *UML'99—The Unified Modeling Language. Beyond the Standard. Second International Conference, Fort Collins, CO, October 28–30. 1999, Proccedings*, Volume 1723, Pages: 416–429. Springer.

93. Okun, V., and Black, P. E. (2003). Issues in software testing with model checkers. citeseer.ist.psu.edu/okun03issues.html.

94. Optimization Department of Cybernetic Institute. OpenOpt. http://openopt.org/.

95. Ostrand, T., Weyuker, E. J., and Bell, R. (2004). Using static analysis to determine where to focus dynamic testing effort. In *WODA'04: Workshop on Dynamic Analysis*.

96. Ostrand, T. J., and Balcer, M. J. (1988). The category-partition method for specifying and generating fuctional tests. *Communications of the ACM*, 31(6):676–686.

97. Owen, D., Desovski, D., and Cukic, B. (2006). Random testing of formal software models and induced coverage. *Random Testing*, Pages: 20–27.

98. Papadimitriou, C. H. (1976). On the complexity of edge traversing. *J. ACM*, 23(3):544–554.

99. Pargas, R. P., Harrold, M. J., and Peck, R. R. (1999). Test-data generation using genetic algorithms. *Software Testing, Verification And Reliability*, 9:263–282.

100. Peleska, J., Löding, H., and Kotas, T. (2007). Test automation meets static analysis. In Koschke, R., Herzog, O., Rödiger, K.-H., and Ronthaler, M., editors, *GI Jahrestagung (2)*, Volume 110 of *Lecture Notes in Informatics*, Pages: 280–290. GI.

101. Peleska, J., and Zahlten, C. (2007). Integrated automated test case generation and static analysis. In *QA+Test 2007: International Conference on QA+Testing Embedded Systems*.

102. Di Pierro, A., and Wiklicky, H. (2002). Probabilistic abstract interpretation and statistical testing. In *PAPM-PROBMIV '02: Proceedings of the Second Joint International Workshop on Process Algebra and Probabilistic Methods, Performance Modeling and Verification*, Pages: 211–212. Springer-Verlag. London, UK.

103. PISATEL LAB. (2002). http://www1.isti.cnr.it/ERI/special.htm.

104. Pretschner, A. (2006). Zur Kosteneffektivität des modellbasierten Testens. *MBEES'06: Modellbasierte Entwicklung eingebetteter Systeme*, 85–94.

105. Ravi Ravindran, A., editor. (2008). *Operations Research and Management Science Handbook*. CRC Press.

106. Rossi, F., van Beek, P., and Walsh, T., editors. (2006). *Handbook of Constraint Programming*. Elsevier.

107. RTCA Inc. (Dec 1992). RTCA/DO-178B, Software Considerations in Airborne Systems and Equipment Certification.

108. Saunders, S. (1999). A comparison of data structures for dijkstra's single source shortest path algorithm.

109. Sharygina, N., Bruttomesso, R., Tsitovich, A., Rollini, S., Tonetta, S., Braghin, C., and Barone-Adesi, K. (2009). OpenSMT. http://verify.inf.unisi.ch/opensmt.

110. Smartesting. Test Designer. http://www.smartesting.com.

111. Sthamer, H., Baresel, A., and Wegener, J. (2001). Evolutionary testing of embedded systems. *QW'01: Proceedings of the 14th International Internet & Software Quality Week*, 1–34.

112. The Choco Team. (2009). Choco Solver 2.1.0. http://choco.emn.fr/.

113. The Mathworks Inc. (1994). Polyspace embedded software verification. http://www.mathworks.com/products/polyspace /index.html.

114. Tip, F., Choi, J.-D., Field, J., and Ramalingam, G. (1996). Slicing class hierarchies in C++. In *OOPSLA'96: Proceedings of the 11th ACM SIGPLAN Conference on Object-Oriented Orogramming, Systems, Languages, and Applications*, Pages: 179–197. ACM Press, New York, NY.

115. Utting, M., and Legeard, B. (2006). *Practical Model-Based Testing: A Tools Approach*. Morgan Kaufmann Publishers Inc., San Francisco, CA.

116. Utting, M., Pretschner, A., and Legeard, B. (2006). A taxonomy of model-based testing. Technical Report 04/2006, Department of Computer Science, The Universiy of Waikato (New Zealand).

117. van Maaren, H., and Franco, J. (2009). The international SAT competitions web page. http://www.satcompetition.org/.

118. Wappler, S., and Lammermann, F. (2005). Using evolutionary algorithms for the unit testing of object-oriented software. In *GECCO'05: Proceedings of the Conference on Genetic and Evolutionary Computation*, Pages: 1053–1060. ACM Press, New York, NY.

119. Wappler, S., and Schieferdecker, I. (2007). Improving evolutionary class testing in the presence of non-public methods. In *ASE'07: Proceedings of the 22nd IEEE/ACM International Conference on Automated Software Engineering*, Pages: 381–384. ACM, New York, NY.

120. Wegener, J., Sthamer, H., and Baresel, A. (2001). Application fields for evolutionary testing. *Eurostar: Proceedings of the 9th European International Conference on Software Testing Analysis & Review*.

121. Weiser, M. D. (1979). *Program Slices: Formal, Psychological, and Practical Investigations of an Automatic Program Abstraction Method*. PhD thesis, University of Michigan, Ann Arbor, MI.

122. Weißleder, S. ParTeG (Partition Test Generator). http://parteg.sourceforge.net.

123. Weißleder, S. (2009). Influencing factors in model-based testing with UML state machines: report on an industrial cooperation. In *Models'09: 12th International Conference on Model Driven Engineering Languages and Systems*.

124. Weißleder, S. (2010). Simulated satisfaction of coverage criteria on UML state machines. *International Conference on Software Testing, Verification, and Validation (ICST)*.

125. Weißleder, S. (2010). Static and dynamic boundary value analysis.

126. Weißleder, S., and Schlingloff, H. (2007). Deriving input partitions from UML models for automatic test generation. In Giese, H., editor, *MoDELS Workshops*, Volume 5002 of *Lecture Notes in Computer Science*, Pages: 151–163. Springer.

127. Weißleder, S. and Schlingloff, H. (2008). Quality of automatically generated test cases based on OCL expressions. In *ICST'08: International Conference on Software Testing, Verification, and Validation*, Pages: 517–520.

128. Weyuker, E. J., and Jeng, B. (1991). Analyzing partition testing strategies. *IEEE Transactions on Software Engineering*, 17(7):703–711.

129. Whitty, R. W. (1991). An exercise in weakest preconditions. *Software Testing, Verification & Reliability*, 1(1):39–43.

5

Automated Statistical Testing for Embedded Systems*

Jesse H. Poore, Lan Lin, Robert Eschbach, and
Thomas Bauer

CONTENTS

---

*This chapter is an updated revision of "Application of Statistical Science to Testing and Evaluating Software Intensive Systems" that appeared in *Statistics, Testing and Defense Acquisition: Background Papers*, Copyright © 1999, the National Academy of Sciences. Courtesy of the National Academy Press, Washington, DC (Cohen, Rolph, and Steffey 1998).

5.1   Introduction

Embedded systems have become quite large over the years, with systems of 10 million lines of code now common. However, many organizations are struggling with the development and testing methods of times gone by when a small team of engineers could retain the modules of an embedded system. Any large, complex, expensive process with myriad ways to do most activities, as is the case with embedded systems development, can have its cost–benefit profile dramatically improved by the use of statistical science. Statistics provide a structure for collecting data and transforming it into information that can improve decision making under uncertainty.

The term "statistical testing" as typically used in the software engineering literature narrowly references randomly generated test cases. The term should be understood, however, as the comprehensive application of statistical science to solving the problems posed by industrial software development. Even when the concept is correctly understood, statistical testing is often dismissed as being impractical because of the high reliability levels called out in industry standards. However, a cost prohibitive standard to demonstrate no more than one failure in 10^5 demands or hours of service (colloquially called "five 9s") makes the standard itself absurd but does not negate the benefits of valid statistical testing. Good standards should acknowledge the value of a legitimate three 9s, for example. Statistical testing enables efficient collection of empirical data that will quantify the behavior of the software intensive system and support economic decisions regarding deployment of dependable systems.

Failures in the field, and the cost (social as well as monetary) of failures in the field, are one motivation behind statistical testing. For many organizations, the collection, classification, and analysis of field failure reports on software products have been standard practice for decades and are now a routine for most software intensive systems, regardless of the maturity of the organization. Field data is analyzed for a variety of reasons, among them the ability to budget support for the next release, to compare with past performance, to compare with competitive systems, and to improve the development process.

Field failure data is unassailable as evidence of need for process improvement. This operational front line is the source of the most compelling statistics. The opportunities to compel process changes move upstream from the field, through system testing, code development, specification writing, and into requirements analysis. Historically, the further one moves upstream, the more difficult it has been to effect a statistically based impact on the software development process that is designed to reduce failures in the field. The methods presented in this chapter facilitate prevention of field failures as well as statistically reasoned and economically beneficial impact on all aspects of the software life cycle.

In general, the concept of "testing in quality" is costly and ineffectual; software quality is achieved in the requirements, architecture, specification, design, code generation, and coding activities (Donohue and Dugan 2003). Statistical testing is not done for the purpose of finding errors (although it will); it is done to demonstrate and document that the system is fit for its intended use. The intended use sets the standard for the demonstration, and even standards that seem low in comparison to other applications of statistics nevertheless require thousands of tests. The problem of doing just enough testing to remove uncertainty regarding critical performance issues and to support a decision that the system is of requisite quality for its mission, environment, or market is amenable to solution by statistical science. The question is not whether to test, but when to test, what to test, and how much to test.

Of course, many embedded products are fielded without benefit of this method. Our emphasis is on the phrase *economical and feasible*. Most product developers see testing as a cost that adds no value; but the peril of failures in the field, product recall, liability, and infamy result in a great deal of expensive testing, much of which adds no valuable information. In the long run, statistical testing will reduce overall costs and shorten the testing phase of the product development cycle. We see a wide range of activities in the field from manual and regression testing to meet certain standards to software-in-the-loop, hardware-in-the-loop, and product testing. As testing moves from the laboratory to the product, each phase becomes more expensive than the previous one. Statistics can be applied to all stages. The goals are lower cost, shorter testing time, and no failures in the field.

There is a substantial body of literature related to the application of statistical science to software testing and evaluation and an even larger literature on testing in general (without the aid of statistics). It is beyond the scope of this chapter to give a comprehensive review of either. However, we will cite key statistical literature that forms the foundation of the Markov Chain Usage Model method presented in this chapter and that can be used in conjunction with the method.

The artifacts of statistics include population models, distributions, parameter estimation, sampling, and inference. Whereas finding the statistics in hardware was seemingly straightforward, as there was variation to study from one copy of a manufactured device to another and always wear, tear, and degradation. Finding the statistics in software has been controversial because all copies of a program are identical and they are not subject to physical wear. However, when the code is changed (to fix a bug) the population of uses while not changing may realize different experiences and outcomes. The relevant statistics for software are found in its use; thus we need statistical artifacts of use.

Testing produces samples of use prior to real use in the field. There is a difference between faults (bugs) in the code and failures in use. Moreover, not all faults have the same contribution to parameters of interest such as reliability and mean-time-to-failure (Boland, Singh, and Cukic 2004). Our application of statistics facilitates a mode of testing that will reveal faults in the order of their contribution to reliability or demonstrate that the highly likely use paths do not fail.

Estimation of software reliability has long been the object of study with twenty or more different published reliability models, that is, different models of failure, repair, estimation, and prediction. Nevertheless, there are many engineers (even company policy) today who insist that reliability is a concept reserved to properties of hardware but not applicable to properties of software. Ironically, for hardware devices with embedded software, this has led to giving the software a free pass, essentially regarding it as perfect. The hardware definition of reliability is the probability that the device has not yet failed at a point in time, but in the fullness of time will surely fail. Software reliability is generally based on the probability of a randomly selected use case executing correctly relative to a specification of correct behavior. The matter is skillfully addressed in Littlewood and Mayne (1989) and Littlewood and Strigini (2000) where it is argued that both use probability theory in a responsible way and are equally entitled to the use of the term "reliability." Of course, our need is for quantification of properties of the embedded system as a whole, whether measuring reliability in time or by demands.

The software testing problem is complex because of the astronomical number of scenarios of use in even the smallest embedded systems. The situation is aggravated when the system is on a communications network, as many products are, because of the variety of network signals that may have to be considered. The domain of testing is large and complex beyond human intuition. Because the software testing problem is so complex, statistical and other mathematical principles should be used to inform and manage the testing strategy.

Most of the methods followed are well within the capability of most test organizations, with a modest amount of training and tool support. Some of the ideas are more advanced and would require the services of a statistician the first few times they were used or until packaged in specialized tool support. Some of the advanced methods would require a resident analyst. However, the methods lend themselves to small and simple beginnings with big payoff and to systematic advancement in small steps with continued good return on the investment.

Section 5.2 establishes the method of modeling the population of uses to be analyzed according to first principles of statistics and introduces a medical device, an embedded system infusion pump, as a running example. Section 5.3 addresses model validation and revision through estimates of long-run statistics of use. Section 5.4 covers many forms of test management for many different testing needs and situation. Section 5.5 covers product certification and process improvement based on test statistics. Section 5.6 is summary and conclusion. Statistical testing requires special tools, and a list of the commands available in the JUMBL (Java Usage Model Builder Library) (Prowell 2003) library of tools developed by UTK SQRL (The University of Tennessee at Knoxville Software Quality Research Laboratory) follows as an appendix to the chapter.

5.2 Understanding the Software Intensive System and Its Use

A software intensive system can be described in terms of how it is going to be used in an operational environment. We consider the software testing problem as a statistical problem where the population contains all possible scenarios of system use. This usually infinite population is characterized with a Markov chain usage model.

5.2.1 First principles

A statistical principle of fundamental importance is that a population to be studied must first be characterized and that characterization must include the infrequent and exceptional as well as the common and typical. It must be possible to represent all questions of interest and all decisions to be made in terms of this characterization. All experimental design methods require such a characterization and representation, in one form or another, at a suitable level of abstraction. When applied to software testing, the population is the set of all possible scenarios of use, with each accurately represented as to frequency of occurrence.

One such method of characterization and representation is the operational usage model. The states of use of the system and the allowable transitions among those states are identified, and the probability of making each allowable transition is determined. These models are then represented in the form of one or more highly structured Markov chains, a type of statistical model (Kemeny and Snell 1960), and the result is called a usage model (Whittaker and Poore 1993, Whittaker and Thomason 1994).

A usage model characterizes the population of usage scenarios for a software intensive system. They are constructed from specifications, user guides, or even existing systems. The "user" might be a human, a hardware device, a network, another software system, or some combination of these. More than one model might be constructed for a single system if there is more than one environment of interest. For example, a medical device might have human, network, and hardware users if it exchanges information with all three. The usage model would be based on the states of use of the system—system off, system on and collecting measurements, system on and transmitting data, and so on—and the allowable transitions among the states. The model could be constructed without regard to whether the supplier will be General Electric or Siemens. It will be irrelevant that one uses a processor made by Intel and the other by Siemens and that they have very different internal hardware states—that one is programmed in C and the other in Ada. It is conceivable that the system would be tested in multiple environments of use, for example, in medical school training or implanted in a human.

When a population is too large for exhaustive study, as is usually the case for all possible uses of a software system, a statistically correct sample must be drawn as a basis for inferences about the population (Kaufman 1996). Figure 5.1 shows the parallel between a classical statistical design and statistical software testing (Poore and Trammell 1998). Under a statistical protocol, the environment of use can be modeled, independent samples taken (Goševa-Popstojanova and Trivedi 2000), and statistically valid statements can be made about a number of matters, including the expected operational performance of the software based on its test performance.

Statistical testing can be initiated at any point in the life cycle of a system, and all of the work products developed along the way become valuable assets that may be used throughout the life of the system. The statistical testing process involves the following six steps:

- Usage model construction

- Model analysis and validation

- Tool chain development

- Test planning

- Testing

- Product and process measurement

Industrial applications require a complete tool chain from requirements analysis to evaluation of testing results in order to process the thousands of test cases needed for standards compliance and company policy. See Bauer et al. (2007) for an example of a tool chain to support statistical testing of an embedded control unit for a car door mirror.

5.2.2 Building usage models

An operational usage model is a formal statistical representation of all possible uses of a system; in the context of this paper, it is always a Markov chain stochastic process. The structure of a usage model may be represented in the familiar form of a directed graph,

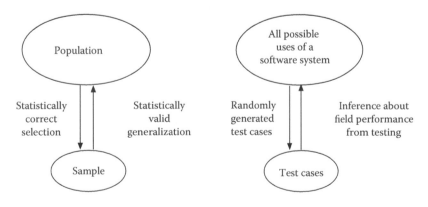

FIGURE 5.1
Parallel between statistical inference and software testing.

where the nodes represent states of system use and the arcs represent possible transitions between states (see Figure 5.2). The structure together with probability distributions over the exit arcs constitutes a Markov chain. (We note that while graphs are used in various software engineering artifacts, they are not statistical models.)

To illustrate the theory we use a model problem, a generic patient controlled analgesia infusion pump controller, as an example throughout the chapter. The original pump controller example is given in Real-Time Systems Group, 2010. It captures the functionality typical of this class of medical device.

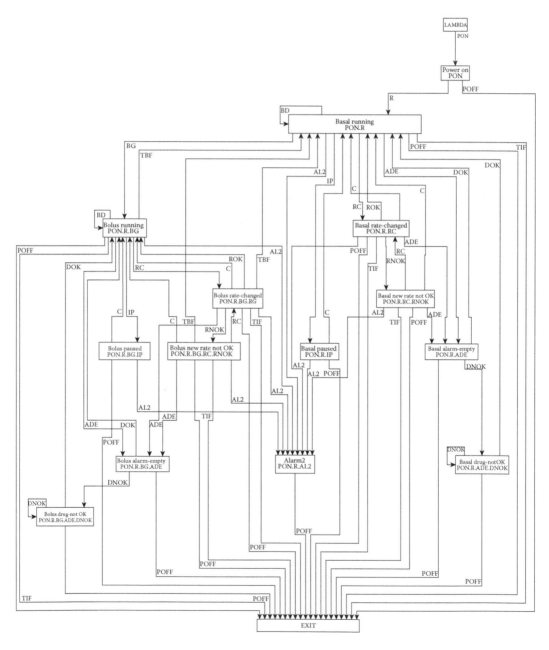

FIGURE 5.2
Usage model structure.

If the graph has any loops or cycles (as is usually the case), then an infinite number of finite sequences through the model are possible, thus an infinite population of usage scenarios. In such graphical form, usage models are easily understood by customers and users, who may participate in model development and validation. As a statistical formalism, a usage model lends itself to statistical analysis that yields quantitative information about system properties.

The basic task in model building (Walton, Poore, and Trammell 1995) is to identify the states of use of the system and the possible transitions among states of use (see Table 5.1). Every possible scenario of use at the chosen level of abstraction must be represented by

TABLE 5.1

Model Structure

From State		To State															
		LAMBDA	Power on	Basal Alarm-Empty	Basal Drug-notOK	Basal New Rate notOK	Basal Paused	Basal Rate-changed	Basal Running	Bolus Alarm-Empty	Bolus Drug-notOK	Bolus New Rate notOK	Bolus Paused	Bolus Rate-changed	Bolus Running	Alarm2	EXIT
		1	2	3	4	5	6	7	8	9	10	11	12	13	14	15	16
1	LAMBDA	0	1	0	0	0	0	0	0	0	0	0	0	0	0	0	0
2	Power on	0	0	0	0	0	0	0	X	0	0	0	0	0	0	0	X
3	Basal Alarm-Empty	0	0	0	X	0	0	0	X	0	0	0	0	0	0	0	X
4	Basal Drug-notOK	0	0	0	X	0	0	0	X	0	0	0	0	0	0	0	X
5	Basal New Rate notOK	0	0	X	0	0	0	X	X	0	0	0	0	0	0	X	X
6	Basal Paused	0	0	0	0	0	0	0	X	0	0	0	0	0	0	X	X
7	Basal Rate-changed	0	0	X	0	X	0	0	X	0	0	0	0	0	0	X	X
8	Basal Running	0	0	X	0	0	X	X	X	0	0	0	0	0	X	X	X
9	Bolus Alarm-Empty	0	0	0	0	0	0	0	0	0	X	0	0	0	X	0	X
10	Bolus Drug-notOK	0	0	0	0	0	0	0	0	0	X	0	0	0	X	0	X
11	Bolus New Rate notOK	0	0	0	0	0	0	0	X	X	0	0	0	X	X	X	X
12	Bolus Paused	0	0	0	0	0	0	0	0	0	0	0	0	0	X	X	X
13	Bolus Rate-changed	0	0	0	0	0	0	0	X	X	0	X	0	0	X	X	X
14	Bolus Running	0	0	0	0	0	0	0	X	X	0	0	X	X	X	X	X
15	Alarm2	0	0	0	0	0	0	0	0	0	0	0	0	0	0	0	1
16	EXIT	1	0	0	0	0	0	0	0	0	0	0	0	0	0	0	0

the model. Thus, every possible scenario of use is represented in the analysis, traceable on the model and potentially generated from the model as a test case. For example, a simple scenario of use is PON–POFF (power on followed immediately by power off), which can be traced down the right hand side of Figure 5.2.

There are both informal methods, such as those associated with "use cases" in an object-oriented approach, and formal methods of discovering the states and transitions.

One formal process developed by Prowell and Poore (2003) drives the discovery of usage states and transitions for discrete systems with a process based on the systematic enumeration of sequences of inputs and leads to a complete, consistent, correct, and traceable usage specification. Requirements analysis and refinement are by-products. The structure of the usage model can be generated in whole or in part directly from the specification (Broadfoot and Broadfoot 2003, Bauer et al. 2007, Carter, Lin, and Poore 2008, Bouwmeester, Broadfoot, and Hopcroft 2009).

The rigorous enumeration process was extended to directly treat matters of fundamental importance to embedded real-time systems: time, continuity, and nondeterminism (Carter 2009). Since many embedded real-time systems are built with MATLAB® Simulink® (Simulink 2010), and similar products, the hybrid automata enumerations were designed with Simulink as the implementation target; moreover, the Simulink system can be used as an oracle for automated testing as described below. As with discrete enumeration, the result is refinement of requirements, precise specifications, traceable decisions, consistency, and completeness. Testing models as well as executable models can be generated directly from the specification. Even when models are developed informally, they should be recast as enumerations to ensure consistency and completeness as an essential aspect of model validation.

Usage models are encoded as finite state, discrete parameter, time homogeneous, and recurrent Markov chains. Inherent in this type of model is the property that the states have no memory; some transitions in an application naturally do not depend on history, whereas others must be made independent of history by state splitting, making the states sufficiently detailed to reflect the relevant history. This leads to calculable growth in the number of states, which must be managed. A usage model is developed in two phases—a structural phase and a statistical phase. The structural phase concerns possible use; the statistical phase expected use. The structure of a model is defined by a set of states and an associated set of directed arcs that define state transitions. When represented as a stochastic matrix, the 0 entries represent the absence of arcs (impossible transitions), the 1 represent the certain transitions, and all other cells have transition probabilities of $0 < x < 1$ (see Table 5.1). This is the structure of the usage model.

The statistical phase is the determination of the transition probabilities, that is, the x's in the structure. There are two basic approaches to this phase, one based on direct assignment of probabilities and the other on deriving the values by analytical methods.

Models should be designed in a standard form consisting of connected submodels with a single entry and single exit. States and arcs can be expanded like macros. Submodels of canonical form can be collapsed to states or arcs. This permits model validation, specification analysis, test planning, and test case generation to occur on various levels of abstraction. The structure of the usage models should be reviewed with the specification writers, real or prospective users, the developers, and the testers. Users and specification writers are essential to represent the application domain and workflow. Developers get an early opportunity to see how the system will be used and can look ahead to implementation strategies that take account of use and workflow. Testers, who are often the model builders, get an early opportunity to define and automate the test environment. In our experience it is in the model development and validation phase that software errors are discovered and prevented, rather than during testing. This is as it must be for the certification of software through demonstration.

5.2.3 Software architecture

The architecture of the software intensive system is an important source of information in building usage models. If the model reflects the architecture of the system, then it will be easier to evolve the usage model as the system evolves. The architecture can be used to directly identify how models should be constructed and how testing should proceed.

A product line of embedded control units for infusion pumps might be based on a common set of objects for measurements, diagnostics analysis, drug dispensing, security, and adjustment by remote control, etc. Each object could be certified independently, and the object interactions as permitted by the supervisor would be certified with the supervisor. A new feature might be added later by developing a new object and modifying the supervisor; this would require a new model for the new object and an update of the model for the supervisor. Importance sampling might be used to emphasize testing of the changed aspects.

Protocols and other standards established by the architecture can also be factors in usage model development. For example, a usage model for the Small Computer System Interface (SCSI) protocol has been developed and used in constructing models of several systems that use it. A protocol for remote communication with medical devices would be similarly versatile.

Architecture and construction using submodels are a key to scalability for producing very large models. Tool support is provided to work with submodels and combining information in order to support product line architectures and scalability (see the Flatten command in the appendix).

5.2.4 Assigning transition probabilities

Transition probabilities among states in a usage model come from historical or projected usage data for the application. Many systems in use today log transaction activity, some to the detail of collection and storage of every keystroke. Because transition probabilities represent classes of users, environments of use, or special usage situations, several sets of probabilities may exist for a single model structure. Moreover, as the system progresses through its life cycle, the probability set may change several times, based on maturation of system use and availability of more information.

When extensive field data for similar or predecessor systems exist, a probability value may be known for every arc of the model (i.e., for every nonzero cell of the stochastic matrix of transition probabilities, as in Column 4 of Table 5.2). For new systems, one might stipulate expected practice based on user interviews, user guides, and training programs. This is a reasonable starting point, but should be open to revision as new information becomes available.

When complete information about system usage is not available, it is advisable to take an analytical approach to generating the transition probabilities, as will be presented in Section 5.3.3. In order to establish defensible plans, it is important that the model builder does not overstate what is known about usage or guess at values.

Embedded systems often present special situations of being in an idle loop waiting for an event to occur or in some steady state of operation that dominates all the special cases leading up to steady state or a change in mode of operation. In these cases, one may want to introduce statistical bias toward the more varied activity and then remove the bias from the analysis.

In the absence of compelling information to the contrary, the mathematically neutral position is to assign uniform probabilities over the transitions from a state in the usage model. Table 5.2 Column 4 represents a model based on Figure 5.2 with uniform transition probabilities across the exit arcs of each state.

TABLE 5.2

Example Usage Models, One Structure, Two Matrices of Transition Probabilities

	From-State		To-State	Stimulus	Uniform Probabilities	Specific Environment
1	LAMBDA	2	Power On	PON	1	1.00
2	Power On	8	Basal Running	R	1/2	0.90
2	Power On	16	EXIT	POFF	1/2	0.10
3	Basal Alarm-Empty	4	Basal Drug-notOK	DNOK	1/3	0.30
3	Basal Alarm-Empty	8	Basal Running	DOK	1/3	0.60
3	Basal Alarm-Empty	16	EXIT	POFF	1/3	0.10
4	Basal Drug-notOK	4	Basal Drug-notOK	DNOK	1/3	0.60
4	Basal Drug-notOK	8	Basal Running	DOK	1/3	0.30
4	Basal Drug-notOK	16	EXIT	POFF	1/3	0.10
5	Basal New Rate notOK	3	Basal Alarm-Empty	ADE	1/6	0.25
5	Basal New Rate notOK	7	Basal Rate-changed	RC	1/6	0.20
5	Basal New Rate notOK	8	Basal Running	C	1/6	0.30
5	Basal New Rate notOK	15	Alarm2	AL2	1/6	0.10
5	Basal New Rate notOK	16	EXIT	POFF	1/6	0.05
5	Basal New Rate notOK	16	EXIT	TIF	1/6	0.10
6	Basal Paused	8	Basal Running	C	1/3	0.60
6	Basal Paused	15	Alarm2	AL2	1/3	0.20
6	Basal Paused	16	EXIT	POFF	1/3	0.20
7	Basal Rate-changed	3	Basal Alarm-Empty	ADE	1/7	0.20
7	Basal Rate-changed	5	Basal New Rate notOK	RNOK	1/7	0.10
7	Basal Rate-changed	8	Basal Running	C	1/7	0.30
7	Basal Rate-changed	8	Basal Running	ROK	1/7	0.20
7	Basal Rate-changed	15	Alarm2	AL2	1/7	0.05
7	Basal Rate-changed	16	EXIT	POFF	1/7	0.05
7	Basal Rate-changed	16	EXIT	TIF	1/7	0.10
8	Basal Running	3	Basal Alarm-Empty	ADE	1/8	0.20
8	Basal Running	6	Based Paused	IP	1/8	0.20
8	Basal Running	7	Basal Rate-changed	RC	1/8	0.15

TABLE 5.2 (Continued)

Example Usage Models, One Structure, Two Matrices of Transition Probabilities

	From-State		To-State	Stimulus	Uniform Probabilities	Specific Environment
8	Basal Running	8	Basal Running	BD	1/8	0.10
8	Basal Running	14	Bolus Running	BG	1/8	0.20
8	Basal Running	15	Alarm2	AL2	1/8	0.05
8	Basal Running	16	EXIT	POFF	1/8	0.05
8	Basal Running	16	EXIT	TIF	1/8	0.05
9	Bolus Alarm-Empty	10	Bolus Drug-notOK	DNOK	1/3	0.30
9	Bolus Alarm-Empty	14	Bolus Running	DOK	1/3	0.60
9	Bolus Alarm-Empty	16	EXIT	POFF	1/3	0.10
10	Bolus Drug-notOK	10	Bolus Drug-notOK	DNOK	1/3	0.30
10	Bolus Drug-notOK	14	Bolus Running	DOK	1/3	0.60
10	Bolus Drug-notOK	16	EXIT	POFF	1/3	0.10
11	Bolus New Rate notOK	8	Basal Running	TBF	1/7	0.10
11	Bolus New Rate notOK	9	Bolus Alarm-Empty	ADE	1/7	0.20
11	Bolus New Rate notOK	13	Bolus Rate-changed	RC	1/7	0.20
11	Bolus New Rate notOK	14	Bolus Running	C	1/7	0.30
11	Bolus New Rate notOK	15	Alarm2	AL2	1/7	0.10
11	Bolus New Rate notOK	16	EXIT	POFF	1/7	0.05
11	Bolus New Rate notOK	16	EXIT	TIF	1/7	0.05
12	Bolus Paused	14	Bolus Running	C	1/3	0.70
12	Bolus Paused	15	Alarm2	AL2	1/3	0.20
12	Bolus Paused	16	EXIT	POFF	1/3	0.10
13	Bolus Rate-changed	8	Basal Running	TBF	1/8	0.10
13	Bolus Rate-changed	9	Bolus Alarm-Empty	ADE	1/8	0.20
13	Bolus Rate-changed	11	Bolus New Rate notOK	RNOK	1/8	0.10
13	Bolus Rate-changed	14	Bolus Running	C	1/8	0.20
13	Bolus Rate-changed	14	Bolus Running	ROK	1/8	0.20
13	Bolus Rate-changed	15	Alarm2	AL2	1/8	0.10
13	Bolus Rate-changed	16	EXIT	POFF	1/8	0.05

continued

TABLE 5.2 (Continued)

Example Usage Models, One Structure, Two Matrices of Transition Probabilities

	From-State		To-State	Stimulus	Uniform Probabilities	Specific Environment
13	Bolus Rate-changed	16	EXIT	TIF	1/8	0.05
14	Bolus Running	8	Basal Running	TBF	1/8	0.20
14	Bolus Running	9	Bolus Alarm-Empty	ADE	1/8	0.20
14	Bolus Running	12	Bolus Paused	IP	1/8	0.10
14	Bolus Running	13	Bolus Rate-changed	RC	1/8	0.10
14	Bolus Running	14	Bolus Running	BD	1/8	0.20
14	Bolus Running	15	Alarm2	AL2	1/8	0.10
14	Bolus Running	16	EXIT	POFF	1/8	0.05
14	Bolus Running	16	EXIT	TIF	1/8	0.05
15	Alarm2	16	EXIT	POFF	1	1.00
16	EXIT	1	LAMBDA		1	1.00

5.3   Model Validation with Product Manager and Customer

A usage model is a readily understandable representation of the system specification that may be reviewed with the customers and users. The following statistics are assured to be available by the mathematical structure of the models and are routinely calculated by the tools (see Tables 5.3 and 5.4 for examples and notice the Analyze command in the appendix).

- Long-run probability. This is the long-run occupancy rate of each state, or the usage profile as a percentage of time spent in each state. These are additive, and sums over certain states may be easier to check for reasonableness than the individual values.

- Probability of occurrence in a single sequence. This is the probability of occurrence of each state in a random use of the software.

- Expected number of occurrences in a single sequence. This is the expected number of times each state will appear in a single random use or test case.

- Expected number of transitions until the first occurrence. For each state, this is the expected number of randomly generated transitions (events of use) before the state will first occur, given that the sequence begins with the initial state (e.g., LAMBDA). This will show the impracticality of visiting some states in random testing without partitioning and stratification.

- Expected sequence length. This is the expected number of state transitions in a random use of the system and may be considered the average length of a use case or test case. Using this value and the transitions until first occurrence, one may estimate the number of test cases until first occurrence.

These statistics should be reviewed for reasonableness in terms of what is known or believed about the application domain and the environment of use. Given the model, these statistics are derived without further assumptions, and if they do not correspond with reality, then the model must be changed. These and other statistics describe the behavior that can

TABLE 5.3

Usage Statistics for the Model with Uniform Probabilities on the Exit Arcs

	State	Long-Run Probability	Probability of Occurrence in a Single Sequence	Expected Occurrences in a Single Sequence	Expected Transitions Until Occurrence
1	LAMBDA	0.22665	1.00000	1.00000	1.0
2	Power on	0.22665	1.00000	1.00000	1.0
3	Basal Alarm-Empty	0.02383	0.09514	0.10514	10.5
4	Basal Drug-notOK	0.01192	0.03386	0.05257	29.5
5	Basal New Rate notOK	0.00298	0.01273	0.01314	78.6
6	Basal Paused	0.02036	0.08473	0.08981	11.8
7	Basal Rate-changed	0.02085	0.08375	0.09200	11.9
8	Basal Running	0.16284	0.50000	0.71846	2.0
9	Bolus Alarm-Empty	0.00401	0.01604	0.01767	62.4
10	Bolus Drug-notOK	0.00200	0.00570	0.00884	175.5
11	Bolus New Rate notOK	0.00045	0.00192	0.00196	521.9
12	Bolus Paused	0.00350	0.01456	0.01543	68.7
13	Bolus Rate-changed	0.00356	0.01447	0.01571	69.1
14	Bolus Running	0.02797	0.08674	0.12342	11.5
15	Alarm2	0.03579	0.15789	0.15789	6.3
16	EXIT	0.22665	1.00000	1.00000	1.0

Number of arcs is 66.

Expected sequence length is approximately 3.412 events.

The log base 2 source entropy is approximately 0.981247281 bits.

The specification complexity index is approximately 4.053 (or 2^4.053 sequences).

be expected in the "long run," that is, in ongoing field use of the software. It may be impractical for enough testing to be done for all aspects of the process to exhibit long-run effects; exceptions can be addressed through special testing situations (as discussed below).

5.3.1 Operational profiles

Operational profiles (Leung 1997, Musa 1998) describe field use. Testing based on an operational profile ensures that the most frequently used features will be tested most thoroughly. When testing schedules and budgets are tightly constrained, profile-based testing yields the highest practical reliability; if failures are seen they would be the high-frequency failures and consequent engineering changes would be those yielding the greatest increase in reliability. Note that critical but infrequently used features, perhaps related to safety, high cost of failure, or high value must receive special attention; for this reason the tools facilitate the attachment of cost and value to arcs of the model and these can be used to drive testing.

One approach to statistical testing is to estimate the operational profiles first and then base random test cases on them. The usage model approach advocated here is to first build

TABLE 5.4

Usage Statistics of Model for Specific Environment

	State	Long-Run Probability	Probability of Occurrence in a Single Sequence	Expected Occurrences in a Single Sequence	Expected Transitions Until Occurrence
1	LAMBDA	0.11914	1.00000	1.00000	1.0
2	Power on	0.11914	1.00000	1.00000	1.0
3	Basal Alarm-Empty	0.06092	0.34386	0.51130	2.9
4	Basal Drug-notOK	0.02611	0.13384	0.21913	7.5
5	Basal New Rate notOK	0.00398	0.03195	0.03338	31.3
6	Basal Paused	0.05197	0.33792	0.43619	3.0
7	Basal Rate-changed	0.03977	0.25798	0.33382	3.9
8	Basal Running	0.25983	0.90000	2.18100	1.1
9	Bolus Alarm-Empty	0.02333	0.13374	0.19584	7.5
10	Bolus Drug-notOK	0.01000	0.05157	0.08393	19.4
11	Bolus New Rate notOK	0.00107	0.00867	0.00898	115.4
12	Bolus Paused	0.01049	0.07522	0.08804	13.3
13	Bolus Rate-changed	0.01070	0.07559	0.08984	13.2
14	Bolus Running	0.10489	0.36150	0.88039	2.8
15	Alarm2	0.03954	0.33184	0.33184	3.0
16	EXIT	0.11914	1.00000	1.00000	1.0

Number of arcs is 66.

Expected sequence length is approximately 7.394 events.

The log base 2 source entropy is approximately 1.458 bits.

The specification complexity index is approximately 11.675 (or 2^11.675 sequences).

a model of system use (describe the stochastic process) based on many decisions as to states of use, allowable transitions, and the probability of those transitions and then calculate the operational profile as the long-run behavior of the stochastic process so described.

A usage model can be designed to simulate any operational condition of interest, such as normal use, nonroutine use, hazardous use, or malicious use. Analytical results are studied during model validation, and surprises are not uncommon. Parts of systems believed to be unimportant may experience surprisingly heavy use while parts that consume a large amount of the development budget may see little use. Since a usage model is based on the software requirements and specifications rather than the code, the model can be constructed early in the life cycle to inform the development process as well as for testing and certification of the code.

5.3.2   Specification complexity

Entropy is defined for a probability distribution or stochastic source as the quantification of uncertainty. The greater the entropy, the more uncertain the outcome or behavior. As new information is incorporated into the source, the behavior of the source generally becomes more predictable and less uncertain. One interpretation of entropy is the minimum average number of "yes or no" questions required to determine one outcome or observation of the random event or process (Ash 1965). As an example, in the pump controller model (Table 5.2) the next state from "Power On" could be either "Basal Running" or "EXIT" with a probability distribution. An entropy can be defined and interpreted as how many binary questions, on average, it takes to determine the next state from "Power On" given that distribution.

Each state of a usage model has a probability distribution across its exit arcs to describe the transitions to other states, which appears as a row of the transition matrix. State entropy gives a measure of the uncertainty in the transition from that state.

Source entropy is by definition the probability-weighted average of the state entropies. Source entropy is an important reference value because the greater the source entropy, the greater the number of sequences (test cases) that it would be necessary to generate from the usage model, on average, to obtain a sample that is representative of usage as defined by the model.

Some systems are untestable in any meaningful sense, even though they might become successful products through "customer testing." Some systems have such a large number of significant paths of use and such a high cost of testing per path that there is insufficient time and budget to perform adequate testing by any criteria, even with the leverage of statistical sampling (Butler and Finelli 1993). Usage models can identify and substantially mitigate such situations early in the process by helping the product manager to reduce features, increase budget, or ultimately decide how to use the available (but inadequate) budget of time and money. Statistical analysis does not create the problem; it simply quantifies the problem.

A usage model represents the capability of the system in an environment of use. All usage steps are probability weighted. Any model with a loop or cycle has an infinite number of paths; however, only a finite number have a large enough probability of occurring to be considered. The *complexity* of a model can be viewed as the number of statistically typical paths (to be thought of as "paths worth considering"). Note that this concept of complexity has nothing to do with the technical challenge posed by the requirements, nor with the intricacies of the ultimate software implementation. It is simply a measure of how many ways the system may be used (how broadly the probability mass is spread over sequences) and therefore a measure of the size of the testing problem.

Complexity analysis can be used to assess the extent to which modification of the specification (and usage model) would reduce the size of the testing problem. By excluding states and arcs from the model, such what-if calculations can be made. For example, modeless display systems that allow the user to switch from any task to any other task are far more expensive to test than modal displays that restrict tasks to categories. It is possible, also, to compare the differences in complexity associated with different environments of use (represented by different sets of transition probabilities as in Tables 5.3 and 5.4). Complexity analysis can be used to assess the impact of changes in the requirements and system implementation on testing. Because the usage model is based on the requirements and specifications, the model can be developed, validated, and analyzed before code is written. An analysis of the complexity of the model may lead to simplification of the specification in various ways, before code development begins.

When the system cannot be changed to reduce complexity and the test budgets cannot be made adequate, usage models can help to focus the budgets on the most important states, arcs, and paths. Certain usage states might be critical to achieve (or to avoid), and the number of pathways by which one might achieve (or avoid) these states could be very important. In a slightly more complex situation, there may be two or more states among which passage should be quick and easy (or virtually impossible). Trajectory entropy provides a measure of the uncertainty in selecting a path from a set of paths. A variation on the techniques of Ekroot and Cover (1993) produces the measure of specification complexity (Walton and Poore 2000). Trajectory entropy is the sum of the uncertainty of the first step in the path plus the conditional uncertainty of the rest of the path, given the first step. This value is the ratio of the source entropy to the stationary probability of the initial state and is used as an index of specification complexity, the minimum average number of yes–no questions one would have to ask to identify the path taken. When 2 is raised to this power, an estimate of the number of paths worth considering is obtained. Many well-posed questions involving states, arcs, and paths can be expressed in a mathematical model with a closed-form solution.

As mentioned above, these statistics and analyses flow from the usage model without further assumptions. If the structure of the model represents the capability of the system, and if the probabilities represent the environment of use, then the conclusions are inescapable. If they do not agree with what is known or believed about the application, then the model must be changed.

Even small models embody a great deal of variation. Consequently, it is not always obvious how to change a model in order to change its statistics. Moreover, small changes in the probabilities can have large and unanticipated side effects (Ou and Dugan 2003). An alternative to the cycle of setting probabilities, analyzing statistics, and revising probabilities is to analytically generate models with stochastic matrices guaranteed to have certain statistics, as described in the next section.

5.3.3   Representing usage models with constraints

An alternative to the direct assignment of transition probabilities discussed in Section 5.2 is to generate transition probabilities with the aid of mathematical programming (specifically, convex programming) (Poore, Walton, and Trammell 2000). Usage models can be represented by a system of constraints, and the matrix of transition probabilities can be generated as the solution to an optimization problem. In general, three forms of constraints are used to define a model: structural, usage, and test management constraints.

Structural constraints define the model structure of states and both the possible and impossible transitions among the usage states.

There are four types of structural constraints:

- $P_{i,j} = 0$ defines an impossible transition between usage state i and usage state j.

- $P_{i,j} = 1$ defines a certain transition between usage state i and usage state j.

- $0 < P_{i,j} < 1$ defines probabilistic transition between usage state i and usage state j.

- Each row of P must sum to one.

If no information about the expected usage of the system is available, one should generate uniform probabilities for the possible transitions from each state. As new information arises, it is recorded in the form of constraints:

- $P_{i,j} = c$ may be used for known usage probabilities, that is, probability values that are exactly known on the basis of historical experience or designed controls.

- $a \leq P_{i,j} \leq b$ defines estimated usage probabilities as a range of values. Defining an estimate as being within a range allows information to be given without being overstated.

- $P_{i,j} = P_{k,m}$ defines equivalent usage probabilities; values that should be the same whether or not one knows what the value should be.

- $P_{i,j} = d\ P_{k,m}$ defines proportional usage probabilities, where one value is a multiple of another.

Probability values can be related to each other by a function to represent what is known about the relationship, without overstating the data and knowledge. More complex constraints may be expressed as follows:

- $P_{i,j} = f(P_{k,m})$, where one value is a function of another.

- $a \leq f(P) \leq b$, where the value of a function of the matrix P is bounded, for example, to constrain the average test case length to a certain range.

Finally, constraints may be used to represent test management controls. Management constraints are of the same forms as usage constraints. A limitation on revisiting previously tested functionality, for example, may be represented in the form of a known usage probability in the section above—a constant that limits the percentage of test cases entering a certain section of the model or a zero to prevent a set of paths from being generated.

For example, certain elements of the right most column of Table 5.2 can be defined by the following constraints:

- $P_{2,8} = 9\ P_{2,16}$.

- $P_{3,8} = P_{4,4} = P_{6,8} = P_{9,14} = P_{10,14} = 0.6$.

5.3.4 Objective functions

Mathematical programming is a technique for determining the values of a finite set of decision variables that optimize an objective function subject to a specified set of mathematical constraints. The general problem of optimizing any function subject to a set of unrestricted constraints can be analytically or computationally intractable. The problem is tractable when it is restricted to convex programming: the minimization of a convex objective function subject to a finite set of convex constraints.

When mathematical programming is used to generate transition probabilities, the solution is optimized for some objective function while satisfying all structural, usage, and management constraints. Theoretically, one could construct a system of constraints for which there is no solution. In practice, if one does not overstate data and knowledge, this is unlikely.

Analysis of a usage model invariably leads to modification of the transition probabilities, in order to incorporate new information or to change focus at different phases of the analysis and testing process. With complex usage models, individual changes in transition probabilities may result in unintended, poorly understood, and unwanted side effects. Better control and understanding is maintained if models are amended through revised or additional constraints and regenerated relative to an optimization objective, rather than by estimation of individual transition probabilities.

Objective functions can be formulated, for example, to minimize cost of testing or to maximize value of testing. Also, entropy measures can be used in objective functions in order to minimize or maximize the uncertainty or variability in the model and, consequently, in the sequences randomly generated from the model.

For example, the structural constraints plus the specific constraints in the section above could be used to generate all other transition probabilities so that the expected test case length is minimized.

There are, in general, many sets of transition probabilities that collectively satisfy a system of constraints. Even when the usage profile (stationary distribution) is fully prescribed, many sets of transition probabilities with the same usage profile are possible for the usage model. Consequently, the certification strategy must be based on carefully reasoned choices among them, in order to support the dependability case. Mathematical programming can be used to make that choice.

Most usage models can be defined with very simple constraints. Again, TML (The Model Language developed by UTK SQRL) and JUMBL support this process.

5.4   Usage Modeling Supports Statistical Testing

As early as possible in the life cycle, one or more usage models are developed and validated. To the best ability of the model developers, with the information available to them, the model represents the operational capability of the system at the desired level of abstraction, and the statistics agree with what is known or believed about the intended environment of use. The following is a summary of the many beneficial uses of the model in planning, managing, and conducting statistical testing.

5.4.1   Testing scripts

A test case is a series of arcs through the usage model from entry to exit. For example, from Figure 5.2 we may encounter the sequence: PON-R-C-ADE-DNOK-DOK-IP-AL2-POFF. A script is associated with each arc of the usage model. Thus, the test case is a series of scripts. These scripts constitute the instructions for testing the transition from one state of use to another as represented by the arc. Scripts should be developed and validated by experienced testers. The scripts are a significant factor in assuring experimental control during testing. Both the TML notation and the JUMBL library support this use of scripts.

In the case of testing performed by humans, the script can tell the tester what to do, what inputs to give the system, and what to look for in deciding whether the transition was made correctly or not. Testing can be a tedious activity that degenerates in effectiveness, unless specific measures are taken to keep the testers focused on what to do and what to look for. Furthermore, testing effectiveness can vary greatly from one person to another, unless steps are taken to assure uniformly effective testing. Every test is a traversal of a series of arcs through the model; if the scripts are granular and are followed, they will form the basis for uniform testing.

In the case of automated testing, the scripts will be commands to software test runners, software-in-the-loop systems, hardware-in-the-loop systems, or other equipment and in most cases will contain the information needed to verify correct performance. Lines of code in various languages have been used as scripts in such a way that the test case literally becomes a program to be compiled and executed by the automated test facility. Obviously, scripting languages such as Python are frequently used to write automated testing scripts.

5.4.2   Recording testing experience

The usage model structure serves as a basis for recording testing experiences, which can be used in assessing test sufficiency and other aspects of the software development process.

Testing experience is recorded in a *testing chain* that is also a Markov chain stochastic process.

A testing chain is started by using just the structure of states and arcs (no transition probabilities) of a usage model. As test sequences are executed, each arc successfully traversed (no failure) is marked on the testing chain, and the relative frequencies of visitation across the exit arcs of each state are calculated. Given enough random sequences from the usage chain, these relative frequencies will converge to the probabilities of the usage chain (model).

Consider tossing a fair coin. We know that in the long run the number of heads should equal the number of tails. But as we begin tossing the coin and recording the outcomes, we might see substantial variation in the early outcomes. Yet, in the long run (which is not too long for such a simple stochastic process) the ratios will converge to 1/2. Because of the immense variation in usage models, it might take thousands of test sequences for the ensemble statistics of testing experience to converge to the statistics of the source—the usage model from which sequences are generated. The measure of similarity between the weights on the usage model (expected activity) and the weights on the testing chain (tested activity) is discussed later as a stopping criterion for testing.

Two types of failures are possible. The first type does not impair or distort the functioning of the system, and the transition to the next state of use can be made. For example, a spelling error may appear in a message or a timer may be off by an insignificant amount. In such cases, a new state is created to represent the failure, and two new arcs are created: one from the departure state to the failure state and one from the failure state to the destination state. Each of the two new arcs receives a mark. Any time in the future that the same failure appears from the same departure state, these two arcs will each be marked again.

A second type of failure is one in which it makes no sense to continue the test case: for example, if the system crashes, it is impossible to continue or if the failure renders further steps meaningless, as in the case of a destroyed file. In such cases, a new state is created to represent the failure, and two new arcs are created, one from the departure state to the failure state and one from the failure state to the termination state (and the remainder of the test case is discarded). Each of the two new arcs receives a mark. Any time in the future that the same failure appears from the same departure state, these two arcs will each be marked again.

Several testing chains can be maintained, each as a separate file with unique identity. One testing chain could be maintained from the beginning of all testing, and another might be maintained for each version of the system, with a new testing chain started each time the code is changed. The cumulative data may be used for process analysis and the data on each version for product analysis. The testing chain can represent all testing experience, special cases as well as random testing, or it can represent only random testing. It is possible, and increasingly frequent, to instrument product code to maintain a "testing chain" based on actual field experience as well. For example, in the case of the infusion pump a record might be kept of every event.

5.4.3 Support for experimental design

Increasingly, statistical experiments are being designed to test software intensive systems (Nair et al. 1998). Although the use of experimental design in software testing is not widespread, the variety of applicable techniques has great potential to transform the testing field.

Designed experiments tell in advance how much testing and what kind of testing will be required to achieve desired results. Indeed, with most of these methods it is possible to influence product design decisions in order to make such testing feasible and more economical. Some characterization of the population under study is necessary for any application of experimental design. The usage model can be of value in all cases.

- *Combinatorial design.* A class of statistical experimental design methods known as combinatorial design is used to generate test sets that cover the n-way combinations of inputs (Cohen 1992, Dalal and Mallows 1998). For certain types of applications, including data entry screens, this approach has been used to minimize the amount of testing required to satisfy use-coverage goals. Combinatorial design deals with test factors, levels within factors, and treatments (combinations of factor levels) but leaves other issues unaddressed; for example, one must choose among many different test cases that cover all pairs of factor levels. Given a usage model, the treatments will appear as visitation of states of use in specific sequences and the likelihood of these sequences arising in use may be taken into account. Both combinatorial design and operational profiles may be used to plan testing.

- *Constraint testing.* There are many situations where certain inputs are mutually exclusive or certain combinations are mandatory. Constraints expressing such situations can be placed on the development or generation of the usage model. This makes the model more efficient, eliminating the generation of impossible or impractical test sequences and improving testing efficiency (Vilkomir, Swain, and Poore 2008).

- *Partition testing.* Partitioning is a standard statistical technique for increasing the efficiency of random sampling (Boland, Singh, and Cukic 2002). It is applicable to increasing the efficiency of random testing as well. Partitions can be identified and defined in terms of the usage model. For example, based on Figure 5.2 test cases might be partitioned into those that include any of the Basal states and those that do not; those that include visiting any of the Bolus states and those that do not. The reliability model of Miller (Miller et al. 1992) can be used since the probability mass of each block of the partition can be calculated from the model, as can the probability mass of test cases run in each block. Similarly, the input space could be partitioned.

- *Rare events and accelerated rate testing.* Some testing must address infrequent but highly critical situations in order to remove uncertainty or estimate reliability that takes rare events into account. Traditional concepts of accelerated testing are applicable (Ehrlich et al. 1998). Experimental design has been used to determine the most efficient approach to testing combinations of factors associated with rare events, and reliability models have been developed for these situations (Alam et al. 1997, Kaufman, Johnson, and Dugan 2002, Tsokos and Nadarajah 2003). Usage models can be built from many different perspectives, including process flow. Critical states, transitions, and subpaths that would have low likelihood of arising in field use (or in a random sample) can be identified from the usage model. The probability of reaching any given state or transition can be calculated directly from the model, as can the traversal of any subpath.

- *Sequential testing.* In some cases each test is so expensive to run or to evaluate that it is important to decide based on the outcome of each test whether or not additional testing is justified. The degree to which the variety and extent of testing are representative of the variety and extent of use expected in the field can be calculated directly from the usage model and the testing record (McDaid and Wilson 2001, Dalal, Poore, and Cohen 2003, Gaver et al. 2003).

- *Economic testing criteria.* Different forms or modes of failures in the field can result in different operational economic loss. Usage models together with mathematical programming methods can be used to design testing to minimize the potential of economic loss from field failure (Sherer 1996).

- *Economic stopping criteria.* Mathematically optimal rules have been developed for supporting decisions to stop testing, based on the known cost of continued testing versus the expected cost of failure in the field (Dalal and Mallows 1988). Quantitative analysis of the usage model can assist in assessing the cost of continued testing and the risk of failure in the field.

5.4.4 Controlling special test situations

Application of statistical science includes creating special, nonrandom test cases. Such testing can remove uncertainty about how the system will perform in specific circumstances of interest, aid in understanding the sources of variation in the population, and contribute to effectiveness and control over all testing. In all instances, however, the usage model is the road map for planning where testing should go and recording where testing has been. A few of the many special situations that can be represented in terms of the usage model are as follows.

- *Model coverage tests.* Using just the structure of the model, a graph-theoretic algorithm generates the minimal sequence of test events (least cost sequence) to cover all arcs (and therefore all states; Gibbons 1985). If it is practical to conduct this test, it is a good first step in that it will confirm that the testers know how to conduct testing and evaluate the results for every state of use and every possible transition. Without test automation, even this compelling testing strategy may not be affordable! Model coverage is a key to smoothly running, long-run sampling strategies that demonstrate a high degree of confidence.

- *Mandatory tests.* Any specific test sequences that are required on contractual, policy, moral, or ethical grounds can be mapped onto the model and run.

- *(Nonrandom) regression tests.* Existing regression test suites can be mapped to the model. This is an effective way to discover the redundancy in the test suite and assess the omissions. One can calculate the probability mass accounted for by the test suite. Of course, one may use the model to create or enhance a regression test set.

- *Most likely use.* The most likely use scenarios can be generated in rank order by probability of occurrence to some number of scenarios or to some cumulative probability mass.

5.4.5 Generating random samples of test cases

Random test cases can be automatically generated from the usage model, constituting a random sample of uses as the basis for statistical estimations about the population. Each test case is a "random walk" through the stochastic matrix, from the initial state to the terminal state. The script associated with each arc of the model is generated at each step of the random walk. One may generate as large a set of test cases as the budget and schedule will bear and establish bounds on test outcomes before incurring the cost of performing the tests.

A random sample of test cases is still a random sample when used multiple times. Thus, it is legitimate to rerun the test set after code changes (regression testing) and to use the results in statistical analysis, provided the code was not changed to specifically execute correctly on the test set. It is not uncommon to see situations where the code always works on the test set but does not work in the field; developers in some organizations literally learn what the testers are testing. Bias in evaluation must also be avoided. Testers may expect correct results because they have always been correct in the past; testers may learn the test set as well. If testing and the random test sets are independent of the developers and maintenance workers, reuse of the random test sets is a valid statistical testing strategy that can facilitate automated testing and substantial reductions in the time and cost of testing.

Some balance must be reached between the amount of test time and money that will be spent in special testing and the amount that will be reserved for testing based on random sampling. Random testing supports inferences about expected operational performance and must dominate all testing when nonrandom tests are included in the analysis.

5.4.6 Importance sampling

As was mentioned above, it is generally the case that many sets of transition probabilities exist that satisfy all known constraints on usage. In other words, there are many usage models (same structure, different transition probabilities) that are consistent with the environment of use.

Objective functions are used to choose the model that satisfies all constraints and is optimal relative to some criterion. By a combination of additional management constraints and objective functions, the resulting model can emphasize aspects of the system or testing process, which are important to testers. The following are among the controls that are possible:

- Costs can be associated with each arc, and one can minimize cost.

- Value can be associated with each arc, and one can maximize value.

- Probabilities associated with exiting arcs that control critical flow can be manipulated.

- Certain long-run effects can be regulated by constraints.

- Some entropy measures can be maximized to increase uncertainty and increase variability in the sequences.

- Some entropy measures can be minimized to reduce variability.

One must be wary of constructing an overly complex model that might be ill-conditioned relative to the numerical methods used in calculating the solution. Too many constraints that are functions of long-run behavior are not advised. (Source entropy of a Markov chain is not a convex function. It becomes convex if the stationary distribution or operational profile is fixed.)

Gutjahr (1997) presents a solution for dynamic revision of probabilities as testing progresses in order to optimize sampling relative to an importance objective.

5.4.7 Test automation

Usage models have led to test automation in almost every situation in which they have been used. Thus, the concept is now usually associated with automated testing. Test automation is attractive because it vastly increases the number of tests that can be run and greatly reduces the unit cost of testing. It is more cost effectively done when planned as a companion to the system development and can also be cost effective for existing systems with an anticipated long-term evolution.

Test automation depends upon three things: (1) generation of test cases in quantity in a form suitable for automated test runners, (2) an oracle or means of evaluating whether or not the system executes the test case correctly, and (3) a test runner that can initiate testing and report results.

The usage model is an excellent means of controlled generation of test cases in any desired quantity. Control is achieved by setting probabilities in order to implement importance sampling. Test cases are produced by walking the graph with a random number generator.

The oracle is the means by which one confirms that each step of the test case does what it is supposed to do (sufficient correctness) and nothing more (complete correctness). This is generally the difficult issue. Some systems have natural and easy oracles, much like double inversion of a matrix or squaring a square root; for example, a disk drive control unit might be tested by writing a file to disk and then reading it back to compare with the original. Sometimes a predecessor system can be used because the behavior of the new system is to be identical to the behavior of the old system.

The JUMBL is a library of command line tools that read and write files in several standard formats; thus, the tools can readily be connected to most commercial and open source test runners. Generally, scripts are sent to the test runners, which will return information for constructing the testing chain and subsequent statistical analysis of testing results.

5.4.8   Testing

Testing is expensive; industry data indicates that about half the software budget is spent on it (Research Triangle Institute 2002). Testing costs are best attacked in the development process, by clarifying and simplifying requirements, providing for testability and test automation, and verifying code against specifications. When high-quality software reaches the test organization, there are two goals: (1) provide the development organization with the most useful information as quickly as possible in order to shorten the overall development cycle and (2) certify the system as quickly and inexpensively as possible. Just "more testing" will certainly add cost, but will not necessarily add new information or significantly improve reliability estimates.

- *Resource and schedule estimation.* Calculations on a usage model provide data for effort, schedule, and cost projections for such goals as covering all states and transitions in the model or demonstrating a target reliability. Estimating the time and cost required to conduct the test associated with each arc of the usage model can lead to estimates for sequences; average sequence lengths can be used to estimate the time and cost of executing test sets.

- *Reliability analysis, with failures.* The testing chain provides the basis for a data-driven estimation of reliability. In the presence of failures, reliability can be assessed without additional mathematical assumptions (in contrast to reliability growth models). The failure states of the testing chain become absorbing, and the reliability of the system is defined as the probability of going from the invocation state to the termination state without being absorbed in a failure state. The failure states in a testing chain can be ranked with respect to their effect on reliability, which is used to help determine the order in which one corrects the code.

- *Reliability analysis, no failures.* In the absence of failures, reliability models based on the binomial are sometimes used (Parnas 1990). Alternatively, the reliability models of Miller (Miller et al. 1992), which are based on partitioning the sample space, can be used to take advantage of the structure of the model in order to improve the confidence in the reliability estimate. The adaptation of the Miller model that is used in the JUMBL is presented in Prowell and Poore (2005). All reliability estimates should be calibrated to field experience.

- *Test sufficiency analysis.* A stopping criterion can be calculated directly from the statistical properties of the usage model and testing chain. The log likelihood ratio (Kullback 1958; Kullback discriminant) can be calculated for these Markov chains and provides evidence for or against the hypothesis that the two stochastic processes are equivalent. The Kullback discriminant is a measure of the difference between expected field usage (usage model) and actual experience in testing (testing chain); it can be monitored during testing because the testing chain is changing with each test event (transition). This is an information-theoretic comparison of the usage and testing chains to assess the degree to which the testing experience has become representative of expected field use. As the testing chain converges to the usage model, it becomes less likely that new information will be gained by further testing generated from the usage model.

As an example, if the following test cases are generated automatically with the JUMBL for the infusion pump case study:

- 27 minimum coverage test cases that cover all the arcs and all the nodes of the model

- 10 test cases with the highest probability

- 1000 random test cases

and we assume that all these 1037 test cases are executed successfully and analyzed against the usage model with uniform transition probabilities across the exit arcs of each state, the JUMBL will generate a test case analysis report as shown below.

Test Case Analysis: Model Pump_Controller

Distribution: (default)
Generated: 5/2/10 4:37 PM

Model Statistics

Node Count	16 nodes
Arc Count	66 arcs
Stimulus Count	16 stimuli
Test Cases Recorded	1,037 cases
Nodes Generated	16 nodes / 16 nodes (1)
Arcs Generated	66 arcs / 66 arcs (1)
Stimuli Generated	16 stimuli / 16 stimuli (1)
Nodes Executed	16 nodes / 16 nodes (1)
Arcs Executed	66 arcs / 66 arcs (1)
Stimuli Executed	16 stimuli / 16 stimuli (1)

Stimulus Statistics

Stimulus	Generated	Executed	Failed	Reliability/ Variance		Optimum Reliability/ Variance		Prior Successes/ Failures	
ADE	119	119	0	0.954	3.31E-04	0.954	3.31E-04	6	6
AL2	168	168	0	0.957	2.25E-04	0.957	2.25E-04	8	8
BD	98	98	0	0.98	1.87E-04	0.98	1.87E-04	2	2
BG	104	104	0	0.991	8.73E-05	0.991	8.73E-05	1	1
C	54	54	0	0.909	1.23E-03	0.909	1.23E-03	6	6
DNOK	62	62	0	0.943	7.59E-04	0.943	7.59E-04	4	4
DOK	57	57	0	0.938	8.75E-04	0.938	8.75E-04	4	4
IP	103	103	0	0.981	1.70E-04	0.981	1.70E-04	2	2
POFF	888	888	0	0.985	1.64E-05	0.985	1.64E-05	14	14
PON	1,037	1,037	0	0.999	9.25E-07	0.999	9.25E-07	1	1
R	530	530	0	0.998	3.52E-06	0.998	3.52E-06	1	1
RC	148	148	0	0.974	1.59E-04	0.974	1.59E-04	4	4
RNOK	34	34	0	0.947	1.28E-03	0.947	1.28E-03	2	2
ROK	16	16	0	0.9	4.29E-03	0.9	4.29E-03	2	2
TBF	26	26	0	0.906	2.57E-03	0.906	2.57E-03	3	3
TIF	149	149	0	0.963	2.21E-04	0.963	2.21E-04	6	6

Arc Statistics

Arc	Probability	Generated	Executed	Failed	Reliability/ Variance		Optimum Reliability/ Variance		Prior Successes/ Failures	
[Alarm2 PON.R.AL2]										
"POFF"	1	168	168	0	0.994	3.42E-05	0.994	3.42E-05	1	1
[Basal Alarm-Empty PON.R.ADE]										
"DNOK"	0.333	32	32	0	0.971	8.16E-04	0.971	8.16E-04	1	1
"DOK"	0.333	33	33	0	0.971	7.71E-04	0.971	7.71E-04	1	1
"POFF"	0.333	34	34	0	0.972	7.30E-04	0.972	7.30E-04	1	1
[Basal Drug-notOK PON.R.ADE.DNOK]										
"DNOK"	0.333	15	15	0	0.941	3.08E-03	0.941	3.08E-03	1	1
"DOK"	0.333	13	13	0	0.933	3.89E-03	0.933	3.89E-03	1	1
"POFF"	0.333	19	19	0	0.952	2.06E-03	0.952	2.06E-03	1	1
[Basal New Rate notOK PON.R.RC.RNOK]										
"ADE"	0.167	4	4	0	0.833	1.98E-02	0.833	1.98E-02	1	1
"AL2"	0.167	1	1	0	0.667	5.56E-02	0.667	5.56E-02	1	1
"C"	0.167	6	6	0	0.875	1.22E-02	0.875	1.22E-02	1	1
"POFF"	0.167	4	4	0	0.833	1.98E-02	0.833	1.98E-02	1	1
"RC"	0.167	4	4	0	0.833	1.98E-02	0.833	1.98E-02	1	1
"TIF"	0.167	5	5	0	0.857	1.53E-02	0.857	1.53E-02	1	1
[Basal Paused PON.R.IP]										
"AL2"	0.333	35	35	0	0.973	6.92E-04	0.973	6.92E-04	1	1
"C"	0.333	31	31	0	0.97	8.64E-04	0.97	8.64E-04	1	1
"POFF"	0.333	26	26	0	0.964	1.19E-03	0.964	1.19E-03	1	1
[Basal Rate-changed PON.R.RC]										
"ADE"	0.143	14	14	0	0.938	3.45E-03	0.938	3.45E-03	1	1
"AL2"	0.143	15	15	0	0.941	3.08E-03	0.941	3.08E-03	1	1
"C"	0.143	12	12	0	0.929	4.42E-03	0.929	4.42E-03	1	1
"POFF"	0.143	18	18	0	0.95	2.26E-03	0.95	2.26E-03	1	1
"RNOK"	0.143	24	24	0	0.962	1.37E-03	0.962	1.37E-03	1	1
"ROK"	0.143	10	10	0	0.917	5.88E-03	0.917	5.88E-03	1	1
"TIF"	0.143	20	20	0	0.955	1.89E-03	0.955	1.89E-03	1	1

continued

[Basal Running PON.R]

"ADE"	0.125	81	81	0	0.988	1.42E-04	0.988	1.42E-04	1	1
"AL2"	0.125	92	92	0	0.989	1.11E-04	0.989	1.11E-04	1	1
"BD"	0.125	86	86	0	0.989	1.26E-04	0.989	1.26E-04	1	1
"BG"	0.125	104	104	0	0.991	8.73E-05	0.991	8.73E-05	1	1
"IP"	0.125	92	92	0	0.989	1.11E-04	0.989	1.11E-04	1	1
"POFF"	0.125	80	80	0	0.988	1.45E-04	0.988	1.45E-04	1	1
"RC"	0.125	109	109	0	0.991	7.97E-05	0.991	7.97E-05	1	1
"TIF"	0.125	103	103	0	0.99	8.90E-05	0.99	8.90E-05	1	1

[Bolus Drug-notOK PON.R.BG.ADE.DNOK]

"DNOK"	0.333	6	6	0	0.875	1.22E-02	0.875	1.22E-02	1	1
"DOK"	0.333	5	5	0	0.857	1.53E-02	0.857	1.53E-02	1	1
"POFF"	0.333	4	4	0	0.833	1.98E-02	0.833	1.98E-02	1	1

[Bolus Alarm-Empty PON.R.BG.ADE]

"DNOK"	0.333	9	9	0	0.909	6.89E-03	0.909	6.89E-03	1	1
"DOK"	0.333	6	6	0	0.875	1.22E-02	0.875	1.22E-02	1	1
"POFF"	0.333	5	5	0	0.857	1.53E-02	0.857	1.53E-02	1	1

[Bolus New Rate notOK PON.R.BG.RC.RNOK]

"ADE"	0.143	1	1	0	0.667	5.56E-02	0.667	5.56E-02	1	1
"AL2"	0.143	2	2	0	0.75	3.75E-02	0.75	3.75E-02	1	1
"C"	0.143	1	1	0	0.667	5.56E-02	0.667	5.56E-02	1	1
"POFF"	0.143	2	2	0	0.75	3.75E-02	0.75	3.75E-02	1	1
"RC"	0.143	1	1	0	0.667	5.56E-02	0.667	5.56E-02	1	1
"TBF"	0.143	2	2	0	0.75	3.75E-02	0.75	3.75E-02	1	1
"TIF"	0.143	1	1	0	0.667	5.56E-02	0.667	5.56E-02	1	1

[Bolus Paused PON.R.BG.IP]

"AL2"	0.333	5	5	0	0.857	1.53E-02	0.857	1.53E-02	1	1
"C"	0.333	3	3	0	0.8	2.67E-02	0.8	2.67E-02	1	1
"POFF"	0.333	3	3	0	0.8	2.67E-02	0.8	2.67E-02	1	1

[Bolus Rate-changed PON.R.BG.RC]

"ADE"	0.125	2	2	0	0.75	3.75E-02	0.75	3.75E-02	1	1
"AL2"	0.125	4	4	0	0.833	1.98E-02	0.833	1.98E-02	1	1
"C"	0.125	1	1	0	0.667	5.56E-02	0.667	5.56E-02	1	1
"POFF"	0.125	4	4	0	0.833	1.98E-02	0.833	1.98E-02	1	1
"RNOK"	0.125	10	10	0	0.917	5.88E-03	0.917	5.88E-03	1	1
"ROK"	0.125	6	6	0	0.875	1.22E-02	0.875	1.22E-02	1	1
"TBF"	0.125	7	7	0	0.889	9.88E-03	0.889	9.88E-03	1	1
"TIF"	0.125	1	1	0	0.667	5.56E-02	0.667	5.56E-02	1	1

[Bolus Running PON.R.BG]

"ADE"	0.125	17	17	0	0.947	2.49E-03	0.947	2.49E-03	1	1
"AL2"	0.125	14	14	0	0.938	3.45E-03	0.938	3.45E-03	1	1
"BD"	0.125	12	12	0	0.929	4.42E-03	0.929	4.42E-03	1	1
"IP"	0.125	11	11	0	0.923	5.07E-03	0.923	5.07E-03	1	1
"POFF"	0.125	14	14	0	0.938	3.45E-03	0.938	3.45E-03	1	1
"RC"	0.125	34	34	0	0.972	7.30E-04	0.972	7.30E-04	1	1
"TBF"	0.125	17	17	0	0.947	2.49E-03	0.947	2.49E-03	1	1
"TIF"	0.125	19	19	0	0.952	2.06E-03	0.952	2.06E-03	1	1

[EXIT]

[LAMBDA]

"PON"	1	1,037	1,037	0	0.999	9.25E-07	0.999	9.25E-07	1	1

[Power on PON]

"POFF"	0.5	507	507	0	0.998	3.84E-06	0.998	3.84E-06	1	1
"R"	0.5	530	530	0	0.998	3.52E-06	0.998	3.52E-06	1	1

Reliabilities

Single Event Reliability	0.987
Single Event Variance	2.87E-06
Single Event Optimum Reliability	0.987
Single Event Optimum Variance	2.87E-06
Single Use Reliability	0.959
Single Use Variance	7.99E-03
Single Use Optimum Reliability	0.959
Single Use Optimum Variance	7.99E-03
Arc Source Entropy	0.981 bits
Kullback Discrimination	7.97E-03 bits
Relative Kullback Discrimination	0.813%
Optimum Kullback Discrimination	7.97E-03 bits
Optimum Relative Kullback Discrimination	0.813%

According to this analysis, given the model and the described testing experience, the system has a single use reliability of 0.959 and a relative Kullback discrimination of 0.813%.

5.5 Product and Process Improvement

Statistical testing supports quantitative certification of software for standards compliance. It can be used across the life cycle supporting incremental development with feedback on the product as well as the development process.

5.5.1 Certification

The certification process involves ongoing evaluation of the merits of continued testing. Stopping criteria are based on reliability, confidence, and remaining uncertainty. Decisions to continue testing are based on an assessment that the goals of testing can still be realized within the schedule and budget remaining.

In most cases, users of statistical testing methods release a version of the software in which no failures have been observed in testing. Reliability estimates such as those in Miller et al. (1992) and Prowell and Poore (2005) are recommended in this case.

Software is sometimes released with known faults. If the test data includes failures, then reliability and confidence may be calculated from the testing chain. The reliability measure computed in this manner reflects all aspects of the sequences tested, including the probability weighting defined by the usage model.

Certification is always relative to a protocol, and the protocol includes the entire testing process and all work products. An independent audit of testing must be possible to confirm correctness of reports. An independent repetition of the protocol should produce the same conclusions, within acceptable statistical variation. Protocols and records of the quality described here provide evidence for dependability assurance cases (Bucchianico et al. 2008, Jackson, Thomas, and Millett 2009).

5.5.2 Incremental development

The Cleanroom software engineering process (Prowell et al. 1998) uses the testing approach described in this chapter. Cleanroom produces software in a stream of increments to be tested, as do some other processes. An increment may be "accepted," indicating that the development process is working well, by different (less stringent) criteria than will be used to certify the final product. If increment certification goals are not met, review of experience may show that changes are needed in the process itself, for example, better verification, changes to the usage model, improved record keeping, more frequent analysis of test data, or rethinking of the entire increment plan. If certification goals are met, the process moves ahead with the next increment or system acceptance.

The historical testing chain and related statistics will reflect consequences of all failures seen and fixed from the very beginning of testing through all versions of the system to the one released. The historical chain may be used to review the development and testing processes across increments. The historical testing chain and the collection of testing chains from version to version can be used to assess reliability growth.

5.5.3   Combining testing information

There are many situations in the life cycle of a software intensive system where it would be beneficial to use existing testing and field use information to identify and minimize the additional testing that is necessary to support a decision regarding the system. These situations can be generally classified as development, reengineering, maintenance, reuse, and porting. Effective configuration control over the software and correct association of the testing records with code or system versions are required to use such information in a statistically valid way. Furthermore, a common basis for describing the testing done and for interpreting the data is necessary. Usage models have the potential to be the common denominator for test planning and evaluation of results in all testing situations in the system life cycle. The specific statistical models for combining information have not been worked out for every situation, but some progress has been made and work continues in this effort to unify testing. The theoretical path seems clear in all cases.

5.5.3.1   Development

With incremental development, each cycle concludes with statistical testing to support the decision to move forward with development of the next increment. In the case of the final increment, the decision is to deploy or accept the system. In each increment some testing is done in the previous section, but most is in the new section. The statistics associated with each testing increment have the same usage model as a basis for combining test information.

An evolutionary procurement is based on the concept of a series of fielded systems with past performance, new requirements, and new technology coming together for each successive version of the system. Previous testing records and field data are available after the first version. The usage model for the fielded version would be a subset of the model for the new version and the starting point for testing of the new development. The field experience from various environments of use could be expressed in terms of the usage model and, together with planned revisions and changes, forms the basis for testing.

5.5.3.2   Reuse

Many systems involve reuse of existing systems or system components, with or without reengineering. Object-oriented reuse ranges from pattern instantiation, to framework integration, to class–subclass hierarchy extensions with polymorphic methods. If a component is to be reused without change, then the usage model originally used to certify the component can be used to assess the testing necessary for the new use. One would have the original usage model and testing records, plus the field use data summarized as an estimate of sequences actually run. A set of transition probabilities would describe the new use. It is straightforward to compare the new use against the records of previous testing and use to determine whether or not the new use requires further testing.

5.5.3.3 Reengineering

Reengineering typically involves changing the technology from which the system is made—for example, one or more of the hardware processor, memory units, power supplies, or even the programming language and data structures—but generally preserves the way the system is used. (Otherwise, it would be new development or maintenance rather than reengineering.) Usage specifications usually survive, with varying degrees of change. The original usage model may change in structure. Usage states and arcs can be associated with underlying changes in the technology. Usage models may be used to assess the extent of change and to guide testing; the greater the change, the harder the testing problem.

5.5.3.4 Maintenance

Maintenance is usually associated with small changes to an operational system. Thus, the developmental testing and field use records are available. Field experience indicates that good understanding of both the usage model (states and arcs) and the architecture and implementation of the system are required in order to map maintenance changes to relevant parts of the usage model. Testing after maintenance must address both known impact areas and the possibility of unanticipated impacts and be planned using records of prior model-based testing and field use.

5.5.3.5 Porting

Porting is the process of moving a system to new or additional "platforms," usually meaning different operating systems for the same hardware, new hardware running the same operating system, or the hardware and operating system of a different vendor. Given a good software architecture and a design that anticipates porting, the changes to the system will be minimal, but the services provided by the hardware and operating systems can be significantly different. Given multiple platforms on which the system must run, what is the optimal amount of testing to be done on each platform in order to support a decision to deploy each? Generating test cases and recording test results based on a common usage model and a common set of statistics make this a tractable problem.

A common framework for planning and recording all testing and field use in the life cycle of a system can lead to substantial cost savings in testing and much better information to support decisions.

5.6 Summary and Conclusion

From a mathematical point of view, the topics in this chapter follow sound problem-solving principles and are direct applications of well-established theory and methodology. The applications of statistical science discussed herein are not in widespread use for software intensive products. Many methods and segments of the process are used in pockets of industry and government, on both experimental and routine bases.

Most usage modeling experience to date is with embedded real-time systems (Oshana 1997), application program interfaces (APIs), and graphical user interfaces (GUIs). Models as small as 20 states and 100 arcs have proven very useful. Typical models are on the order of 500 states and 2000 arcs; large models of more than 2000 states and 20,000 arcs are in use. Even the largest models developed to date (20,000 states) are small in comparison to similar mathematical models used in industrial operations research and econometrics, and are manageable with available tool support. Large models must be accepted as appropriate to many software systems and to the testing problem they pose. The size is not to be lamented because the larger and more complex the testing problem, the greater the need for the assistance that modeling and simulation afford.

Since 1992, the IBM Storage Systems Division has applied Markov chain usage models for certification of tape drives, tape controllers, tape libraries, disk drives, and disk controllers. Some products are tested with several different usage models, including models of customer use, a data communication protocol model, a model keyed to the injection of hardware and media failures, and a stress model. Many of these models are reused from product to product because only the technology of the product changes and not the architecture of the product, the way it is used, or the standards to which it is built. Transition probabilities have been determined by instrumentation measurements collected during internal use and external customer command traces originally collected for performance analysis. The test facility is highly automated and employs compiler-writing technology to automatically compile executable test cases from abstract arc labels, which permits testing of a large number of scripts. Stopping criteria are based on both reliability estimates and substantial agreement between testing experience and expected field experience. Use of this technology has significantly reduced the testing effort and improved field reliability.

A project with the Oak Ridge National Laboratory created test models for approximately 40 programs in a library to support theoretical physics calculations (Sayre and Poore 2007). A project is currently underway to use the methods of Carter (2009) for a weigh-in-motion hybrid system used in loading vehicles onto ships and airplanes.

Verum reported a large-scale industrial application in application to medical devices (Bouwmeester, Broadfoot, and Hopcroft 2009). The methods discussed here have been integrated into the Verum Compliance Test Framework for the certification of industrial software.

The collaboration between UTK SQRL and the Fraunhofer IESE has resulted in wider use and improved tools (Fraunhofer IESE).

Many industrial experiments have been conducted as proof of concept demonstrations in applications for fuel injection, car door mirror controls, and automatic transmissions, for example. These required integration with existing test facilities such as Rational Test RealTime®, PROVEtech: TA®, and MATLAB Simulink.

The process is well supported by tools and documentation (Prowell 2003). The JUMBL contains software tools to support all aspects of statistical testing based on Markov chain usage models. JUMBL has been made freely available by The University of Tennessee for several years. There have been several thousand downloads of the library, and we know of several commercial products around the world that make use of the library. The appendix to this paper lists the capabilities of the library.

Some activities of statistical testing are computationally intensive, with run-time for analyses a function of the number of states or the number of arcs in a usage model. While the computations would seem routine to an operations research analyst, they might seem prohibitive to some software engineers. Ironically, software engineering environments tend to be computationally starved.

Automated statistical testing is an *economical and feasible* way to demonstrate that an embedded system meets standards and criteria for moving to the next stage of development or for releasing it as a product. By "automated" we mean that the test cases are automatically generated, automatically executed, automatically evaluated as to pass or fail, and that the experimental record is automatically recorded. As the term "experimental record" suggests, one approaches statistical testing as experiments to demonstrate and document that the system under test satisfies various criteria. This form of testing provides supporting evidence for dependability assurance cases (Gutjahr 2000, Jackson, Thomas, and Millett 2009).

Statistical testing based on usage models can be applied to large and complex systems because the modeling can be done at various levels of abstraction and because the models effectively allow analysis and simulation of *use* of the application rather than the application itself.

References

Alam, M.S. et al. (1997). Assessing software reliability performance under highly critical but infrequent event occurrences. In *Proc. 8th Int. Symp. on Reliability Eng.*, Pages: 294–307. Albuquerque, NM.

Ash, R. (1965). *Information Theory*. John Wiley and Sons, New York, NY.

Bauer, T. et al. (2007). From requirements to statistical testing of embedded systems. In *Proc. 4th Int. Workshop on Software Eng. for Automotive Sys.*, Pages: 3–9. Minneapolis, MN.

Boland, P.J., Singh, H., and Cukic, B. (2002). Stochastic orders in partition and random testing of software. *J. Appl. Prob.* 39(3): 555–565.

Boland, P.J., Singh, H., and Cukic, B. (2004). The stochastic precedence ordering with applications in sampling and testing. *J. Appl. Prob.* 41(1): 73–82.

Bouwmeester, L., Broadfoot, G.H., and Hopcroft, P.J. (2009). Compliance test framework. In *Proc. 2nd Workshop on Model-Based Testing in Practice*, Pages: 97–106. Enscede, NL.

Broadfoot, G.H. and Broadfoot, P.J. (2003). Academia and industry meet: Some experiences of formal methods in practice. In *Proc. IEEE Computer Soc. 10th Asia-Pacific Software Eng. Conf.*, Pages: 49–59. Chiangmai, Thailand.

Bucchianico, A.D. et al. (2008). Statistical certification of software systems. *Comm. in Statistics —Simulation and Computation* 37(2): 346–359.

Butler, R.W. and Finelli, G.B. (1993). The infeasibility of quantifying the reliability of life-critical real-time software. *IEEE Trans. on Software Eng.* 19(1): 3–12.

Carter, J.M. (2009). Sequence-based specification of real-time embedded systems. Ph. D. dissertation, The University of Tennessee, Knoxville. http://sqrl.eecs.utk.edu/btw/files/jd.pdf (accessed August 30, 2010).

Carter, J.M., Lin, L., and Poore, J.H. (2008). Automated functional testing of Simulink control models. In *Proc. 1st Workshop on Model-Based Testing in Practice*, Pages: 41–50. Berlin, Germany.

Cohen, D.M. (1992). The AETG system: An approach to testing based on combinatorial design. *IEEE Trans. on Software Eng.* 23(7): 437–444.

Cohen, M.L., Rolph, J.E., and Steffey, D.L., eds. (1998). *Statistics, Testing, and Defense Acquisition: New Approaches and Methodological Improvements*. The National Academies Press, Washington, DC.

Dalal, S.R., and Mallows, C.L. (1988). When should one stop testing software? *J. Am. Statistical Assoc.* 83(403): 872–879.

Dalal, S.R., and Mallows, C.L. (1998). Factor-covering designs for testing software. *Technometrics* 40(3): 234–243.

Dalal, S.R., Poore, J.H., and Cohen, M.L., eds. (2003). *Innovations in Software Engineering for Defense Systems*. The National Academies Press, Washington, DC.

Donohue, S.K., and Dugan, J.B. (2003). Modeling the "good enough to release" decision using V& V preference structures and Bayesian belief networks. In *Proc. Annual Reliability and Maintainability Symp.*, Pages: 568–573. Tampa, FL.

Ehrlich, W.K. et al. (1998). Software reliability assessment using accelerated testing methods. *Appl. Statist.* 47(1): 15–30.

Ekroot, L., and Cover, T.M. (1993). The entropy of Markov trajectories. *IEEE Trans. on Information Theory* 39(4): 1418–1421.

Fraunhofer IESE, Department of Testing and Inspections. http://www.iese.fraunhofer.de/ competence/quality/tai/index.jsp (accessed August 30, 2010).

Gaver, D.P. et al. (2003). Probability models for sequential-stage system reliability growth via failure mode removal. *Int. J. of Reliability, Quality and Safety Eng.* 10(1): 15–40.

Gibbons, A.M. (1985). *Algorithmic Graph Theory*. Cambridge University Press.

Goševa-Popstojanova, K. and Trivedi, K.S. (2000). Failure correlation in software reliability models. *IEEE Trans. on Reliability* 49(1): 37–48.

Gutjahr, W.J. (1997). Importance sampling of test cases in Markovian software usage models. *Probability in the Eng. and Informational Sci.* 11(19): 2–6.

Gutjahr, W.J. (2000). Software dependability evaluation based on Markov usage models. *Performance Evaluation* 40(4): 199–222.

Jackson, D., Thomas, M., and Millett, L.I., eds. (2009). *Software for Dependable Systems: Sufficient Evidence?* The National Academies Press, Washington, DC.

Kaufman, G.M. (1996). Successive sampling and software reliability. *J. Statistical Planning and Inference* 49(3): 343–369.

Kaufman, L.M., Johnson, B.W., and Dugan, J.B. (2002). Coverage estimation using statistics of the extremes for when testing reveals no failures. *IEEE Trans. on Computers* 51(1): 3–12.

Kemeny, J.G., and Snell, J.L. (1960). *Finite Markov Chains*. D. Van Nostrand Company, Inc.

Kullback, S. (1958). *Information Theory and Statistics*. John Wiley and Sons, New York, NY.

Leung, Y.-W. (1997). Software reliability allocation under an uncertain operational profile. *Journal of the Operational Research Society* 48(4): 401–411.

Littlewood, B., and Mayne, A.J. (1989). Predicting software reliability and discussion. *Phil. Trans. R. Soc. Lond. A* 327(1596): 513–527.

Littlewood, B., and Strigini, L. (2000). Software reliability and dependability: A roadmap. In *Proc. Conf. on the Future of Software Eng.*, Pages: 175–188. Limerick, Ireland.

McDaid, K., and Wilson, S.P. (2001). Deciding how long to test software. *The Statistician* 50(2): 117–134.

Miller, K. et al. (1992). Estimating the probability of failure when testing reveals no failures. *IEEE Trans. on Software Eng.* 18(1): 33–43.

Musa, J. (1998). *Software Reliability Engineering*. McGraw-Hill.

Nair, V.N. et al. (1998). A statistical assessment of some software testing strategies and applications of experimental design techniques. *Statistica Sinica* 8(1): 165–184.

Oshana, R. (1997). Software testing with statistical usage based models. *Embedded Systems Programming* 10(1): 40–55.

Ou, Y., and Dugan, J.B. (2003). Approximate sensitivity analysis for acyclic Markov reliability models. *IEEE Trans. on Reliability* 52(2): 220–230.

Parnas, D. (1990). An evaluation of safety critical software. *Comm. Assoc. Computing Machin.* 23(6): 636–648.

Poore, J.H., and Trammell, C.J. (1998). Engineering practices for statistical testing. *Crosstalk* (DoD software engineering journal-newsletter), April 1998, 24–28.

Poore, J.H., Walton, G.H., and Trammell, C.J. (2000). A constraint-based approach to the representation of software usage models. *Information and Software Technol.* 42(12): 825–833.

Prowell, S.J. (2003). JUMBL: A tool for model-based statistical testing. In *Proc. 36th Ann. Hawaii Int. Conf. on System Sci.*, Pages: 337–345. Big Island, HI.

Prowell, S.J., and Poore, J.H. (2003). Foundations of sequence-based software specification. *IEEE Trans. on Software Eng.* 29(5): 417–429.

Prowell, S.J., and Poore, J.H. (2005). Reliability computation for usage based testing. In *Mod. Stat. and Mathematical Methods in Reliability*, ed. Wilson, A. et al., chap. 27. World Science.

Prowell, S.J. et al. (1998). *Cleanroom Software Engineering: Technology and Process*. Addison-Wesley.

Real-Time Systems Group, University of Pennsylvania (2010). Documentation of a generic infusion pump. http://rtg.cis.upenn.edu/gip-docs/GPCA% 20Pump% 20Model.doc (accessed August 30, 2010).

Real-Time Systems Group, University of Pennsylvania (2010). Simulink model of the generic infusion pump. http://rtg.cis.upenn.edu/gip-docs/GIP-model.tgz (accessed August 30, 2010).

Research Triangle Institute (2002). The economic impacts of inadequate infrastructure for software testing. National Institute of Standards and Technology RTI Project.

Sayre, K., and Poore, J.H. (2007). Automated testing of generic computational science libraries. In *Proc. 40th Ann. Hawaii Int. Conf. on System Sci.*, Pages: 277–285. Big Island, HI.

Sherer, S.A. (1996). Statistical software testing using economic exposure assessments. *Software Eng. J.* 11(5): 293–298.

Simulink® 7 User's Guide (March 2010). http://www.mathworks.com/access/helpdesk/help/pdf_doc/simulink/sl_using.pdf (accessed August 30, 2010).

Tsokos, C., and Nadarajah, S. (2003). Extreme value models for software reliability. *Stochastic Analysis and Applications* 21(3): 719–735.

Vilkomir, S., Swain, T., and Poore, J.H. (2008). Combinatorial test case selection with Markovian usage models. In *Proc. IEEE Computer Soc. 5th Int. Conf. on Information Technol.: New Generations*, Pages: 3–8. Las Vegas, NE.

Walton, G.H., and Poore, J.H. (2000). Measuring complexity and coverage of software specifications. *Information and Software Technol.* 42(12): 859–872.

Walton, G.H., Poore, J.H., and Trammell, C.J. (1995). Statistical testing of software based on a usage model. *Software—Practice and Experience* 25(1): 97–108.

Whittaker, J.A., and Poore, J.H. (1993). Markov analysis of software specifications. *ACM Trans. on Software Eng. and Methodol.* 2(1): 93–106.

Whittaker, J.A., and Thomason, M.G. (1994). A Markov chain model for statistical software testing. *IEEE Trans. on Software Eng.* 30(10): 812–824.

Appendix: A Summary of the JUMBL Commands Supporting Statistical Testing

Testing Process	JUMBL Command	Use in the Testing Process
Usage modeling	jumbl Write <usage model file in old format> [--type=<new format type>] [--suffix=<suffix>]	Convert a constructed usage model from one format to another format (SM by default). The model formats supported by the JUMBL include SM, TML, MML, EMML, GML, CSV, MOD, DOT, GDL, and HTML.
	jumbl Check <usage model file>	Check if the usage model has correct structure. If so, it also reports some overall model statistics.
	jumbl Prune <usage model file>	Prune a bad usage model. Remove unreachable nodes (from the source) and trapped nodes (that cannot reach the sink).

continued

Testing Process	JUMBL Command	Use in the Testing Process
	jumbl Flatten <usage model file that contains references> [--collapse]	Flatten a usage model that contains references to component models by either collapsing or instantiating (the default). The result is a single "flat" model.
Model analysis and validation	jumbl Analyze [--key=<distribution key>] [--suffix=<suffix>] [--model_engine=<model engine>] <usage model file>	Analyze a usage model with the specified distribution key and model analysis engine, and generate a comprehensive report of model statistics in HTML. Supported model analysis engines include Quick (the default), Fast, Simple, and Simulation).
Test planning	jumbl GenTest --min [--key=<cost distribution key>] <usage model file>	Generate minimum coverage test cases (test cases that cover all the arcs of the model with the minimum cost or by default the minimum number of test steps).
	jumbl GenTest [--num=<test case number>] [--key=<distribution key>] <usage model file>	Generate a specified number of random test cases (by default a single random test case) from the model with the specified distribution key.
	jumbl GenTest --weight [--num=<test case number>] [--key=<distribution key>] [--sum] <usage model file>	Generate a specified number of weighted test cases (by default a single weighted test case) from the model with the specified distribution key in either decreasing order of probability (by default) or increasing order of weight (arc weights are summed).
	jumbl CraftTest	Create test cases by hand, or edit existing test cases.
	jumbl ManageTest List <test record file>	Display a directory of the content of a test record.
	jumbl ManageTest Add <target test record file> (<individual test case file> | <test record file with selector>)+	Add test cases to a test record. The test cases are added at the end of the test record.
	jumbl ManageTest Insert <target test record file> <starting index for added test cases> (<individual test case file> | <test record file with selector>)+	Add test cases to a test record. The test cases are added starting at a given index.

continued

Testing Process	JUMBL Command	Use in the Testing Process
	jumbl ManageTest Delete <test record file with selector>	Remove selected test cases from a test record.
	jumbl ManageTest Export <test record file with selector> [--type=<exported test case format>] [--suffix=<suffix>] [--extension=.<extension>]	Write individual test cases in a test record to separate files (usually used to write test cases in executable form for automated testing). By default the individual test cases are written in TXT files.
Testing	jumbl ManageTest ReadResults <test record file> <test result file>+	Read one or more test result files containing execution information and apply the information to the test record.
	jumbl ManageTest WriteResults <test record file> <test result file>	Write the test execution information stored in a test record to a test result file in XML format.
	jumbl RecordResults <test record file> <selector>* jumbl RecordResults --file=<test result file same as the command line>	Record the results of executing one or more test cases in a test record. Record failure steps and indicate whether testing stopped after the last failure step for failed test cases.
	jumbl ManageTest ShowResults <test record file>	Display a directory of the content of a test record with results of test execution shown, along with some simple reliability measures.
Product and process measurement	jumbl Analyze [--key=<distribution key>] [--suffix=<suffix>] [--test_engine=<test engine>] <test record file>	Analyze a test record with the specified distribution key and test analysis engine, and generate a comprehensive report of use statistics in HTML, including reliabilities and measures of test sufficiency. Supported test analysis engines include Simple.

6

How to Design Extended Finite State Machine Test Models in Java

Mark Utting

CONTENTS

6.1 Introduction

Above all others, the key skill that is needed for model-based testing (MBT) is the ability to write good test models that capture just the essential aspects of your *system under test* (SUT). This chapter focuses on developing the skill of modeling for MBT. After this introduction, which gives an overview of MBT and its pros and cons, Section 6.2 compares two of the most common styles of models used for MBT—SUT input models and finite state models (FSM)—and discusses their suitability for embedded systems. Then, in Section 6.3, we develop some simple graphical FSM test models for testing a well-known kind of Java collection (Set<E>) and show how this model can be expressed as an extended finite state machine (EFSM) model in Java.

Section 6.4 illustrates how the ModelJUnit tool (ModelJUnit 2010) can be used to generate a test suite from this model and discusses several different kinds of test generation algorithms. Section 6.5 describes one of the simplest test generation algorithms possible and shows how you can implement a complete MBT tool in just a couple of dozen lines of code, using Java reflection. Section 6.6 turns to the practical issues of connecting the generated tests to some implementation of `Set<E>` and reports on what happens when we execute those tests on a `HashSet<String>` object and on an implementation of `Set<String>` that has an off-by-one bug. It also describes how we can estimate the strength of the generated test suite using SUT code coverage metrics and the Jumble mutation analysis tool (Jumble 2010). As well as illustrating general EFSM testing techniques, Sections 6.3 through 6.6 are also useful as a brief tutorial introduction to using ModelJUnit.

Section 6.7 discusses the modeling and testing of a larger, embedded system example—a subset of the GSM 11-11 protocol used within mobile phones, Section 6.8 discusses related work and tools, and Section 6.9 draws some brief conclusions.

6.1.1 What is model-based testing?

The basic idea of MBT is that instead of designing dozens or hundreds of test cases manually, we design a small model of the desired behavior of the SUT and then select an algorithm to automatically generate some tests from that model (El-Far and Whittaker 2002, Utting and Legeard 2007). In this chapter, most of the models that we write will be state machines, which have some internal state that represents the current state of the SUT, and some actions that represent the behaviors of the SUT. We will express these state machine models in the Java programming language, so some programming skills will be required when designing the models. The open-source ModelJUnit tool can then take one of these models, use reflection to automatically explore the model, visualize the model, and generate however many test cases you want. It can also measure how well the generated tests cover the various aspects of the model, which can give us some idea of how comprehensive the test suite is.

6.1.2 What are the pros and cons?

Like all test automation techniques, MBT has advantages and disadvantages. One of the advantages is that generating the tests automatically can save large amounts of time, compared to designing tests by hand. However, this is partially offset by the time taken to design the test model. Most published case studies show that MBT reduces overall costs (Dalal et al. 1999, Farchi, Hartman, and Pinter 2002, Bernard et al. 2004, Horstmann, Prenninger, and El-Ramly 2005, Jard et al. 2005), typically by 20% –30% , but sometimes more dramatically—up to 90% (Clark 1998).

Another advantage of MBT is that it is easy to generate lots of tests, far more than could be designed by hand. For example, it can be useful to generate and execute thousands of tests overnight, with everything automated. Of course, having more tests does not necessarily mean that we have better tests. But MBT can produce a test suite that systematically covers all the combinations of behavior in the model, and this is likely to be less ad hoc than a manually design test suite where it is easy to miss some cases. Case studies have shown that model-based test suites are often as good at fault detection as manually designed test suites (Dalal et al. 1999, Farchi, Hartman, and Pinter 2002, Bernard et al. 2004, Pretschner et al. 2005). In addition, model-based test suites can be better at detecting *requirements* errors than manually designed test suites because typically half or more of all the faults found by a model-based test suite are because of errors in the model (Stobie 2005). Detecting these model errors is very useful since they often point to requirements issues and misunderstandings about the expected behavior of the SUT. The process of modeling the SUT exposes requirements issues as well.

The main disadvantage of MBT is the time and expertise necessary to design the model. A test model has to give an accurate description of the expected SUT behavior, so precise executable models are needed. They may be expressed in some programming language, in a precise subset of UML with detailed state machines, or using some finite-state machine notation such as graphs. So the person designing the model needs to have some programming or modeling skills as well as SUT expertise. It takes some experience to be able to design a test model at a good level of abstraction so that it is not overly detailed and large, but it still captures the essence of the SUT that we want to test. This chapter will give examples of how to develop such models for several different kinds of SUT.

One last advantage that we must mention is evolution. When requirements change, updating a large manually designed test suite can be a lot of work. But with MBT, it is not necessary to update the tests—we can just update the test model and regenerate a new test suite. Since a good test model is much smaller than the generated test suite, this can result in faster response to changing requirements.

6.2   Different Kinds of Models

The term "model-based testing" can be used to describe many different kinds of test generation (Utting and Legeard 2007, page 7). This is because different kinds of models are appropriate for different kinds of SUT. Two of the most widely used kinds of models for MBT are *input models* and *finite-state models*, so we shall start with a brief overview and comparison of these two kinds.

If your SUT is batch oriented (it takes a collection of input values, processes them, and then produces some output), then one simple kind of model is to just define a small set of test values for each input variable. For example, if we are testing a print function that must print several different kinds of documents onto several different kinds of printers and work on several different operating systems, we might define an *input model* that simply defines several important test values for each input variable:

document: {plain text, rich text+images, html+images, PDF}
printer: {color inkjet printer, black&white laser, postscript printer}
op.system: {Windows XP, Windows Vista, Linux, Mac OS X}

Given this input model, we could then choose between several different algorithms to generate a test suite. If we want to test *all combinations* of these test inputs, our test suite would contain $4 \times 3 \times 4 = 48$ test cases. If we want to test *all pairs* of test input values (Czerwonka 2008), then 16 test cases would suffice. If we are happy with the dangerous strategy of testing *all input values* but ignoring any interactions between different choices, then four test cases could cover all the test input values. This is an example of how we can model the possible inputs of our SUT in a very simple way and then choose a test generation strategy/algorithm to generate a test suite from that model of the inputs. This kind of input-only model is useful for generating test inputs in a systematic way, but it does not help us to know what the expected output is or to determine whether the test passes or fails. Another example of input-only models is grammar-based testing (Coppit and Lian 2005), where various random generation algorithms are used to generate complex input values (such as sample programs to test a compiler or SQL queries to test a database system) from a regular expression or a context free grammar.

In this chapter, we focus on testing state-based SUTs, where the behavior of the SUT varies depending upon what state it is in. For such systems, our test cases usually contain a sequence of actions that interact with the SUT, sending it a sequence of input commands and values, as well as specifying the expected outputs of the SUT. The output of the SUT depends on the current state of the SUT, as well as upon the current input value. For example, if we call the `isEmpty()` method of a Java collection object, it will sometimes return `true` and sometimes `false`, depending on whether the internal state of the collection object is empty or not. Similarly, if we send a "TurnLeft" command to a wheelchair controller, it may respond differently depending on whether the wheelchair is currently moving or stationary. Embedded systems that contain software are usually best modeled as state-based systems.

For these state-based systems, it is important to use a richer state-based model of the expected behavior of the SUT that keeps track of the current state of the SUT. This means that the model cannot only be used to generate input values to send to the SUT, but it can also tell us the expected response of the SUT because the model knows roughly what state the SUT is in. For modeling state-based systems, it is common to use *finite-state machines* or *UML state machines* (Lee and Yannakakis 1996, Binder 1999, Utting and Legeard 2007, Jacky et al. 2008). In this chapter, we will see how one style of *extended finite-state machine* can be written in Java and used to generate test sequences that send input values and actions to the SUT as well as checking the expected SUT outputs. By using this kind of rich model of the SUT, we can generate test cases from the model automatically, and those test cases can automate the verdict assignment problem of deciding whether each test has passed or failed when it is executed.

> If you want the generated tests to automate the pass/fail verdict, your model must capture the current state or expected outputs of the SUT, so use a finite-state model, not an input-only model.

6.3   How to Design a Test Model

We will start by designing a test model for a very small system that we want to test. We will model the Java `Set<E>` interface, which is an interface to a collection of objects of type E. In later sections, we will generate tests from this model and execute those tests on a couple of different implementations of sets.

Here is a summary of the main methods defined in the `Set<E>` interface. We divide them into two groups: the *mutator* methods that can change the state of the set and the *query* methods that return information about the set but do not change its state.

	Result Type	Set<E>Method	Description
Mutator Methods	boolean	add(E obj)	adds obj to this set
	boolean	remove(Object obj)	removes obj from this set
	void	clear()	removes all elements from this set
Query Methods	boolean	contains(Object obj)	true if this set contains obj
	boolean	equals(Object obj)	compares this set with obj
	boolean	isEmpty()	true if this set contains no elements
	int	size()	the number of elements in this set
	Iterator<E>	iterator()	iterates over the elements in this set

The first step of modeling any embedded system is the same: identify the input commands that change the state of the SUT and the query/observation points that allow us to observe the state of the SUT without changing its state.

6.3.1 Designing an FSM model

To understand the idea of a state-based model, let us start by drawing a diagram of the states that a set may go through as we call some of its mutator methods. Starting from a newly constructed empty set, imagine that we add some string called s1, then add a second string s2, then remove s2, then remove s1 to get an empty set again. If we draw a diagram of this sequence of states, we get Figure 6.1.

Each circle represents one state of the set (a snapshot of what we would see if we could look inside the set object), with the contents of the set written inside the circle. Each arrow represents an action (a call to a mutator method) that changes the set from one state to another state. Of course, a moments thought makes us realize that the first and last states are both empty and are actually indistinguishable. All our query methods give exactly the same results for a newly constructed empty set as they do for a set that has just had all its members removed. So we should redraw our state diagram to merge these two states into one. Similarly for the two states that contain just the s1 string. They are indistinguishable, so should be merged. This gives us a smaller diagram, where some of the arrows form loops (Figure 6.2).

This state diagram is a big improvement over our first state diagram because it has several loops, and these loops give us more ways of going through the diagram and generating tests. Note that any path through the state diagram defines a sequence of method calls, and we can view any sequence as a test sequence.

> The more loops, choices, and alternative paths we have in our model, the better because they enable us to generate a wider variety of test sequences.

For example, the leftmost loop tells us that the remove(s1) method undoes the effect of the add(s1) method because it returns to the same empty state. So no matter how many times we go around the add(s1);remove(s1) loop, the set should still be empty. Similarly, the rightmost loop shows that remove(s2) undoes the effect of the add(s2) method.

FIGURE 6.1
Example states of a Set<E> object.

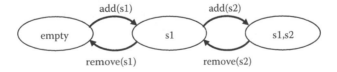

FIGURE 6.2
States from Figure 6.1, with identical states merged.

We do not want to just generate lots of test sequences; we also want to be able to execute each test sequence on a SUT and automatically determine whether the test has passed or failed. There are two ways we can do this. For methods that return results, we can annotate each transition of our state diagram with the expected result of each method call. For example, the `add(s1)` transition from the empty state should return true because the `s1` string was not a member of the empty set—so we could write this transition as `add(s1)/true` to indicate the expected result.

The other way of checking whether a test sequence has passed or failed is to check that the internal state of the SUT agrees with the expected state of the model. It is not always possible to do this because if the internal state of the SUT is private, we may not be able to observe it. But most SUTs provide a few query methods that give us some information about the current state of the SUT, and this allows us to check if that state agrees with our model. For our `Set<E>` example, we can use the `size()` method to check that a SUT contains the expected number of strings, and we can use the `contains(String)` method to check if each of the expected strings is in the set. In fact, it is a good strategy to call as many of the query methods as possible after each state transition since this helps to test all the query methods (checking that they are consistent with each other) and also verifies that the SUT state is correct. We could explicitly show every query method as a self-transition in our state diagram, but this would clutter the state diagram too much. So we will show only the mutator methods in our state diagrams here, but we will see later how the query methods can be added into the model after each transition.

Our state diagram is already a useful little test model that captures some of the expected behavior of a `Set<E>` implementation, but it does not really test the full functionality yet. The `clear()` method is never used, and we are testing only two strings so far. We need to add some more transitions and states to obtain a more comprehensive model. This raises the most important question of MBT:

How big does our model have to be?

The answer usually is, the smaller the better. A small model is quicker to write, easier to understand, and will not give an excessive number of tests. A good model will have a high level of *abstraction*, which means that it will omit all details that are not essential for describing the behavior that we want to test. However, we still want to meet our test goals, which in this case is to test all the mutator methods. So we will add some `clear()` transitions into our model. Also, it is often a good goal to ensure that the model is *complete*, which means that we have modeled the behavior of every mutator method call in every state. Our state diagram above calls `add(s1)` from the empty state, but not from the other states, so it is currently incomplete. If we expand it to include all five actions (`clear()`, `add(s1)`, `add(s2)`, `remove(s1)`, `remove(s2)`) in every state, we get the state diagram shown in Figure 6.3.

Note how our goal of having a complete model forced us to consider several additional cases that we might not have considered if we were designing test sequences in a more ad hoc fashion. For example, the `add(s1)` transition out of the s1 state models the behavior of `add(s1)` when the string `s1` is already in the set and checks that we do not end up with *two* copies of `s1` in the set. Similarly, the `remove(s1)` transition out of the s2 state models what should happen when the member to be removed is not in the set—the remove method should leave the set unchanged and should return false. The `clear()` transition out of the empty state might not have occurred to a manual test designer, but it serves the useful purpose of ensuring that `clear()` can be called multiple times in a row without crashing. The point is that designing a model (especially a complete model) leads us to consider all

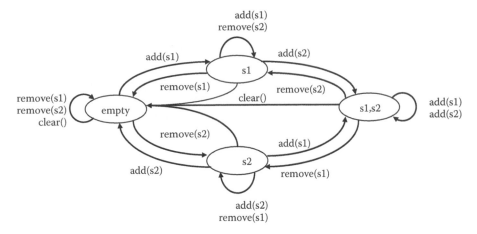

FIGURE 6.3
Finite-state diagram for Set<E> with two strings, s1 and s2.

the possible cases in a very systematic way, which can improve the quality of our testing, and is a good way of finding omissions and errors in the requirements (Stobie 2005).

The remaining question about our model that we should discuss is how many different string values should we test? Why have we tested just two strings? A real implementation can handle hundreds or millions of strings, so should we not test large numbers of strings, too? This is another question about how abstract our model should be. To keep our model small, we want to model as few strings as possible, but still exercise the essential features of sets. Zero strings would be rather uninteresting since the set would always be empty. One string would mean that the set is either empty or contains just that one string. This would allow us to check that the set ignores duplicate adds and duplicate removes, but it would not allow us to test that adding a string leaves all other strings in the set unchanged. Two is the minimum number of strings that covers the main behaviors of a set, so this is the best number of strings to use in our model. If we expanded our model to three different strings, it would have 8 states and 7 actions, with a total of 56 transitions. This would be significantly more time consuming to design, but it would give little additional testing power.

One of the key skills of developing good test models is finding a good level of abstraction, to minimize the size of the model, while still covering the essential SUT features that you want to test.

6.3.2   From FSM to EFSM: Writing models in Java

Embedded systems often have quite complex behavior, so they require reasonably large models to accurately summarize their behavior. As models become larger, it quickly becomes tedious to draw them graphically. Instead, we will write them as Java classes, following an EFSM style. An extended finite-state machine is basically a finite-state machine with some *state variables* added to the model to keep track of more details about the current SUT state and *actions* (code that updates the state variables) added to the transitions. These features can make models much more concise because the state variables can define many different states, and one Java method can define many similar transitions in the model. We will use the ModelJUnit style of writing the models because it is simple and effective.

ModelJUnit is an open-source tool that aims to be the simplest possible introduction to MBT for Java programmers (Utting and Legeard 2007). The models are written in Java so that you do not have to learn some new modeling language. In fact, a model is just a Java class that implements a certain interface (FsmModel). The state variables of the class are used to define all the possible states of the state machine model, and the "Action" methods of the Java class define the transitions of the state machine model. Figure 6.4 shows some Java code that defines our two-string test model of the Set<E> interface—we model just the three mutator operations at this stage.

We now discuss each feature of this class, showing how it defines our two-string test model. Line 02 defines a class called SimpleSet and says that it implements the FsmModel interface defined by ModelJUnit. This tells us that the class can be used for MBT and means that it must define the getState and reset methods. Line 04 defines a Boolean variable for each of the two strings that we are interested in. The programmer realized that the two strings can be treated independently and that all we need to know about each string is whether it is in the set or not. So when the variable s1 is true, it means that the first string is in the set, and when the variable s2 is true, it means that the second string is in the set. (We will decide on the precise contents of the two strings later). Choosing the state variables of the model is the step that requires the most insight and creativity from the programmer.

Lines 06–07 define the getState() method, which allows ModelJUnit to read the current state of the model at any time. It returns a string that shows the values of the two Boolean variables, with each Boolean converted to a single "T" or "F" character to make the state names shorter. Lines 09–10 define the reset method, which is called each time a new test sequence is started. It sets both Boolean variables to false, meaning that the set is empty.

The remaining lines of the model give five *action* methods. These define the transitions of the state machine because the code inside these methods changes the state variables of the model. For example, the addS1 method models the action of adding the first string into the model, so it sets the s1 flag to true to indicate that the first string should now be in

```
01: /** A model of a set with two elements: s1 and s2. */
02: public class SimpleSet implements FsmModel
03: {
04:     protected boolean s1, s2;
05:
06:     public Object getState()
07:     { return (s1 ? "T" : "F") + (s2 ? "T" : "F"); }
08:
09:     public void reset(boolean testing)
10:     { s1 = false; s2 = false; }
11:
12:     @Action public void addS1() { s1 = true;}
13:     @Action public void addS2() { s2 = true;}
14:     @Action public void removeS1() { s1 = false;}
15:     @Action public void removeS2() { s2 = false;}
16:     @Action public void clear() { s1 = false; s2 = false;}
17: }
```

FIGURE 6.4
Java code for the SimpleSet model.

the set. These action methods are marked with a `@Action` annotation, to distinguish them from other auxiliary methods that are not intended to define transitions of the model.

6.4 How to Generate Tests with ModelJUnit

ModelJUnit provides a graphical user interface (GUI) that can load a model class, explore that model interactively or automatically, visualize the state diagram that the model produces, generate any number of tests from the model, and analyze how well the generated tests cover the model. If we compile our `SimpleSet` model (using a standard Java compiler) and then load it into the ModelJUnit GUI, we see something like Figure 6.5, where the "Edit Configuration" panel shows several test generation options that we can choose between.

If we accept the default options and use the "Random Walk" test generation algorithm to generate the default size test suite of 10 tests, the small test sequence shown in the left panel of Figure 6.5 is generated. Each triple (*Sa, Action, Sb*) indicates one step of the test sequence, where *Action* is the test method that is being executed, starting in state *Sa* and finishing in state *Sb*. For example, the first line tells us to start with an empty set (state = "FF"), add the second string, and then check that the set corresponds to state FT (i.e., it contains the second string but not the first string). Then, the second and third lines check that adding then removing the first string brings us back to the same FT state.

Since this test sequence is generated by a purely random walk, it is not very smart (it tests the `addS2` action on the full set four times!). However, even such a naive algorithm as this will test every transition (i.e., every action going out of every state) if we generate a long enough test sequence. On average, the random walk algorithm will cover every transition of this small model if we generate a test sequence of about 125 steps. More sophisticated

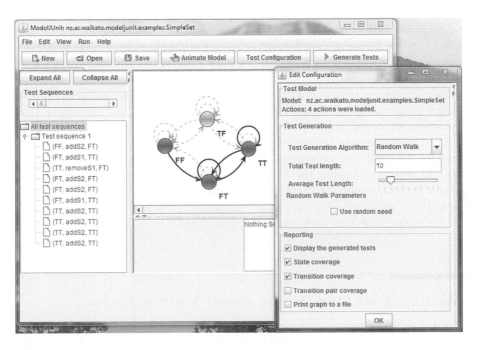

FIGURE 6.5
Screenshot of ModelJUnit GUI and Test Configuration Panel.

```
01: /** An example of generating tests from the set model. */
02: public static void main(String[] args)
03: {
04:    Tester tester = new RandomTester(new SimpleSet());
05:    tester.addListener(new VerboseListener()); // print the tests
06:    tester.generate(1000); // generate a long sequence of tests
07: }
```

FIGURE 6.6
ModelJUnit code to generate tests by a random traversal of a model.

algorithms can cover every transition more quickly. For example, ModelJUnit also has a "Greedy Random Walk" algorithm that gives priority to unexplored paths, and this takes about 55 steps on average to cover every transition. There is also the "Lookahead Walk" algorithm that does a lookahead of several transitions (three by default) to find unexplored paths, and this takes only 25 steps to test all 25 transitions. This happens to be the shortest possible test sequence that ensures *all-transitions coverage* of this model. Such minimum-length test sequences are called *Chinese Postman Tours* (Kwan 1962, Thimbleby 2003) because postmen also have the goal of finding the shortest closed circuit that takes them down every street in their delivery area.

The ModelJUnit GUI is convenient but not necessary. We can also write code that automates the generation of a test suite from our model. For example, the code shown in Figure 6.6 will generate and print a random sequence of 1000 tests. We put the test generation code inside a main method so that we can execute it from the command line. Another common approach is to put it inside a JUnit test method so that it can be executed as part of a larger suite of tests.

The code in Figure 6.6 generates a random sequence of 1000 add, remove, and clear calls. First, we create a "tester" object and initialize it to use a RandomTester object, which implements a "Random Walk" algorithm. We pass an instance of our SimpleSet model to the RandomTester object, and it uses Java reflection facilities to determine what actions our model provides. The next line (Line 05) adds a VerboseListener object to the tester so that some information about each test step will be printed to standard output as the tests are generated. The final line asks the tester to generate a sequence of 1000 test steps. This will include a random mixture of add, remove, and clear actions and will also perform a reset action occasionally, which models the action of creating a new instance of the Set<E> class that starts off in the empty state. The reset actions also mean that we generate lots of short understandable test sequences, rather than one long sequence.

Although the usual way of generating tests is via the ModelJUnit API, as in Figure 6.6, for simple testing scenarios, the ModelJUnit GUI can write this kind of test generation code for you. As you modify the test configuration options, it displays the Java code that implements the currently chosen options so that you can see how to use the API or cut and paste the code into your Java test generation programs.

6.5   Writing Your Own Model-Based Testing Tool

ModelJUnit provides a variety of useful test generation algorithms, model visualization features, model coverage statistics, and other features. However, its core idea of using reflection and randomness to generate tests from a Java model is very simple and can easily be implemented in other languages or in application-specific ways. Figure 6.7 shows the

```java
public class SimpleMBT {
  public static final double RESET_PROBABILITY = 0.01;
  protected FsmModel model_;
  protected List<Method> methods_ = new ArrayList<Method>();
  protected Random rand_ = new Random(42L); // use a fixed seed

  SimpleMBT(FsmModel model) {
    this.model_ = model;
    for (Method m : model.getClass().getMethods()) {
      if (m.getAnnotation(Action.class) != null) {
        methods_.add(m);
      }
    }
  }

  /** Generate a random test sequence of length 1.
   *  @return the name of the action done, or "reset".
   */
  public String generate() throws Exception {
    if (rand_.nextDouble() < RESET_PROBABILITY) {
      model_.reset(true);
      return "reset";
    } else {
      int i = rand_.nextInt(methods_.size());
      methods_.get(i).invoke(model_, new Object[0]);
      return methods_.get(i).getName();
    }
  }

  public static void main(String[] args) throws Exception {
    FsmModel model = new SimpleSet();
    SimpleMBT tester = new SimpleMBT(model);
    for (int length = 0; length < 100; length++) {
      System.out.println(tester.generate() + ": " + model.getState());
    }
  }
}
```

FIGURE 6.7
A simple MBT tool.

code for a simple MBT tool that just performs random walks of all the @Action methods in a given Java model, with a 1% probability of doing a reset at each step instead of an action. This occasional reset helps to prevent the test generation from getting stuck within one part of the model when the model contains irreversible actions. Many variations and improvements of this basic strategy are possible, but this illustrates how easy it can be to develop a simple MBT tool that is tailored to your testing environment.

6.6 Automating the Execution of Tests

We have now seen how we can *generate* tests automatically from a model of the expected behavior of the SUT. The generated test sequences have been printed in a human-readable format. If our SUT has a physical interface or a GUI, we could *manually* execute these generated test sequences by pushing buttons and looking to see if the current state of the SUT seems to be correct. This can be quite a useful approach for embedded systems that are difficult to connect to a computer.

But it would be nice to automate the *execution* of the tests, as well as the generation, if possible. This section discusses two alternative ways of automating the test execution: *offline* and *online* testing. Both approaches can be used on embedded systems. They both require an API connection to the SUT so that commands can be sent to the SUT and its current state can be observed. For embedded SUTs, this API often connects with some hardware, such as digital to analog converters, which connect to the SUT.

6.6.1 Offline testing

One simple, low-tech approach to executing the tests is to write a separate *adaptor* program that reads the generated test sequence, converts each action in a call to a Set<E> implementation, and then checks the new state of that implementation after the call to ensure that it agrees with the expected state, and report a test failure when they disagree. This adaptor program is essentially a little interpreter of the generated test commands, sending commands to the SUT via the API and checking the results. It plays the same role as a human who interfaces to the SUT and executes the tests manually.

This approach is called *offline* testing because the test generation and the test execution are done independently, at separate times and perhaps on separate computers. Offline testing can be useful if you need to execute the generated tests in many different environments or on a different computer to the test generator, or you want to use your existing test management tool to manage and execute the generated tests.

6.6.2 Online testing

Online testing is when the tests are being executed on the SUT at the same time as they are being generated from the model. This gives immediate feedback and even allows a test generation algorithm to observe the actual SUT output and adapt its test generation strategy accordingly, which is useful if the model or the SUT is *nondeterministic* (Hierons 2004, Miller et al. 2005). Online testing creates a tighter, faster connection between the test generator and the SUT, which can permit better error reporting and fast execution of much larger test suites, so it is generally the best approach for embedded systems, unless there are clear reasons why offline testing is preferable.

In this section, we shall extend our SimpleSet model so that it performs online testing of a Java SUT object that implements the Set<E> interface. Figure 6.8 shows a Java Model class that is similar to SimpleSet, but also has a pointer (called sut) to a Set<String> implementation that we want to test. Each of the @Action methods is extended so that as well as updating the state of the model (the s1 and s2 variables), it also calls one of the SUT methods. For example, after the addS1 action sets s1 to true (to indicate that string s1 should be in the set after this action), it calls sut.add(s1) to make the corresponding change to the SUT object. Then, it calls various query methods to check that the updated SUT state is the same as the state of the model (since all of the @Action methods do the same state checks in this example, we move those checks into a method called checkSUT() and call this at the end of each @Action method).

We have written this online testing class as a standalone class so that you can see the model updating code and the SUT updating code next to each other. An alternative style is to use inheritance to extend a model class (like SimpleSet) by creating a subclass that overrides each action method and adds the SUT actions and the checking code.

```
01: public class SimpleSetWithAdaptor implements FsmModel
02: {
03:   protected boolean s1, s2;
04:   protected Set<String> sut; // the implementation we are testing
05:
06:   // our test data for the SUT
07:   protected String str1 = "some string";
08:   protected String str2 = "";  // empty string
09:
10:   /** Tests a StringSet implementation. */
11:   public SimpleSetWithAdaptor(Set<String> systemUnderTest)
12:   { this.sut = systemUnderTest; }
13:
14:   public Object getState()
15:   { return (s1 ? "T" : "F") + (s2 ? "T" : "F"); }
16:
17:   public void reset(boolean testing)
18:   { s1 = false; s2 = false; sut.clear(); checkSUT(); }
19:
20:   @Action public void addS1()
21:   { s1 = true; sut.add(str1); checkSUT(); }
22:
23:   @Action public void addS2()
24:   { s2 = true; sut.add(str2); checkSUT(); }
25:
26:   @Action public void removeS1()
27:   { s1 = false; sut.remove(str1); checkSUT(); }
28:
29:   @Action public void removeS2()
30:   { s2 = false; sut.remove(str2); checkSUT(); }
31:
32:   /** Check that the SUT is in the expected state. */
33:   protected void checkSUT()
34:   {
35:     Assert.assertEquals(s1, sut.contains(str1));
36:     Assert.assertEquals(s2, sut.contains(str2));
37:     int size = (s1 ? 1 : 0) + (s2 ? 1 : 0);
38:     Assert.assertEquals(size, sut.size());
39:     Assert.assertEquals(!s1 && !s2, sut.isEmpty());
40:     Assert.assertEquals(!s1 && s2,
41:       sut.equals(Collections.singleton(str2)));
42:   }
43: }
```

FIGURE 6.8
An extension of SimpleSet that performs online testing.

A checking method such as `checkSUT()` typically calls one or more of the SUT query methods to see if the expected state (of the model) and the actual state of the SUT agree. For this example, we have decided to test a set of strings, using the two sample strings defined as `str1` and `str2` in Figure 6.8. So we can see if the first string is in the set by calling `sut.contains(str1)`, and we expect that this should be true exactly when our model has set the Boolean variable `s1` to true. So we use standard JUnit methods to check that `s1` is equal to `sut.contains(str1)`. We check that relationship between `str1` and `s2` in the same way. We add several additional checks on the `size()`, `isEmpty()`, and `equals(_)` methods of the SUT, partly to gain more confidence that the SUT state is correct, and partly so that we test those SUT query methods. They will be called many times, in every SUT state that our model allows, so they will be well tested. Finally, note that in Figure 6.8, we are not checking the return value of `sut.add(_)`, but we can easily do this by checking that the return value equals the initial value of the `s1` flag.

   Note how each action method updates the model, then updates the SUT in a similar way, then checks that the model state agrees with the SUT state. So as we execute a sequence of these action methods, the model and the SUT are evolving in parallel, each making the same changes, and the checkSUT() method is checking that they agree about what the next state should be. This nicely illustrates the essential idea behind MBT:

> Implement your system twice and run the two implementations in parallel to check them against each other.

But of course, no one really wants to implement a system twice! The trick that makes MBT useful is that those two "implementations" have very different goals:

1. The SUT implementation needs to be efficient, robust, scale to large data sets, and it must implement all the functionality in the requirements.

2. The model "implementation" can be a vastly simplified system that implements only one or two key requirements, handles only a few small data values chosen for testing purposes, and does not need to be efficient or scalable.

This difference means that it is usually practical to "implement" (design and code) a model in a few hours or a few days, whereas the real SUT takes months of careful planning and coding. We repeat: the key to cost-effective modeling is finding a good level of *abstraction* for the model.

> *Abstraction:* Deciding which requirements are the key ones that must be tested and which ones can be ignored or simplified for the purposes of testing.

6.6.3   Test execution results

We can use this model to test the `HashSet` class from the standard Java library simply by passing `new HashSet<String>()` to the constructor of our `SimpleSetWithAdaptor` class and then using that to generate any number of tests, either by using the ModelJUnit GUI or by executing some test generation code similar to Figure 6.6. When we do this, no errors are detected. This is not surprising since the standard Java library classes are widely used and thoroughly tested. If we write our own simple implementation of `Set<String>` and insert an off-by-one bug into its `equals` method (see the `StringSetBuggy` class in the ModelJUnit distribution for details), we get the following output when we try to generate a test sequence of length 60 using the Greedy Random Walk algorithm.

```
done (FF, addS2, FT)
done (FT, addS1, TT)
done (TT, removeS1, FT)
done (FT, removeS2, FF)
done (FF, removeS2, FF)

FAILURE: failure in action addS1 from state FF due to
  AssertionFailedError: expected:<false> but was:<true>
...
Caused by: AssertionFailedError: expected:<false> but was:<true>
  ...
  at junit.framework.Assert.assertEquals(Assert.java:149)
  at SimpleSetWithAdaptor.checkSUT(SimpleSetWithAdaptor.java:123)
  at SimpleSetWithAdaptor.addS1(SimpleSetWithAdaptor.java:87)
  ... 10 more
```

This pinpoints the failure as being detected by the `sut.equals` call on line 41 of Figure 6.8, when the `checkSUT` method was called from the `addS1` action with the set being empty. Interestingly, the test sequence shows us that `checkSUT` had tested the `equals` method on an empty set several times previously, but the failure did not occur then—it required a `removeS1` followed by an `addS2` to detect the failure. A manually designed JUnit test suite may not have tested that particular combination, but the random automatic generation will always eventually generate such combinations and detect such failures, if we let it generate long enough sequences.

If we fix our off-by-one error, then all the tests pass, and ModelJUnit reports that 100% of the transitions of the model have been tested. If we measure the code coverage of this `StringSet` implementation, which just implements a set as an ArrayList with no duplicate entries, we find that the generated test suite has covered 93.3% of the code (111 out of 119 JVM instructions, as measured by the EclEmma plugin for Eclipse [Emma 2009]). The untested code is the `iterator()` method, which we did not call in our `checkSUT()` method, and one exception case to do with null strings.

6.6.4 Mutation analysis of the effectiveness of our testing

It is also interesting to use the Jumble mutation analysis tool (Jumble 2010) to measure the effectiveness of our automatically generated test suite. Jumble analyzes the Java bytecode of a SUT class, creates lots of *mutants* (minor modifications that cause the program to have different behavior), and then runs our tests on each mutant to see if they detect the error that has been introduced. On this SUT class, `StringSet.java`, Jumble creates 37 different mutants and reports that our automatically generated tests detect 94% (35 out of 37) of those mutants.

```
Mutating modeljunit.examples.StringSet
Tests: modeljunit.examples.StringSetTest
Mutation points = 37, unit test time limit 2.58s
M FAIL: modeljunit.examples.StringSet:44: changed return value
.M FAIL: modeljunit.examples.StringSet:56: 0 -> 1
..................................
Score: 94%
```

This is a high level of error detection, which indicates that our automatically generated tests are testing our simple set implementation quite thoroughly and that our model accurately captures most of the behavior of the `Set<E>` interface. One of the two mutants that were not detected is in the `iterator()` method, which we did not test in our model. The other mutant indicates that we are not testing the case where the argument to the `equals` method is a different type of object (not a set). This is a low-priority case that could be ignored or could easily be covered by a manually written JUnit test.

6.6.5   Testing with large amounts of data

What if we wanted to do some performance testing to test that sets can handle hundreds or thousands of elements? For example, we might know that a SUT like `HashSet<E>` expands its internal data structures after a certain number of elements have been added, so we suspect that testing a set with only two elements is inadequate.

One approach would be to expand our model so that it uses a bit vector to keep track of hundreds of different strings and knows exactly when each string is in or out of the set. It is not difficult to write such a model, but when we start to generate tests, we quickly find that the model has so many states to explore that it will be impossible to test all the states or all the transitions. For example, with 100 strings, the model would have 2^{100} states and even more transitions. Many of these states would be similar, so many of the tests that we generate would be repetitive and uninteresting.

A more productive style is to keep our model small (e.g., two or three Boolean flags), but change our interpretation of one of those Boolean flags `s2` so that instead of meaning that "*str2 is in the set,*" it now means "*all the strings 'x1', 'x2' ... 'x999' are in the set.*" This leaves the behavior of our model unchanged and means that all we need to change is the code that updates the SUT. For example, the `addS2()` action becomes

```
23:   @Action public void addS2()
24a:  {
24b:     s2 = true;
24c:     for (int i=1; i<1000; i++)
24d:       { sut.add("x"+i); }
24e:     checkSUT();
24f:  }
```

With this approach, we can generate the same short test sequence as earlier and easily cover all the states and transitions of our small model, while the tests can scale up to any size of set that we want. This is another good example of using *abstraction* when we design the model—we decided that even though we want to test a thousand strings, it is probably not necessary to test them all independently—testing two groups of strings should give the same fault-finding power.

> When possible, it is good to keep the model small and abstract and make the adaptor code do the donkey work.

6.7   Testing an Embedded System

In this section, we shall briefly see how this same MBT approach can be used to model and test an embedded system such as the *Subscriber Identification Module* (SIM) card embedded in GSM mobile phones. The SIM card stores various data files that contain private data of the user and of the Telecom provider, so it protects these files via a system of access permissions and PIN codes. When a SIM card is inserted into a mobile phone, the phone communicates with the SIM card by sending small packets of bytes that follow the GSM 11.11 standard protocol (Bernard et al. 2004). A summary of some key features of this GSM 11.11 protocol is given in Utting and Legeard (2007, Chapter 9), together with use cases, UML class diagrams and UML state machine models, plus examples of generating tests from those UML models. In this section, we give a brief overview of how the same system can be modeled in Java and show how we can generate tests that send packets of bytes to the SIM and check the correctness of its responses. We execute the generated tests on a simulator of the SIM card so that we can measure the error detection power using Jumble. The generated tests could equally well be executed on real hardware, if we have a test execution platform with the hardware to connect to the physical SIM and send and receive the low-level packets produced by the tests.

6.7.1   The SIM card model

Our test model of the SIM card is defined in a Java class called `SimCard` (420 source lines of code), which contains 6 enumerations, 12 data variables, and 15 actions, plus the usual `reset` and `getState` methods. There is also a small supporting class called `SimFile` (24 source lines of code) that models the relevant aspects of the File objects stored within the SIM—we do not model the full contents of each file—a couple of bytes of data is sufficient to test that the correct file contents are being retrieved. The full source code of this SIM card model is included as one of the example models in the ModelJUnit distribution.

Figure 6.9 shows all the data variables of the model. The `files` map models the contents of all the files and directories on the SIM—these are constant throughout testing since this model does not include any write operations. The `DF` and `EF` variables model the currently selected *directory file* and the currently selected *elementary file* within that directory, respectively. The PIN variable corresponds to the correct PIN number, which is set to 11 by the reset method of the model, then may be set to 12 or back to 11 by `changePIN` actions during testing (two PIN numbers are sufficient for testing purposes).

The next four variables (`status_en`, `counter_PIN_try`, `perm_session,` and `status_PIN_block`) model all the PIN-related aspects of the SIM security, and the following two variables (`counter_PUK_try` and `status_blocked`) model the Personal Unblocking Key (PUK) checking—entry of a correct PUK code allows a user to unblock a card that had `status_PIN_block` set to `Blocked` because of three incorrect PIN attempts. However, after 10 incorrect PUK attempts, `status_PUK_block` will be set to `Blocked`, which means that all future attempts to unblock the SIM by entering a correct PUK will fail. When testing a real SIM chip this effectively destroys the SIM chip since there is no way of resetting the SIM to normal functionality once it has blocked PUK entry.

Figure 6.10 shows one of the more interesting methods in the model, `Unblock_PIN`. This model the user trying to enter a PUK code in order to set the SIM to use a new PIN number, which is typically done after the old PIN is blocked due to three incorrect PIN attempts. The `Unblock_PIN` method takes the PUK code and the new PIN code as inputs, and these are typically eight digit and four digit integers, respectively. If we chose input values at random, there would be $10^8 \times 10^4 = 10^{12}$ possible combinations of inputs, most of which would have the same effect. So to focus the test generation on the most interesting cases, we decide to define just two actions that call Unblock_PIN—one with the correct PUK number and a new PIN code of 12, and one with an incorrect PUK number. Repeated applications of the latter action will test the PUK blocking features of the SIM. This illustrates a widely used

```
public class SimCard implements FsmModel
{
  public enum E_Status {Enabled, Disabled};
  public enum B_Status {Blocked, Unblocked};
  public enum Status_Word {sw_9000, sw_9404, sw_9405, sw_9804,
                           sw_9840, sw_9808, sw_9400};
  public enum File_Type {Type_DF, Type_EF};
  public enum Permission {Always, CHV, Never, Adm, None};
  public enum F_Name {MF, DF_GSM, EF_LP, EF_IMSI, DF_Roaming, EF_FR, EF_UK};

  // These variables model the attributes within each Sim Card.
  protected static final int GOOD_PUK = 1223; // the correct PUK code
  public static final int Max_Pin_Try = 3;
  public static final int Max_Puk_Try = 10;
  /** This models all the files on the SIM and their contents */
  protected Map<F_Name,SimFile> files = new HashMap<F_Name,SimFile>();
  /** The currently-selected directory (never null) */
  protected SimFile DF;
  /** The current elementary file, or null if none is selected */
  protected SimFile EF;
  /** The correct PIN (can be 11 or 12) */
  protected int PIN;
  /** Say whether PIN-checking is Enabled or Disabled */
  protected E_Status status_en;
  /** Number of bad PIN attempts: 0 .. Max_Pin_Try */
  protected int counter_PIN_try;
  /** True means a correct PIN has been entered in this session */
  protected boolean perm_session;
  /** Set to Blocked after too many incorrect PIN attempts */
  protected B_Status status_PIN_block;
  /** Number of bad PUK attempts: 0 .. Max_Puk_Try */
  protected int counter_PUK_try;
  /** Set to Blocked after too many incorrect PUK attempts */
  protected B_Status status_PUK_block;
  /** The status word returned by each command */
  protected Status_Word result;
  /** The data returned by the Read_Binary command */
  protected String read_data;
  /** The adaptor object that interacts with the SIM card */
  protected SimCardAdaptor sut = null;
```

FIGURE 6.9
Data variables of the SimCard model.

test design strategy called *equivalence classes* (Copeland 2004): when testing a general-purpose method that has many possible combinations of input values, we manually choose just a strategic few of those input combinations—one for each different kind of behavior that is possible. In our Unblock_PIN model method, the choice of a good or bad PUK code determines the outcome of the second if condition (puk == GOOD_PUK), and since all the other if conditions are determined by the state variables of the model, these two PUK values are sufficient for us to test all the possible behaviors of this model method. This style of having several @Action methods that all call the same method, with carefully chosen different parameter values, is often used in ModelJUnit to reduce the size of the state space

```
@Action public void unblockPINGood12() { Unblock_PIN(GOOD_PUK,12);}
@Action public void unblockPINBad()  { Unblock_PIN(12233446,11);}
public void Unblock_PIN(int puk, int newPin)
{
 if (status_block == B_Status.Blocked) {
  result = Status_Word.sw_9840; /*@REQ: Unblock_CHV1 @*/
 } else if (puk == GOOD_PUK) {
  PIN = newPin;
  counter_PIN_try = 0;
  counter_PUK_try = 0;
  perm_session = true;
  status_PIN_block = B_Status.Unblocked;
  result = Status_Word.sw_9000;
  if (status_en == E_Status.Disabled) {
     status_en = E_Status.Enabled; /*@REQ: Unblock5 @*/
  } else {
    // leave status_en unchanged
  } /*@REQ: Unblock7,Unblock2 @*/
 } else if (counter_PUK_try == Max_Puk_Try - 1) {
   System.out.println("BLOCKED PUK!!! PUK try counter="+counter_PUK_try);
   counter_PUK_try = Max_Puk_Try;
   status_block = B_Status.Blocked;
   perm_session = false;
   result = Status_Word.sw_9840; /*@REQ: REQ7, Unblock4 @*/
 } else {
   counter_PUK_try = counter_PUK_try + 1;
   result = Status_Word.sw_9804; /*@REQ: Unblock3 @*/
 }

 if (sut != null) {
  sut.Unblock_PIN(puk, newPin, result);
 }
}
```

FIGURE 6.10
The Unblock_PIN actions of the SimCard model.

that is explored during testing, while still ensuring that the important different behaviors are tested.

6.7.2   Connecting the test model to an embedded SUT

The last two lines of the Unblock_PIN method in Figure 6.10 show how we can connect the model to an implementation of the SIM, via some adapter code (shown in Figure 6.11) that handles the low-level details of assembling, sending, and receiving packets of bytes.

The SimCard model defines quite a large finite-state machine. If we analyze the state variables of the model and think about which combinations of values are possible, we find that there are 10 possible directory/file settings (DF and EF), 2 PIN values, 4 values for counter_PIN_try, and 11 values for counter_PUK_try, plus several other flags (but their values are generally correlated with other data values), so there are likely to be around $10 \times 2 \times 4 \times 11 = 880$ states in the model and up to 15 times that number of transitions (since there are 15 @Action methods in the model). This is too large for us to want to test exhaustively, but it is easy to use the various random walk test generation algorithms of

```
public class SimCardAdaptor
{
 protected byte[] apdu = new byte[258];
 protected byte[] response = null;
 protected GSM11Impl sut = new GSM11Impl();

 /** Sets up the first few bytes of the APDU, ready to send to the SIM. */
 protected void initCmd(int cmdnum, int p1, int p2, int p3)
 {
  for (int i=0; i<apdu.length; i++)
  { apdu[i] = 0; }
  apdu[0] = (byte)0xA0;
  apdu[1] = (byte)(cmdnum & 0xFF);
  apdu[2] = (byte)(p1 & 0xFF);
  apdu[3] = (byte)(p2 & 0xFF);
  apdu[4] = (byte)(p3 & 0xFF);
 }

 public void Unblock_PIN(int puk, int newPin, SimCard.Status_Word result)
 {
  initCmd(0x2C, 0x00, 0x00, 0x10);
  setChv(5, puk); // pack the PUK into bytes 5..12
  setChv(13, newPin); // pack the PIN code into bytes 13..20
  response = sut.cmd(apdu);
  checkStatus(result, 0);
 }
...
```

FIGURE 6.11
Adapter class that connects SimCard model to a SIM implementation.

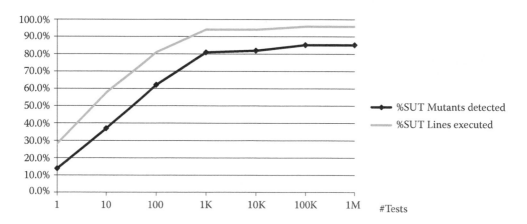

FIGURE 6.12
How SUT coverage and error detection increase with test sequence length.

ModelJUnit to generate test suites of any desired length. If we are doing online testing, we can just keep generating and executing tests until we find an error, or until a certain number of seconds or hours has elapsed. The randomness aspect of the generation means that the longer we test, the more thoroughly we cover all the possible sequences of actions. Figure 6.12 shows how effective a simple random walk test generation algorithm can be at finding errors. To measure the error detection power of the generated tests, we wrote a software simulation of a SIM chip in Java and used Jumble to generate 298 different mutants of that simulation. Each mutant is like a potential bug in the SUT. Then, we generate different size test suites and measure what percentage of the mutants/bugs is detected by the test suite and what percentage of the lines of code in the SUT is executed by the test suite. Figure 6.12 shows that the bug-detection rate rises rapidly up to a test length of 1000, and then more slowly to the maximum of 85% of mutants detected by a test suite with 100,000 steps (the remaining mutants were mostly modifying data in parts of the sample SIM files that were outside the scope of the model, so the mutations were not detectable with this model). The SUT code coverage follows a similar pattern and reaches a maximum of 95.8% of lines executed by the generated tests. It might seem impractical to execute test suites this large, but the total generation and online execution time (using the simulated Java SIM as the SUT) of a million test steps takes less than 7 s on an Intel Core 2 Duo 2.5GHz, and 100,000 tests takes less than 1 s, so test suites this large are quite practical. Our GSM model is reasonably large (more than 570 states and 8000 transitions), so it is worthwhile to generate lots of tests so that we cover lots of different scenarios. For example, an average test suite of length 100,000 tests the blocked PIN situation about 3000 times, uses the PUK about 6000 times, and tests the blocked-PUK situation (which destroys the SIM card) about three times.

6.8 Related Work and Tools

There are many other languages and tools that can also be used for EFSM MBT. In this chapter, we have written EFSM models in Java, but the design strategies are transferable to most kinds of state-based models and tools. For example, the NModel tool (Jacky et al. 2008) uses a similar style of model, but in C#. Spec Explorer (Veanes et al. 2008) is a MBT tool from Microsoft that uses EFSM models written in C#, but it can also combine these with scenarios written in a regular-expression style. Recent versions of Spec Explorer are integrated with Visual Studio and have GUI facilities for visualizing the EFSMs and the generated tests. There are also several MBT tools that use EFSM models written as UML state machines, with the actions written in Java (Conformiq 2009) or in OCL (Smartesting 2009). The EFSM modeling principles are similar across all these tools, but they differ in their algorithms for generating tests and their facilities for visualizing models and tests. ModelJUnit differs by taking a very simple random-exploration approach to test generation (and Section 6.5 shows how you can build a similar test generation tool in any language that supports reflection) and by providing an API for generating tests, whereas the other tools use a GUI or command line program to generate tests.

6.9 Conclusions

We have explored the key ideas of MBT: creating a small *model* of the system that you want to test, and then using various kinds of tools to automatically generate a test suite from that model. The ModelJUnit philosophy is to use Java as the modeling language because it is familiar, and to use reflection plus some simple random choice algorithms to generate the test suites. This is a simple approach that can be implemented quite easily in any language that supports reflection. It can be used to generate offline test suites, or for online testing.

MBT allows you to automatically generate and execute a large number of tests. With careful design of the model, it can give good coverage of the SUT behavior and code. The problem of maintaining the test suite disappears since it can be regenerated at any time. Instead, you must maintain the test model, but this is typically smaller and less repetitive than a test suite.

References

Bernard, E., Legeard, B. Luck, X., and Peureux, F. (2004). Generation of test sequences from formal specifications: GSM 11.11 standard case-study. *Software: Practice and Experience*, 34(10), 915–948.

Binder, R. V. (1999). *Testing Object-Oriented Systems: Models, Patterns, and Tools.* The Addison-Wesley Object Technology Series. Addison-Wesley, Boston, MA.

Clarke, J. (1998). Automated test generation from behavioral models. *Proceedings of the 11th Software Quality Week (QW'98)*, Software Research Inc., San Francisco, CA.

Conformiq Inc. Web site. http://www.conformiq.com. Accessed on September 2009).

Copeland, L. (2004). *A Practitioner's Guide to Software Test Design.* Artech House Publishers, Norwood, MA.

Coppit, D., and Lian, J. (2005). Yagg: an easy-to-use generator for structured test inputs. *Proceedings of the 20th IEEE/ACM International Conference on Automated Software Engineering (ASE'05), Long Beach, CA*, Pages: 356–359. ACM, New York, NY.

Czerwonka, J. (2008). http://www.pairwise.org (accessed Aug. 2008).

Dalal, S. R., Jain, A. Karunanithi, N. et al. (1999). Model-based testing in practice. *Proceedings of the 21st International Conference on Software Engineering (ICSE '99), Los Alamitos, CA, USA*, Pages: 285–294. ACM, New York.

El-Far, I. K., and Whittaker, J.A. (2002). Model-based software testing. *Encyclopedia of Software Engineering, Volume 1*, ed. Marciniak, J. J., Pages: 825–837. Wiley-InterScience, New York.

Emma. (2009). The Emma Eclipse plugin and the Emma command line Java coverage tool are available from http://www.eclemma.org and http://emma.sourceforge.net, respectively (accessed May 2010).

Farchi, E., Hartman, A., and Pinter, S. S. (2002). Using a model-based test generator to test for standard conformance. *IBM Systems Journal*, 41(1), 89–110.

Kwan M.-K. (1962). Graphic programming using odd or even points. *Chinese Mathematics*, 1:273–277.

Hierons, R. M. (2004). Testing from a nondeterministic finite state machine using adaptive state counting. *IEEE Transactions on Computers*, 53(10):1330–1342.

Horstmann, M., Prenninger, W., and El-Ramly, M. (2005). Case studies. *Model-Based Testing of Reactive Systems*, eds. Broy, M., et al. Springer LNCS 3472, Pages: 439–461. Springer-Verlag, Heidelberg.

Jacky, J., Veanes, M., Campbell, C., and Schulte, W. (2008). *Model-based Software Testing and Analysis with C#* , Cambridge University Press. For details about the NModel tool see http://www.codeplex.com/NModel. Accessed on 9 June, 2011.

Jard, C., and Thierry Jéron, T. (2005). TGV: Theory, principles and algorithms. *International Journal on Software Tools for Technology Transfer (STTT)*, 7(4):297–315.

Jumble web site. (2010). http://jumble.sourceforge.net (accessed Aug 2010).

Lee, D., and Yannakakis, M. (1996). Principles and methods of testing finite state machines—a survey. *Proceedings of the IEEE*, 84(2):1090–1126.

Miller, R. E., Chen, D., Lee, D., and Hao, R. (2005). Coping with nondeterminism in network protocol testing. *Testing of Communicating Systems: 17th IFIP TD 6/WG 6.1 International Conference, TESTCOM 2005, Montreal, Canada, May 31–June 2, 2005, Proceedings*, Volume 3502 of LNCS, Pages: 129–145. Springer-Verlag, Heidelberg.

ModelJUnit web site. (2010). http://modeljunit.sourceforge.net (accessed Sep. 2010).

Pretschner, A., Prenninger, W., Wagner, S., et al. (2005). One evaluation of model-based testing and its automation. *Proceedings of the 27th International Conference on Software Engineering (ICSE'05), St. Louis, May 2005*, 392–401. ACM Press, New York, NY.

Smartesting Inc. Web site, http://www.smartesting.com. Accessed on September 2009).

Stobie, K. (2005). Model-based testing in practice at Microsoft. *Proceedings of the Workshop on Model Based Testing (MBT 2004)*, eds. Gurevich et al. Volume 111 of Electronic Notes in Theoretical Computer Science, Pages: 5–12. Elsevier, January 2005.

Thimbleby, H. (2003). The directed Chinese Postman Problem. *Software: Practice and Experience*, 33(11), 1081–1096.

Utting, M., and Legeard, B. (2007). *Practical Model-Based Testing: A Tools Approach*. Morgan Kaufmann, San Francisco, CA.

Veanes, M., Campbell, C., Grieskamp, W., et al. (2008). Model-based testing of object-oriented reactive systems with Spec Explorer. *Formal Methods and Testing*. LNCS 4949, Pages: 39–76. Springer-Verlag, Heidelberg.

7

Automatic Testing of Lustre/Scade Programs

Virginia Papailiopoulou, Besnik Seljimi, and Ioannis Parissis

CONTENTS

SCADE (Safety Critical Application Development Environment)* is a tool-suite dedicated to the development of critical embedded systems in many domains in industry (avionics, nuclear energy, train transportation). It is mainly used in the design of major applications in the aerospace field. It provides facilities for the hierarchical definition of the system components into a graphical editor, for their simulation and verification as well as for automatic code generation.

The Lustre language [3] is the backbone of SCADE; it is a synchronous declarative, data-flow language. Its deterministic nature and formal semantics make it suitable for programming the control part of reactive synchronous systems.

In industrial Lustre/Scade applications, safety is a factor of high importance since a possible failure could cost human lives or severe damage to equipment or environment. Therefore, the verification and validation of such programs are major issues. The fact that

This work has been partially supported by SIESTA (www.siesta-project.com), a project of the French National Research Agency (ANR).
*www.esterel-technologies.com.

LUSTRE is both a programming language and a temporal logic [23] makes it possible to specify the required properties of a program in the same formalism used for the implementation. Formal verification of LUSTRE programs is then easy to carry out [10], but well-known state explosion problems remain the major limitation of such an approach and testing remains the main verification technique.

Test objectives are extracted from the functional requirements of the system (e.g., a requirement of DO-178B [7] standard) and test scenarios must be designed accordingly. The actual practices of the test professionals still involve mainly manual test construction. Hence, the test activities are very expensive, and automating some of them is an important concern for both industry and academia.

This chapter addresses two issues related to the automation of the testing activities related to LUSTRE/SCADE:

Test generation.   Usually, test design is manual and must ensure that the system requirements are adequately implemented. Investigations on the automation of the test generation process for LUSTRE programs have mainly focused on automatic test data generation for which promising contributions and tools have been proposed. They generally process formal specifications to extract test data [16] or to build test data generators [21, 22, 25]. In this chapter, we focus on LUTESS V2 [27] and on a model-based testing methodology using this tool. LUTESS V2 enables automatic construction of test input generators from a test model, written in a LUSTRE-like language. Several test models can be built for the same program simulating normal behaviors and system failures randomly or in specific situations.

Test coverage assessment.   Manual requirement-based and automated model-based test generation aim at ensuring that all the requirements have been adequately implemented but do not ensure that the entire program has been tested: code coverage metrics are mainly used to provide a measure of test coverage of a program. In the current state of practice, a SCADE program is compiled into an equivalent C code according to various compilation options and coverage is assessed on C code that is automatically generated. The resulting C code, however, depends on the compiler and because there is no standardized compiler, it is difficult to establish a formal relation between the C code and the original SCADE program. Hence, although possible, assessing the test coverage of the generated C code does not provide meaningful information about the corresponding SCADE program coverage. Moreover, SCADE users are not necessarily familiar with C or others often used programming languages, so C code coverage criteria are not relevant to them. A coverage criteria family is presented in this chapter, defined directly on the LUSTRE/SCADE specifications, enabling the coverage assessment of a SCADE model. Since LUSTRE is a data-flow language, common criteria defined on the program control flow graph (CFG) do not directly apply to a LUSTRE program. The proposed approach formally defines a hierarchy of coverage criteria [14] implemented in a prototype tool, LUSTRUCTU [12]. LUSTRUCTU analyzes LUSTRE programs and extracts the conditions that a test input sequence must verify in order to meet a criterion and computes the coverage ratio achieved after the execution of a test data sequence. The proposed criteria are inspired from several past investigations. For instance, Woodward, Hedley, and Hennell [31] defines LCSAJs, intermediate constructions that can be concatenated to build arbitrary long program subpaths. The associated test adequacy criteria called *Test Effectiveness Ratio* (TER_i) correspond to various levels of specified path coverage, in that it can be tailored to meet whatever path coverage is specified. Adequacy criteria focusing on the data flow of a program and defined on CFG have been proposed in [6, 26, 18, 15]. Finally, adequacy criteria focusing on Boolean expression coverage have been defined in [29, 30, 5, 4].

The impact of the proposed contributions on a common development process is shown in Figure 7.1 by the rectangles with dashed outline and rounded corners. The contributions

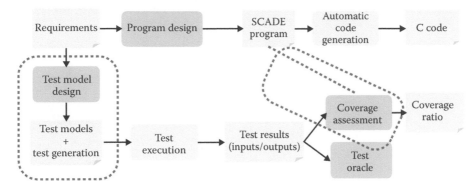

FIGURE 7.1
Scope of the proposed approaches with respect to the development process.

pertain to two approaches, the test data generation process and the coverage evaluation, respectively. These approaches are tailored to industrial needs and could have a positive impact in effectively testing real-world applications. Case studies are used to demonstrate the application of these approaches as well as to empirically evaluate their performance and complexity.

In our approach, the requirement-oriented-test generation is considered independently of the specification-based (or model-based) coverage assessment. This choice stems from the current industrial practices mentioned above. Some recent investigations [24] suggested measuring the extent to which requirements have been covered by applying a test suite. For this, requirements must be first transformed into temporal logic formulas. Coverage metrics are then defined directly on the formalized requirements in order to determine how well a test suite has exercised the requirements set. In addition, these coverage criteria can be used to automatically generate sets of test cases for requirements testing.

After an overview of the LUSTRE language, the two main sections of this chapter concentrate on the automation of the test coverage assessment and of the test generation, respectively.

7.1 LUSTRE Overview

LUSTRE [9] is a data-flow language. Contrary to imperative languages that describe the control flow of a program, LUSTRE describes how outputs are computed from inputs. Any variable or expression is represented by an infinite sequence of values and take the n-th value at the n-th cycle of the program execution, as it is shown in Figure 7.2. At each tick of a global clock, all inputs are read and processed simultaneously and all outputs are emitted, according to the *synchrony* hypothesis.[*]

A LUSTRE program is structured into nodes. A node is a set of equations that define the node outputs as a function of its inputs. Each variable can be defined only once within a node and the order of equations is of no matter. Specifically, when an expression E is assigned to a variable X, $X{=}E$ that indicates that the respective sequences of values are

---

[*]The synchrony hypothesis states that the software reaction is sufficiently fast so that every change in the external environment is taken into account.

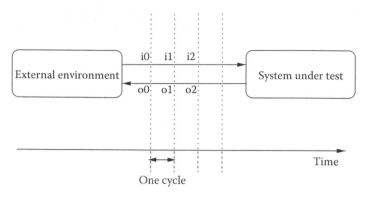

FIGURE 7.2
Synchronous software operation.

```
node Never(A: bool) returns (never_A: bool);
let
  never_A = not(A) -> not(A) and pre(never_A);
tel;
```

	c_1	c_2	c_3	c_4	...
A	*false*	*false*	*true*	*false*	...
never_A	*true*	*true*	*false*	*false*	...

FIGURE 7.3
Example of a LUSTRE node.

identical throughout the program execution; at any cycle, X and E have the same value. Once a node is defined, it can be used inside other nodes similar to any other operator.

The operators supported by LUSTRE are the common arithmetic and logical operators (+, -, *, /, and, or, not) as well as two specific temporal operators: the *precedence* (pre) and the *initialization* (->). The pre operator introduces to a sequence a delay of one time unit, while the -> operator –also called *followed by* (fby)– allows the initialization of a sequence. Let $X = (x_0, x_1, x_2, x_3, \ldots)$ and $E = (e_0, e_1, e_2, e_3, \ldots)$ be two LUSTRE expressions. Then, pre(X) denotes the sequence $(nil, x_0, x_1, x_2, x_3, \ldots)$, where *nil* is an undefined value, while X->E denotes the sequence $(x_0, e_1, e_2, e_3, \ldots)$.

LUSTRE neither supports loops (constructs such as for and while) nor recursive calls. Consequently, the execution time of a LUSTRE program can be statically computed and the satisfaction of the synchrony hypothesis can be checked.

A simple LUSTRE program is given in Figure 7.3, followed by an instance of its execution. This program has a single input Boolean variable and a single output Boolean variable. The output is *true* if and only if the input has never been *true* since the beginning of the program execution.

7.1.1 Operator network

The transformation of the inputs into the outputs in a LUSTRE program is done via a set of operators. Therefore, it can be represented by a directed graph, the so-called *operator network*. An operator network is a graph with a set of N operators that are connected to each other by a set of $E \subseteq N \times N$ directed edges. Each operator represents a logical or a numerical computation. With regard to the corresponding LUSTRE program, an operator network has as many input and output edges as the program input and output variables, respectively.

Figure 7.4 shows the corresponding operator network for the node of Figure 7.3. At the first execution cycle, the output `never_A` is the negation of the input `A`; for the rest of the execution, the output equals to the result of the conjunction of its previous value and the negation of `A`.

An operator represents a data transformation from an input edge into an output edge. There are two types of operators:

1. The basic operators that correspond to a basic computation.

2. The compound operators that correspond to the case where in a program, a node calls another node.

A basic operator is denoted as $\langle e_i, s \rangle$, where e_i, $i = 1, 2, 3, \ldots$, stands for its inputs edges and s stands for the output edge.

7.1.2 Clocks in Lustre

In Lustre, any variable or expression denotes a flow that is each infinite sequence of values is defined on a clock, which represents a sequence of time. Thus, a flow is a pair that consists of a sequence of values and a clock.

The clock serves to indicate when a value is assigned to the flow. This means that a flow takes the n-th value of its sequence of values at the n-th instant of its clock. Any program has a cyclic behavior and that cycle defines a sequence of times, a clock, which is the *basic clock* of a program. A flow on the basic clock takes its n-th value at the n-th execution cycle of the program. Slower clocks can be defined through flows of Boolean values. The clock defined by a Boolean flow is the sequence of times at which the flow takes the value *true*.

Two operators affect the clock of a flow: `when` and `current`.

1. `When` is used to **sample** an expression on a slower clock. Let E be an expression and B a Boolean expression with the same clock. Then, `X=E when B` is an expression whose clock is defined by B and its values are the same as those of E's only when B is *true*. This means that the resulting flow X has not the same clock with E or, alternatively, when B is *false*, X is not defined at all.

2. `Current` operates on expressions with different clocks and is used to **project** an expression on the immediately faster clock. Let E be an expression with the clock defined by the Boolean flow B, which is not the basic clock. Then, `Y=current(E)` has the same clock as B and its value is the value of E at the last time that B was *true*. Note that until B is *true* for the first time, the value of Y will be *nil*.

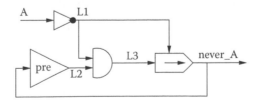

FIGURE 7.4
The operator network for the node `Never`.

TABLE 7.1

The Use of the Operators **when** and **current**

E	e_0	e_1	e_2	e_3	e_4	e_5	e_6	e_7	e_8	...
B	*false*	*false*	*true*	*false*	*true*	*false*	*false*	*true*	*true*	...
X=E when B			$x_0 = e_2$		$x_1 = e_4$			$x_2 = e_7$	$x_3 = e_8$	...
Y=current(X)	$y_0 = nil$	$y_1 = nil$	$y_2 = e_2$	$y_3 = e_2$	$y_4 = e_4$	$y_5 = e_4$	$y_6 = e_4$	$y_7 = e_7$	$y_8 = e_8$	...

```
node ex2cks(m:int) returns (c:bool; y:int);
var (x:int) when c;
let
  y = if c then current(x) else pre(y)-1;
  c = true -> (pre(y)=0);
  x = m when c;
tel;
```

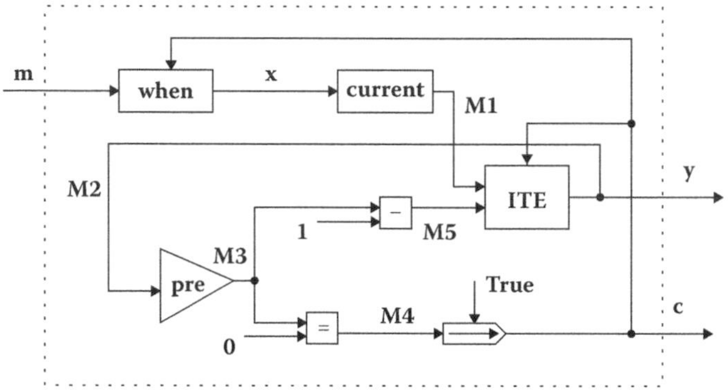

FIGURE 7.5

The **ex2cks** example and the corresponding operator network; two clocks are used, the basic clock and the flow c.

The sampling and the projection are two complementary operations: a projection changes the clock of a flow to the clock that the flow had before its last sampling operation. Trying to project a flow that was not sampled produces an error. Table 7.1 provides further detail on the use of the two temporal LUSTRE operators.

An example [8] of the use of clocks in LUSTRE is given in Figure 7.5. The LUSTRE node **ex2cks**, as indicated by the rectangle with a dashed outline, receives as input the signal m. Starting from this input value when the clock c is true, the program counts backwards until zero; from this moment, it restarts from the current input value and so on.

7.2 Automating the Coverage Assessment

The development of safety-critical software, such as deployed in aircraft control systems, requires a thorough validation process ensuring that the requirements have been exhaustively checked and the program code has been adequately exercised. In particular, according to the DO-178B standard, at least one test case must be executed for each requirement; the achieved code coverage is assessed on the generated C program. Although it is possible to apply many of the adequacy criteria to the CFG of the C program, this is not an interesting option for many reasons. First, the translation from LUSTRE to C depends on the used compiler and compilation options. For instance, the C code may implement a sophisticated automaton minimizing the execution time, but it can also be a "single loop" without explicit representation of the program states. Second, it is difficult if not impossible to formally establish a relation between the generated C code and the original LUSTRE program. As a result, usual adequacy criteria applied to the generated C code do not provide meaningful information on the LUSTRE program coverage. For these reasons, specific coverage criteria have been defined for LUSTRE applications.

More precisely, in this section, we describe a coverage assessment approach that conforms to the synchronous data-flow paradigm on which LUSTRE/SCADE applications are based. After a brief presentation of the LUSTRE language and its basic features, we provide the formal definitions of the structural coverage metrics. Then, we introduce some extensions of these metrics that help adequately handle actual industrial-size applications; such applications are usually composed of several distinct components that constantly interact with each other and some functions may use more than one clock. Therefore, the proposed extensions allow efficiently applying the coverage metrics to complex major applications, taking into account the complete set of the LUSTRE language.

7.2.1 Coverage criteria for LUSTRE programs

The following paragraphs present the basic concepts and definitions of the coverage criteria for LUSTRE programs.

7.2.1.1 Activation conditions

Given an operator network N, paths can be defined in the program, that is, the possible directions of flows from the inputs through the outputs. More formally, a path is a finite sequence of edges $\langle e_0, e_1, \ldots, e_n \rangle$, such that for $\forall i \epsilon [0, n-1]$, e_{i+1} is a successor of e_i in N. A *unit path* is a path with two edges (thus, with only one successive edge). For instance, in the operator network of Figure 7.4, the following complete paths can be found.

$$p_1 = \langle A, L_1, never\_A \rangle$$
$$p_2 = \langle A, L_1, L_3, never\_A \rangle$$
$$p_3 = \langle A, L_1, never\_A, L_2, L_3, never\_A \rangle$$
$$p_4 = \langle A, L_1, L_3, never\_A, L_2, L_3, never\_A \rangle$$

Obviously, one could discover infinitely many paths in an operator network depending on the number of cycles repeated in the path (i.e., the number of **pre** operators in the path). However, we only consider paths of finite length by limiting the number of cycles. That is, a path of length n is obtained by concatenating a path of length $n-1$ with a unit path (of length 2). Thus, beginning from unit paths, longer paths can be built. A path is then finite, if it contains no cycles or if the number of cycles is limited.

A Boolean LUSTRE expression is associated with each pair $\langle e, s \rangle$, denoting the condition on which the data flows from the input edge e through the output s. This condition is called *activation condition*. The evaluation of the activation condition depends on what type of operators the paths is composed of. Informally, the notion of the activation of a path is strongly related to the propagation of the effect of the input edge through the output edge. More precisely, a path activation condition shows the dependencies between the path inputs and outputs. Therefore, the selection of a test set satisfying the activation conditions of the paths in an operator network leads to a notion for program coverage. Since covering all paths in an operator network could be impossible because of their potentially infinite number and length, in our approach, coverage is defined with regard to a given path length that is actually determined by the number of cycles included in the path.

TABLE 7.2

Activation Conditions for All LUSTRE Operators

Operator	Activation Condition
$s = NOT(e)$	$AC(e,s) = true$
$s = AND(a,b)$	$AC(a,s) = not(a)\, or\, b$
	$AC(b,s) = not(b)\, or\, a$
$s = OR(a,b)$	$AC(a,s) = a\, or\, not(b)$
	$AC(b,s) = b\, or\, not(a)$
$s = ITE(c,a,b)$	$AC(c,s) = true$
	$AC(a,s) = c$
	$AC(b,s) = not(c)$
relational operator	$AC(e,s) = true$
$s = FBY(a,b)$	$AC(a,s) = true$ -> $false$
	$AC(b,s) = false$ -> $true$
$s = PRE(e)$	$AC(e,s) = false$ -> $pre(true)$

Table 7.2 summarizes the formal expressions of the activation conditions for all LUSTRE operators (except for **when** and **current** for the moment). In this table, each operator **op**, with the input e and the output s, is paired with the respective activation condition $AC(e,s)$ for the unit path $\langle e,s\rangle$. Note that some operators may define several paths through their output, so the activation conditions are listed according to the path inputs.

Let us consider the path $p_2 = \langle A, L_1, L_3, never\_A\rangle$ in the corresponding operator network for the node **Never** (Figure 7.4). The condition under which that path is activated is represented by a Boolean expression showing the propagation of the input A through the output $never\_A$. To calculate its activation condition, we progressively apply the rules for the activation conditions of the corresponding operators according to Table 7.2.* Starting from the end of the path, we reach the beginning, moving one step at a time along the unit paths. Therefore, the necessary steps would be the following:

$AC(p_2) = false$ -> $AC(p')$, where $p' = \langle A, L_1, L_3\rangle$

$AC(p') = not(L_1)\, or\, L_2\, and\, AC(p'') = A\, or\, pre(never\_A)\, and\, AC(p'')$, where $p'' = \langle A, L_1\rangle$

$AC(p'') = true$

After backward substitutions, the Boolean expression for the activation condition of the selected path is:

$$AC(p_2) = false\ \text{->}\ A\, or\, pre(never\_A).$$

In practice, in order for the path output to be dependent on the input, either the input has to be *true* at the current execution cycle or the output at the previous cycle has to be *true*. Note that at the first cycle of the execution, the path is not activated.

7.2.1.2   Coverage criteria

A LUSTRE/SCADE program is compiled into an equivalent C program. Given that the format of the generated C code depends on the compiler, it is difficult to establish a formal relation between the original LUSTRE program and the final C one. In addition, major

---

*In the general case (path of length n), the path p containing the **pre** operator is activated if its prefix p' is activated at the previous cycle of execution, that is $AC(p) = false$ -> $pre(AC(p'))$. Similarly, in the case of the initialization operator **fby**, the given activation conditions are respectively generalized in the forms: $AC(p) = AC(p')$ -> $false$ (i.e., the path p is activated if its prefix p' is activated at the initial cycle of execution) and $AC(p) = false$ -> $AC(p')$ (i.e., the path p is activated if its prefix p' is always activated except for the initial cycle of execution).

industrial standards, such as DO-178B in the avionics field, demand coverage to be measured on the generated C code. In order to tackle these problems, three coverage criteria specifically defined for LUSTRE programs have been proposed [14]. They are specified on the operator network according to the length of the paths and the input variable values.

Let $\mathcal{T}$ be the set of test sets (input vectors) and $P_n = \{p | length(p) \leq n\}$ the set of all complete paths in the operator network whose length is less than or equal to n. Then, the following families of criteria are defined for a given and finite order $n \geq 2$. The input of a path p is denoted as $in(p)$, whereas a path edge is denoted as e.

1. *Basic Coverage Criterion (BC).* This criterion is satisfied if there is a set of test input sequences, $\mathcal{T}$, that activates at least once the set P_n. Formally, $\forall p \in P_n$, $\exists t \in \mathcal{T} \colon AC(p) = true$. The aim of this criterion is basically to ensure that all the dependencies between inputs and outputs have been exercised at least once. In case that a path is not activated, certain errors, such as a missing or misplaced operator, could not be detected.

2. *Elementary Conditions Criterion (ECC).* In order for an input sequence to satisfy this criterion, it is required that the path p is activated for both input values, *true* and *false* (taking into account that only Boolean variables are considered). Formally, $\forall p \in P_n$, $\exists t \in \mathcal{T} \colon in(p) \land AC(p) = true$ and $not(in(p)) \land AC(p) = true$. This criterion is stronger than the previous one in the sense that it also takes into account the impact that the input value variations have on the path output.

3. *Multiple Conditions Criterion (MCC).* In this criterion, the path output depends on all the combinations of the path edges, including the internal ones. A test input sequence is satisfied if and only if the path activation condition is satisfied for each edge value along the path. Formally, $\forall p \in P_n$, $\forall e \in p$, $\exists t \in \mathcal{T} \colon e \land AC(p) = true$ and $not(e) \land AC(p) = true$.

The above criteria form a hierarchical relation: MCC satisfies all the conditions that ECC does, which also subsumes BC.

The path length is a fundamental parameter of the criteria definition. It is mainly determined by the number of cycles that a complete path contains. In fact, as this number increases, so does the path length as well as the number of the required execution cycles for its activation. Moreover, the coverage of cyclic paths strongly depends on the number of execution cycles and, consequently, on the test input sequences length. In practice, professionals are usually interested in measuring the coverage for a set of paths of a given number of cycles $(c \geq 0)^*$ rather than a given path length. Therefore, it is usually more convenient to consider various sets of complete paths in an operator network according to the number of cycles c contained in them and hence determine the path length n in relation to c.

7.2.2 Extension of coverage criteria to when and current operators

The above criteria have been extended in order to support the two temporal LUSTRE operators when and current. These operators allow to handle the case where multiple clocks are present, which is a common case in many industrial applications.

The use of multiple clocks implies the filtering of some program expressions. It consists of changing their execution cycle, activating the latter only at certain cycles of the basic clock. Consequently, the associated paths are activated only if the respective clock is true. As a result, the tester must adjust this filtered path activation rate according to the global *timing*.

---

*Note that $c = 0$ denotes the set of complete cycle-free paths.

7.2.2.1   Activation conditions for `when` and `current`

Informally, the activation conditions associated with the `when` and `current` operators are based on their intrinsic definition. Since the output values are defined according to a condition (i.e., the *true* value of the clock), these operators can be represented by means of the conditional operator `if-then-else`. For the expression E and the Boolean expression B with the same clock,

1. `X=E when B` could be interpreted as `X=if B then E else NON_DEFINED` and similarly,

2. `Y=current(X)` could be interpreted as `Y=if B then X else pre(X)`.

Hence, the formal definitions of the activation conditions result as follows:

Definition 1. Let e and s be the input and output edges, respectively, of a `when` operator and let b be its clock. The activation conditions for the paths $p_1 = \langle e,s \rangle$ and $p_2 = \langle b,s \rangle$ are

$$AC(p_1) = b$$
$$AC(p_2) = true \qquad \qquad \square$$

Definition 2. Let e and s be the input and output edges, respectively, of a `current` operator and let b be the clock on which it operates. The activation condition for the path $p = \langle e,s \rangle$ is

$$AC(p) = b. \qquad \qquad \square$$

As a result, to compute the paths and the associated activation conditions of a LUSTRE node involving several clocks, one has to just replace the `when` and `current` operators by the corresponding conditional operator (see Figure 7.6). At this point, two basic issues must be further clarified. The first one concerns the `when` case. Actually, there is no definition of the value of the expression X, when the clock B is not *true* (branch *NON_DEF* in Figure 7.6a). By default, at these instants, X does not occur and such paths (beginning with a nondefined value) are infeasible.* In the `current` case, the operator implicitly refers to the clock parameter B, without using a separate input variable (see Figure 7.6b). This indicates

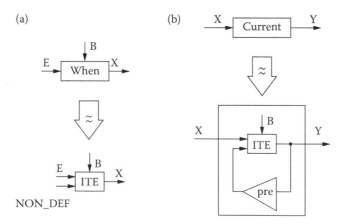

FIGURE 7.6
Modeling the `when` and `current` operators using `if-then-else`.

---

*An infeasible path is a path that is never executed by any test cases, hence it is never covered.

that `current` always operates on an already sampled expression, so the clock that determines its output activation should be the one on which the input is sampled.

Let us assume the path $p = \langle m, x, M_1, M_2, M_3, M_4, c \rangle$ in the example of Section 7.1.2, displayed in bold in Figure 7.5. Following the same procedure for the activation condition computation and starting from the last path edge, the activation conditions for the intermediate unit paths are

$$AC(p) = false\ \text{->}\ AC(p_1), \text{ where } p_1 = \langle m, x, M_1, M_2, M_3, M_4 \rangle$$
$$AC(p_1) = true\ and\ AC(p_2), \text{ where } p_2 = \langle m, x, M_1, M_2, M_3 \rangle$$
$$AC(p_2) = false\ \text{->}\ pre(AC(p_3)), \text{ where } p_3 = \langle m, x, M_1, M_2 \rangle$$
$$AC(p_3) = c\ and\ AC(p_4), \text{ where } p_4 = \langle m, x, M_1 \rangle$$
$$AC(p_4) = c\ and\ AC(p_5), \text{ where } p_5 = \langle m, x \rangle$$
$$AC(p_5) = c$$

After backward substitutions, the activation condition of the selected path is

$$AC(p) = false\ \text{->}\ pre(c).$$

This condition corresponds to the expected result and is consistent with the above definitions, according to which the clock must be true to activate the paths with `when` and `current` operators.

In order to evaluate the impact of these temporal operators on the coverage assessment, we consider the operator network of Figure 7.5 and the paths

$$p_1 = \langle m, x, M_1, y \rangle$$
$$p_2 = \langle m, x, M_1, M_2, M_3, M_4, c \rangle$$
$$p_3 = \langle m, x, M_1, M_2, M_3, M_5, y \rangle$$

Intuitively, if the clock c holds true, any change of the path input is propagated through the output, hence the above paths are activated. Formally, the associated activation conditions to be satisfied by a test set are

$$AC(p_1) = c$$
$$AC(p_2) = false\ \text{->}\ pre(c)$$
$$AC(p_3) = not(c)\ and\ false\ \text{->}\ pre(c) \qquad \qquad \square$$

Eventually, the input test sequences satisfy the BC. Indeed, as soon as the input m causes the clock c to take the required values, the activation conditions are satisfied since the latter depend only on the clock. In particular, in case the value of m at the first cycle is an integer different from zero (for sake of simplicity, let us consider $m = 2$), the BC is satisfied in two steps since the corresponding values for c are $c=true$, $c=false$. On the contrary, if at the first execution cycle m is equal to zero, the basic criterion is satisfied after three steps with the corresponding values for c: $c=true$, $c=true$, $c=false$. These two samples of input test sequences and the corresponding outputs are shown in Table 7.3.

Admittedly, the difficulty to meet the criteria is strongly related to the complexity of the system under test as well as to the test case generation effort. Moreover, activation conditions covered with short input sequences are easy to be satisfied, as opposed to long test sets that correspond to complex execution instances of the system under test. Experimental evaluation on more complex case studies, including industrial software components, is necessary and part of our future work in order to address these problems. Nonetheless, the enhanced definitions of the structural criteria presented above complete the coverage assessment issue for LUSTRE programs, as all the language operators are supported. In addition, the complexity of the criteria is not further affected, because, in essence, we use nothing but `if-then-else` operators.

TABLE 7.3

Test Cases Samples for the Input m

	c_1	c_2	c_3	c_4	...		c_1	c_2	c_3	c_4
m	$i_1 (\neq 0)$	i_2	i_3	i_4	...	m	$i_1 (= 0)$	i_2	i_3	...
c	*true*	*false*	*false*	*true*	...	c	*true*	*true*	*false*	...
y	i_1	$i_1 - 1$	0	i_4	...	y	0	i_2	$i_2 - 1$	...

It should be noted that the presented coverage criteria are limited to LUSTRE specifications that exclusively handle Boolean variables. The definition of the criteria implies that the path activation is examined in relation to the possible values that path inputs can take on, that is *true* and *false*. This means that, in case of integer inputs, the criteria would be inapplicable. Since in practice, applications deal with variables of different types, the criteria extension to more variable types appears to be a significant task and must be further studied.

7.2.3 LUSTRUCTU

LUSTRUCTU [13] is an academic tool that integrates the above criteria and automatically measures the structural coverage of LUSTRE/SCADE programs. It requires three inputs: the LUSTRE program under test, the required path length and the maximum number of loops in a path, and finally the criterion to satisfy. The tool analyzes the program and constructs its operator network. It then finds the paths that satisfy the input parameters and extracts the conditions that a test input sequence must satisfy in order to meet the given criterion. This information is recorded in a separate LUSTRE file, the so-called *coverage node*. This node receives as inputs; the inputs of the program under test and computes the coverage ratio at the output. The program outputs become the node local variables. For each path of length lower or equal to the value indicated in the input, its activation condition and the accumulated coverage ratio are calculated. These coverage nodes are compiled and executed (similar to any other regular LUSTRE program) over a given test data set[*] and the total coverage ratio[†] is computed.

An important remark is that the proposed coverage assessment technique is independent of the method used for test data generation. In other words, LUSTRUCTU simply considers a given test data set and computes the achieved coverage ratio according to the given criterion. In theory, any test data generation technique may be used. However, in our tests, we generally employ randomly generated test cases in order to obtain unbiased results, independent of any functional or structural requirements.

7.2.4 SCADE MTC

In SCADE, coverage is measured through the Model Test Coverage (MTC) module, in which the user can define custom criteria by defining the conditions to be activated during testing. Indeed, MTC measures the coverage of low-level requirements (LLR coverage), with regard to the demands and objectives of DO-178B standard, by assessing how thoroughly the SCADE model (i.e., system specification) has been exercised. In particular, each elementary SCADE operator is associated with a set of features concerning the possible behaviors of the operator. Therefore, structural coverage of the SCADE model is determined by the activation ratio of the features of each operator. Thus, the coverage approach previously presented could be easily integrated in SCADE in the sense that activation conditions

---

[*]Test input sequences are given in a .xml file.

[†]Coverage ratio $= \frac{\text{Number of satisfied activation conditions}}{\text{Number of activation conditions}}$.

corresponding to the defined criteria (BC, ECC, MCC) could be assessed once they are transformed into suitable MTC expressions.

7.2.5 Integration testing

So far, the existing coverage criteria are defined on a unit-testing basis and cannot be applied to LUSTRE nodes that locally employ user-defined operators (compound operators). The cost of computing the program coverage is affordable as long as the system size remains small. However, large or complex nodes must be locally expanded and code coverage must be globally computed. As a result, the number and the length of the paths to be covered increase substantially, which renders these coverage metrics impracticable when the system size becomes large.

In particular, as far as relatively simple LUSTRE programs are concerned, the required time for coverage computation is rather short. This holds particularly in the case of basic and of elementary condition coverage [11] for which paths are relatively short and the corresponding activation conditions are simple, respectively. As long as the path length remains low, the number of the activation conditions to be satisfied is computationally affordable. However, coverage analysis of complex LUSTRE nodes (Figure 7.7) may involve a huge number of paths and the coverage cost may become prohibitive and, consequently, the criteria inapplicable.

This is particularly true for the MCC criterion, where the number of the activation conditions to be satisfied increases dramatically when the length and the number of paths are high. In fact, in order to measure the coverage of a node that contains several other nodes (compound operators), the internal nodes are unfolded, the paths and the corresponding activation conditions are locally computed, and then they are combined with the global node coverage. This may result in a huge number of paths and activation conditions. Indeed, covering a path of length k requires $2(k-1)$ activation conditions to be satisfied. Consequently, satisfying a criterion for the set of paths P_n, r_i being the number of paths of length equal to i, requires the satisfaction of $2(r_2 + 2r_3 + \cdots + (n-1)r_n)$ activation conditions.

We are currently investigating an integration testing technique for the coverage measurement of large-scale LUSTRE programs that involve several internal nodes. This coverage assessment technique involves an approximation for the coverage of the called nodes by extending the definition of the activation conditions for these nodes. Coverage criteria are redefined, not only according to the length of paths but also with respect to the level of

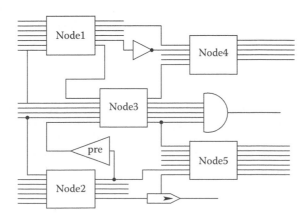

FIGURE 7.7
Example of the operator network of a complex LUSTRE program.

integration. This extension reduces the total number of paths at the system level and hence, the overall complexity of the coverage computation.

To empirically evaluate the proposed coverage approach, the extended criteria were applied to an alarm management component developed for embedded software used in the field of avionics. This component involves several LUSTRE nodes and it is representative of typical components in the avionics application area.

The module on which we focused during the experiment contains 148 lines of LUSTRE code with 10 input variables and 3 output variables, forming two levels of integration. The associated operator network comprises 32 basic operators linked to each other by 52 edges. Tests were performed on a Linux Fedora 9, Intel Pentium 2GHz and 1GB of memory. We are interested in complexity issues in terms of the prospective gain in the number of paths with reference to the coverage criteria that do not require full node expansion, the relative difficulty to meet the criteria, as well as the fault detection ability of the criteria.* For complete paths with at most three cycles, the preliminary results show a remarkable decrease in the number of paths and activation conditions, particularly for the MCC, which suggests that the extended criteria are useful for measuring the coverage of large-scale programs. The required time to calculate the activation conditions is relatively negligible; a few seconds (maximum 2 minutes) were necessary to calculate complete paths with maximum of 10 cycles and the associated activation conditions. Even for the MCC, this calculation remains minor, considering that the number of paths to be analyzed is computationally affordable.

For a complete presentation of the extended criteria as well as their experimental evaluation, the reader is advised to refer to [20].

7.3   Automating the Test Data Generation

This section introduces a technique for automated, functional test data generation, based on formal specifications. The general approach used by LUTESS to automatically generate test data for synchronous programs is first presented. It uses a specification language, based on LUSTRE, including specific operators applied to specify test models. Recent research extended this approach so that programs with integer parameters can be included [27]. Furthermore, existing test operators were adapted to the new context and new operators were added. These extensions are implemented into a new version of the tool, called LUTESS V2 [28]. After presenting the specification language and its usage, a general methodology to apply while testing with LUTESS V2 is proposed. The application of this methodology in a well-known case study [19] showed that it allows for an efficient specification and testing of industrial programs.

7.3.1   LUTESS

LUTESS is a tool transforming a formal specification into a test data generator. The dynamic generation of test data requires three components to be provided by the user: the software environment specification (Δ), the system under test (Σ), and a test oracle (Ω) describing the system requirements, as shown in Figure 7.8. The system under test and the oracle are both synchronous executable programs.

---

*Mutation testing [2] was used to simulate various faults in the program. In particular, a set of mutation operators was defined and several mutants were automatically generated. Then, the mutants and the coverage nodes were executed over the same test input data and the mutation score (ratio of killed mutants) was compared with the coverage ratio.

LUTESS builds a test input generator from the test specification and links it to the system under test and the oracle. It coordinates their execution and records the input and output sequences as well as the associated oracle verdicts using a trace collector.

A test is a sequence of single action–reaction cycles:

1. The generator produces an input vector.

2. It sends this input vector to the system under test.

3. The system reacts with an output vector that is sent back to the generator.

The generator produces a new input vector, and this sequence is repeated. At each cycle, the oracle observes the produced inputs and outputs to detect failures.

7.3.1.1 LUTESS V2 testnodes

A test specification is defined in a special node, called *testnode*, written in a language that is a superset of LUSTRE. The inputs and outputs of the software under test are the outputs and inputs for a testnode, respectively. The general form of a testnode is given in Figure 7.9.

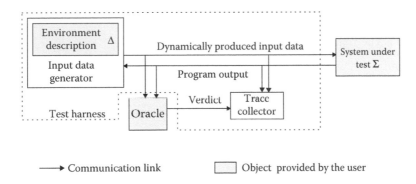

FIGURE 7.8
The LUTESS testing environment.

```
testnode Env(<SUT outputs>) returns (<SUT inputs>);
var <local variables>;
let
    environment(Ec₁,Ec₂,....,Ecₙ);
    prob(C₁,E₁,P₁);
    ...
    prob(Cₘ,Eₘ,Pₘ);
    safeprop(Sp₁,Sp₂,....,Spₖ);
    hypothesis(H₁,H₂,....,Hₗ);
    <definition of local variables>;
tel;
```

FIGURE 7.9
Testnode syntax.

There are four operators specifically introduced for testing purposes:

1. The `environment` operator makes it possible to specify invariant properties of the program environment.

2. The `prob` operator is used to define conditional probabilities. The expression `prob(C,E,P)` means that if the condition `C` holds, then the probability of the expression `E` to be `true` is equal to `P`.

3. The `safeprop` operator is exploited by LUTESS to guide the test generation toward situations that could violate the program safety properties (see safety-property-guided testing).

4. The `hypothesis` operator introduces knowledge or assumptions in the test generation process targeting to improve the fault-detection ability of safety-property-guided testing.

These operators are illustrated in a simple example and explained in detail in the next sections.

7.3.1.2   An air-conditioner example

Figure 7.10 shows the signature of a simple air conditioner controller.

The program has three inputs:

1. `OnOff` is true when the On/Off button is pressed by the user and false otherwise:

2. `Tamb` is the ambient temperature expressed in Celsius degrees,

3. `Tuser` is the temperature selected by the user,

and two outputs:

1. `IsOn` indicates that the air conditioner is on,

2. `Tout` is the temperature of the air emitted by the air conditioner.

This program is supposed to compute, according to the difference between the ambient and the user-selected temperature, the temperature of the air to be emitted by the air conditioner.

7.3.2   Using LUTESS V2

The following paragraphs describe the basic steps in the specification of the external environment of a system using LUTESS V2.

7.3.2.1   The `environment` operator

Figure 7.11 shows a trivial use of the `environment` operator. This specification would result in a test data generator issuing random values for `OnOff`, `Tamb`, and `Tuser`.

Obviously, the behavior of the actual software environment, although not completely deterministic, is not random. For instance, the temperature variation depends on the respective values of the ambient temperature and of the issued air temperature. This can

```
node AC(OnOff: bool; Tamb, Tuser : int)
returns (IsOn: bool; Tout: int)
```

FIGURE 7.10
The interface of the air conditioner.

be expressed by means of two properties, stating that if the air emitted by the air conditioner is either hotter or colder than the ambient temperature, the latter cannot decrease or increase, respectively. Moreover, we can specify that the ambient temperature remains in some realistic interval. We can write such properties with usual relational and arithmetical operators available in the LUSTRE language. To allow a test generator for producing the test data consistent with such constraints, they are specified in the `environment` operator, as shown in Figure 7.12.

Each property in the `environment` operator is a LUSTRE expression that can refer to the present or past values of the inputs and only to past values of the outputs. Therefore, the resulting test generator issues at any instant a random input satisfying the environment properties.

Table 7.4 shows an instance of the generated test sequence corresponding to the testnode of Figure 7.12.

7.3.2.2 The `prob` operator

The `prob` operator enables defining conditional probabilities that are helpful in guiding the test data selection. These probabilities are used to specify advanced execution scenarios such

```
testnode EnvAC(IsOn: bool; Tout: bool)
returns (OnOff: bool; Tamb, Tuser : int)
let
   environment(true);
tel;
```

FIGURE 7.11
Unconstrained environment.

```
testnode EnvAC(IsOn: bool; Tout: bool)
returns (OnOff: bool; Tamb, Tuser : int)
let
   environment(
      -- the user can choose a
      -- temperature between 10° and 40°
      Tuser >= 10 and Tuser <= 40,
      -- the ambient temperature
      -- should be between -20° and 60°
      Tamb >= -20 and Tamb <= 60,
      -- the temperature cannot decrease
      -- if hot air is emitted
      true -> implies(pre IsOn and
                      pre (Tout - Tamb) > 0,
        not(Tamb < pre Tamb)) ,
      -- the temperature cannot increase
      -- if cold air is emitted
      true -> implies(pre IsOn and
                      pre (Tout - Tamb) < 0,
        not(Tamb > preTamb))
   );
tel;
```

FIGURE 7.12
Constrained environment for the air conditioner.

as operational profiles [17] or fault simulation. Let us consider Figure 7.13. The previous example of the air-conditioner environment specification has been modified with some of the invariant properties now specified as expressions that hold with some probability. Also, probabilities have been added that specify low and high probability to push the OnOff button when the air conditioner is on and off, respectively. This leads to longer sub-sequences with a working air conditioner (IsOn = true).

Note that any invariant property included in the environment operator has an occurrence probability equal to 1.0. In other words, environment(E)⇔prob(true,E,1.0).

No static check of consistency on the probability definitions is performed, so the user can, in fact, specify a set of conditional probabilities that are impossible to satisfy at a given situation. If the generator encounters such a situation, different options to allow a satisfiable solution, such as partial satisfaction, can be specified.

Table 7.5 shows an instance of a test sequence after the execution of the generator corresponding to the testnode of Figure 7.13.

7.3.2.3   The safeprop operator

Safety properties express that the system cannot reach highly undesirable situations. They must always hold during the system operation. The safeprop operator automates the searching for the test data according to their ability to violate the safety properties.

The basic idea is to ignore such test sequences that cannot violate a given safety property. Consider a simple property $i \Rightarrow o$, where i is an input and o an output of the software. In this case, the input $i = false$ should not be generated since the property could not be violated regardless of the value of the produced output o. Of course, even after ignoring such sequences, it is not guaranteed that the program under test will reach a faulty situation since outputs are not known.

Table 7.6 shows a sequence produced when the following operator is added to the testnode of Figure 7.13:

```
safeprop(implies(IsOn and Tamb<Tuser,Tout>Tuser));
```

TABLE 7.4
Generated Test Data—Version 1

	t_0	t_1	t_2	t_3	t_4	t_5	t_6	t_7	t_8	t_9	t_{10}	t_{11}	t_{12}	t_{13}
OnOff	0	1	1	0	1	0	0	1	0	0	1	0	0	0
Tamb	-5	8	31	38	40	4	10	30	43	10	23	28	21	10
Tutil	20	21	14	35	13	17	24	22	20	11	36	17	20	40
IsOn	0	1	0	0	1	1	1	0	0	0	1	1	1	1
Tout	30	25	9	34	4	21	28	20	13	11	40	14	20	50

TABLE 7.5
Generated Test Data—Version 2

	t_0	t_1	t_2	t_3	t_4	t_5	t_6	t_7	t_8	t_9	t_{10}	t_{11}	t_{12}	t_{13}
OnOff	0	1	0	0	1	1	0	0	0	0	0	0	0	0
Tamb	28	31	27	29	4	41	18	52	59	7	13	5	57	-2
Tutil	36	26	32	29	40	32	32	19	22	36	10	19	12	18
IsOn	0	1	1	1	0	1	1	1	1	1	1	1	1	1
Tout	38	25	33	29	52	29	36	8	10	45	9	23	-3	24

```
testnode EnvAC(IsOn: bool; Tout: bool)
returns (OnOff: bool; Tamb, Tuser : int)
let
  environment(
    -- the user can choose a
    -- temperature between 10° and 40°
    Tuser >= 10 and Tuser <= 40,
    -- the ambient temperature
    -- should be between -20° and 60°
    Tamb >= -20 and Tamb <= 60
  );
  -- if hot air is emitted,
  -- the ambient temperature can hardly decrease
  prob( false -> pre IsOn and pre (Tout-Tamb)>0,
      true -> Tamb < pre Tamb, 0.1 );
  -- if cold air is emitted, t
  -- the ambient temperature hardly increases
  prob( false -> pre IsOn and pre (Tout-Tamb)<0,
      true -> Tamb > pre Tamb, 0.1 );
  -- High probability to press the OnOff button
  --   when the air-conditioner is not On
  prob( false -> not(pre IsOn), OnOff, 0.9 );
  -- Low probability to press the OnOff button
  --   when the air-conditioner is On
  prob( false -> pre IsOn, OnOff, 0.1 );
tel;
```

FIGURE 7.13
Using occurrence probabilities for expressions.

TABLE 7.6
Safety-Property-Guided Testing

	t_0	t_1	t_2	t_3	t_4	t_5	t_6	t_7	t_8	t_9	t_{10}	t_{11}	t_{12}	t_{13}
OnOff	0	0	1	0	0	1	1	1	0	1	0	1	0	1
Tamb	-9	17	27	-20	7	10	6	10	-20	7	-14	14	0	14
Tuser	36	26	32	29	40	32	32	19	40	36	10	19	12	18
IsOn	0	0	1	1	1	0	1	0	0	1	1	0	0	1
Tout	51	29	33	45	51	39	40	22	60	45	18	20	16	19

Note that the generated values satisfy `Tamb < Tuser`, which is a necessary condition to violate this property.

As a rule, a safety property can refer to past values of inputs that are already assigned. Thus, the generator must anticipate values of the present inputs that allow the property to be violated in the future. Given a number of steps k, chosen by the user, `safeprop(P)` means that such test inputs should be generated that can lead to a violation of P in the next k execution cycles. In order to do so, LUTESS posts the property constraints for each cycle, according to three strategies:

1. The *Union strategy* would select inputs able to lead to a violation of P at *any* of the next k execution cycles: $\neg P_t \vee \neg P_{t+1} \vee ... \vee \neg P_{t+k-1}$.

2. The *Intersection strategy* would select inputs able to lead to a violation of P at *each* of the next k execution cycles: $\neg P_t \wedge \neg P_{t+1} \wedge ... \wedge \neg P_{t+k-1}$.

3. The *Lazy strategy* would select inputs able to lead to a violation of P *as soon as possible* within the next k execution cycles: $\neg P_t \vee (P_t \wedge \neg P_{t+1}) \vee \ldots \vee ((P_t \wedge \ldots \wedge P_{t+k-2}) \wedge P_{t+k-1})$.

Depending on the type of the expression inside the `safeprop` operator, each of these strategies produces different results. In most cases, as the value of k increases, the *union strategy* is too weak (input values are not constrained) and the *intersection strategy* too strong (unsatisfiable). The *lazy strategy* is a trade-off between these two extremes. To illustrate this, consider the safety property $i_{t-1} \wedge \neg i_t \Rightarrow o_t$. In this case, with $k = 2$, we obtain with the following:

1. Using the *union strategy*, we only impose $i_t = true$ when $i_{t-1} = false$, otherwise any value of i_t is admitted.

2. Using the *intersection strategy*, there is no solution at all.

3. Using the *lazy strategy*, we impose always $i_t = \neg i_{t-1}$, resulting in a sequence alternating the value of i at each step.

7.3.2.4   The `hypothesis` operator

The generation mode guided by the safety properties has an important drawback. Since the program under test is considered as a black-box, the input computation is made assuming that any reaction of the program is possible. In practice, the program would prevent many of the chosen test inputs from leading to a state where a property violation is possible.

Taking into account hypotheses on the program could be an answer to this problem. Such hypotheses could result from the program analysis or could be properties that have been successfully tested before. They can provide information, even incomplete, on the manner how outputs are computed and hence provide better inputs for safety-property-guided testing.

By adding to the testnode of Figure 7.13 the following two statements:

```
hypothesis( true -> OnOff = IsOn<>pre(IsOn) );
safeprop( implies(IsOn and Tamb<Tuser, Tout>Tuser) ),
```

we introduce a hypothesis stating that the `OnOff` button turns the air conditioner on or off. The condition `IsOn=true` is necessary to violate the safety property, but since `IsOn` is an output of the software, we cannot directly set it to true. The hypothesis provides information about the values to be given to the `OnOff` input in order to obtain `IsOn=true` as output. The violation of the safety property then depends only on the `Tout` output.

Table 7.7 shows a sequence produced by the test generator corresponding to the above specification. We can remark that the `OnOff` button is pressed only once when the air conditioner was off (`pre IsOn = false`).

TABLE 7.7

Using Hypotheses in Safety-Property-Guided Testing

	t_0	t_1	t_2	t_3	t_4	t_5	t_6	t_7	t_8	t_9	t_{10}	t_{11}	t_{12}	t_{13}
OnOff	0	1	0	0	0	0	0	0	0	0	0	0	0	0
Tamb	-9	17	27	-20	7	10	6	10	-20	7	-14	14	0	14
Tuser	36	26	32	29	40	32	32	19	40	36	10	19	12	18
IsOn	0	1	1	1	1	1	1	1	1	1	1	1	1	1
Tout	51	29	33	45	51	39	40	22	60	45	18	20	16	19

7.3.3 Toward a test modeling methodology

The above operators enable the test engineer to build test models according to a methodology that has been defined and applied in several case studies. One of such case studies is a steam boiler control system [19], a system that operates on a significantly large set of input/output variables and internal functions. In previous work, it has been used to assess the applicability of several formal methods [1]. The primary function of the boiler controller is to keep the water level between the given limits, based on inputs received from different boiler devices.

The modeling and testing methodology consists of the following incremental approach:

1. *Domain definition*: Definition of the domain for integer inputs. For example, the water level cannot be negative or exceed the boiler capacity.

2. *Environment dynamics*: Specification of different temporal relations between the current inputs and past inputs/outputs. These relations often include, but are not limited to, the physical constraints of the environment. For example, we could specify that when the boiler valve opens, the water level can only decrease.

   The above specifications are introduced in the testnode by means of the `environment` operator. Simple random test sequences can be generated, without a particular test objective, but considering all and only inputs allowed by the environment.

3. *Scenarios*: Having in mind a specific test objective, the test engineer can specify more precise scenarios, by providing additional invariant properties or conditional probabilities (applying the `prob` operator). As a simple example, consider the *stop* input that stops the controller when true; a completely random value will stop the controller prematurely and thus prevent the testing of all the following behaviors. In this case, lowering the probability of *stop* being true keeps the controller running.

4. *Property-based testing*: This step uses formally specified safety properties in order to guide the generation toward the violation of such a property. Test hypotheses can also be introduced and possibly make this guidance more effective.

Applying this methodology to the steam boiler case study showed that relevant test models for the steam boiler controller were not difficult to build. Modeling the steam boiler environment required a few days of work. Of course, the effort required for a complete test operation is not easy to assess as it depends on the desired thoroughness of the test sequences, which may lead the tester to write several conditional probabilities corresponding to different situations (and resulting in different testnodes). Building a new testnode to generate a new set of test sequences usually requires a slight modification of a previous *testnode*. Each of these *testnodes* can then be used to generate a large number of test sequences with little effort. Thus, when compared to manual test data construction, which is still a current practice by many test professionals, such an automatic generation of test cases could certainly facilitate the testing process.

The steam boiler problem requires exchanging a given number of messages between the system controller and the physical system. The main program handles 38 inputs and 34 outputs, Boolean or integer, and it is composed of 30 internal functions. The main node is comprised, when unfolded, of 686 lines of LUSTRE code. Each testnode consists of about 20 invariant properties modeling the boiler environment to which various conditional probabilities or safety properties are added. The average size of a testnode, together with the auxiliary nodes, approximates 200 lines of LUSTRE code. It takes less than 30 seconds to generate a sequence of hundred steps, for any of the test models we used (tests performed on a Linux Fedora 9, Intel Pentium 2GHz, and 1GB of memory).

References

1. Abrial, J.-R. (1995). Steam-boiler control specification problem. *Formal Methods for Industrial Applications*, Volume 1165 of *LNCS*, 500–509.

2. Budd, T. A., DeMillo, R. A., Lipton, R. J., and Sayward, F.G. (1980). Theoretical and empirical studies on using program mutation to test the functional correctness of programs. In *ACM Symposium on Principles of Programming Languages*, Las Vegas, Nevada.

3. Caspi, P., Pilaud, D., Halbwachs, N., and Plaice, J. (1987). Lustre: A declarative language for programming synchronous systems. *POPL*, 178–188.

4. Chen, T.Y., and Lau, M. F. (2001). Test case selection strategies based on boolean specifications. *Software Testing, Verification and Reliability, 11*(3), 165–180.

5. Chilenski, J.J., and Miller, S.P. (1994). Applicability of modified condition/decision coverage to software testing. *Software Engineering Journal, 9*(5), 193–200.

6. Clarke, L. A., Podgurski, A., Richardson, D. J., and Zeil, S. J. (1989). A formal evaluation of data flow path selection criteria. *IEEE Transactions on Software Engineering, 15*(11), 1318–1332.

7. DO-178B (1992). Software Considerations in Airborne Systems and Equipment Certification. Technical report, RTCA, Inc., www.rtca.org.

8. Girault, A., and Nicollin, X. (2003). Clock-driven automatic distribution of lustre programs. In *3rd International Conference on Embedded Software, EMSOFT'03*, Volume 2855 of *LNCS*, Pages: 206–222. Springer-Verlag, Philadelphia.

9. Halbwachs, N., Caspi, P., Raymond, P., and Pilaud, D. (1991). The synchronous data flow programming language lustre. *Proceedings of the IEEE, 79*(9), 1305–1320.

10. Halbwachs, N., Lagnier, F., and Ratel, C., (1992). Programming and verifying real-time systems by means of the synchronous data-flow language lustre. *Transactions on Software Engineering, 18*(9), 785–793.

11. Lakehal, A., and Parissis, I. (2007). Automated measure of structural coverage for lustre programs: A case study. In *proceedings of the 2nd IEEE International Workshop on Automated Software Testing (AST'2007), a joint event of the 29th ICSE*. Minneapolis, MN.

12. Lakehal, A., and Parissis, I. (2005). Lustructu: A tool for the automatic coverage assessment of lustre programs. In *Proceedings of the 16th IEEE International Symposium on Software Reliability Engineering*, Pages: 301–310. Chicago, IL.

13. Lakehal, A., and Parissis, I. (2005). Lustructu: A tool for the automatic coverage assessment of lustre programs. In *IEEE International Symposium on Software Reliability Engineering*, Pages: 301–310. Chicago, IL.

14. Lakehal, A., and Parissis, I. (2005). Structural test coverage criteria for lustre programs. In *Proceedings of the 10th International Workshop on Formal Methods for Industrial Critical Systems: a satellite event of the ESEC/FSE'05*, Pages: 35–43, Lisbon, Portugal.

15. Laski, J. W., and Korel, B. (1983). A data flow oriented program testing strategy. *IEEE Transactions on Software Engineering 9*(3), 347–354.

16. Marre, B. and Arnould, A. (2000). Test sequences generation from lustre descriptions: Gatel. *Proceedings of the 15th IEEE Conference on Automated Software Engineering, Grenoble, France*, 229–237.

17. Musa, J. D. (1993). Operational profiles in software-reliability engineering. *IEEE Software, 10*(2), 14–32.

18. Ntafos, S. C. (1984). An evaluation of required element testing strategies. In *International Conference on Software Engineering*, Pages: 250 256. Orlando, FL.

19. Papailiopoulou, V., Seljimi, B., and Parissis, I. (2009). Revisiting the steam-boiler case study with lutess: modeling for automatic test generation. In *12th European Workshop on Dependable Computing*, Toulouse, France.

20. Papailiopoulou, V. (2010). Test automatique de programmes lustre/scade. Phd thesis, Université de Grenoble, France.

21. Parissis, I., and Ouabdesselam, F. (1996). Specification-based testing of synchronous software. *ACM-SIGSOFT Foundations of Software Engineering*, 127–134.

22. Parissis, I., and Vassy, J. (2003). Thoroughness of specification-based testing of synchronous programs. In *Proceedings of the 14th IEEE International Symposium on Software Reliability Engineering*, 191–202.

23. Pilaud, D., and Halbwachs, N. (1988). From a synchronous declarative language to a temporal logic dealing with multiform time. *Proceedings of Formal Techniques in Real-Time and Fault-Tolerant Systems*, Warwick, United Kingdom, Volume 331 of *Lecture Notes in Computer Science*, 99–110.

24. Rajan, A. (2008). Coverage metrics for requirements-based testing. Phd thesis, University of Minnesota, Minneapolis.

25. Raymond, P., Nicollin, X. Halbwachs, N., and Weber, D. (1998). Automatic testing of reactive systems. *Proceedings of the 19th IEEE Real-Time Systems Symposium*, Madrid, Spain, 200–209.

26. Richardson, D., and Clarke, L. (1985). Partition analysis: a method combining testing and verification. *IEEE Transactions on Software Engineering, 11*(12), 1477–1490.

27. Seljimi, B., and Parissis, I. (2006). Using CLP to automatically generate test sequences for synchronous programs with numeric inputs and outputs. In *17th International Symposium on Software Reliability Engineering*, Pages: 105–116. Raleigh, North Carolina.

28. Seljimi, B., and Parissis, I. (2007). Automatic generation of test data generators for synchronous programs: Lutess V2. In *Workshop on Domain Specific Approaches to Software Test Automation*, Pages: 8–12. Dubrovnik, Croatia.

29. Vilkomir, S. A., and Bowen, J. P. (2001). Formalization of software testing criteria using the Z notation. In *International Computer Software and Applications Conference (COMPSAC)*, Pages: 351–356. Chicago, IL.

30. Vilkomir, S. A., and Bowen, J. P. (2002). Reinforced condition/decision coverage (RC/DC): A new criterion for software testing. In *International Conference of B and Z Users*, Pages: 291–308. Grenoble, France.

31. Woodward, M. R., Hedley, D., and Hennell, M. A. (1980). Experience with path analysis and testing of programs. *IEEE Transactions on Software Engineering, 6*(3), 278–286.

8

Test Generation Using Symbolic Animation
of Models

Frédéric Dadeau, Fabien Peureux, Bruno Legeard, Régis Tissot, Jacques Julliand, Pierre-Alain Masson, and Fabrice Bouquet

CONTENTS

In the domain of embedded systems, models are often used either to generate code, possibly after refinement steps, but they also provide a functional view of the modeled system that can be used to produce black-box test cases, without considering the actual details of implementation of this system. In this process, the tests are generated by appling given test selection criteria on the model. These test cases are then played on the system and the results obtained are compared with the results predicted by the model, in order to ensure the conformance between the concrete system and its abstract representation. Test selection criteria aim at achieving a reasonable coverage of the functionalities or requirements of the system, without involving a heavyweight human intervention.

We present in this chapter work on the B notation to support model design, intermediate verification, and test generation. In B machines, the data model is described using abstract data types (such as sets, functions, and relations) and the operations are written in a code-like notation based on generalized substitutions. Using a customized animation tool, it is possible to animate the model, that is, to simulate its execution, in order to ensure that the operations behave as expected w.r.t. the initial informal requirements. Furthermore, this animation process is also used for the generation of test cases, with more or less automation. More precisely, our work focuses on symbolic animation that improves classical model animation by avoiding the enumeration of operation parameters. Parameter values become abstract variables whose values are handled by dedicated tools (provers or solvers). This process has been tool supported with the BZ-Testing-Tools framework that has been industrialized and commercialized by the company Smartesting (Jaffuel and Legeard 2007). We present in this chapter the techniques used to perform the symbolic animation of B models using underlying set-theoretical constraint solvers, and we describe two test generation processes based on this approach.

The first process employs animation in a fully automated manner, as a means for building test cases that reach specific test targets computed so as to satisfy a structural coverage criterion over the operations of the model, also called static test selection criterion. In contrast, the second one is a Scenario-Based Testing (SBT) approach, also said to satisfy dynamic test selection criteria, in which manually designed scenarios are described as sequences of operations, possibly targeting specific states. These scenarios are then animated in order to produce the test cases. The goals of such automation are twofold. First, it makes it possible to reduce the effort in test design, especially on large and complex systems. Second, the application of model coverage criteria improves the confidence in the efficiency of the testing phase in detecting functional errors. We illustrate the use and the complementarity of these two techniques on the industrial case of a smart card application named IAS—Identification Authentication Signature—an electronic platform for loading applications on latest-generation smart cards.

8.1 Motivations and Overall Approach

In the domain of embedded systems, a model-based approach for design, verification, or validation is often required, mainly because these kinds of systems are often of a safety-critical nature (Beizer 1995). In that sense, a defect can be relatively costly in terms of money or human lives. The key idea is thus to detect the possible malfunctions as soon as possible. The use of formal models, on which mathematical reasoning can be performed, is therefore an interesting solution. In the context of software testing, the use of formal models makes it possible to achieve an interesting automation of the process, the model being used as a basis from which the test cases are computed. In addition, the model predicts the expected results, named the oracle, that describe the response that the System Under Test (SUT) should provide (modulo data abstraction). The conformance of the SUT w.r.t. the initial model is based on this oracle.

We rely on the use of behavioral models, which are models describing an abstraction of the system, using state variables, and operations that may be executed, representing a transition function described using generalized substitutions. The idea for generating tests from these models is to animate them, that is, to simulate their execution by invoking their operations. The sequences obtained represent abstract test cases that have to be concretized to be run on the SUT. Our approach considers two complementary test generation techniques that use model animation in order to generate the tests. The first one is based on a structural coverage of the operations of the model, and the second is based on dynamic selection criteria using user-defined scenarios.

Before going further into the details of our approach, let us define the perimeter of the embedded systems we target. We consider embedded systems that do not present concurrency, or strong real-time constraints (i.e., time constraints that cannot be discretized). Indeed, our approach is suitable for validating the functional behaviors of electronic transaction applications, such as smart cards applets or discrete automotive systems, such as front wipers or cruise controllers.

8.1.1 Context: The B abstract machines notation

Our work focuses on the use of the B notation (Abrial 1996) for the design of the model to be used for testing an embedded system. Several reasons motivate this choice. B is a very convenient notation for modeling embedded systems, grounded on a well-defined semantics. It makes it possible to easily express the operations of the system using a functional approach. Thus, each command of the SUT can be modeled by a B operation that acts as a function updating the state variables. Moreover, the operations syntax displays conditional structures (IF...THEN...ELSE...END) that are similar to any programming language. One of the advantages of B is that it does not require the user to know the complete topology of the system (compared to automata-based formal notations), which simplifies its use in the industry. Notice that we do not consider the entire development process described by the B method. Indeed, this latter starts from an abstract machine and involves successive refinements that would be useless for test generation purposes (i.e., if the code is generated from the model, there is no need to test the code). Here, we focus on abstract machines; this does not restrict the expressiveness of the language since a set of refinements can naturally be flattened into a single abstract machine.

B is based on a set-theoretical data model that makes it possible to describe complex structures using sets, relations (set of pairs), and a large variety of functions (total/partial functions, injections, surjections, bijections), along with numerous set/relational operators. The dynamics of the model, namely the initialization and the operations, are expressed using *Generalized Substitutions* that describe the possible atomic evolution of the state variables including simple assignments ($x := E$), multiple assignments ($x, y := E, F$ also written $x := E \parallel y := F$), conditional assignments (IF Cond THEN $Subst_1$ ELSE $Subst_2$ END), bounded choice substitutions (CHOICE $Subst_1$ OR .... OR $Subst_N$ END), or unbounded choice substitutions (ANY z WHERE Predicate(z) THEN Subst END) (see Abrial 1996, p. 227 for a complete list of generalized substitutions).

An abstract machine is organized in clauses that describe (1) the constants of the system and their associated properties, (2) the state variables and the invariant (containing the data typing information) (3) the actual invariant (properties that one wants to see preserved through the possible execution of the machine), (4) the initial state, and (5) the atomic state evolution described by the operations.

Figure 8.1 gives an example of a B abstract machine that will be used to illustrate the various concepts presented in this chapter. This machine models an electronic purse, similar as those embedded on smart cards, managing a given amount of money (variable `balance`). A PIN code is also used to identify the card holder (variable `pin`). The holder may try to authenticate using operation `VERIFY_PIN`. Boolean variable `auth` states whether or not the holder is authenticated. A limited number of tries is given for the holder to authenticate (three in the model). When the user fails to authenticate, the number of tries decreases until reaching zero, corresponding to a state in which the card is definitely blocked (i.e., no command can be successfully invoked). The model provides a small number of operations that make it possible: to set the value of the PIN code (`SET_PIN` operation), to authenticate the holder (`VERIFY_PIN` operation), and to credit the purse (`CREDIT` operation) or to pay a purchase (`DEBIT` operation).

```
MACHINE
    purse
CONSTANTS
    max_tries
PROPERTIES
    max_tries ∈ ℕ ∧ max_tries = 3
VARIABLES
    balance, pin, tries, auth
INVARIANT
    balance ∈ ℕ ∧ balance ≥ 0 ∧ pin ∈ -1..9999 ∧
    tries ∈ 0..max_tries ∧ auth ∈ BOOLEAN ∧  ...
INITIALIZATION
    balance := 0 ‖ pin := -1 ‖ tries := max_tries ‖ auth := false
OPERATIONS
    sw ← SET_PIN(p) ≙ ...
    sw ← VERIFY_PIN(p) ≙ ...
    sw ← CREDIT(a) ≙ ...
    sw ← DEBIT(a) ≙ ...
END
```

FIGURE 8.1
B abstract machine of a simplified electronic purse.

8.1.2   Model-based testing process

We present in this part the use of B as a formal notation that makes it possible to describe the behavior of the SUT. In order to produce the test cases from the model, the B model is animated using constraint solving techniques. We propose to develop two test generation techniques based on this principle, as depicted in Figure 8.2.

The first technique is fully automated and aims at applying structural coverage criteria on the operations of the machine so as to derive test cases that are supposed to exercise all the operations of the system, involving decision coverage and data coverage as a boundary analysis of the state variables. Unfortunately, this automated process shows some limitations, which we will illustrate. This leads us to consider a guided technique based on the design of scenarios. Both techniques rely on the use of animation, either to compute the test sequences by a customized state exploration algorithm or to animate the user-defined scenarios.

These two processes compute test cases that are said to be abstract since they are expressed at the model level. These tests thus need to be concretized to be run on the SUT. To achieve that, the validation engineer has to write an *adaptation layer* that will be in charge of bridging the gap between the abstract and the concrete level (basically model operations are mapped to SUT commands, and abstract data values are translated into concrete data values).

8.1.3   Plan of the chapter

The chapter is organized as follows. Section 8.2 describes the principle of symbolic animation that will be used in the subsequent sections. The automated boundary test generation technique is presented in Section 8.3, whereas the SBT approach is described

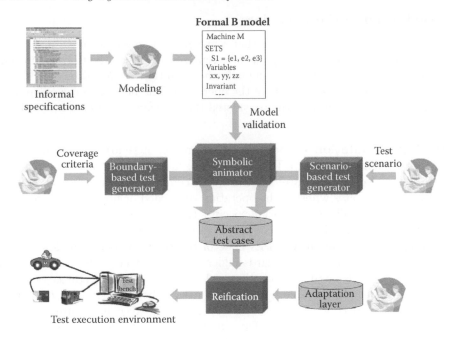

FIGURE 8.2
Test generation processes based on symbolic animation.

in Section 8.4. The usefulness and complementarity of these two approaches are illustrated in Section 8.5 on industrial case studies on smart card applets. Finally, Section 8.6 presents the related works, and Section 8.7 concludes and gives an overview of the open issues.

8.2   Principles of Symbolic Animation

For the test generation approaches to be relevant, it is mandatory to ensure that the model behaves as expected since the system will be checked against the model. Model animation is thus used for ensuring that the model behaves as described in the initial requirements. This step is done in a semi-automated way, by using a dedicated tool—a model animator—with which the validation engineer interacts. Concretely, the user chooses which operation he wants to invoke. Depending on the current state of the system and the values of the parameters, the animator computes and displays the resulting states that can be obtained. By comparing these states with the informal specification, the user can evaluate its model and correct it if necessary. This process is complementary to the verification that involves properties that have to be formally verified on the model.

The symbolic animation improves the "classical" model animation by giving the possibility to abstract the operation parameters. Once a parameter is abstracted, it is replaced by a symbolic variable that is handled by dedicated constraints solvers. Abstracting all the parameter values turns out to consider each operation as a set of "behaviors" that are the basis from which symbolic animation can be performed (Bouquet et al. 2004).

8.2.1   Definition of the behaviors

A *behavior* is a subpart of an operation that represents one possible effect of the operation. Each behavior can be defined as a predicate, representing its activation condition, and a substitution that represents its effect, namely the evolution of the state variables and the instantiation of the return parameters of the operation. The behaviors are computed as the paths in the control flow graph of the considered B operation, represented as a before–after predicate.*

Example 1 (Computation of behaviors). Consider a smart card command, named VERIFY_PIN aimed at checking a PIN code proposed as parameter against the PIN code of the card. As for every smart card command, this command returns a code, named `sw` for status word, that indicates whether the operation succeeded or not and possibly indicating the cause of the failure. The precondition specifies the typing information on the parameter `p` (a four-digit number). First, the command cannot succeed if there are no remaining tries on the card and if the current PIN code of the card has been previously set. If the digits of the PIN code match, the card holder is authentified, otherwise there are two cases: either there are enough tries on the card, and the returned status word indicates that the PIN is wrong, or the holder has performed his/her last try, and the status word indicates that the card is now blocked. This operation is given in Figure 8.3, along with its control flow graph representation. This command presents four behaviors, which are made of the conjunction of the predicates on the edges of a given path, that are denoted by the sequence of nodes from 1 to 0. For example, behavior [1,2,3,4,0], defined by predicate $p \in 0..9999 \land \mathtt{tries} > 0 \land \mathtt{pin} \neq -1 \land p = \mathtt{pin} \land \mathtt{auth}' = \mathtt{true} \land \mathtt{tries}' = \mathtt{max\_tries} \land \mathtt{sw} = \mathtt{ok}$ represents a successful authentication of the card holder. In this predicate, X' designates the value of variable X after the execution of the operation. $\square$

8.2.2   Use of the behaviors for the symbolic animation

When performing the symbolic animation of a B model, the operation parameters are abstracted and the operations are considered through their behaviors. Each parameter is thus replaced by a symbolic variable whose value is managed by a constraint solver.

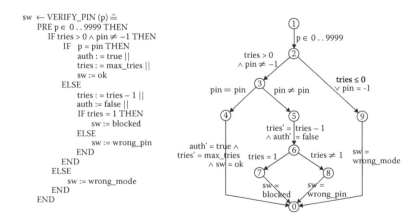

FIGURE 8.3
B code and control flow graph of the VERIFY_PIN command.

---

*A before–after predicate is a predicate involving state variables before the operation and after, using a primed notation.

Definition 1 (Constraint Satisfaction Problem [CSP]). A CSP is a triplet $\langle X, D, C \rangle$ in which

- $X = \{X_1, \ldots, X_N\}$ is a set of N variables,

- $D = \{D_1, \ldots, D_N\}$ is a set of domains associated to each variable $(X_i \in D_i)$,

- C is a set of constraints that relate variable values altogether.

A CSP is said to be *consistent* if there exists at least one valuation of the variables in X that satisfies the constraints of C. It is *inconsistent* otherwise. $\square$

Activating a transition from a given state is equivalent to solving a CSP whose variables X are given by the state variables of the current state (i.e., the state from which the transition is activated), the state variables of the after state (i.e., the state reached by the activation of the transition), and the parameters of the operation. According to the B semantics, the domains D of the state variables and the operation parameters can be found in the invariant of the machine and in the precondition of the operation, respectively. The constraints C are the predicates composing the behavior that is being activated, enriched with equalities between the before and after variables that are not assigned within the considered behavior.

The feasibility of a transition is defined by the consistency of the CSP associated to the activation of the transition from a given state. The iteration over the possible activable behaviors is done by performing a depth-first exploration of the behavior graph.

Example 2 (Behavior activation). Consider the activation of the VERIFY_PIN operation given in Example 1. Suppose the activation of this operation from the state s_1 defined by: `tries = 2, auth = false, pin = 1234`. Two behaviors can be activated. The first one corresponds to an invocation `ok ← VERIFY_PIN(1234)` that covers path [1,2,3,4,0], and produces the following consistent CSP (notice that data domains have been reduced so as to give the most human-readable representation of the corresponding states):

$$CSP_1 = \langle \{tries, auth, pin, p, tries', auth', pin', sw\},$$
$$\{\{2\}, \{false\}, \{1234\}, \{1234\}, \{3\}, \{true\}, \{1234\}, \{ok\}\},$$
$$\{Inv, Inv', tries > 0, pin \neq -1, p = pin, tries' = 3,$$
$$auth' = true, pin' = pin, sw = ok\}\rangle \tag{8.1}$$

where Inv and Inv' designate the constraints from the machine invariant that apply on the variables before and after the activation of the behavior, respectively. The second behavior that can be activated corresponds to an invocation `wrong_pin ← VERIFY_PIN(p)` that covers path [1,2,3,5,6,8,0] and produces the following consistent CSP:

$$CSP_2 = \langle \{tries, auth, pin, p, tries', auth', pin', sw\},$$
$$\{\{2\}, \{false\}, \{1234\}, 0..1233 \cup 1235..9999, \{1\}, \{false\}, \{1234\}, \{wrong\_pin\}\},$$
$$\{Inv, Inv', tries > 0, pin \neq -1, p \neq pin, tries' = tries - 1,$$
$$auth' = false, tries \neq 1, pin' = pin, sw = wrong\_pin\}\rangle \tag{8.2}$$

State variables may also become symbolic variables, if their after value is related to the value of a symbolic parameter. A variable is said to be symbolic if the domain of the variable contains more than one value. A system state that contains at least one symbolic state variable is said to be a *symbolic state* (as opposed to a concrete state).   □

Example 3 (Computation of Symbolic States). Consider the SET_PIN operation that sets the value of the PIN on a smart card:

```
sw ← SET_PIN(p) ≙
    PRE p ∈ 0..9999 THEN
        IF pin = -1 THEN  pin := p ∥ sw := ok
        ELSE sw := wrong_mode
        END
    END
```

From the initial state, in which auth = false, tries = 3, and pin = -1, the SET_PIN operation can be activated to produce a symbolic state associated with the following CSP:

$$CSP_0 = \langle \{tries, auth, pin, p, tries', auth', pin', sw\}, \\ \{\{3\}, \{false\}, \{-1\}, 0..9999, \{3\}, \{false\}, 0..9999, \{ok\}\}, \\ \{Inv, Inv', pin = -1, pin' = p, sw = ok\}\rangle \tag{8.3}$$

□

The symbolic animation process works by exploring the successive behaviors of the considered operations. When two operations have to be chained, this process acts as an exploration of the possible combinations of successive behaviors for each operation.

In practice, the selection of the behaviors to be activated is done in a transparent manner and the enumeration of the possible combinations of behaviors chaining is explored using backtracking mechanisms. For animating B models, we use CLPS-BZ (Bouquet, Legeard, and Peureux 2004), a set-theoretical constraint solver written in SICStus Prolog (SIC 2004) that is able to handle a large subset of the data structures existing in the B machines (sets, relations, functions, integers, atoms, etc.).

Once the sequence has been played, the remaining symbolic parameters can be instantiated by a simple labeling procedure, which consists of solving the constraints system and producing an instantiation of the symbolic variables, obtaining an abstract test case.

It is important to notice that constraint solvers work with an internal representation of constraints (involving constraint graphs and/or polyhedra calculi for relating variable values). Nevertheless, consistency algorithms used to acquire and propagate constraints are insufficient to ensure the consistency of a set of constraints, and a labeling procedure always has to be employed to guarantee the existence of solutions in a CSP associated to a symbolic state.

The use of symbolic techniques avoids the complete enumeration of the concrete states when animating the model. It thus makes it possible to deal with large models, which represent billions of concrete states, by gathering them into symbolic states. As illustrated in the experimental section, such techniques ensure the scalability of the overall approach.

The next two sections will now describe the use of symbolic animation for the generation of test cases.

8.3   Automated Boundary Test Generation

We present in this section, the use of the symbolic animation for automating the generation of model-based test cases. This technique aims at a structural coverage of the transitions of the system. To make it simple, each behavior of each operation of the B machine is targeted; the test cases thus aim at covering all the behaviors. In addition, a symbolic representation of the system states makes it possible to perform a boundary analysis from which the test targets will result (Legeard, Peureux, and Utting 2002, Ambert, Bouquet, Legeard, and Peureux 2003). This technique is recognized as a pertinent heuristics for generating test data (Beizer 1995).

The tests that we propose comprise four parts, as illustrated in Figure 8.4. The first part, called *preamble*, is a sequence of operations that brings the system from the initial state to a state in which the test target, namely a state from which the considered behavior can be activated, is reached. The *body* is the activation of the behavior itself. Then, the *identification* phase is made of user-defined calls to observation operations that are supposed to retrieve internal values of the system so that they can be compared to model data in order to establish the conformance verdict of the test. Finally, the *postamble* phase is similar to the preamble, but it brings the system back to the initial state or to another state that reaches another test target. The latter part is important to chain the test cases. It is particularly useful when testing embedded systems since the execution of the tests on the system is very costly and such systems take usually much time to be reset by hand.

This automated test generation technique requires some testability hypotheses to be employed. First, the operations of the B machine have to represent the control points of the system to be tested, so as to ease the concretization of the test cases. Second, it is mandatory that the concrete data of the SUT can be compared to the abstract data of the model, so as to be able to compare the results produced by the execution of the test cases with the results predicted by the model. Third, the SUT has to provide observation points that can be modeled in the B machine (either by return values of operations, such as the status words in the smart cards or by observation operations).

We will now describe how the test cases can be automatically computed, namely how the test targets are extracted from the B machine and how the test preambles and postambles are computed.

8.3.1 Extraction of the test targets

The goal of the tests is to verify that the behaviors described in the model exist in the SUT and produce the same result. To achieve that, each test will focus on one specific behavior of an operation. Test targets are defined as the states from which a given behavior can be activated. These test targets are computed so as to satisfy a structural coverage of the machine operations.

Definition 2 (Test Target). Let $OP = \langle (Act_1, \mathit{Eff}_1)[] \dots [](Act_N, \mathit{Eff}_N) \rangle$ be the set of behaviors extracted from operation OP, in which Act_i denotes the activation condition of behavior i, Eff_i denotes its effect, and $[]$ is an operator of choice between behaviors. Let

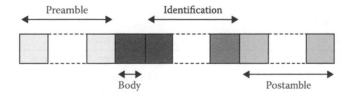

FIGURE 8.4
Composition of a test case.

Inv be the machine invariant. A test target is defined by a predicate that characterizes the states of the invariant from which a behavior i can be activated: $Inv \wedge Act_i$. □

The use of underlying constraint solving techniques makes it possible to provide interesting possibilities for data coverage criteria. In particular, we are able to perform a boundary analysis of the behaviors of the model. Concretely, we will consider *boundary goals* that are states of the model for which at least one of the state variable is at an extremum (minimum or maximum) of its current domain.

Definition 3 (Boundary Goal). Let $minimize(V,C)$ and $maximize(V,C)$ be functions that instantiate a symbolic variable V to its minimal and maximal value, respectively, under the constraints given in C. Let Act_i be the activation condition of behavior i, let $\vec{P}$ be the parameters of the corresponding operation, and let $\vec{V}$ be the set of state variables that occur in behavior i, the boundary goals for the variables $\vec{V}$ are computed by

$$BG^{min} = minimize(f(\vec{V}), Inv \wedge \exists \vec{P}.Act_i)$$
$$BG^{max} = maximize(f(\vec{V}), Inv \wedge \exists \vec{P}.Act_i)$$

in which f is an optimization function that depends on the type of the variable:

if $\vec{X}$ is a set of integers, $f(\vec{X}) = \sum_{x \in \vec{X}} x$
if $\vec{X}$ is a set of sets, $f(\vec{X}) = \sum_{x \in \vec{X}} card(x)$
otherwise, $f(\vec{X}) = 1$ □

Example 4 (Boundary Test Targets). Consider behavior [1,2,3,4,5,0] from the VERIFY_PIN operation presented in Figure 8.3. The machine invariant gives the following typing informations:

$$Inv \mathrel{\hat{=}} tries \in 0..3 \wedge pin \in -1..9999 \wedge auth \in \{true, false\}$$

The boundary test targets are computed using the minimization/maximization formulas:

$$BG^{min} = minimize(tries + pin, Inv \wedge \exists p \in 0..9999.(tries > 0 \wedge pin \neq -1 \wedge pin = p))$$
$$\rightsquigarrow tries = 1, \; pin = 0$$
$$BG^{max} = maximize(tries + pin, Inv \wedge \exists p \in 0..9999.(tries > 0 \wedge pin \neq -1 \wedge pin = p))$$
$$\rightsquigarrow tries = 3, \; pin = 9999$$ □

In order to improve the coverage of the operations, a predicate coverage criterion (Offutt, Xiong, and Liu 1999) can be applied by the validation engineer. This criterion acts as a rewriting of the disjunctions in the decisions of the B machine. Four rewritings are possible, which enables satisfying different specification coverage criteria, as given in Table 8.1.

Rewriting 1 leaves the disjunction unmodified. Thus, the Decision Coverage criterion will be satisfied if a test target satisfies either P_1 or P_2 indifferently (also satisfying the Condition

TABLE 8.1
Decision Coverage Criteria Depending on Rewritings

N	Rewriting of $P_1 \vee P_2$	Coverage Criterion
1	$P_1 \vee P_2$	Decision Coverage (DC)
2	$P_1 \; [] \; P_2$	Condition/Decision Coverage (C/DC)
3	$P_1 \wedge \neg P_2 \; [] \; \neg P_1 \wedge P_2$	Full Predicate Coverage (FPC)
4	$P_1 \wedge P_2 \; [] \; P_1 \wedge \neg P_2 \; [] \; \neg P_1 \wedge P_2$	Multiple Condition Coverage (MCC)

Coverage criterion). Rewriting 2 produces two test targets, one considering the satisfaction of P_1, and the other the satisfaction of P_2. Rewriting 3 will also produce two test targets, considering an exclusive satisfaction of P_1 without P_2 and vice versa. Finally, Rewriting 4 produces three test targets that will cover all the possibilities to satisfy the disjunctions.

Notice that the consistency of the resulting test targets is checked so as to eliminate inconsistent test targets.

Example 5 (Decision coverage). Consider behavior [1,2,9,0] from operation `VERIFY_PIN` presented in Figure 8.3. The selection of the Multiple Condition Coverage criterion will produce the following test targets:

1. $Inv \land \exists p \in 0..9999 . (tries \leq 0 \land pin = -1)$
2. $Inv \land \exists p \in 0..9999 . (tries > 0 \land pin = -1)$
3. $Inv \land \exists p \in 0..9999 . (tries \leq 0 \land pin \neq -1)$

providing contexts from which boundary goals will then be computed. □

We now describe how symbolic animation reaches these targets by computation of the test preamble.

8.3.2 Computation of the test cases

Once the test targets and boundary goals are defined, the idea is to employ symbolic animation in an automated manner that will aim at reaching each target. To achieve that, a state exploration algorithm that is a variant of the A* path-finding algorithm and based on a Best-First exploration of the system states has been developed.

This algorithm aims at finding automatically a path, from the initial state, that will reach a given set of states characterized by a predicate. A sketch of the algorithm is given in Figure 8.5. From a given state, the symbolic successors, through each behavior, are computed using symbolic animation (procedure `compute_successors`). Each of these successors is then evaluated to compute the distance to the target. This latter is based on a heuristics that considers the "distance" between the current state and the targeted states (procedure `compute_distance`). To do that, the sum of the distances between each state variable is considered; if the domains of the two variables intersect, then the distance for these variables is 0, otherwise a customized formula, involving the type of the variable and the size of the domains, computes the distance (see Colin, Legeard, and Peureux 2004 for more details). The computation of the sequence restarts from the most relevant state, that is, the one presenting the smallest distance to the target (procedure `remove_minimal_distance` returning the most interesting triplet ⟨state, sequence of behaviors, distance⟩ and removing it from the list of visited states). The algorithm starts with the initial state (denoted by *s_init* and obtained by initializing the variables according to the INITIALIZATION clause of the machine denoted by the `initialize` function). It ends if a zero-distance state is reached by the current sequence, or if all sequences have been explored for a given depth.

Since reachability of the test targets cannot be decided, this algorithm is bounded in depth. Its worst-case complexity is $O(n^d)$, where n is the number of behaviors in all the operations of the machine and d is the depth of the exploration (maximal length of test sequence). Nevertheless, the heuristics consisting in computing the distance between the states explored and the targeted states to select the most relevant states improves the practical results of the algorithm.

The computation of the preamble ends for three possible reasons. It may have found the target, and thus, the path is returned as a sequence of behaviors. Notice that, in practice, this path is often the shortest from the initial state, but it is not always the case because

```
SeqOp ← compute_preamble(Depth, Target)
 begin
     s_init   ←   initialize ;
     Seq_curr ←   [init] ;
     dist_init ← compute_distance(Target, s_init) ;
     visited ←   [⟨ s_init, Seq_curr, dist_init ⟩] ;
      while visited ≠ [] do
         ⟨ s_curr, Seq_curr, MinDist ⟩ ← remove_minimal_distance(visited) ;
         if length(Seq_curr) < Depth then
            [(s_1, Seq_1),...,(s_N, Seq_N)] ← compute_successors((s_curr, Seq_curr)) ;
             for each (s_i, Seq_i) ∈ [(s_1, Seq_1),...,(s_N, Seq_N)] do
                 dist_i ← compute_distance(Target, s_i) ;
                 if dist_i = 0 then
                     return Seq_i;
                 else
                     visited ← visited ∪ (s_i, Seq_i, dist_i) ;
                 end if
             done
         end if
     done
     return [];
 end
```

FIGURE 8.5
State exploration algorithm.

of the heuristics used during the search. The algorithm may also end by stating that the target has not been reached. This can be because the exploration depth was too low, but it may also be because of the unreachability of the target.

Example 6 (Reachability of the test targets). Consider the three targets given in Example 5. The last two can easily be reached. Target 2 can be reached by setting the value of the PIN, and Target 3 can be reached by setting the value of the PIN, followed by three successive authentication failures.

Nevertheless, the first target will never be reached since the decrementation of the tries can only be done if pin $\neq -1$. In order to avoid considering unreachable targets, the machine invariant has to be complete enough to catch at best the reachable states of the system, or, at least, to exclude unreachable states. In the example, completing the invariant by: $pin = -1 \Rightarrow tries = 3$ makes Target 1 inconsistent, and thus removes it from the test generation process. □

The sequence returned by the algorithm represents the preamble, to which the invocation of the considered behavior (representing the test body) is concatenated. If operation parameters are still constrained, they are also instantiated to their minimal or maximal value. The observation operations are specified by hand, and the (optional) postamble is computed on the same principle as the preamble.

8.3.3 Leirios test generator for B

This technique has been industrialized by the company Smartesting,[*] a startup created from the research work done at the university of Franche-Comté in 2003, in a toolset named

---
[*]www.smartesting.com.

Leirios* Test Generator for B machines (Jaffuel and Legeard 2007) (LTG-B for short). This tool presents features of animation, test generation, and publication of the tests. In a perspective of industrial use, the tool brings out the possibility of requirements traceability. Requirements can be tagged in the model by simple markers that will make it possible to relate them to the corresponding tests that have been generated (see Bouquet et al. 2005 for more details). The tool also presents test generation reports that show the coverage of the test targets and/or the coverage of the requirements, as illustrated in the screenshot shown in Figure 8.6.

8.3.4 Limitations of the automated approach

Even though this automated approach has been successfully used in various industrial case studies on embedded systems (as will be described in Section 8.5), the feedback from the field experience has shown some limitations.

The first issue is the problem of reachability of the test targets. Even if the set of system states is well defined by the machine invariant, the experience shows that some test targets require an important exploration depth to be reached automatically, which may strongly increase the test generation time. Second, the lack of observations on the SUT may weaken the conformance relationship. As explained before, it is mandatory to dispose of a large number of observations points on the SUT to improve the accuracy of the conformance verdict. Nevertheless, if a limited number of observation is provided by the test bench (e.g., in smart cards only status words can be observed), it is mandatory to be able to check that the system has actually and correctly evolved. Finally, an important issue is the coverage of the dynamics of the system (e.g., ensure that a given sequence of commands cannot be executed successfully if the sequence is broken). Knowing the test-generation driving possibilites of the LTG-B tool, it is possible to encode the dynamics of the system by additional (ghost) variables on which a specific coverage criterion will be applied. This

FIGURE 8.6
A screenshot of the LTG-B user interface.

---

*Former name of the company Smartesting.

solution is not recommended because it requires a good knowledge of how the tool works to be employed, which is not necessarily the case of any validation engineer. Again, if limited observation points are provided, this task is all the more complicated. This weakness is amplified by the fact that the preambles are restricted to a single path from the initial state and do not cover possibly interesting situations that would have required different sequences of operation to be computed (e.g., increasing their length, involving repetitions of specific sequences of operations, etc.).

These reasons led us to consider a complementary approach, also based on model animation, that would overcome the limitations described previously. This solution is based on user-defined scenarios that will capture the know-how of the validation engineer and assist him in the design of his/her test campaigns.

8.4   Scenario-Based Test Generation

SBT is a concept according to which the validation engineer describes scenarios of use cases of the system, thus defining the test cases. In the context of software testing, it consists of describing sequences of actions that exercise the functionalities of the system. We have chosen to express scenarios as regular expressions representing sequences of operations, possibly presenting intermediate states that have to be reached.

Such an approach is related to combinatorial testing, which uses combinations of operations and parameter values, as done in the TOBIAS tool (Ledru et al. 2004). Nevertheless, combinatorial approaches can be seen as input-only, meaning that they do not produce the oracle of the test and only provide a syntactical means for generating tests, without checking the adequacy of the selected combinations w.r.t. a given specification. Thus, the numerous combinations of operations calls that can be produced may turn out to be not executable in practice. In order to improve this principle, we have proposed to rely on symbolic animation of formal models of the system in order to free the validation engineer from providing the parameters of the operations (Dadeau and Tissot 2009). This makes it possible to only focus on the description of the successive operations, possibly punctuated with checkpoints, as intermediate states, that guide the steps of the scenario. The animation engine is then in charge of computing the feasibility of the sequence at unfolding-time and to instantiate the operation parameters values. One of the advantages of our SBT approach is that it helps the production of test cases by considering symbolic values for the parameters of the operations. Thus, the user may force the animation to reach specific states, defined by predicates, that add constraints to the state variables values. Another advantage is that it provides a direct requirement traceability of the tests, considering that each scenario addresses a specific requirement.

8.4.1   Scenario description language

We present here the language that we use for designing the scenarios, first introduced in Julliand, Masson, and Tissot 2008*a*. As its core are regular expressions that are then unfolded and played by the symbolic animation engine. The language is structured in three layers: the *sequence* layer, the *model* layer, and the *directive* layer, which are described in the following.

8.4.1.1   Sequence and model layers

The *sequence* layer (Figure 8.7) is based on regular expressions that make it possible to define test scenarios as operation sequences (repeated or alternated) that may possibly

```
SEQ   ::=   OP1 | "(" SEQ ")"
       |   SEQ "." SEQ
       |   SEQ REPEAT (ALL_or_ONE)?
       |   SEQ CHOICE SEQ
       |   SEQ "↝(" SP ")"

REPEAT  ::=  "?" | n | n..m
```

FIGURE 8.7
Syntax of the sequence layer.

```
OP     ::=   operation_name
        |   "$OP"
        |   "$OP \ {" OPLIST "}"

OPLIST  ::=   operation_name
         |   operation_name "," OPLIST

SP     ::=   state_predicate
```

FIGURE 8.8
Syntax of the model layer.

lead to specific states. The *model* layer (Figure 8.8) describes the operation calls and the state predicates at the model level and constitutes the interface between the model and the scenario. A set of rules specifies the language.

Rule **SEQ** (axiom of the grammar) describes a sequence of operation calls as a regular expression. A step in the sequence is either a simple operation call, denoted by **OP1**, or a sequence of operation calls that leads to a state satisfying a state predicate, denoted by **SEQ** ↝(**SP**). This latter represents an improvement w.r.t. usual scenarios description languages since it makes it possible to define the target of an operation sequence, without necessarily having to enumerate all the operations that compose the sequence. Scenarios can be composed by the concatenation of two sequences, the repetition of a sequence, and the choice between two or more sequences. In practice, we use bounded repetition operators: 0 or 1, exactly n times, at most m times, and between n and m times. Rule **SP** describes a state predicate, whereas **OP** is used to describe the operation calls that can be (1) an operation name, (2) the $OP keyword, meaning "any operation," or (3) $OP\{OPLIST} meaning "any operation except those of **OPLIST**."

8.4.1.2 Test generation directive layer

This layer makes it possible to drive the step of test generation, when the tests are unfolded. We propose three kinds of directives that aim at reducing the search for the instantiation of a test scenario. This part of the language is given in Figure 8.9.

Rule **CHOICE** introduces two operators denoted | and ⊗, for covering the branches of a choice. For example, if S_1 and S_2 are two sequences, $S_1 | S_2$ specifies that the test generator has to produce tests that will cover S_1 and other tests that will cover sequence S_2, whereas $S_1 ⊗ S_2$ specifies that the test generator has to produce test cases covering either S_1 or S_2.

$$\text{CHOICE} \quad ::= \quad "|"$$
$$\qquad\qquad\qquad | \quad "\otimes"$$

$$\text{ALL\_or\_ONE} \quad ::= \quad "\_one"$$

$$\text{OP1} \quad ::= \quad \text{OP} \ | \ "[" \text{OP} "]"$$
$$\qquad\qquad | \quad "[" \ \text{OP} \ "/w" \ \text{BHRLIST} \ "]"$$
$$\qquad\qquad | \quad "[" \ \text{OP} \ "/e" \ \text{BHRLIST} \ "]"$$

$$\text{BHRLIST} \quad ::= \quad \underline{bhr\_label} \ \ ("," \ \underline{bhr\_label})*$$

FIGURE 8.9
Syntax of the test generation directive layer.

Rule ALL_or_ONE makes it possible to specify if all the solutions of the iteration will be returned (when not present) or if only one will be selected (_one).

Rule OP1 indicates to the test generator that it has to cover one of the behaviors of the OP operation (default option). The test engineer may also require all the behaviors to be covered by surrounding the operation with brackets. Two variants make it possible to select the behaviors that will be applied, by specifying which behaviors are authorized (/w) or refused (/e) using labels that have to tag the operations of the model.

Example 7 (An example of a scenario). Consider again the VERIFY_PIN operation from the previous example. A piece of scenario that expresses the invocation of this operation until the card is blocked, whatever the number of remaining tries might be, is expressed by (VERIFY_PIN$^{0..3}$ _one) $\rightsquigarrow$ (tries=0).

8.4.2   Unfolding and instantiation of scenarios

The scenarios are unfolded and animated on the model at the same time, in order to produce the test cases. To do that, each scenario is translated into a Prolog file, directly interpreted by the symbolic animation engine of BZ-Testing-Tools framework. Each solution provides an instantiated test case. The internal backtracking mechanism of Prolog is used to iterate on the different solutions. The instantiation mechanism involved in this part of the process aims at computing the values of the parameters of the operations composing the test case so that the sequence is *feasible* (Abrial 1996, p. 290). If a given scenario step cannot be activated (e.g., because of an unsatisfiable activation condition), the subpart of the execution tree related to the subsequence steps of the sequence is pruned and will not be explored.

Example 8 (Unfolding and instantiation). When unfolded, scenario (VERIFY_PIN$^{0..3}$ *_one*) $\rightsquigarrow$ (tries=0) will produce the following sequences:

(1) $\epsilon \rightsquigarrow$ (tries=0)
(2)   VERIFY_PIN(P_1) $\rightsquigarrow$ (tries=0)
(3)   VERIFY_PIN(P_1) . VERIFY_PIN(P_2) $\rightsquigarrow$ (tries=0)
(4)   VERIFY_PIN(P_1) . VERIFY_PIN(P_2) . VERIFY_PIN(P_3) $\rightsquigarrow$ (tries=0)

where P_1, P_2, P_3 are variables that will have to be instantiated afterwards. Suppose that the current system state gives tries=2 (remaining tries) and pin=1234. Sequence (1) can not be satisfied, (2) does not make it possible to block the card after a single authentication failure, sequence (3) and (4) are feasible, leading to a state in which the card is blocked. According to the selected directive (_one), only one sequence will be kept (here, (3) since it represents the lowest number of iterations).

The solver then instantiates parameters P_1 and P_2 for sequence (3). This sequence activates behavior $[1, 2, 3, 5, 6, 8, 0]$ of VERIFY_PIN followed by behavior $[1, 2, 3, 5, 6, 7, 0]$ that blocks the card (cf. Figure 8.3). The constraints associated with the variables representing

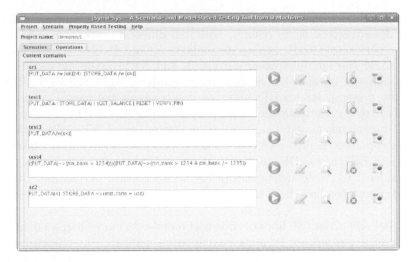

FIGURE 8.10
The jSynoPSys SBT tool.

the parameters are thus $P_1 \neq 1234$ and $P_2 \neq 1234$. A basic instantiation will then return $P_1 = P_2 = 0$, resulting in sequence: `VERIFY_PIN(0)` ; `VERIFY_PIN(0)`. ☐

These principles have been implemented into a tool named jSynoPSys (Dadeau and Tissot 2009), a SBT tool working on B Machines. A screenshot of the tool is displayed in Figure 8.10. The tool makes it possible to design and play the scenarios. Resulting tests can be displayed in the interface or exported to be concretized. Notice that this latter makes it possible to reuse existing concretization layers that would have been developed for LTG-B.

8.5 Experimental Results

This section relates the experimental results obtained during various industrial collaborations in the domain of embedded systems: smart card applets (Bernard 2004) or operating systems (Bouquet et al. 2002), ticketing applications, automotive controllers (Bouquet, Lebeau, and Legeard 2004), and space on-board software (Chevalley, Legeard, and Orsat 2005). We first illustrate the relevance of the automated test generation approach compared to manual test design. Then, we show the complementary of the two test generation techniques presented in this chapter.

8.5.1 Automated versus manual testing—The GSM 11.11 case study

In the context of an industrial partnership with the smart card division* of the Schlumberger company, a comparison has been done between a manual and an automated approach for the generation of test cases. The selected case study was the GSM 11.11 standard (European Telecommunications Standards Institute 1999) that defines, on mobile phones, the interface between the Subscriber Identification Module (SIM) and the Mobile Equipment (ME).

The part of the standard that was modeled consisted of the structure of the SIM, namely its organization in directories (called Dedicated Files—DF) or files (called Elementary

---

*Now Parkeon – www.parkeon.com.

Files—EF), and the security aspects of the SIM, namely the access control policies applied to the files. Files are accessible for reading, with four different access levels: ALWays (access can always be performed), CHV (access depends on a Card Holder Verification performed previously), ADM (for administration purposes), and NEVer (the file cannot be directly accessed through the interface). The commands modeled were SELECT_FILE (used to explore the file system), READ_BINARY (used to read in the files if permitted), VER-IFY_CHV (used to authenticate the holder), and UNBLOCK_CHV (used to unblock the CHV when too many unsuccessful authentication attempts with VERIFY_CHV happened). In addition, a command named STATUS makes it possible to retrieve the internal state of the card (current EF, current DF, and current values of tries counters). Notice that no command was modeled to create/delete files or set access control permission: the file system structure and permission have been modeled as constants and manually created on the test bench.

The B model was about 500 lines of code and represents more than a milion of concrete states. Although it was written by our research team members, the model did not involve complicated B structures and thus did not require a high level of expertise in B modeling. A total of 42 boundary goals have been computed, leading to the automated computation of 1008 test cases. These tests have been compared to the existing test suite, which had been handwritten by the Schlumberger validation team and covering the same subset of the GSM 11.11 standard. This team performed the comparison.

It showed that the automated test suite included 80% of the manual tests. More precisely, since automated test cases cover behaviors atomically, a single manual test may usually exercise the SUT in the same way that several automated tests would do. On the opposite end of the spectrum, 50% of the automated tests were absent from the manual test suite. Among them, for 20% of tests that were not produced automatically, three reasons appear. Some of the missing tests (5%) considered boundary goals that have not been generated. Other tests (35%) considered the activation of several operations from the boundary state that is not considered by the automated approach. Whereas these two issues are not crucial, and do not put the process into question; it appeared that the rest of the tests (60%) covered parts of the informal requirements that were not expressed in the B model. To overcome this limitation, a first attempt of SBT has been proposed, asking the validation engineer to provide tests designed independently, with the help of the animation tool.

The study also compared the efforts for designing the test cases. As shown in Table 8.2, the automated process reduces test implementation time, but adds time for the design of the B model. On the example, the overall effort is reduced by 30%.

8.5.2 Completing functional tests with scenarios—The IAS case study

The SBT process has been designed during the French National project POSE* that involved the leader of smart cards manufacturers, Gemalto, and that aimed at the validation of security policies for the IAS platform.

TABLE 8.2
Comparison in Terms of Time Spent on the Testing Phase in Persons/Day

Manual Design		Automated Process	
Design of the test plan	6 p/d	Modeling in B	12 p/d
		Test generation	Automated
Implementation and test execution	24 p/d	Test execution	6 p/d
Total	30 p/d	Total	18 p/d

---

*http://www.rntl-pose.info.

IAS stands for Identification, Authentication, and electronic Signature. It is a standard for Smart Cards developed as a common platform for e-Administration in France and specified by GIXEL. IAS provides identification, authentication, and signature services to the other applications running on the card. Smart cards, such as the French identity card or the "Sesame Vitale 2" health card, are expected to conform to IAS. Being based on the GSM 11.11 interface, the models present similarities. This platform presents a file system containing DFs and EFs. In addition, DFs host Security Data Objects (SDO) that are objects of an application containing highly sensitive data such as PIN codes or cryptographic keys. The access to an object by an operation in IAS is protected by security rules based on the security attributes of the object. The access rules can possibly be expressed as a conjunction of elementary access conditions, such as Never (which is the rule by default, stating that the command can never access the object), Always (the command can always access the object), or User (user authentication: the user must be authenticated by means of a PIN code). The application of a given command to an object can then depend on the state of some other SDOs, which complicates the access control rules.

The B model for IAS is 15,500 lines long. The complete IAS commands have been modeled as a set of 60 B operations manipulating 150 state variables. A first automated test generation campaign was experimented with and produced about 7000 tests. A close examination of the tests found the same weakness as for the GSM 11.11 case study, namely, interesting security properties were not covered at best, and manual testing would be necessary to overcome this weakness.

The idea of the experiment was to relate to the Common Criteria (C.C.) norm (CC 2006), a standard for the security of Information Technology products that provides a set of assurances w.r.t. the evaluation of the security implemented by the product. When a product is delivered, it can be evaluated w.r.t. the C.C. that ensure the conformance of the product w.r.t. security guidelines related to the software design, verification, and validation of the standard. In order to pass the current threshold of acceptance, the C.C. require the use of a formal model and evidences of the validation of the given security properties of the system. Nevertheless, tracing the properties in the model in order to identify dedicated tests was not possible since some of the properties were not directly expressed in the original B model.

For the experimentation, we started by designing a simplified model called Security Policy Model (SPM) that focuses on access control features. This model is 1100 lines long with 12 operations manipulating 20 state variables and represents the files management with authentications on their associated SDOs.

In order to complete the tests that are generated automatically from the complete model, three scenarios have been designed for exercising specific security properties that could not be covered previously. The scenarios and their associated tests provide direct evidences of the validation of given properties. Each scenario is associated with a *test need* that informally expresses the intention of the scenario w.r.t. the property and provides documentation on the test campaign.

- The first scenario exercises a security property stating that the access to an object protected by a PIN code requires authentication by means of the PIN code. The tests produced automatically exercise this property in a case where the authentication is obtained, and in a case where it is not. The scenario completes these tests by considering the case in which the authentication has first been obtained, but lost afterwards. The unfolding of this scenario provided 35 instantiated sequences, illustrating the possible ways of losing an authentication.

- The second scenario exercises the case of homonym PIN files located in different DFs, and their involvement in the access control conditions. In particular, it aimed at ensuring that an authenticated PIN in a specific DF is not mistakenly considered in an access control condition that involves another PIN displaying the same name but located in another DF. The unfolding of this scenario resulted in 66 tests.

- The third and last scenario exercises a property specifying that the authentication obtained by means of a PIN code not only depends on the location of the PIN but also on the life cycle state of the DF where a command protected by the PIN is applied. This scenario aimed at testing situations where the life cycle state of the directory is not always activated (which was not covered by the first campaign). The unfolding of this scenario produced 82 tests.

In the end, the three scenarios produced 183 tests that were run on the SUT. Even if this approach did not reveal any errors, the execution of these tests helps increasing the confidence in the system w.r.t. the considered security properties. In addition, the scenarios could provide direct evidence of the validation of these properties, which were useful for the C.C. evaluation of the IAS.

Notice that, when replaying the scenarios on the complete IAS model, the SBT approach detected a nonconformance between the SPM and the complete model because of a different interpretation of the informal requirements in the two models.

8.5.3 Complementarity of the two approaches

These two case studies illustrate the complementarity of the approaches. The automated boundary test generation approach is efficient at replacing most of the manual design of the functional tests, saving efforts in the design of the test campaigns. Nevertheless, it is mandatory to complete the test suite to exercise properties related to the dynamics of the system to be tested. To this end, the SBT approach provides an interesting way to assist the validation engineer in the design of complementary tests. In both cases, the use of symbolic techniques ensures the scalability of the approach.

Finally, it is important to notice that the effort of model design is made beneficial by the automated computation of the oracle and the possibility to script the execution of the tests and the verdict assignment. Notice also that, if changes appear in the specifications, a manual approach would require the complete test suite to be inspected and updated, whereas our approaches would only require to propagate these changes in the model and let the test generation tool recompute the new test suites, saving time and efforts of test suite maintenance.

8.6 Related Work

This section is divided into two subsections The first subsection is dedicated to automated test generation using model coverage criteria. The second compares our SBT process with similar approaches.

8.6.1 Model-based testing approaches using coverage criteria

Many model-based testing approaches rely on the use of a Labeled Transition System or a Finite-State Machine from which the tests are generated using dedicated graph exploration algorithms (Lee and Yannakakis 1996). Tools such as TorX (Tretmans and Brinksma 2003) and TGV (Jard and Jéron 2004) use a formal representation of the system written as Input–Output Labeled Transition Systems, on which test purposes are applied to select the relevant test cases to be produced. In addition, TorX proposes the use of test heuristics that help filtering the resulting tests according to various criteria (test length, cycle coverage, etc.). The conformance is established using the ioco (Tretmans 1996) relationship. The major differences with our automated approach is that, first, we do not know the topology of the system. As a consequence, the treatment of the model differs. Second, these processes are based on the online (or on-the-fly) testing paradigm in which the model program and the implementation are considered together. On the contrary, our approach is amenable to offline testing that requires a concretization step for the tests to be run on the SUT and the conformance to be established. Notice that the online testing approaches described previously may also be employed offline (Jéron 2009).

The STG tool (Clarke et al. 2001) improves the TGV approach by considering Input–Output Symbolic Transitions Systems, on which deductive reasoning applies, involving constraint solvers or theorem provers. Nevertheless, the kind of data manipulated are often restricted to integers and Booleans, whereas our approach manipulates additional data types, such as collections (sets, relation, functions, etc.) that may be useful for the modeling step. Similarly, AGATHA (Bigot et al. 2003) is a test generation tool based on constraint solving techniques that works by building a symbolic execution graph of systems modeled by communicating automata. Tests are then generated using dedicated algorithms in charge of covering all the transitions of the symbolic execution graph.

The CASTING (van Aertryck, Benveniste, and Le Metayer 1997) testing method is also based on the use of operations written in DNF for extracting the test cases (Dick and Faivre 1993). In addition, CASTING considers decomposition rules that have to be selected by the validation engineer so as to refine the test targets. CASTING has been implemented for B machines. Test targets are computed as constraints applying on the before and after states of the system. These constraints define states that have to be reached by the test generation process. To achieve that, the concrete state graph is built and explored. Our approach improves this technique by considering symbolic techniques that perform a boundary analysis for the test data, potentially improving the test targets. Moreover, the on-the-fly exploration of the state graph avoids the complete enumeration of all the states of the model.

Also based on B specifications, ProTest (Satpathy, Leuschel, and Butler 2005) is an automated test generator coupled with the ProB model checker (Leuschel and Butler 2003). ProTest works by first building the concrete system state graph through model animation that is then explored for covering states and transitions using classical algorithms. One point in favor of ProTest/ProB is that it covers a larger subset of the B notation as our approach, notably supporting sequences. Nevertheless, the major drawback is the exhaustive exploration of all the concrete states that complicates the industrial use of the tool on large models. In particular, the IAS model used in the experiment reported in Section 8.5.2 can not be handled by the tool.

8.6.2 Scenario-based testing approaches

In the literature, a lot of SBT work focuses on extracting scenarios from UML diagrams, such as the SCENTOR approach (Wittevrongel and Maurer 2001) or SCENT (Ryser and Glinz 1999), both using statecharts. The SOOFT approach (Tsai et al. 2003) proposes an object-oriented framework for performing SBT. In Binder (1999), Binder proposes the notion of round-trip scenario test that covers all event-response path of a UML sequence diagram. Nevertheless, the scenarios have to be completely described, contrary to our approach that abstracts the difficult task of finding well-suited parameter values.

In the study by Auguston, Michael, and Shing (2005), the authors propose an approach for the automated scenario generation from environment models for testing of real-time reactive systems. The behavior of the system is defined as a set of events. The process relies on an attributed event grammar (AEG) that specifies possible event traces. Even if the targeted applications are different, the AEG can be seen as a generalization of regular expressions that we consider.

Indirectly, the test purposes of the STG (Clarke et al. 2001) tool, described as IOSTS (Input/Output Symbolic Transition Systems), can be seen as scenarios. Indeed, the test purposes are combined with an IOSTS of the SUT by an automata product. This product restricts the possible executions of the system to those evidencing the test purpose. Such an approach has also been adapted to the B machines in (Julliand, Masson, and Tissot 2008*b*).

A similar approach is the test by model checking, where test purposes can be expressed in the shape of temporal logic properties, as is the case in Amman, Ding, and Xu (2001) or Tan, Sokolsky, and Lee (2004). The model checker computes witness traces of the properties by a synchronized product of the automata of the property and of a state/transition model of the sytem under test. These traces are then used as test cases. An input/output temporal logic has also been described in Rapin (2009) to express temporal properties w.r.t. IOSTS. The authors use an extension of the AGATHA tool to process such properties.

As explained in the beginning of this chapter, we were inspired by the TOBIAS tool (Ledru et al. 2004) that works with scenarios expressed using regular expressions representing the combinations of operations and parameters. Our approach improves this principle by avoiding the enumeration of the combinations of input parameters. In addition, our tool provides test driving possibilities that may be used to easily tackle the combinatorial explosion, inherent to such an approach. Nevertheless, on some points, the TOBIAS input language is more expressive than ours and a combination of these two approaches, which would employ the TOBIAS tool for describing the test cases, is currently under study. Notice that an experiment has been done in Maury, Ledru, and du Bousquet (2003) for coupling TOBIAS with UCASTING, the UML version of the CASTING tool (van Aertryck, Benveniste, and Le Metayer 1997). This work made it possible to use UCASTING for (1) filtering the large tests sequences combinatorially produced by TOBIAS, by removing traces that were not feasible on the model or (2) to instantiate operation parameters. Even if the outcome is similar, our approach differs since the inconsistency of the test cases is detected without having to completely unfold the test sequences. Moreover, the coupling of these tools did not include as many test driving options, to reduce the number of test cases, as we propose.

The technique for specifying scenarios can be related to Microsoft Parameterized Unit Tests (PUT for short) (Tillmann and Schulte 2005), in which the user writes skeletons of test cases involving parameterized data that will be instantiated automatically using constraint solving techniques. Moreover, the test cases may contain basic structures such as conditions and iterations, which will be unfolded during the process, so as to produce test cases. Our approach is very similar in its essence, but some differences exist. First, our scenarios do not contain data parameters. Second, we express them on the model, whereas the PUT approach aims at producing test cases that will be directly executed on the code, leaving the question of the oracle not addressed. Nevertheless, the question of refining the scenario description language so as to propagate some symbolic parameterized data along the scenario is under study.

8.7   Conclusion and Open Issues

This chapter has presented two test generation techniques using the symbolic animation of formal models, written in B, used for automating test design in the context of embedded systems such as smart cards. The first technique relies on the computation of boundary goals that define tests targets. These are then automatically reached by a customized state exploration algorithm. This technique has been industrialized by the company Smartesting and applied on various case studies in the domain of embedded systems, in particular in the domain of electronic transactions. The second technique considers user-defined scenarios, expressed as regular expressions on the operations of the model and intermediate states, that are unfolded and animated on the model so as to filter the inconsistent test cases. This technique has been designed and experimented with during an industrial partnership. This SBT approach has shown to be very convenient, firstly with the use of a dedicated scenario description language that is easy to put into practice. Moreover, the connection between the tests, the scenarios, and the properties from which they originate can be directly established, providing a means for ensuring the traceability of the tests, which is useful in the context of high-level evaluation of C.C, that requires evidences of the validation of specific properties of the considered software.

The work presented here has been applied to B models, but it is not restricted to this formalism, and the adaptation to UML, in partnership with Smartesting, is currently being studied.

Even if the SBT technique overcomes the limitations of the automated approach, in terms of relevance of the preambles, reachability of the test targets, and observations, the design of the scenario is still a manual step that requires the validation engineer to intervene. One interesting lead would be to automate the generation of the scenarios, in particular using high-level formal properties that they would exercise. Another approach is to use model abstraction (Ball 2005) for generating the tests cases, based on dynamic test selection criteria, expressed by the scenarios.

Finally, we have noticed that a key issue in the process is the ability to deal with changes and evolutions of the software at the model level. We are now working on integrating changes in the Model-based Testing process. The goal is twofold. First, it would avoid the complete recomputation of the test suites, thus saving computation time. Second, and more importantly, it would make it possible to classify tests into specific test suites dedicated to the validation of software evolutions by ensuring nonregression and nonstagnation of the parts of system.

References

Abrial, J. (1996). *The B-Book*, Cambridge University Press, Cambridge, United Kindgom.

Ambert, F., Bouquet, F., Legeard, B., and Peureux, F. (2003). Automated boundary-value test generation from specifications—method and tools. In *4th Int. Conf. on Software Testing, ICSTEST 2003*, Pages: 52–68. Cologne, Allemagne.

Amman, P., Ding, W., and Xu, D. (2001). Using a model checker to test safety properties. In *ICECCS'01, 7th Int. Conf. on Engineering of Complex Computer Systems*, Page: 212. IEEE Computer Society, Washington, DC.

Auguston, M., Michael, J., and Shing, M.-T. (2005). Environment behavior models for scenario generation and testing automation. In *A-MOST '05: Proceedings of the 1st International Workshop on Advances in Model-Based Testing*, Pages: 1–6. ACM, New York, NY.

Ball, T. (2005). A theory of predicate-complete test coverage and generation. In de Boer, F., Bonsangue, M., Graf, S., and de Roever, W.-P., eds, *FMCO'04*, Volume 3657, of *LNCS*, Pages: 1–22. Springer-Verlag, Berlin, Germany.

Beizer, B. (1995). *Black-Box Testing: Techniques for Functional Testing of Software and Systems.* John Wiley & Sons, New York, NY.

Bernard, E., Legeard, B., Luck, X., and Peureux, F. (2004). Generation of test sequences from formal specifications: GSM 11-11 standard case study. *International Journal of Software Practice and Experience* **34**(10), 915–948.

Bigot, C., Faivre, A., Gallois, J.-P., Lapitre, A., Lugato, D., Pierron, J.-Y., and Rapin, N. (2003). Automatic test generation with AGATHA. In Garavel, H. and Hatcliff, J., eds, *Tools and Algorithms for the Construction and Analysis of Systems, 9th International Conference, TACAS 2003*, Volume 2619, *Lecture Notes in Computer Science*, Pages: 591–596. Springer-Verlag, Berlin, Germany.

Binder, R.V. (1999). *Testing Object-oriented Systems: Models, Patterns, and Tools.* Addison-Wesley Longman Publishing Co., Inc., Boston, MA.

Bouquet, F., Jaffuel, E., Legeard, B., Peureux, F., and Utting, M. (2005). Requirement traceability in automated test generation—application to smart card software validation. In *Procs. of the ICSE Int. Workshop on Advances in Model-Based Software Testing (A-MOST'05)*. ACM Press, St. Louis, MO.

Bouquet, F., Julliand, J., Legeard, B., and Peureux, F. (2002). Automatic reconstruction and generation of functional test patterns—application to the Java Card Transaction Mechanism (confidential). Technical Report TR-01/02, LIFC—University of Franche-Comté and Schlumberger Montrouge Product Center.

Bouquet, F., Lebeau, F., and Legeard, B. (2004). Test case and test driver generation for automotive embedded systems. In *5th Int. Conf. on Software Testing, ICS-Test 2004*, Pages: 37–53. Düsseldorf, Germany.

Bouquet, F., Legeard, B., and Peureux, F. (2004). CLPS-B: A constraint solver to animate a B specification. *International Journal on Software Tools for Technology Transfer, STTT* **6**(2), 143–157.

Bouquet, F., Legeard, B., Utting, M., and Vacelet, N. (2004). Faster analysis of formal specifications. In Davies, J., Schulte, W., and Barnett, M., eds, *6th Int. Conf. on Formal Engineering Methods (ICFEM'04)*, Volume 3308, of *LNCS*, Pages: 239–258. Springer-Verlag, Seattle, WA.

CC (2006). Common Criteria for Information Technology Security Evaluation, version 3.1, Technical Report CCMB-2006-09-001.

Chevalley, P., Legeard, B., and Orsat, J. (2005). Automated test case generation for space on-board software. In Eurospace, ed, *DASIA 2005, Data Systems In Aerospace Int. Conf.*, Pages: 153–159. Edinburgh, UK.

Clarke, D., Jéron, T., Rusu, V., and Zinovieva, E. (2001). Stg: A tool for generating symbolic test programs and oracles from operational specifications. In *ESEC/FSE-9: Proceedings of the 8th European software engineering conference held jointly with 9th ACM SIG-SOFT international symposium on Foundations of software engineering*, Pages: 301–302. ACM, New York, NY.

Colin, S., Legeard, B., and Peureux, F. (2004). Preamble computation in automated test case generation using constraint logic programming. *The Journal of Software Testing, Verification and Reliability* **14**(3), 213–235.

Dadeau, F. and Tissot, R. (2009). jSynoPSys—a scenario-based testing tool based on the symbolic animation of B machines. *ENTCS, Electronic Notes in Theoretical Computer Science, MBT'09 proceedings* **253**(2), 117–132.

Dick, J. and Faivre, A. (1993). Automating the generation and sequencing of test cases from model-based specifications. In Woodcock, J. and Gorm Larsen, P. eds, *FME '93: First International Symposium of Formal Methods Europe*, Volume 670 of *LNCS*, Pages: 268–284. Springer, Odense, Denmark.

European Telecommunications Standards Institute (1999). *GSM 11-11 V7.2.0 Technical Specifications.*

Jaffuel, E. and Legeard, B. (2007). LEIRIOS test generator: Automated test generation from B models. In *B'2007, the 7th Int. B Conference—Industrial Tool Session*, Volume 4355 of *LNCS*, Pages: 277–280. Springer, Besancon, France.

Jard, C. and Jéron, T. (2004). Tgv: Theory, principles and algorithms, a tool for the automatic synthesis of conformance test cases for non-deterministic reactive systems. *Software Tools for Technology Transfer (STTT)* **6**.

Jéron, T. (2009). Symbolic model-based test selection. *Electronical Notes Theoritical Computer Science* **240**, 167–184.

Julliand, J., Masson, P.-A., and Tissot, R. (2008*a*). Generating security tests in addition to functional tests. In *AST'08, 3rd Int. workshop on Automation of Software Test*, Pages: 41–44. ACM Press, Leipzig, Germany.

Julliand, J., Masson, P.-A., and Tissot, R. (2008*b*). Generating tests from B specifications and test purposes. In *ABZ'2008, Int. Conf. on ASM, B and Z*, Volume 5238 of *LNCS*, Pages: 139–152. Springer, London, UK.

Ledru, Y., du Bousquet, L., Maury, O., and Bontron, P. (2004). Filtering TOBIAS combinatorial test suites. In Wermelinger, M. and Margaria, T., eds, *Fundamental Approaches to Software Engineering, 7th Int. Conf., FASE 2004*, Volume 2984 of *LNCS*, Pages: 281–294. Springer, Barcelona, Spain.

Lee, D. and Yannakakis, M. (1996). Principles and methods of testing finite state machines—a survey. In *Proceedings of the IEEE*, Pages: 1090–1123.

Legeard, B., Peureux, F., and Utting, M. (2002). Automated boundary testing from Z and B. In *Proc. of the Int. Conf. on Formal Methods Europe, FME'02*, Volume 2391 of *LNCS*, Pages: 21–40. Springer, Copenhaguen, Denmark.

Leuschel, M. and Butler, M. (2003). ProB: A model checker for B. In Araki, K., Gnesi, S., and Mandrioli, D., eds, *FME 2003: Formal Methods*, Volume 2805 of *LNCS*, Pages: 855–874. Springer.

Maury, O., Ledru, Y., and du Bousquet, L. (2003). Intgration de TOBIAS et UCASTING pour la gnration de tests. In *16th International Conference Software and Systems and their applications-ICSSEA*, Paris.

Offutt, A., Xiong, Y., and Liu, S. (1999). Criteria for generating specification-based tests. In *Proceedings of the 5th IEEE International Conference on Engineering of Complex Computer Systems (ICECCS'99)*, Pages: 119–131. IEEE Computer Society Press, Las Vegas, Nevada.

Rapin, N. (2009). Symbolic execution based model checking of open systems with unbounded variables. In *TAP '09: Proceedings of the 3rd International Conference on Tests and Proofs*, Pages: 137–152. Springer-Verlag, Berlin, Heidelberg.

Ryser, J. and Glinz, M. (1999). A practical approach to validating and testing software systems using scenarios.

Satpathy, M., Leuschel, M., and Butler, M. (2005). ProTest: an automatic test environment for B specifications. *Electronic Notes in Theroretical Computer Science* **111**, 113–136.

SIC (2004). *SICStus Prolog 3.11.2 manual documents.* http://www.sics.se/sicstus.html.

Tan, L., Sokolsky, O., and Lee, I. (2004). Specification-based testing with linear temporal logic. In *IRI'2004, IEEE Int. Conf. on Information Reuse and Integration*, Pages: 413–498.

Tillmann, N. and Schulte, W. (2005). Parameterized unit tests. *SIGSOFT Softw. Eng. Notes* **30**(5), 253–262.

Tretmans, G.J. and Brinksma, H. (2003). Torx: automated model-based testing. In Hartman, A. and Dussa-Ziegler, K., eds, *First European Conference on Model-Driven Software Engineering*, Pages: 31–43, Nuremberg, Germany.

Tretmans, J. (1996). Conformance testing with labelled transition systems: implementation relations and test generation. *Computer Networks and ISDN Systems*, **29**(1), 49–79.

Tsai, W. T., Saimi, A., Yu, L., and Paul, R. (2003). Scenario-based object-oriented testing framework. *qsic* **00**, 410.

van Aertryck, L., Benveniste, M., and Le Metayer, D. (1997). Casting: a formally based software test generation method. *Formal Engineering Methods, International Conference on* **0**, 101.

Wittevrongel, J. and Maurer, F. (2001). Scentor: scenario-based testing of e-business applications. In *WETICE '01: Proceedings of the 10th IEEE International Workshops on Enabling Technologies*, Pages: 41–48. IEEE Computer Society, Washington, DC.

Part III

Integration and Multilevel Testing

9

Model-Based Integration Testing with Communication Sequence Graphs

Fevzi Belli, Axel Hollmann, and Sascha Padberg

CONTENTS

While unit testing is supposed to guarantee the proper function of single units, integration testing (ITest) is intended to validate the communication and cooperation between different components. ITest is important because many events are caused by integration-related faults such as, for example, failures during money transfer, air- and spacecraft crashes, and many more that are not detectable during unit testing. This chapter introduces an approach to model-based integration testing. After a brief review of existing work (1) communication sequence graphs (CSG) are introduced for representing the communication between software components on a meta-level and (2) based on CSG and other introduced notions test coverage criteria are defined. A case study based on a robot-controlling application illustrates and validates the approach.

9.1 Introduction and Related Work

Testing is the validation method of choice applied during different stages of software production. In practice, testing is often still carried out at the very end of the software development process. It is encouraging, however, that some companies, for example, in the aircraft industry, follow a systematic approach using phase-wise verification and validation while developing, for example, embedded systems. Disadvantages of a "Big-Bang-Testing" (Myers 1979) that is carried out at the end of development are obvious. Sources of errors interfere with each other resulting in late detection, localization, and correction of faults. This, in turn, becomes very costly and time consuming.

Several approaches to ITest have been proposed in the past. Binder (1999) gives different examples of ITest techniques, for example, top-down and bottom-up ITest. Hartmann et al. (Hartmann, Imoberdorf, and Meisinger 2000) use UML statecharts specialized for *object-oriented programming* (OOP). Delamaro et al. (Delamaro, Maldonado, and Mathur 2001) introduced a communication-oriented ITest approach that mutates the interfaces of software units. An overview of mutation analysis results is given by Offutt (1992). Saglietti et al. (Saglietti, Oster, and Pinte 2007) introduced an interaction-oriented, higher-level approach and several test coverage criteria.

In addition, many approaches to ITest of object-oriented software (OOS) have been proposed. Buy et al. (Buy, Orso, and Pezze 2000) defined method sequence trees for representing the call structure of methods. Daniels et al. (Daniels and Tai 1999) introduced different test coverage criteria for method sequences. Martena et al. (Martena, DiMilano, Orso, and Pezzè 2002) defined interclass testing for OOS. Zhao and Lin (2006) extended the approach of Hartmann, Imoberdorf, and Meisinger (2000) by using the *method message paths* for ITest, illustrating the communication between objects of classes. A *method message path* is defined as a sequence of method execution paths linked by messages, indicating the interactions between methods in OOS. Hu, Ding, and Pu (2009) introduced a path-based approach focused on OOS in which a forward slicing technique is used to identify the call statements of a unit and by connecting the units via interface net and mapping tables. The path-based approach considers units as nodes; the interface nets are input and output ports of the nodes representing the parameters of the unit, and the mapping tables describe the internal mapping from the in-ports to the out-ports of a node. Furthermore, Sen (2007) introduced a *concolic* testing approach that integrates conditions into graphs for concrete and symbolic unit testing. Hong, Hall, and May (1997) detailed test termination criteria and test adequacy for ITest and unit testing.

In this chapter, CSG are introduced to represent source code at different levels of abstraction. Software systems with discrete behavior are considered. In contrast to existing, mostly state-based approaches described above, CSG-based models are stateless, that is, they do not concentrate on internal states of the software components,* but rather focus on events.

CSGs are directed graphs enriched with some semantics to adopt them for ITest. This enables the direct application of well-known algorithms from graph theory, automata theory, operation research, etc. for test generation and test minimization. Of course, UML diagrams could also be used for ITest, done by Hartmann et al. (2000); in this case, however, some intermediate steps would be necessary to enable the application of formal methods. The approach presented in this chapter is applicable to both OOP and non-OOP programming.

The syntax of CSG is based on *event sequence graphs* (ESGs) (Belli, Budnik, and White 2006). ESGs are used to generate test cases for user-centered black-box testing of human-machine systems.

ITest makes use of the results of unit testing. Therefore, a uniform modeling for both unit testing and ITest is aimed at by using the same modeling techniques for both levels. Section 9.2 explains how CSG are deployed for unit testing (Section 9.2.2) and ITest (Section 9.2.3), after a short introduction to fault modeling on ITest (Section 9.2.1). This section also introduces a straightforward strategy for generating test cases and mutation testing to the CSG (Section 9.2.4). A case study in Section 9.3 exemplifies and validates the approach.

---

*Note that "software component" and "unit" are used interchangeably.

For the case study, a robot-control application is chosen that performs a typical assembly process. Using different coverage criteria, test sets are generated from CSG models of the system under consideration (SUC). Mutation analysis is applied to SUC for evaluating the adequacy of the generated test sets. Section 9.4 gives a summary of the approach and concludes the chapter referring to future research work.

9.2 Communication Sequence Graphs for Modeling and Testing

Depending on the applied programming language, a software component represents a set of functions including variables forming data structures. Classes contain methods and variables in the object-oriented paradigm. In the following, it is assumed that unit tests have already been conducted and ITest is to be started. In case that no model exists, the first step of ITest is supposed to model the components c_i of the SUC, represented as $C = \{c_1, \ldots, c_n\}$.

9.2.1 Fault modeling

Figure 9.1 shows the communication between a calling software component, $c_i \in C$, and an invoked component, $c_j \in C$. Messages to realize this communication are represented as tuples M of parameter values and global variables and can be transmitted correctly (*valid*) or faultily (*invalid*), leading to the following combinations:

- M_{ci} (c_i, c_j): correct input from c_i to c_j, (valid case)

- M_{co} (c_j, c_i): correct output from c_j back to c_i, (valid case)

- M_{fi} (c_i, c_j): faulty input from c_i to c_j, (invalid case)

- M_{fo} (c_j, c_i): faulty output from c_j back to c_i (invalid case).

Figure 9.1 illustrates the communication process. Two components, c_i, $c_j \in C$ of a software system C communicate with each other by sending a message from c_i to c_j, that is, the communication is directed from c_i to c_j.

We assume that either M_{ci} (c_i, c_j) or M_{fi} (c_i, c_j) is the initial invocation. As the reaction of this invocation, c_j sends its response back to c_i. The response is M_{co} (c_j, c_i) or

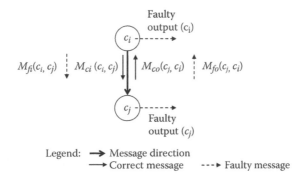

FIGURE 9.1

Message-oriented model of integration faults between two software components c_i and c_j.

M_{fo} (c_j, c_i). We assume further that the tuples of *faulty* message M_{fi} (c_i,c_j) and M_{fo} (c_j, c_i) cause faulty outputs of c_i and c_j as follows:

- Component c_j produces faulty results based on

  - Faulty parameters transmitted from c_i in M_{fi} (c_i,c_j), or
  - Correct parameters transmitted from c_i in M_{ci} (c_i,c_j), but perturbed during transmission resulting in a faulty message.

- Component c_i produces faulty results based on

  - Faulty parameters transmitted from c_i to c_j, causing c_j to send a faulty output back to c_i in M_{fo} (c_j, c_i), or
  - Correct, but perturbed parameters transmitted from c_i to c_j, causing c_j to send a faulty output back to c_i in M_{co} (c_j, c_i) resulting in a faulty message.

The message direction in this example indicates that c_j is reactive and c_i is pro-active. If c_j sends the message first, the message will be transmitted in the opposite direction.

This fault model helps to consider potential system integration faults and thus, to generate tests to detect them. Perturbation during transmission arises if either

- The message is being corrupted, or

- The messages are re-ordered, or

- The message is lost.

A message is corrupted when its content is corrupted during transmission. When the order of messages is corrupted, the receiving unit uses faulty data. When a message is lost, the receiving unit does not generate an output.

The terminology in this chapter is used in such a manner that *faulty* and *invalid*, and *correct* and *valid* are interchangeable. A faulty message results from a perturbation of the message content. If a faulty message is sent to a correct software unit, this message can result in a correct output of the unit, but the output deviates from the specified output that corresponds to the input. This is also defined as a faulty output.

9.2.2 Communication sequence graphs for unit testing

In the following, the term *actor* is used to generalize notions that are specific to the great variety of programming languages, for example, functions, methods, procedures, basic blocks, and so on. An elementary *actor* is the smallest, logically complete unit of a software component that can be activated by or activate other actors of the same or other components (Section 9.2.3). A software component, $c \in C$, can be represented by a CSG as follows.

Definition 1. A CSG for a software component, $c \in C$ is a directed graph $CSG = (\Phi, E, \Xi, \Gamma)$, where

- The set of *nodes* Φ comprises all actors of component c, where a node/actor is defined as an abstract node/actor ϕ^{a} in case it can be refined to elementary actors $\phi^{\mathrm{a}}_{1,2,3,\ldots,\mathrm{n}}$.

- The set of *edges* E describes all pairs of valid concluding invocations (calls) within the component, an edge (ϕ, ϕ')$\in E$ denotes that actor ϕ' is invoked after the invocation of actor ϕ ($\phi \to \phi'$).

- $\Xi \subseteq \Phi$ and $\Gamma \subseteq \Phi$ represent initial/final invocations (nodes). $\square$

Figure 9.2 shows a CSG for Definition 1 including an abstract actor ϕ_2^a that is refined using a CSG. This helps to simplify large CSGs. In this case, ϕ_2^a is an abstract actor encapsulating the actor sequence $\phi_{2.1}^a$, $\phi_{2.2}^a$, and $\phi_{2.3}^a$.

To identify the initial and final invocations of a CSG, all $\phi \in \Xi$ are preceded by a pseudo vertex "[" $\notin \Phi$ (*entry*) and all $\phi \in \Gamma$ are followed by another pseudo vertex "]" $\notin \Phi$ (*exit*). In OOS, these nodes typically represent invocations of a constructor and destructor of a class.

CSG is a derivate of ESG (Belli, Budnik, and White 2006) differing in the following features.

- In an ESG, a node represents an event that can be a user input or a system response, both of which lead interactively to a succession of user inputs and expected system outputs.

- In a CSG, a node represents an actor invoking another actor of the same or another software component.

- In an ESG, an edge represents a sequence of immediately neighboring events.

- In a CSG, an edge represents an invocation (call) of the successor node by the preceding node.

Readers familiar with ITest modeling will recognize similarity of CSG with *call graphs* (Grove et al. 1997). However, they differ in many aspects as summarized below.

- CSGs have explicit boundaries (entry [begin] and exit [end] in form of initial/final nodes) that enable the representation of not only the activation structure but also the functional structure of the components, such as the initialization and the destruction of a software unit (for example, the call of the constructor and destructor method in OOP).

- CSGs are directed graphs for systematic ITest that enable the application of rich notions and algorithms of graph theory. The latter are useful not only for generation of test cases based on criteria for graph coverage but also for optimization of test sets.

- CSGs can easily be extended not only to represent the control flow but also to precisely consider the data flow, for example, by using Boolean algebra to represent constraints (see Section 9.4).

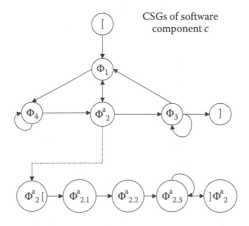

FIGURE 9.2
A *CSG* including a refinement of the abstract actor ϕ_2^a.

Definition 2. Let Φ and E be the finite set of nodes and arcs of CSG. Any sequence of nodes $(\phi_1, \ldots, \phi_k)$ is called a *communication sequence* (CS) if $(\phi_i, \phi_{i+1}) \in E$, for $i = 1, \ldots, k-1$. $\square$

The function l (*length*) determines the number of nodes of a CS. In particular, if $l\ (CS) = 1$, then it is a CS of length 1, which denotes a single node of CSG. Let α and ω be the functions to determine the initial and final invocation of a CS. For example, given a sequence $CS = (\phi_1, \ldots, \phi_k)$, the initial and final invocation are $\alpha\ (CS) = \phi_1$ and $\omega\ (CS) = \phi_k$, respectively. A $CS = (\phi, \phi')$ of length 2 is called a *communication pair* (CP).

Definition 3. A CS is a *complete communication sequence* (CCS) if $\alpha\ (CS)$ is an initial invocation and $\omega\ (CS)$ is a final invocation.

Now, based on Definitions 2 and 3, the ***i-sequence coverage criterion*** can be introduced that requires generation of CCSs that sequentially invoke all CSs of length $i \in \mathbb{N}$.

At first glance, i-sequence coverage, also called *sequence coverage criterion*, is similar to All-n-Transitions coverage (Binder 1999). However, i-sequence coverage focuses on CSs. CSG does not have state transitions, but it visualizes CSs of different length $(2, 3, 4, \ldots, n)$ that are to be covered by tests cases. Section 9.4 will further discuss this aspect.

The i-sequence coverage criterion is fulfilled by covering all sequences of nodes and arcs of a CSG of length i. It can also be used as a test termination criterion (Hong, Hall, and May 1997). All CSs of a given length i of a *CSG* are to be covered by means of CCSs that represent test cases. Thus, test case generation is a derivation of *Chinese Postman Problem*, understood as finding the shortest path or circuit in a graph by visiting each arc. Polynomial algorithms supporting this test generation process have been published in previous works (Aho et al. 1991, Belli, Budnik, and White 2006).

The coverage criteria introduced in this chapter are named in accordance with the length of the CS to be covered. The coverage criterion of CSs of length 1 is called *1-sequence coverage criterion* or *actor coverage criterion*, where every actor is visited at least once. The coverage criterion of CSs of length 2 is called *2-sequence coverage criterion* or *communication pair criterion*, etc. Finally, coverage criteria of CSs of length *len* are called *len-sequence coverage criterion* or *communication len-tuple criterion*. Algorithm 1 sketches the test case generation process for unit testing. $\square$

Algorithm 1 Test Case Generation Algorithm for Unit Testing

Input: *CSG*

len := maximum length of communication
sequences (CS) to be covered

Output: Test report of succeeded and failed test cases

FOR $i := 1$ **TO** *len* **Do**
  Cover all CS of *CSG* by means of CCS
  Apply test cases to SUC and observe system outputs

9.2.3 Communication sequence graphs for integration testing

For ITest, the communication between software components has to be tested thoroughly. This approach is based on the communication between pairs of components including the study of the control flow.

Definition 4. Communication between actors of two different software components, $CSG_i = (\Phi_i, E_i, \Xi_i, \Gamma_i)$ and $CSG_j = (\Phi_j, E_j, \Xi_j, \Gamma_j)$ is defined as an *invocation relation* $IR(CSG_i, CSG_j) = \{(\phi, \phi')\ |\phi \in \Phi_i \text{ and } \phi' \in \Phi_j, \text{ where } \phi \text{ activates } \phi'\}$.

A $\phi \in \Phi$ may invoke an additional $\phi' \in \Phi'$ of another component. Without losing generality, the notion is restricted to communication between two units. If ϕ causes an invocation of a third unit, this can also be represented by a second invocation considering the third one. □

Definition 5. Given a set of $CSG_1, \ldots, CSG_n$ describing n components of a system C a set of invocation relations $IR_1, \ldots, IR_m$, the *composed* CSG_C is defined as $CSG_C = (\{\Phi_1 \cup \cdots \cup \Phi_n\}, \{E_1 \cup \cdots \cup E_n \cup IR_1 \cup \cdots \cup IR_m\}, \{\Xi_1 \cup \cdots \cup \Xi_n\}, \{\Gamma_1 \cup \cdots \cup \Gamma_n\})$. □

An example of a composed CSG built of a $CSG_1 = (\{\phi_1, \phi_2, \phi_3, \phi_4\}, \{(\phi_1, \phi_4), (\phi_1, \phi_2), (\phi_2, \phi_1), (\phi_2, \phi_3), (\phi_3, \phi_1), (\phi_3, \phi_3), (\phi_4, \phi_2), (\phi_4, \phi_4)\}, \{\phi_1\}, \{\phi_3\})$ and $CSG_2 = (\{\phi_1', \phi_2', \phi_3'\}, \{(\phi_1', \phi_2'), (\phi_2', \phi_1'), (\phi_2', \phi_2'), (\phi_2', \phi_3'), (\phi_3', \phi_1')\}, \{\phi_2'\}, \{\phi_3'\})$ for two software components c_1 and c_2 is given by Figure 9.3. Invocation of ϕ_1' by ϕ_2 is denoted by a dashed line, that is $IR(CSG_1, CSG_2) = \{(\phi_2, \phi_1')\}$.

Based on the *i*-sequence coverage criterion, Algorithm 2 represents a test case generation procedure. For each software component $c_i \in C$, a CSG_i and invocation relations IR serve as input. As a first step, the composed CSG_C is to be constructed. The nodes of CSG_C consist of the nodes of $CSG_1, \ldots, CSG_n$. The edges of CSG_C are given by the edges of $CSG_1, \ldots, CSG_n$ and the invocation relations IRs among these graphs.

The coverage criteria applied for ITest are called in the same fashion as those for unit testing.

Algorithm 2 Test Case Generation Algorithm for Integration Testing

---

Input: $CSG_1, \ldots, CSG_n$
$IR_1, \ldots, IR_m$
$len :=$ maximum length of communication
sequences (CS) to be covered

Output: Test report of succeeded and failed test cases

$CSG_C = (\{\Phi_1 \cup \cdots \cup \Phi_n\}, \{E_1 \cup \cdots \cup E_n \cup IR_1 \cup \cdots \cup IR_m\},$
$\{\Xi_1 \cup \cdots \cup \Xi_n\}, \{\Gamma_1 \cup \cdots \cup \Gamma_n\})$

Use Algorithm 1 for test case generation.

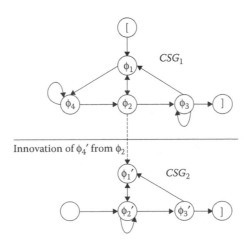

FIGURE 9.3
Composed CSG_C consisting of CSG_1 and CSG_2 and an invocation between them.

9.2.4   Mutation analysis for CSG

The previous sections, 9.2.2 and 9.2.3, defined CSG and introduced algorithms for test case generation with regards to unit testing and ITest. In the following, mutation analysis is used to assess the adequacy of the test cases with respect to their fault detection effectiveness.

Mutation analysis was introduced by DeMillo et al. in 1978 (DeMillo, Lipton, and Sayward 1978). A set of mutation operators syntactically manipulates the original software and thus, seeds semantic faults leading to a set of mutants that represent faulty versions of the software given. A test set is said to be *mutation adequate* with respect to the program and mutation operators if at least one test case of the test set detects the seeded faults for each mutant. In this case, the mutant is said to be *distinguished* or *killed*. Otherwise, the mutant remains *live* and the test set is marked as *mutation inadequate*. The set of *live* mutants may also contain *equivalent* mutants that must be excluded from analysis. Equivalent mutants differ from the original program in their syntax, but they have the same semantics. A major problem of mutation testing is that in general *equivalent* mutants cannot be detected automatically. Thus, the *mutation score* $\mathcal{MS}$ for a given program P and a given test set T is:

$$\mathcal{MS}(\mathcal{P}, \mathcal{T}) = \frac{Number\ of\ killed\ mutants}{Number\ of\ all\ mutants - Number\ of\ equivalent\ mutants}.$$

The ideal situation results in the score 1, that is, all mutants are killed.

Applying a mutation operator only once to a program yields a *first-order* mutant. Multiple applications of mutation operators to generate a mutant are known as *higher-order* mutants. An important assumption in mutation analysis is the *coupling effect*, that is, assuming that test cases that are capable of distinguishing first-order mutants will also most likely kill higher-order mutants. Therefore, it is common to consider only first-order mutants. A second assumption is the *competent programmer hypothesis*, which assumes that the SUC is close to being correct (Offutt 1992).

The procedure of integration and mutation testing a system or program P based on CSG is illustrated in Figure 9.4.

Algorithms 1 and 2 generate the test sets (see Figure 9.4, arc (1)) to be executed on the system or program P (see arc (2)). If ITest does not reveal any faults, this could mean that

- SUC is fault-free or, more likely,

- The generated test sets are not adequate to detect the remaining faults in SUC.

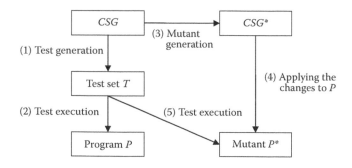

FIGURE 9.4
Software ITest and mutation analysis with CSG.

TABLE 9.1

List of Mutation Operators for CSGs

Name	Description
AddNod	Inserts a new node into the CSG
DelNod	Deletes a node from the CSG
AddEdg	Inserts a new edge into the CSG (also applicable for self-loops)
DelEdg	Deletes an edge from the CSG by deactivating the destination of the edge (also applicable for self-loops)
AddInvoc	Inserts an invocation between actor ϕ of software component c to ϕ' of component c'
DelInvoc	Deletes an invocation between actor ϕ of component c to ϕ' of component c'

Therefore, in order to check the adequacy of the test sets, a set of mutation operators modifies the CSG for generating first-order mutants (see Figure 9.4, arc (3)). Based on CSG, six *basic operators* are defined that realize insertion and/or deletion of nodes or edges of the CSG (compare to Belli, Budnik, and Wong 2006). The operators are listed in Table 9.1.

After applying the mutation operators to P (see Figure 9.4, arc (4)) and so producing the mutants P^*, the generated test sets are executed on P^* (see arc (5)). If some mutants are not killed, the test set is not adequate. In this case, the length of the CSs has to be increased. If all mutants are now killed, the test set is adequate for ITest.

The operator AddNod in Table 9.1 adds a new node to the CSG, generating a new CS from one node via the new node to another node of the same software unit, that is, a new call is inserted in the source code of a component between two calls.

DelNod deletes a node from the CSG and connects the former ingoing edges to all successor nodes of the deleted node, that is, an invocation is removed from the source code of a software unit.

The mutation operator AddEdg inserts a new edge from one node to another node of the same software component that had no connection before applying the operator. Alternatively, it inserts a self-loop at one node that had no self-loop before applying the operator to the CSG. In other words, after a call it is possible to execute another invocation of the same component that was not a successor before the mutation.

If a self-loop is added, a call is repeated using different message data. Similarly, DelEdg deletes an edge from one node to another node of the same software component that had a connection before applying the operator or deletes a self-loop at one node that had a self-loop before applying the operator on the CSG. In this manner, the order of the calls is changed. It is not possible to execute a call that was a successor of another invocation of the same unit before the mutation. In case of removing a self-loop, a call cannot be repeated.

While AddInvoc inserts an invocation between actors of different software units, DelInvoc deletes it. In other words, a call is inserted in or removed from another component.

9.3 Case Study

To validate and demonstrate the approach, a case study was performed using a robot (that is, RV–M1 manufactured by Mitsubishi Electronics). Figure 9.5 shows the robot in its working area within a control cabinet.

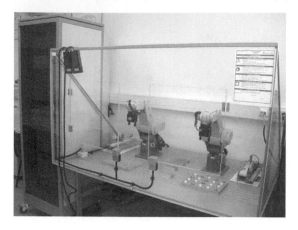

FIGURE 9.5
Robot System RV-M1 (refer to Belli, Hollmann, and Padberg 2009).

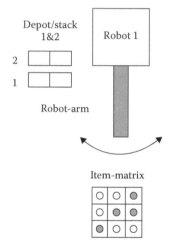

FIGURE 9.6
Working area/buffers of the robot (Belli, Hollmann, and Padberg 2009).

9.3.1 System under consideration

The SUC is a part of the software system implemented in C++ that controls the robot RV-M1. The robot consists of two mechanical subsystems, an arm and a hand. The arm of RV-M1 can move items within the working area. These items can also be stored in two buffers as sketched in Figure 9.6. The robot can grab items contained in the item matrix and transport them to a depot. For this purpose, its arm is moved to the appropriate position, the hand is closed, moved to the stacking position, and the hand releases the item.

9.3.2 Modeling the SUC

The mechanical subsystems of the robot are controlled by 14 software units listed in Table 9.2.

An example of the CSG of the software unit, `RC_constructor/init`, including the corresponding implementation, is given in a brief format in Figures 9.7, 9.8, and 9.9.

TABLE 9.2

List of Software Units

Name	Description
StackConstruct	The main application constructing a stack of items in the depot taken from the item matrix.
SC_control	The main control block determines the destination of the items.
RobotControl	The RobotControl class controls the other software units from SerialInterface to RoboPosPair.
RC_constructor/init	The RobotControl constructor method starts the units SerialInterface to RoboPosPair.
RC_moveMatrixMatrix	The RobotControl method `moveMatrixMatrix` moves an item from one matrix position to another free position.
RC_moveMatrixDepot	The RobotControl method `moveMatrixDepot` moves an item from a matrix position to a free depot position.
RC_moveDepotMatrix	The RobotControl method `moveDepotMatrix` moves an item from a depot position to a free matrix position.
RC_moveDepotDepot	The RobotControl method `moveMatrixMatrix` moves an item from one depot position to another free position.
SerialInterface	The SerialInterface class controls the interface of a PC to robot controlling units.
MoveHistory	The MoveHistory class saves all executed movements of the robot.
MH_Add	The MoveHistory method `Add` adds a movement to the history.
MH_Undo MH_UndoAll	The MoveHistory method `Undo` removes a movement from the history und reversing the movement. If the history is empty, it corresponds to `MH_UndoAll`.
RoboPos	The RoboPos class provides the source and destination positions of the robot.
RoboPosPair	The RoboPosPair class combines two stack positions of one depot destination position.

The dashed lines between these graphs represent communication (method invocations) between the components. The *RobotControl* unit is initialized by its constructor method call `RobotControl::RobotControl`, which activates the `SerialInterface`, `MoveHistory`, `RoboPos`, and the RoboPosPair software units of the robot system. Figures 9.8 and 9.9 provide commentaries explaining the CSG structure of 9.7.

Figure 9.10 shows the StackConstruct application of the robot system. The corresponding source code of the application is given in Figure 9.11.

The StackConstruct application builds a stack on Depot 1/2 by moving matrix items. The robot system is initialized by `rc->init()` and the matrix items, 00,01,02,10,11,12,20,21 are moved to the depot positions D1LH (depot1 left hand) floor, D1LH second stack position, D1RH (depot1 right hand) floor position, D1RH second stack position, D2LH floor position, D2LH second stack position, D2RH floor position, or D2RH second stack position. Finally, all items are put back to their initial positions by `rc->stop` and the robot system is shut down.

9.3.3 Test generation

Five test sets were generated by using Algorithm 2. Test sets T_i, consisting of *CCSs*, were generated to cover all sequences of length $i \in \{1,2,3,4,5\}$. Test set T_1 achieves the actor coverage criterion and the test cases of T_1 are constructed to cover all actors of the software

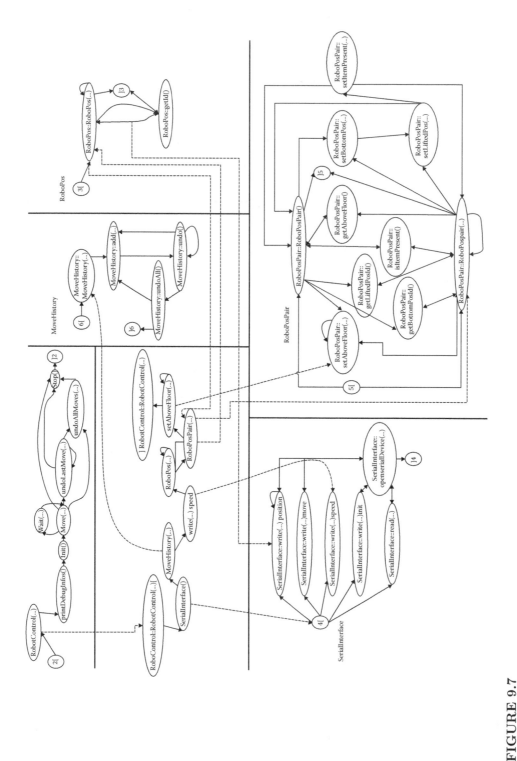

FIGURE 9.7

CSG for initializing the robot system (dashed lines represent calls between components).

```
RoboControl :: RoboControl(int id, int speed)
{

  if(speed < 0 || speed > 9) {
  }
  /* The invocation new SerialInterface() prepares the component SerialInterface, calling
   * its constructor method SerialInterface : : SerialInterface and new MoveHistory(this)
   * the unit MoveHistory saving all movements of the robot system */

  si = new SerialInterface();
  mh = new MoveHistory(this);

  this -> id = id;
  this -> speed = speed;

  /* si -> write(SP,this speed) calls the actor SerialInterface::write(int cmd, int id)
   * setting the speed of the robot */

  si -> write(SP, this speed);    // send desired speed to robot

  /* For defining the start and the destination of intermediate positions of the system new RoboPos(...)
   * invokes RoboPos::RoboPos(int id, float x, float y, float z, float slope, float roll)
   * which transfers the position information calling
   * void SerialInterface::write(int cmd, int id, float x, float y, float z, float slope, float roll).
   * As there are several positions the actors new RoboPos(...),
   * RoboPos::RoboPos(int id, float x, float y, float z, float slope, float roll)
   * and void SerialInterface::write(int cmd, int id, float x, float y, float z, float slope, float roll)
   * have self-loops */

  idlePos =             new RoboPos(1, -2.7, +287.4, +333.3, -90.2, -3.1);
  intermediatePosMatrix = new RoboPos(2, +14.4, +314.7, +102.7, -91.6, +.4);
  intermediatePosDepot1 = new RoboPos(3, +359.5, +23.1, +106.3, -90.5, -3.8);
  intermediatePosDepot2 = new RoboPos(4, +359.5, -73.1, +106.3, -90.5, +10.1);
```

FIGURE 9.8

Extract of source code of robot component RoboControl::RoboControl including its invocations (part 1).

```
/* All placing positions for the items are set by calling new RoboPosPair(...), which comprises nested invocations.
 * It calls the constructor method calls RoboPosPair:: RoboPosPair and
 * RoboPosPair:: RoboPosPair(RoboPos* bottom, RoboPos* lifted, bool itemPresent) of the unit RoboPosPair.
 * This component defines a stack of items which can be constructed at the placing positions. */

matrixPositions[0][0] = new RoboPosPair( new RoboPos(10, . . . ), new RoboPos(20, . . . ), true);
matrixPositions[1][0] = new RoboPosPair( new RoboPos(11, . . . ), new RoboPos(21, . . . ), true);
matrixPositions[2][0] = new RoboPosPair( new RoboPos(12, . . . ), new RoboPos(22, . . . ), true);

matrixPositons[0][1] = new RoboPosPair( new RoboPos(13, . . . ), new RoboPos(23, . . . ), true);
matrixPositions[1][1] = new RoboPosPair( new RoboPos(14, . . . ), new RoboPos(24, . . . ), true);
matrixPositions[2][1] = new RoboPosPair( new RoboPos(15, . . . ), new RoboPos(25, . . . ), true);

matrixPositions[0][2] = new RoboPosPair( new RoboPos(16, . . . ), new RoboPos(26, . . . ), true);
matrixPositions[1][2] = new RoboPosPair( new RoboPos(17, . . . ), new RoboPos(27, . . . ), true);
matrixPositions[2][2] = new RoboPosPair( new RoboPos(18, . . . ), new RoboPos(28, . . . ), true);

/* Additionally new RoboPos(...) is called twice for setting two stack-postions on each depot-position,
 * again new RoboPos(...) invokes
 * void SerialInterface::write(int cmd, int id, float x, float y, float z, float slope, float roll) */

depotPositons[0][0] = new RoboPosPair( new RoboPos(30, . . . ), new RoboPos(32, . . . ), false);
depotPositons[0][1] = new RoboPosPair( new RoboPos(31, . . . ), new RoboPos(33, . . . ), false);
depotPositons[1][0] = new RoboPosPair( new RoboPos(34, . . . ), new RoboPos(36, . . . ), false);
depotPositons[1][1] = new RoboPosPair( new RoboPos(35, . . . ), new RoboPos(37, . . . ), false);

RoboPosPair* depot1Floor1RH = new RoboPosPair( new RoboPos(40, . . . ), new RoboPos(42, . . . ), false);
RoboPosPair* depot1Floor1RH = new RoboPosPair( new RoboPos(41, . . . ), new RoboPos(43, . . . ), false);
RoboPosPair* depot2Floor1RH = new RoboPosPair( new RoboPos(44, . . . ), new RoboPos(46, . . . ), false);
RoboPosPair* depot2Floor1RH = new RoboPosPair( new RoboPos(45, . . . ), new RoboPos(47, . . . ), false);

/* The initialization is finished by defining the four depot positions as well as the actual
 * stack-position by calling depotPositions[0][0] -> setAboveFloor (depot1Floor1RH), where this
 * actor invokes void RoboPosPair::setAboveFloor (RoboPosPair* aboveFloor)
 * several times indicated by the self-loops on both actors. */

depotPositons[0][0] -> setAboveFloor(depot1Floor1RH);
depotPositons[0][1] -> setAboveFloor(depot1Floor1LH);
depotPositons[1][0] -> setAboveFloor(depot2Floor1RH);
depotPositons[1][1] -> setAboveFloor(depot2Floor1LH);
}
```

FIGURE 9.9

Extract of source code of robot component RoboControl::RoboControl including its invocations (part 2).

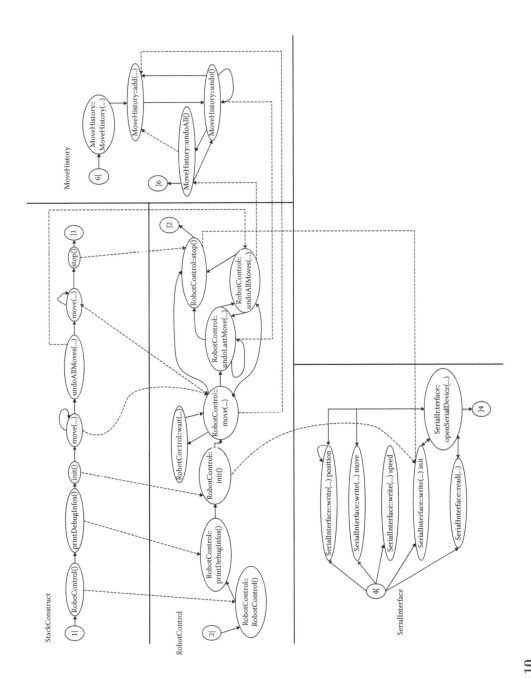

FIGURE 9.10

CSG of StackConstruct application of robot system (dashed lines represent calls between components).

```
int main( int argc, char *argv[])
{
    RoboControl* rc = new RoboControl(1,4);
    rc->printDebugInfos();

    rc->init();

    rc->move(M00,D1LH);
    rc->move(M01,D1LH);
    rc->move(M02,D1RH);
    rc->move(M10,D1LH);
    rc->move(M11,D2LH);
    rc->move(M12,D2LH);
    rc->move(M20,D2RH);
    rc->move(M21,D2RH);
    rc->move(M22,M22);
    rc->undoAllMoves();

    rc->stop();

    printf("shutting down...\n");
    return 0;
}
```

FIGURE 9.11
Source code of StackConstruct application.

components including connected invocation relations. Test set T_2 attains the coverage of the communication pair criterion. Test cases of T_2 are generated to cover sequences of length 2. This means that every CP of all units and IRs are covered. Test set T_3 fulfills the communication triple criterion. Test cases of T_3 are constructed to cover each communication triple, that is, sequences of length 3 of the robot system. The test sets T_4 and T_5 achieve the communication quadruple and quintuple criterion. The test cases are constructed to cover the robot system sequences of length 4 or 5.

The mutation adequacy of these test sets is evaluated by a mutation analysis in the following section.

9.3.4   Mutation analysis and results

Each of the six basic mutation operators of Section 9.2.4 was used to construct one mutant of each unit of the SUC (14 software units and 6 mutants each, plus an additional two for adding or deleting self-loops). These 112 mutants were then applied to test sets T_1, T_2, T_3, T_4, and T_5. Test generation is terminated if a higher coverage does not result in an increased mutation score. After the execution of the test cases of the test set T_5, all faults injected were revealed. Figure 9.12 summarizes the complete analysis by calculating the mutation score.

As a result of the case study, the mutation scores for the test sets T_1, T_2, T_3 , T_4, and T_5 improved with respect to their length of CS. Test set T_1 only detects the faults injected in the software unit *StackConstruct* (unit 1[) and its invocations, so this criterion is only applicable for systems having a simple invocation structure. While the length of the CSs increases, the CSs kill more mutants. T_2 detects all mutants of T_1 and faults injected in the *RobotControl* unit (unit 2[), including its invocations invoked by unit

Applied test sets based on coverage criteria	6 Mutation operators applied => 8 mutants generated of 14 units each	
		Mutation Score (MS)
T_1 Actor coverage criterion	112 mutants	0.0680
T_2 Communication pair criterion	112 mutants	0.2233
T_3 Communication triple criterion	112 mutants	0.6699
T_4 Communication quadruple criterion	112 mutants	0.9223
T_5 Communication quintuple criterion	112 mutants	1.0000

FIGURE 9.12
Results of mutation analysis.

StackConstruct (unit 1[). This continues through T_5 that then kills all mutants of the robot system.

Figure 9.13 shows the CS for detecting one mutant that is detectable by a test case of T_5. The test case revealed a perturbed invocation in the MH_Add software unit inserted by the mutation operator AddNod. The CS has the length of five to reach the mutation via the actors RobotControl::undoAllMoves(...), MoveHistory::undoAll(), MoveHistory::Add(...), t_moveCmd(), and insert(...).

Only the test case of T_5 detected this mutant because every call of the sequence provided a message to the next call. They were not given in T_4 to T_1.

9.3.5 Lessons learned

Modeling the SUC with CSG and analyzing the generated test cases using mutation analysis revealed some results that are summarized below.

Lesson 1. Use different abstraction levels for modeling SUC

As methods in classes contain several invocations, the overview of the system becomes unavailable when all invocations are drawn in one CSG of the system. The solution is to focus on the invocations of one software unit to all other units and to build several CSGs. Using abstract actors and their refinement in abstraction helps to keep a manageable view on the system.

Lesson 2. Use mutation analysis to determine the maximum length of the communication sequences for generating the test cases

Section 9.3.4 showed that all mutants were killed using the test cases of the set T_5. Consequently, this SUC needs at least the entire T_5 set to test the system thoroughly. In case when no faults can be found in the SUC by traditional testing, mutation analysis can also be used to find the maximum length of the CSs. The maximum length will be achieved if the mutation score reaches 100% by executing the test cases of the last generated test set.

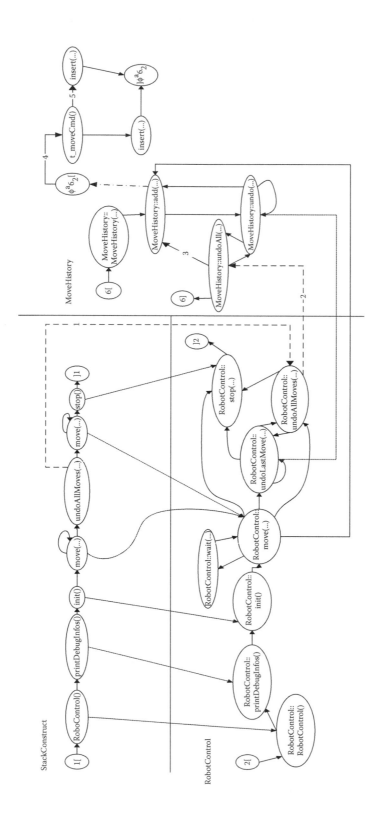

FIGURE 9.13

Communication sequence for detecting mutant *insert* (...) in component *MoveHistory*.

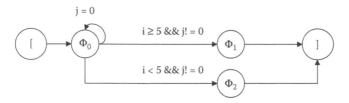

FIGURE 9.14
CSG augmented by Boolean expression.

9.4 Conclusions, Extension of the Approach, and Future Work

This chapter introduced CSG and CSG-related notions, which are used in this approach to ITest. Mutation analysis is used to evaluate the adequacy of generated test sets. The CSG of a robot system was modeled and the corresponding implementation was exemplified. The generated test sets for testing the SUC were applied to all mutants of the system according to Figure 9.4. The results are the following, (1) ITest can be performed in a communication-oriented manner by analyzing the communication between the different units, (2) CSG can be used to model the SUC for generating test cases, (3) different abstraction levels of the CSG help to keep the testing process manageable, and (4) mutation analysis helps to determine the maximum length of the CSs.

Ongoing research work includes augmenting CSG by labeling the arcs with Boolean expressions. This enables the consideration of *guards*, that are, conditions that must be fulfilled to invoke ϕ' after ϕ. This extension requires appropriate expansion of the selected test case generation algorithms (Algorithms 1 and 2). An example of a fragment of a CSG augmented by Boolean expressions is given in Figure 9.14.

Compared to *concolic unit testing* (Sen 2007), this approach is easier to apply to integration testing.

Similar to the *All-transitions criterion* specified on state-based models (Binder 1999), an *All-Invocations-criterion* for generating test cases could be introduced to the CSG that covers all the invocations directly as a testing goal. The test cases generated by Algorithm 2, however, already include these invocations. Therefore, a special *All-Invocations-criterion* is not needed.

At present, mutation operators reflect insertion and/or deletion of entities of the CSG. Apart from combining these basic operations in order to form operations of higher order, Boolean expressions should be included in the CSG concept. This also enables the consideration of further mutation operators.

References

Aho, A.V., Dahbura, A., Lee, D., and Uyar, M. (1991). An optimization technique for protocol conformance test generation based on UIO sequences and rural chinese postman tours. *IEEE Transactions on Communications*, Volume 39, Number 11, Pages: 1604–1615.

Belli, F., Budnik, C.J., and White, L. (2006). Event-based modelling, analysis and testing of user interactions: approach and case study. *Software Testing, Verification & Reliability*, Pages: 3–32.

Belli, F., Budnik, C.J., and Wong, W.E. (2006). Basic operations for generating behavioral mutants. *MUTATION '06: Proceedings of the Second Workshop on Mutation Analysis*, Page: 9. IEEE Computer Society, Los Alamitos, CA.

Belli, F., Hollmann, A., and Padberg, S. (2009). Communication sequence graphs for mutation-oriented integration testing, *Proceedings of the Workshop on Model-Based Verification & Validation*, Pages: 373–378. IEEE Computer Press, Washington, DC.

Binder, R.V. (1999). *Testing object-oriented systems: models, patterns, and tools.* Addison-Wesley Longman Publishing Co., Inc., Boston, MA.

Buy, U., Orso, A., and Pezze, M. (2000). Automated testing of classes. *ACM SIG-SOFT Software Engineering Notes*, Volume 25, Number 5, Pages: 39–48. ACM, New York, NY.

Daniels, F.J. and Tai, K.C. (1999). Measuring the effectiveness of method test sequences derived from sequencing constraints. *International Conference on Technology of Object-Oriented Languages*, Page: 74. IEEE Computer Society, Los Alamitos, CA.

Delamaro, M.E., Maldonado, J.C., and Mathur, A.P. (2001). Interface mutation: an approach for integration testing. *Transactions on Software Engineering, IEEE*, Volume 27, Number 3, Pages: 228–247. IEEE Press, Piscataway, NJ.

DeMillo, R.A., Lipton, R.J., and Sayward, F.G. (1978). Hints on test data selection: help for the practicing programmer. *IEEE Computer*, Volume 11, Number 4, Pages: 34–41.

Grove, D., DeFouw, G., Dean, J., and Chambers, C. (1997). Call graph construction in object-oriented languages. *Proceedings of the 12th ACM SIGPLAN Conference on Object-oriented Programming, Systems, Languages, and Applications*, Volume 32, Number 10, Pages: 108–124. ACM, New York, NY.

Hartmann, J., Imoberdorf, C., and Meisinger, M. (2000). UML-based integration testing. *ISSTA '00: Proceedings of the 2000 ACM SIGSOFT International Symposium on Software Testing and Analysis*, Pages: 60–70. ACM, New York, NY.

Hong, Z., Hall, P.A.V., and May, J.H.R. (1997) Software unit test coverage and adequacy. *ACM Computing Surveys*, Volume 29, Number 4, Pages: 366–427.

Hu, J., Ding, Z., and Pu, G. (2009). Path-based Approach to Integration Testing. *Proceedings of the Third IEEE International Conference on Secure Software Integration and Reliability Improvement*, Pages: 431–432. IEEE Computer Press, Washington, DC.

Martena, V., DiMilano, P., Orso, A., and Pezzè, M. (2002). Interclass testing of object oriented software. *Proceedings of the IEEE International Conference on Engineering of Complex Computer System*, Pages: 135–144. Georgia Institute of Technology, Washington, DC.

Myers, G.J. (1979). *Art of Software Testing.* John Wiley & Sons, Inc., New York, NY.

Offutt, A.J. (1992). Investigations of the software testing coupling effect. *ACM Transactions on Software Engineering and Methodology*, Pages: 5–20. ACM, New York, NY.

Saglietti, F., Oster, N., and Pinte, F. (2007). Interface coverage criteria supporting model-based integration testing. *Workshop Proceedings of 20th International Conference on*

Architecture of Computing Systems (ARCS 2007), Pages: 85–93. Berlin/Offenbach: VDE Verlag, University of Erlangen-Nuremberg, Erlangen, Germany.

Sen, K. (2007). Concolic testing. *Proceedings of the 22nd IEEE/ACM International Conference on Automated Software Engineering (ASE '07)*, Pages: 571–572. ACM, New York, NY.

Zhao, R. and Lin, L. (2006). An UML statechart diagram-based MM-path generation approach for object-oriented integration testing. *International Journal of Applied Mathematics and Computer Sciences*, Pages: 22–27.

10

A Model-Based View onto Testing: Criteria for the Derivation of Entry Tests for Integration Testing

Manfred Broy and Alexander Pretschner

CONTENTS

10.1 Introduction

In many application domains, organization, cost, and risk considerations continue to lead to increasingly distributed system and software development processes. In these contexts, suppliers provide components, or entire subsystems, that are assembled by system integrators. One prominent, and tangible, example for such a development paradigm is the automotive domain where multiple embedded systems are integrated into a car (see Pretschner et al. 2007, Reiter 2010). For reasons of economy, suppliers aim at selling their subsystems to as many car manufacturers (usually and somewhat counterintuitively called original equipment manufacturers, or OEMs, in this context) as possible. This requires that their components work correctly in a multitude of different environments, which motivates thorough testing (and specification) of the components of a car under development. Each OEM, on the other hand, wants to make sure that the external components work as expected in its particular cars.

To reduce the cost of integration, the OEM subjects the external component to a separate set of component tests before integrating the component with the rest of the car, and subsequently performing integration tests. This process comes by the name of *entry testing* for integration testing, and the rational management of this process is the subject of this chapter.

We tackle the following main problem. Assume an OEM orders some external component, to be integrated with the rest of its system, say a "residual" car that lacks this component (or several variants of such a "residual" car). Can we find criteria for test derivation that allows the OEM to reduce the overall cost of testing by pushing effort from the integration test for *the residual car that is composed with the component* to entry tests for *the component only*? In other words, is it possible to find circumstances and test criteria for the component that generalize to test criteria for the combination of the residual car and the component?

In response to this question, we present three contributions:

- First, we provide a formalized conceptual model that captures several testing concepts in the context of reactive systems, including the fundamental notions of module, integration, and system tests. In particular, we investigate the nature of test drivers and stubs for integrated embedded systems. As far as we know, no such set of precise definitions existed before.

- Second, we relate these concepts to the activities of the systems development process, thus yielding a comprehensive view of a development process for distributed embedded systems. This comprehensive view relies on the formal framework supporting both architecture and component specifications.

- Third, we show the usefulness of the formalized conceptual model by providing criteria for shifting effort from integration testing to component entry tests. We also investigate the benefits for the suppliers that have an interest in defining tests such that their components work correctly in all anticipated environments.

Our contributions provide further arguments for the attractiveness of model-based development processes. Moreover, the results generalize to other application domains. For instance, we see an analogous fundamental structure in service-oriented architectures, or the cloud: for a provider P (the integrator, or OEM) to provide a service S (the car), P relies on a set of different services $S_1, \ldots, S_n$ (components provided by the suppliers $H_1, \ldots, H_m$). Obviously, P wants to make sure that the supplied services perform as expected while only caring about the own service S (and its variants). The suppliers H_i, on the other hand, want to sell their services S_j to as many other parties as possible. They must hence find principles as to how to optimize the selection of their component tests.

This chapter consists of a conceptual and a methodological part and is structured as follows. We introduce the fundamental concepts of systems, interfaces, behaviors, composition, and architectures in Section 10.2. These are necessary to precisely define a model-based development process and the fundamental notions of architecture and component faults in Section 10.3. Since the focus of this paper is on testing, we use the framework of Sections 10.2 and 10.3 to introduce a few essential testing concepts in Section 10.4. In Section 10.5, we continue the discussion on the model-based development process of Section 10.3 by focusing on the integration testing phase and by explaining how to select component tests on the grounds of the architecture. Because these tests are essentially derived from a simulation of the subsystem to be tested, the tests are likely to reflect behaviors that usually are verified at integration time and are hence likely to identify faults that would otherwise surface only at integration testing time. We put our work in context and conclude in Section 10.6.

10.2 Background: Systems, Specifications, and Architectures

In this section, we briefly introduce the syntactic and semantic notion of a *system*, its *interface,* and that of a *component*. This theoretical framework is in line with earlier work (Broy and Stølen 2001). While this chapter is self-contained, knowledge of this reference work may help with the intuition behind some of the formalizations.

The fundamental concepts of system interfaces and system behaviors are introduced in Section 10.2.1. In Section 10.2.2, we show how to describe system behaviors by means of state machines. Section 10.2.3 introduces the notion of architectures that essentially prescribe how to compose subsystems. The formal machinery is necessary for the definition of a model-based development process in Section 10.2.3.4 and, in particular, for the precise definition of architecture and component faults.

10.2.1 Interfaces and behaviors

We start by shortly recalling the most important foundations on which we will base our model-based development process for multifunctional systems. We are dealing with models of *discrete systems*. A discrete system is a technical or organizational unit with a clear boundary. A discrete system interacts with its environment across this boundary by exchanging messages that represent discrete events. We assume that messages are exchanged via channels. Each instance of sending or receiving a message is a discrete event.

We closely follow the FOCUS approach described in Broy and Stølen (2001). Communication between components takes place via input and output channels over which streams of messages are exchanged. The messages in the streams received over the input channels represent the input events. The messages in the streams sent over the output channels represent the output events.

Systems have *syntactic interfaces* that are described by their sets of input and output channels. Channels are used for communication by transmitting messages and to connect systems. Channels have a type that indicates which messages are communicated over the channels. Hence, the syntactic interfaces describe the set of actions for a system that are possible at its interface. Each action consists in the sending or receiving of an instance of a message on a particular channel.

It is helpful to work with messages of different types. A type is a name for a data set, a channel is a name for a communication line, and a stream is a finite or an infinite sequence of data messages. Let TYPE be the set of all types. With each type $T \in$ TYPE, we associate the set CAR(T) of its data elements. CAR(T) is called the *carrier set* for the type T. A set of typed channels is a set of channels where a type is given for each of its channels.

Definition 1 (Syntactic Interface). Let I be a set of typed input channels and O be the set of typed output channels. The pair $(I \blacktriangleright O)$ denotes the *syntactic interface* of this system. For each channel $c \in I$ with type T_1 and each message $m \in$ CAR(T_1), the pair (m, c) is called an input message for the syntactic interface $(I \blacktriangleright O)$. For each channel $c \in O$ with type T_2 and each message $m \in$ CAR(T_2), the pair (m, c) is called an output message for the syntactic interface $(I \blacktriangleright O)$. $\square$

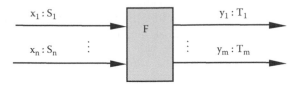

FIGURE 10.1
Graphical representation of a system F as a data flow node with its syntactic interface. The x_i are input channels of type S_i, and the y_j are output channels of type T_j. Channels x_i and y_i need not be ordered.

Figure 10.1 shows the system F with its syntactic interface in a graphical representation by a data flow node.

In FOCUS, a system encapsulates a state and is connected to its environment exclusively by its input and output channels. Streams of messages (see below) of the specified type are transmitted over channels.

A discrete system has a semantic interface represented by its interactive behavior. The behavior is modeled by a function mapping the streams of messages given on its input channels to streams of messages provided on its output channels. We call this the *black box behavior* or the *interface behavior* of discrete systems.

Definition 2 ([Nontimed] Streams). Let IN denote the natural numbers. Given a set M, by M*, we denote the set of finite sequences of elements from M. By M^∞, we denote the set of infinite sequences of elements of M that can be represented by functions $IN\backslash\{0\} \to M$. By M^ω, we denote the set $M^* \cup M^\infty$, called the set of finite and infinite (nontimed) *streams*. □

In the following, we work with streams that include discrete timing information. Such streams represent histories of communications of data messages transmitted within a time frame. To keep the time model simple, we choose a model of discrete time where time is structured into an infinite sequence of finite time intervals of equal length.

Definition 3 (Timed Streams). Given a message set M of data elements, we represent a *timed stream* s by a function

$$s : IN\backslash\{0\} \to M^*,$$

where M* is the set of finite sequences over the set M (which is the carrier set of the type of the stream). By $(M^*)^\infty$, we denote the set of *timed streams*. □

Intuitively, a timed stream maps abstract time intervals to finite sequences of messages. For a timed stream $s \in (M^*)^\infty$ and an abstract time interval $t \in IN\backslash\{0\}$, the sequence $s(t)$ of messages denotes the sequence of messages communicated within time interval t as part of the stream s.

We will later work with one simple basic operator on streams: $x{\downarrow}t$ denotes the prefix of length $t \in IN$ of the stream x (which is a sequence of length t carrying finite sequences as its elements; $x{\downarrow}0$ is the empty sequence).

A (timed) channel history for a set of typed channels C (which is a set of typed identifiers) assigns to each channel $c \in C$ a timed stream of messages communicated over that channel.

Definition 4 (Channel History). Let C be a set of typed channels. A (total) *channel history* is a mapping (let IM be the universe of all messages)

$$x : C \to (IN\backslash\{0\} \to IM^*)$$

such that x(c) is a stream of type Type(c) for each channel $c \in C$. We denote the set of all channel histories for the channel set C by $\vec{C}$ A finite (also called partial) channel history is a mapping

$$x : C \rightarrow (\{1, \ldots, t\} \rightarrow IM^*)$$

for some number $t \in IN$. $\qquad\square$

For each history $z \in \vec{C}$ and each time $t \in IN$, $z{\downarrow}t$ yields a finite history for each of the channels in C represented by a mapping of the type $C \rightarrow (\{1, \ldots, t\} \rightarrow IM^*)$.

For a given syntactic interface $(I \blacktriangleright O)$, the behavior of a system is defined by a relation that relates the input histories in $\vec{I}$ with the output histories in $\vec{O}$. This way, we get a (nondeterministic) functional model of a system behavior.

For reasons of compositionality, we require behavior functions to be causal. Causality assures a consistent time flow between input and output histories in the following sense: in a causal function, input messages received at time t do only influence output at times $\geq$t (in the case of strong causality at times $\geq t+1$, which indicates that there is a delay of at least one time interval before input has effect on output). A detailed discussion is contained in earlier work (Broy and Stølen 2001).

Definition 5 (I/O-Behavior). Let $\wp(X)$ denote the powerset of set X. A strongly causal function F: $\vec{I} \rightarrow \wp(\vec{O})$ is called *I/O-behavior*. By IF[I $\blacktriangleright$ O], we denote the set of all (total and partial) I/O-behaviors with syntactic interface $(I \blacktriangleright O)$, and by IF, the set of all I/O-behaviors. $\qquad\square$

Definition 6 (Refinement, Correctness). The black box behavior, also called interface behavior of a system with syntactic interface $(I \blacktriangleright O)$ is given by an I/O-behavior F from IF[I $\blacktriangleright$ O]. Every behavior F' in IF[I $\blacktriangleright$ O] with

$$F'(x) \subseteq F(x)$$

for all $x \in \vec{I}$ is called a refinement of F. A system implementation is correct w.r.t. the specified behavior F if its interface behavior is a refinement of F. $\qquad\square$

10.2.2 State machines and interface abstractions

A system is any syntactic artifact; the semantics of which are defined as or can be mapped to an interface behavior as described above. Examples for systems include those specified by state machines or FOCUS formulae. Systems interact with their environment via their interfaces. Each system can be used as a component of a larger system, and each component is a system by itself. Components and systems can be composed to form larger systems. The composition of systems consists of connecting output channels of one component to one or more input channels of another component. In case of feedback to the same component, causality problems may arise that can be solved by adding delay, or latch, components (Broy and Stølen 2001). While components can of course be broken down hierarchically in a top-down development approach, it is sensible to speak of *atomic components* when a bottom-up development approach is favored: atomic components are those that are not the result of composing two or more existing components.

It is sometimes more convenient to specify atomic components as state machines rather than by relations on streams. However, by virtue of interface abstractions, the former can directly be transformed into the latter.

10.2.3 Describing systems by state machines

In this section, we introduce the concept of a state machine with input and output that relates well to the introduced concept of interface. It will be used as model representing implementations of systems.

Definition 7 (State Machine with Input and Output). Given a state space Σ, a state machine (Δ, Λ) with input and output according to the syntactic interface $(I \blacktriangleright O)$ with messages over some set M consists of a set $\Lambda \subseteq \Sigma$ of initial states as well as of a state transition function

$$\Delta \colon (\Sigma \times (I \to M^*)) \to \wp(\Sigma \times (O \to M^*))$$

By $\mathrm{SM}[I \blacktriangleright O]$, we denote the set of all state machines.   □

For each state $\sigma \in \Sigma$ and each valuation $a \colon I \to M^*$ of the input channels in I by sequences, we obtain a set of state transitions. Every pair $(\sigma', b) \in \Delta(\sigma, a)$ represents a successor state σ' and a valuation $b \colon O \to M^*$ of the output channels. The channel valuation b consists of the sequences produced by the state transition as output. (Δ, Λ) is a state machine with possibly infinite state space.

As shown in Broy (2007a) and Broy (2007b), every such state machine describes an I/O-behavior for each state of its state space. Conversely, every I/O-behavior can be modeled by a state machine with input and output. Partial machines describe services that are partial I/O-behaviors. As shown in Broy (2007a) and (Broy 2007b), there is a duality between state transition machines with input and output and I/O-behaviors. Every state machine specifies an I/O-behavior and every I/O-behavior represents and can be represented by a state machine. Therefore, from a theoretical point of view, there is no difference between state machines and I/O-behaviors. I/O-behaviors specify the set of state machines with identical interface behaviors.

10.2.4 From state machines to interface behaviors

Given a state machine, we may perform an interface abstraction. It is given by the step from the state machine to its interface behavior.

Definition 8 (Black Box Behavior and Specifying Assertion). Given a state machine $A = (\Delta, \Lambda)$, we define a behavior F_A as follows (let Σ be the state space for A)

$$F_A(x) = \{y \in \overrightarrow{O}\} : \exists\, \sigma : \mathrm{IN} \to \Sigma : \sigma(0) \in \Lambda \wedge \forall\, t \in \mathrm{IN} : (\sigma(t+1), y.(t+1)) \in \Delta(\sigma(t), x.(t+1))\}.$$

Here for $t \in \mathrm{IN} \backslash \{0\}$, we write x.t for the mapping in $I \to M^*$ with

$$(x.t)(c) = (x(c))(t)$$

for $c \in I$. F_A is called the black box behavior for A and the logical expression that is equivalent to the proposition $y \in F_A(x)$ is called the *specifying assertion*.   □

F_A is causal by construction. If A is a Moore machine (i.e., the output depends on the state only), then F_A is strongly causal.

State machines can be described by state transition diagrams or by state transition tables.

10.2.5 Architectures and composition

In this section, we describe how to form architectures from subsystems, called the components of the architecture. Architectures are concepts to build systems. Architectures contain precise descriptions of how the composition of their subsystems takes place. In other words, architectures are described by the sets of systems forming their components together with mappings from output to input channels that describe internal communication. In the following, we assume that each system used in architecture as a component, which has a unique identifier k. Let K be the set of names for the components of an architecture.

Definition 9 (Set of Composable Interfaces). A set of component names K with a finite set of interfaces $(I_k \blacktriangleright O_k)$ for each $k \in K$ is called *composable*, if

1. the sets of input channels I_k, $k \in K$, are pairwise disjoint,

2. the sets of output channels O_k, $k \in K$, are pairwise disjoint,

the channels in $\{c \in I_k : k \in K\} \cap \{c \in O_k : k \in K\}$ have the same channel types in $\{c \in I_k : k \in K\}$ and $\{c \in O_k : k \in K\}$. □

If channel names are not consistent for a set of systems to be used as components, we simply may rename the channels to make them consistent.

Definition 10 (Syntactic Architecture). A syntactic architecture $A = (K, \xi)$ with interface $(I_A \blacktriangleright O_A)$ is given by a set K of component names with composable syntactic interfaces $\xi(k) = (I_k \blacktriangleright O_k)$ for $k \in K$.

1. $I_A = \{c \in I_k : k \in K\} \backslash \{c \in O_k : k \in K\}$ denotes the set of *input* channels of the architecture,

2. $D_A = \{c \in O_k : k \in K\}$ denotes the set of *generated* channels of the architecture,

3. $O_A = D_A \backslash \{c \in I_k : k \in K\}$ denotes the set of *output* channels of the architecture,

4. $D_A \backslash O_A$ denotes the set of *internal* channels of the architecture,

5. $C_A = \{c \in I_k : k \in K\} \cup \{c \in O_k : k \in K\}$ the set of all channels.

By $(I_A \blacktriangleright D_A)$, we denote the *syntactic internal interface* and by $(I_A \blacktriangleright O_A)$, we denote the *syntactic external interface* of the architecture. □

A syntactic architecture forms a directed graph with its components as its nodes and its channels as directed arcs. The input channels in I_A are ingoing arcs and the output channels in O_A are outgoing arcs.

Definition 11 (Interpreted Architecture). An interpreted architecture (K, ψ) for a syntactic architecture (K, ξ) associates an interface behavior $\psi(k) \in IF[I_k \blacktriangleright O_k]$ with every component $k \in K$, where $\xi(k) = (I_k \blacktriangleright O_k)$. □

In the following sections, we define an interface behavior for interpreted architectures by composing the behaviors of the components.

10.2.6 Glass box views onto interpreted architectures

We first define composition of composable systems. It is the basis for giving semantic meaning to architectures.

Definition 12 (Composition of Systems—Glass Box View). For an interpreted architecture A with syntactic internal interface $(I_A \blacktriangleright D_A)$, we define the glass box interface behavior $[\times]A \in IF[I_A \blacktriangleright D_A]$ by the equation (let $\psi(k) = F_k$):

$$([\times]A)(x) = \{y \in \overrightarrow{D}_A : \exists z \in \overrightarrow{C}_A : x = z|I_A \wedge y = z|D_A \wedge \forall k \in K : z|O_k \in F_k(z|I_k)\},$$

where | denotes the usual restriction operator. Internal channels are not hidden by this composition, but the streams on them are part of the output. The formula defines the result of the composition of the k behaviors F_k by defining the output y of the architecture $[\times]$ A with the channel valuation z of all channels. The valuation z carries the input provided by x expressed by $x = z|I_A$ and fulfills all the input/output relations for the components expressed by $z|O_k \in F_k(z|I_k)$. The output of the composite system is given by y which the restriction $z|D_A$ of z to the set D_A of output channels of the architecture $[\times]$ A.   □

For two composable systems $F_k \in IF[I_k \blacktriangleright O_k]$, k = 1, 2, we write

$$F_1 \times F_2$$

for $[\times]$ $\{F_k: k = 1, 2\}$. Composition of composable systems is commutative

$$F_1 \times F_2 = F_2 \times F_1$$

and associative

$$(F_1 \times F_2) \times F_3 = F_1 \times (F_2 \times F_3).$$

The proof of this equation is straightforward. We also write therefore with $K = \{1, 2, 3, \ldots\}$

$$[\times]\{F_k \in IF[I_k \blacktriangleright O_k] : k \in K\} = F_1 \times F_2 \times F_3 \times \cdots.$$

From the glass box view, we can derive the black box view as demonstrated in the following chapter.

10.2.7 Black box views onto architectures

The black box view of the interface behavior of an architecture is an abstraction of the glass box view.

Definition 13 (Composition of Systems—Black Box View). Given an interpreted architecture with syntactic external interface $(I_A \blacktriangleright O_A)$ and glass box interface behavior $[\times]$ $A \in IF[I_A \blacktriangleright D_A]$, we define the black box interface behavior $F_A \in IF[I_A \blacktriangleright O_A]$ by

$$F_A(x) = (F(x))|O_A$$

Internal channels are hidden by this composition and in contrast to the glass box view not part of the output.   □

For an interpreted architecture with syntactic external interface $(I_A \blacktriangleright O_A)$, we obtain the black box interface behavior $F_A \in IF[I_A \blacktriangleright O_A]$ specified by

$$F_A(x) = \{y \in \overrightarrow{O}_A : \exists\, z \in \overrightarrow{C}_A : x = z|I_A \wedge y = z|O_A \wedge \forall\, k \in K : z|O_k \in F_k(z|I_k)\}$$

and write

$$F_A = \otimes\{F_k \in IF[I_k \blacktriangleright O_k] : k \in K\}.$$

For two composable systems $F_k \in IF[I_k \blacktriangleright O_k]$, k = 1, 2, we write

$$F_1 \otimes F_2$$

for $\otimes\{F_1, F_2\}$ Composition of composable systems is commutative

$$F_1 \otimes F_2 = F_2 \otimes F_1$$

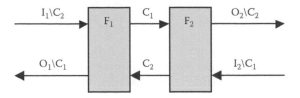

FIGURE 10.2
Composition $F_1 \otimes F_2$.

and associative

$$(F_1 \otimes F_2) \otimes F_3 = F_1 \otimes (F_2 \otimes F_3).$$

The proof of this equation is straightforward. We also write therefore with $K = \{1, 2, 3, \ldots\}$

$$\otimes \{F_k \in IF[I_k \blacktriangleright O_k] : k \in K\} = F_1 \otimes F_2 \otimes F_3 \otimes \cdots.$$

The idea of the composition of systems as defined above is shown in Figure 10.2 with $C_1 = I_2 \cap O_1$ and $C_2 = I_1 \cap O_2$. For properties of the algebra, we refer the reader to Broy and Stølen (2001) and Broy (2006).

In a composed system, the internal channels are used for internal communication.

Given a syntactic architecture $A = (K, \xi)$ and specifying assertions S_k for the systems $k \in K$, the specifying assertion for the glass box behavior is given by $\forall k \in K: S_k$, and for the black box behavior by $\exists c_1, \ldots, c_j: \forall k \in K: S_k$, where $\{c_1, \ldots, c_j\}$ denotes the set of internal channels.

The set of systems together with the introduced composition operators form an algebra. The composition of systems (strongly causal stream processing functions) yields systems and the composition of services yields services.

Composition is a partial function on the set of all systems. It is only defined if the syntactic interfaces fit together. Syntactic interfaces fit together if there are no contradictions in the channel names and types.

Since it ignores internal communication, the black box view is an abstraction of the glass box view of composition.

10.2.8   Renaming

So far, we defined the composition using the names of components to connect them only for sets of components that are composable in the sense that their channel names and types fit together. Often, the names of the components may not fit. Then, renaming may help.

Definition 14 (Renaming Components' Channels). Given a component $F \in IF$ $[I \blacktriangleright O]$, a renaming is a pair of mappings $\alpha\colon I' \to I$ and $\beta\colon O \to O'$, where the types of the channels coincide in the sense that c and $\alpha(c)$ as well as e and $\beta(e)$ have the same types for all $c \in I$ and all $e \in O$. By a renaming $\rho = (\alpha, \beta)$ of F, we obtain a component $\rho[F] \in IF$ $[I' \blacktriangleright O']$ such that for $x \in \overrightarrow{I}'$

$$\rho[F](x) = \beta(F(\alpha(x))),$$

where for $x \in \overrightarrow{I}'$ the history $\alpha(x) \in \overrightarrow{I} \, prime$ is defined by

$$\alpha(x)(c) = x(\alpha(c))$$

for $c \in I$. $\square$

Note that by a renaming, a channel in I' or O may be used in several copies in I or O'. Given an interpreted architecture $A = (K, \psi)$ with a set of components $\psi(k) = F_k \in [I_k \blacktriangleright O_k]$ for $k \in K\}$ and a set of renamings $R = \{\rho_k : k \in K\}$, where ρ_k is a renaming of F_k for all $k \in K$, we call (A, R, ψ) an interpreted architecture with renaming if the set $\{\rho_k[F_k] : k \in K\}$ is well defined and composable. The renamings R define the connections that make A an architecture.

10.2.9   Composing state machines

A syntactic architecture forms a directed graph with its components as its nodes and its channels as directed arcs. The input channels in I_A are ingoing arcs and the output channels in O_A are outgoing arcs.

Definition 15 (Architecture Implemented by State Machines). An implemented architecture (K, ζ) of a syntactic architecture (K, ξ) associates a state machine $\zeta(k) = (\Delta_k, \Lambda_k) \in SM[I_k \blacktriangleright O_k]$ with every $k \in K$, where $\xi(k) = (I_k \blacktriangleright O_k)$.   □

In the following sections, we define an interface behavior for interpreted architectures by composing the behaviors of the components.

Next, we define the composition of a family of state machines $R_k = (\Delta_k, \Lambda_k) \in SM[I_k \blacktriangleright O_k]$ for the syntactic architecture $A = (K, \xi)$ with interface $(I_A \blacktriangleright O_A)$ with $\xi(k) = (I_k \blacktriangleright O_k)$. It is the basis for giving semantic meaning to implementations of architectures.

Definition 16 (Composition of State Machines—Glass Box View). For an implemented architecture $R = (K, \zeta)$ for a syntactic architecture $A = (K, \xi)$, we define the composition $(\Delta_R, \Lambda_R) \in SM[I_A \blacktriangleright D_A]$ by the equation (let $\zeta(k) = (\Delta_k, \Lambda_k)$ with state space Σ_k): The state Σ_R is defined by the direct product (let for simplicity $K = \{1, 2, 3, \dots\}$)

$$\Sigma_R = \Sigma_1 \times \Sigma_2 \times \Sigma_3 \times \cdots,$$

the initial state is defined by

$$\Lambda_R = \Lambda_1 \times \Lambda_2 \times \Lambda_3 \times \cdots,$$

and the state transition function Δ is defined by

$$\Delta R(\sigma, a) = \{(\sigma', b) : \exists z : C \to M^* : b = z|DA \wedge a = z|IA \wedge \forall k \in K :$$
$$(\sigma'k, z|Ok) \in \Delta k(\sigma k, z|Ik)\}.$$

Internal channels are not hidden by this composition, but their messages on them are part of the output.   □

Based on the implementation, we can talk about tests in the following section.

10.3   Model-Based Development: Specification and Implementation

In the previous sections, we have introduced a comprehensive set of modeling concepts for systems. We can now put them together in an integrated system description approach.

When building a system, in the ideal case, we carry out the following steps that we will be able to cast in our formal framework:

1. System specification
2. Architecture design
   a. Decomposition of the system into a syntactic architecture
   b. Component specification (enhancing the syntactic to an interpreted architecture)
   c. Architecture verification
3. Implementation of the components
   a. (Ideally) code generation
   b. Component (module) test and verification
4. Integration
   a. System integration
   b. Component entry test
   c. Integration test and verification
5. System test and verification

A *system specification* is given by a syntactic interface $(I \blacktriangleright O)$ and a specifying assertion S (i.e., a set of properties), which specifies a system interface behavior $F \in IF[I \blacktriangleright O]$.

An *architecture specification* is given by a composable set of syntactic interfaces $(I_k \blacktriangleright O_k)$ for component identifiers $k \in K$ and a component specification S_k for each $k \in K$. Each specification S_k specifies a behavior $F_k \in IF[I_k \blacktriangleright O_k]$. In this manner, we obtain an interpreted architecture.

The *architecture specification is correct* w.r.t. the system specification F if the composition of all components results in a behavior that refines the system specification F. Formally, the architecture is correct if for all input histories $x \in \overrightarrow{I}$,

$$\otimes\{F_k : k \in K\}(x) \subseteq F(x).$$

Given an *implementation* R_k for each component identifier $k \in K$, the implementation R_k with interface abstraction F'_k is correct if for all $x \in \overrightarrow{I}_k$ we have:

$$F'_k(x) \subseteq F_k(x)$$

(note that it does not matter if F'_k was generated or implemented manually). Then, we can integrate the implemented components into an implemented architecture

$$F' = \otimes\{F'_k : k \in K\}.$$

The following basic theorem of modularity is easily proved by the construction of composition (for details see Broy and Stølen 2001).

Theorem 1. Modularity. If the architecture is correct (i.e., if $\otimes\{F_k: k \in K\}(x) \subseteq F(x)$) and if the components are correct (i.e., $F'_k(x) \subseteq F_k(x)$ for all k), then the *implemented system is correct*:

$$F'(x) \subseteq F(x) \quad \text{for all } x \in \overrightarrow{I}. \qquad \square$$

A system (and also a subsystem) is hence called *correct* if the interface abstraction of its implementation is a refinement of its interface specification.

Before we consider the missing steps (4) and (5) of the development process in more detail in Sections 10.4 and 10.5, it is worthwhile to stress that we clearly distinguish between

1. the architectural design of a system, and
2. the implementation of the components of an architectural design.

An architectural design consists in the identification of components, their specification, and the way they interact and form the architecture. If the architectural design and the specification of the constituting components are sufficiently precise, then we are able to determine the result of the composition of the components of the architecture, according to their specification, even without providing an implementation of all components! If the specifications address behavior of the components and the design is modular, then the behavior of the architecture can be derived from the behavior of the components and the way they are connected. In other words, in this case, the architecture has a specification and a—derived—specified behavior. This specified behavior can be put in relation with the requirements specification for the system, and, as we will discuss later, also with component implementations.

The above process includes two steps of verification, *component verification* and *architecture verification*. These possibly reveal *component faults* (of a component/subsystem w.r.t. its specification) and *architecture faults* (of an architecture w.r.t. the system specification). If both verification steps are performed sufficiently carefully and the theory is modular, which holds here (see Broy and Stolen 2001), then correctness of the system follows from both verification steps.

The crucial point here is that architecture verification w.r.t. the system specification is enabled *without the need for actual implementations of the components*. In other words, it becomes possible before the implemented system exists. The precise implementation of the verification of the architecture depends of course on how its components are specified. If the specification consists of state machines, then the architecture can be simulated, and simulation results compared to the system specification. In contrast, if the component specifications are given by descriptive specifications in predicate logic, then deductive verification becomes possible.

Furthermore, if we have a hierarchical system, then the scheme of specification, design, and implementation can be iterated for each subhierarchy. An idealized top-down development process then proceeds as follows. We obtain a requirement specification for the system and from this, we derive an architectural design and specification. This results in specifications for components that we can take as requirements specifications for the subsequent step in which the components are designed and implemented. Given a specified architecture, test cases can be derived for integration test.

Given component specifications, we implement the components with the specifications in mind and then verify them with respect to their specifications. This of course entails some methodological problems if the code for the components has been generated from the specification in which case only the code generator and/or environment assumptions can be checked, as described in earlier work (Pretschner and Philipps 2005).

Now, if we have an implemented system for a specification, we can have either errors in the architecture design—in which case the architecture verification would fail—or we can have errors in the component implementation. An obvious question is that of the *root cause* of an architecture. Examples of architecture errors include

1. Connecting an output port to an incorrect input port and to forget about such a connection.
2. To have a mismatch in provided and expected sampling frequency of signals.

3. To have a mismatch in the encoding.

4. To have a mismatch in expected and provided units (e.g., km/h instead of m/s).

One fundamental difference between architecture errors and component errors of course is liability: in the first case, the integrator is responsible, while in the second case, responsibility is with the supplier.*

Assume a specified architecture to be given. Then, a *component fault* is a mismatch between the component specification, which is provided as part of the architecture, and the component implementation. An *architecture fault* is a mismatch between the behavior as defined by the architecture and the overall system specification. In an integrated system, we are hence able to distinguish between component faults and architecture faults.

With the outlined approach, we gain a number of interesting options to make the entire development process more precise and controllable. First of all, we can provide the architecture specification by a model, called the *architecture model*, where we provide a possibly nondeterministic state machine for each of the components. In this case, we can even simulate and test the architecture before actually implementing it. A more advanced and ambitious idea would be to provide formal specifications for each of the components. This would allow us to verify the architecture by logical techniques since the component specifications can be kept very abstract at the level of what we call a logical architecture. Such a verification could be less involved than it would be, if it were performed at a concrete implementation level.

Moreover, by providing state machines for each of the components, we may simulate the architecture. Thus, we can on the one hand test the architecture by integration tests in an early stage, and we can moreover generate integration tests from the architecture model to be used for the integration of the implemented system, as discussed below. The same is possible for each of the components with given state machine descriptions from which we can generate tests. We can, in fact, logically verify the components. Given state machines for the components, we can automatically generate hundreds of test cases as has been shown in Pretschner et al. (2005). For slightly different development scenarios, this leads to a fully automatic test case generation procedure for the component implementations.

10.4 Testing Systems: Preliminaries

We are now ready to formally define central testing notions and concepts. In Section 10.4.1, we define tests and related concepts as such. In Section 10.4.2, we show how to formally relate requirements to test cases.

10.4.1 System tests

A system test describes an instance of a finite system behavior. A system test case is given by a pair of finite histories. Such a pair is also called a *scenario*.

Definition 17 (System Test Case). Given a syntactic interface $(I \blacktriangleright O)$, a system test case till time $t \in IN$ is a pair $(x \downarrow t, \{y_1 \downarrow t, y_2 \downarrow t, \ldots, y_n \downarrow t\})$ for histories $x \in \overrightarrow{I}$ and $y_1, y_2, \ldots, y_n \in \overrightarrow{O}$. The finite history $x \downarrow t$ is called the *stimulus* and set $\{y_1 \downarrow t, y_2 \downarrow t, \ldots, y_n \downarrow t\}$ is called the *anticipation* that is used as oracle for the test. □

---

*Both architecture and component errors can be a result of an invalid specification and an incorrect implementation. This distinction touches the difference between validation and verification. We may safely ignore the case of invalid specifications (i.e., validation) in this chapter.

The anticipation specifies the set of correct outputs.

Definition 18 (Test Suite). A test suite is a set of test cases.   □

Before we turn our attention to the definition of nondeterministic tests, we define what it means for a system to pass a test.

Definition 19 (Passing and Failing Tests). A positive test is a test where we expect the system behavior to match the anticipation. A negative test is a test where we expect the system behavior *not* to match the anticipation. Given a system with behavior F ∈ IF [I ▶ O] and a system test (a, B) till time t ∈ IN, we say that the system behavior F *passes* a (positive) test if there *exist* histories x ∈ $\overrightarrow{I}$ and y ∈ $\overrightarrow{O}$ with y ∈ F(x) and a = x↓t and y↓t ∈ B. Then, we write

$$\text{pt}(F, (a, B)).$$

Otherwise, we say that F fails the test. The system F passes the test *universally* if for all histories x ∈ $\overrightarrow{I}$ and *all* y ∈ $\overrightarrow{O}$ with y ∈ F(x) and a = x↓t, we get y↓t ∈ B. Then we write

$$\text{ptu}(F, (a, B))$$

We say that the system *passes a negative* test (a, B) if there exist x ∈ $\overrightarrow{I}$ and y ∈ $\overrightarrow{O}$ with y ∈ F(x) and a = x↓t such that y↓t ∉ B. It passes a negative test (a, B) universally if for all y ∈ $\overrightarrow{I}$ and y ∈ $\overrightarrow{O}$ with y ∈ F(x) and a = x↓t, it holds that y↓t ∉ B.   □

In general, we test, of course, not interface behaviors but implementations. An implementation of a system with syntactic interface (I ▶ O) is given by a state machine A = (Δ, Λ) ∈ SM[I ▶ O]. The state machine passes a test if its interface abstraction F_A passes the test.

The decision to define anticipations as sets rather than as singletons is grounded in two observations, one relating to abstraction and one relating to nondeterminism.

In terms of abstraction, it is not always feasible or desirable to specify the expected outcome in full detail (Utting, Pretschner, and Legeard 2006)—otherwise, the oracle of the system would become a full-fledged fully detailed model of the system under test. In most cases, this is unrealistic because of cost considerations. Hence, rather than precisely specifying one *specific value*, test engineers specify *sets of values*. This is witnessed by most assertion statements in xUnit frameworks, for instance, where the assertions usually consider only a subset of the state variables, and then usually specify sets of possible values for these variables (e.g., greater or smaller than a specific value). Hence, one reason for anticipations being set is cost effectiveness: to see if a system operates correctly, it is sufficient to see if the expected outcome is in a given range.

The second reason is related to nondeterministic systems. Most distributed systems are nondeterministic—events happen in different orders and at slightly varying moments in time; asynchronous bus systems nondeterministically mix up the order of signals—and the same holds for continuous systems—trajectories exhibit jitter in the time and value domains.

Testing nondeterministic systems of course is notoriously difficult. Even if a system passes a test case, there may exist runs that produce output that is not specified in the anticipation. Vice versa, if we run the system with input a = x↓t and it produces some y ∈ F(x) with y↓t ∉ B, we cannot conclude that the system does not pass the test (but we know that it does not pass it universally). Hence, from a practical perspective, the guarantees that are provided by a test suite are rather weak in the nondeterministic case (but this is the nature of the beast, not of our conceptualization).

However, from a practical perspective, in order to cater to jitter in the time and value domains as well as to valid permutations of events, it is usually safe to assume that the actual testing infrastructure takes care of this (Prenninger and Pretschner 2004): at the model level, test cases assume deterministic systems, whereas at the implementation level, systems can be nondeterministic as far as jitter and specific event permutations are concerned.

A deterministic system that passes a test always passes the test universally. Moreover, if a system passes a test suite (a set of tests) universally, this does not mean that the system is deterministic—it is only deterministic as far as the stimuli in the test suite are concerned.

10.4.2 Requirements-based tests

Often it is recommended to produce test cases when documenting requirements. This calls for a consideration of the coverage of requirements.

A functional requirement specifies the expected outcome for *some* input streams in the domain of a system behavior. Hence a functional requirement for a system (a set of which can form the system specification) with a given syntactic interface is a predicate

$$R : (\overrightarrow{I} \to \wp(\overrightarrow{O})) \to \{\text{true}, \text{false}\}.$$

A test (a, B) is called *positively relevant* for a requirement if every system behavior F that does not pass the test universally does not fulfill the requirement R. Or expressed positively, if F fulfills requirement R, then it passes the test universally. This is formally expressed by

$$R(F) \Rightarrow \text{ptu}(F, (a, B)).$$

A test (a, B) is called *negatively relevant* for a requirement if every system behavior F that does pass the test universally does not fulfill the requirement R. Or expressed positively, if F fulfills requirement R, then it does not pass the test universally. This is formally expressed by

$$R(F) \Rightarrow \neg \text{ptu}(F, (a, B)).$$

Two comments are in order here. First, note that in this context, F denotes the set of possible executions of an *implemented system*. This is different from the *specification* of a system. In the context of a nondeterministic system, F is a "virtual" artifact as it can be obtained concretely. It is nevertheless necessary to define relevant concepts in the context of testing.

Second, the intuition behind these definitions becomes apparent when considering their contraposition, as stated in the definitions. Positive relevance means that if a test does not pass (universally), then the requirement is not satisfied. This seems like a very natural requirement on "useful" test cases, and it will usually come with a *positive* test case. Negative relevance, in contrast, means that if a test passes universally, then the requirement is not satisfied which, as a logical consequence, is applicable in situations where negative tests are considered.

At least from a theoretical perspective, it is perfectly possible to consider dual versions of relevance that we will call significance.

A test (a, B) is called *positively significant* for a requirement if every system behavior F that does not fulfill the requirement R does not pass the test universally. Or expressed positively, if F passes the test universally, then it fulfills requirement R. This is formally expressed by

$$\text{ptu}(F, (a, B)) \Rightarrow R(F).$$

Again, by contraposition, significance stipulates that if a requirement is not satisfied, then the test does not pass. The fact that this essentially means that correctness of a system w.r.t. a stated requirement can be proved by testing demonstrates the limited practical applicability of the notion of significance, except maybe for specifications that come in the form of (existentially interpreted [Krüger 2000]) sequence diagrams.

For symmetry, a test (a, B) is called _negatively significant_ for a requirement if every system behavior F that does not pass the test universally fulfills the requirement R. This is formally expressed by

$$\neg ptu(F, (a, B)) \Rightarrow R(F).$$

Of course, in practice, a significant test is only achievable for very simple requirements.

Among other things, testing can be driven by fault models rather than by requirements. Fault-based tests can be designed and run whenever there is knowledge of a class of systems. Typical examples include limit-value testing or stuck-at-1 tests. The idea is to identify those situations that are typically incorrectly developed. These "situations" can be of a syntactic nature (limit tests), can be related to a specific functionality ("we always get this wrong"), etc. In our conceptual model, fault-based tests correspond to tests for requirements where the requirement stipulates that the system is brought into a specific "situation."

10.5 Model-Based Integration Testing

We can now continue our discussion of the model-based development process sketched in Section 10.2.3.4. In Section 10.5.1, we use our formal framework to describe integration tests. In Section 10.5.2, we highlight the beneficial role of executable specifications that, in addition to being specifications, can be used for test case generation and also as stubs. In Section 10.5.3, we argue that these models, when used as environment model for a component to be tested, can be used to guide the derivation of tests that reflect the integration scenario. In Section 10.5.4, we propose a resulting testing methodology that we discuss in Section 10.5.5.

10.5.1 Integration tests

The steps from syntactic architecture $A = (K, \xi)$ with interface $(I_A \blacktriangleright O_A)$ and an implemented architecture $R = (K, \zeta)$ to a system (Δ_R, Λ_R) are called _integration_. The result of integration is an interpreted architecture $B = (K, \psi)$ with $\psi(k) = F_{\zeta(k)}$.

Integration can be performed in a single step (called big-bang) by composing all components at once. It can also be performed incrementally by choosing an initial subset of components to be tested and then add some more components to be tested, etc., until the desired system is obtained.

In practice, incremental integration requires the implementation of stubs and drivers. Traditionally, drivers are components that provide input to the system to be tested (mainly used in bottom-up integration). Stubs are components that provide interfaces and serve as dummies for the functionality of those components that are required for the system under test calls but that are not implemented yet.

In our context, the distinction between stubs and drivers is immaterial. At the level of abstraction that we consider here, we do not have a notion of functions that are called. In addition, the purpose of a stub is to provide input (i.e., a return value) to the calling function. In other words, all that is necessary is a component _that provides input_ (perhaps depending on some output received) to: (1) the top-level interface of a system, (2) all those channels that have not yet been connected to components (because these components are not part of the current incomplete architecture). Hence, the only part that matters is the input part—which we can easily encode in a test case! Assuming that developers have access to all channels in the system, we simply move the internal channels for which input must be provided to the external interface of the system.

An integration test consists of two parts:

1. A strategy that determines in which order sets of components are integrated with the current partial system under test (a single-stage big-bang, top-down, bottom-up, ...).

2. A set of finite histories for the (external) input channels and all those internal (output) channels of the glass box view of the architecture that are directly connected to the current partial system under test.

Definition 20 (Integration Strategy). Let S be a system built by an architecture with the set of components that constitute the final glass box architecture with the set of component identifiers K. Any set $\{K_1, \ldots, K_j\}$ with $K_1 \subset K_2 \cdots \subseteq K_j = K$ is called an *incremental integration strategy*. □

Given a syntactic architecture $A = (K, \xi)$ with interface $(I_A \blacktriangleright O_A)$ and an implemented architecture $R = (K, \zeta)$, an integration strategy determines a family of syntactic architectures $(K_i, \xi|K_i)$ with implemented architectures $R_i = (K_i, \zeta|K_i)$. We arrive at a family of interpreted architectures $B_i = (K_i, \psi|K_i)$ with $\psi(k) = \Gamma_{\zeta(k)}$ and interface behaviors F_i. An interesting question is what the relations between the F_i are. In general, these are not refinement relations.

Definition 21 (Integration Test Set). Let all definitions be as above. We define behaviors $S_j = [\times] \{F_k \in IF[I_k \blacktriangleright O_k]: k \in K_i \}$ with external interface $(I_j \blacktriangleright O_j)$ be a glass box architecture that is to be tested in iteration i, where $1 \leq i \leq j$. A set of system test cases for S_i is an integration test set for stage i of the integration. □

Definition 22 (Integration Test). For a glass box architecture consisting of a set of components, K, and a set of internal and external channels I and O, an integration test is a mapping $\{1, \ldots, j\} \to \wp(K) \times \wp(IF[I \blacktriangleright O])$ that stipulates which components are to be tested at stage i of the integration, and by which tests. □

Note that the notions of failing and passing tests carry over to integration tests unchanged.

10.5.2 The crucial role of models for testing

Usually, when we compose subsystems into architectures, the resulting system shows a quite different functionality compared to its subsystems. In particular, properties that hold for a particular subsystem do not hold any longer for the composed system (at least not for the external channels). The inverse direction also is true. This is because in order to go to the black box behavior on the one hand, a projection onto the output channels visible for the system is provided. On the other hand, some of the input of a component is no longer provided by the environment, but instead is now produced inside the systems on the internal channels. If this is the case, the behavior of the overall system is likely different from the behavior of its component and every test case of a component does not correspond to a test case, also not to an integration test case, for the overall system.

Next, we study a special case. In this case, we assume that we have a subsystem of a larger architecture which receives its input mainly from the system's environment, even within the overall architecture and produces output for the rest of the system only to a small extent and receives input from the rest of the system only to a small extent. This is typical for systems in the automotive domain, where suppliers develop components that carry a certain subfunctionality of the overall system.

One of the main issues is now to separate system and integration tests in a manner such that many of the system and integration test can be performed at the component test level already while only a subset of the system and the integration test remains for later phases. What we are interested in is finding appropriate methods to decompose system and integration tests such that the system and integration test can be performed as much as possible during the component test phases.

The advantage of this is that we do not have more expensive debugging at the integration test and system test level since we can have early partial integration tests and early partial system tests. As a result, the development process is accelerated and moved one step closer to concurrent engineering is done.

If the behavior specifications of the components happen to be executable—as, for instance, in the form of executable machines—we are in a particularly advantageous position. Declarative specifications enable us to derive meaningful tests, both the input part and the expected output part called the oracle (remember that an actual implementation is to be tested against this model). Operational specifications, in addition, *allow us to directly use them as stubs when actual testing is performed*, in the sense of model-in-the-loop testing (Sax, Willibald, and Müller-Glaser 2002). Hence, in addition to using them as specifications, we can use the models for two further purposes: for deriving tests that include the input and expected output parts, and as simulation components, or stubs, when it comes to performing integration tests.* This of course requires runtime driver components that bridge the methodologically necessary gap between the abstraction levels of the actual system under test and the model that serves as stub (Pretschner and Philipps 2005, Utting, Pretschner, and Legeard 2006).

10.5.3 Using the architecture to derive entry-level component tests

In the model-based system description sketched in Section 10.2.3, we have access to both a system specification and an architecture. With models of the single components that are connected in the architecture, however, we are ready for testing.

We have defined architecture faults to be mismatches between system specification and the interpreted architecture. These mismatches can happen at two levels: at the model level (architecture against system specification) and at the level of the implementation (implemented system against either architecture behavior or system specification). Architecture faults of the latter kind can, of course, only be detected at integration testing time. Component faults, in contrast, can be detected both at integration and module testing time. In the following, we will assume that it is beneficial to detect component faults at component testing time rather than at integration testing time, the simple reason being (a) that fault localization is simpler when the currently tested system is small and (b) that (almost) all integration tests have to be performed again by regression test once the faulty component has been shipped back to the supplier, fixed, and reintegrated with the system.

One natural goal is then to find as many component faults as early as possible. With our approach of using both component and architecture models at the same time, we can—rather easily, in fact—shift testing effort from the integration testing phase to the component testing phase.

The idea is simple. Assume we want to test component C in isolation and ensure that as many as possible faults that are likely to evidence during the integration testing phase are tested for during the component testing phase. Assume, then, that integration stage

---

*We deliberately do not consider using models for automatic code generation here.

j is the first to contain component C (and all stages after j also contain C). Assuming a suitable model-based testing technology to exist, we can then derive tests for the *subsystem* of integration phase $j+n$ and project these tests to the input and output channels of C. It is precisely the structure of this composed subsystem at stage $j+n$ that we exploit for testing C in its actual context—without this structure, we would have to assume any environment, that is, no constraints on the possible behaviors. These projections are, without any further changes, tests for component C. By definition, they are relevant to the integration with all those components that are integrated with C at stage $j+n$. Faults in C that, without a model of the architecture and the other components, would only have been found when performing integration testing, are now found at the component testing stage. In theory, of course, this argument implies that no integration testing would have to be performed. By the very nature of models being abstractions, this is unfortunately not always the case, however.

More formally, if we study a system architecture

$$F = \otimes\{F_k : k \in K\} \in IF[I \blacktriangleright O]$$

with the interface $(I \blacktriangleright O)$, we can distinguish internal components from those that interact with the environment. Actually, we distinguish three classes of system components:

1. Internal components $k \in K$: for them there is no overlap in their input and output channels with the channels of the overall system F: If $F_k \in IF[I_k \blacktriangleright O_k]$, then $I \cap I_k = \varnothing$ and $O \cap O_k = \varnothing$.

2. External output providing components $k \in K$: $O \cap O_k \neq \varnothing$.

3. External input accepting components $k \in K$: $I \cap I_k \neq \varnothing$.

4. Both external output providing and external input accepting components $k \in K$: $I \cap I_k \neq \varnothing$ and $O \cap O_k \neq \varnothing$.

In the case of an external input and output providing component $k \in K$, we can separate the channels of $F_k \in IF[I_k \blacktriangleright O_k]$ as follows:

$$I'_k = I_k \cap I, \quad I''_k = I_k \setminus I$$

$$O'_k = O_k \cap O, \quad O''_k = O_k \setminus O$$

This leads to the diagram presented in Figure 10.3 that depicts the component F_k as a part of a system's architecture.

Following Broy (2010a), we specify projections of behaviors for systems.

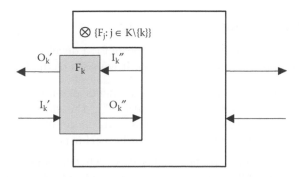

FIGURE 10.3
Component F_k as part of an architecture.

Definition 23 (Projection of Behaviors). Given syntactic interfaces $(I_1 \blacktriangleright O_1)$ and $(I \blacktriangleright O)$, where $(I_1 \blacktriangleright O_1)$ is a syntactic subinterface of $(I \blacktriangleright O)$, we define for a behavior function $F \in IF[I \blacktriangleright O]$ its *projection* $F^\dagger$ $(I_1 \blacktriangleright O_1) \in IF[I_1 \blacktriangleright O_1]$ to the syntactic interface $(I_1 \blacktriangleright O_1)$ by the following equation (for all input histories $x \in \overrightarrow{I}_1$):

$$F^\dagger(I_1 \blacktriangleright O_1)(x) = \{y|O_1 : \exists x'\overrightarrow{I} \in \overrightarrow{I} : x = x'|I_1 \wedge y \in F(x')\}. \qquad \square$$

When doing component tests for component k, we consider the behavior F_k with interface $(I'_k \cup I''_k \, O'_k \cup O''_k)$. The idea essentially is to use a simulation of the environment of k, $\otimes\{F_j : j \in K\backslash\{k\}\}$, which, via I''_k, provides input to k, to restrict the set of possible traces of k. This directly reduces the set of possible traces that can be used as tests. If tests for $F^\dagger_k$ $(I''_k \blacktriangleright O''_k)$ are not included in the set of traces given by $F^\dagger$ $(I''_k \blacktriangleright O''_k)$ in the sense that the behavior of k is undefined for the respective input, then the respective component test is useless because it corresponds to a behavior that will never be executed. Using the environment of k, $\otimes\{F_j : j \in K\backslash\{k\}\}$ allows us to eliminate such useless tests. Note, of course, that these tests are useless only from the system integrator's perspective. They are certainly not useless for the supplier of the component who, in contrast, wants to see the component used in as many different contexts as possible. This, of course, is the problem that providers of libraries face as well.

In other words, we must consider the question of what the relationship and essential difference between the behavior

$$F_k\dagger[I'_k \blacktriangleright O'_k]$$

and the system behavior

$$F\dagger[I'_k \blacktriangleright O'_k]$$

is. Only if there are test cases (a, B) that can be used for both views, we can push some system tests for system F to the component testing phase. To achieve this, we compose a model (a simulation) of the component's environment with the model of the component. We now use this composed model rather than the model of the component only for the derivation of tests. Using a projection of the resulting composed behavior to the I/O channels of the component only yields precisely the set of possible traces of the component when composed with the respective environment, and test selection can be restricted to this set.

Note that there are no objections to exploiting the inverse of this idea and use tests for one component D, or rather the output part of these tests, as input parts of tests for all those components the input ports of which are connected to the output ports of D. In fact, this idea has been successfully investigated in earlier work (Pretschner 2003), but in this cited work, we failed to see the apparently more relevant opposite direction.

10.5.4   A resulting testing methodology

The above considerations naturally lead to a proposal for a development and testing strategy for integrators and component suppliers. Briefly, it consists of the following steps.

1. Build (executable) models for all those components or subsystems that are to be integrated. These models serve as specification for the suppliers, as basis for test case generation at the integrator's site, and as stubs or simulation components at the integrator's site when integration tests are to be performed.

2. Build an architecture that specifies precisely which components are connected in which sense. Together with the models of the single components, this provides a behavior specification for each conceivable subsystem that may be relevant in the context of integration testing.

3. Derive component-level tests for each supplied component from the respective models in isolation.

4. Decide on an integration testing strategy. In other words, decide on subsystems in the architecture that form fundamental blocks. Examples include strongly connected subgraphs, more or less hierarchic substructures, etc.

5. Compose the models, according to the architecture, that correspond to the components at integration stage j. Since by compositionality, this is a model as well, derive tests for this composed model, and project the test cases to the I/O channels of each single component in the set. These projections are test cases for each single component. Collecting these components tests for all components at all stages yields the component tests that are relevant for integration testing.

6. Execute the generated tests from steps (3) and (5) for each component C by possibly using the executable models of the other components as stubs or simulation components.

This outlines a methodology for development and testing for integrators and suppliers to save test efforts at the integration test level.

10.5.5   Discussion

What we have shown is just an example of applying a strictly model-based theory for discussing different approaches to carry out tests, in this case integration tests. We conent that modeling techniques are useful not only when applied directly in system development: they are certainly useful when working out dedicated methodologies (e.g., for testing). We did not fully evaluate the possibilities of model-based development, for instance for test case generation, but rather designed and assessed certain test strategies on the basis of a model-based system description.

10.6   Summary and Outlook

Based on the FOCUS modeling theory, we have worked out formal descriptions of fundamental notions for testing complex systems. In particular, we have precisely defined the difference between component and architecture errors that naturally leads to the requirement of *architecture models* in addition to *models of components*. In a second step, we have shown how to use these architecture models to reduce the set of possible test cases for a component by considering only those traces that occur in the integrated system. By definition, this allows a system integrator to reduce the number of entry-level component tests to those that really matter. Our results are particularly relevant in the automotive domain but generalize to distributed systems as found in service-oriented architectures or when COTS components are to be integrated into a system.

The ideas in this chapter clearly advocate the use of model-based engineering processes. While most of these processes relate to the automatic generation of code, we carefully look into the advantages of models for testing, regardless of whether or not code is generated. To our knowledge, this has been done extensively for single components; we are not aware, however, of a systematic and well-founded treatment of model-based testing for distributed systems.

We are aware that there are plenty of well-known obstacles to implementing model-driven engineering processes on a large scale. We see the pressing need for further research into understanding which general and which domain-specific abstractions can be used, into systematic treatments of bridging the different levels of abstraction when system tests are executed, into whether or not these abstractions discard too much information so as to be useful for testing, and into whether or not model-based testing is indeed a cost-effective technology.

References

Broy, M. and Stølen, K. (2001). *Specification and Development of Interactive Systems: Focus on Streams, Interfaces, and Refinement.* Springer, New York.

Broy, M. (2006). The 'grand challenge' in informatics: engineering software-intensive systems. IEEE Computer. 72–80, 39, issue 10.

Broy, M., Krüger, I., and Meisinger, C.M. (2007). A formal model of services. TOSEM — ACM *Trans. Softw. Eng. Methodol.* 16, 1, article no. 5.

Broy, M. (2010a). Model-driven architecture-centric engineering of (embedded) software intensive systems: Modeling theories and architectural milestones. *Innovations Syst. Softw. Eng.*

Broy, M. (2010b). Multifunctional software systems: structured modelling and specification of functional requirements. Science of Computer Programming, accepted for publication.

Krüger, I. (2000). Distributed System Design with Message Sequence Charts, Ph.D. dissertation, Technische Universität München.

Philipps, J., Pretschner, A., Slotosch, O., Aiglstorfer, E., Kriebel, S., and Scholl, K. (2003). Model-based test case generation for smart cards. Proc. Formal Methods for Industrial Critical Systems, Trondheim, Pages: 168–182. *Electronic Notes in Theoretical Computer Science*, 80.

Prenninger, W. and Pretschner, A. (2005). Abstractions for model-based testing. Proc. 2nd Intl. Workshop on Test and Analysis of Component Based Systems (TACoS'04), Barcelona, March 2004. *Electronic Notes in Theoretical Computer Science* 116:59–71.

Pretschner, A. (2003). Compositional generation of MC/DC integration test suites. *ENTCS* 82(6):1–11.

Pretschner, A. and Philipps, J. (2005). Methodological issues in model-based testing. In Broy, M., Jonsson, B., Katoen, J.-P., Leucker, M., and Pretschner, A. *Model-Based Testing of Reactive Systems*, Volume 3472 of Springer LNCS, Pages: 281–291.

Pretschner, A., Prenninger, W., Wagner, S., Kühnel, C., Baumgartner, M., Sostawa, B., Zölch, R., and Stauner, T. (2005). One evaluation of model-based testing and its automation. *Proc. 27th Intl. Conf. on Software Engineering* (ICSE'05), Pages: 392–401. St. Louis.

Pretschner, A., Broy, M., Krüger, I., and Stauner, T. (2007). Software engineering for automotive systems: A roadmap. *Proc. Future of Software Engineering*, 55–71.

Reiter, H. (2010). Reduktion von Integrationsproblemen für Software im Automobil durch frühzeitige Erkennung und Vermeidung von Architekturfehlern. Ph. D. Thesis, Technische Universität München, Fakultät für Informatik, forthcoming.

Sax, E., Willibald, J., and Müller-Glaser, K. (2002). Seamless testing of embedded control systems. In *Proc. 3rd IEEE Latin American Test Workshop*, S. 151–153.

Utting, M., Pretschner, A., and Legeard, B. (2006). A taxonomy of model-based testing. Technical report 04/2006, Department of Computer Science, The University of Waikato, New Zealand.

11

Multilevel Testing for Embedded Systems

Abel Marrero Pérez and Stefan Kaiser

CONTENTS

11.1 Introduction

Multilevel testing constitutes an evolving methodology that aims at reducing the effort required for functional testing of large systems, where the test process is divided into a set of subsequent test levels. This is basically achieved by exploiting the full test reuse potential across test levels. For this purpose, we analyze the commonality shared between test levels as well as the variability and design a test reuse strategy that takes maximum advantage of the commonality while minimizing the effects of the variability. With this practice, we achieve reductions in test effort for testing system functions across test levels, which are characterized by high commonality and low variability.

We focus on large embedded systems such as those present in modern automobiles. Those embedded systems are mainly driven by software (Broy 2006) and consist of a large number of electronic components. The system's complexity increases continuously as a consequence of new functionality and a higher level of functional distribution. Regarding testing, this implies a necessity to continuously increase the efficiency as well—something we can only achieve by enhancing our testing methods and tools.

Increasing the testing efficiency constitutes the fundamental challenge for novel testing methodologies because of the large cost of testing, an activity that consumes around 50% of development cost (Beizer 1990).

The separation of the testing process into independent test levels contributes to establishing different methods and tools across test levels. This heterogeneity helps counter any efforts toward test level integration. It also results in a higher effort being required for updating many different methods and tools to the state of the art. Thus, a higher level of homogeneity is desired and in practice often necessary.

A further significant problem in the field of large embedded systems is the level of redundancy that the test automation advances of the past decade have produced. Merely repeating test executions or developing additional test cases across test levels does not automatically lead to a higher test quality. The reduction in effort brought about by test automation should never obstruct the view on testing cost. The creation of new test cases and the assessment of new test results (especially for failed test cases) are costly activities that cannot be efficiently automated. We thus need to avoid the execution of similar or even identical test cases at different test levels whenever this repeated execution is redundant. In order to systematically define a test execution at a specific test level as redundant, appropriate test strategies must be applied. They should define what must be tested at the different test levels and should take the entire test process into consideration—instead of defining the testing scope at each test level independently.

Such an integrated test strategy will consider the execution of numerous similar or even identical test cases at different test levels. This results from the refinement/abstraction relation between consecutive test levels and does not represent any form of redundancy. Hence, on the one hand, there is evidence of the existence of significant commonalities between test cases across test levels.

However, the strict separation of the test process into independent test levels indirectly leads to an underestimation of the potential synergies and commonalities shared by the different test levels. In consequence, multiple test implementations of very similar test artifacts coexist in practice at different test levels. Great effort was necessary for their creation—and is further necessary for their maintenance.

Our objective is thus to reduce this design and maintenance effort by reusing test cases across test levels. We take advantage of previous work on reusing test specifications and focus on reusing test implementations. Our work is mainly based on multilevel test models and multilevel test cases, which are test design concepts supporting an integrative methodology for all test levels. These concepts are presented and discussed in-depth in this contribution, especially highlighting the potential benefits for the entire test process.

This introduction is followed by a summary of related work that provides insight into previous work regarding test case reuse across test levels. The subsequent sections introduce the different test levels for embedded systems, analyze their commonality and variability, and describe our initial solution for multilevel testing: multilevel test cases. In the main segment, we describe strategies for test level integration and introduce multilevel test models as our model-based approach in this context. The contributions are validated using an automated light control (ALC) example before concluding with a summary and a brief discussion of the practical relevance of multilevel testing.

11.2   Related Work

Partial solutions for the problems mentioned in the introduction are currently available. Research is in progress in many of these areas. In this section, we follow the argumentation pattern of the introduction in order to provide an overview of related work in our research field.

11.2.1 Facing complexity

Manual testing nowadays appears to be a relict of the past: expensive, not reproducible, and error-prone. The automation of the test execution has significantly contributed to increasing testing efficiency. Since the full potential in terms of efficiency gains has already been achieved, research on test automation does not focus on test execution anymore, but on other test activities such as automatic test case generation. As an example, search-based testing uses optimization algorithms for automatically generating test cases that fulfill some optimization criteria, for example, worst-case scenarios. Such algorithms are also applicable to functional testing (Bühler and Wegener 2008). Automatically searching for the *best* representatives within data equivalence classes using evolutionary algorithms is proposed in (Lindlar and Marrero Pérez 2009), which leads to an optimization of the test data selection within equivalence classes.

Automatic test case generation is the main objective of model-based testing approaches, which take advantage of test models. Generally speaking, models are the result of an abstraction (Prenninger and Pretschner 2005). In this context, model-based testing increases the testing efficiency as it benefits from the loss of information provided by the abstraction. Later on, the missing details are introduced automatically in order to provide concrete test cases. Providing additional details is not necessary when testing abstract test objects such as models (Prenninger and Pretschner 2005). Zander-Nowicka described such an approach for models from the automotive domain (Zander-Nowicka 2008).

However, most test objects are not that abstract. The additional details necessary for test execution are provided by test adapters, test case generators, and compilers. While test adapters represent separate instances that operate at the test model interfaces, test case generators and compilers perform a transformation of the abstract test model into executable test cases. The utilized approach is typically closely related to the kind of test models used. In our contribution, we apply a combination of both approaches.

Basically, we differentiate between test abstraction and interface abstraction. As a consequence, low-level test cases, for instance written in C, can possess a particularly abstract interface and vice versa, an abstract test model can feature a very concrete interface. We use time partition testing (TPT) (Lehmann 2003) for test modeling, which employs a compiler to generate executable test cases from the abstract test models. Additionally, we use test adapters for adapting the abstract test model interface to the concrete test object interface.

Abstraction principles go beyond our differentiation in test and interface abstraction. Prenninger and Pretschner describe four different abstraction principles: functional, data, communication, and temporal abstraction (Prenninger and Pretschner 2005).

Functional abstraction refers to omitting functional aspects that are not relevant to the current test. It plays the key role in this contribution because multilevel testing addresses test objects at different test levels and hence at different abstraction levels. In this context, selecting the appropriate abstraction level for the test models represents a crucial decision. Data abstraction considers the mapping to concrete values, whereas temporal abstraction typically addresses the description of time in the form of events. Both principles will be considered in the context of the test adapters in this contribution. Only communication abstraction, from our point of view a combination of data and temporal abstraction, is beyond the scope of the contribution. Data abstraction and temporal abstraction are widely used within the model-based testing domain, but in this contribution, we will consider them in the context of what we have previously called interface abstraction.

Most approaches using test adapters mainly consider data abstraction. A recently published report (Aichernig et al. 2008) generically describes test adapters as functions that map abstract test data to concrete values. Temporal abstraction typically represents an additional requirement when time plays a central role for test execution. Larsen et al. present an approach for testing real-time embedded systems online using UPPAAL-TRON. In their work, they use test adapters to map abstract signals and events to concrete physical signals in order to stimulate the test object (Larsen et al. 2005).

The concept of test adapters refers to the adapter concept in component-based design introduced by Yellin and Strom (1997). Adapters are placed between components and are responsible for assuring the correct interaction between two functionally compatible components. Adapters are further responsible for what they call interface mapping, typically data type conversion (Yellin and Strom 1997). Thus, clear differences exist between the test adapter concept and the original adapter concept from component-based design. In the latter, adapters are namely not specifically supposed to help bridge abstraction differences between the interfaces they map.

11.2.2   Methods and tools heterogeneity

The lack of homogeneity along the test process has been addressed by industry in recent years. Wiese et al. (2008) describe a set of means for test homogenization within their company. One of their central ideas is making testing technologies portable across test levels.

There are several testing technologies supporting multiple test platforms, that is, specific test environments at a specific test level. In the field of embedded systems, the main representatives are TPT (Lehmann 2003) and TTCN-3 (European Telecommunications Standards Institute 2009-06). TPT's platform independence is based on the TPT virtual machine, which is capable of executing test cases on almost any platform. For test execution, the TPT virtual machine is directly nested in the test platform. TTCN-3 test cases are also executed close to the test platform using a platform and a system adapter. For a more detailed technology overview, please consult Marrero Pérez and Kaiser (2009).

Such technologies are reuse friendly and ease homogenization attempts in the industry. For homogenization, however, the test interface represents the central problem, as denoted by Burmester and Lamberg (2008). Implementing abstract interfaces in combination with test adapters (also called *mapping layer* in this context) rapidly leads to platform independence and thus reusability (Burmester and Lamberg 2008, Wiese et al. 2008). Obviously, any model-based approach providing the appropriate test case generators and/or test adapters can lead to test cases that can be executed at different platforms. All published approaches for homogenization strategies are based on data abstraction.

Reuse across test levels is not a great challenge technologically, but methods for implementing test cases capable of testing test objects at different abstraction levels sometimes featuring strongly differing interfaces have not been developed to date. This means that while in theory we can already reuse test cases across test levels today, we do not exactly know what the crucial issues are that have to be taken into account in order to be successful in this practice.

11.2.3   Integrated test specifications

Reducing redundancy across test levels implies reducing their independence, that is, establishing relations between them. Hiller et al. (2008) have reported their experience in creating a central test specification for all test levels. A common test specification contributes to test level integration by establishing a central artifact for the different testing teams. The test levels for which each test case must be executed are declared as an additional attribute in the test specification. The systematic selection of test levels for each test case contributes to avoiding the execution of the same test cases at multiple test levels where this is unreasonable. Hiller et al. argue that the test efficiency can be increased by using tailored test management technologies. For instance, a test case that failed at a specific test level in the current release should temporarily not be executed at any higher test level until the fault has been found and fixed (Hiller et al. 2008).

Our approach will benefit from such an integrated test specification for different reasons. Firstly, we can take advantage of the additional attribute in the test specification providing the test levels where the test case should be specified. Secondly, we benefit from particularly abstract test cases that were specifically designed for being executable at different test levels. Lastly, the common test specification constitutes a further artifact featuring an integrative function for the different test levels, which represents additional support for our methodology.

11.2.4 Test reuse across test levels

Schätz and Pfaller have proposed an approach for executing component test cases at the system level (Schätz and Pfaller 2010). Their goal is to test particular components from the system's interface, that is, with at least partially limited component interface visibility. For this purpose, they automatically transform component test cases into system test cases using formal descriptions of all other system components. Note that these test cases do not aim at testing the system, but rather a single component that they designate component under test. Thus, their motivation for taking multiple test levels into consideration clearly differs from ours.

The transformation performed by Schätz and Pfaller results in a test case that is very similar or even identical to the result of appending a test adapter to the original component test case. Consequently, we can state that their work (Schätz and Pfaller 2010) shows that test adapters can be generated automatically, provided that the behavior of all other system components is exactly known and formally specified. This assumption can be made for software integration testing, where the considered software components may be formally specified. However, when analog hardware parts are considered, the complexity of their physical behavior—including tolerances—often makes a formal description of their behavior with the required precision impossible.

Another approach that considers test level integration was presented by Benz (2007). He proposes a methodology for taking advantage of component test models for integration testing. More concretely, Benz uses task models for modeling typically error-prone component interactions at an abstract level. From the abstract test cases generated using the task model, executable test cases that can stimulate the integrated components are generated based on a mapping between the tasks and the component test models. Hence, Benz takes advantage of the test models of another test level in order to refine abstract test cases without actually reusing them.

Mäki-Asiala (2005) introduced the concept of vertical reuse for designating test case reuse across test levels. This concept had been used before in the component-based design for addressing the reuse of components within a well-defined domain. In this context, vertical reuse is also known as domain-specific reuse (Gisi and Sacchi 1993).

As in every reuse approach, commonality and variability are decisive for vertical reuse. Mäki-Asiala (2005) states that similarities between the test levels must be identified, as well as the tests possessing the potential to be reused and to reveal errors at different test levels. His work, however, lacks instructions for these identification processes. There is no description of how to identify reuse potentials and error revelation potentials.

Mäki-Asiala provides a set of guidelines for test case reuse in TTCN-3, discussing their benefits for vertical reuse. The guidelines were designed for reusing tests with little effort without considering test adapters so that interface visibility becomes a major issue (Mäki-Asiala 2005). Lehmann (2003) also addresses the interface visibility problem in his thesis, highlighting the impossibility of test reuse for different test object interfaces. As mentioned above, we address this problem by indirect observation, similar to Schätz and Pfaller. Analogous to component-based design, we require functional compatibility between test and test object in order to design our test adapters (Yellin and Strom 1997).

Hence, we can conclude that there are only a few approaches to test reuse across test levels—a fact that demonstrates the novelty of our approach. While Mäki-Asiala presents generic approaches to test reuse, Schätz and Pfaller address cross-level test reuse at consecutive test levels only. There is a lack of an integrative approach that considers the entire test process, the test levels of which are described in the subsequent section.

11.3   Test Levels for Embedded Systems

In the domain of embedded systems the V model (Spillner et al. 2007, Gruszczynski 2006) constitutes the reference life cycle model for development and testing, especially in the automotive domain (Schäuffele and Zurawka 2006). It consists of a left-hand branch representing the development process that is characterized by refinement. The result of each development level is artifacts that—in terms of functionality—specify what must be tested at the corresponding test levels in the V model's right-hand branch (Deutsche Gesellschaft für Qualität e.V. 1992).

A complete right-hand branch of the V model for embedded systems is shown in Figure 11.1. It starts at the bottom of the V with two branches representing software and hardware. After both hardware and software are integrated and tested, these branches merge at the system component integration test level. Before system integration, the system components are tested separately. The only task remaining after the system has been tested is acceptance testing.

Note that in Figure 11.1 each integration test level is followed by a test level at which the integrated components are functionally tested as a whole before a new integration test level is approached. Hence, the V model's right-hand branch consists of pairs of integration and *integrated* test levels. From software component testing up to acceptance testing, the V model features three different integration test levels for embedded systems: component integration (either software or hardware), software/hardware integration, and system integration. This makes it different from the V model for software systems which features only a single integration test level (cf. Spillner et al. 2007). But in analogy to that model, a test level comprising the completely integrated unit follows each integration test level, as mentioned before.

Since our goal is functional testing, we will exclude all integration test levels from our consideration. In doing so, we assume that integration testing specifically focuses on testing the component interfaces and their interaction, leaving functional aspects to the test level that follows. In fact, integration and system testing are often used equivalently in the automotive domain (see for instance, Schätz and Pfaller 2010).

Acceptance testing, which is typically considered as not belonging to development (Binder 1999), is out of the scope of our approach. By excluding this test level and the integration test levels, we focus on the remaining four test levels: software (hardware) component testing, software (hardware) testing, system component testing, and system testing.

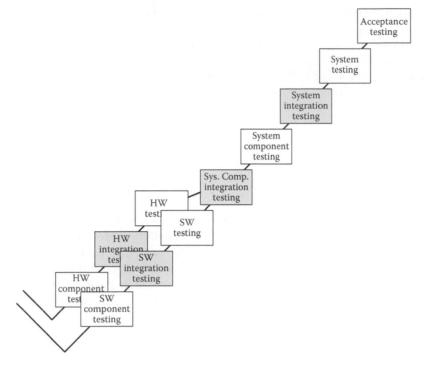

FIGURE 11.1
Right-hand branch of the V model for embedded systems, featuring all test levels.

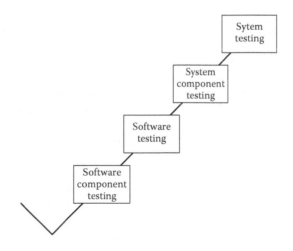

FIGURE 11.2
Test levels considered in this chapter. (Reprinted from *The Journal of Systems and Software*, 83, no. 12, Marrero Pérez, A., and Kaiser, S., Bottom-up reuse for multi-level testing, 2392–2415. Copyright 2010, with permission from Elsevier.)

As most of the functionality in modern vehicles is implemented in software, we will not consider the hardware branch in Figure 11.1 any further.

Hence, the test levels we cover are depicted in Figure 11.2. Our V model's right-hand branch thus starts with software component testing, which typically addresses a single software function (for instance, a C function). Different software components are integrated to constitute the complete software of a control unit (software testing). After software/hardware integration, the so-called electronic control unit (ECU) is tested at the system component test level. This test typically includes testing the real sensors and actuators connected to the ECU under test. Our test process concludes with the test of the entire system, consisting of a set of ECUs, sensors, and actuators, most of which may be physically available in the laboratory. Any components not available will be simulated, typically in real time.

11.4   Commonality and Variability Across Test Levels

A very basic test case for a vehicle's headlights could look as follows: *From the OFF state, turn the lights on. After* 10 *s turn the lights off.* Manually performing this test case in a car is not a big issue. But what happens when the switch is rotated? There is no wire directly supplying the lamps through the switch. Instead, there will be some kind of control unit receiving the information about the actual switch position. Some logic implemented in software then decides if the conditions for turning the headlights on are given, and should this be the case, a driver provides the current needed by the light bulbs. In addition, this functionality may be distributed across different control units within the vehicle.

At this point, we can identify the simplest version of the typical embedded control system pattern consisting of a unidirectional data flow from sensors toward actuators:

$$sensor \rightarrow hardware \rightarrow software \rightarrow hardware \rightarrow actuator$$

Dependencies between components, systems, and functions make reality look less neat. However, a key lesson from this pattern remains valid: as in the headlights example, the software *decides* and the remaining parts of the system—such as hardware—are gradually taking on secondary roles in terms of system functionality.

For the basic headlights test case described above, this implies that it can be employed at the software component test level for testing the component in charge of deciding whether the enabling conditions are met or not. In fact, this test case is valid for all test levels. It is possible to test at every test level whether the headlights will switch on in the final system.

Not only is it possible to repeat this particular test case at every test level, it is also reasonable and even efficient. The reason for this is that the testing of the functionality of the system has to begin as soon as the first software components are available. There is no point in waiting until the first car prototype is built, because the earlier a fault is detected, the less expensive the fault correction process becomes.

Software components are the first test objects available for testing. Along the right-hand branch of the V model, further test objects will become available successively till a completely integrated system makes testing at the top test level possible. Because of this temporal availability order, it appears reasonable to at least perform basic functional tests at each test level in order to ensure, for instance, that the headlights will work in the first prototype car. In addition to the benefits reported by testing earlier on in development, less effort is required for revealing and identifying faults at lower than at upper test levels. This keeps the lower test levels attractive for the last part of development where all test levels are already available.

The headlights example demonstrates that there are significant similarities between the functional test cases across test levels. Consequently, a large set of functional test cases are execution candidates for different test levels, provided that they are specified at a reasonable functional abstraction level. For example, because of the abstraction level, a software tester as well as a hardware tester will know how to perform the test case *turn the headlights on and off*.

The key for the commonality is thus the functionality, and the key for having a low variability is the level of abstraction. When test cases address functional details, which are often test level specific, the variability becomes higher and there is less commonality to benefit from. For instance, if the headlights test case described above would utilize specific signal and parameter names, it would be more difficult to reuse.

The variability is thus primarily given by the differences between the test objects. Both a software component and an ECU may implement the same function, for example, the headlights control, but their interfaces are completely different. We will address this issue in the following sections in combination with the already mentioned interface abstraction.

A further variability aspect concerns the test abstraction level, which increases along the V model's right-hand branch. Abstract test levels require testing less functional details than concrete test levels. In fact, many functional details are not even testable at abstract test levels because these details are not observable. In other cases, the details are observable but doing so requires a high effort that makes reuse unaffordable. As a consequence, there is no point in trying to reuse every test case at every test level, even if it were technically possible.

Our solution to address the variability that originates from the different abstraction levels is to separate test cases into different groups depending on their level of abstraction. This approach will be described in Section 11.6, after the introduction of multilevel test cases.

11.5 Multilevel Test Cases

We introduced multilevel test cases in (Marrero Pérez and Kaiser 2009) as a modularization concept for structuring test cases that permits reusing major parts of test cases across test levels. Multilevel test cases reflect the commonality and variability across test levels. As shown in Figure 11.3, they consist of an abstract test case core (TCC) representing the commonality and test level-specific test adapters that encapsulate the variability.

The only form of variability accepted in the TCC is parameterization. The parameters cover both test case variability and test object variability. Within the test case variability, those parameters addressing differences across test levels are of particular interest here. With parameterization, we can rely on invariant test behavior across test levels. The test behavior provides an interface consisting of a set of signals $\vec{TE}(t)$ for evaluation and another set of signals $\vec{TS}(t)$ for stimulation. Thus, the interface of the TCC does not take data or temporal abstraction into consideration, but operates at a technical abstraction level using discrete signals, that is, value sequences that are equidistant in time. Without this practice, it would not be possible to consider complex signals because the loss of information caused by abstraction would be prohibitive.

Test adapters are divided into three different modules: input test adapter (ITA), output test adapter (OTA), and parameter test adapter (PTA). As their names suggest, they are in charge of observation, stimulation, and parameterization of the different test objects at each test level. Within the embedded systems domain, both ITAs and OTAs are functions relating the TCC interface signals $\vec{TE}(t)$ and $\vec{TS}(t)$ to the test object interface signals

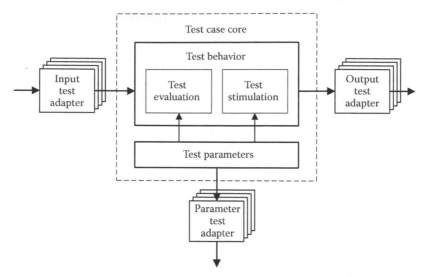

FIGURE 11.3
Structure of multilevel test cases. (Reprinted from *The Journal of Systems and Software*, 83, no. 12, Marrero Pérez, A., and Kaiser, S., Bottom-up reuse for multi-level testing, 2392–2415. Copyright 2010, with permission from Elsevier.)

$T\vec{E}'(t)$ and $T\vec{S}'(t)$:

$$T\vec{E}(t) = ITA(T\vec{E}') \tag{11.1}$$

$$T\vec{S}'(t) = OTA(T\vec{S}) \tag{11.2}$$

The ITA function typically provides functional abstraction as well as temporal and data abstraction, if necessary. In contrast, the OTA function will refine the test stimulation (TS) functionally but also add temporal and data details if required. The same applies to interface abstraction. Here, the ITA will abstract the test object interface while the OTA will perform a refinement. These differences should be kept in mind, even though Equations 11.1 and 11.2 demonstrate that both functions $ITA(T\vec{E}')$ and $OTA(T\vec{S})$ technically represent a mapping between interfaces.

Multilevel test cases primarily focus on reducing the test case implementation effort through reuse. For this reason, it is crucial that the design of test adapters and their validation require less effort than creating and validating new test cases. In addition, the maintenance effort necessary in both cases has to be taken into consideration.

The resulting test case quality will mainly depend on the test developers, however. There is no evidence indicating that multilevel test cases directly increase the test case quality. Indirectly, our aim is to reduce the test implementation effort so that fewer resources are necessary for reaching the same quality level.

11.6   Test Level Integration

This chapter addresses our methodological approach to test reuse across test levels, which primarily consists of departing from the strict separation of test levels stipulated by conventional test processes. The objective is to show how to take advantage of the common functionality that must be tested across multiple test levels while taking the cross-level differences into account. These differences concern the abstraction levels and the interfaces, as discussed in Section 11.4.

11.6.1 Top-down refinement versus top-down reuse

Test level integration implies relating test levels and hence analyzing and establishing dependencies between them. A straightforward approach for test level integration thus consists of reflecting the abstraction/refinement relationship between consecutive test levels by introducing such a relationship between test cases at the different test levels. Following this approach, abstract test cases at the top test level are refined stepwise towards more concrete test cases at the test levels below (top-down refinement). Figure 11.4 schematically shows the refinement process across test levels. The differences in the rectangles' size point out the increasing amount of details present in the test cases across the test levels. Top-down refinement can be performed parallel to the refinement process in the V model's left-hand branch.

A further integration approach consists of reusing—not refining—test cases from the top test level down towards the lowest test level (top-down reuse). Instead of adding details with refinement, the details are introduced in form of additional test cases for each test level here. These test cases will be reused at the test levels below, as well, but never at the test levels above since they test details that are out of the scope of more abstract test levels. The principle of top-down reuse is depicted in Figure 11.5. The vertical arrows relating test cases at different test levels indicate reuse. With top-down reuse four different test case groups are created, one at each test level (A through D in Figure 11.5).

Top-down refinement best fits the development process in the left-hand branch of the V model. Each development artifact possesses a specific testing counterpart so that traceability is given. Assuming a change in requirements, this change will affect one or more test cases at the top test level. Analogous to the update in requirements causing changes in the artifacts

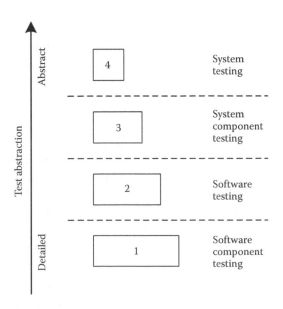

FIGURE 11.4
Top-down refinement.

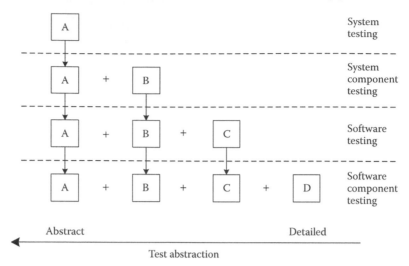

FIGURE 11.5
Top-down reuse.

at all development levels below, the change in the abstract test case will affect all test cases that refine it. Without appropriate linking mechanisms between the different test levels, a large set of test cases would then have to be updated manually. In fact, the manual effort for updating all test cases in case the refinement cannot be automatically performed makes this approach impracticable nowadays.

The second approach—top-down reuse—overrides the described update problem by avoiding test refinement along the V model while preserving the traceability between development and testing as well as across test levels. The key idea of this approach consists of describing test cases at the highest possible level of functional abstraction and then reusing these test cases at every lower test level. Thus, top-down reuse does not imitate the refinement paradigm of development on the V model's right-hand branch. Instead, it provides better support for the testing requirements, as we argue on three accounts.

Firstly, testing requires simplicity. The test objects are sufficiently complex—testing has to be kept simple in order to allow focusing on the test object and fault detection and not on the test artifacts themselves. Top-down reuse supports the simple creation of simple test cases at single abstraction levels without taking other abstraction levels into consideration. Additionally, the relationship between test levels is kept simple and clear. Refining test cases towards simple test cases is not trivial, however.

Secondly, automatic testing requires executable artifacts—test cases—that must be implemented at each test level. Therefore, top-down reuse has a great effort reduction potential in terms of test implementation, which is not featured by top-down refinement.

Thirdly, the current test process lacks collaboration between test teams at different test levels. Team collaboration is clearly improved by both approaches, but top-down reuse goes the extra mile in test level integration by assuring that identical test cases are shared across test levels. Team members at different test levels can thus discuss on the basis of a common specification and implementation.

The advantages of top-down reuse are clear and thus lead us to selecting it as our preferred test level integration approach. So far, we only focused on test case specification. The reuse approach also features advantages for test case design and test case implementation, though, as we already alluded to when arguing for top-down reuse.

11.6.2 Multilevel test design strategy

The main difference between test specification and test design is that we must take the test interface into account for test design, that is, we must decide which information channels to use for TS and which for test object observation and hence test evaluation (TE).

The test interface differs across test levels. In fact, different test objects at different abstraction levels possess different interfaces. However, while designing and implementing test cases, we must define a test interface as a means of communication (stimulation and observation) with the test object. The variability of the test interface across test levels in conjunction with the necessity of defining a test case interface common to all test levels makes it difficult to completely reuse entire test designs and test implementations along the V model.

The alternative is introducing interface refinement through test adapters supporting test case reuse across test levels by taking the differences in the test interface into consideration. With this practice, we can design and implement TCCs with abstract interfaces exactly following the top-down specification described in the previous section. Later on, these TCCs will be reused at test levels with more detailed interfaces using test adapters that perform an interface refinement.

The described design strategy utilizes the multilevel test case concepts described in Section 11.5. In fact, the combination of top-down reuse and interface refinement represents a reasonable approach for multilevel test case design, our solution for test level integration.

A central question that has not yet been addressed concerns the concept of an abstract test interface. We can approach this term from the perspective of data, temporal, or functional abstraction.

In this contribution, we assume that the test interface consists of two signal sets: a set of test inputs for TE and a set of test outputs for TS. The exclusive use of signals implies a low—and basically constant—level of temporal abstraction. It represents, however, a fundamental decision for supporting complex signals, as already mentioned in Section 11.5. Temporal abstraction differences between signals may exist, but they will be insignificant. For instance, a signal featuring (a few) piecewise constant phases separated by steps may seem to simply consist of a set of events. This could be considered to imply temporal abstraction. However, such a signal will not contain less temporal information than a real-world sensor measurement signal featuring the same sampling rate.

In contrast to temporal abstraction, the data abstraction level of a signal can vary substantially. These variations are covered by the typing information of the signals. For example, a Boolean signal contains fewer details than a floating point signal and is thus more abstract.

Following the definition from Prenninger and Pretschner (2005) given in Section 11.2.1, we consider an interface to be abstract in terms of functionality if it omits any functional aspects that are not directly relevant for the actual test case. This includes both omitting any irrelevant signals and within each of these also omitting any details without a central significance to the test. This fuzzy definition complicates the exact determination of the functional abstraction level of an interface. However, as we will show below, such accuracy is also unnecessary.

Considering the V model's right-hand branch, starting at the bottom with software components and gradually moving closer to the physical world, we can state that all three aspects of interface abstraction reveal decreasing interface abstraction levels. On the one hand, the closer we are to the real world, the lower the temporal and data abstraction levels of the test object signals tend to be. On the other hand, software functions will not take more arguments or return more values than strictly necessary for performing their functionality. Consequently, the functional abstraction level of the test interface will also be highest at the lowest test level. At the test levels, above additional components are incrementally integrated, typically causing the test interface to include more signals and additional details not significant to the very core of the function being tested but supporting other integrated components.

Hence, we propose a bottom-up approach for test design and test implementation, as shown in Figure 11.6. In addition to the basic schematic from Figure 11.5, test adapters performing interface abstraction are depicted here. They adapt the narrow interface of the test cases to the more detailed—and thus wider—interface of the corresponding test objects. The figure also proposes the concatenation of test adapters for reducing the effort of interface refinement at each test level to a refinement of the interface of the previous test level.

11.6.3 Discussion

In summary, our test level integration approach consists of two parts. Firstly, a top-down reuse approach for test specification that considers the increasing test abstraction* along the V model's right-hand branch (see Figure 11.5). In analogy to Hiller et al. (2008), this implies introducing a central test specification covering all test levels. Secondly, a bottom-up approach for test design and test implementation that takes the decreasing interface abstraction along the test branch of the V model into consideration using interface refinement (see Figure 11.6). Multilevel test cases constitute our candidates for this part.

As far as reuse is concerned, Wartik and Davis state that the major barriers to reuse are not technical, but organizational (Wartik and Davis 1999). Our approach covers different abstraction levels mainly taking advantage of reuse but also using refinement. In this

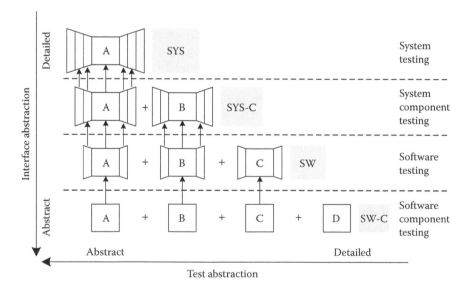

FIGURE 11.6
Bottom-up test design and test implementation. (Reprinted from *The Journal of Systems and Software*, 83, no. 12, Marrero Pérez, A., and Kaiser, S., Bottom-up reuse for multilevel testing, 2392–2415. Copyright 2010, with permission from Elsevier.)

---

*Cf. Section 11.2.1.

context, our technical solution achieves a substantially higher relevance than in pure reuse situations without reducing the importance of organizational aspects.

One of the most important disadvantages of test level integration is the propagation of faults within test cases across test levels. This represents a straightforward consequence of the reuse approach, which requires high-quality test designs and test implementations in order to minimize the risk of such fault propagation. In effect, not only faults are propagated across test levels but also quality. Reusing high-quality test cases ensures high-quality results.

Test level integration introduces several advantages in comparison to fully independent test levels including better traceability for test artifacts, lower update effort, better conditions for collaboration across test levels, and above all, a noticeable reduction of the test specification, test design, and test implementation effort.

Although multilevel testing does not constitute a universal approach, and test levels cannot be entirely integrated because of the lack of commonality in many cases, the relevance of test reuse across test levels is significant for large embedded systems. As mentioned before, there is clear evidence that the conventional test process lacks efficiency with respect to functional testing, and multilevel testing represents a tailored solution to this problem.

Note that the ensuing lack of universality does not represent a problem, since both multilevel test cases and multilevel test models—which are presented in the subsequent section—are compatible with conventional test approaches. In fact, multilevel testing can be considered a complement of the latter, which supports increasing the testing efficiency with respect to selected functions.

The reduction in effort is particularly evident for test specifications and has even been successfully validated in practice (Hiller et al. 2008). For test designs and test implementations, the successful application of multilevel testing depends on the effort required for interface refinement. Only if creating a new test case causes greater effort than refining the interface of an existing one, multilevel testing should be applied.

In order to reduce the interface refinement effort, we already proposed the concatenation of test adapters and the reuse of test adapters within a test suite in (Marrero Pérez and Kaiser 2009). Another possible effort optimization for multilevel test cases consists in automatically configuring test adapters from a signal mapping library. In the next section, we will present an approach for a further optimization, consisting of designing multilevel test models instead of multilevel test cases.

11.7 Multilevel Test Models

Multilevel test models constitute an extension of the multilevel test case concept. Instead of designing single scenarios that exercise some part of a system's function, multilevel test models focus on the entire scope and extent of functional tests that are necessary for a system's function. In other words, multilevel test models aim at abstractly representing the entire test behavior for a particular function at all test levels and hence at different functional abstraction levels.

Note that the *entire test behavior* does not imply the overall test object behavior. The test behavior modeled by multilevel test models does not necessarily have to represent a generalization of all possible test cases. Instead, a partial representation is also possible and often more appropriate than a complete one. Multilevel test models are tied to the simplicity principle of testing, too.

The main objective of multilevel test models is the reduction of the design and implementation effort in relation to multilevel test cases by achieving gains in functional abstraction and consequently taking advantage of the commonalities of the different test cases for a specific system's function. In this context, multilevel test models can also be seen as artifacts that can be *reused* for providing all functional multilevel test cases for a single function. The more commonalities these test cases share, the more the required effort can be reduced by test modeling.

For the design of multilevel test models, we propose the structure shown in Figure 11.7, which basically corresponds to Figure 11.3 and hence best stands for the extension of the multi-level test case concept. The essential concept is to design a test model core (TMC) that constitutes an abstract test behavior model featuring two abstract interfaces: one for stimulation and the other for evaluation. On the one hand, the refinement of the stimulation interface toward the test object input interface at each test level will be the responsibility of the output test adapter model (OTAM). On the other hand, the input test adapter model (ITAM) will abstract the test object outputs at each test level toward the more abstract TMC input interface. With this practice, we can derive a multilevel test case from the multilevel test model through functional refinement of both TMC and test adapter model.

The blocks within the TMC in Figure 11.7 reveal that we continue to separate test case stimulation and test case evaluation, as well as data and behavior. With respect to the data, test parameters are used whenever data has to be considered within the test model. These parameters may be related to the test object (test object parameters) or exclusively regard the test model and the test cases (test parameters).

In addition to extensive structural similarities, a large set of differences exists between test cases and test models in terms of reuse across test levels. These differences start with the design and implementation effort. Furthermore, multilevel test models will typically offer better readability and maintainability, which are important premises for efficiently sharing implementations across test teams. The advantages of gains in abstraction apply to the differences between test adapters and test adapter models, as well. Test adapter models will cover any interface refinement (TMC output) or interface abstraction (TMC input) for a

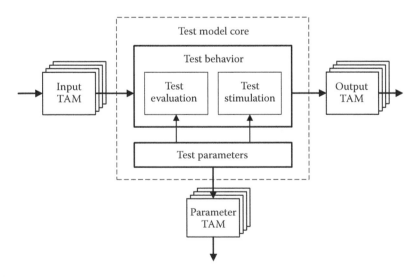

FIGURE 11.7

Structure of multilevel test models. (Adapted from Marrero Pérez, A., and Kaiser, S., Multi-level test models for embedded systems, *Software Engineering*, pages 213–224, © 2010b/GI.)

single test level. A test adapter capable of performing this could also be implemented without modeling, but because of the missing functional abstraction, it would clearly possess inferior qualities in terms of design and implementation effort, readability, and maintainability. Additionally, we can concatenate test adapter models in exactly the same way we proposed in Marrero Pérez and Kaiser (2009) for test adapters.

Multilevel test cases are always designed at a specific abstraction level. Some multilevel test cases will be more abstract than others, but there is a clear separation of abstraction levels. Multilevel test models cannot rely on this separation, however. They have to integrate test behavior at different functional abstraction levels into a single model. This is one of the most important consequences of the extension of multilevel test cases to multilevel test models. This issue is addressed next.

11.7.1   Test model core interface

The test interface plays a central role for test cases because there is no better way of keeping test cases simple than to focus on the stimulation and evaluation interfaces. However, this is only half the story for test models. Through abstraction and (at least partial) generalization, the interfaces are typically less important for test models than they are for test cases. Other than common test models, though, TMCs integrate various functional abstraction levels. For this reason, we claim that the TMC interface is of particular relevance for the TMC behavior.

As described in Section 11.6, the test cases for a specific test level are obtained top-down through reuse from all higher test levels and by adding new test cases considering additional details. The test interface of these new test cases will consist of two groups of signals:

- Signals shared with the test cases reused from above.

- Test-level specific signals that are used for stimulating or observing details and are thus not used by any test level above.

With this classification, we differentiate between as many signal groups as test levels are available. Test cases at the lowest test level could include signals from all groups. Furthermore, the signals of each group represent test behavior associated with a specific functional abstraction level.

We can transfer this insight to multilevel test models. If we consider multilevel test models as compositions of a (finite) set of multilevel test cases designed at different abstraction levels following the design strategy described in Section 11.6.2, we can assume that the TMC interface will be a superset of the test case interfaces and thus include signals from all groups. Note that the signal groups introduced above refer to the abstraction level of the functionality being tested and not to interface abstraction.

11.7.2   Test model behavior

We take an interface-driven approach for analyzing the central aspects of the test model behavior concerning multiple test levels. By doing so, we expect to identify a set of generic requirements that the TMC has to meet in order to be reusable across test levels.

In this context, we propose dividing the test behavior into different parts according to the signal groups defined in the previous section. Each part is associated with a specific model region within a multilevel test model. Since there are as many signal groups as test levels, there will also be as many model regions as test levels.

Figure 11.8 schematically shows the structure of a multilevel test model consisting of four regions. Each region possesses its own TS and TE and is responsible for a group of

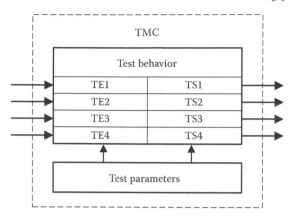

FIGURE 11.8
Division of the test behavior into regions. (Adapted from Marrero Pérez, A., and Kaiser, S., Multi-level test models for embedded systems, *Software Engineering*, pages 213–224, © 2010b/GI.)

input and output signals. Note that each signal is used by a single region within the test model.

Relating test behavior and test interface in this manner implies separating the different abstraction levels within the test behavior—a direct consequence of the signal group definition provided in the previous section. This is not equivalent to creating separate test models for each functional abstraction level, however. Our proposition considers designing in each region of the TMC only the behavior required by the test level-specific signals of the corresponding group, omitting the behavior of other shared signals (cf. Section 11.7.1), which are provided by other regions.

With this approach, the functionally most abstract test cases are derived from the abstract model parts, while the least abstract test cases may be modeled by one or more parts of the model, depending on the signals needed. A link between the different model parts will only be required for the synchronization of the test behavior across abstraction levels. Such a synchronization will be necessary for modeling dependencies between signals belonging to different groups, for instance.

The separation of test behavior into different parts concerning different abstraction levels permits avoiding the design of TMCs mixing abstract and detailed behavior. This aspect is of particular significance for test case derivation. Multilevel test cases testing at a specific functional abstraction level cannot evaluate more detailed behavior for the following three reasons.

Firstly, the corresponding signals may not be provided by the test object. In this case, the signals providing detailed information about the test object behavior will be internal signals within the SUT that are not visible at the SUT interface.

Secondly, even in case these internal signals were observable, there would be no point in observing them at the current abstraction level. Functional testing is a black-box technique and so the only interesting observation signals are those at the SUT interface.

Thirdly, the ITA will not be able to create these internal signals because the SUT interface will not provide the details that are necessary for reconstructing those internal signals featuring a lower abstraction level.

In summary, we have presented an approach for extending multi-level test cases to multilevel test models, which possess a similar structure at a higher functional abstraction level. For the design and implementation of multilevel test models, we basically follow the strategy for multilevel test cases presented in Section 11.6 with some extensions such as test behavior parts. The resulting test models will cover all abstraction levels and all interface signal groups for a single system's function. Furthermore, multilevel test cases derived from multilevel test models feature a separation of the test behavior into stimulation and evaluation, as well as into different abstraction levels.

11.8 Case Study: Automated Light Control

This section presents a case study of a multilevel test model for the vehicle function ALC. With this example, we aim at validating the proposed approach. This function is similar to the ALC function presented in Schieferdecker et al. (2006). As a proof of concept, we will follow the design strategy proposed in this contribution for the ALC. Hence, we will start specifying a set of test cases using the top-down reuse approach. We will then proceed with the design of the multilevel test model using Simulink® 6.5.* The design will also include a set of test adapter models.

11.8.1 Test specification

The ALC controls the state of the headlights automatically by observing the actual outside illumination and switching them on in the darkness. The automated control can be overridden using the headlight rotary switch, which has three positions: *ON*, *OFF*, and *AUTO*. For this functionality at the very top level, we create two exemplary test cases TC_1 and TC_2. TC_1 switches between *OFF* and *ON*, while TC_2 brings the car into darkness and back to light while the switch is in the *AUTO* position.

Three ECUs contribute to the ALC, namely the driver panel (DP), the light sensor control (LSC), and the ALC. Figure 11.9 depicts how these ECUs are related. All three control units are connected via a CAN-Bus. The rotary light switch is connected to the DP, the two light sensors report to the LSC, and the headlights are directly driven from the ALC.

At the system component test level we reuse both test cases TC_1 and TC_2 for testing the ALC, but in addition create a new test case TC_3 in which the headlights are switched on because of a timeout in the switch position signal in the CAN bus. The existence of

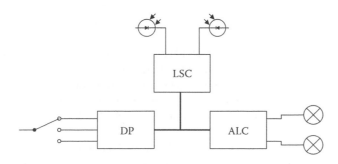

FIGURE 11.9
ALC system.

---

*The MathWorks, Inc.—MATLAB®/Simulink®/Stateflow®, 2006, http://www.mathworks.com/.

timeouts in the CAN bus is not within the scope of the system integration abstraction level, so that this test case belongs to the second test case group. The effect will be that the headlights are ON as long as the current switch position is not received via the CAN bus.

The ALC ECU includes two software components: the darkness detector (DD) and the headlight control (HLC). The DD receives an illumination signal from the LSC ECU and returns *false* if it is light outside and *true* if it is dark. The HLC uses this information as well as the switch position for deciding whether to switch the headlights on. At the software test level, we reuse all three test cases from the test levels above without adding any new test cases. This means that there are no functional details to test that were not already included in the test cases of the test level above.

Finally, at the software component level, we reuse the test cases once again but add two new test cases TC_4 and TC_5. TC_4 tests that the headlights are ON when the value 3 is received at the switch position input. This value is invalid, because only the integer values $0 = OFF$, $1 = ON$, and $2 = AUTO$ are permitted. When detecting an invalid value, the headlights should be switched on for safety reasons. TC_5 checks that variations in the darkness input do not affect the headlights when the switch is in the ON position. This test case is not interesting at any higher test level, although it would be executable. In terms of functional abstraction, it is too detailed for reuse.

All five specified test cases are included in Table 11.1. Each test case consists of a set of test steps for each of which there are indications regarding the actions that must be performed and the expected results. In analogy to Hiller et al. (2008), we have also specified at which test levels each test case must be executed.

Using a significantly reduced number of test cases and a simple example, this test specification demonstrates how the top-down reuse approach works. We obtain test cases that must be reused at different test levels. Instead of specifying these test cases four times (once at each test level) each time analyzing a different artifact on the V model's left-hand branch, our approach only requires a single specification. The reduction in effort attained by this identification of common test cases is not only limited to test case creation but also applies to other activities such as reviews or updates. These activities benefit from test specifications that do not include multiple test cases being similar or even identical.

11.8.2   Test model core design

We design the core of our multilevel test model in Simulink® as shown in Figure 11.10. The TMC generates input values for the HLC software component in the stimulation subsystem, evaluates the headlights state in the evaluation subsystem, and includes an additional test control subsystem that provides the current test step for synchronizing stimulation and evaluation. The test control concept used in this contribution is similar to the one discussed in Zander-Nowicka et al. (2007).

The test design follows a bottom-up strategy. The HLC component features two inputs (Switch and Darkness) and a single output (Headlights). All these inputs are utilized by the top-level test cases in Table 11.1 so that there is a single interface group in this example and all test cases will share these interface signals. As a consequence, there will be only one test behavior part within the TMC, even though we will be designing behavior at different abstraction levels.

Both TS and TE are designed in a similar way (see Figure 11.11). The behavior of each test step is described separately and later merged. Figure 11.12 provides insight into the stimuli generation for the last test step (cf. Table 11.1). Note that the variables are not declared in the model but as test parameters in MATLAB's® workspace. Figure 11.13 presents the TE for the last test step. The test verdict for this step is computed as specified in Table 11.1. Finally, the test control shown in Figure 11.14 mainly consists of a Stateflow

TABLE 11.1

Test specification for the automated light control function

TC_1: Headlights ON/OFF switching
Test levels: 1, 2, 3, 4

Step	Actions	Pass Conditions
1	Set switch to *ON*	headlights = *ON*
2	Set switch to *OFF*	headlights = *OFF*

TC_2: Headlights automatic ON/OFF
Test levels: 1, 2, 3, 4

Step	Actions	Pass Conditions
1	Set switch to *AUTO*	-
2	Set illumination to *LIGHT*	headlights = *OFF*
3	Set illumination to *DARK*	headlights = *ON*
4	Set illumination to *LIGHT*	headlights = *OFF*

TC_3: Headlights ON—switch CAN timeout
Test levels: 1, 2, 3

Step	Actions	Pass Conditions
1	Set switch to *OFF*	headlights = *OFF*
2	CAN timeout	headlights = *ON*
3	Remove timeout	headlights = *OFF*

TC_4: Headlights *ON* for invalid switch position input
Test levels: 1

Step	Actions	Pass Conditions
1	Set switch to *AUTO*	-
2	Set illumination to *LIGHT*	headlights = *OFF*
3	Set switch to INVALID	headlights = *ON*
4	Set switch to AUTO	headlights = *OFF*

TC_5: Darkness variations with switch *ON*
Test levels: 1

Step	Actions	Pass Conditions
1	Set switch to *ON*	headlights = *ON*
2	Set darkness to *false*	headlights = *ON*
3	Set darkness to *true*	headlights = *ON*
4	Set darkness to *false*	headlights = *ON*

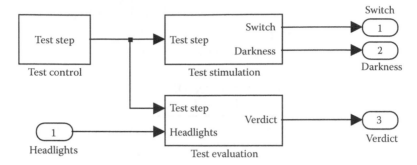

FIGURE 11.10
Test model core.

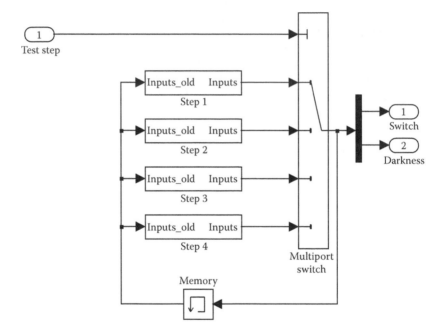

FIGURE 11.11
Test stimulation.

diagram, where each state represents a test step. Within the diagram, the variable t represents the local time, that is, the time the test case has spent in the current state.

Since the test interface exclusively contains shared signals, both stimulation and evaluation models take advantage of this similarity, making the model simpler (cf. Figures 11.12 and 11.13). Besides this modeling effect, the interface abstraction is also evident, particularly for the darkness signal, which is actually a simple Boolean signal instead of the signal coming from the sensor.

11.8.3 Test adapter models

We implement two test adapter models (input and output) for every test level in Simulink®. These models describe the relation between interfaces across test levels. The aim in this case

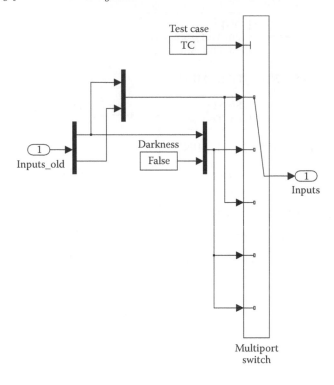

FIGURE 11.12
Test stimulation (test step 4).

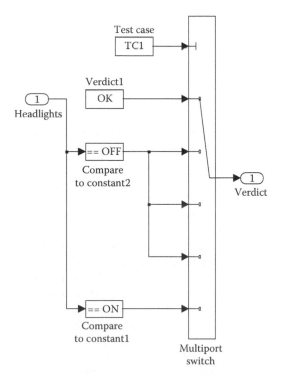

FIGURE 11.13
Test evaluation (test step 4).

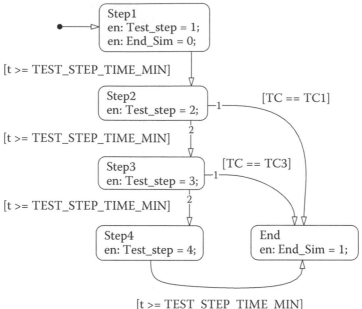

FIGURE 11.14
Test control.

FIGURE 11.15
Input test adapter model for system testing.

is not to model partial but complete behavior so that the test adapter models are valid for any additionally created scenarios without requiring updates.

The ITAMs are in charge of abstracting the test object interface toward the TMC inputs. In the ALC example, there is only a single observation signal in the TMC, namely the Boolean signal *headlights*. In fact, there will not be much to abstract from. Within the software, we will be able to access this signal, and for the hardware, we have a test platform that is capable of measuring the state of the driver for the relay that is executing on the ECU. Hence, the test adapters basically have to pass the input signals to the outputs while only the name may change (see for instance Figure 11.15).

At the system component test level the ITAM is of more interest, as shown in Figure 11.16. Here, the test platform provides inverted information on the relay status, that is, *Headlights_TP = true* implies that the headlights are *off*. Thus, the test adapter must logically negate the relay signal in order to adapt it to the test model.

For the ALC, we applied test adapter model concatenation in order to optimize the modeling. With this practice, it is not necessary to logically negate the relay signal at the system test level anymore (cf. Figure 11.15), even though we observe exactly the same signal. The test team at the system test level uses a test model including a test adapter chain from all test levels below. Hence, the test model already includes the negation.

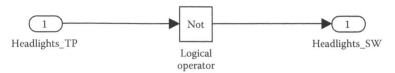

FIGURE 11.16
Input test adapter model for system component testing.

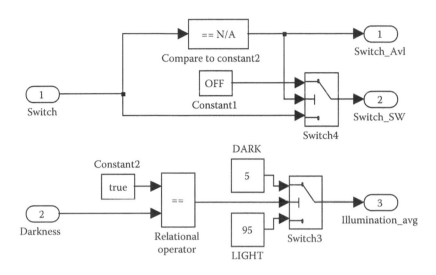

FIGURE 11.17
Output test adapter model for software testing.

The OTAMs do not abstract but refine the test model interfaces introducing additional details. A good example is the OTAM for the software test level that is depicted in Figure 11.17.

The first signal to be refined is the switch position. There is another software component within the ALC ECU that merges the signals *Switch_Avl* and *Switch_SW*.* As a consequence, the test adapter has to invert this behavior by splitting the switch position. When the position is *N/A* (not available), the *Switch_SW* signal is set to *OFF*.

The darkness signal also has to be refined toward the *Illumination_avg*, which is a continuous signal containing the average value of the two light sensors and whose range goes from 0 (dark) to 100 (light). In this case, we invert the function of the DD by setting an illumination of 5% for darkness and 95% for daylight.

The OTAM for system component testing includes a data type conversion block in which a Boolean signal is refined to an 8-bit integer in order to match the data type at the test platform (cf. Figure 11.18). The test platform allows us to filter all CAN messages including the switch position information in order to stimulate a timeout.

At the system test level, the status of the switch signal on the CAN bus is out of the abstraction level's scope (see OTAM in Figure 11.19). Had we actually used this signal for

---

Switch_Avl is a Boolean signal created by the CAN bus driver that indicates a timeout when active, that is, the information on the switch position has not been received for some specified period of time and is hence not available. *Switch_SW* represents the signal coming from the CAN bus driver into the software, containing the information on the switch position.

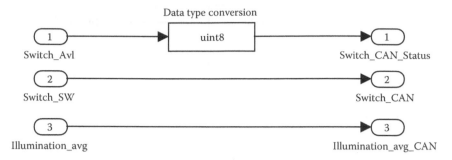

FIGURE 11.18
Output test adapter model for system component testing.

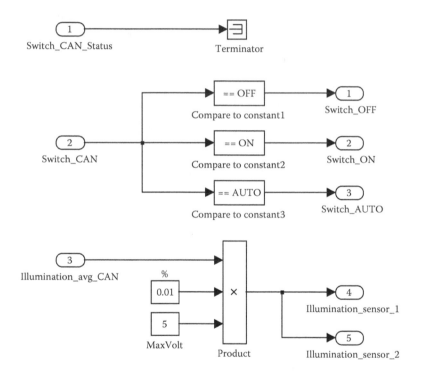

FIGURE 11.19
Output test adapter model for system testing.

the TMC interface instead of mapping it to the software component input, this signal would have belonged to the signal group corresponding to the system component test level.

The switch delivers three Boolean signals, one for each position. We have to split this signal here so that the DP can merge it again. For the illumination, we must refine the average into two sensor signals delivering a voltage between 0 and 5 V. We opt to provide identical sensor signals through the test adapter for simplicity's sake.

In analogy to the ITAMs, the advantages of test adapter concatenation are clearly visible. Additionally, several interface refinements have been presented.

The ALC case study provides insight into the concepts presented in this chapter, particularly for multilevel test models. It is a concise example that demonstrates that the proposed design strategies are feasible and practicable. Even though the test modeling techniques applied in Simulink® feature a rather low abstraction level, the advantages of reuse, particularly in terms of reductions in effort, have become clear.

11.9 Conclusion

Multilevel testing is an integrative approach to testing across test levels that is based on test reuse and interface refinement. In this chapter, we presented test specification and test design strategies both for test cases and test models that aim at integrating test levels. We have identified test level integration as an approach promising a major potential for reduction of test effort. The strategies described in this chapter are applicable to both test models and test cases based on the two key instruments of the developed methodology: multilevel test models and multilevel test cases. We paid particular attention to describing the idiosyncrasies of the model-based approach in comparison to using test cases while aiming at formulating a generic and flexible methodology that is applicable to different kinds of models. There are no comparable approaches in literature for test level integration, particularly none that are this extensive.

The benefits of multilevel testing include significant reductions in effort, especially if the multilevel test models are utilized only when they are more efficient than conventional approaches. Apart from the efficiency gains, we have discussed further reuse benefits in this contribution. Among the new possibilities provided by our test level integration, these benefits include improving cross-level collaboration between test teams for more systematic testing, supporting test management methods and tools, and automatically establishing a vertical traceability between test cases across test levels.

We have evaluated our multilevel testing approach in four industrial projects within production environments. Details of this effort are documented in other work (Marrero Pérez and Kaiser 2010a). As a summary of the results, we compared the test cases at two different test levels and found that around 60% of the test cases were shared by both test levels, which is a very high rate considering that the test levels were not consecutive.

We applied our multilevel testing approach to reuse the test cases from the lower test level applying interface refinement and noticed substantial reductions in test effort with respect to test design and test management. The only exception in our analysis was the validation of the test adapter models, which caused greater effort than expected because of our manual approach. However, we expect to noticeably reduce this effort by automating the creation of the test adapter models, an approach whose feasibility has been demonstrated in Schätz and Pfaller (2010).

Lastly, our evaluation confirmed that our test level integration approach scales. We validated our approach in projects of different sizes including both small and large test suites and obtained comparable results in both cases.

Summing up, we offer an optimized approach tailored to embedded systems, but applicable to other domains whose test process is characterized by multiple test levels. This contribution advocates focusing on systematic methodologies more intensively, after years of concentration on technical aspects within the testing domain.

Acknowledgments

We would like to thank Oliver Heerde for reviewing this manuscript.

References

Aichernig, B., Krenn, W., Eriksson, H., and Vinter, J. (2008). State of the art survey—Part a: Model-based test case generation.

Beizer, B. (1990). *Software Testing Techniques*. International Thomson Computer Press, London, UK, 2 edition.

Benz, S. (2007). Combining test case generation for component and integration testing. In *3rd International Workshop on Advances in Model-based Testing (A-MOST 2007)*, Pages: 23–33.

Binder, R.V. (1999). *Testing Object-oriented Systems*. Addison-Wesley, Reading, MA.

Broy, M. (2006). Challenges in automotive software engineering. In *28th International Conference on Software Engineering (ICSE 2006)*, Pages: 33–42.

Bühler, O. and Wegener, J. (2008). Evolutionary functional testing. *Computers & Operations Research*, 35(10):3144–3160.

Burmester, S. and Lamberg, K. (2008). Aktuelle Trends beim automatisierten Steuergerätetest. In Gühmann, C., editor, *Simulation und Test in der Funktions- und Softwareentwicklung für die Automobilelektronik II*, Pages: 102–111.

Deutsche Gesellschaft für Qualität e.V. (1992). *Methoden und Verfahren der Software-Qualitätssicherung*, Volume 12-52 of *DGQ-ITG-Schrift*. Beuth, Berlin, Germany, 1 edition.

European Telecommunications Standards Institute. The Testing and Test Control Notation version 3; Part 1: TTCN-3 Core Language, 2009-06.

Gisi, M.A. and Sacchi, C. (1993). A positive experience with software reuse supported by a software bus framework. In *Advances in software reuse: Selected papers from the 2nd International Workshop on Software Reusability (IWSR-2)*, Pages: 196–203.

Gruszczynski, B. (2006). An overview of the current state of software engineering in embedded automotive electronics. In *IEEE International Conference on Electro/Information Technology*, Pages: 377–381.

Hiller, S., Nowak, S., Paulus, H., and Schmitfranz, B.-H. (2008). Durchgängige Testmethode in der Entwicklung von Motorkomponenten zum Nachweis der Funktionsanforderungen im Lastenheft. In Reuss, H.-C., editor, *AutoTest 2008*.

Larsen, K.G., Mikucionis, M., Nielsen, B., and Skou, A. (2005). Testing real-time embedded software using UPPAAL-TRON. In *5th ACM International Conference on Embedded Software (EMSOFT 2005)*, Pages: 299–306.

Lehmann, E. (2003). *Time partition testing*. PhD thesis, Technische Universität Berlin.

Lindlar, F. and Marrero Pérez, A. (2009). Using evolutionary algorithms to select parameters from equivalence classes. In Schlingloff, H., Vos, T.E.J., and Wegener, J., editors, *Evolutionary test generation*, Volume 08351 of *Dagstuhl Seminar Proceedings*.

Mäki-Asiala, P. (2005). *Reuse of TTCN-3 code*, Volume 557 of *VTT Publications*. VTT, Espoo, Finland.

Marrero Pérez, A. and Kaiser, S. (2009). Integrating test levels for embedded systems. In *Testing: Academic & Industrial Conference – Practice and Research Techniques (TAIC PART 2009)*, Pages: 184–193.

Marrero Pérez, A. and Kaiser, S. (2010a). Bottom-up reuse for multi-level testing. *The Journal of Systems and Software*, 83(12): 2392–2415.

Marrero Pérez, A. and Kaiser, S. (2010b). Multi-level test models for embedded systems. In *Software Engineering (SE 2010)*, Pages: 213–224.

Prenninger, W. and Pretschner, A. (2005). Abstractions for model-based testing. *Electronic Notes in Theoretical Computer Science*, 116:59–71.

Schätz, B. and Pfaller, C. (2010). Integrating component tests to system tests. *Electronic Notes in Theoretical Computer Science*, 260:225–241.

Schäuffele, J. and Zurawka, T. (2006). *Automotive software engineering*. Vieweg, Wiesbaden, Germany, 3 edition.

Schieferdecker, I., Bringmann, E., and Großmann, J. (2006). Continuous TTCN-3: Testing of embedded control systems. In *Workshop on Software Egineering for Automotive Systems (SEAS 2006)*, Pages: 29–36.

Spillner, A., Linz, T., and Schaefer, H. (2007). *Software testing foundations*. Rockynook, Santa Barbara, CA, 2nd edition.

Wartik, S. and Davis, T. (1999). A phased reuse adoption model. *The Journal of Systems and Software*, 46(1):13–23.

Wiese, M., Hetzel, G., and Reuss, H.-C. (2008). Optimierung von E/E-Funktionstests durch Homogenisierung und Frontloading. In Reuss, H.-C., editor, *AutoTest 2008*.

Yellin, D.M. and Strom, R.E. (1997). Protocol specifications and component adaptors. *ACM Transactions on Programming Languages and Systems*, 19(2):292–333.

Zander-Nowicka, J. (2008). *Model-based testing of real-time embedded systems in the automotive domain*. PhD thesis, Technische Universität Berlin.

Zander-Nowicka, J., Marrero Pérez, A., Schieferdecker, I., and Dai, Z. R. (2007). Test design patterns for embedded systems. In Schieferdecker, I. and Goericke, S., editors, *Business Process Engineering. 10th International Conference on Quality Engineering in Software Technology (CONQUEST 2007)*, Pages: 183–200.

12

Model-Based X-in-the-Loop Testing

**Jürgen Großmann, Philip Makedonski, Hans-Werner Wiesbrock,
Jaroslav Svacina, Ina Schieferdecker, and Jens Grabowski**

CONTENTS

Software-driven electronic control units (ECUs) are increasingly adopted in the creation of more secure, comfortable, and flexible systems. Unlike conventional software applications, ECUs are real-time systems that may be affected directly by the physical environment they operate in. Whereas for software applications testing with specified inputs and checking whether the outputs match the expectations are in many cases sufficient, such an approach is no longer adequate for the testing of ECUs. Because of the real-time requirements and the close interrelation with the physical environment, proper testing of ECUs must directly consider the feedback from the environment, as well as the feedback from the system under test (SUT) to generate adequate test input data and calculate the test verdict. Such simulation and testing approaches dedicated to verify feedback control systems are normally realized using so called closed loop architectures (Montenegro, Jhnichen, and Maibaum 2006,

Lu et al. 2002, Kendall and Jones 1999), where the part of the feedback control system that is being verified is said to be "in the loop." During the respective stages in the development lifecycle of ECUs, models, software, and hardware are commonly placed in the loop for testing purposes.

Currently, often proprietary technologies are used to set up closed loop testing environments and there is no methodology that allows the technology-independent specification and systematic reuse of testing artifacts, such as tests, environment models, etc. for closed loop testing. In this chapter, we propose such a methodology, namely "X-in-the-Loop testing," which encompasses the testing activities and the involved artifacts during the different development stages. This work is based on the results from the TEMEA project*. Our approach starts with a systematic differentiation of the individual artifacts and architectural elements that are involved in "X-in-the-loop" testing. Apart from the SUT and the tests, the environment models, in particular, must be considered as a subject of systematic design, development, and reuse. Similar to test cases, they shall be designed to be independent from test platform-specific functionalities and thus be reusable on different testing levels.

This chapter introduces a generic approach for the specification of reusable "X-in-the-loop" tests on the basis of established modeling and testing technologies. Environment modeling in our context will be based on Simulink® (The MathWorks 2010b). For the specification and realization of the tests, we propose the use of TTCN-3 embedded (TEMEA Project 2010), an extended version of the standardized test specification language TTCN-3 (ETSI 2009b, ETSI 2009a). The chapter starts with a short motivation in Section 12.1 and provides some generic information about artifact reuse in Section 12.2. In Section 12.3, we describe an overall test architecture for reusable closed loop tests. Section 12.4 introduces TTCN-3 embedded, Section 12.5 provides examples on how vertical and horizontal reuse can be applied to test artifacts, and Section 12.6 presents reuse as a test quality issue. Section 12.7 concludes the chapter.

12.1 Motivation

An ECU usually interacts directly with its environment, using sensors and actuators in the case of a physical environment, and with network systems in the case of an environment that consists of other ECUs. To be able to run and test such systems, the feedback from the environment is essential and must usually be simulated. Normally, such a simulation is defined by so-called *environment models* that are directly linked with either the ECU itself during Hardware-in-the-Loop (HiL) tests, the software of the ECU during Software-in-the-Loop (SiL) tests, or in the case of Model-in-the-Loop (MiL) tests, with an executable model of the ECU's software. Apart from the technical differences that are caused by the different execution objects (an ECU, the ECU's software, or a model of it), the three scenarios are based on a common architecture, the so-called *closed loop architecture*.

Following this approach, a test architecture can be structurally defined by *generic environment models* and specific *functionality-related test stimuli* that are applied to the closed loop. The environment model and the SUT constitute a self-contained functional entity, which is executable without applying any test stimuli. To accommodate such an architecture, test scenarios in this context apply a systematic interference with the intention to disrupt the functionality of the SUT and the environment model. The specification of

---

*The project TEMEA "**Te**sting Specification Technology and **M**ethodology for **E**mbedded Real-Time Systems in **A**utomobiles" (TEMEA 2010) is co-financed by the European Union. The funds are originated from the European Regional Development Fund (ERDF).

such test scenarios has to consider certain architectural requirements. We need *reactive stimulus components* for the generation of test input signals that depend on the SUT's outcome, *assessment capabilities* for the analysis of the SUT's reaction, and a *verdict setting mechanism* to propagate the test results. Furthermore, we recommend a *test specification and execution language*, which is expressive enough to deal with reactive control systems.

Because of the application of model-based software development strategies in the automotive domain, the design and development of reusable models are well known and belong to the state of the art (Conrad and Dörr 2006, Fey et al. 2007, Harrison et al. 2009). These approved development strategies and methods can be directly ported to develop highly reusable environment models for testing and simulation and thus provide a basis for a generic test architecture that is dedicated to the reuse of test artifacts. Meanwhile, there are a number of methods and tools available for the specification and realization of environment models, such as Simulink® and Modelica (Modelica Association 2010). Simulink® in particular is supported by various testing tools and is already well established in the automotive industry. Both Modelica and Simulink® provide a solid technical basis for the realization of environment models, which can be used either as self-contained simulation nodes, or, in combination with other simulation tools, as part of a co-simulation environment.

12.2 Reusability Pattern for Testing Artifacts

Software reuse (Karlsson 1995) has been an important topic in software engineering in both research and industry for quite a while now. It is gaining a new momentum with emerging research fields such as software evolution. Reuse of existing solutions for complex problems minimizes extra work and the opportunity to make mistakes. The reuse of test specifications, however, has only recently been actively investigated. Notably, the reusability of TTCN-3 tests has been studied in detail as a part of the Tests & Testing Methodologies with Advanced Languages (TT-Medal) (TT-Medal 2010) project, but the issue has been investigated also in Karinsalo and Abrahamsson 2004, Mäki-Asiala 2004. Reuse has been studied on three levels within TT-Medal—TTCN-3 language level (Mäki-Asiala, Kärki, and Vouffo 2006, Mäki-Asiala et al. 2005), test process level (Mäntyniemi et al. 2005), and test system level (Kärki et al. 2005). In the following, focus will be mainly on reusability on the TTCN-3 language level and how the identified concepts transfer to TTCN-3 embedded.

Means for the development of reusable assets generally include establishing and maintaining a good and consistent structure of the assets, the definition of and adherence to standards, norms, and conventions. It is furthermore necessary to establish well-defined interfaces and to decouple the assets from the environment. In order to make the reusable assets also usable, detailed documentation is necessary, but also the proper management of the reusable assets, which involves collection and classification for easy locating and retrieval. Additionally, the desired granularity of reuse has to be established upfront so that focus can be put on a particular level of reuse, for example, on a component level or on a function level.

On the other hand, there are the three viewpoints on test reuse as identified in Kärki et al. 2005, Mäki-Asiala 2004, Mäki-Asiala et al. 2005:

Vertical - which is concerned with the reuse between testing levels or types (e.g., component and integration testing, functional and performance testing);

Horizontal - which is concerned with the reuse between products in the same domain or family (e.g., standardized test suites, tests for product families, or tests for product lines);

Historical - which is concerned with the reuse between product generations (e.g., regression testing).

While the horizontal and historical viewpoints have been long recognized in software reuse, vertical reuse is predominantly applicable only to test assets. "X-in-the-Loop" testing is closest to the vertical viewpoint on reuse, although it cannot be entirely mapped to any of the viewpoints since it also addresses the horizontal and historical viewpoints as described in the subsequent sections.

Nevertheless, the reuse of real-time test assets can be problematic, since, similar to real-time software, context-specific time, and synchronization constraints are often embedded in the reusable entities. Close relations between the base functionality and the real-time constraints often cause interdependencies that reduce the reusability potential. Thus, emphasis shall be placed on context-independent design from the onset of development, identifying possible unwanted dependencies on the desired level of abstraction and trying to avoid them whenever a feasible alternative can be used instead. This approach to reuse, referred to as "revolutionary" (Mäntyniemi et al. 2005), or "reuse by design" involves more upfront planning and a sweeping transformation on the organizational level, which requires significant experience in reuse. The main benefit is that multiple implementations are not necessary. In contrast, the "evolutionary" (Mäntyniemi et al. 2005) approach to reuse involves a gradual transition toward reusable assets, throughout the development, by means of adaptation and refactoring to suit reusability needs as they emerge. The evolutionary approach requires less upfront investment in and knowledge of reuse and involves less associated risks, but in turn may also yield less benefits. Knowledge is accumulated during the development process and the reusability potential is identified on site. Such an approach is better suited for vertical reuse in systems where requirements are still changing often.

Despite the many enabling factors from a technological perspective, a number of organizational factors inhibiting the adoption of reuse, as well as the risks involved, have been identified in Lynex and Layzell 1997, Lynex and Layzell 1998, Tripathi and Gupta 2006. Such organizational considerations are concerned primarily with the uncertainties related to the potential for reusability of production code and its realization.

The basic principle of this approach is to develop usable assets first, then turn them into reusable ones. In the context of "X-in-the-Loop" testing, the aim is to establish reusability as a design principle, by providing a framework, an architecture, and support at the language level. Consequently, a revolutionary approach to the development of the test assets is necessary.

12.3   A Generic Closed Loop Architecture for Testing

A closed loop architecture describes a feedback control system. In contrast to open loop architectures and simple feedforward controls, models, especially the environment model, form a central part of the architecture. The input data for the execution object is calculated directly by the environment model, which itself is influenced by the output of the execution object. Thus, both execution object and environment model form a self-contained entity. In terms of testing, closed loop architectures are more difficult to handle than open loop architectures. Instead of defining a set of input data and assessing the related output data, tests in a closed loop scenario have to be integrated with the environment model. Usually, neither environment modeling nor the integration with the test system and the individual tests are carried out in a generic way. Thus, it would be rather difficult to properly define and describe test cases, to manage them, and even to reuse them partially.

In contrast, we propose a systematic approach on how to design reusable environment models and test cases. We think of an environment model as defining a generic test scenario. The individual test cases are defined as perturbations of the closed loop runs in a controlled manner. A test case in this sense is defined as a generic environment model (basic test scenario) together with the description of its intended perturbation. Depending on the underlying test strategies and intentions, the relevant perturbations can be designed based on functional requirements (e.g., as black box tests) or derived by manipulating standard inputs stochastically to test robustness. They can also be determined by analyzing limit values or checking interfaces. In all cases, we obtain a well-defined setting for closed loop test specifications.

To achieve better maintenance and reusability, we rely on two basic principles. First, we introduce a proper architecture, which allows the reuse of parts of an environment model. Secondly, we propose a standard test specification language, which is used to model the input perturbations and assessments and allows the reuse of parts of the test specification in other test environments.

12.3.1 A generic architecture for the environment model

For the description of the environment model, we propose a three layer architecture, consisting of a computation layer, a pre- and postprocessing layer, and a mapping (Figure 12.1). The computation layer contains the components responsible for the stimulation of the SUT (i.e., the abstract environment model and the perturbation component) and for the assessment of the SUT's output (i.e., the assessment component). The pre- and postprocessing

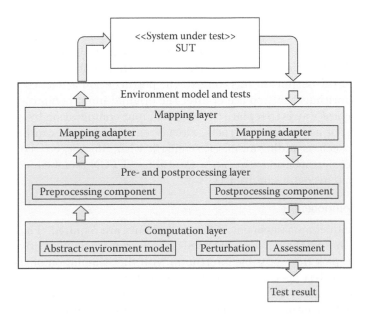

FIGURE 12.1
Closed loop architecture.

layer contains the preprocessing and postprocessing components, which are responsible for signal transformation. The mapping layer contains the so-called mapping adapters that provide the direct connectivity between the SUT and the environment model.

Based on this architecture, we will present a method for the testing of controls with feedback. In what follows, we present a more detailed description of the individual entities of the environment model.

12.3.1.1   The computation layer

An essential part in closed loop test systems is the calculation of the environmental reaction, that is, for each simulation step, the input values for the execution object are computed by means of the output data and the abstract environment model. Moreover, special components are dedicated to testing. The perturbation component is responsible for the production of test stimuli by means of test perturbations and the assessment component is responsible for the verdict setting.

- Abstract environment model

  The abstract environment model is a generic model that simulates the environment where the ECU operates in. The level of technical abstraction is the same as that of the test cases we intend to specify. In a model-based approach, such a model can be designed as a reusable entity and realized using commercial modeling tools like Simulink, Stateflow® (The MathWorks 2010c), or Modelica. Hardware entities that are connected to the test environment may be used as well. Moreover, the character of the abstract environment model directly depends on the chosen test level. For a module test, such a test model can be designed using Simulink. In an integration test, performed with CANoe (Vector Informatics 2010), for example, the model can be replaced by a Communication Application Programming Language (CAPL) Node. And last but not least, in a HiL scenario, this virtual node can be replaced by other ECUs that interoperate with the SUT via Controller Area Network (CAN) bus communication.

- Test perturbation

  As sketched above, a test stimulus specification is defined as a perturbation of the closed loop run. Thus, the closed loop is impaired and the calculation of the input data is partially dissociated from its original basis, that is, from the environment model output or the SUT output. In this context, the perturbation is defined by means of data generation algorithms that replace or alter the original input computation of the closed loop. The simplest way is to replace the original input computation by a synthetic signal. With the help of advanced constructs, already existing signals can be manipulated in various ways, such as adding an offset, scaling it with a factor, etc. For the algorithmic specification of perturbation sequences, we will use TTCN-3 embedded, the hybrid and continuous systems extension to TTCN-3.

- Assessment component

  In order to conduct an assessment, different concepts are required. The assessment component must observe the outcome of the SUT and set verdicts respectively, but it should not interfere (apart from possible interrupts) with the stimulation of the SUT. TTCN-3 is a standardized test specification and assessment language and its hybrid and continuous systems extension TTCN-33e provides just the proper concepts for the definition of assessments for continuous and message-based communication. Furthermore, by relying on a standard, assessment specifications can be reused in other environments, for example, conformance testing, functional testing, or interoperability testing.

12.3.1.2   The pre- and postprocessing layer and the mapping layer

The pre- and postprocessing layer (consisting of pre- and postprocessing components) and the mapping layer (consisting of mapping adapters) provide the intended level of abstraction between the SUT and the computation layer. Please note that we have chosen the following perspective here—preprocessing refers to the preparation of the data that is emitted by the SUT and fed into the testing environment and postprocessing refers to the preparation of the data that is emitted by the test perturbation or the abstract environment model and sent to the SUT.

- Preprocessing component

  The preprocessing component is responsible for the measurement and preparation of the outcome of the SUT for later use in the computation layer. Usually, it is neither intended nor possible to assess the data from the SUT without preprocessing. We need an abstract or condensed version of the output data. For example, the control of a lamp may be performed by pulse-width modulated current. To perform a proper assessment of the signal, it would suffice to know the duty cycle. The preprocessing component serves as such an abstraction layer. This component can be easily designed and developed using modeling tools, such as Simulink, Stateflow, or Modelica.

- Postprocessing component

  The postprocessing component is responsible for the generation of the concrete input data for the SUT. It adapts the low-level interfaces of the SUT to the interfaces of the computation layer, which are usually more abstract. This component is best modeled with the help of Simulink, Stateflow, Modelica, or other tools and programming languages, which are available in the underlying test and simulation infrastructure.

- Mapping adapter

  The mapping adapter is responsible for the syntactic decoupling of the environment model and the execution object, which in our case is the SUT. Its main purpose is to relate (map) the input ports of the SUT to the output ports of the environment model and vice versa. Thus, changing the names of the interfaces and ports of the SUT would only lead to slight changes in the mapping adapter.

12.3.2   Requirements on the development of the generic test model

Another essential part of a closed loop test is the modeling of the environment feedback. Such a model is an abstraction of the environment the SUT operates in. Because this model is only suited for testing and simulation, generally we need not be concerned with completeness and performance issues in any case. However, closed loop testing always depends on the quality of the test model and we should carefully develop this model to get reliable results when using real-time environments as well as when using simpler software-based simulation environments. Besides general quality issues that are well-known from model-based development, the preparation of reusable environment models must address some additional aspects.

Because of the signal transformations in the pre- and postprocessing components, we can design the test model at a high level of abstraction. This eases the reuse in different environments. Moreover, we are not bound to a certain processor architecture. When we use processors with floating point arithmetic in our test system, we do not have to bother with scalings. The possibility to upgrade the performance of our test system, for example, by adding more random access memory or a faster processor helps mitigate performance problems. Thus, when we develop a generic test model, we are primarily interested in the correct functionality and less in the performance or the structural correctness. We will therefore focus on functional tests using the aforementioned model and disregard structural tests.

To support the reuse in other projects or in later phases of the same project, we should carefully document the features of the model, their abstractions, their limits, and the meaning of the model parameters. We will need full version management for test models in order to be able to reproducibly check the correctness of the SUT behavior with the help of closed loop tests.

In order to develop proper environment models, high skills in the application domain and good capabilities in modeling techniques are required. The test engineer, along with the application or system engineer shall therefore consult the environment modeler to achieve sufficient testability and the most appropriate level of abstraction for the model.

12.3.3 Running example

As a guiding example, we will consider an engine controller that controls the engine speed by opening and closing the throttle. It is based on a demonstration example (Engine Timing Model with Closed Loop Control) provided by the MATLAB® (The MathWorks 2010a) and Simulink® tool suites.

The engine controller model (Figure 12.2) has three input values, namely, the valve trigger (`Valve_Trig_`), the engine speed (`Eng_Speed_`), and the set point for the engine speed (`Set_Point_`). Its objective is the control of the air flow to the engine by means of the throttle angle (`Throttle_Angle`). In contrast to the controller model, the controller software is specifically designed and implemented to run on a real ECU with fixed-point arithmetic, thus we must be concerned with scaling and rounding when testing the software.

The environment model (Figure 12.3) is a rather abstract model of a four-cylinder spark ignition engine. It processes the engine angular velocity (shortened: engine speed [rad/s]) and the crank angular velocity (shortened: crank speed [rad/s]) controlled by the throttle angle. The throttle angle is the control variable, the engine speed the observable. The angle of the throttle controls the air flow into the intake manifold. Depending on the engine speed, the air is forced from the intake manifold into the compression subsystem and periodically pumped into the combustion subsystem. By changing the period for pumping, the air mass provided for the combustion subsystem is regulated, and the torque of the engine, as well as the engine speed, is controlled. Technically, the throttle angle should be controlled at any valve trigger in such a way that the engine speed approximately reaches the set point speed.

To model this kind of environment model, we use floating point arithmetic. It is a rather small model, but it is reasonable enough to test basic features of our engine controller (e.g., the proper functionality, robustness, and stability). Moreover, we can potentially extend the scope of the model by providing calibration parameters to adapt the behavior of the model to different situations and controllers later.

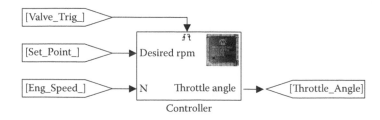

FIGURE 12.2
ECU model for a simple throttle control.

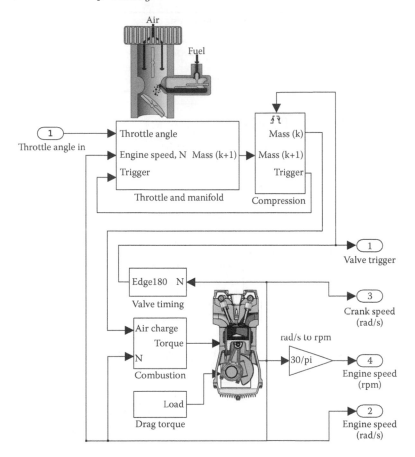

FIGURE 12.3
Environment model for a simple throttle control.

To test the example system, we will use the test architecture proposed in Section 12.3.1. The perturbation and assessment component is a TTCN-3 embedded component that compiles to a Simulink® S-Function. Figure 12.4 shows the complete architecture and identifies the individual components.

The resulting test interface, that is, the values that can be assessed and controlled by the perturbation and assessment component are depicted in Table 12.1. Note that we use a test system centric perspective for the test interface, that is, system inputs are declared as outputs and system or environment model outputs as inputs. The mapping between the test system-specific names and the system and environment model-specific names is defined in the first and second column of the table.

On the basis of the test architecture and the test interface, we are now able to identify typical test scenarios:

- The set point speed `to_Set_Point` jumps from low to high. How long will it take till the engine speed `ti_Engine_Speed` reaches the set point speed?

- The set point speed `to_Set_Point` falls from high to low. How long will it take till the engine speed `ti_Engine_Speed` reaches the set point speed? Are there any over-regularizations?

- The engine speed sensor retrieves perturbed values `to_Engine_Perturbation`. How does the controller behave?

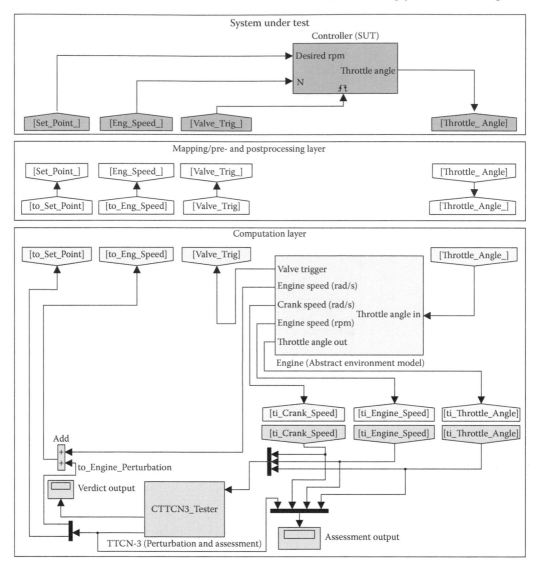

FIGURE 12.4
Test architecture in Simulink®.

TABLE 12.1
Engine Controller Test Interface

Test System Symbol	System Symbol	Direction	Unit	Data Type
ti_Crank_Speed	Crank Speed (rad/s)	In	rpm	Double
ti_Engine_Speed	Engine Speed (rpm)	In	rpm	Double
ti_Throttle_Angle	Throttle Angle Out	In	rad	Double
to_Engine_Perturbation	Eng_Speed_	Out	rpm	Double
to_Set_Point	Set_Point_	Out	rpm	Double

12.4   TTCN-3 Embedded for Closed Loop Tests

For the specification of the tests, we rely on a formal testing language, which provides dedicated means to specify the stimulation of the system and the assessment of the system's reaction. To emphasize the reusability, the language should provide at least sufficient support for modularization as well as support for the specification of reusable entities, such as functions, operations, and parameterization.

In general, a testing language for software-driven hybrid control systems should provide suitable abstractions to define and assess analog and sampled signals. This is necessary in order to be able to simulate the SUT's physical environment and to interact with dedicated environment models that show continuous input and output signals. On the other hand, modern control systems consist of distributed entities (e.g., controllers, sensors, actuators) that are interlinked by network infrastructures (e.g., CAN or FlexRay buses in the automotive domain). These distributed entities communicate with each other by exchanging complex messages using different communication paradigms such as asynchronous event-based communication or synchronous client server communication.* Typically, this kind of behavior is tested using testing languages, which provide support for event- or message-based communication and provide means to assess complex data structures.

In recent years, there have been many efforts to define and standardize formal testing languages. In the telecommunications industry, the Testing and Test Control Notation (TTCN-3) (ETSI 2009b, ETSI 2009a) is well established and widely proliferated. The language is a complete redefinition of the Tree and Tabular Combination Notation (TTCN-2) (ISO/IEC 1998). Both notations are standardized by the European Telecommunications Standards Institute (ETSI) and the International Telecommunication Union (ITU). Additional testing and simulation languages, especially ones devoted to continuous systems in particular, are available in the field of hardware testing or control system testing. The Very High Speed Integrated Circuit Hardware Description Language (VHDL) (IEEE 1993) and its derivative for analog and mixed signals (VHDL-AMS) (IEEE 1999) is useful in the simulation of discrete and analogue hardware systems. However, both languages were not specifically designed to be testing languages. The Boundary Scan Description Language (BSDL) (Parker and Oresjo 1991) and its derivative the Analog Boundary Scan Description Language (ABSDL) (Suparjo et al. 2006) are testing languages that directly support the testing of chips using the boundary scan architecture (IEEE 2001) defined by the Institute of Electrical and Electronics Engineers (IEEE). The Time Partition Testing Method (TPT) (Bringmann and Kraemer 2006) and the Test Markup Language (TestML) (Grossmann and Mueller 2006) are approaches that have been developed in the automotive industry about 10 years ago, but are not yet standardized. The Abbreviated Test Language for All Systems (ATLAS) (IEEE 1995) and its supplement, the Signal and Method Modeling Language (SMML) (IEEE 1998), define a language set that was mainly used to test control systems for military purposes. Moreover, the IEEE currently finalizes the standardization of an XML-based test exchange format, namely the Automatic Test Mark-up Language (ATML) (SCC20 ATML Group 2006), which is dedicated to exchanging information on test environments, test setups, and test results in a common way. The European Space Agency (ESA) defines requirements on a language used for the development of automated test and operation procedures and standardized a reference language called Test and Operations Procedure Language (ESA-ESTEC 2008). Last, but not

---

*Please refer to AUTOSAR (AUTOSAR Consortium 2010), which yields a good example of an innovative industry-grade approach to designing complex control system architectures for distributed environments.

least, there exist a huge number of proprietary test control languages that are designed and made available by commercial test system manufacturers or are developed and used in-house only.

Most of the languages mentioned above are neither able to deal with complex discrete data that are exhaustively used in network interaction, nor with distributed systems. On the other hand, TTCN-3, which is primarily specializing in testing distributed network systems, lacks support for discretized or analogue signals to stimulate or assess sensors and actuators. ATML, which potentially supports both, is only an exchange format, yet to be established, and still lacking user-friendly representation formats.

The TTCN-3 standard provides a formal testing language that has the power and expressiveness of a normal programming language with formal semantics and a user-friendly textual representation. It also provides strong concepts for the stimulation, control, and assessment of message-based and procedure-based communication in distributed environments. Our anticipation is that these kinds of communication will become much more important for distributed control systems in the future. Additionally, some of these concepts can be reused to define signal generators and assessors for continuous systems and thus provide a solid basis for the definition of analogue and discretized signals. Finally, the overall test system architecture proposed in the TTCN-3 standard (ETSI 2009c) shows abstractions that are similar to the ones we defined in Section 12.3.1. The TTCN-3 system adapter and the flexible codec entities provide abstraction mechanisms that mediate the differences between the technical SUT interface and the specification level interfaces of the test cases. This corresponds to the pre- and postprocessing components from Section 12.3.1. Moreover, the TTCN-3 map statement allows the flexible specification of the mappings between so-called ports at runtime. Ports in TTCN-3 denote the communication-related interface entities of the SUT and the test system. Hence, the map statement directly corresponds to the mapping components from Section 12.3.1. In addition, the TTCN-3 standard defines a set of generic interfaces (i.e., Test Runtime Interface (TRI) (ETSI 2009c), Test Control Interface (TCI) (ETSI 2009d)) that precisely specify the interactions between the test executable, the adapters, and the codecs, and show a generalizable approach for a common test system architecture. Last, but not least, the TTCN-3 standard is actually one of the major European testing standards with a large number of contributors.

To overcome its limitations and to open TTCN-3 for embedded systems in general and for continuous real-time systems in particular, the standard must be extended. A proposal for such an extension, namely TTCN-3 embedded (TEMEA Project 2010), was developed within the TEMEA (TEMEA 2010) research project and integrates former attempts to resolve this issue (Schieferdecker and Grossmann 2007, Schieferdecker, Bringmann, and Grossmann 2006). Next, we will outline the basic constructs of TTCN-3 and TTCN-3 embedded and show how the underlying concepts fit to our approach for closed loop testing.

12.4.1   Basic concepts of TTCN-3 embedded

TTCN-3 is a procedural testing language. Test behavior is defined by algorithms that typically assign messages to ports and evaluate messages from ports. For the assessment of different alternatives of expected messages, or timeouts, the port queues and the timeout queues are frozen when the assessment starts. This kind of snapshot semantics guarantees a consistent view on the test system input during an individual assessment step. Whereas the snapshot semantics provides means for a pseudo parallel evaluation of messages from several ports, there is no notion of simultaneous stimulation and time-triggered evaluation. To enhance the core language for the requirements of continuous and hybrid behavior, we introduce the following:

- The notions of time and sampling.

- The notions of streams, stream ports, and stream variables.

- The definition of statements to model a control flow structure similar to that of hybrid automata.

We will not present a complete and exhaustive overview of TTCN-3 embedded.[*] Instead, we will highlight some basic concepts, in part by providing examples and show the applicability of the language constructs to the closed loop architecture defined in the sections above.

12.4.1.1   Time

TTCN-3 embedded provides dedicated support for time measurement and time-triggered control of the test system's actions. Time is measured using a global clock, which starts at the beginning of each test case. The actual value of the clock is given as a float value that represents seconds and which is accessible in TTCN-3 embedded using the keyword **now**. The clock is sampled, thus it is periodically updated and has a maximum precision defined by the sampling step size. The step size can be specified by means of annotations to the overall TTCN-3 embedded module. Listing 12.1 shows the definition of a TTCN-3 embedded module that demands a periodic sampling with a step size of 1 msec.

Listing 12.1
Time

---

module myModule { ... } **with** {**stepsize** "0.001"}   1

12.4.1.2   Streams

TTCN-3 supports a component-based test architecture. On a conceptual level, test components are the executable entities of a test program. To realize a test setup, at least one test component and an SUT are required. Test components and the SUT are communicating by means of dedicated interfaces called ports. While in standard TTCN-3 interactions between the test components and the SUT are realized by sending and receiving messages through ports, the interaction between continuous systems can be represented by means of so-called streams. In contrast to scalar values, a stream represents the entire allocation history applied to a port. In computer science, streams are widely used to describe finite or infinite data flows. To represent the relation to time, so-called timed streams (Broy 1997, Lehmann 2004) are used. Timed streams additionally provide timing information for each stream value and thus enable the traceability of timed behavior. TTCN-3 embedded provides timed streams. In the following, we will use the term (measurement) record to denote the unity of a stream value and the related timing in timed streams. Thus, concerning the recording of continuous data, a record represents an individual measurement, consisting of a stream value that represents the data and timing information that represents the temporal perspective of such a measurement.

TTCN-3 embedded sends and receives stream values via ports. The properties of a port are described by means of port types. Listing 12.2 shows the definition of a port type for incoming and outgoing streams of the scalar type float and the definition of a component type that defines instances of these port types (`ti_Crank_Speed`, `ti_Engine_Speed`,

---

[*]For further information on TTCN-3 embedded, please refer to (TEMEA 2010).

ti_Throttle_Angle, to_Set_Point, to_Engine_Perturbation) with the characteristics defined
by the related port-type specifications (Listing 12.2, Lines 5 and 7).

Listing 12.2
Ports

```
type port FloatInPortType stream {in float};                                    1
type port FloatOutPortType stream {out float};                                  2
                                                                                3
type component EngineTester{                                                     4
  port FloatInPortType ti_Engine_Speed, ti_Crank_Speed, ti_Throttle_Angle;      5
  port FloatOutPort to_Engine_Perturbation, to_Set_Point;                       6
}                                                                               7
```

With the help of TTCN-3 embedded language constructs, it is possible to modify, access,
and assess stream values at ports. Listing 12.3 shows how stream values can be written to
an outgoing stream and read from an incoming stream.

Listing 12.3
Stream Value Access

```
to_Set_Point.value := 5000.0;                                                   1
to_Set_Point.value := ti_Engine_Speed.value + 200.0;                            2
```

Apart from the access to an actual stream value, TTCN-3 embedded provides access
to the history of stream values by means of index operators. We provide time-based
indices and sample-based indices. The time-based index operator **at** interprets time parameters as the time that has expired since the test has started. Thus, the expression
ti_Engine_Speed.at(10.0).value yields the value that has been available at the stream 10 s
after the test case has started. The sample-based index operator **prev** interprets the index
parameter as the number of sampling steps that have passed between the actual valuation and the one that will be returned by the operator. Thus, t_engine.prev(12).value
returns the valuation of the stream 12 sampling steps in the past. The expression
t_engine.prev.value is a short form of t_engine.prev(1).value. Listing 12.4 shows some
additional expressions based on the index operators.

Listing 12.4
Stream History Access

```
to_Set_Point.value := ti_Engine_Speed.at(10.0).value;                           1
to_Set_Point.value := ti_Engine_Speed.prev.value                                2
                    + ti_Engine_Speed_perturb.prev(2).value;                    3
```

Using the **assert** statement, we can assess the outcome of the SUT. The assert statement specifies the expected behavior on the SUT by means of relational expressions. Hence,
we can use simple relational operators that are already available in standard TTCN-3 and
apply them to the stream valuation described above to express predicates on the behavior
of the system. If any of the predicates specified by an active assert statement is violated,
the test verdict is automatically set to fail and the test fails. Listing 12.5 shows the specification of an assertion that checks whether the engine speed is in the range between 1000
and 3000 rpm.

Listing 12.5
Assert

```
assert ( ti_Engine_Speed . value > 1000.0 ,                           1
         ti_Engine_Speed . value < 3000.0) ;                          2
// the values must be in the range ]1000.0, 3000.0[.                 3
```

12.4.1.3   Access to time stamps and sampling-related information

To complement the data of a stream, TTCN-3 embedded additionally provides access to sampling-related information, such as time stamps and step sizes, so as to provide access to all the necessary information related to measurement records of a stream. The time stamp of a measurement is obtained by means of the **timestamp** operator. The timestamp operator yields the exact measurement time for a certain stream value. The exact measurement time denotes the moment when a stream value has been made available to the test system's input and thus strongly depends on the sampling rate. Listing 12.6 shows the retrieval of measurement time stamps for three different measurement records. Line 3 shows the retrieval of the measurement time for the actual measurement record of the engine speed, Line 4 shows the same for the previous record, and Line 5 shows the time stamp of the last measurement record that has been measured before or at 10 s after the start of the test case.

Listing 12.6
Time Stamp Access

```
var float myTimePoint1 , myTimePoint2 , myTimePoint3 ;               1
...                                                                  2
myTimePoint1 := ti_Engine_Speed . timestamp ;                        3
myTimePoint2 := ti_Engine_Speed . prev () . timestamp ;              4
myTimePoint3 := ti_Engine_Speed . at (10.0) . timestamp ;            5
```

As already noted, the result of the timestamp operator directly relates to the sampling rate. The result of `ti_Engine_Speed.timestamp` need not be equal to `now`, when we consider different sampling rates at ports. The same applies to the expression `ti_Engine_Speed.at(10.0).timestamp`. Dependent on the sampling rate, it may yield 10.0, or possibly an earlier time (e.g., when the sampling rate is 3.0, we will have measurement records for the time points 0.0, 3.0, 6.0, and 9.0 and the result of the expression will be 9.0).

In addition to the timestamp operator, TTCN-3 embedded enables one to obtain the step size that has been used to measure a certain value. This information is provided by the **delta** operator, which can be used in a similar way as the **value** and the **timestamp** operators. The delta operator returns the size of the sampling step (in seconds) that precedes the measurement of the respective measurement record. Thus, `ti_Engine_Speed.delta` returns:

```
ti_Engine_Speed.timestamp - ti_Engine_Speed.prev.timestamp
```

Please note, TTCN-3 embedded envisions dynamic sampling rates at ports. The delta and timestamp operators are motivated by the implementation of dynamic sampling strategies and thus can only develop their full potential in such contexts. Because of the space limitations, the corresponding, concepts are not explained here.

Listing 12.7 shows the retrieval of the step size for different measurement records.

Listing 12.7
Sampling Access

```
var float myStepSize1, myStepSize2, myStepSize3;                1
...                                                              2
myStepSize1 := ti_Engine_Speed.delta;                           3
myStepSize2 := ti_Engine_Speed.prev().delta;                    4
myStepSize3 := ti_Engine_Speed.at(10.0).delta;                  5
```

12.4.1.4 Integration of streams with existing TTCN-3 data structures

To enable the processing and assessment of stream values by means of existing TTCN-3 statements, we provide a mapping of streams, stream values, and the respective measurement records to standard TTCN-3 data structures, namely records and record-of structures. Thus, each measurement record, which is available at a stream port, can be represented by an ordinary TTCN-3 record with the structure defined in Listing 12.8. Such a record contains fields, which provide access to all value and sampling-related information described in the sections above. Thus, it includes the measurement value (`value_`) and its type* (`T`), its relation to absolute time by means of the `timestamp_` field as well as the time distance to its predecessor by means of the `delta_` field. Moreover, a complete stream or a stream segment maps to a record-of structure, which arranges subsequent measurement records (see Listing 12.8, Line 4).

Listing 12.8
Mapping to TTCN-3 Data Structures

```
Measurement <in type T> {T value_, float delta_, float timestamp_}   1
type record of Measurement<float> Float_Stream_Records;              2
```

To obtain stream data in accordance to the structure in Listing 12.8, TTCN-3 embedded provides an operation called **history**. The history operation extracts a segment of a stream from a given stream port and yields a record-of structure (stream record), which complies to the definitions stated above. Please note, the data type `T` depends on the data type of the stream port and is set automatically for each operation call.

The history operation has two parameters that characterize the segment by means of absolute time values. The first parameter defines the lower temporal limit and the second parameter defines the upper temporal limit of the segment to be returned. Listing 12.9 illustrates the usage of the history operation. We start with the definition of a record-of structure that is intended to hold measurement records with float values. In this context, the application of the history operation in Line 2 yields a stream record that represents the first ten values at `ti_Engine_Speed`. Please note, the overall size of the record of structure that is the number of individual measurement elements depends on the time interval defined by the parameters of the history operation, as well as on the given sampling rate (see Section 12.4.1).

Listing 12.9
The History Operation

```
type record of Measurement<float> Float_Stream_Records;                 1
var Float_Stream_Records speed := ti_Engine_Speed.history(0.0,10.0);    2
```

---

*The type in this case is passed as a type parameter which is possible with the new TTCN-3 advanced parameterization extension (ETSI 2009e).

We can use the record-of representation of streams to assess complete stream segments. This can be achieved by means of an assessment function, which iterates over the individual measurement records of the stream record, or by means of so-called stream templates, which characterize a sequence of measurement records as a whole. While such assessment functions are in fact only small TTCN-3 programs, which conceptually do not differ from similar solutions in any other programming language, the template concepts are worth explaining in more detail here.

A template is a specific data structure that is used to specify the expectations on the SUT not only by means of distinct values but also by means of data-type specific patterns. These patterns allow, among other things, the definition of ranges (e.g., `field := (lowerValue .. upperValue)`), lists (e.g., `field := (1.0, 10.0, 20.0)`), and wildcards (e.g., `field := ?` for any value or `field := *` for any or no value). Moreover, templates can be applied to structured data types and record-of structures. Thus, we are able to define structured templates that have fields with template values. Last but not least, templates are parameterizable so that they can be instantiated with different value sets.*

Values received from the SUT are checked against templates by means of certain statements.

- The **match** operation already exists in standard TTCN-3. It tests whether an arbitrary template matches a given value completely. The operation returns true if the template matches and false otherwise. In case of a record-of representation of a stream, we can use the match operation to check the individual stream values with templates that conform to the type definitions in Section 12.8.

- The **find** operation has been newly introduced in TTCN-3 embedded. It scans a record-structured stream for the existence of a structure that matches the given template. If such a structure exists, the operation returns the index value of the matching occurrence. Otherwise, it returns −1. In case of a record-of representation of a stream, we can use the find statement to search the stream for the first occurrence of a distinct pattern.

- The **count** operation has been newly introduced in TTCN-3 embedded as well. It scans a record-structured stream and counts the occurrences of structures that match a given template. The operation returns the number of occurrences. Please note, the application of the count-operation is not greedy and checks the templates iteratively starting with each measurement record in a given record-of structure.

Listing 12.10 shows a usage scenario for stream templates. It starts with the definition of a record template. The template specifies a signal pattern with the following characteristics:

Listing 12.10
Using Templates to Specify Signal Shapes

```
template Float_Stream_Record toTest := {                                    1
    {value_ := ?                 ,    delta_ := 0.0,    timestamp_:= ?},      2
    {value_ := (1900.0 .. 2100.0) , delta_ := 2.0,    timestamp_:= ?},      3
    {value_ := (2900.0 .. 3100.0) , delta_ := 2.0,    timestamp_:= ?},      4
    {value_ := (2950.0 .. 3050.0) , delta_ := 2.0,    timestamp_:= ?},      5
    {value_ := ?                 ,    delta_ := 2.0,    timestamp_:= ?}      6
}                                                                            7
```

---

*Please note that the TTCN-3 template mechanism is a very powerful concept, which cannot be explained in full detail here.

```
// checks, whether a distinct segment conforms to the template toTest      10
match( ti_Engine_Speed . history (2.0 , 10.0), toTest );                    11
// finds the first occurrence of a stream segment that conforms to toTest   12
find ( ti_Engine_Speed . history (0.0 , 100.0), toTest );                   13
// counts all occurrences of stream segments that conform to toTest         14
count( ti_Engine_Speed . history (0.0 , 100.0), toTest );                   15
```

the signal starts with an arbitrary value; after 2 s, the signal value is between 1900 and 2100; after the next 2 s, the signal value reaches a value between 2900 and 3100, thereafter (i.e., 2 s later), it reaches a value between 2950 and 3050, and finally ends with an arbitrary value. Please note, in this case, we are not interested in the absolute time values and thus allow arbitrary values for the `timestamp_` field.

One should note that a successful match requires that the stream segment and the template have the same length. If this is not the case, the match operation fails.

Such a definition of stream templates (see Listing 12.10) can be cumbersome and time consuming. To support the specification of more complex patterns, we propose the use of generation techniques and automation, which can easily be realized by means of TTCN-3 functions and parameterized templates.* Listing 12.11 shows such a function that generates a template record out of a given record. The resulting template allows checking a stream against another stream (represented by means of the `Float_Stream_Record myStreamR`) and additionally allows a parameterizable absolute tolerance for the value side of the stream.

Listing 12.11
Template Generation Function

```
                                                                           1
function generateTemplate(in Float_Stream_Record myStreamR, in float tolVal)   2
return template Float_Stream_Record{                                        3
  var integer i;                                                           4
  var template Float_Stream_Record toGenerate                              5
  template Measurement<float> tolerancePattern(in float delta,             6
                                  in float value,                          7
                                  in float tol) := {                       8
    delta_ := delta,                                                       9
    value_ := ((value − (tol / 2.0)) .. (value + (tol / 2.0))),           10
    timestamp_:= ?                                                        11
  }                                                                        12
                                                                          13
  for (i := 0, i < sizeof(myStreamR), i := i + 1){                        14
    toGenerate[i] := tolerancePattern(myStreamR[i].delta,                 15
                              myStreamR[i].value, tolVal);                 16
  }                                                                        17
  return toGenerate;                                                      18
}                                                                         19
```

---

*Future work concentrates on the extensions for the TTCN-3 template language to describe repetitive and optional template groups. This will yield a regular expression like calculus, which provides a much more powerful means to describe the assessment for stream records.

These kind of functions are not intended to be specified by the test designer, but rather provided as part of a library. The example function presented here is neither completely elaborated nor does it provide the sufficient flexibility of a state-of-the-art library function. It is only intended to illustrate the expressive power and the potential of TTCN-3 and TTCN-3 embedded.

12.4.1.5 Control flow

So far, we have only reflected on the construction, application, and assessment of single streams. For more advanced test behavior, such as concurrent applications, assessment of multiple streams, and detection of complex events (e.g., zero crossings or flag changes), richer capabilities are necessary.

For this purpose, we combine the concepts defined in the previous section with state machine-like specification concepts, called modes. Modes are well known from the theory of hybrid automata (Alur, Henzinger, and Sontag 1996, Lynch et al. 1995, Alur et al. 1992). A mode is characterized by its internal behavior and a set of predicates, which dictate the mode activity. Thus, a simple mode specification in TTCN-3 embedded consists of three syntactical compartments: a mandatory body to specify the mode's internal behavior; an invariant block that defines predicates that must not be violated while the mode is active; and a transition block that defines the exit condition to end the mode's activity.

Listing 12.12
Atomic Mode

```
cont { //body                                                          1
        // ramp, the value increases at any time step by 3             2
        to_Set_Point.value := 3.0 * now;                               3
        // constant signal                                             4
        to_Engine_Perturbation.value := 0.0;                           5
}                                                                      6
inv {  // invariants                                                   7
        // stops when the set point exceeds a value of 20000.0         8
        to_Set_Point.value > 20000.0;                                  9
}                                                                     10
until { //transition                                                 11
        [ti_Engine_Speed.value > 2000.0]{to_Engine_Perturbation.value := 2.0;}  12
}                                                                    13
```

In the example in Listing 12.12, the set point value `to_Set_Point` increases linearly in time and the engine perturbation `to_Engine_Perturbation` is set constantly to 0.0. This holds as long as the invariant holds and the **until** condition does not fire. If the invariant is violated, that is the set point speed exceeds 20000.0, an error verdict is set and the body action stops. If the **until** condition yields true, that is the value of `ti_Engine_Speed` exceeds 2000.0, the `to_Engine_Perturbation` value is set to 2.0 and the body action stops.

To combine different modes into larger constructs, we provide parallel and sequential composition of individual modes and of composite modes. The composition is realized by the **par** operator (for parallel composition) and the **seq** operator (for sequential composition). Listing 12.13 shows two sequences, one for the perturbation actions and other for the assessment actions, which are themselves composed in parallel.

Listing 12.13
Composite Modes

```
par { // overall perturbation and assessment                           1
  seq{// perturbation sequence                                         2
    cont{// perturbation action 1}                                     3
    cont{// perturbation action 2}                                     4
    ...}                                                               5
  seq{// assessment sequence                                          6
    cont{// assessment action 1}                                       7
    cont{// assessment action 1}                                       8
    ...}                                                               9
}                                                                     10
```

In general, composite modes show the same structure and behavior as atomic modes as far as invariants and transitions are concerned. Hence, while being active, each invariant of a composite mode must hold. Additionally, each transition of a composite mode ends the activity of the mode when it fires. Moreover, a sequential composition ends when the last contained mode has finished and a parallel composition ends when all contained modes have finished. Furthermore, each mode provides access to an individual local clock that returns the time that has passed since the mode has been activated. The value of the local clock can be obtained by means of the keyword **duration**.

Listing 12.14
Relative Time

```
seq{// perturbation sequence                                                            1
    cont{to_Set_Point.value := 20000.0;} until (duration > 3.0)                          2
    cont{to_Set_Point.value := 40000.0 + 100.0 * duration;} until (duration > 2.0)       3
} until (duration > 4.0)                                                                 4
```

Listing 12.14 shows the definition of three modes, each of which has a restricted execution duration. The value of `to_Set_Point` is increased continuously by means of the duration property. The duration property is defined locally for each mode. Thus, the valuation of the property would yield different results in different modes.

12.4.2   Specification of reusable entities

This section presents more advanced concepts of TTCN-3 embedded that are especially dedicated to specifying reusable entities. We aim to achieve a higher degree of reusability by modularizing the test specifications and supporting the specification of abstract and modifiable entities. Some of these concepts are well known in computer language design and already available in standard TTCN-3. However, they are only partially available in state-of-the-art test design tools for continuous and hybrid systems and they must be adapted to the concepts we introduced in Section 12.4.1. The concepts dedicated to support modularization and modification are the following:

- Branches and jumps to specify repetitions and conditional mode execution.

- Symbol substitution and referencing mechanisms.

- Parameterization.

12.4.2.1 Conditions and jumps

Apart from the simple sequential and parallel composition of modes, stronger concepts to specify more advanced control flow arrangements, such as conditional execution and repetitions are necessary. TTCN-3 already provides a set of control structures for structured programming. These control structures, such as if statements, while loops, and for loops are applicable to the basic TTCN-3 embedded concepts as well. Hence, the definition of mode repetitions by means of loops, as well as the conditional execution of assertions and assignments inside of modes are allowed. Listing 12.15 shows two different use cases for the application of TTCN-3 control flow structures that directly interact with TTCN-3 embedded constructs. In the first part of the listing (Lines 4 and 6), an if statement is used to specify the conditional execution of assignments inside a mode. In the second part of the listing (Lines 10–14), a while loop is used to repeat the execution of a mode multiple times.

Listing 12.15

Conditional Execution and Loops

```
cont {  //body                                                     1
   // ramp until duration >= 4.0                                   2
   if (duration < 4.0) {to_Set_Point.value := 3.0 * now;}          3
   // afterwards the value remains constant                        4
   else {to_Set_Point.value := to_Set_Point.prev.value;}           5
}                                                                   6
                                                                    7
// saw tooth signal for 3 minutes with a period of 5.0 seconds     8
while (now < 180.0) {                                               9
   cont {                                                          10
      to_Set_Point.value := 3.0 * duration;                       11
   } until (duration > 5.0)                                       12
}                                                                  13
```

For full compatibility with the concepts of hybrid automata, definition of so-called transitions must be possible as well. A transition specifies the change of activity from one mode to another mode. In TTCN-3 embedded, we adopt these concepts and provide a syntax, which seamlessly integrates with already existing TTCN-3 and TTCN-3 embedded concepts. As already introduced in the previous section, transitions are specified by means of the `until` block of a mode. In the following, we will show how a mode can refer to multiple consecutive modes by means of multiple transitions and how the control flow is realized.

A transition starts with a conditional expression, which controls the activation of the transition. The control flow of transitions resembles the control flow of the already existing (albeit antiquated) TTCN-3 `label` and `goto` statements. These statements have been determined sufficiently suitable for specifying the exact control flow after a transition has fired. Thus, there is no need to introduce additional constructs here. A transition may optionally contain arbitrary TTCN-3 statements to be executed when the transition fires.

Listing 12.16 illustrates the definition and application of transitions by means of pseudo code elements. The predicate `<activation_predicate>` is an arbitrary predicate expression that may relate to time values or stream values or both. The `<optional_statement_list>` may contain arbitrary TTCN-3 or TTCN-3 embedded statements except blocking or time-consuming statements (alt statements and modes). Each goto statement relates to a label definition that specifies the place where the execution is continued.

Listing 12.16

Transitions

```
label labelsymbol_1                                                              1
cont {} until {                                                                  2
 [<activation_predicate>] {<optional_statement_list>} goto labelsymbol_2;       3
 [<activation_predicate>] {<optional_statement_list>} goto labelsymbol_3;       4
}                                                                                5
label labelsymbol_2; cont {} goto labelsymbol_4;                                6
label labelsymbol_3; cont {} goto labelsymbol_1;                                7
label labelsymbol_4;                                                            8
```

Listing 12.17 shows a more concrete example that relates to our example from Section 12.3.3. We define a sequence of three modes that specify the continuous valuation of the engine's set point (`to_Set_Point`), depending on the engine's speed (`engine_speed`). When the engine speed exceeds 2000.0 rpm, the set point is decreased (`goto decrease`), otherwise it is increased (`goto increase`).

Listing 12.17

Condition and Jumps

```
testcase myTestcase() runs on EngineTester {                                     1
                                                                                 2
  // reusable mode application                                                   3
  cont {to_Set_Point.value := 3.0 * now;}                                        4
  until {                                                                        5
   [duration > 10.0 and engine_speed.value > 2000.0] goto increase;             6
   [duration > 10.0 and engine_speed.value <= 2000.0] goto decrease             7
  }                                                                              8
  label increase;                                                               9
  cont {to_Set_Point.value := 3 * now;} until  {[duration > 10.0 ] goto end;}   10
  label decrease;                                                              11
  cont {to_Set_Point.value := 3 * now;} until  (duration > 10.0 )              12
  label end;                                                                   13
}                                                                              14
```

12.4.2.2 Symbol substitution and referencing

Similar to the definition of functions and functions calls, it is possible to declare named modes, which can then be referenced from any context that would allow the explicit declaration of modes. Listing 12.18 shows the declaration of a mode type (Line 1), a named mode* (Line 4) and a reference to it within a composite mode definition (Line 12).

---

*Named modes, similar to other TTCN-3 elements that define test behavior, can be declared with a **runs on** clause, in order to have access to the ports (or other local fields) of a test component type.

Listing 12.18
Symbol Substitution

```
type mode ModeType();                                                          1
                                                                               2
// reusable mode declaration                                                   3
mode ModeType pert_seq() runs on EngineTester seq {                            4
  cont {to_Set_Point.value := 2000.0} until (duration >= 2.0)                  5
  cont {to_Set_Point.value := 2000.0 + duration / to_Set_Point.delta * 10.0}  6
  until (duration >= 5.0)                                                      7
}                                                                              8
                                                                               9
testcase myTestcase() runs on EngineTester {                                   10
  par {                                                                        11
    pert_seq(); // reusable mode application                                   12
    cont {assert(engine_speed.value >= 500.0)}                                 13
  } until (duration > 10.0)                                                    14
}                                                                              15
```

12.4.2.3 Mode parameterization

To provide a higher degree of flexibility, it is possible to specify parameterizable modes. Values, templates, ports, and modes can be used as mode parameters. Listing 12.19 shows the definition of a mode type, which allows the application of two float parameters and the application of one mode parameter of the mode type `ModeType`.

Listing 12.19
Mode Parameterization

```
type mode ModeType2(in float startVal, in float increase, in ModeType assertion);  1
                                                                                    2
// reusable mode declaration                                                        3
mode ModeType assert_mode() runs on EngineTester :=                                  4
  cont {assert(engine_speed.value >= 500.0)}                                         5
                                                                                    6
mode ModeType2 pert_seq_2(in float startVal,                                        7
                          in float increase,                                        8
                          in ModeType assertion)                                     9
runs on EngineTester par {                                                          10
  seq{// perturbation sequence                                                      11
    cont{to_Set_Point.value := startVal} until (duration >= 2.0)                    12
    cont{to_Set_Point.value := startVal + duration / to_Set_Point.delta * increase} 13
    until (duration >= 5.0)                                                         14
  }                                                                                 15
  assertion();                                                                      16
}                                                                                   17
                                                                                    18
testcase myTestcase() runs on EngineTester {                                         19
  // reusable mode application                                                       20
  pert_seq_2(1000.0, 10.0, assert_mode);                                            21
  pert_seq_2(5000.0, 1.0, cont{assert(engine_speed.value >= 0.0)});                 22
}                                                                                   23
```

Lines 23 and 24 illustrate the application of parameterizable reusable modes. Line 23 applies `pert_seq_2` and sets the parameter values for the initial set point to 1000.0 and the parameter for the increase to 10.0, and the `assert_mode` is passed as a parameter to be applied within the `pert_seq_2` mode. Line 24 shows a more or less similar application of `pert_seq_2`, where an inline mode declaration is passed as the mode parameter.

12.5 Reuse of Closed Loop Test Artifacts

Thus far, we have discussed reusability on the level of test specification language elements (referencing mechanisms, modifications operators, parameterization), in this section, we additionally focus on the reuse of the high-level artifacts of our generic closed loop architecture for testing.

In this context, however, different aspects must be taken into account. Three-tier development and closed loop testing (MiL, SiL, HiL) methodologies have an increased need for traceability and consistency between the different testing levels.

Consider a scenario in which one subcontractor delivers test artifacts that are to be used on components from various suppliers. These components in turn may often utilize the same basic infrastructure and parts of a generic environment, which also contributes to increased reuse potential. Furthermore, there may be conformance tests for certain components provided by standardization organizations, which should again be reusable at the same testing level across different implementations, both of the components, but potentially also of the environment they shall operate in. Therefore, despite impeding factors of an organizational nature when production code is concerned, in the scope the current context, there are many organizational factors that not only facilitate the reuse of test artifacts but also increase the need and potential for reuse.

In the following, a general approach for vertical and horizontal test specification and test model reuse at different closed loop test levels will be presented and accompanied by an example that outlines the practical reuse of TTCN-3 embedded specifications and the environment model from Section 12.3.3 in a MiL and a HiL scenario. Since closed loop tests, in general, require a complex test environment, an appropriate test management process and tool support throughout the life cycle of the SUT is required. This subject is addressed in the last subsection.

12.5.1 Horizontal reuse of closed loop test artifacts

Reusable test suites can be developed using concepts such as symbol substitution, referencing, and parameterization. Environment modeling based on Simulink® provides modularization concepts such as subsystems, libraries, and model references, which facilitate the reusability of environment models. Since the generic closed loop architecture for testing clearly separates the heterogeneous artifacts by using well-defined interfaces, the notation-specific modularization and reuse concepts can be applied without interfering with each other. By lifting reuse to the architecture level, at least the environment model and the specification of the perturbation and assessment functionality can be (re-)used with different SUTs. The SUTs may differ in type and in version, but as long as they are built on common interfaces and share common characteristics, then this kind of reuse is applicable.

In general, the reuse of test specifications across different products is nowadays often used for testing products or components, which are based on a common standard (conformance testing). The emerging standardization efforts in embedded systems development (e.g., AUTOSAR [AUTOSAR Consortium 2010]) indicate the emerging need for such an approach.

12.5.2 Vertical reuse of environment models

Depending on the type of the SUT, varying testing methods on different test levels may be applied, each suited for a distinct purpose. With the MiL test, the functional aspects of the model are validated. SiL testing is used to detect errors that result from software-specific issues, for instance, the usage of fixed-point arithmetic. MiL and SiL tests are used in the early design and development phases, primarily to discover functional errors within the software components. These types of test can be processed on a common PC hardware, for instance through co-simulation, and, therefore, are not suitable for addressing real-time matters. To validate the types of issues that result from the usage of a specific hardware, HiL tests must be used.

In principle, the SUTs exhibit the same logical functionality through all testing levels. However, they are implemented with different technologies and show different integration levels with other components, including the hardware. In the case of different implementation technologies, which often result in interfaces with the same semantics but different technological constraints and access methods, the reuse of the environment model and the perturbation and assessment specifications is straightforward. The technological adaptation is realized by means of the mapping components that bridge the technological as well as the abstraction gap between the SUT and the environment (Figure 12.5).

12.5.3 Test reuse with TTCN-3 embedded, Simulink®, and CANoe

In Section 12.3.3, we provided an example that shows the application of our ideas within a simple MiL scenario. This section demonstrates the execution of test cases in a MiL Scenario and outlines the reuse of some of the same artifacts in a HiL scenario. It provides a proof of concept illustration of the applicability of our ideas. The main artifacts for reuse are the environment model (i.e., generic test scenarios) and the test specifications. The reuse of the test specifications depends on their level of abstraction (i.e., the semantics of the specification must fit the test levels we focus on) and on some technological issues (e.g., the availability of a TTCN-3 test engine for the respective test platform). Within this example,

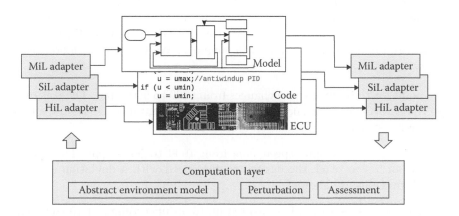

FIGURE 12.5
Vertical reuse of environment models.

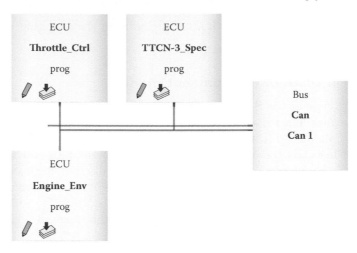

FIGURE 12.6
Integration with Vector CANoe.

we show that we are able to find the most appropriate level of abstraction for the test specifications and the environment model. The technological issues are not discussed here.

Tests on MiL level are usually executed in the underlying MATLAB® simulation. The environment and the controller have a common time base (the simulation time). Based on the well-tested controller model, the object code will be generated, compiled, and deployed. On PiL level, the target hardware provides its own operating system and thus its own time base. Furthermore, it will be connected to other controllers, actuators, and sensors via a bus system (e.g., CAN bus, FlexRay Bus, etc.) using a HiL environment.

In principle, we use pre- and postprocessing components to link our test system to the bus system and/or to analog devices and thus to the controller under test. The well-established toolset CANoe (Vector Informatics 2010) supports such kind of integration between hardware-based controllers, Simulink®-based environment models and test executables by means of software-driven network simulations (see Figure 12.6).

To reuse the TTCN-3 embedded test specification, we only need a one time effort to build a TTCN-3 embedded test adapter for Simulink® and CANoe. Both being standard tools, most of the effort can be shared.

We refer to a model-based development process and specify tests for the controller that was already introduced in Section 12.3.3. The controller regulates the air flow and the crankshaft of a four-cylinder spark ignition engine with an internal combustion engine (Figure 12.3). From the requirements, we deduce abstract test cases. They can be defined semiformally and understandably in natural language. We will use the following test cases as guiding examples.

- The set point for the engine speed changes from 2000 to 5000 rpm. The controller should change the throttle angle in such a way that the system takes less than 5 s to reach the desired engine speed with a deviation of 100 rpm.

- The set point for the engine speed falls from 7000 to 5000 rpm. The system should take less than 5 s to reach the desired engine speed with a deviation of 100 rpm. No over-regularizations are allowed.

- The engine speed sensor data is measured up to an uncertainty of 10 rpm. The control must be stable and robust for that range. Given the manipulated variable, the throttle angle, the deviation caused by some disturbance should be less than 100 rpm.

In the next step, we implement the test cases, that is, concretize them in TTCN-3 embedded so that they can be executed by a test engine. The ports of our system are as defined in Listing 12.2 (based on the test architecture in Figure 12.4).

The first and the second abstract test cases analyze the test behavior in two similar situations. We look into the system when the engine speed changes. In the first test case, it jumps from 2000 to 5000 rpm and in the second from 7000 to 5000 rpm. In order to do the specification job only once, we define a parameterizable mode that realizes a step function (Listing 12.20). The engine speed is provided by an environment model. The same applies for the other data. To test the robustness (see abstract test case 3), we must perturb the engine speed. This is realized by means of the output `to_Engine_Perturbation`.

Listing 12.20
Speed Jump

```
type mode Set_Value_Jump(in float startVal, in float endVal);               1
                                                                            2
mode Set_Value_Jump pert_seq(in float startVal, in float endVal)            3
runs on ThrottleControlTester seq{                                          4
  cont {to_Set_Point.value := startVal} until (duration >= 5.0)             5
  // the first 5 seconds the set point is given as startVal rpm             6
  cont {to_Set_Point.value := endVal } until (duration >= 10.0)             7
  // the next 10 seconds the set point should be endVal rpm                 8
}                                                                           9
                                                                           10
testcase TC_Speed_Jump() runs on ThrottleControlTester {                   11
  // reusable mode application                                             12
    pert_seq (2000.0, 5000.0);                                             13
}                                                                          14
```

A refined and more complex version of the mode depicted above, which uses linear interpolation and flexible durations, can easily be developed by using the ideas depicted in Listing 12.17.

In order to assess the tests, we have to check whether the controller reacts in time. For this purpose, we have to check whether the values of a stream are in a certain range. This can be modeled by means of a reusable mode too (Listing 12.21).

Listing 12.21
Guiding Example Assertion

```
type mode Range_Check(in float startTime, in float endTime, in float setValue,   1
                      in float Dev, out FloatInPortType measuredStream);          2
                                                                                  3
// reusable mode declaration                                                      4
mode Range_Check range_check (in float startTime,                                 5
                      in float endTime, in float setValue,                        6
                      in float Dev, out FloatInPortType measuredStream)           7
seq {                                                                             8
    // wait until the startTime                                                   9
    cont {} until (duration >= startTime);                                       10
    // check the engine speed until the endTime was reached                      11
    cont {assert(measuredStream.value >= (setValue − Dev) &&                     12
```

```
      measuredStream.value <= (setValue + Dev))}    13
      until (duration >= endTime)                   14
}                                                    15
```

The executable test specification of the complete abstract test case `TC_Speed_Jump` is given in Listing 12.22.

Listing 12.22
Guiding Example Test Specification

```
                                                             1
  testcase TC_Speed_Jump() runs on ThrottleControlTester{    2
    par{                                                      3
      // set the jump                                         4
      pert_seq (2000.0, 5000.0);                              5
      // check the control signal, where Desired_Speed is     6
      // the assumed desired value                            7
      range_check(10.0, 10.0, Desired_Speed, 100.0, ti_Engine_Speed );  8
    }                                                         9
  }                                                          10
```

In order to check for a possible overshoot, we will use the maximum value of a stream over a distinct time interval. This can be easily realized by using the constructs introduced in Section 12.4, thus we will not elaborate further on this here. To test the robustness, a random perturbation of the engine speed is necessary. It is specified by means of the random value function **function rnd(float seed) return float** of TTCN-3. The function returns a random value between 0.0 and 1.0. Listing 12.23 shows the application of the random function to the engine perturbation port.

Listing 12.23
Random Perturbation

```
// rnd(float seed) retrieves random values between [0,1]    1
to_Engine_Perturbation := rnd(0.2) * 20.0 - 10.0;           2
```

This random function can be used to define the perturbation resulting from uncertain measurement. To formulate a proper assessment for the third abstract test case, the parameterized mode `range_check` can be reused with the stream `ti_Throttle_Angle`. We will leave this exercise for the reader.

By using proper TTCN-3 embedded test adapter, we can perform the tests in a MATLAB® simulation, and as outlined in a CANoe HiL environment as well. Figure 12.7 shows the result of a test run of the first two test cases defined above with TTCN-3 embedded and Simulink®. While the upper graph shows the progress of the set point value (i.e., `to_Set_Point`) and the engine speed value (i.e., `ti_Engine_Speed`), the lower graph represents the time course of the throttle angle (i.e., `ti_Throttle_Angle`).

This example shows the reusability of TTCN-3 embedded constructs and together with co-simulation, we establish an integrated test process over vertical testing levels.

12.5.4 Test asset management for closed loop tests

Closed loop tests require extended test asset management with respect to reusability because of the fact that several different kinds of reusable test artifacts are involved. Managing test data systematically relies on a finite data set representation. For open loop tests, this often

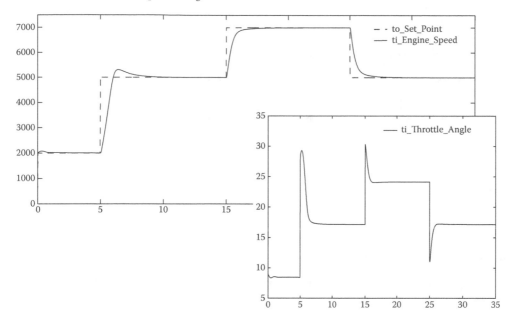

FIGURE 12.7
Test run with Simulink® and TTCN-3 embedded.

consists of input data descriptions that are defined by finitely many support points and interpolation prescriptions, such as step functions, ramps, or splines. The expectation is usually described by the expected values or ranges at specific points in time or time spans.

This no longer works for closed loop tests. In such a context, an essential part of a test case specification is defined by a generic abstract environment model, which may contain rather complex algorithms. At a different test level, this software model may be substituted by compiled code or a hardware node. In contrast to open loop testing, there is a need for a distinct development and management process incorporating all additional assets into the management of the test process. In order to achieve reproducibility and reusability, the asset management process must be carefully designed to fit into this context. The following artifacts that uniquely characterize closed loop test specifications must be taken into consideration:

- Abstract environment model (the basic generic test scenario).

- Designed perturbation and assessment specification.

- Corresponding pre- and postprocessing components.

All artifacts that are relevant for testing must be versioned and managed in a systematic way. Figure 12.8 outlines the different underlying supplemental processes. The constructive development process that produces the artifacts to be tested is illustrated on the right-hand side. Parallel to it, an analytic process takes place. Whereas in open loop testing, the tests are planned, defined, and performed within this analytic process; in closed loop architectures, there is a need for an additional complementary development process for the environment models. The skills required to develop such environment models and the points of interest they have to meet distinguish this complementary process from the development processes of the tests and the corresponding system.

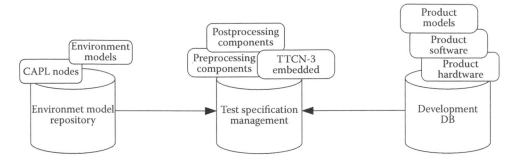

FIGURE 12.8
Management of test specifications.

12.6   Quality Assurance and Guidelines for the Specification of Reusable Assets

As outlined above, TTCN-3 embedded, which is similar to standard TTCN-3, has been designed with reusability in mind, providing a multitude of variability mechanisms. This implies that similar aspects shall be taken into consideration for the specification of reusable assets using TTCN-3 embedded as well. Reusability is inevitably connected to quality, especially when the development of reusable test assets is concerned. Reusability was even identified as one of the main quality characteristics in the proposed quality model for test specifications (Zeiß et al. 2007). On the other hand, quality is also critically important for reusability. Reusable assets must be of particularly high quality since deficiencies in such assets will have a much larger impact on the systems they are reused in. Defects within reusable assets may or may not affect any and every system they are used in. Furthermore, modifications to remedy an issue in one target system, may affect the other target systems, both positively and negatively. Thus, special care should be taken to make sure that the reusable assets are of the necessary quality level. Quality may further affect the reusability in terms of adaptability and maintainability. The assets may have to be adapted in some contexts and they must be maintained to accommodate others, extend or improve functionality, correct issues, or simply be reorganized for even better reusability. If the effort for maintenance or adaptation is too high, it will offset (part of) the benefits of having reusable test assets. Hence, quality is even more important to reuse than reuse is to quality, and thus quality assurance is necessary for the effective development of reusable assets with TTCN-3 embedded.

Furthermore, if an established validation process is employed, the use of validated reusable libraries would increase the percentage of validated test code in the testware, as noted in (Schulz 2008). Establishing such a process on the other hand, will increase the confidence in the reusable assets. A validation process will again involve quality assurance.

In addition to the validation of reusable assets, assessment of the actual reusability of different assets may be necessary to determine the possible candidates for validation and possible candidates for further improvement. This can be achieved by establishing reusability goals and means to determine whether these goals are met (e.g., by defining metrics models or through testability analysis). If they are not met, either the asset is not suitable for reuse, or its implementation does not adhere to the reusability specifications for that asset. In Mäki-Asiala 2004, two metrics for quantitative evaluation of reusability are illustrated in a small case study. The metrics themselves were taken from Caruso (1993). Further metrics for software reuse are described in (Frakes and Terry 1996). Additional metrics specifically concerning the reuse of test assets may have to be defined. There are ongoing studies that use advanced approaches to assess the reusability of software components (Sharma, Grover, and Kumar 2009). Such approaches could be adapted to suit the test domain.

While there is a significant body of work on quality assurance for standard TTCN-3 (Bisanz 2006, Neukirchen, Zeiß, and Grabowski 2008, Neukirchen et al. 2008, Zeiß 2006), quality assurance measures for TTCN-3 embedded remain to be studied, as TTCN-3 embedded is still in the draft stage. Similar to standard TTCN-3, metrics, patterns, code smells, guidelines, and refactorings should be defined to assess the quality of test specifications in TTCN-3 embedded, detect issues, and correct them efficiently. Based on a survey of existing methodologies, a few examples for quality assurance items for TTCN-3 embedded that are related to reusability will be briefly outlined below.

The main difficulty in the design of TTCN-3 libraries, as identified by Schulz 2008, is to anticipate the evolution of use of libraries. Therefore, modularization, in the form of separation of concepts and improved selective usage, and a layered structure of library organization are suggested as general guiding principles when developing libraries of reusable assets. Furthermore, in Schulz 2008, it is also recommended to avoid component variables and timers, as well as local verdicts and stop operations (unless on the highest level, that is, not within a library), when designing reusable behavior entities. The inability to pass functions as parameters and the lack of an effective intermediate verdict mechanism are identified as major drawbacks of the language. Meanwhile, the first issue has been resolved within the TTCN-3 standard by means of an extension package that enables so-called behavior types to be passed as parameters to functions, testcases, and altsteps (ETSI 2010). Similarly, in TTCN-3 embedded modes can be passed as parameters.

Generic style guidelines that may affect the reusability potential of assets are largely transferable to TTCN-3 embedded, for example,

- Restricting the nesting levels of modes.

- Avoiding duplicated segments in modes.

- Restricting the use of magic values (explicit literal or numerical values) or if possible avoiding them altogether.

- Avoiding the use of over-specific runs on statements.

- Proper grouping of certain closely related constructs.

- Proper ordering of constructs with certain semantics.

In Mäki-Asiala (2004), 10 guidelines for the specification of reusable assets in TTCN-3 were defined. These are concerned with the reusability of testers in concurrent and nonconcurrent contexts, the use and reuse of preambles and postambles, the use of high-level functions, parameterization, the use of selection structures, common types, template modifications, wildcards, and modularization based on components and on features. The guidelines are also related to the reusability factors that contributed to their development. The guidelines are rather generic and as such also fully applicable to TTCN-3 embedded.

In Mäki-Asiala, Kärki, and Vouffo (2006), four additional guidelines specific to the vertical reuse viewpoint are defined. They involve the separation of test configuration from test behavior, the exclusive use of the main test component for coordination and synchronization, redefinition of existing types to address new testing objectives, and the specification of system- and configuration- related data as parameterized templates. Again, these guidelines are valid for TTCN-3 embedded as well. In addition, they can be adapted to the specific features of TTCN-3 embedded. The functional description should be ideally separated from the real-time constraints, and continuous behavior specifications shall be separated from noncontinuous behavior.

At this stage, only guidelines based on theoretical assumptions and analogies from similar domains can be proposed. The ultimate test for any guideline is putting it into practice. Apart from validating the effectiveness of guidelines, practice also helps for the improvement and extension of existing guidelines, as well as for the definition of new guidelines.

When discussing guidelines for the development of reusable real-time components, often cited in the literature are the conflicts between performance requirements on one side and reusability and maintainability on the other (Häggander and Lundberg 1998). TTCN-3 embedded, however, abstracts from the specific test platform and thus issues associated with test performance can be largely neglected at the test specification level. Thus, the guidelines shall disregard performance. Ideally, it should be the task of the compilation and adaptation layers to ensure that real-time requirements are met.

The quality issues that may occur in test specifications implemented in TTCN-3 embedded (and particularly those that affect the reusability) and the means for their detection and removal remain to be studied in more detail. There is ongoing research within the TEMEA project concerning the quality assurance of test specifications implemented in TTCN-3 embedded. As of this writing, there are no published materials on the subject. Once available, approaches to the quality assurance can be ultimately integrated in a test development process and supported by tools to make the development of high-quality reusable test specifications a seamless process. Other future prospects include approaches and tool support for determining the reusability potential of assets both during design and during implementation to support both the revolutionary and evolutionary approaches to reuse.

12.7   Summary

The "X-in-the-Loop" testing approach both suggests and presupposes enormous reusability potential. During the development cycle of embedded systems, software models are reused directly (for code generation) or indirectly (for documentation purposes) for the development of the software. The developed software is then integrated into the hardware (with or without modifications). Thus, it makes sense to reuse tests through all of the development phases. Another hidden benefit is that tests extended in the SiL and HiL levels can be reused back in earlier levels (if new test cases are identified at later levels that may as well be applicable to earlier levels). If on the other hand a strict cycle is followed, where changes are only done at the model level and always propagated onward, this would still reduce the effort significantly, as those changes will have to be made only once. For original equipment manufacturers (OEMs) and suppliers this will also add more transparency and transferability to different suppliers as well (on all levels, meaning reusable tests can be applied to models from one supplier, software from another, hardware from yet another).

The proposed test architecture supports the definition of environment models and test specification on a level of abstraction, which allows the reuse of the artifacts on different test systems and test levels. For the needs of the present domain, we introduced TTCN-3 embedded, an extension of the standardized test specification language TTCN-3, which provides the capabilities to describe test perturbations and assessments for continuous and hybrid systems. Whereas TTCN-3 is a standard already, we propose the introduced extensions for standardization as well. Thus, the language does not only promise to solve the reusability issues on technical level but also addresses organizational issues, such as long-term availability, standardized tool support, education, and training.

The ideas presented in this chapter are substantial results of the project "Testing Specification Technology and Methodology for Embedded Real-Time Systems in Automobiles" (TEMEA). The project is co-financed by the European Union. The funds are originated from the European Regional Development Fund (ERDF).

References

Alur, R., Courcoubetis, C., Henzinger, T. A., and Ho, P.-H. (1992). Hybrid automata: an algorithmic approach to the specification and verification of hybrid systems. In *Hybrid Systems*, Pages: 209–229.

Alur, R., Henzinger, T. A., and Sontag, E. D. (Eds.) (1996). *Hybrid Systems III: Verification and Control, Proceedings of the DIMACS/SYCON Workshop, October 22-25, 1995, Ruttgers University, New Brunswick, NJ, USA*, Volume 1066 of *Lecture Notes in Computer Science*. Springer, New York, NY.

AUTOSAR Consortium (2010). Web site of the AUTOSAR (AUTomotive Open System ARchitecture) consortium. URL: http://www.autosar.org.

Bisanz, M. (2006). Pattern-based smell detection in TTCN-3 test suites. Master's thesis, ZFI-BM-2006-44, ISSN 1612-6793, Institute of Computer Science, Georg-August-Universität Göttingen (Accessed on 2010).

Bringmann, E. and Kraemer, A. (2006). Systematic testing of the continuous behavior of automotive systems. In *SEAS '06: Proceedings of the 2006 International Workshop on Software Engineering for Automotive Systems*, Pages: 13–20. ACM Press, New York, NY.

Broy, M. (1997). Refinement of time. In Bertran, M. and Rus, T. (Eds.), *Transformation-Based Reactive System Development, ARTS'97*, Number 1231 in Lecture Notes on Computer Science (LNCS), Pages: 44–63. TCS Springer, New York, NY.

Conrad, M. and Dörr, H. (2006). Model-based development of in-vehicle software. In Gielen, G. G. E. (Ed.), *DATE*, Pages: 89–90. European Design and Automation Association, Leuven, Belgium.

ESA-ESTEC (2008). Space engineering: test and operations procedure language, standard ECSS-E-ST-70-32C.

ETSI (2009a). Methods for Testing and Specification (MTS). The Testing and Test Control Notation Version 3, Part 1: TTCN-3 Core Language (ETSI Std. ES 201 873-1 V4.1.1).

ETSI (2009b). Methods for Testing and Specification (MTS). The Testing and Test Control Notation Version 3, Part 4: TTCN-3 Operational Semantics (ETSI Std. ES 201 873-4 V4.1.1).

ETSI (2009c). Methods for Testing and Specification (MTS). The Testing and Test Control Notation Version 3, Part 5: TTCN-3 Runtime Interfaces (ETSI Std. ES 201 873-5 V4.1.1).

ETSI (2009d). Methods for Testing and Specification (MTS). The Testing and Test Control Notation Version 3, Part 6: TTCN-3 Control Interface (ETSI Std. ES 201 873-6 V4.1.1).

ETSI (2009e). Methods for Testing and Specification (MTS). The Testing and Test Control Notation Version 3, TTCN-3 Language Extensions: Advanced Parameterization (ETSI Std. : ES 202 784 V1.1.1).

ETSI (2010). Methods for Testing and Specification (MTS). The Testing and Test Control Notation Version 3, TTCN-3 Language Extensions: Behaviour Types (ETSI Std. : ES 202 785 V1.1.1).

Fey, I., Kleinwechter, H., Leicher, A., and Müller, J. (2007). Lessons Learned beim Übergang von Funktionsmodellierung mit Verhaltensmodellen zu Modellbasierter Software-Entwicklung mit Implementierungsmodellen. In Koschke, R., Herzog, O., Rödiger, K.-H., and Ronthaler, M. (Eds.), *GI Jahrestagung (2)*, Volume 110 of *LNI*, Pages: 557–563. GI.

Frakes, W. and Terry, C. (1996). Software reuse: metrics and models. *ACM Comput. Surv. 28*(2), 415–435.

Grossmann, J. and Mueller, W. (2006). A formal behavioral semantics for TestML. In *Proc. of IEEE ISoLA 06, Paphos Cyprus*, Pages: 453–460.

Häggander, D. and Lundberg, L. (1998). Optimizing dynamic memory management in a multithreaded application executing on a multiprocessor. In *ICPP '98: Proceedings of the 1998 International Conference on Parallel Processing*, Pages: 262–269. IEEE Computer Society. Washington, DC.

Harrison, N., Gilbert, B., Lauzon, M., Jeffrey, A. Lalancette, C., Lestage, D. R., and Morin, A. (2009). A M&S process to achieve reusability and interoperability. URL: ftp://ftp.rta.nato.int/PubFullText/RTO/MP/RTO-MP-094/MP-094-11.pdf.

IEEE (1993). *IEEE Standard VHDL (IEEE Std.1076-1993.)*. The Institute of Electrical and Electronics Engineers, Inc, New York, NY.

IEEE (1995). *IEEE Standard Test Language for all Systems–Common/Abbreviated Test Language for All Systems (C/ATLAS) (IEEE Std.716-1995.)*. The Institute of Electrical and Electronics Engineers, Inc, New York, NY.

IEEE (1998). User's manual for the signal and method modeling language. URL: http://grouper.ieee.org/groups/scc20/atlas/SMMLusers_manual.doc.

IEEE (1999). *IEEE Standard VHDL Analog and Mixed-Signal Extensions (IEEE Std. 1076.1-1999.)*. The Institute of Electrical and Electronics Engineers, Inc, New York, NY.

IEEE (2001). *IEEE Standard Test Access Port and Boundary-Scan Architecture (IEEE Std.1149.1 -2001)*. The Institute of Electrical and Electronics Engineers, Inc, New York, NY.

ISO/IEC (1998). Information technology - open systems interconnection - conformance testing methodology and framework - part 3: The tree and tabular combined notation (second edition). International Standard 9646-3.

Karinsalo, M. and Abrahamsson, P. (2004). Software reuse and the test development process: a combined approach. In *ICSR*, Volume 3107 of *Lecture Notes in Computer Science*, Pages: 59–68. Springer.

Kärki, M., Karinsalo, M., Pulkkinen, P., Mäki-Asiala, P., Mäntyniemi, A., and Vouffo, A. (2005). Requirements specification of test system supporting reuse (2.0). Technical report, Tests & Testing Methodologies with Advanced Languages (TT-Medal).

Karlsson, E.-A. (Ed.) (1995). *Software Reuse: A Holistic Approach*. John Wiley & Sons, Inc., New York, NY.

Kendall, I. R. and Jones, R. P. (1999). An investigation into the use of hardware-in-the-loop simulation testing for automotive electronic control systems. *Control Engineering Practice 7*(11), 1343–1356.

Lehmann, E. (2004). *Time Partition Testing Systematischer Test des kontinuierlichen Verhaltens von eingebetteten Systemen*. Ph. D. thesis, TU-Berlin, Berlin.

Lu, B., McKay, W., Lentijo, S., Monti, X. W. A., and Dougal, R. (2002). The real time extension of the virtual test bed. In *Huntsville Simulation Conference*. Huntsville, AL.

Lynch, N. A., Segala, R., Vaandrager, F. W., and Weinberg, H. B. (1995). Hybrid i/o automata. See Alur, Henzinger, and Sontag (1996), Pages: 496–510.

Lynex, A. and Layzell, P. J. (1997). Understanding resistance to software reuse. In *Proceedings of the 8th International Workshop on Software Technology and Engineering Practice (STEP '97) (including CASE '97)*, Pages: 339. IEEE Computer Society.

Lynex, A. and Layzell, P. J. (1998). Organisational considerations for software reuse. *Ann. Softw. Eng. 5*, 105–124.

Mäki-Asiala, P. (2004). Reuse of TTCN-3 code. Master's thesis, University of Oulu, Department of Electrical and Information Engineering, Finland.

Mäki-Asiala, P., Kärki, M., and Vouffo, A. (2006). Guidelines and patterns for reusable TTCN-3 tests (1.0). Technical report, Tests & Testing Methodologies with Advanced Languages (TT-Medal).

Mäki-Asiala, P., Mäntyniemi, A., Kärki, M., and Lehtonen, D. (2005). General requirements of reusable TTCN-3 tests (1.0). Technical report, Tests & Testing Methodologies with Advanced Languages (TT-Medal).

Mäntyniemi, A., Mäki-Asiala, P., Karinsalo, M., and Kärki, M. (2005). A process model for developing and utilizing reusable test assets (2.0). Technical report, Tests & Testing Methodologies with Advanced Languages (TT-Medal).

The MathWorks (2010a). MATLAB® - the language of technical computing. URL: http://www.mathworks.com/products/matlab/.

The MathWorks (2010b). Web site of the Simulink® tool - simulation and model-based design. URL: http://www.mathworks.com/products/simulink/.

The MathWorks (2010c). Web site of the Stateflow® tool - design and simulate state machines and control logic. URL: http://www.mathworks.com/products/stateflow/.

Modelica Association (2010). Modelica - a unified object-oriented language for physical systems modeling. URL: http://www.modelica.org/documents/ModelicaSpec30.pdf.

Montenegro, S., Jähnichen, S., and Maibaum, O. (2006). Simulation-based testing of embedded software in space applications. In Hommel, G. and Huanye, S. (Eds.), *Embedded Systems - Modeling, Technology, and Applications*, Pages: 73–82. Springer Netherlands. 10.1007/1-4020-4933-1_8.

Neukirchen, H., Zeiß, B., and Grabowski, J. (2008, August). An Approach to Quality Engineering of TTCN-3 Test Specifications. *International Journal on Software Tools for Technology Transfer (STTT)*, Volume 10, Issue 4. (ISSN 1433-2779), Pages: 309–326.

Neukirchen, H., Zeiß, B., Grabowski, J., Baker, P., and Evans, D. (2008, June). Quality assurance for TTCN-3 test specifications. *Software Testing, Verification and Reliability (STVR)*, Volume 18, Issue 2. (ISSN 0960-0833), Pages: 71–97.

Parker, K. P. and Oresjo, S. (1991). A language for describing boundary scan devices. *J. Electron. Test. 2*(1), 43–75.

Poulin, J. and Caruso, J. (1993). A reuse metrics and return on investment model. In *Proceedings of the Second International Workshop on Software Reusability*, Pages: 152–166.

SCC20 ATML Group (2006). IEEE ATML specification drafts and IEEE ATML status reports.

Schieferdecker, I., Bringmann, E., and Grossmann, J. (2006). Continuous TTCN-3: testing of embedded control systems. In *SEAS '06: Proceedings of the 2006 international workshop on Software engineering for automotive systems*, Pages: 29–36. ACM Press, New York, NY.

Schieferdecker, I. and Grossmann, J. (2007). Testing embedded control systems with TTCN-3. In Obermaisser, R., Nah, Y., Puschner, P., and Rammig, F. (Eds.), *Software Technologies for Embedded and Ubiquitous Systems*, Volume 4761 of *Lecture Notes in Computer Science*, Pages: 125–136. Springer Berlin / Heidelberg.

Schulz, S. (2008). Test suite development with TTCN-3 libraries. *Int. J. Softw. Tools Technol. Transf. 10*(4), 327–336.

Sharma, A., Grover, P. S., and Kumar, R. (2009). Reusability assessment for software components. *SIGSOFT Softw. Eng. Notes 34*(2), 1–6.

Suparjo, B., Ley, A., Cron, A., and Ehrenberg, H. (2006). Analog boundary-scan description language (ABSDL) for mixed-signal board test. In *International Test Conference*, Pages: 152–160.

TEMEA (2010). Web site of the TEMEA project (Testing Methods for Embedded Systems of the Automotive Industry), founded by the European Community (EFRE). URL: http://www.temea.org.

TEMEA Project (2010). Concepts for the specification of tests for systems with continuous or hybrid behaviour, TEMEA Deliverable. URL: http://www.temea.org/deliverables/D2.4.pdf.

Tripathi, A. K. and Gupta, M. (2006). Risk analysis in reuse-oriented software development. *Int. J. Inf. Technol. Manage. 5*(1), 52–65.

TT-Medal (2010). Web site of the TT-Medal project - tests & testing methodologies with advanced languages. URL: http://www.tt-medal.org/.

Vector Informatics (2010). Web site of the CANoe tool - the development and test tool for can, lin, most, flexray, ethernet and j1708. URL: http://www.vector.com/vi_canoe\ _en.html.

Zeiß, B. (2006). A Refactoring Tool for TTCN-3. Master's thesis, ZFI-BM-2006-05, ISSN 1612-6793, Institute of Computer Science, Georg-August-Universität Göttingen.

Zeiß, B., Vega, D., Schieferdecker, I., Neukirchen, H., and Grabowski, J. (2007). Applying the ISO 9126 quality model to test specifications – exemplified for TTCN-3 test specifications. In *Software Engineering 2007 (SE 2007). Lecture Notes in Informatics (LNI) 105. Copyright Gesellschaft für Informatik*, Pages: 231–242. Köllen Verlag, Bonn.

Part IV

Specific Approaches

13

A Survey of Model-Based Software Product
Lines Testing

Sebastian Oster, Andreas Wübbeke, Gregor Engels, and Andy Schürr

CONTENTS

13.1 Introduction

Software product line (SPL) engineering is an approach to improve reusability of software within a range of products that share a common set of features [Bos00, CN01, PBvdL05]. Because of the systematic reuse, the time-to-market and costs for development and maintenance decrease, while the quality of the individual products increases. In this way, SPLs enable developers to provide rapid development of customized products. The concepts behind the product line paradigm are not new. Domains such as the automotive industry have successfully applied product line development for several years. The software developing industry has recently adopted the idea of SPLs. Especially when analyzing the development of embedded systems, it is evident that the product line paradigm has gained increasing importance, while developing products for particular domains, such as control units in the automotive domain [TH02, GKPR08].

In recent years, the development of software for embedded systems has changed to model-based approaches. These approaches are frequently used to realize and to implement SPLs. The development of mobile phone software [Bos05] or automotive system electronic controller units [vdM07] is an example of embedded systems, which are developed using an SPL approach.

However, every concept of development is as sufficient and reliable as it is supported by concepts for testing. In single system engineering, testing often consumes up to 25% or even

50% of the development costs [LL05]. Because of the variability within an SPL, the testing of SPLs is more challenging than single system testing. If these challenges are solved by adequate approaches, the benefits outweigh the higher complexity. For example, the testing of a component, which is to be reused in different products, illustrates one challenge in this context. The component must be tested accurately, as a fault would be introduced into every product including it. It is not sufficient to test this component only once in an arbitrary configuration because the behavior of the component varies depending on the corresponding product. Identifying and solving the SPL-specific test challenges are important to achieve the benefits the product line paradigm provides.

This chapter deals with the challenges in SPL testing by initially examining the requirements for model-based testing (MBT) SPLs by defining a conceptional MBT approach. Afterwards, a summary and a discussion of various existing MBT approaches for SPLs is presented. We thereby focus on the best-known and most discussed approaches and provide a comparison of these. The different approaches are compared using the conceptional MBT approach as previously defined. In concluding, we will explore open research topics on the basis of the summary and comparison. Summarizing current approaches for model-based SPL testing supports further studies and helps developers identify possible methodologies for their development process. Thus, our contribution encourages researchers to observe open research objectives and the software industry to choose proper concepts for MBT SPLs.

The remainder of this chapter is structured as follows: In Section 13.2, we explain all relevant fundamentals concerning SPL Engineering and the variability concept. Section 13.3 concentrates on the problem of testing an SPL and outlines the requirements that every testing approach should fulfill. To exemplify the comparison, we introduce a running example in Section 13.4. In Section 13.5, we then introduce the criteria used to compare different MBT approaches for SPLs. In Section 13.6, we describe different approaches for MBT of SPLs. For each methodology, its functionality and scope are explained by means of our running example. In addition, we provide a brief summary of each approach. In Section 13.7, we discuss the different approaches, according to the criteria for comparison, introduced in Section 13.5. Finally, we conclude this chapter in Section 13.8.

13.2   Software Product Line Fundamentals

The basic idea within the SPL approach is to develop software products from common parts. Indeed, the SPL development paradigm addresses the problem of developing similar functionalities for different products. The goal is to reuse these functionalities in all products rather than repeatedly developing them.

To achieve the advantages stated in Section 13.1, the SPL must address a certain domain [vdM07] such as an engine control unit or software for mobile phones. This assures that similar functional requirements are shared by different products of the SPL. The reuse of these common parts achieves an increased development speed. Moreover, the parts are quality assured when they are reused in additional applications because they have already been tested in the SPL context. This reduces the test effort for new products, but it does not make testing redundant. Examples such as the *Ariane V* accident [Lyo96] demonstrate the importance of testing components already in use, when introducing them to new products, in order to assume correct behavior of the entire system.

Furthermore, the development of an SPL affects all activities of the development process. In contrast to single system development, the development process is separated into two levels: domain engineering (often also called platform development) and application engineering. The former allows the development of the common and variable parts of the product line, while the latter allows developing an application (also called product) considering the use of individual parts. The two levels, domain and application engineering, again are separated into five and four activities, respectively (see Figure 13.1):

- Product management

- Requirements engineering

- Design

- Realization

- Testing

The activity product management is only part of the domain engineering, supporting the sizing and the evolution of the entire product line. The product management controls the other four activities in domain engineering, starting with the development of the common and variable requirements, their design, realization, and finally testing. Domain testing provides testing of common and variable parts without deriving a product from the SPL. The output of every activity is artifacts including variability.

Each activity in application engineering is supported by the corresponding activity in domain engineering; the development artifacts of the domain level are the basis for the

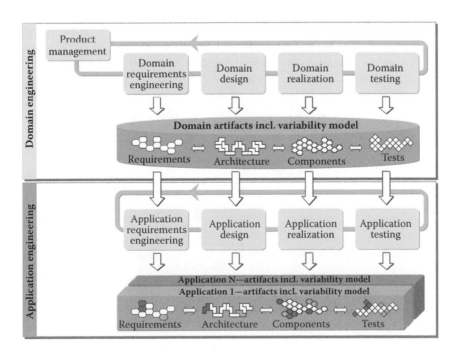

FIGURE 13.1
Software product line development process. (Adapted from K. Pohl, G. Böckle, and F. van der Linden, *Software Product Line Engineering: Foundations, Principles and Techniques*, Springer, New York, 2005.)

development in the application engineering level. This is enabled by deriving the desired common and variable parts from the particular domain engineering activity for the corresponding application engineering activity. Deriving refers to the process of binding the variability within the particular artifact to receive application-specific artifacts. This derivation is done for every activity in application engineering. The next step, in each application engineering activity, is the development of application individual parts. The common and variable parts are illustrated in Figure 13.1 by the unshaded symbols, while the product individual parts are depicted by the shaded symbols. The derivation of the common and variable parts for each activity is illustrated by a long unshaded arrow. Further information and variants of the SPL development process can be found in [PBvdL05], [CN01], and [Gom04].

In addition to the SPL development process, the central aspect concerning the reuse of artifacts is the concept of variability. This concept provides the possibility to define particular artifacts of the domain engineering as not necessarily being part of each application of the product line. Variability appears within artifacts and is defined by variation points, and variants. The former defines where variations may occur and the latter, which characteristics are to be selected for the particular variation point. A variation point can be of different types, namely *optional, alternative, or,* and in the case of a feature diagram, the type *mandatory.* Details on variation point types can be found in [vdM07]. Figure 13.2 shows the connection between parts in domain engineering and variation points as a meta model in UML syntax. Parts having at least one variation point are called variable parts. Each part of an artifact can contain a variation point. A variation point itself has at least one variant, which is again a part. Depending on the type of variation point, one to every variant can be bound in product derivation. To model variation points and their corresponding variants, there are different types of variability models known in research. One of the most familiar is the feature model published by [KCH+90].

In this chapter, we focus on the activities domain and application testing of the SPL development process. Furthermore, we concentrate in particular on MBT approaches of embedded systems, as we mentioned in Section 13.1. In the next section, the conceptual MBT approach and the SPL-specific challenges of this are presented.

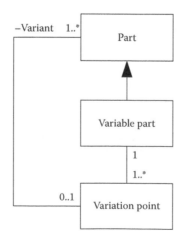

FIGURE 13.2
Relation between variation point and variants.

13.3   Testing Software Product Lines

Because of the variability, testing is even more challenging in SPL engineering than for single systems. A summary of general approaches for SPL testing is given in [TTK04]. The authors distinguish between the following approaches for testing.

Product-by-product testing

In the majority of cases, each instance of an SPL is tested individually—the so-called product-by-product testing approach. As modern product lines can comprise thousands of products, it is impracticable to test each product individually. For example, nowadays there exist millions of possible configurations of a car. Testing the manifold configurations of each potential new car is both time and cost prohibitive. Product-by-product testing can be improved by identifying a minimal set of products that for testing purposes are representative of all other products. Instead of testing all products, the representative set is tested. However, finding a minimal test set is an NP-complete problem [Sch07]. Different heuristics are used to approximate a minimal test set. Promising procedures are mentioned in [Sch07, OMR10] and we refer to this work for further details.

Incremental testing

This methodology stipulates a regression testing technique. One product is chosen to be the first product derived from the SPL and to be tested individually. All other products are tested using regression testing techniques with respect to the commonalities between the different products [McG01]. To identify those parts of a product that remain unchanged and those that vary is the challenge within this test approach. Additionally, the question arises as to whether it is sufficient that only the modified and added parts have to be tested. Again, we like to quote the *Ariane V* accident [Lyo96] to emphasize the importance of testing components already in use.

Reusable test assets

A very promising approach is the generation of reusable test assets. These assets are created during domain engineering and customized during application engineering [TTK04]. All model-based approaches belong to this category since test models are initially created during domain engineering and reused in application engineering.

Division of responsibility

In [TTK04], testing is partitioned according to the levels of the development process, for instance the V-model. For example, Unit testing is performed during domain engineering and the other levels of the V-model are carried out later on during the application engineering activities.

All these approaches can be realized using model-based techniques. The approaches summarized in this chapter comply with at least one of the general approaches in TTK04. MBT is derived from the idea of developing software based on models. This means to explicitly model the structure and behavior of the system by using models on the basis of (semi)formal modeling approaches. The difference to nonmodel-based approaches is made by replacing the informal model by an explicit representation [PP04]. Afterwards, these models are used to generate code for the system implementation. For MBT, an additional test model representing system requirements is used to derive test cases. In Figure 13.3, the

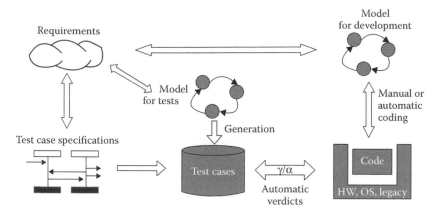

FIGURE 13.3

Model-based testing overview. (Adapted from Pretschner, A., and Philipps, J., Methodological issues in model-based testing, in *Model-Based Testing of Reactive Systems*, Broy, M. et al. (eds.), 281–291, Springer, New York, 2004.)

basic idea of model-based development and testing is depicted. On the top-left corner, the informal customer requirements are illustrated as a cloud. From these requirements, three engineering artifacts are derived: first, the development model (top-right), the test model (center), and the test case specification. The last describes the test cases that have to be written based on the test strategy. The test case specification, together with the test model, is used to generate test cases. This generation step can be automatic or semiautomatic. The development model is used to implement the code that we must test. The test cases are used to validate the code against the test model, in an automatic or semiautomatic way.

Different starting points for the model-based generation of test cases exist. In our example (Figure 13.3), there are two independent models used to either generate test cases (test model), or to develop the code (development model). These models are both derived from the informal user requirements. Another version of the MBT approach uses only one model to derive both the code and the test cases. This version has a drawback concerning the significance of the test results: Errors in the implementation model are not found because it is also the basis for the test generation. Further information on MBT can for example be found in [PP04] and [Rob00].

We summarize different approaches to transferring the concept of MBT to SPLs. To facilitate the comparison of the different approaches, we introduce a conceptional process model comprising all MBT procedures for SPLs. This model is depicted in Figure 13.4 and forms a superset of all test approaches to be introduced in this chapter. Thus, the idiosyncrasies and properties of each model-based approach are captured subsequently by specializing the process model introduced here.

Figure 13.4 is based on the following two principles.

1. According to the development process for SPLs depicted in Figure 13.1, the testing process is subdivided into application and domain engineering.

2. Each phase can be subdivided according to the levels of the V-model introduced in DW00. For each step in the development process (left-hand side), a corresponding test level exists (right-hand side). Therefore, we can visualize the different levels of testing: Unit tests, Integration tests, and System tests. Each test phase contains a test model and a test case specification to generate test cases.

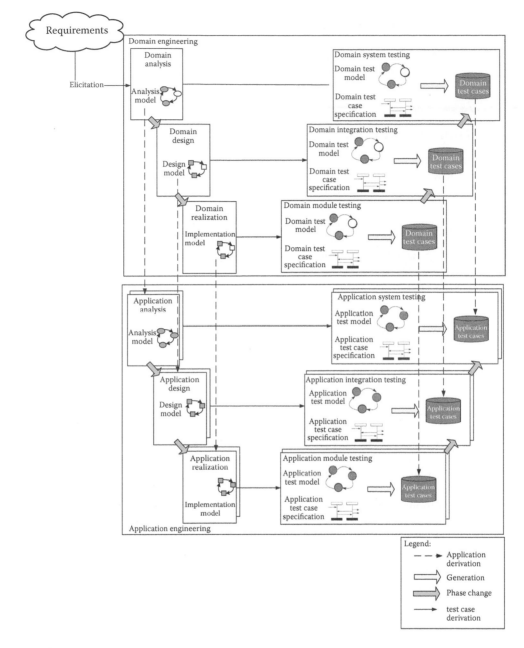

FIGURE 13.4
Conceptional process model for software product line testing.

Additional vertical edges connecting domain and application engineering indicate that artifacts developed in domain engineering are reused in application engineering. Further details about this model are explained in Section 13.6, in which each approach is explained by means of this process model.

Using this process model, we can discuss:

- Whether the approaches in our scope differentiate between domain testing and application testing,

- Whether they consider unit, integration, and system testing, and

- How the test case generation is accomplished.

13.4   Running Example

In this section, we introduce a running example to illustrate the different SPL testing approaches in a consistent way. In our simplified example scenario, a company develops mobile phone games. The game portfolio of the company contains a game called *Bomberman*. The company wants to offer this game to different mobile phone producers, who will sell their mobile phones bundled together with the *Bomberman* game. Every mobile phone the game should be compatible with has some different features. This means the mobile phones have different hardware configurations, which are important for the game software. In our simplified example, the relevant hardware and software components are

- A Java mobile edition runtime environment on every mobile phone.

- Bluetooth on some mobile phones.

- Internet connection on some mobile phones.

- Photo camera on some mobile phones.

- A touch screen on exactly one mobile phone.

The different hardware and software components of mobile phones result in the need for an individual game configuration for every different configurable mobile phone type. These game configurations have commonalities such as the usage of the Java-based runtime environment, which are in part included in mobile phone configurations. The goal is to develop the commonalities of the different mobile phone games only once and to reuse them in the different products. To achieve this goal, the development paradigm SPL will be used. The domain (see Section 13.2, domain engineering) of the *mobile phone game* product line contains features, which are part in all or some game versions of *Bomberman*.

The connection between the domain engineering and three possible game configurations is depicted in Figure 13.5. The three games A, B, and C are illustrated as bullets. The intersection of all bullets displays the commonalities. In our case, it is the Java game engine. Beyond that, game version A requires a mobile phone with an internet connection and a photo camera, version B relies on a photo camera and a bluetooth connection, and version C is designed for an internet connection, a bluetooth connection, and a touch screen interface. These features are all depicted in the intersections of only two of the circles. Consequently, they are not part of all possible game versions, but of some. From a domain point of view, these features are variable. As depicted in Figure 13.5, the feature *camera game profile* is part of game version A and B. This feature enables the game player to add a photo to his/her game profile by using the photo camera of the mobile phone. The internet high score (version A and C) provides the functionality to publish the high score of the player via an internet connection. In version B and C, a bluetooth multiplayer functionality is integrated. Beyond that, only version C contains a touch screen interface. Figure 13.5 shows how this feature is individual to game version C only as it is in part of the C circle that is not overlapping. Feature-oriented modeling like in KCH$^+$90 offers the possibility to model the mandatory and variable features of an SPL. In Figure 13.6, the features of our

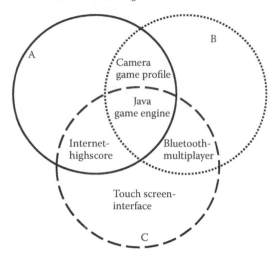

FIGURE 13.5
Mobile phone game product line Bomberman.

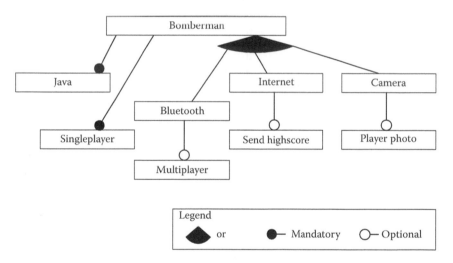

FIGURE 13.6
Feature model of running example.

running example are explicitly depicted. The *Bomberman* game has mandatory features *Java* and *Singleplayer*. Beyond this, the customer can choose from the features *Bluetooth*, *Internet*, and *Camera*. Each of these features has another optional feature as a child feature, namely *Multiplayer*, *Send highscore*, and *Player photo*. As most of the sections that follow where a comparison between approaches is made rely on UML models, we now provide the appropriate product-specific use cases for *Gameplay Bomberman* by means of three activity diagrams (Figure 13.7).

The activity diagram of Product *A* is depicted on the left-hand side of Figure 13.7. After game start and watching the intro, the player may take a photo for the player profile, as the mobile phone is equipped with a camera. Afterwards, the main menu is shown to the player. The player can now start the game, show the highscore list, or exit the game. If the player starts the game, it runs till the player exits the game or the player loses the game.

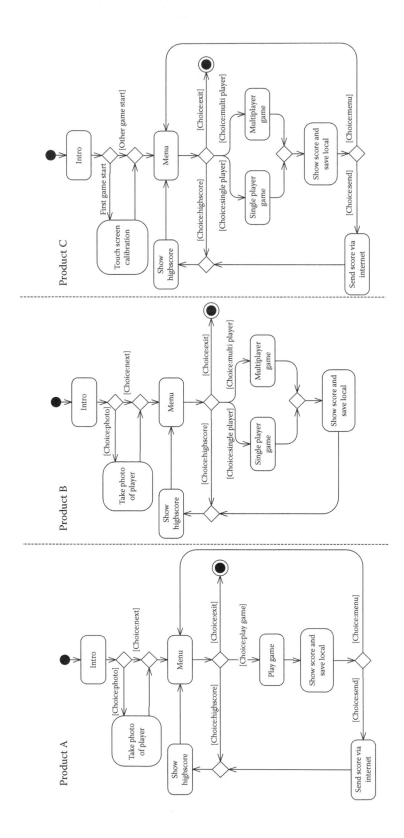

FIGURE 13.7

Activity diagrams of game play from the three different products.

The score is then shown to the player and it is saved to the mobile phone's memory. As the mobile phone is equipped with an Internet connection, the player can choose to send the score to the common highscore list. Otherwise, the player can return to the menu. In the first case, after sending the score, the highscore list is shown to the player and he/she can then return to the menu.

The second Product *B* realizes a slightly different use case. As the mobile phone is not equipped with an Internet connection, it is not possible for the player to send the achieved score to the highscore list. As a point of difference with Product *A*, it is possible to play multiplayer games, as the mobile phone does come equipped with the *Bluetooth* feature.

The third Product *C* is equipped with a product-specific *Touchscreen* functionality. This feature is calibrated after finishing the game intro, but only if the game is executed for the first time. In contrast to Products *A* and *B*, this product is not equipped with a camera and, consequently, the player cannot take a photo for the player profile.

The three activity diagrams depicted in Figure 13.7 illustrate the commonalities and variability of the use case scenarios. As a consequence, the development of these products using the paradigm SPL provides reuse concerning the use cases and corresponding test cases. Note that, the three illustrated examples *A*, *B*, and *C* are only sample products. On the basis of the existing commonalities and the variability of the domain engineering, other configurations can be constructed, too. We restrict our example to the three products already introduced. Later on, the example is used to explain the different MBT approaches.

13.5 Criteria for Comparison

To examine the different SPL MBT approaches, we need comparison criteria to identify significant differences. We discuss the approaches with respect to six main categories that we have chosen according to the best of our knowledge: (1) *Input* of the test methodology, (2) *Output* of the test methodology, (3) *Test Levels*, (4) *Traceability* between test cases and requirements, architecture, and the feature model, (5) integration of the test process within the *Development Process*, (6) and *Application* of the test approach. The different criteria are depicted in Figure 13.8. Each category contains corresponding criteria for comparison.

Input

In this category, we examine the inputs required for the test approaches. Since we concentrate on model-based approaches, we determine what kind of test model is used. We analyze every approach in this context and examine how the test model is created. Furthermore, additional test case creation inputs are considered, for example, models of the development process or further system information.

Output

The *Output* describes the output of the test approaches. We usually expect test cases or a description of them to be generated. Regarding the generation process, we additionally examine the degree of automation, for example, whether the test case generation is executed automatically or semi-automatically. Another important approach to evaluating MBT approaches is to measure the coverage of its tests. Therefore, we include the existence and type of coverage criteria belonging to a certain approach as a criterion to compare the different approaches.

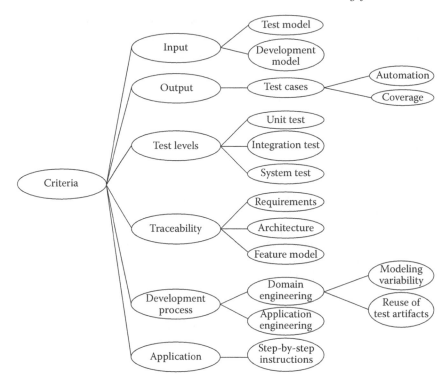

FIGURE 13.8
Criteria tree for comparison of SPL approaches.

Test Levels

We examine whether a test approach covers a specific test level, for example, unit, integration, or system testing.

Traceability

An important property of SPL testing approaches is the mapping between test cases and requirements, architecture, and features of the feature model. It offers the possibility to trace the errors found back to the responsible component and even back to the requirements specification. Besides that, if a requirement is changed, we know exactly which test must be adapted. Another point on traceability is the derivation of concrete product-specific test cases. By choosing requirements for a concrete product, the corresponding test cases should be determined by tracing from the selected variants of the requirements to the test cases covering this requirement. Therefore, we check if traceability is supported and to what extent.

Development Process

The question arises whether the testing approach can be integrated within the development process of SPLs. A key attribute is the differentiation between domain and application engineering. If an approach differentiates between domain and application engineering, we must examine how variability within the different activities is modeled and handled. Then, we determine how reuse and variability interact for testing.

Application

Finally, we introduce a criterion focusing on the application of the test approaches. This criterion is mainly important for the software industry. It addresses the question of whether a particular approach can be integrated into a company's development process. To be able to do this, a step-by-step description is important.

The presented criteria will be used to compare the existing MBT approaches for SPL. Before starting the comparison, the running example is presented in the next section. This example is used to compare the approaches. This provides transparency of the comparison to the reader.

13.6   Model-Based Test Approaches

Various frameworks and approaches to SPL engineering exist. Among these are COPA [OAMvO00], FODA [KCH+90], FORM [KKL+98], PLUS [Gom04], PuLSE [IES], QADA [MND02], and the "Generative Software Development" approach introduced by Czarnecki [CE00]. There is, however, a shortage when it comes to testing in nearly all of these approaches. In TTk04, Tevanlinna, et al. present open questions on testing SPLs. The subdomain test case design and specification are explored in Wüb08 and shortcomings are elicitated.

In this chapter, we only focus on approaches that include a test methodology that is based on the MBT paradigm. Furthermore, we explain other MBT approaches that are used in combination with SPL engineering. Each approach is described according to the following scheme. First, we briefly describe the approach and how it aligns with the engineering process. Subsequently, we apply the approach to our running example to clarify the test development process. At this point, we have to restrict ourselves to some extent and only focus on prominent characteristics. Finally, we summarize each approach individually, in preparation for the discussion in Section 13.7.

13.6.1   CADeT

CADeT (Customizable Activity Diagrams, Decision Tables, and Test specifications) is a MBT approach developed by Erika Olimpiew [Oli08]. The approach is based on the PLUS (Product Line UML based Software engineering) method by Gomaa [Gom04]. The approach focuses on deriving test cases from textual use cases by converting them into activity diagrams.

The method defines the following steps on the platform engineering level:

1. Create activity diagrams from use cases

2. Create decision tables and test specifications from activity diagrams

3. Build a feature-based test plan

4. Apply a variability mechanism to customize decision tables

In application engineering, the following steps are defined:

1. Select and customize test specifications for a given application

2. Select test data for the application

3. Test application

The approach is illustrated in Figure 13.9. The top-left rectangle depicts the activities in SPL Engineering. Here the activity diagrams, decision tables, and test specifications are created. All artifacts are stored in an SPL Repository. Now by using a feature diagram a feature-based test plan is built to define test cases (feature-based test derivation) illustrated in the bottom-right rectangle. The test models are then enriched with test data and further information to render them test ready. The final step is to test the product by using the created test cases.

In the following paragraph, we are going to apply the CADeT approach to our running example introduced in 13.4.

Application

First, we transform the feature diagram of our running example, taken from Section 13.4, into the PLUS-specific notation (see Figure 13.10).

The type of feature (e.g., *common, optional, zero or more of*) is depicted as a stereotype within the feature. The features are linked by *requires* and *mutually includes* associations, depending on the type of the features to be linked. The feature diagram is the basis for deriving products from the SPL. In CADeT, the more specific requirements are defined using textual use cases with variation points. Using our running example from Section 13.4, we specified a use case diagram by using the CADeT notation. The result is depicted in

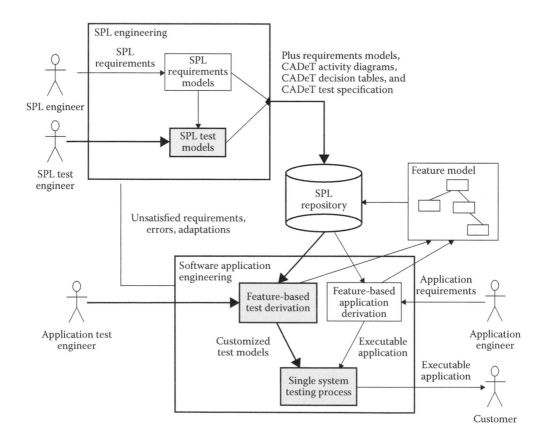

FIGURE 13.9
Overview of the CADeT approach (Adapted from Olimpiew, E. M., Model-Based Testing for Software Product Lines, PhD thesis, George Mason University, 2008.)

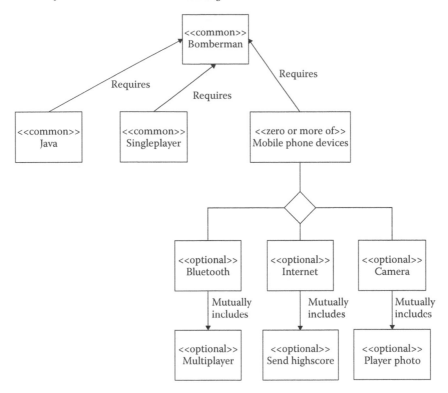

FIGURE 13.10
Feature model of running example in PLUS notation.

Figure 13.11. The use cases are stereotyped by either <<*kernel*>> for a common use case or <<*optional*>> for use cases depending on variation points. The stereotype <<*selection*>> models the selection rule for two or more optional use cases being associated to a certain use case (for further information, see Gom04). The variation points use cases are depicted in the corresponding use case and on the association between linked use cases. Each use case has a textual representation. Because of the space restrictions, we do not represent the use case representation of our running example. More information about designing such use cases, with variability can be found in [Oli08].

On the basis of the textual use cases, we now manually create activity diagrams with variability definitions. One activity diagram for our example is illustrated in Figure 13.12 (We left out the pre- and postconditions). In the diagram, the use cases from Figure 13.11 are depicted as rectangles without curved edges. In every rectangle, the single steps of the use case are modeled as activities. Extended use cases have the stereotype <<*extension use case*>> and the corresponding variation point as a condition on the control flow arc. This activity diagram together with the feature diagram from Figure 13.10 is now the basis for deriving a decision table containing test cases for application testing.

The decision table shown in Table 13.1 contains a test case in every column. Each test case has a feature condition that enables us to select the appropriate test cases for a specific product derivation. Each test case specification consists of a pre- and postcondition, taken from the use case and the test steps taken from the activity diagram. Dependent on a specific feature condition, the test case does not contain variability anymore. If the feature condition is fulfilled, the corresponding test case is part of the derived product.

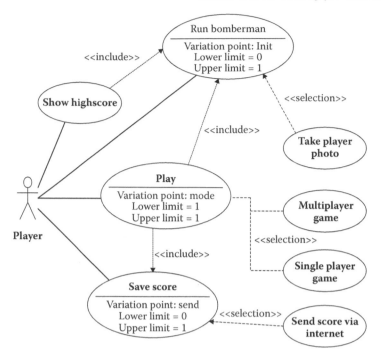

FIGURE 13.11
Use case diagram in CADeT notation.

After deriving the product, the test cases have to be enriched by test data and detailed pre- and postconditions. Besides, CADeT provides a strategy, based on the feature diagram, to test sample configurations of the corresponding SPL (feature-based test plan).

Summary

The CADeT method provides a detailed description of how to apply it to an SPL project. It aims at the system test level and defines use cases for requirements modeling based on PLUS. The approach belongs to the category "reusable test assets" defined in Section 13.3. Furthermore, a feature diagram provides an overview of the possible features that derived products. The test cases are derived automatically based on activity diagrams, which are the test models containing variability in domain engineering. The activity diagrams are manually derived from the use cases defined in domain engineering. The activity diagrams together with the feature model define the test case specification on the domain engineering level. By choosing a particular feature configuration, product test cases can be derived from the test case specification. CADeT offers different feature coverage criteria to test example product configurations. Test coverage, in a classical point of view, is not supported by the approach. In addition, test data definition in domain engineering is not supported by CADeT. Figure 13.13 shows the CADeT approach according to the conceptional process model.

13.6.2 ScenTED

ScenTED (Scenario-based TEst case Derivation) is a MBT procedure introduced by Reuys et al. [RKPR05]. This approach can be integrated in the SPL development process, depicted in Figure 13.1, since it distinguishes between domain testing and application testing. It supports the derivation of test cases for system testing and mainly focuses on the question:

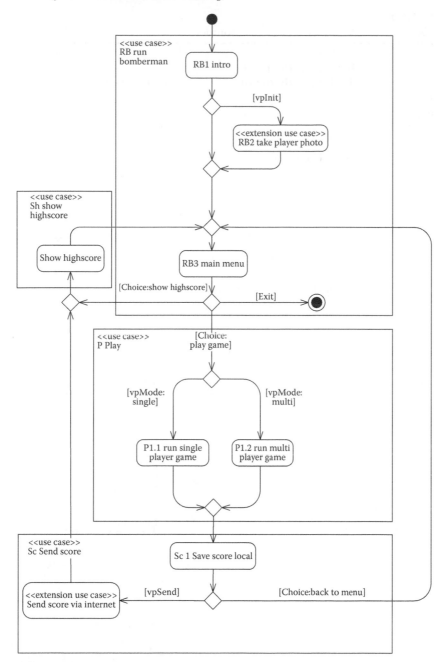

FIGURE 13.12
CADeT activity diagram of Bomberman.

What to test during domain and what during application engineering and how to reuse test artifacts, for example, test cases, test data, and test scripts?

To overcome these challenges, ScenTED generates domain test cases, including variability, and reuses these test cases to derive test cases for each individual product in application engineering. Complete testing is impossible during domain engineering due to the variability: testing can only be performed after variability is bound.

TABLE 13.1

CADeT Decision Table of Use Case Play Bomberman

RB	<<use case>> Run Bomberman Test Specification	1 <<reuse as is>>	2 <<adaptable>>
Feature condition	Player photo	T	F
Precondition		Mobile phone is running	Mobile phone is running
Execution condition	Start Bomberman		
Actions			
RB1	Mobile phone shows intro (out intro)	X	X
RB2.1	<<optional input step>> User takes photo (in photo)		X
RB2.2	<<optional output step>> Photo is stored and shown to the user. (out show photo)		X
RB3	User enters main menu (out menu)	X	X
Postcondition		Main menu	Main menu

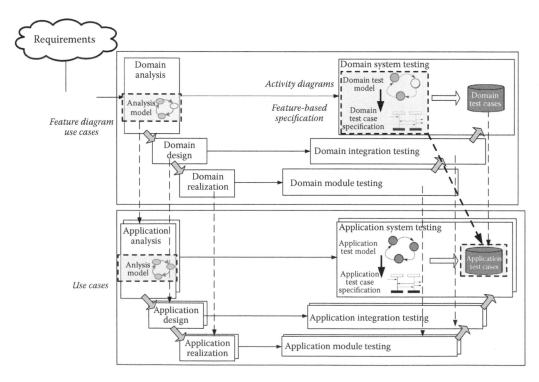

FIGURE 13.13

CADeT according to our conceptional process model.

ScenTED requires that the representation of the system requirements are described by use case diagrams. In this chapter, we focus on the system testing approach. Figure 13.14 depicts the ScenTED process. The main activities within ScenTED are numbered and described in the following.

1. Each Use Case is translated into an activity diagram representing the test model. The variability within a Use Case—mirroring variability in functional requirements—has to be incorporated into the corresponding activity diagram. There, the possible variants are described using alternative paths. Variation points within the activity diagram control the selection of possible variants and express differences between applications. Each variation point has a name and contains the cardinality of the variants it encompasses.

2. Domain test scenarios are derived satisfying a certain branch coverage criterion within the activity diagrams. This coverage criterion ensures that each branch within an activity diagram is covered by at least one test case. Each variation point is used as a placeholder for each possible variant. Each possible variant is represented by a domain test scenario. A domain test scenario contains test information, for example, pre- and postcondition, input parameter, and the expected output.

3. At the time applications are derived, variability is resolved. Application test cases are derived based on the domain test scenarios. Each product must be tested individually, using the corresponding application test scenarios.

The domain test scenarios for system testing can be reused for various applications.

Application

To demonstrate the system testing procedure, described in ScenTED, we apply it to our running example introduced in Section 13.4. ScenTED expects that requirements are described as Use Case diagrams. Figure 13.15 depicts the Use Cases of our running example.

We choose the two use cases, *Start Game* and *Play*, to demonstrate ScenTED. In the first step of ScenTED, the use cases are translated into activity diagrams depicted in Figure 13.16. The left-hand use case describes the starting of a game. The variability within this use case is denoted as the variation point, (stereotype $<<VP>>$). According to this variation point, two variants are possible: (1) The player may choose a photo or (2) The player starts the game without choosing a photo. Choosing the photo is an optional variant that can be selected. Comments are used to describe further details about the variation point and the alternative paths. Using this activity diagram to derive domain test scenarios,

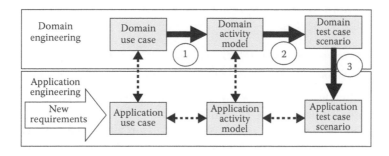

FIGURE 13.14
ScenTED concept [RKPR05].

Domain: Handy game

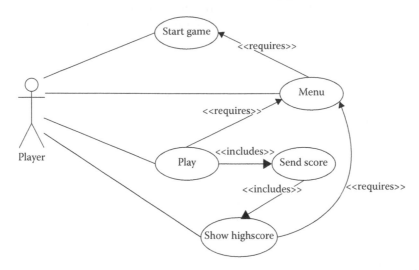

FIGURE 13.15
Use case diagram of the running example.

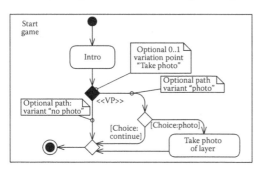
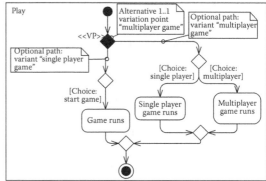

FIGURE 13.16
Two use cases: *Start Game* and *Play*.

we first use the variation point as a placeholder for all possible variants. $\{Intro, \{\}_{photo}\}$. Intro is the first state to be reached and $\{\}_{photo}$ represents the placeholder for the variation point. Now, we have to consider all possible paths that can replace the variation point. We obtain the following sequences:

$$\{Intro, \{\{\}_{Variant"without\_photo"}, \{\}_{Variant\_"photo"}\}_{photo}\}$$
$$\{Intro, \{\{\}_{Variant"without\_photo"}, \{Take\_photo\_of\_player\}_{Variant\_"photo"}\}_{photo}\}$$

In the first scenario, the player wishes to play without taking a photo, and in the second scenario, the photo, is taken before the game starts.

The use case on the right-hand side describes the Domain use case *Play*, which contains a variation point *multiplayer game*. The user may either play the game as a single player or in a multiplayer game, where several other players can join the game. In comparison to the variation point mentioned in the left-hand use case, this variation point describes two alternative variants. Exactly one of the alternative paths must to be chosen. As a consequence, we obtain the following test sequences for the variation point:

$$\{\{\}_{multiplayer\_game}\}$$

either:

$$\{\{\{Game\_runs\}_{single\ player}, \{single\ player\_Game\_runs\}_{multiplayer}\}_{multiplayer\_game}\}$$

or:

$$\{\{\{Game\_runs\}_{single\ player}, \{multiplayer\_Game\_runs\}_{multiplayer}\}_{multiplayer\_game}\}$$

After deriving Domain use cases, we can generate domain test cases. For that, we must model each scenario using sequence diagrams and add additional information about the test. In fact, several sequence diagrams are derived from a single use case depending on the degree of variability. In our running example, we only generate one sequence diagram per use case, both depicted in Figure 13.17. Using these sequence diagrams, test cases can be generated. Pre- and postconditions support the derivation of test sequences and additional information for testing. They are added by using comments. Pre- and postconditions can be used as test criteria to check whether the result of each step of the testing process is as expected. The final step of the ScenTED methodology is the generation of test cases for a certain application. According to the requirements for a certain product, a product use case is derived. For this purpose, domain use cases may be reused. To generate a test case for a

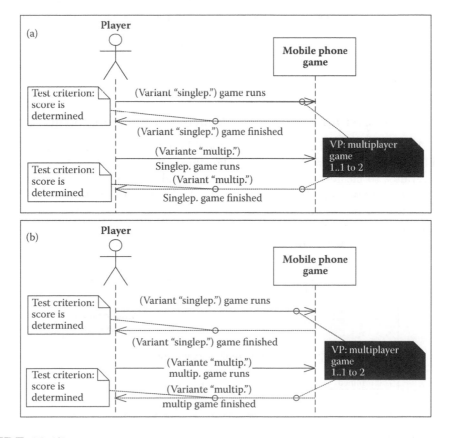

FIGURE 13.17
Domain test cases as sequence diagrams.

certain application using a domain test case, all variants that do not belong to it have to be removed. The result is a sequence diagram for a certain application. Figure 13.18 depicts a test case scenario for a product realizing a single player game. In comparison to the domain test case in Figure 13.17, we only have to remove the multiplayer part. Indeed, it is possible that further requirements during application engineering must be taken into account when deriving a product. Therefore, existing domain test cases need to be adapted or new use cases must be added and new test cases must be generated. This would be the case for product C in Figure 13.7 in Section 13.4, as it would require touchscreen calibration.

Summary

ScenTED preserves variability within the entire process of deriving domain and application test cases for system testing. Test cases for different products can be derived easily from the domain test cases and variability is preserved within the domain test cases. Integration tests and Unit tests are not considered in ScenTED. Furthermore, the entire process may be automated and provides an adapted coverage metric [RKPR05]. However, the metrics only take the activity diagrams into account, offering a very rough description of system properties and functionalities. We suggest to consider further test data to ensure sufficient coverage. Besides that, it is not possible to relate variation points with each other and a variation point can only represent one specific use case. Figure 13.19 depicts those parts of our conceptional process model that are covered by the ScenTED approach.

The Analysis Models in the Domain and Application Engineering are described using use case diagrams. On the basis of these use case diagrams, domain test models by means of activity diagrams are generated. Subsequently, domain test case specifications are set up as sequence diagrams. The domain test models and the domain test case specifications serve as a basis for application test models and application test case specifications. With regard to the classification of test approaches for SPLs, ScenTED realizes a combination of product-by-product and reusable test assets.

13.6.3   UML-based approach for validating software product lines

In HVR04, Hartmann et al. describe a UML model-based approach to validate SPLs in the context of a project conducted for Siemens. The approach extends the UML, and especially activity diagrams, to include variability annotations and use this. The approach addresses the test level *system testing* and is integrated with the IBM/Rational Rose tool. The goal of this approach is to formalize and automate the test case design, generation, and execution.

As mentioned before, the approach uses activity diagrams to generate test cases. These diagrams are extended by *swimlanes*, which decide between the success path and

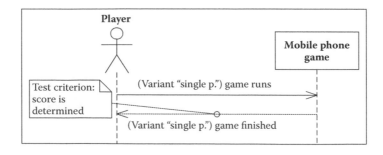

FIGURE 13.18
Application test case for a singleplayer game.

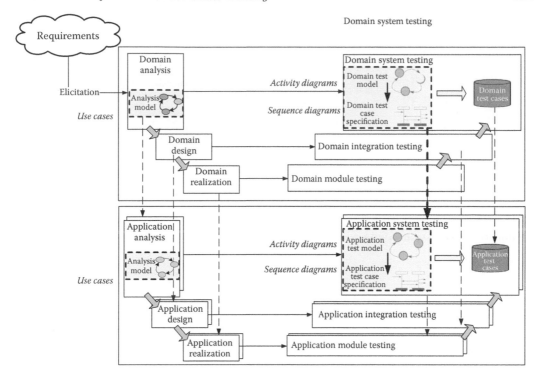

FIGURE 13.19
ScenTED according to our conceptional process model.

```
<<Product Line>>
    Product A
    Product B
    Product C
```

FIGURE 13.20
Names of product variants.

paths with all exceptions of the activity diagram. Furthermore, the activities (depicted as rectangles with rounded corners) are annotated by the stereotypes <<UserAction>>, <<SystemResponse>>, and <<Include>>. The first two stereotypes indicate whether the activity is a user or system activity, the latter enables us to include another activity diagram into the original one.

To model variability within an activity diagram, another stereotype named <<Variance>> is used. In the following, we apply the approach to our running example.

Application

The first step of the approach is to determine all possible product variants and define them, using the stereotype <<ProductLine>> in the root activity diagram, using a note (see Figure 13.20). The note contains the three product instances we defined in Section 13.4.

Variability is then defined by using the existing product variants to decide which variants belong to each product (Figure 13.21). In our example, the optional activity *Pre Game*

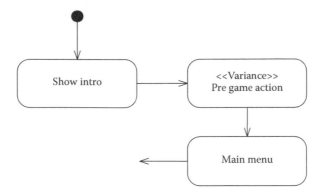

FIGURE 13.21
Definition of variation points with UML activity diagrams.

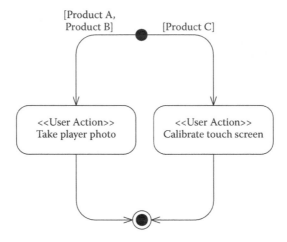

FIGURE 13.22
Subdiagram identifying the product variants.

Action is defined as a variable step within the main activity diagram. The particular variants are modeled in subdiagrams of the corresponding activity.

Figure 13.22 shows the subdiagram for the activity *Pre Game Action*. Depending on the selected product, a specific path through the diagram is selected (see conditions on the arcs). If *Product A* or *Product B* is chosen, the user action *Take a player photo* is selected. For *Product C*, the calibration of the touch screen action (*Calibrate Touch Screen*) is chosen.

In another layer of activity diagrams, the modeling of the detailed behavior of each variant is depicted. Figure 13.23 shows the detailed modeling of the activity *Take player photo*. Here, two swimlanes are used to sort the activities by user actions and system responses.

After defining the variability by means of the product line activity diagrams, the designer is able to add other test relevant attributes such as decision nodes and nodes with information on test case derivation. These notes hold the stereotype <<define>>. The test model is now used to generate test cases by using well-known category partitioning methods. The user chooses a specific product from the product line (e.g., *Product A*). The variability within the activity diagrams is then bound and the categories concerning the choices and input data are built.

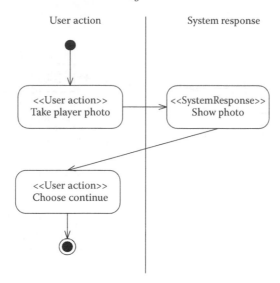

FIGURE 13.23
The *Take Player Photo* activity.

Summary

Hartmann et al. provide a UML model-based approach to generate test cases for system testing. Activity diagrams represent the system specification and test model. As only one representation is used for both the system specification as well as the test model, this approach suffers from the drawback discussed in Section 13.3 where the common bias prevents independent detection of errors. The test model is used to derive product-specific test cases. The approach belongs to the category "reusable test assets" defined in Section 13.3. The approach by Hartmann et al. does not explicitly model variability within activity diagrams. They use a tree structure of activity diagrams to build up the appropriate activity diagram for the selected product. As the selection criteria for the subactivity diagrams depend on concrete product variants, the approach does not apply to new product variants, that is, those that do not already exist in the model. This means that all possible or desired product variants must be known or anticipated from the beginning. Furthermore, the approach does not describe in detail how to generate test cases. From our point of view, the generation is not a product line specific problem, as the generation works on activity diagrams without variability. Finally, the approach leaves out detailed information about coverage and test data and the additional stereotypes complicate the automation of this approach. Figure 13.24 shows the approach according to the conceptional process model.

13.6.4   Model-checking approach

Kishi et al. focus on the fact that the majority of faults in software engineering originates during requirements analysis and design phases [KN04]. Therefore, they propose checking the design of each product derived by the application engineering before actually building these products. Thus, they examine whether a certain product realizes the required features. To check the design, a design model, consisting of a state machine in combination with model checking, is used. Former approaches to design testing, such as reviewing and testing, are not capable of checking the design in its entirety [KN04]. Therefore, the authors apply model checking for design testing in SPL development. Their approach is part of the application

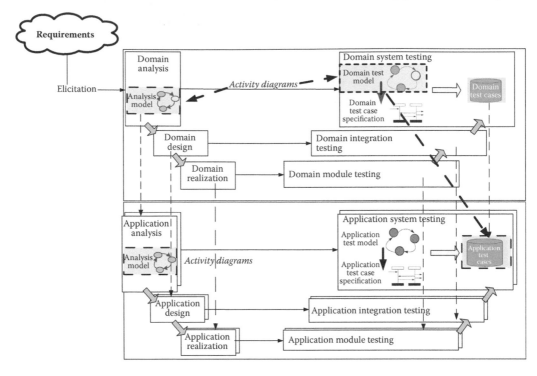

FIGURE 13.24
UML-based approach according to our conceptional process model.

engineering phase. Because the authors use model checking to generate test sequences for the design model, they state that they are not performing design verification but design testing.

Model checking always works on the basis of two models [KN04]. On the one hand, there is the product itself, described by means of a state machine. This model is referred to as *target* model representing the behavior of the product. On the other hand, an additional state machine is set up sending event sequences to the *target* model. Kishi et al. use this model in combination with logical formulae, which serve as *environment* models to emulate event sequences. These event sequences then serve as test sequences to test the design of the product. If the required behavior is met, it can be checked by applying every possible event sequence to the design model. For each event sequence, a correct design reaches the expected state. The model checker generates test cases using the event sequences from the *environment* model. We apply our running example to describe the reuse technique proposed by Kishi et al. We focus on the overall process and leave out details about the model-checking procedure.

Application

To demonstrate the reuse technique, we first describe two possible products of our running example by way of an *environment* and a *target* model. Then, we introduce a reusable *environment* model, capable of describing both products in this single model by way of a reusable *environment* model. Kishi et al. use state machines with logical expressions to describe the behavior of the product, as well as class diagrams to express variability. Please note that we can only describe the theoretical application of this approach. A detailed description of the model-checking technique goes beyond the scope of this survey.

Figure 13.25 depicts a rather simple product of our running example, only allowing the user to select a single player game and to show the highscore. The *environment* model consists of a state machine, realizing the necessary transitions: the choice of a single player game (*Choice: single player*), the choice to exit (*Choice: Exit*), and the choice to view the highscore (*Choice: Highscore*). The *target* model uses these transitions to realize the behavior of this simple product.

The second product, as shown in Figure 13.26, includes more functionality. Its class diagram consists of three classes: *Operator*, *Bomberman*, and *Internet*. The classes *Operator* and *Internet* belong to the *environment* model of this product, whereas *Bomberman* as the *target* model, describes the mobile phone game. The *target* model should be self-explanatory. It offers the user the particular multiplayer and highscore sending functionality that is only available when a connection to the internet is possible. These functionalities are multiplayer and sending highscore.

The two corresponding *environment* models describe the possible input sequences for the *target* model. The *Operator* model includes the additional transition *Choice:multiplayer* in comparison to the simple product, enabling a change of state to the multiplayer mode. The *Internet* model includes the transitions to send the highscore, or to return to the menu without sending. According to Kishi et al., it is possible to set up reusable *environment* models including variability, capable of describing both products. The reusable set still consists of the two *environment* models, *Operator* and *Internet*, and the *target* model *Bomberman*. Please note that the *target* model for this reusable set is the one depicted in Figure 13.27.

In comparison to the previous products, the class *Internet* is marked as optional. This means that the transitions described by the *environment* model *Internet* can only be used when this class is chosen. In addition to that, the state machine describing the *environment* model *Operator* is adapted. A placeholder transition is the only transition remaining in the *Operator* state machine. This placeholder transition is linked to an additional class diagram, containing all choices we require for the two products. Since the *Choice: multiplayer* and *Choice: highscore* are only necessary if the optional *environment* model *Internet* is included, these classes representing the *Choice: multiplayer* and *Choice: highscore* transitions are

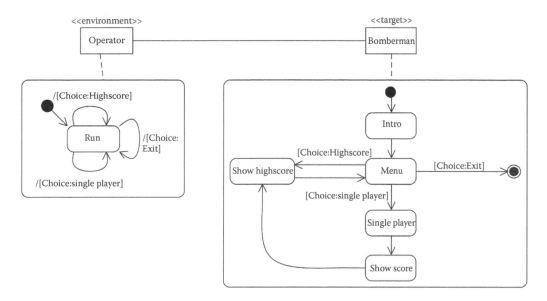

FIGURE 13.25
Simple product.

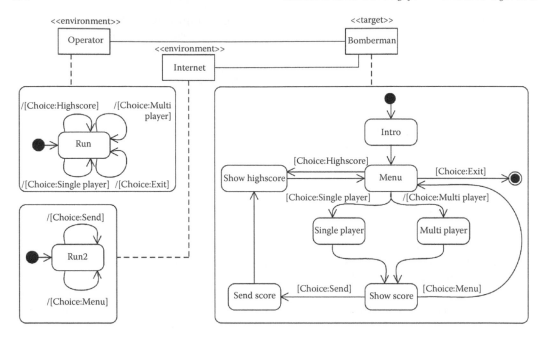

FIGURE 13.26
Complex product.

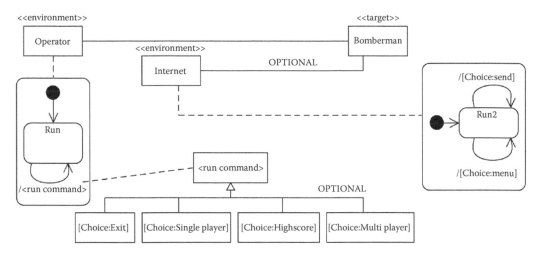

FIGURE 13.27
Reusable environment model.

optional as well. All other transitions can be used as in Figure 13.25. These *environment* models can therefore be then used to describe the *target* models of both products.

Summary

The work of Kishi et al. introduces an approach to test the design of a product using model checking and reusable state machines. The authors expect to test the design more exhaustively than when other, more traditional, approaches are applied such as reviewing. The complete verification of a product is, however, not achieved. Rather than applying

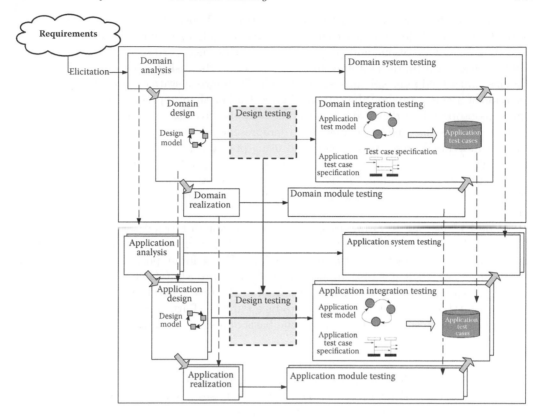

FIGURE 13.28
Model-checking approach according to our conceptional process model.

model checking on individual state machines of products of an SPL, a reusable *environment* model is introduced, using a single *target* model. However, a well-known problem for model checking is state explosion [CGP99]. This problem also arises in this approach.

To realize traceability, Kishi et al. group features into different groups in a feature model. Actors and components described in the design model are associated with these groups. For a detailed description of grouping the features, we refer to [KN04].

Figure 13.28 depicts those parts of our conceptional process model that are covered by the model checking approach. The approach introduced by Kishi et al. applies model checking for design testing [KN04]. Since variability is preserved, the design testing approach is applied to the domain engineering and application engineering process. Regarding the different test methodologies, Kishi et al. realize a combination of product-by-product and reusable test assets.

13.6.5 Reusing state machines for automatic test case generation

In [WSS08] the authors describe a methodology to reuse state machines as test models for automatic test case generation in product line engineering. To foster the reuse of these test models, the authors explain two different approaches to reuse state machines. One approach was introduced by the OMG [OMG09] and the other approach was developed by the authors themselves. We do not discuss the OMG approach in this chapter, but refer to [OMG09]. The authors use a state machine as a test model representing the behavior of the SPL to generate test suites for each product. Therefore, variability has to be realized within the state machine. Rather than using the generalization and specialization technique proposed by the OMG, Weißleder et al. leave the state machine unchanged but change its context—its class [OMG09]. Operations and properties within this class correspond to guards, events, and transitions within the state machine. According to the Liskov's principle, all properties of a class must hold in its subclasses, too [Lis87]. Therefore, a state machine that models the behavior of a class also describes all subclasses. Rather than applying inheritance between states, as proposed by the OMG, inheritance is applied to classes as a whole.

The authors use one state machine to represent an entire product line. Differences between individual products are expressed using pre- and postconditions, or values within the attributes in the context classes. Therefore, each product derived from the SPL is described by a corresponding subclass with corresponding properties and OCL constraints. These classes configure the state machine so that it describes the behavior of the derived products.

To derive test suites automatically, OCL expressions are used to generate test input partitions. For that, the OCL constraints of the subclass of a specific product are translated into input parameters of the state machine. Using input partitions is a well-known approach for testing and is also applied in the CTE approach [GG93]. Because of the complexity, we cannot provide detailed information of the test generation process and refer to WS08 for further details. The authors introduce ParTeG (Partition Test Generator), which automatically derives boundary tests on the basis of a state machine [Wei09].

Application

Figure 13.29 depicts the state machine representing the **mobile phone game** product line and its context class. Furthermore, it shows two subclasses describing two possible products of the running example SPL. Each of those products has different attributes and OCL constraints. The product *Bomber1* consists of the features: bluetooth, internet connection, and touchscreen. *Bomber2* contains only: bluetooth and camera. The OCL constraint in *Bomber1* determines that the highscore is shown if *show highscore* is selected and if the previous state is either *menu* or *send highscore*. The OCL constraint in *Bomber2* is slightly different. It does not include the state *send highscore* because an internet connection is not available in that product.

These relatively abstract and simple class diagrams are actually sufficient to derive Unit tests using ParTeG. ParTeG derives test cases for both products on the basis of the single state machine and the pre- and postconditions. We refer to Wei09 for further details about the test case generation process.

Summary

Weißleder et al. propose to reuse test models for MBT of SPLs by means of reusing state machines for context classes [Wei09]. A state machine has to be set up representing the entire SPL. However, one must decide whether this is really more time consuming than creating a state machine for each product individually, as described in Section 13.6.4. When it comes to coverage, this approach seems very promising because well-known coverage criteria are applied and combined. In contrast to the other approaches, this method preserves variability within a single test model. It is not necessary to generate individual models including variability for each feature, feature combination, or use case. However, there are still open questions that must be discussed. For instance, one important question is how to derive class hierarchies from product lines automatically. The authors assume that aspect-oriented approaches within the development process may offer a solution. Each feature may be represented by a certain aspect. Then, a hierarchical structure of aspects can be set up.

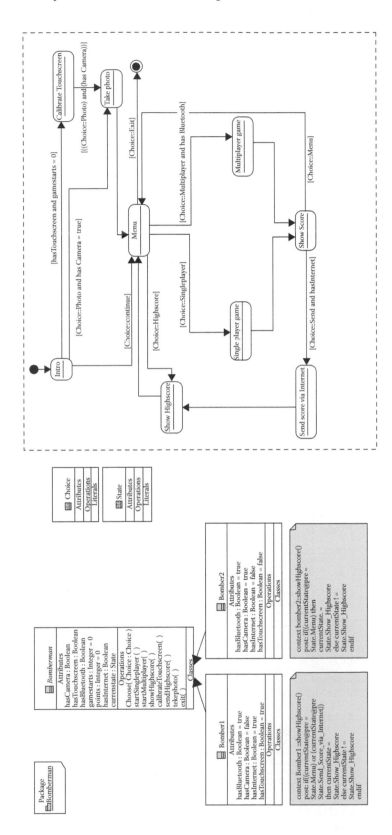

FIGURE 13.29

State machine with its context class.

Figure 13.30 depicts those parts of our conceptional process model that are covered by the Reusing State Machine approach. Weißleder et al. also focus on system testing, preserving variability within a domain test model. Thus, they test according to the reusable test asset method. One single state machine is used as test model describing the behavior of the entire SPL. Test models for the individual applications are represented by subclasses. The test case specifications are described using OCL constraints. Application test cases are derived using ParTeG.

13.6.6   Framework for product line test development

Kang et al. [KLKL07] emphasize that it is important to be equipped with a development method for SPLs, before describing a test methodology. Therefore, a formal framework for product line development, including test derivation, is introduced. After introducing such a framework, systematic methods are sketched for

* deriving product line tests,

* adapting product line tests for a product instance, and

* deriving product specific tests.

The aim of these methods is to link basic product line development artifacts (according to the author's definition: features, architecture, use case scenarios) and concepts such as variability, to product line test concepts (test architecture, test scenarios including variability). The main artifacts for product line engineering are the feature model, use cases,

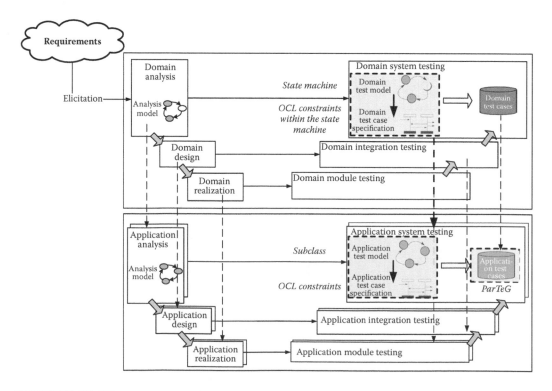

FIGURE 13.30
Reusing state machine approach, according to our conceptional process model.

and use case scenarios representing system requirements, and components representing the architecture of the implementation.

This procedure is very similar to the ScenTED approach, described in Section 13.6.2, as it also focuses on system testing. Figure 13.31 depicts the test development process. To shorten the description, we do not describe the entire procedure. We begin at the point where use case scenarios and product line test architectures are taken into account, to derive test scenarios. Our scope is marked with a rectangle in Figure 13.31.

To enrich the use case scenarios with variability, Kang et al. use an extended sequence diagram notation, which they call "variability extended sequence diagram." Require and exclude dependencies are added to the notation of sequence diagrams. Kang et al. use the orthogonal variability model (OVM) notation presented in [PM08] to depict the product line architecture. Therefore, they can use variation points to describe variable parts within the architecture. Once the product line architecture is selected, its test architecture can be generated. The test architecture contains components required for testing and test components. The so-called test components determine whether the components behave as expected.

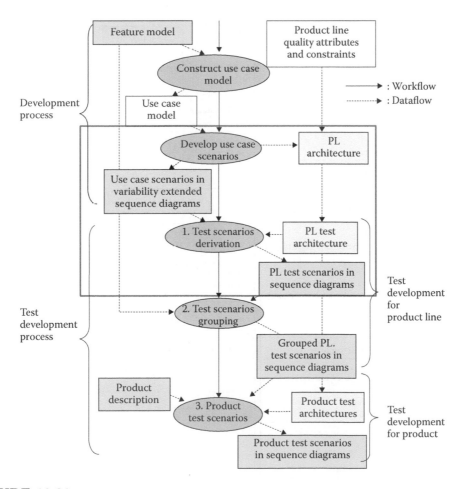

FIGURE 13.31

SPL test development process. (According to S. Kang, J. Lee, M. Kim, and W. Lee, Towards a Formal Framework for Product Line Test Development, in *Proceedings of the 7th IEEE International Conference on Computer and Information Technology (CIT2007)*, 921–926, 2007.)

Subsequently, test scenarios are generated on the basis of the use case scenarios, along with the product line test architecture. These test scenarios are described using sequence diagrams, which are extended to express variability. The authors offer an extension to depict *optional*, *exclusive*, and *inclusive* variability. The product line test scenarios can be reused for test derivation for certain products. Since it is known from the development process which use case belongs to which feature, all relevant test scenarios are selected when deriving a product.

Application

We apply our running example to the process introduced by Kang et al. Figure 13.32 depicts a sequence diagram of our running example, by way of a sequence diagram. It shows the use case differentiating between single player and multiplayer mode.

The variability is preserved by utilizing *selection* statements and *optional*, *exclusive*, and *inclusive* constructs within the UML sequence diagram.

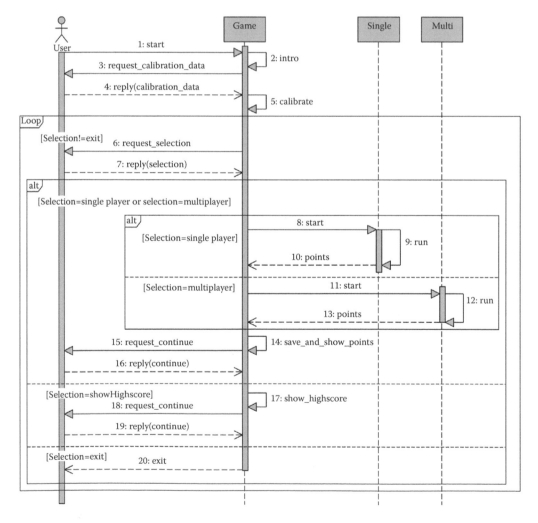

FIGURE 13.32

Sequence diagram describing our running example.

To generate a test sequence, according to the Kang approach, we must involve a test architecture. A test architecture is derived from the SPL architecture. Figure 13.33 depicts an extract of the running example architecture (marked as (b)) and the corresponding *variation point* (marked as (a)). The corresponding test architecture is shown on the right-hand side of Figure 13.33 (marked as (c)). There, we introduce the so-called test components (*TC1, TC2*). Each of these *test components* can either observe (passive), or observe and inject (active) a certain message or phenomenon. In this case, *TC1* is an active tester, injecting messages into the *Game* component. *TC2* is a passive tester, observing the incoming messages sent from the game component.

Kang et al. use the test architecture and the use case scenario (cf. Figure 13.34) to derive a test scenario in the form of sequence diagrams. This is actually the same sequence as with the additional *test components*. To generate tests for a product, all test sequences belonging to a certain product are grouped. The architectures of the products are taken into account in identifying the corresponding test sequences. We refer to [KLKL07] for further details.

Summary

Kang et al. introduce a complete procedure for product line testing and for generating test cases for products derived from SPLs [KLKL07]. For testing purposes, the authors use use case scenarios and test architectures.

Kang et al. state that sequence diagrams describing the use cases must be complemented to depict *optional*, *exclusive*, and *inclusive* variability. We believe, however, that the original notation of the sequence diagram is feasible to describe the required variability. In contrast to ScenTED, the approach by Kang et al. does not use activity diagrams, however, present a step-by-step procedure for SPL testing. Some open questions still remain:

- How are the *test components* implemented?

- How is the test architecture derived from the SPL architecture?

Figure 13.35 depicts the parts of our conceptional process model that are covered by the approach introduced by Kang et al.

The domain analysis model is formed by sequence diagrams describing use cases. Rather than using an application analysis model, Kang et al. group the sequences of the domain

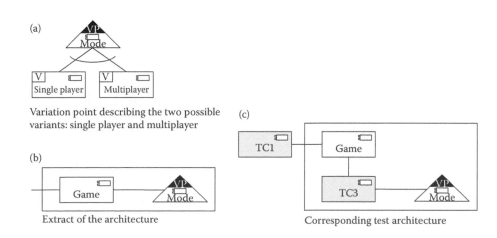

FIGURE 13.33
Extract of the test architecture of our running example.

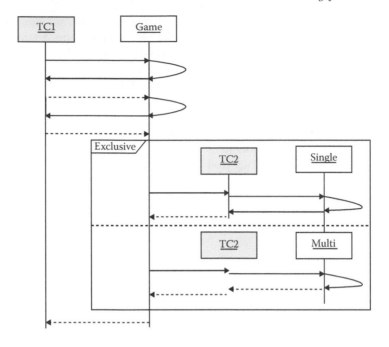

FIGURE 13.34
Test development process.

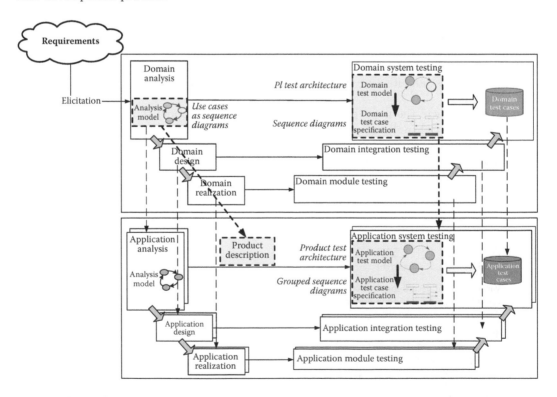

FIGURE 13.35
Framework for SPL test development according to our conceptional process model.

analysis model to form product-specific use cases. Domain and application test models are described using test architectures and the test cases are specified using sequence diagrams. In both cases, the domain artifacts are reused by grouping them to obtain application specific test models and application-specific test case specifications. Thus, Kang et al. apply the reusable test asset approach in combination with product-by-product testing.

13.7   Discussion

In this section, we compare the different approaches, discuss the results of the comparison, and highlight interesting findings regarding the possibility to combine different approaches. To compare the approaches, we use the criteria for comparison introduced in Section 13.5. Figure 13.36 shows the tree of criteria together with their individual values, which are marked as dotted nodes. Furthermore, we discuss the results of our comparison, which are depicted in Table 13.2. The criteria in Figure 13.36 are marked with bold lines. Each node with a bold line has a corresponding row in Table 13.2.

Table 13.2 offers an overview of all approaches introduced in this chapter. There is no approach covering all criteria for comparison. The comparison in Table 13.2 illustrates that all approaches have different advantages and scopes. Similarities and unique characteristics can be easily identified using this table. Some of these unique characteristics are as follows:

- Design Testing in the Model Checking approach.

- Using test architectures in the Framework-based approach.

- Fully automated test case derivation using ParTeG in the Reusing State Machine approach.

Regarding the unique characteristics and different scopes, it appears to be promising to combine different approaches. As the model-checking approach is situated on a different level than the other approaches and focuses on design testing, it can be combined with every other testing approach. Another promising combination is the integration of CADeT and the Reusing State Machine approach. CADeT offers a good integration into the development process and a step-by-step procedure to apply it to any SPL. However, it fails to consider test coverage and variability. Reusing State Machines offer coverage criteria and variability but do not offer a detailed step-by-step process or integration in the development process. The framework approach introduced by Kang et al. can also be combined with an approach offering a coverage criteria. In addition to these presented, many other combinations are feasible.

It is interesting to observer that most rely on activity diagrams as test models. Model-based test approaches for embedded systems mainly rely on state machines as test models. However, no disadvantage was found in using activity diagrams as test models within SPLs of embedded systems. Even though activity diagrams are generally not suitable to describe the behavior of a reactive system very well, they can be used to describe their use cases for test case derivation. All approaches can be applied to embedded systems. For instance, ScenTED was applied and evaluated by the authors using an embedded SPL, by means of a medical product line. A more far reaching discussion about handling, for example, reactive systems is desirable. Please note that the effort to create a test model based on use cases and the process of doing so is not well described by any approach. The authors rely on the well-known approaches introduced in the MBT community.

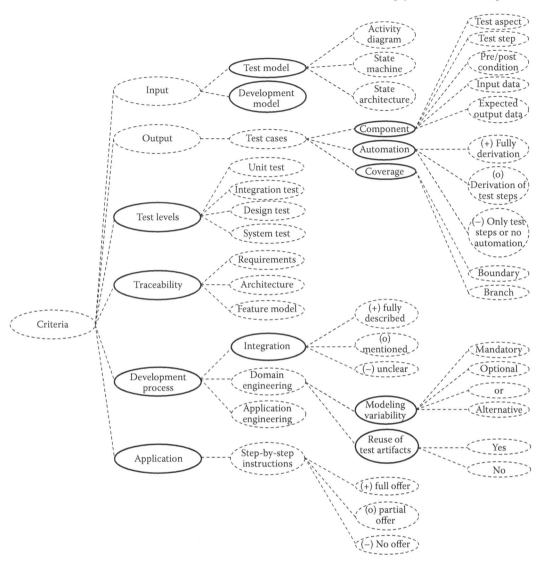

FIGURE 13.36
Criteria tree with individual metrics.

We discuss some criteria not covered by the majority of approaches to motivate further research activities in this area. A list of general research questions is provided in our conclusion in Section 13.8. First observation is that only system and design testing is supported using model-based approaches. Furthermore, most approaches generate test cases, which are not directly executable or do not consider parameterization. The most costly parts of the definition of an executable test case, like building the test model and defining the test data, are not covered by any approach. Measuring the coverage of the test cases is often not possible because only a few approaches introduce coverage criteria. Last but not least, we discuss how effectively the described approaches may be applied to SPLs. CADeT and the framework for product line test development offer a step-by-step procedure. However, the framework is described very briefly in the literature found. All other approaches either start at a certain point in the development process, without describing how to get there, or stop at a certain point, without describing the actual test derivation or test execution.

TABLE 13.2
Comparison of Model-Based Test Approaches for SPL

	CADeT—Section 13.6.1	ScenTED—Section 13.6.2	UML Based Approach—Section 13.6.3	Model Checking—Section 13.6.4	Reusing State Machines—Section 13.6.5	Framework Approach—Section 13.6.6
Input—Test Model	Activity diagram	Activity diagram	Activity diagram	State machine	State machine	Test architecture
Input—Dev. Model	Use cases	Use cases	Activity diagrams	Design model	Not mentioned	Sequence diagrams
Output—TC—Component	Test steps pre-, postcondition	Test steps pre-, and postcondition input data expected output data	Test steps	Test steps input data	Test steps input data pre- and postcondition	Not mentioned
Output—TC—Automation	(o)	(−)	(−)	(o)	(o)/(+)	(−)
Output—TC—Coverage	Not mentioned	Branch coverage with variability	Not mentioned	Not mentioned	Boundary based coverage	Not mentioned
Test Levels	System test	System test	System test	Design test	System test	System test
Traceability	Yes	No	No	No	Yes	Yes
Dev. Process	Domain & application eng.	Application eng.	Application eng.	Domain & application eng.	Application eng.	Application eng.
Dev. Process Integration	(+)	(+)	(−)	(o)	(o)	(+)
Dev. Process—Variability	Mandatory optional	Mandatory, optional, or, alternative	Optional	Mandatory, optional, or, alternative	Mandatory, optional, or, alternative	Mandatory, optional, or, alternative
Reuse of Artifacts	Yes	Yes	Yes	Yes	Yes	Yes
Application—Instructions	(+)	(o)	(o)	(o)	(o)	(+)

13.8   Conclusion

In this chapter, the advantages of the SPL approach were exposed, focusing on the testing activity. First, some fundamentals on SPL engineering and variability were discussed. Next, the special character concerning the testing of product lines was described, finishing with a conceptual process model for SPL testing. After this, criteria for comparing different testing approaches and a running example to make this comparison plausible were presented. Subsequently, the different approaches were introduced by using the running example. Each approach is summarized by comparing it to the conceptual process model for SPL testing. Finally to sum up the outputs of the different approach descriptions, a comparison on the basis of the defined criteria was made.

This shows the SPL approach to be well suited to develop product lines with a high amount of derived, customer individual products. The conclusion provided an overview of approaches to test an SPL by comparing them using the same running example. The approaches define the important elements of a product line testing approach, such as variability definition, test case generation, and product derivation. All of these elements encourage the SPL developers to gain the SPL-specific advantages such as reduction of costs or a higher product quality (see Section 13.1). Beside this, a detailed look at the comparison, reveals some interesting facts concerning the compared approaches. Every approach concentrates on particular elements. For example, the CADeT approach provides traceability from features to test case specifications. The ScenTED approach integrates test and variability coverage for activity diagrams. All but one of the approaches concentrate on the system test level and none of the approaches considers test data in domain engineering.

It becomes clear that the combination of different approaches could improve the SPL test process. Some possible approach-combinations are proposed in the discussion section. Beyond this, some elements of SPL testing remain uncovered. This is also revealed in the comparison of the different approaches with the conceptional process model for SPL testing, which we introduced in Section 13.3. Some parts of the process model, such as Unit Testing, have not been considered up until now. Based on the comparison, the following research objectives can be exposed sorted by relevance:

- Integration of all Test Levels in one testing approach. The variability makes it interesting to think of different test times (domain and application time). The test time should depend on the variability contained in the System Under Test.

- Provide concrete test cases with their test data definitions in domain engineering to foster reusabiliy.

- Implement traceability between development activities and test level as well as within development activities or test levels.

- Automate more elements of the test case generation by using combinations of different development and test models.

The first research objective asks for an integrated testing approach for SPLs recognizing all test levels. If an integrated testing approach is available reuse becomes supported both between different test levels and for different products of an SPL, which extends the possibilities for reuse. The second research objective aims at reusing all parts of an executable test case [Wüb08]. Especially the definition of test data is a very time-consuming part of the test case development process. The reuse of test data could also provide a reduction of testing effort. Traceability between development and test artifacts supports for example the impact analysis for changes to the SPL. This is part of product line evolution. Evolution of software intensive systems is not new at all, but the task is more delicate recognizing variability [Pus02]. Traceability also supports the last research objective. Automatic test case generation can help reducing the testing effort. This is not an SPL but a test-specific objective.

References

Bos00   Bosch, J. (2000). *Design and Use of Software Architectures—Adopting and Evolving a Product Line Approach*. Addison-Wesley Longman, Amsterdam.

Bos05   Bosch, J. (2005). Software product families in Nokia. In Obbink, J. and Pohl, K. editors, *Proc. of the 9th International Conference on Software Product Line Engineering (SPLC2009)*, Pages: 2–6.

CE00   Czarnecki, K. and Eisenecker, U. (2000). *Generative Programming: Methods, Tools, and Applications*. Addison-Wesley Professional, New York.

CGP99   Clarke, E. M., Grumberg, O., and Peled, D. A. (1999). *Model Checking*. MIT Press.

CN01   Clements, P. and Northrop, L. (2001). *Software product lines: practices and patterns*. Addison-Wesley Longman Publishing Co., Inc., Boston, MA.

DW00   Dröschel, W. and Wiemers, M. (2000). *Das V-Modell 97*. Oldenbourg, Berlin, Germany.

GG93   Grochtmann, M. and Grimm, K. (1993) Classification trees for partition testing. *Software Testing, Verification and Reliability*, 3:63–82.

GKPR08   Grönniger, H., Krahn, H., Pinkernell, C., and Rumpe, B. (2008). Modeling variants of automotive systems using views. In *Tagungsband Modellierungs-Workshop MBEFF: Modellbasierte Entwicklung von eingebetteten Fahrzeugfunktionen*, Volume Informatik-Bericht 2008-01. CFG-Fakultt, TU Braunschweig, Germany.

Gom04   Gomaa, H. (2004) *Designing Software Product Lines with UML: From Use Cases to Pattern-Based Software Architectures (The Addison-Wesley Object Technology Series)*. Addison-Wesley Professional, Redwood City, CA.

HVR04   Hartmann, J., Vieira, M., and Ruder, A. (2004). A UML-based approach for validating product lines. In Geppert, B., Krueger, C., and Li, J., editors, *Proceedings of the International Workshop on Software Product Line Testing (SPLiT 2004)*, Pages: 58–65.

IES   Fraunhofer IESE. PuLSE Product Lines for Software Systems, http://www.iese.fraunhofer.de/Products_Services/pulse/, last visit 30.09.2009.

KCH+90   Kang, K. C., Cohen, S. G., Hess, J. A., Novak, W. E., and Peterson, A. S. (1990). Feature-oriented domain analysis (FODA) feasibility study. Technical report, Carnegie-Mellon University Software Engineering Institute.

KKL+98   Kang, K. C., Kim, S., Lee, J., Kim, K., Shin, E., and Huh, M. (1998). FORM: a feature-oriented reuse method with domain-specific reference architectures. *Ann. Softw. Eng.*, 5:143–168.

KLKL07   Kang, S., Lee, J., Kim, M., and Lee, W. (2007). Towards a formal framework for product line test development. In *Proceedings of the 7th IEEE International Conference on Computer and Information Technology (CIT2007)*, Pages: 921–926.

KN04   Kishi, T. and Noda, N. (2004). Design testing for product line development based on test scenarios. In Geppert, B., Krueger, C., and Li, J., editors, *Proceedings of the International Workshop on Software Product Line Testing (SPLiT 2004)*, Pages: 19–26.

Lis87   Liskov, B. (1987). Keynote address – data abstraction and hierarchy. In *OOPSLA '87: Addendum to the proceedings on Object-oriented programming systems, languages and applications (Addendum)*, Pages: 17–34. ACM, New York, NY.

LL05   Ludewig, J. and Lichter, H. (2005). *Software Engineering*. Dpunkt.Verlag GmbH, Heidelberg, Germany.

Lyo96   Lyons, J. L. (1996) Ariane 5: flight 501 failure. Technical report, The Inquiry Board.

McG01   McGregor, J. D. (2001). Testing a software product line. Technical Report CMU/SEI-2001-TR-022.

MND02   Matinlassi, M., Niemelä, E., and Dobrica, L. (2002). Quality-driven architecture design and quality analysis method: a revolutionary initiation approach to a product line architecture. *VTT PUBLICATIONS, Technical Research Centre of Finland, Espoo*, 456.

OAMvO00   Obbink, H., America, P., Müller, J., and van Ommering, R. (2000). COPA: a component-oriented platform architecting method for families of software-intensive electronic products. In *Proceedings of the 1st Internatinal Conference on Software Product Line Engineering (SPLC2000)*, Pages: 167–180.

Oli08   Olimpiew, E. M. (2008). *Model-based testing for software product lines*. PhD thesis, George Mason University.

OMG09   OMG. (2009). Unified Modeling Language (UML). Technical report, version2.2.

OMR10   Oster, S., Markert, F., and Ritter, P. (2010). Automated incremental pairwise testing of software product lines. In *Proceedings of the 14th International Software Product Line Conference*, Pages: 196–210.

PBvdL05   Pohl, K., Böckle, G., and van der Linden, F. (2005). *Software Product Line Engineering: Foundations, Principles and Techniques*. Springer, Secaucus, NJ.

PM08   Pohl, K. and Metzger, A. (2008). Variabilitätsmanagement in Software-Produktlinien. In Herrmann, K. and Bruegge, B., editors, *Proceedings Software Engineering 2008*, Pages: 28–41.

PP04   Pretschner, A. and Philipps, J. (2004). Methodological issues in model-based testing. In *Model-Based Testing of Reactive Systems*, Pages: 281–291.

Pus02   Pussinen, M. (2002). A survey on software product-line evolution. Technical report, Institute of Software Systems, Tampere University of Technology, Tampere, Finland.

RKPR05   Reuys, A., Kamsties, E., Pohl, K., and Reis, S. (2005). Model-based system testing of software product families. In *CAiSE*, Pages: 519–534.

Rob00   Robinson, H. (2000). Intelligent test automation. *Software Testing & Quality Engineering*, 5:24–32.

Sch07   Scheidemann, K. (2007). *Verifying Families of System Configurations*. PhD thesis, Technical Univerity of Munich.

TH02   Thiel, S. and Hein, A. (2002). Modeling and using product line variability in automotive systems. *IEEE Softw.*, 19(4):66–72.

TTK04   Tevanlinna, A., Taina, J., and Kauppinen, R. (2004). Product family testing: a survey. *ACM SIGSOFT Software Engineering Notes*, 29:12–12.

vdM07   von der Maßen, T. (2007). *Feature-basierte Modellierung und Analyse von Variabilität in Produktlinienanforderungen*. PhD thesis, Rheinisch-Westfälische Technische Hochschule Aachen.

Wei09   Weißleder, S. (2009). ParTeG (Partition Test Generator), http://parteg. sourceforge.net, last visit 30.09.2009.

WS08   Weißleder, S. and Schlingloff, B. H. (2008). Deriving input partitions from UML models for automatic test generation. *Lecture Notes in Computer Science*, 5002:151–163.

WSS08   Weißleder, S., Sokenou, D., and Schlingloff, B. H. (2008). Reusing state machines for automatic test generation in product lines. In *Proceedings of the 1st Workshop on Model-based Testing in Practice (MoTiP2008)*.

Wüb08   Wübbeke, A. (2008). Towards an efficient reuse of test cases for software product lines. In Thiel, S. and Pohl, K., editors, *Proceedings of the 12th International Software Product Line Conference (2008) Second Volume*, Pages: 361–368.

14

Model-Based Testing of Hybrid Systems

Thao Dang

CONTENTS

14.1 Introduction

Hybrid systems, that is, systems exhibiting both continuous and discrete dynamics, have proven to be a useful mathematical model for various physical phenomena and engineering systems. Due to the safety critical features of many such applications, much effort has been devoted to the development of automatic analysis methods and tools for hybrid systems, based on formal verification. Although these methods and tools have been successfully applied to a number of interesting case studies, their applicability is still limited to systems of small size because of the complexity of formal verification. It is thus clear that for systems of industrial size, one needs more lightweight methods. Testing is another validation approach, which can be used for much larger systems and is a standard tool in industry, although it can only reveal an error but does not permit proving its absence. A question of great interest is thus to bridge the gap between the verification and testing approaches, by defining a formal framework for testing of hybrid systems and developing methods and tools that help automate the testing process.

In this work, we adopt a model-based testing approach. This approach allows the engineer to perform validation during the design, where detecting and correcting errors on a model are less expensive than on an implementation.

We first briefly review related work in model-based testing. The development of the first model-based testing frameworks was motivated by digital circuit testing and is based on Mealy machines [30]. More recently, frameworks based on other models, such as finite labeled transition systems, were proposed (e.g., see [39]). These models are of asynchronous nature and appropriate for the applications in communication protocols. Another important application area is software testing for which models, such as flow graphs, and coverage techniques have been used [20]. Recently, model-based testing has been extended to real-time systems. Timed automata have become a popular model for modeling and verifying real-time systems during the past decade, and a number of methods for testing real-time systems based on variants of this model or other similar models (such as timed Petri nets) have been proposed (e.g., see [26, 34, 8]. Although the practice of testing, especially in industry, is still empirical and ad hoc, formal testing has become progressively accepted [17]. This is, on the one hand, due to the success of the formal techniques in a number of domains (such as model checking of digital circuits) and, on the other hand, due to the development of commercial tools for automatic test generation. Among these tools, we can mention: Telelogic TestComposer (http://www.telelogic.com) for SDL models, Reactis Tester (http://www.reactive-systems.com) for Simulink® models (http://www.mathworks.com), Conformiq Test Generator (http://www.conformiq.com) for UML State-Chart models.

Concerning hybrid systems, model-based testing is still a new research domain. We defer a discussion on related work on hybrid systems testing to Section 14.14. A number of special characteristics of hybrid systems make their testing particularly challenging, in particular:

- *Combination of the complexity in both discrete and continuous aspects.* While continuous systems have been well studied in control theory and continuous mathematics, and discrete systems have been investigated in computer science, the interaction between continuous and discrete dynamics leads to fundamental problems (such as undecidability), which are not yet well understood or for which a general solution is often impossible.

- *Infiniteness of the state space of a hybrid system and of the input space.* In general, in order to test an open system, one first must feed an input signal to the system and then check whether the behavior of the system induced by this input signal is as expected. When there is an infinite number of possible input signals, it is important to choose the ones that lead to interesting scenarios (with respect to the property/functionality to test).

To deal with these issues, we take an approach that draws on ideas from two domains: the algorithmic analysis methodology from computer science and methods from control theory. To model hybrid systems, we use hybrid automata [22]. This model, which can be roughly described as an extension of automata with continuous variables evolving according to differential equations, is a mathematical model largely used by computer scientists and control engineers to reason about problems related to hybrid systems. In addition, this model is expressive enough to describe complex hybrid phenomena arising in numerous applications, and its well-defined semantics permits accurate interpretation of testing results.

The main results we present in this chapter can be summarized as follows.

- *Formal framework for conformance testing of hybrid systems.* This framework uses the commonly accepted hybrid automaton model and allows, on the one hand, to formally reason about the relation between a system under test (SUT) and a specification, and on the other hand, to develop test generation algorithms.

- *Novel test coverage measure.* This is a challenging problem in testing. Intuitively, test coverage is a way to characterize the relation between the number and the type of tests to execute and the portion of the system's behavior effectively tested. The classical notions of coverage, introduced mainly for software testing (such as statement coverage, if-then-else branch coverage, path coverage), are unsuitable for the behaviors of a hybrid system defined as solutions of some differential equations. We thus propose a novel coverage measure, which on the one hand reflects our goal of testing safety and reachability properties and, on the other hand, can be efficiently computed. This measure is based on the equidistribution degree of a set of states over the state space and furthermore can be used to guide the test generation process.

- *Coverage-guided test generation.* We first propose a test generation algorithm that is based on the Rapidly-exploring Random Tree (RRT) algorithm [28], a probabilistic motion planning technique in robotics. This RRT algorithm has been successful in finding feasible trajectories in motion planning. We then include in this algorithm a procedure for guiding the test generation process using the above-mentioned coverage measure. Furthermore, we introduce a new notion of disparity between two point sets, in order to tackle "blocking" situations the test generation algorithm may enter. Indeed, in order to increase the coverage, the algorithm may try to explore the directions that are not reachable by the system's dynamics. We can detect such situations by comparing the distribution of the goal states and the visited states, using their disparity. If the disparity is large, it means that the visited states do not follow the goal states. This indicates that the goal states may not be reachable and we should change the goal state sampling strategy.

- *Actuator and sensor imprecision.* Because of the limitations of practical actuators and sensors, the tester cannot realize exactly an input value specified as a real-valued vector as well as measure exactly the state of the system. We handle this using sensitivity analysis.

- *Tool development.* We have implemented a tool for conformance testing of hybrid systems, called HTG. The core of the tool is the implementation of the coverage-guided test case generation algorithm and the methods for estimating the coverage and disparity measures.

- *Applications.* Besides traditional applications of hybrid systems, we explore a new domain, which is analog and mixed signal circuits. Indeed, hybrid systems provide a mathematical model appropriate for the modeling and analysis of these circuits. The choice of this application domain is motivated by the need in automatic tools to facilitate the design of these circuits which, for various reasons, is still lagging behind the digital circuit design. Besides hybrid automata described using a textual language, the tool HTG can accept as input electrical circuits specified using SPICE netlists. We have treated a number of case studies from control applications as well as from analog and mixed signal circuits. The experimental results obtained using the tool HTG show its applicability to systems with complex dynamics and its scalability to high-dimensional systems (with up to 200 continuous variables).

Before presenting these results, we first describe our conformance testing framework and test coverage measure.

14.2   Model

Conformance testing provides a means to assess the correctness of an implementation with respect to a specification by performing experiments on the implementation and observing its responses. When the specification is described by a formal model, the international standard "Formal Methods in Conformance Testing" (FMCT) [38] provides a framework of conformance testing, which includes abstract concepts (such as conformance, test cases, test execution, test generation) and the requirements on these concepts. A testing approach that is based on a formal model is called a *model-based testing* approach. Depending on the type of formal models, various frameworks can be developed for conformance testing. In this work, following the standard FMCT, we are interested in developing a conformance testing framework for embedded systems, using the hybrid automaton model [4]. Intuitively, a hybrid automaton is an automaton augmented with continuous variables that evolve according to some differential equations.

Definition 1 (Hybrid automaton). A *hybrid automaton* is a tuple $\mathcal{A} = (\mathcal{X}, Q, E, F, \mathcal{I}, \mathcal{G}, \mathcal{R})$, where

- $\mathcal{X}$ is the continuous state space and is a bounded subset of $\mathbb{R}^n$;

- Q is a (finite) set of locations (or discrete states);

- $E \subseteq Q \times Q$ is a set of discrete transitions;

- $F = \{F_q \mid q \in Q\}$ such that for each $q \in Q$, $F_q = (f_q, U_q)$ defines a differential equation:

$$\dot{x}(t) = f_q(x(t), u(t)),$$

  where $x \in \mathcal{X}$ is the continuous state, $u(\cdot) \in \mathcal{U}_q$ is the input of the form $u : \mathbb{R}^+ \to U_q \subset \mathbb{R}^m$. The set $\mathcal{U}_q$ is the set admissible inputs and consists of piecewise continuous functions. We assume that all f_q are Lipschitz continuous.[*] In location q, the evolution of the continuous variables is governed by $\dot{x}(t) = f_q(x(t), u(t))$.

- $\mathcal{I} = \{\mathcal{I}_q \subseteq \mathcal{X} \mid q \in Q\}$ is a set of staying conditions;

- $\mathcal{G} = \{\mathcal{G}_e \mid e \in E\}$ is a set of guards such that for each discrete transition $e = (q, q') \in E$, $\mathcal{G}_e \subseteq \mathcal{I}_q$;

- $\mathcal{R} = \{\mathcal{R}_e \mid e \in E\}$ is a set of reset maps. For each $e = (q, q') \in E$, $\mathcal{R}_e : \mathcal{G}_e \to 2^{\mathcal{I}_{q'}}$ defines how x may change when $\mathcal{A}$ switches from q to q';

- The initial state of the automaton is denoted by (q_{init}, x_{init}).   □

---

[*]The function f_q is Lipschitz continuous if there exists a constant K such that $\forall x, y : ||f_q(x) - f_q(y)|| \leq K||x - y||$, where $|| \cdot ||$ is some norm of $\mathbb{R}^n$. This condition ensures the existence and uniqueness of solutions of the differential equations.

A *hybrid state* is a pair (q, x), where $q \in Q$ and $x \in \mathcal{X}$. The hybrid state space is $\mathcal{S} = Q \times \mathcal{X}$. In the rest of the paper, for brevity, we often use "state" to refer to a hybrid state.

A state (q, x) of $\mathcal{A}$ can change in two ways as follows: (1) by a *continuous evolution*, the continuous state x evolves according to the dynamics f_q, while the location q remains constant; (2) by a *discrete evolution*, x satisfies the guard condition of an outgoing transition, the system changes the location by taking this transition and possibly changing the values of x according to the associated reset map. More formally, continuous and discrete evolutions are defined as follows.

Given a real number $h > 0$ and an admissible input function $u(\cdot) \in \mathcal{U}_q$, $(q, x) \overset{u(\cdot),h}{\rightarrow} (q, x')$ is a *continuous evolution* at the location q from the hybrid state (q, x) to (q, x'), iff $x' = \xi_{x,u(\cdot)}(h)$ and for all $t \in [0, h] : \xi_{x,u(\cdot)}(t) \in \mathcal{I}_q$, where $\xi_{x,u(\cdot)}(t)$ is the solution of the differential equation at the location q with the initial condition x and under the input $u(\cdot)$. In other words, x' is reached from x under the input $u(\cdot)$ after exactly h time, and we say that $u(\cdot)$ is *admissible* starting at (q, x) for h time.

Given a transition $e = (q, q') \in E$, $(q, x) \overset{e}{\rightarrow} (q', x')$ is a discrete evolution iff $x \in \mathcal{G}_e$ and $x' \in \mathcal{R}_e(x)$. We say that (q', x') is reachable from (q, x) and the discrete transition e is admissible at (q, x). Unlike *continuous evolutions*, *discrete evolutions* are instantaneous, which means that they do not take time.

It is important to note that this model allows capturing *nondeterminism* in both continuous and discrete dynamics. The nondeterminism in continuous dynamics is caused by the uncertainty in the input function. For example, when the input is used to model some external disturbances or modeling errors, we do not know the exact input fucntion but only its range. The nondeterminism in discrete dynamics is caused by the fact that at some states it is possible for the system to stay at the current location or to switch to another one. In addition, multiple transitions can be enabled at some states. This nondeterminism is useful for describing disturbances from the environment and imprecision in modeling and implementation. We assume that the hybrid automata we consider are non-Zeno.*

Figure 14.1 shows a hybrid automaton that has three locations q_1, q_2, q_3. From each location q_i, there is a discrete transition (q_i, q_j) such that $j \neq i$.

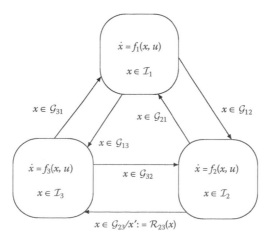

FIGURE 14.1
A hybrid automaton.

---

*A Zeno behavior can be described informally as the system making an infinite number of discrete transitions in a finite amount of time.

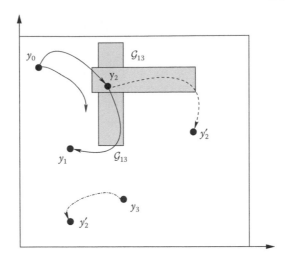

FIGURE 14.2
Hybrid automaton evolution.

Figure 14.2 illustrates the evolution of this hybrid automaton. We denote by $\mathcal{G}_{ij}$ the guard of the transition (q_i, q_j). In Figure 14.2, only the guards $\mathcal{G}_{12}$ and $\mathcal{G}_{23}$ are the shaded regions. All the locations have the same staying set, which is the bounding box.

As we can see from the figure, starting from the initial state (q_1, y_0), the system can generate a set of infinite number of trajectories. For instance, we consider the following trajectories. Starting from (q_1, y_0), under two different continuous input functions, from the system generates two different trajectories, which can reach the guards $\mathcal{G}_{12}$ and $\mathcal{G}_{13}$. The points y_1, y_2', and y_3' correspond to three different choices: (1) remaining at location q_1 and continuing with the dynamics of f_1; (2) taking the transition (q_1, q_2) at the point y_2 and evolving under the dynamics of f_2 to y_2'; (3) taking the transition (q_1, q_3) at the point y_2, jumping then to y_3 according to the reset $\mathcal{R}_{13}$ and from there evolving under the dynamics of f_3 to y_3'.

14.3   Conformance Testing

In the first part of this section, we define the notions of inputs and observations. In the second part, we define conformance relation, test cases, and test executions.

Our testing goal is to make statements about the conformance relation between the behaviors of an implementation or, more generally, a SUT and a specification. The specification is formal and modeled by a hybrid automaton. The conformance is defined as a relation $\preccurlyeq \subseteq \Xi \times \mathcal{H}$, where Ξ is a set of SUTs of interest, and $\mathcal{H}$ is a set of hybrid automata modeling the specifications of interest. The SUT are physical systems, but it can be assumed that all the SUTs in Ξ can be described by a class of formal models, which is a set $\mathcal{H}_s$ of hybrid automata. It is important to note that we assume that a model for each SUT in Ξ exists* but do not assume that we know it. This assumption enables us to include the SUT in our formal framework and to express formally the conformance relation $\preccurlyeq$ between

---

*Concerning embedded systems design, this assumption is not restrictive since, as mentioned in the introduction, hybrid systems can be used to model various physical systems with mixed continuous-discrete dynamics and are particularly appropriate for embedded systems.

the models of the SUTs and the specifications, that is $\preccurlyeq \subseteq \mathcal{H}_s \times \mathcal{H}$. Note that here we use the same notation $\preccurlyeq$ for the relation between the physical SUT and the specification and the relation between the model of the SUT and the specification. A SUT $S_t \in \Xi$ is said to *conform* to a specification $\mathcal{A} \in \mathcal{H}$ if and only if the model $\mathcal{A}_s \in \mathcal{H}_s$ of the SUT is related to $\mathcal{A}$ by $\preccurlyeq$, that is, $\mathcal{A}_s \preccurlyeq \mathcal{A}$.

The SUT often operates within some environment. In our testing framework, a tester plays the role of the environment and it performs experiments on the SUT in order to study the conformance relation between the SUT and the specification. Such an experiment is called a *test*, and its specification is called a *test case*. A set of test cases is called a *test suite*, and the process of applying a test to a SUT is called a *test execution*. The tester works as follows. It emits the control inputs to the SUT and measures the observation sequences in order to produce a verdict $\nu \in \{P, F\}$, where P means "pass" (the observed behavior is allowed by the specification), F means "fail" (the observed behavior is not allowed by the specification). We continue by giving a detailed description of conformance relation. The problem of how to perform test executions and derive verdicts is discussed at the end of this section.

14.3.1 Conformance relation

Recall that the specification is modeled by a hybrid automaton $\mathcal{A}$ and the SUT by another hybrid automaton $\mathcal{A}_s$. For brevity, when the context is clear, we often say "the SUT" to mean the automaton $\mathcal{A}_s$. To define the conformance relation, the notions of *inputs and observations* are necessary.

An input of the system that is controllable by the tester is called a *control input*; otherwise, it is called a *disturbance input*. We consider the following input actions. Control inputs are realized by actuators and observations by sensors. In practical systems, because of actuator and sensor imprecision, control inputs and observations are subject to errors. In the rest of this section, we assume that the tester can realize exact input values and observations are exactly measured. The testing problem with actuator and sensor imprecision is addressed in Section 14.11. It is worth noting that in this work, we use the trace inclusion to define conformance relation. In the literature of conformance testing for discrete systems, more complex relations are considered, for example, input–output conformance relation (see [39]).

Continuous input action. All the continuous inputs are assumed to be controllable by the tester. Since we want to implement the tester as a computer program, we are interested in piecewise-constant input functions; indeed, a computer cannot generate a function from reals to reals. Hence, a *continuous control action* $(\bar{u}_q, h)$, where $\bar{u}_q$ is the value of the input and h is the *duration*, specifies that the automaton continues with the continuous dynamics at the location q under the input $u(t) = \bar{u}_q$ for exactly h time. We say that $(\bar{u}_q, h)$ is *admissible* at (q, x) if the input function $u(t) = \bar{u}_q$ for all $t \in [0, h]$ is admissible starting at (q, x) for h time.

Discrete input actions. The discrete transitions are partitioned into *controllable* corresponding to discrete control actions and *uncontrollable* corresponding to discrete disturbance actions. The tester emits a discrete control action to specify whether the system should take a controllable transition (among the enabled ones) or continue with the same continuous dynamics. In the former case, it can also control the values assigned to the continuous variables by the associated reset map. Although nondeterminism caused by the reset maps can be expressed in this framework (and considered in the proposed test generation algorithms), for simplicity of notation and clarity of explanation, we omit this type of non determinism in the subsequent definitions. Hence, we denote a discrete control action by the corresponding transition, such as (q, q').

We use the following assumption about the inputs: continuous control actions are of higher priority than discrete actions. This means that after a continuous control action $(\bar{u}_q, h)$ is applied, no discrete transitions can occur during h time, that is until the end of that continuous control action. This assumption is not restrictive from a modeling point of view. Indeed, by considering all the possible values of h, we can capture the cases where a discrete transition can occur before the termination of a continuous control action.

In this work, we are only interested in testing *nonblocking behaviors*, we thus need the notion of *admissible input sequences*. We write $(q, x) \overset{\iota}{\rightarrow} (q', x')$ to indicate that (q', x') is reached after applying the input action ι to the state (q, x).

Definition 2 (Admissible input sequence). For a state (q, x), a sequence of input actions $\omega = \iota_0, \iota_1, \ldots, \iota_k$ is admissible at (q, x) if

- ι_0 is admissible at (q, x), and

- for each $i = 1, \ldots, k$, let (q_i, x_i) be the state such that $(q_{i-1}, x_{i-1}) \overset{\iota_{i-1}}{\rightarrow} (q_i, x_i)$, then ι_i is admissible at (q_i, x_i).

The sequence $(q, x), (q_1, x_1), \ldots, (q_k, x_k)$ is called the *trace* starting at (q, x) under ω and is denoted by $\tau((q, x), \omega)$.   □

We write $(q, x) \overset{\omega}{\rightarrow} (q', x')$ to indicate that (q', x') is reached from (q, x) after ω. We also say that (q', x') is forward reachable from (q, x) and (q, x) is backward reachable from (q', x'). In the rest of this chapter, we simply say "reachable" to mean "forward reachable"; "backward reachable" is explicitly stated.

By the assumption about the inputs, uncontrollable discrete transitions cannot occur during a continuous control action. However, they can occur between control actions. Hence, the result of applying a control action is nondeterministic. To determine all possible traces that can be generated by applying a sequence of control actions, we must define an *admissible sequence of control actions*.

Given a state (q, x) and a control action c, let σ be a disturbance input sequence such that $c \oplus \sigma$, where $\oplus$ is the concatenation operator, is an admissible input sequence at (q, x). The sequence σ is called a disturbance input sequence *admissible after the control action* c. We denote by $\Lambda(c, (q, x))$, the set of all such disturbance input sequences.

To know whether a sequence of control actions is admissible, we must know which disturbance inputs are admissible after each control action. This means that we must know the successors after each control action. We first consider a sequence of two control actions $\omega_c = c_0 c_1$. After accepting the control action c_0 and all the disturbance input sequences admissible after c_0, the set of all possible successors of (q, x) is

$$\Upsilon(c_0, (q, x)) = \{(q', x') \mid \exists \sigma \in \Lambda(c_0, (q, x)) : (q, x) \overset{c_0 \oplus \sigma}{\rightarrow} (q', x')\}.$$

It should be noted that if the first c_0, is admissible at (q, x), then $\Upsilon(c_0, (q, x))$ is not empty. We use the same notation Λ for the *set of all disturbance input sequences admissible after the control action sequence* $\omega_c = c_0 c_1$:

$$\Lambda(\omega_c, (q, x)) = \bigcup_{(q', x') \in \Upsilon(c_0, (q, x))} \Lambda(c_1, (q', x')).$$

Therefore, we can now determine the set of all input sequences that can occur when we apply the control sequence $\omega_c = c_0 c_1$. We denote this set by $\Sigma(\omega_c, (q, x))$, which can be defined as follows:

$$\Sigma(\omega_c,(q,x)) = \{c_0 \oplus \sigma_0 \oplus c_1 \oplus \sigma_1 \mid \sigma_0 \in \Lambda(c_0,(q,x))$$
$$\wedge \; \exists (q',x') \in \Upsilon(c_0,(q,x)) : \sigma_1 \in \Lambda(c_1,(q',x'))\}.$$

For a sequence ω_c of more than two control actions, the set $\Sigma(\omega_c,(q,x))$ can be defined similarly.

Definition 3 (Admissible control action sequence). A control action sequence ω_c is admissible starting at (q,x) iff $\Sigma(\omega_c,(q,x))$ is not empty. The set of traces starting at (q,x) after an admissible control action sequence ω_c is $Tr((q,x),\omega_c) = \{\tau((q,x),\sigma) \mid \sigma \in \Sigma(\omega_c,(q,x))\}$. ☐

Intuitively, this means that an admissible control action sequence, when being applied to the automaton, does not cause it to be blocked. We denote by $S_{\mathcal{C}}(\mathcal{A})$ the *set of all admissible control action sequences* for the hybrid automaton $\mathcal{A}$ starting at the initial state (q_{init}, x_{init}).

14.3.1.1 Observations

We use the following assumptions about the *observability* of the hybrid automata $\mathcal{A}$ and $\mathcal{A}_s$:

- The locations of the hybrid automata $\mathcal{A}$ and $\mathcal{A}_s$ are observable.

- We assume a subset $V_o(\mathcal{A})$ and $V_o(\mathcal{A}_s)$ of observable continuous variables of $\mathcal{A}$ and $\mathcal{A}_s$, respectively. In addition, we assume that $V_o(\mathcal{A}) \subseteq V_o(\mathcal{A}_s)$, which means that an observable continuous variable of $\mathcal{A}$ is also an observable variable of $\mathcal{A}_s$.

Systems with more general partial observability are not considered in this chapter and is part of our current research. Since not all the continuous variables are observable, the following projection operator is necessary. The projection of a continuous state x of $\mathcal{A}$ on the observable variables $V_o(\mathcal{A})$ is denoted by $\pi(x, V_o(\mathcal{A}))$. The projection can be then defined for a trace as follows. The projection of a trace $\tau = (q_0, x_0), (q_1, x_1), (q_2, x_2) \ldots$ on $V_o(\mathcal{A})$ is

$$\pi(\tau, V_o(\mathcal{A})) = (q_0, \pi(x_0, V_o(\mathcal{A}))), (q_1, \pi(x_1, V_o(\mathcal{A}))), (q_2, \pi(x_2, V_o(\mathcal{A}))) \ldots.$$

A pair $(q, \pi(x, V_o(\mathcal{A}))$ where q is a location and x is the continuous state of the automation $\mathcal{A}$ is called an *observation*.

Definition 4 (Observation sequence). Let ω be an admissible control action sequence starting at the initial state (q_{init}, x_{init}) of $\mathcal{A}$. The set of *observation sequences* associated with ω is $S_{\mathcal{O}}(\mathcal{A}, \omega) = \{\pi(\tau, V_o(\mathcal{A})) \mid \tau \in Tr((q_{init}, x_{init}), \omega)\}$. ☐

Before continuing, we remark that it is straightforward to extend the framework to observations described by (q, y), where q is a location, $y = g(x)$, and g is an output map.

14.3.1.2 Conformance relation

In the definition of the conformance relation between a SUT $\mathcal{A}_s$ and a specification $\mathcal{A}$, we assume that the set of all admissible control action sequences of $\mathcal{A}$ is a subset of that of $\mathcal{A}_s$, that is $S_{\mathcal{C}}(\mathcal{A}) \subseteq S_{\mathcal{C}}(\mathcal{A}_s)$. This assumption assures that the SUT can admit all the control action sequences that are admissible by the specification. Detecting the cases where the physical SUT does not admit some inputs that are allowed by the specification requires the ability to identify the states of the system from the observations. We do not consider this problem in this work.

Definition 5 (Conformance). The SUT $\mathcal{A}_s$ is conform to the specification $\mathcal{A}$, denoted by $\mathcal{A} \preceq \mathcal{A}_s$, iff

$$\forall \omega \in S_C(\mathcal{A}): \pi(S_O(\mathcal{A}_s, \omega), V_o(\mathcal{A})) \subseteq S_O(\mathcal{A}, \omega). \qquad \square$$

Intuitively, the SUT $\mathcal{A}_s$ is conform to the specification $\mathcal{A}$ if under every admissible control action sequence, the set of observation sequences of $\mathcal{A}_s$ is included in that of $\mathcal{A}$.

14.3.2 Test cases and test executions

In our framework, a *test case* is represented by a tree where each node is associated with an observation and each path from the root with an observation sequence. Each edge of the tree is associated with a control action. A physical *test execution* can be described as follows:

- The tester applies a test ζ to the SUT S_t.

- It measures and records a number of observations.

- The observations are measured at the end of *each* continuous control action and after *each* discrete (disturbance or control) action.

This conformance testing procedure is denoted by $exec(\zeta, S_t)$, which leads to an observation sequence, or a set of observation sequences if multiple runs of ζ are possible because of nondeterminism. The above test execution process uses a number of implicit assumptions. First, observation measurements take zero time, and in addition, no measurement error is considered. Second, the tester is able to realize exactly the continuous input functions, which is often impossible in practice because of actuator imprecision, as mentioned earlier. Under these assumptions, one can only test the conformance of *a model of the SUT* to the specification in discrete time. These issues must be considered in order to address the actual testing of real systems and this is discussed in Section 14.11.

We focus on the case where each test execution involves a single run of a test case. The remaining question is how to interpret the observation sequences in order to produce a verdict. Let Ω denote the observation sequence domain. We thus define a verdict function: $\boldsymbol{v}: \Omega \to \{\mathbf{pass}, \mathbf{fail}\}$. Note that an observation sequence must cause a unique verdict. The observation sequences in Ω are grouped into two disjoint sets: the set O_p of observation sequences that cause a "pass" verdict, the set O_f that cause a "fail" verdict. Therefore, saying "The SUT S_t passes the test ζ" formally means $\boldsymbol{v}(exec(\zeta, S_t)) = \mathbf{pass}$. This can then be extended to a test suite.

We now discuss some important requirements for a test suite. A test suite T_s is called *complete* if for a specification specified as a hybrid automaton $\mathcal{A}$:

$$S_t \preceq \mathcal{A} \Longleftrightarrow S_t \text{ passes } T_s. \qquad (14.1)$$

This means that a complete test suite can distinguish exactly between all conforming and nonconforming systems. In practice, it is generally impossible to fulfill this requirement, which often involves executing an infinite test suite. A weaker requirement is *soundness*. A test suite is sound if a system does not pass the test suite, then the system is nonconforming. We can see that this requirement is weaker than completeness since it corresponds only to the left-to-right implication in 14.1.

After defining all the important concepts, it now remains to tackle the problem of generating test cases from a specification model. In particular, we want the test suites to satisfy the *soundness requirement*. A hybrid automaton may have an infinite number of infinite traces; however, the tester can only perform a finite number of test cases in finite time. Therefore, we must select a finite portion of the input space of the specification $\mathcal{A}$ and test the conformance of the SUT $\mathcal{A}_s$ with respect to this portion. The selection is done using a coverage criterion that we formally define in the next chapter. Hence, our testing problem is formulated as to automatically generate a set of test cases from the specification automaton to satisfy this coverage criterion.

14.4 Test Coverage

Test coverage is a way to evaluate testing quality. More precisely, it is a way to relate the number of tests to carry out with the fraction of the system's behaviors effectively explored.

As mentioned in the introduction, a major problem with extending the "classic" testing approach to hybrid system is the infiniteness of the input signal space and of the state space. Indeed, in practice, it is only possible to test the system with a finite number of input functions, for a bounded time horizon and, furthermore, the results are only in the form of a finite number of finite sequences of points on trajectories of the system. In other words, a continuous/hybrid tester cannot produce in practice the output signals that are functions from reals to reals but only their approximation in discrete time. Given an analysis objective, such as to verify a safety property, the question that arises is thus how to choose appropriate input signals so as to fulfill the analysis objective as best as possible.

Since it is impossible to enumerate all the admissible external inputs to the hybrid system in question, much effort has been invested in defining and implementing notions of *coverage* that guarantee, to some extent, that the finite set of input stimuli against which the system is tested is sufficient for validating correctness. Test coverage criteria are indeed a way to evaluate the testing quality, or the degree of fulfilling the desired analysis objective. More precisely, it is a way to relate the number of simulations to carry out with the fraction of the system's behaviors effectively explored.

For discrete systems, specified using programming languages or hardware design languages, some syntactic coverage measures can be defined, such as statement coverage and if-then-else branch coverage, path coverage, *etc.* (e.g., see [20, 39]). In this work, we treat continuous and hybrid systems that operate in a metric space (typically $\mathbb{R}^n$) and where not much inspiration for the coverage issues derives from the syntax. On the other hand, the metric nature of the state space encourages more *semantic* notions of coverage, namely that all system trajectories generated by the input test patterns form a kind of dense network in the reachable state space without too many big unexplored "holes."

Two main challenges in defining a test coverage measure are the following. First, it should be meaningful to reflect testing quality with respect to a given analysis objective. Second, one must be able to compute this measure. The above-mentioned classic coverage notions mainly used in software testing are not appropriate for the trajectories of continuous and hybrid systems defined by differential equations. However, geometric properties of the hybrid state space can be exploited to define a coverage measure which, on the one hand, has a close relationship with the properties to verify and, on the other hand, can be efficiently computed or estimated. In this work, we are interested in *state coverage* and focus on a measure that describes how "well" the visited states represent the reachable set of the system. This measure is defined using the *star discrepancy* notion in statistics, which characterizes the uniformity of the distribution of a point set within a region. Note that the reachable sets of hybrid systems are often nonconvex with complex geometric form, therefore considering only corner cases does not always cover the behaviors that are important for reachabilily properties, especially in high dimensions. Hence, for a fixed number of visited states (which reflects the computation cost to produce a test suite), we want the visited states to be equidistributed over the reachable set as much as possible since this provides a good representation of all possible reachable states.

14.4.1 Star discrepancy

We first briefly recall the star discrepancy. The star discrepancy is an important notion in equidistribution theory as well as in quasi-Monte Carlo techniques (e.g., see [7]). Recently, it was also used in probabilistic motion planning to enhance the sampling uniformity [29].

Let P be a set of k points inside $\mathcal{B} = [l_1, L_1] \times \ldots \times [l_n, L_n] \subset \mathbb{R}^n$. Let $\mathcal{J}$ be the set of all sub-boxes J of the form $J = \prod_{i=1}^{n}[l_i, \beta_i]$ with $\beta_i \in [l_i, L_i]$ (see Figure 14.3). The local discrepancy of the point set P with respect to the sub-box J is defined as follows:

$$D(P, J) = \left| \frac{A(P, J)}{k} - \frac{vol(J)}{vol(\mathcal{B})} \right|,$$

where $A(P, J)$ is the number of points of P that are inside J, and $vol(J)$ is the volume of the box J.

Definition 6 (Star discrepancy). The star discrepancy of a point set P with respect to the box $\mathcal{B}$ is defined as:

$$D^*(P, \mathcal{B}) = sup_{J \in \mathcal{J}} D(P, J). \tag{14.2}$$

It is not hard to prove the following property of the star discrepancy [37]. □

Proposition 1. The star discrepancy of a point set P with respect to a box $\mathcal{B}$ satisfies $0 < D^*(P, \mathcal{B}) \leq 1$. □

Intuitively, the star discrepancy is a measure for the irregularity of a set of points. A large value $D^*(P, \mathcal{B})$ means that the points in P are not much equidistributed over $\mathcal{B}$. When the region is a box, the star discrepancy measures how badly the point set estimates the volume of the box.

Example. To show an intuitive meaning of the star discrepancy, we use some sequences of 100 points inside a two-dimensional unit box. The first example is the Faure sequence [16],

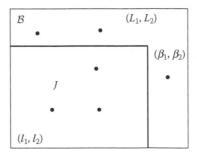

FIGURE 14.3
Illustration of the star discrepancy notion.

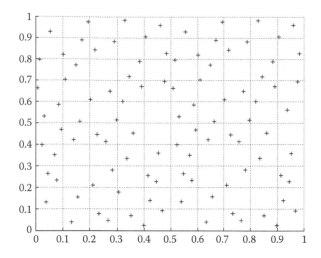

FIGURE 14.4

Faure sequence of 100 points. Its star discrepancy value is 0.048.

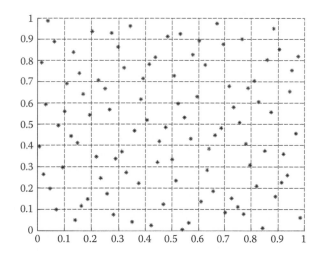

FIGURE 14.5

Halton sequence of 100 points. The star discrepancy value is 0.05.

a well-known low-discrepancy sequence (see Figure 14.4). As we can observe from the figure, this set of points 'covers well' the box, in the sense that the points are well equidistributed over the box. Its star discrepancy value is 0.048.

The second example is the Halton sequence [41] shown in Figure 14.5, which is also a well-known low-discrepancy sequence. The value of the star discrepancy of the Halton sequence is about 0.050, indicating that the Faure sequence is more equidistributed than the Halton sequence. The star discrepancy values of these two sequences are however close, and indeed visually it is difficult to see from the figures which one is better equidistributed.

We now give another example, which is a sequence of 100 points generated by a pseudo-random function provided by the C library system. This sequence is shown in Figure 14.6, from which we can observe that this sequence is not well equidistributed over the box. This is confirmed by its star discrepancy value 0.1. The star discrepancy is thus a meaningful measure that can characterize the uniformity quality of a point set distribution.

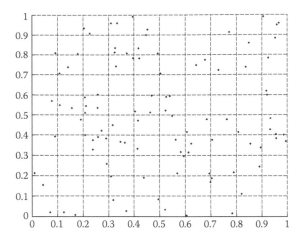

FIGURE 14.6
A sequence of 100 points generated by a pseudo-random function in the C library. Its star discrepancy value is 0.1.

14.4.2 Coverage estimation

To evaluate the coverage of a set of states, we must compute the star discrepancy of a point set, which is not an easy problem (e.g., see [14]). Many theoretical results for one-dimensional point sets are not generalizable to higher dimensions, and among the fastest algorithms, the one proposed in [14] has time complexity $\mathcal{O}(k^{1+n/2})$, where n is the dimension. In this work, we do not try to compute the star discrepancy but approximate it by estimating a lower and an upper bound. These bounds, as well as the information obtained from their estimation, are then used to decide which parts of the state space have been "well explored" and which parts must be explored more. This estimation is carried out using a method published in [37]. Let us briefly describe this method for computing the star discrepancy $D^*(P,\mathcal{B})$ of a point set P w.r.t. a box $\mathcal{B}$. Although in [37], the box $\mathcal{B}$ is $[0,1]^n$, we extended it to the case where $\mathcal{B}$ can be any full-dimensional box. Intuitively, the main idea of this estimation method is to consider a finite box partition of the box $\mathcal{B}$, instead of considering an infinite number of all sub-boxes as in the definition of the star discrepancy. Let $\mathcal{B} = [l_1, L_1] \times \ldots \times [l_n, L_n]$. In what follows, we often call this box $\mathcal{B}$ the *bounding box*. We define a box partition of $\mathcal{B}$ as a set of boxes $\Pi = \{\boldsymbol{b}^1, \ldots, \boldsymbol{b}^m\}$ such that $\cup_{i=1}^m \boldsymbol{b}^i = \mathcal{B}$ and the interiors of the boxes $\boldsymbol{b}^i$ do not intersect. Each such box is called an *elementary box*. Given a box $\boldsymbol{b} = [\alpha_1, \beta_1] \times \ldots \times [\alpha_n, \beta_n] \in \Pi$, we define $\boldsymbol{b}^+ = [l_1, \beta_1] \times \ldots \times [l_n, \beta_n]$ and $\boldsymbol{b}^- = [l_1, \alpha_1] \times \ldots \times [l_n, \alpha_n]$ (see Figure 14.7 for an illustration).

For any finite box partition Π of $\mathcal{B}$, the star discrepancy $D^*(P,\mathcal{B})$ of the point set P with respect to $\mathcal{B}$ satisfies: $C(P,\Pi) \leq D^*(P,\mathcal{B}) \leq B(P,\Pi)$, where the upper and lower bounds are

$$B(P,\Pi) \quad = \quad \max_{\boldsymbol{b}\in\Pi} \max \left\{ \frac{A(P,\boldsymbol{b}^+)}{k} - \frac{vol(\boldsymbol{b}^-)}{vol(\mathcal{B})}, \frac{vol(\boldsymbol{b}^+)}{vol(\mathcal{B})} - \frac{A(P,\boldsymbol{b}^-)}{k} \right\}. \quad (14.3)$$

$$C(P,\Pi) \quad = \quad \max_{\boldsymbol{b}\in\Pi} \max \left\{ \left| \frac{A(P,\boldsymbol{b}^-)}{k} - \frac{vol(\boldsymbol{b}^-)}{vol(\mathcal{B})} \right|, \left| \frac{A(P,\boldsymbol{b}^+)}{k} - \frac{vol(\boldsymbol{b}^+)}{vol(\mathcal{B})} \right| \right\}. \quad (14.4)$$

The imprecision of this approximation is the difference between the upper and lower bounds, which can be bounded by $B(P,\Pi) - C(P,\Pi) \leq W(\Pi)$, where

$$W(\Pi) = \max_{\boldsymbol{b}\in\Pi}(vol(\boldsymbol{b}^+) - vol(\boldsymbol{b}^-))/vol(\mathcal{B}). \quad (14.5)$$

Thus, one must find a partition Π such that this difference is small.

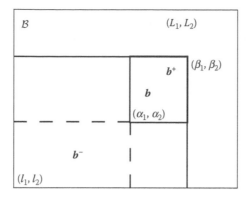

FIGURE 14.7
Illustration of the boxes b^- and b^+.

14.4.3 Hybrid systems test coverage

Since a hybrid system can only evolve within the staying sets of the locations, we are interested in the coverage with respect to these sets. We assume that all the staying sets are boxes since the star discrepancy is defined for points inside a box. If a staying set $\mathcal{I}_q$ is not a box, we can take the smallest oriented (or nonaxis-aligned) box that encloses it and apply the star discrepancy definition in 14.2 to that box after an appropriate coordinate transformation. Another possible solution, which is part of our current work and reported in this chapter, is to consider the discrepancy notion for sets with more general geometric forms [7].

Definition 7 (Test coverage). Let $\mathcal{P} = \{(q, P_q) \mid q \in Q \ \wedge \ P_q \subset \mathcal{I}_q\}$ be the set of states. The coverage of $\mathcal{P}$ is defined as:

$$Cov(\mathcal{P}) = \frac{1}{\|Q\|} \sum_{q \in Q} 1 - D^*(P_q, \mathcal{I}_q),$$

where $\|Q\|$ is the number of locations in Q. $\qquad \square$

We can see that a large value of $Cov(\mathcal{P})$ indicates a good space-covering quality. If $\mathcal{P}$ is the set of states visited by a test suite, our objective is to maximize $Cov(\mathcal{P})$.

14.5 Test Generation

Our test generation is based on a randomized exploration of the reachable state space of the system. It is inspired by the RRT algorithm, which is a successful motion planning technique for finding feasible trajectories of robots in an environment with obstacles (see [27] for a survey). More precisely, we extend the RRT algorithm to hybrid systems. Furthermore, we combine it with a guiding tool in order to achieve a good coverage of the system's behaviors we want to test. To this end, we use the coverage measure defined in the previous section. Before continuing, we mention that related work on test generation for hybrid models is described in Section 14.14.

In this section, we describe the extension of the RRT algorithm to hybrid system, which we call the hRRT algorithm. The combination of the hRRT algorithm with the guiding tool is explained in the next section.

The algorithm stores the visited states in a tree, the root of which corresponds to the initial state. The construction of the tree is summarized in Algorithm 1.

Algorithm 1 Test generation algorithm hRRT

$k = 1$
$\mathcal{T}^k.init(s_{init})$ $\triangleright$ s_{init}: *initial state*
repeat
     $s_{goal} = \text{SAMPLING}(\mathcal{S})$ $\triangleright$ $\mathcal{S}$: *hybrid state space*
     $s_{near}^k = \text{NEIGHBOR}(\mathcal{T}^k, s_{goal}^k)$
     $(s_{new}^k, u_{q_{near}}^k) = \text{CONTINUOUSSUCC}(s_{near}^k, h)$ $\triangleright$ h: *time step*
     $\text{DISCRETESUCC}(\mathcal{T}^k, s_{new}^k)$
     $k{+}{+}$
until $k \geq k_{max}$

The tree constructed at the kth iteration is denoted by $\mathcal{T}^k$. The function SAMPLING samples a hybrid state $s_{goal}^k = (q_{goal}^k, x_{goal}^k)$ to indicate the direction toward which the tree is expected to evolve. Then, a starting state $s_{near}^k = (q_{near}^k, x_{near}^k)$ is determined as a neighbor of s_{goal}^k using a hybrid distance, which is described later. Expanding the tree from s_{near}^k toward s_{goal}^k is done as follows:

- The function CONTINUOUSSUCC tries to find the input $u_{q_{near}}^k$ such that, after one time step h, the current continuous dynamics at q_{near}^k takes the system from s_{near}^k toward s_{goal}, and this results in a new continuous state x_{new}^k. A new edge from s_{near} to $s_{new}^k = (q_{near}^k, x_{new}^k)$, labeled with the associated input $u_{q_{near}}^k$, is then added to the tree. To find s_{new}^k, when the set U is not finite it can be sampled, or one can solve a local optimal control problem.

- Then, from s_{new}^k, the function DISCRETESUCC computes its successors by all possible discrete transitions and add them in the tree. A discrete successor by a transition is computed by testing whether s_{new}^k satisfies its guard and if so applying the associated reset function to s_{new}^k.

The algorithm terminates after some maximal number of iterations. Another possible termination criterion is that a satisfactory coverage value is reached. In the classic RRT algorithms, which work in a continuous setting, only x_{goal} must be sampled, and a commonly used sampling distribution of x_{goal} is uniform over $\mathcal{X}$. In addition, the point x_{near} is defined as a nearest neighbor of x_{goal} in some usual distance, such as the Euclidian distance. In our hRRT algorithm, the goal state sampling is not uniform and the function SAMPLING plays the role of guiding the exploration via a biased sampling of x_{goal}.

14.5.1   Hybrid distance and computation of neighbors

Defining a metric for the hybrid state space is difficult because of the discrete component of a hybrid state. We propose an approximate distance between two hybrid states, which is used in the function NEIGHBOR of the hRRT algorithm.

Given two hybrid states $s = (q, x)$ and $s' = (q', x')$, if they have the same discrete component, that is, $q = q'$, we can use some usual metric in $\mathbb{R}^n$, such as the Euclidian metric. When $q \neq q'$, it is natural to use the average length of the trajectories from one to another.

Figure 14.8 illustrates this definition. We consider a trajectory corresponding to a sequence of transitions $\gamma = e_1, e_2$, where $e_1 = (q, q_1)$ and $e_2 = (q_1, q')$. We call γ a discrete path from $s = (q, x)$ to $s' = (q', x')$.

- The average length of the discrete path γ is simply the distance between the image of the first guard $\mathcal{G}_{(q,q_1)}$ by the first reset function $\mathcal{R}_{(q,q_1)}$ and the second guard $\mathcal{G}_{(q_1,q')}$. This distance is shown in the middle figure. The average distance between two sets is defined as the Euclidean distance between their geometric centroids.

- The average length of trajectories from $s = (q, x)$ to $s = (q', x')$ following the discrete path γ is the sum of three distances (shown in Figure 14.8 from left to right): the distance between x and the first guard $\mathcal{G}_{(q,q_1)}$, the average length of the path, and the distance between $\mathcal{R}_{(q_1,q')}(\mathcal{G}_{(q_1,q')})$ and x'.

Now, we are ready to define the *hybrid distance* from s to s'. We denote by $\Gamma(q, q')$ the set of all discrete paths from q to q' in the hybrid automaton $\mathcal{A}$.

Definition 8 (Hybrid distance). Given two hybrid states $s = (q, x)$ and $s' = (q', x')$, the *hybrid distance* from s to s', denoted by $d_H(s, s')$, is defined as follows:

- If $q = q'$, then $d_H(s, s') = ||x - x'||$, where $|| \cdot ||$ is the Euclidean distance in $\mathbb{R}^n$.

- If $q \neq q'$, there are two cases:

  - If $\Gamma(q, q') \neq \emptyset$, then $d_H(s, s') = \min_{\gamma \in \Gamma(q,q')} len_\gamma(s, s')$. The path γ that minimizes $len_\gamma(s, s')$ is called the *shortest path* from s to s'.

  - Otherwise, $d_H(s, s') = \infty$. $\qquad\square$

It is easy to see that the hybrid distance d_H is only a pseudo metric since it does not satisfy the symmetry requirement. Indeed, the underlying discrete structure of a hybrid automaton is a directed graph. In the above definition, instead of the Euclidian distance, we can use any metric in $\mathbb{R}^n$.

Then, in each iteration of hRRT, the function NEIGHBOR can be computed as follows. A neighbor of the goal state s_{goal} is:

$$s_{near} \in arg\,min_{s \in V}\, d_H(s, s_{goal}),$$

where V is the set of all the states stored at the vertices of the tree.

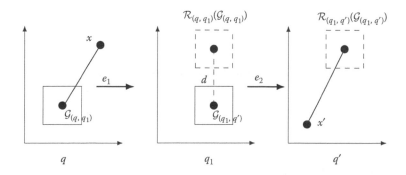

FIGURE 14.8
Illustration of the average length of a trajectory.

14.5.2   Computing continuous and discrete successors

We first describe the function CONTINUOUSSUCC. If the states s_{near} and s_{goal} have the same location component, we want to expand the tree from x_{near} toward x_{goal} as closely as possible, using the continuous dynamics at that location.

When the states s_{near} and s_{goal} are at different locations, let γ be the shortest path from s_{near} to s_{goal}. It is natural to make the system follow this path. Therefore, we want to steer the system from x_{near} toward the first guard of γ. In both of the two cases, one must solve an optimal control problem with the objective of minimizing the distance to some target point. This problem is difficult especially for systems with nonlinear continuous dynamics. Thus, we can trade some optimality for computational efficiency. When the input set U is not finite, we sample a finite number of inputs and pick from this set a best input that makes the system approach the boundary of the guard of γ as much as possible. In addition, we can prove that by appropriately sampling the input set, the completeness property of our algorithm is preserved [13]). It is important to emphasize that the function CONTINUOUSSUCC must assure that the trajectory segment from x_{near} remains in the staying set of the current location.

The computation of discrete successors in DISCRETESUCC, which involves testing a guard condition and applying a reset map, is rather straightforward.

14.5.3   Test cases and verdicts

The tree constructed by the hRRT algorithm can be used to extract a test suite. In addition, when applying such test cases to the SUT, the tree can be used to compare the observations from the physical systems and the expected observations in the tree, which allows a decision whether the system satisfies the conformance relation.

14.6   Coverage-Guided Test Generation

In this section, we propose a tool for guiding the test generation algorithm. This tool is based on the coverage measure defined using the star discrepancy. The goal of the guiding tool is to use the sampling process to bias the evolution of the tree toward the interesting region of the state space, in order to rapidly achieve a good coverage quality. In each iteration, we use the information of the current coverage to improve it. Indeed, the coverage estimation provides not only an approximate value of the current coverage but also the information about which regions need to be explored more.

Sampling a goal state $s_{goal} = (q_{goal}, x_{goal})$ in the hybrid state space $\mathcal{S}$ consists of two steps:

1. Sample a goal location q_{goal} from the set Q of all the locations, according to some probability distribution.

2. Sample a continuous goal state x_{goal} inside the staying set $\mathcal{I}_{q_{goal}}$ of the location q_{goal}.

14.6.1   Location sampling

Recall that we want to achieve a good testing coverage quality, which is equivalent to a small value of the star discrepancy of the points visited at each location. More concretely, in each iteration, we want to bias the goal state sampling distribution according to the current coverage of the visited states. To do so, we first sample a location and then a continuous state. Let $\mathcal{P} = \{(q, P_q) \mid q \in Q \land P_q \subset \mathcal{I}_q\}$ be the current set of visited states. The location sampling distribution depends on the current continuous state coverage of each location:

$$Pr[q_{goal} = q] = \frac{D^*(P_q, \mathcal{I}_q)}{\sum_{q' \in Q} D^*(P_{q'}, \mathcal{I}_{q'})},$$

where the notation Pr is used for probabilities. Intuitively, we assign a higher sampling probability to the locations with lower continuous state coverage. As we have shown earlier, the star discrepancy is approximated by a lower bound and an upper bound. We thus compute the above probability $Pr[q_{goal} = q]$ using these bounds and then taking the mean of the results.

14.6.2 Continuous state sampling

We now show how to sample x_{goal}, assuming that we have already sampled a location $q_{goal} = q$. In the remainder of this chapter, to give geometric intuitions, we often call a continuous state a point. In addition, since all the staying sets are assumed to be boxes, we denote the staying set $\mathcal{I}_q$ by the box $\mathcal{B}$ and denote the current set of visited points at the location q simply by P instead of P_q. Let k be the number of points in P. Let Π be a finite box partition of $\mathcal{B}$ that is used to estimate the star discrepancy of P. The sampling process consists of two steps. In the first step, we sample an elementary box $\boldsymbol{b}_{goal}$ from the set Π; in the second step, we sample a point x_{goal} in $\boldsymbol{b}_{goal}$ uniformly.

The elementary box sampling distribution in the first step is biased in order to optimize the coverage. Guiding is thus done via the goal box sampling process. Let Π be the box partition used in the coverage estimation, and we denote by P the current set of visited states. The objective is to define a probability distribution over the set of elementary boxes of Π. This probability distribution is defined at each iteration of the test generation algorithm. Essentially, we favor the selection of a box if adding a new state in this box allows to improve the coverage of the visited states. This is captured by a potential influence function, which assigns to each elementary box $\boldsymbol{b}$ in the partition a real number that reflects the change in the coverage if a new state is added in $\boldsymbol{b}$. The current coverage is given in form of a lower and an upper bound. In order to improve the coverage, we aim at reducing both of the bounds. More details on the method can be found in [13].

Let us summarize the developments so far. We have shown how to sample a goal hybrid state. This sampling method is not uniform but biased in order to achieve a good coverage of the visited states. From now on, the algorithm hRRT in which the function SAMPLING uses this coverage-guided method is called the gRRT algorithm, which means "guided hRRT."

We can prove that the gRRT algorithm preserves the *probabilistic completeness* of RRT [13]. Roughly speaking, the probabilistic completeness property [25] states that if the trace we search for is feasible, then the probability that the algorithm finds it approaches 1 as the number k of iterations approaches infinity.

To demonstrate the performance of gRRT, we use two illustrative examples. For brevity, we call the classical RRT algorithm using uniform sampling and the Euclidian metric hRRT. The reason we choose these examples is that they differ in the reachability property. In the first example, the system is "controllable" in the sense that the entire state space is reachable from the initial states (by using appropriate inputs), but in the second example, the reachable set is only a small part of the state space. These examples will also be used to validate the efficiency of the new guiding method that we propose.

Example 1. This is a two-dimensional continuous system where the state space $\mathcal{X}$ is a box $\mathcal{B} = [-3, -3] \times [3, 3]$. The continuous dynamics is $f(x,t) = u(t)$, where the input set is $U = \{u \in \mathbb{R}^2 \mid ||u|| \leq 0.2\}$. $\qquad\qquad\qquad\square$

We use 100 input values resulting from a discretization of the set U. The initial state is $(-2.9, -2.9)$. The time step is 0.002. Figure 14.9 shows the result obtained using gRRT, and Figure 14.10 shows the evolution after each iteration of the coverage of the states generated by gRRT (solid curve) and by hRRT (dashed curve), which indicates that gRRT achieved a better coverage quality especially in convergence rate.

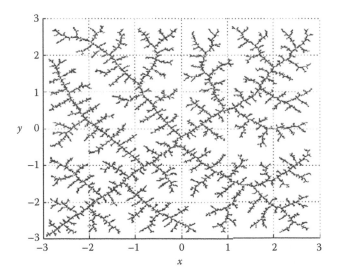

FIGURE 14.9
The gRRT exploration result for Example 1.

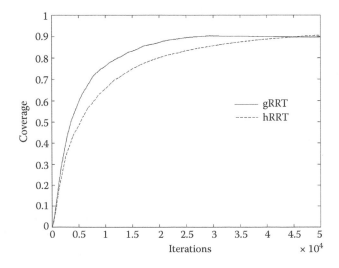

FIGURE 14.10
The test coverage evolution using hRRT and gRRT.

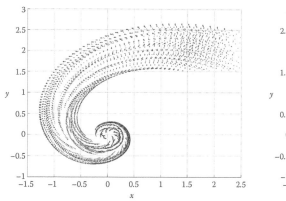 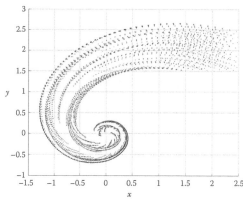

FIGURE 14.11
Results obtained using the guided sampling method (left) and using the uniform sampling method (right).

Example 2. This example is a linear system with a stable focus at the origin. Its dynamics is

$$\begin{pmatrix} \dot{x} \\ \dot{y} \end{pmatrix} = \begin{pmatrix} -1 & -1.9 \\ 1.9 & -1 \end{pmatrix} \times \begin{pmatrix} x \\ y \end{pmatrix} + \begin{pmatrix} u_1 \\ u_2 \end{pmatrix}.$$

We let the dynamics be slightly perturbed by an additive input u. The state space is the box $\mathcal{B} = [-3, -3] \times [3, 3]$. The input set $U = \{u \in \mathbb{R}^2 \mid ||u|| \leq 0.2\}$. Figure 14.11 shows the results obtained after 50,000 iterations. We can see that again the guided sampling method achieved a better coverage result for the same number of points since the points visited by the guided sampling method are more equidistributed over the state space. $\square$

14.7   Controllability Issue

From different experiments with Example 2, we observe that the coverage performance of gRRT is not satisfying when the reachable space is only a small part of the entire state space. To illustrate this, we increase the state space from $\mathcal{B} = [-3, -3] \times [3, 3]$ to $\mathcal{B}' = [-5, -5] \times [5, 5]$. For the larger state space, the coverage quality is poorer (see Figure 14.12). This can be explained as follows. There are boxes, such as those near the bottom right vertex of the bounding box, which have a high potential of reducing the bounds of the star discrepancy. Thus, the sampler frequently selects these boxes. However, these boxes are not reachable from the initial states, and all attempts to reach them do not expand the tree beyond the boundary of the reachable set. This results in a large number of points concentrated near this part of the boundary, while other parts of the reachable set are not well explored.

It is important to emphasize that this problem is not specific to gRRT. The RRT algorithm using the uniform sampling method and, more generally, any algorithm that does not take into account the differential constraints of the system, may suffer from this phenomenon. This phenomenon can however be captured by the evolution of the disparity between the set of goal states and the set of visited states. This notion is formally defined in the next section. Roughly speaking, it describes how different their distributions are. When the disparity does not decrease after a certain number of iterations, this often indicates that the system cannot approach the goal states, and it is better not to favor an

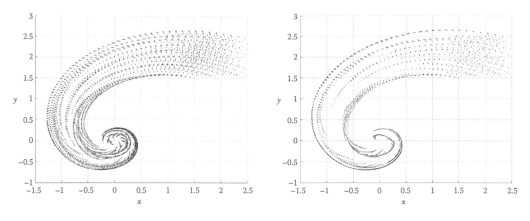

FIGURE 14.12
Results for the state spaces $\mathcal{B}$ (left) and $\mathcal{B}'$ (right).

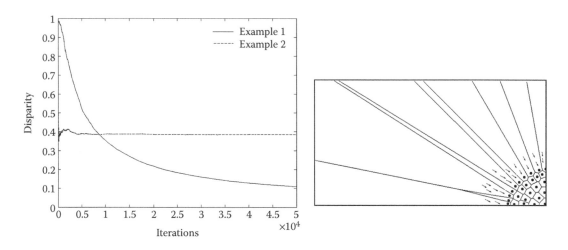

FIGURE 14.13
Left: The evolution of the disparity between the set P^k of visited states and the set G^k of goal states. Right: Illustration of the controllability issue.

expansion toward the exterior but a *refinement*, that is an exploration in the interior of the already visited regions.

Figure 14.13 shows the evolution of the disparity between the set P^k of visited states at the kth iteration and the set G^k of goal states for the two examples. We observe that for the system of Example 1 that can reach any state in the state space (by choosing appropriate admissible inputs), the visited states follow the goal states, and thus the disparity gets stabilized over time. However, in Example 2, where the system cannot reach everywhere, the disparity does not decrease for a long period of time, during which most of the goal states indicate a direction toward which the tree cannot be expanded further.

Figure 14.13 shows the Voronoi diagram* of a set of visited states. In this example, the boundary of the reachable set can be seen as an "obstacle" that prevents the system from

---

*The Voronoi diagram of a set V of points in $\mathbb{R}^n$ is the partition of $\mathbb{R}^n$ into k polyhedral regions. Each region corresponds to a point $v \in V$, called the Voronoi cell of v, is defined as the set of points in $\mathbb{R}^n$, which are closer to v than to any other points in V.

crossing it. Note that the Voronoi cells of the states on the boundary are large (because they are near the large unvisited part of the state space). Hence, if the goal states are uniformly sampled over the whole state space, these large Voronoi cells have higher probabilities of containing the goal states, and thus the exploration is "stuck" near the boundary, while the interior of the reachable set is not well explored.

To tackle this problem, we introduce the notion of disparity to describe the "difference" in the distributions of two sets of points. The controllability problem can be detected by a large value of the disparity between the goal states and the visited states. We can thus combine gRRT with a disparity-based sampling method, in order to better adapt to the dynamics of the system. This is the topic of the next section.

14.8 Disparity

The notion of disparity between two point sets that we develop here is inspired by the star discrepancy. Indeed, by definition, the star discrepancy of a set P w.r.t. the box $\mathcal{B}$ can be seen as a comparison between P and an "ideal" infinite set of points distributed all over $\mathcal{B}$.

Let P and Q be two sets of points inside $\mathcal{B}$. Let J be a sub-box of $\mathcal{B}$, which has the same bottom-left vertex as $\mathcal{B}$ and the top-right vertex of which is a point inside $\mathcal{B}$. Let Γ be the set of all such sub-boxes. We define the local disparity between P and Q with respect to the sub-box J as: $\gamma(P,Q,J) = |A(P,J)/||P|| - A(Q,J)/||Q|||$, where $A(P,J)$ is the number of points of P inside J and $||P||$ is the total number of points of P.

Definition 9 (Disparity). The disparity between P and Q with respect to the bounding box $\mathcal{B}$ is defined as: $\gamma^*(P,Q,\mathcal{B}) = sup_{J\in\Gamma}\gamma(P,Q,J)$. $\square$

The disparity satisfies $0 < \gamma^*(P,Q,\mathcal{B}) \leq 1$. This property is a direct consequence of the above definition. A small value $\gamma^*(P,Q,\mathcal{B})$ means that the distributions of the sets P and Q over the box $\mathcal{B}$ are "similar".

To illustrate our notion of disparity, we consider two well-known sequences of points: the Faure sequence [16] and the Halton sequence [41], which are shown in Figure 14.14. Their disparity is 0.06, indicating that they have similar distributions. We then compare the Faure sequence with a sequence generated by the C library. Figure 14.15 displays these two sequences, each of which has 100 points. The star discrepancy coverage of the Faure sequence is much better than that of the C sequence, and in fact the disparity between them (which is 0.12) is twice larger than that between the Faure and Halton sequences. The last example is used to compare the Faure sequence and a set of 100 points concentrated in some small rectangle inside the bounding box. Their disparity is 0.54.

14.8.1 Disparity estimation

The exact computation of the disparity is as difficult as the exact computation of the star discrepancy, which is because of the infinite number of the sub-boxes. We propose a method for estimating a lower and an upper bound for this new measure. Let Π be a box partition of $\mathcal{B}$. Let P, Q be two sets of points inside $\mathcal{B}$. For each elementary box $\boldsymbol{b}\in\Pi$, we denote $\mu_m(\boldsymbol{b}) = \max\{\mu_c(\boldsymbol{b}),\mu_o(\boldsymbol{b})\}$, where $\mu_c(\boldsymbol{b}) = A(P,\boldsymbol{b}^+)/||P|| - A(Q,\boldsymbol{b}^-)/||Q||$, $\mu_o(\boldsymbol{b}) = A(Q,\boldsymbol{b}^+)/||Q|| - A(P,\boldsymbol{b}^-)/||P||$. We also denote

$$c(\boldsymbol{b}) = \max\left\{\left|\frac{A(P,\boldsymbol{b}^-)}{||P||} - \frac{A(Q,\boldsymbol{b}^-)}{||Q||}\right|, \left|\frac{A(P,\boldsymbol{b}^+)}{||P||} - \frac{A(Q,\boldsymbol{b}^+)}{||Q||}\right|\right\}.$$

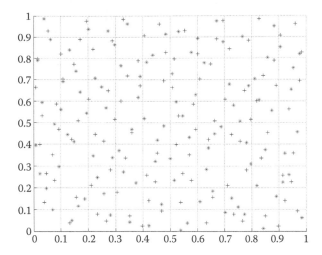

FIGURE 14.14
The disparity between the Faure sequence (drawn using the + signs) and the Halton
sequence (drawn using the * signs) is 0.06.

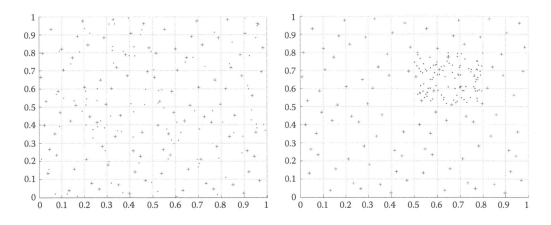

FIGURE 14.15
Left: The disparity between the Faure sequence (+ signs) and a C pseudo-random sequence
(* signs) is 0.12. Right: The disparity between the Faure sequence (+ signs) and another C
pseudo-random sequence (* signs) is 0.54.

Theorem 1 (Upper and lower bounds). An upper bound $B_d(P, Q, \Pi)$ and a lower
bound $C_d(P, Q, \Pi)$ of the disparity between P and Q are: $B_d(P, Q, \Pi) = \max_{\boldsymbol{b} \in \Pi} \{\mu_m(\boldsymbol{b})\}$
and $C_d(P, Q, \Pi) = \max_{\boldsymbol{b} \in \Pi} \{c(\boldsymbol{b})\}$. $\square$

The proof of the theorem can be found in [12].

14.8.2   Estimation error

We now give a bound on the estimation error. For each elementary box $\boldsymbol{b} = [\alpha_1, \beta_1] \times \ldots \times
[\alpha_n, \beta_n] \in \Pi$, we define the $\mathcal{W}$-zone, denoted by $\mathcal{W}(\boldsymbol{b})$, as follows: $\mathcal{W}(\boldsymbol{b}) = \boldsymbol{b}^+ \setminus \boldsymbol{b}^-$. We recall
that $\boldsymbol{b}^+ = [l_1, \beta_1] \times \ldots \times [l_n, \beta_n]$ and $\boldsymbol{b}^- = [l_1, \alpha_1] \times \ldots \times [l_n, \alpha_n]$.

We can prove the following bound on the error of the estimation, defined as the difference between the upper and lower bounds [12].

Theorem 2 (Error bounds). Let $B_d(P, Q, \Pi)$ and $C_d(P, Q, \Pi)$ be the upper and lower bounds of the disparity between P and Q. Then,

$$B_d(P, Q, \Pi) - C_d(P, Q, \Pi) \le \max_{\boldsymbol{b} \in \Pi} \max \left\{ \frac{A(P, \mathcal{W}(\boldsymbol{b}))}{||P||}, \frac{A(Q, \mathcal{W}(\boldsymbol{b}))}{||Q||} \right\}. \qquad \square$$

The above error bounds can be used to dynamically refine the partition.

14.9 Disparity-Guided Sampling

The essential idea of our disparity-based sampling method is to detect when the dynamics of the system does not allow the tree to expand toward the goal states and then to avoid such situations by favoring a refinement, that is an exploration near the already visited states.

A simple way to bias the sampling toward the set P^k of already visited states is to reduce the sampling space. Indeed, we can make a bounding box of the set P^k and give more probability of sampling inside this box than outside it. Alternatively, we can guide the sampling using the disparity information as follows. The objective now is to reduce the disparity between the set G^k of goal states and the set P^k of visited states. Like the guiding method using the star discrepancy, we define for each elementary box $\boldsymbol{b}$ of the partition a function $\eta(\boldsymbol{b})$ reflecting the potential for reduction of the lower and upper bounds of the disparity between P^k and G^k. This means that we favor the selection of the boxes that make the distribution of goal states G^k approach that of the visited states P^k. Choosing such boxes can improve the quality of refinement. The formulation of the potential influence function for the disparity-based sampling method is similar to that for the coverage-guided sampling.

A combination of the coverage-guided and the disparity-guided sampling methods is done as follows. We fix a time window T_s and a threshold ϵ. When using the coverage guide method, if the algorithm detects that the disparity between the G^k and P^k does not decrease by ϵ after T_s time, it switches to the disparity-guided method till the disparity is reduced more significantly and switches back to the coverage-guided method. Note that a nondecreasing evolution of the disparity is an indication of the inability of the system to approach the goal states. In an interactive exploration mode, it is possible to let the user manually decide when to switch. As mentioned earlier, we call the resulting algorithm agRRT (the letter "a" in this acronym stands for "adaptive").

Examples. We use the examples in the previous section to demonstrate the coverage improvement of agRRT. Figure 14.16 shows that the final result for Example 1 obtained using agRRT has a better coverage than that obtained using gRRT. The solid curve represents the coverage of the set P^k of visited states and the dashed one the coverage of the set G^k of goal states. The dash-dot curve represents the disparity between G^k and P^k.

The result obtained using agRRT for Example 2 is shown in Figure 14.17, which also indicates an improvement in coverage quality. The figure on the right shows the set of generated goal states. The states are drawn in dark color. In this example, we can observe the adaptivity of the combination of gRRT and agRRT. Indeed, in the beginning, the gRRT algorithm was used to rapidly expand the tree. After some time, the goal states sampled from the outside of the exact reachable space do not improve the coverage since they only

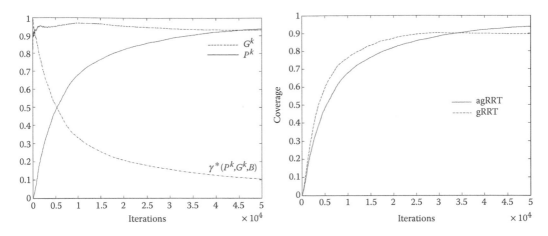

FIGURE 14.16
Left: Test coverage of the result obtained using agRRT for Example 1. Right: Comparing
gRRT and agRRT.

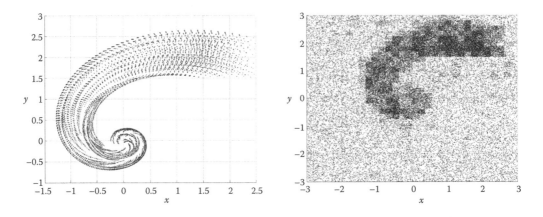

FIGURE 14.17
Result after $k = 50,000$ iterations, obtained using agRRT (left: the set of visited states P^k,
right: the set of goal states G^k).

create more states near the boundary of the reachable set. In this case, the disparity between
P^k and G^k does not decrease, and the agRRT is thus used to enable an exploration in the
interior of the reachable set. The interior of the reachable set thus has a higher density of
sampled goal states than the outside, as one can see in the figure.

14.10   Termination Criterion Using Disparity

We first explain why it is not suitable to use the star discrepancy coverage to decide the
termination of the test generation algorithm. Indeed, the star discrepancy allows to compare the spatial coverage quality between the point sets of the same cardinality; it however
may give misleading comparison between the sets of different cardinalities. To illustrate

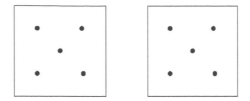

FIGURE 14.18
The point set P (left) and the point set $P \cup Q$ (right).

this, we consider a set of two-dimensional points: $P = \{(0.25, 0.25), (0.5, 0.5), (0.75, 0.25),$ $(0.25, 0.75), (0.75, 0.75)\}$ inside the bounding box $\mathcal{B} = [0, 1] \times [0, 1]$, shown in Figure 14.18 (left). The star discrepancy estimation gives: $D^*(P, \mathcal{B}) \simeq 0.347$. An arising question is how the star discrepancy changes when we add more points in P. We now consider another nonempty set Q of points inside $\mathcal{B}$ that does not contain any points of P. Different experiments show that the star discrepancy of the union $P \cup Q$ may be larger or smaller than that of P. For example, for the following set $Q = \{(0.06, 0.06), (0.12, 0.12), (0.06, 0.12), (0.12, 0.06),$ the value of $D^*(P \cup Q, \mathcal{B}) \simeq 0.493$ (see Figure 14.18-right). In other words, adding this set Q in P increases the star discrepancy, which can be easily understood because the set Q is not well equidistributed over the box $\mathcal{B}$. However, from a verification point of view, the union $P \cup Q$ provides more information about the correctness of the system than the set P. In addition, geometrically speaking, the set $P \cup Q$ "covers more space" than the set P.

Therefore, we do not use the star discrepancy coverage measure to decide when the test generation algorithm can terminate. Instead, to detect when the coverage reaches a "saturation," we use the disparity between two consecutive sets of visited states. If its value remains below some predefined threshold Δ_γ after a predefined number K of iterations, we can stop the algorithm because the distribution of the sets of visited states is not much changed and the coverage is not significantly improved. This criterion can be used together with a maximal number of iterations.

14.11 Actuator and Sensor Imprecision

Because of the limitations of practical actuators and sensors, the tester cannot realize exactly an input value specified as a real-valued vector as well as measure exactly the state of the system. We first explain how actuator imprecision influences the testing process.

Actuator imprecision. We consider the following continuous dynamics of a location: $\dot{x}(t) = f(x(t), u(t))$. Given an initial state x and an input value u, let $\xi_{x,u}$ denote the unique trajectory during the time interval $[0, h]$.

Because of actuator imprecision, when the tester emits an input u to the SUT, indeed some input $v = u + \delta_u$ with $|\delta_u| \leq \epsilon_u$ is applied. We call ϵ_u the actuator imprecision bound. After one step, this uncertainty causes the new state $y = \xi_{x,v}(h)$ to deviate from the exact state $y' = \xi_{x,u}(h)$. In the next step, this deviation can be considered as an uncertainty in the initial condition, namely $(y' - y)$, which causes further deviation.

Therefore, only when the observation measured by the tester lies outside some admissible deviation neighborhood, the conformance property is considered violated. Therefore, we must compute the admissible deviation neighborhoods.

Let $M_x = \partial \xi_{x,u} / \partial x$ and $M_u = \partial \xi_{x,u} / \partial u$ denote the partial derivatives of $\xi_{x,u}$ with respect to the initial condition x and to the input u. They are called the *sensitivity matrices*. Given a neighboring initial state y and a neighboring input value v, we can get an estimation of $\xi_{y,v}$ by dropping higher order terms in the Taylor expansion of $\xi_{x,u}(t)$ seen as a function of x and of u:

$$\hat{\xi}_{y,v}(h) = \xi_{x,u}(h) + M_x(h)(y-x) + M_u(h)(v-u). \tag{14.6}$$

The deviation is thus $\delta = |M_x(h)(y-x) + M_u(h)\epsilon_u|$. The deviation neighborhood is defined as a ball centered at $\xi_{x,u}(h)$ with radius δ. To compute the sensitivity matrices, we solve the following differential equations, which result from taking the derivative of the solution:

$$\dot{x} = f(x,u), \tag{14.7}$$

$$\dot{M}_x = \frac{\partial f(x,u)}{\partial x} M_x, \tag{14.8}$$

$$\dot{M}_u = \frac{\partial f(x,u)}{\partial x} M_u + \frac{\partial f(x,u)}{\partial u} \tag{14.9}$$

with the initial condition $\xi_x(0) = x$ and $M_x(0) = \mathbf{I}_n$ (the identity matrix) and $M_u(0) = \mathbf{0}_n$ (the zero matrix). These are $n+m+1$ ordinary differential equations of order n. Note that the cost can be made less than that of the resolution of $n+m+1$ different systems since the Jacobian $\partial f(x,u)/\partial x$ can be extensively reused in the computation.

Therefore, in the RRT tree construction, when we compute a new state from a given state x and input u, we additionally compute the corresponding sensitivity matrix $M_x(h)$ and $M_u(h)$ and associate these matrices to the new node created to store the new state. We also propagate the neighborhoods in order to detect possible executions of uncontrollable discrete transitions. If the current state does not enable an uncontrollable transition but the associated neighborhood intersects with its guard, we compute the image of the intersection by the reset map and take its centroid to define a new visited state. The diameter of the image is the radius of the associated neighborhood.

Verdict. During the test, let $\tilde{s}^i = (\tilde{q}^i, \tilde{x}^i)$ be the observation at the ith step and $s^i = (q^i, x^i)$ be the corresponding expected state in the tree. Let M_x^i and M_u^i be the sensitivity matrices associated with the node (q^i, x^i). The SUT is assumed to satisfy the conformance property at the initial state. The verdict is then decided as follows. We suppose that the conformance property is not violated until the ith step. There are two cases:

1. (C1) $s^{(i+1)}$ is reached by the continuous dynamics. Then, if $|\tilde{x}^{(i+1)} - x^{(i+1)}| > M_x^i |\tilde{x}^i - x^i| + M_u^i \epsilon_u$, we conclude that the conformance property is violated and stop the test. Otherwise, we continue the test.

2. (C2) $s^{(i+1)}$ is reached by a discrete transition. Then, if $\tilde{q}^i = q^i$ and $|\tilde{x}^{(i+1)} - x^{(i+1)}| > \delta^{(i+1)}$, where $\delta^{(i+1)}$ is the deviation radius, we conclude that the conformance property is violated and stop the test. Otherwise, we continue the test.

Sensor imprecision. Let the sensor imprecision be modeled by an upper bound ϵ_x on the distance between the actual continuous state and the observation measured by the tester. The distance between the observation and the expected states is now tested against a bound larger by ϵ_x, that is $|\tilde{x}^{(i+1)} - x^{(i+1)}| > M_x^i(\tilde{x}^i - x^i) + M_u^i \epsilon_u + \epsilon_x$ for the above (C1) and $|\tilde{x}^{(i+1)} - x^{(i+1)}| > \delta^{(i+1)} + \epsilon_x$ for (C2).

14.12   Tool HTG

We have implemented the above algorithms in a test generation tool, called HTG. Its implementation uses a data structure similar to a k-d tree, which facilitates the required operations, such as updating the tree, finding neighbors, updating the star discrepancy and disparity. Indeed, using this data structure, we can efficiently encode a box partition for storing and accessing the visited states as well as for the star discrepancy and disparity estimations. In the following, we briefly describe some important functions. A more detailed description of the implementation can be found in [12].

14.12.1   Numerical simulation

In most classic versions of hybrid automata, continuous dynamics are defined using ODEs. To analyze analog and mixed-signal circuits, the behavior of which are described using differential algebraic equations (DAEs), we adapt the model to capture this particularity and use the well-known RADAU algorithm for solving DAEs [21]. In addition, with view to applications in analog and mixed-signal circuits, an efficient and reliable simulation method is key. The state-of-the-art SPICE simulator is prone to convergence problems when dealing with circuit components with stiff characteristics. We also integrated in our test generation tool a connection to the platform SICONOS, developed at INRIA Rhônes-Alpes, that contains a number of simulation, algorithms based on *the nonsmooth approach* [2], which have proved to be efficient for systems with stiff dynamics.

14.12.2   SPICE netlists as input systems

An important new feature of the tool is its ability to deal with circuits describing the use of SPICE, in addition to textual descriptions of a hybrid automata. This facilitates the application of the tool to practical circuits. Note that our test generation method works on the differential equations of a hybrid automaton, which is an appropriate mathematical model to describe circuit dynamics. However, the circuit equations must be generated from a SPICE netlist. Common SPICE parsers do not directly provide a succinct form of circuit equations but generate complex equations for the numerical resolution of each simulation step. Furthermore, using SPICE descriptions, we cannot specify uncertain inputs. The solution we propose to address this can be described as follows:

- The inputs (controllable by tester) can be source currents or voltages in SPICE. Their uncertainty is described in an auxiliary file.

- We use the tool ACEF [1] to generate the numerical simulation equations in each step. The values of the controllable inputs, computed by the test generation algorithm, are communicated to ACEF [1] before the generation in each step. The generation can be optimized by only updating some terms involving the modified input in the equations of the previous steps.

The test cases generated by the tool can be visualized by different standard viewers such as MATLAB® and GNU plots.

The tool can also be used as a systematic simulator to verify properties of a model. For this problem, the tester can manipulate not only control inputs but also disturbance inputs. Systematic simulation can be seen as a good compromise between exhaustive verification and ad hoc simulation.

14.13   Applications

The tool has been tested on various examples, which proved its scalability to high-dimensional systems (up to a few hundred continuous variables) [13]. In addition, we have also successfully applied the tool to a number of case studies in control systems and in analog and mixed signal circuits. In the following, to provide an overview of the applicability of the tool, we briefly report some of these applications.

14.13.1   Aircraft collision avoidance system

To illustrate the application of our algorithm to hybrid systems, we use the aircraft collision avoidance problem [32], which is a well-known benchmark in the hybrid systems literature. In [32], the authors treated the problem of collision avoidance of two aircraft. To show the scalability of our approach, we consider the same model with N aircraft.

As shown in Figure 14.19, all the aircraft are at a fixed altitude. Each aircraft i has three states (x_i, y_i, θ_i), where x_i and y_i describe the position and θ_i is the relative heading of the aircraft. Each aircraft begins in straight flight at a fixed relative heading (mode 1).

As soon as two aircrafts are within the distance R between each other, they enter mode 2. In this mode, each aircraft makes an instantaneous heading change of 90 degrees, and begins a circular flight for π time units. After that, they switch to mode 3 and make another instantaneous heading change of 90 degrees and resume their original headings from mode 1.

The dynamics of the system are shown in Figure 14.20. The guard transition between modes 1 and 2 is given by $Dij < R$, which means that the aircraft i is at R distance from the aircraft j. The dynamics of each aircraft is as follows:

$$\dot{x}_i = vcos(\theta_i) + dx_i,$$
$$\dot{y}_i = vsin(\theta_i) + dy_i$$
$$\dot{\theta}_i = \omega.$$

The continuous inputs are dx_i and dx_i describing the external disturbances on the aircraft (such as wind):

$$dx_i = d_1 sin(\theta_i) + d_2 cos(\theta_i),$$
$$dy_i = -d_1 cos(\theta_i) + d_2 sin(\theta_i),$$

and $-\delta \le d_1, d_2 \le \delta$.

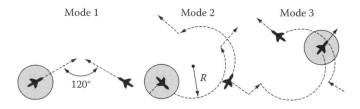

FIGURE 14.19
Aircraft behavior in the three modes. (Adapted from Mitchell, I., and Tomlin, C., Level set method for computation in hybrid systems, in Krogh, B., and Lynch, N. (eds.), *Hybrid Systems: Computation and Control*, 311–323, Springer, New York, 2000.)

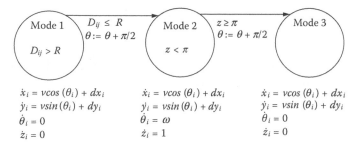

FIGURE 14.20
System dynamics for the three modes.

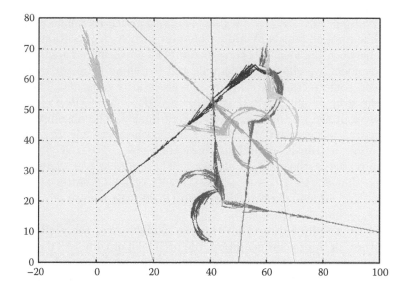

FIGURE 14.21
Eight-aircraft collision avoidance (50,000 visited states, computation time: 10 min.)

Results. For N aircraft, the system has $3N + 1$ continuous variables (one for modeling a clock). For the case of $N = 2$ aircraft, when the collision distance is 5 (that is a collision occurs if the distance between the aircraft is not greater than 5), no collision was detected after visiting 10,000 visited states, and the computation time was 0.9 min. The result for $N = 8$ aircraft with the disturbance bound $\delta = 0.06$ is shown in Figure 14.21, where we show the projected positions of the eight aircraft on a two-dimensional space. For this example, the computation time for 50,000 visited states was 10 min and a collision was found. For a similar example with $N = 10$ aircraft, the computation time was 14 min and a collision was also found.

14.13.2 Robotic vehicle benchmark

Another example is adapted from the robotic navigation system benchmark [35]. We consider a car with the following continuous dynamics with five variables: $\dot{x} = vcos(\theta)$, $\dot{y} = vsin(\theta)$, $\dot{\theta} = vtan(\phi)/L$, $\dot{v} = u_0$, $\dot{\phi} = u_1$, where x, y, θ describe the position and heading of the car, v is its speed, and ϕ is its steering angle. The car can be in one of three car modes (smooth car, smooth unicycle, smooth differential drive). In this work, we consider only the smooth car mode.

The inputs of the system are u_0 and u_1, which are respectively the acceleration and steering control. The system uses a hybrid control law with three driver modes. In the first driver mode, called *RandomDriver*, the control inputs are selected uniformly at random between their lower and upper bounds. In the second driver mode, called *StudentDrive*, when the speed is low, u_0 is randomly chosen as in first mode; otherwise, the strategy is to reduce the speed. In the third driver mode, called *HighwayDrive*, the strategy is to reduce the speed when it is high and increase it when it is low. A detailed description of this control law can be found in [35].

Rather than to analyze a realistic navigation system model, we use this example to test the efficiency of our algorithms on a hybrid system with a larger number of locations. To this end, we created from this system two models. The terrain is partitioned into K rectangles using a regular grid $\mathcal{G} = \{0, \ldots, K_x - 1\} \times \{0, \ldots, K_y - 1\}$. Each rectangle is associated with a driver mode. The first model is a hybrid automaton with $K_x K_y$ locations and the system can only switch between the locations corresponding to adjacent rectangles.

In the second model, we allow more complicated switching behavior by letting the system jump between some rectangles that are not necessarily adjacent. The rectangle corresponding to the grid point $(i, j) \in \mathcal{G}$ is $R_{ij} = [il_x, jl_y] \times [(i + 1)l_x, (j + 1)l_y]$, where l_x and l_y are the sizes of the grid in the x and y coordinates. The absolute index of R_{ij} is an integer defined as follows: $\iota(R_{ij}) = iK_y + j$. From the rectangle R_{ij} with even absolute index, we allow a transition to R_{mn} such that $\iota(R_{mn}) = (\iota(R_{ij}) + J) \bmod (K_x K_y)$ (where $J > 0$ and *mod* is the modulo division). The guard set at R_{ij} is the rightmost band of width ϵ_g, that is $[(i + 1)l_x - \epsilon_g, jl_y] \times [(i + 1)l_x, (j + 1)l_y]$. After switching to R_{mn}, the car position (x, y) is reset to a random point inside the square of size ϵ_r defined as $[ml_x, (n + 1)l_y - \epsilon_r] \times [ml_x + \epsilon_r, (n + 1)l_y]$.

We compared the results obtained for the two models using the gRRT algorithm and the hRRT algorithm. In this experimentation, the hRRT algorithm uses a uniform sampling (both over the discrete and continuous state space). Since we want to focus on the performance of the guiding tool, the two algorithms use the same hybrid distance definition and implementation. The parameters used in this experimentation are: $l_x = l_y = 20$, $l_x = l_y = 20$, the car position $(x, y) \in [-100, 100] \times [-100, 100]$, $\epsilon_g = \epsilon_r = 6$, $J = 6$. The number of locations in the hybrid automata is 100. For the first model without jumps, in terms of coverage efficiency, the algorithms are comparable. For the model with jumps, gRRT systematically produced better coverage results. However, gRRT is not always better than hRRT in terms of the number of covered locations. This is because of our coverage definition using the average of the continuous-state coverages of all the locations.

In terms of time efficiency, we now report the computation time of gRRT for the experimentations with various maximal visited states. For the first model, the computation times of gRRT are: 4.7 s for 10,000 states in the tree, 1 min 26 s for 50,000 states, 6 min 7 s for 100,000 states. For the second model, the computation times of gRRT are: 4.2 s for 10,000 states in the tree, 2 min 5 s for 50,000 states, 4 min 40 s for 100,000 states, and 20 min 22 s for 150,000 states.

14.13.3 Analog and mixed-signal circuits

Using the above results, we implemented a test generation tool and tested it on a number of control applications, which proved its scalability to high-dimensional systems [33]. In this implementation, all the sets encountered in the hybrid automaton definition are convex polyhedra. For circuit applications, we use the well-known RADAU algorithm for solving

DAEs [21]. We recall that solving high index* and stiff DAEs is computationally expensive, and in order to evaluate the efficiency of the test generation algorithm, we have chosen two practical circuits with DAEs of this type. The three circuits we treated are a transistor amplifier, a voltage-controlled oscillator (VCO), and a Delta–Sigma modulator circuit.

Transistor amplifier. The first example is a transistor amplifier model [21], shown in Figure 14.22, where the variable y_i is the voltage at node i; U_e is the input signal and $y_8 = U_8$ is the output voltage. The circuit equations are a system of DAEs of index 1 with eight continuous variables: $M\dot{y} = f(y, u)$. The function f is given by

$$
\begin{pmatrix}
-U_e/R_0 + y_1/R_0 \\
-U_b/R_2 + y_2(1/R_1 + 1/R_2) - (\alpha - 1)g(y_2 - y_3) \\
-g(y_2 - y_3) + y_3/R_3 \\
-U_b/R_4 + y_4/R_4 + \alpha g(y_2 - y_3) \\
-U_b/R_6 + y_5(1/R_5 + 1/R - 6) - (\alpha - 1)g(y_5 - y_6) \\
-g(y_5 - y_6) + y_6/R_7 \\
-U_b/R_8 + y_7/R_8 + \alpha g(y_5 - y_6) \\
y_8/R_9
\end{pmatrix}.
$$

The circuit parameters are: $U_b = 6$; $U_F = 0.026$; $R_0 = 1000$; $R_k = 9000$, $k = 1, \ldots, 9$; $C_k = k10^{-6}$; $\alpha = 0.99$; $\beta = 10^{-6}$. The initial state $y_{init} = (0, U_b/(R_2/R_1 + 1),$ $U_b/(R_2/R_1+1), U_b, U_b/(R_6/R_5 + 1), U_b/(R_6/R_5 + 1), U_b, 0)$. To study the influence of circuit parameter uncertainty, we consider a perturbation in the relation between the current through the source of the two transistors and the voltages at the gate and source $I_S = g(U_G - U_S) = \beta(e^{U_G - U_S/U_F} - 1) + \epsilon$, with $\epsilon \in [\epsilon_{min}, \epsilon_{max}] = [-5e - 5, 5e - 5]$. The input signal $U_e(t) = 0.1sin(200\pi t)$. The acceptable interval of U_8 in the nonperturbed circuit is $[-3.01, 1.42]$. Once the initial transient period has settled down, the test case indicates the presence of traces with overshoots after $18,222$ iterations (corresponding to 1.1 min of computation time). The total computation time for generating $50,000$ states was 3 min. Figure 14.22 shows the generated states projected on U_8 over the first 0.03 s.

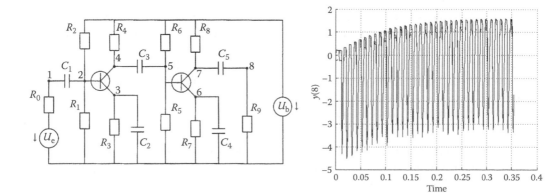

FIGURE 14.22
Test generation result for the transistor amplifier.

---

*The differential index of a DAE is a measure of the singularity of the DAE. It characterizes the difficulty in numerically solving the equation.

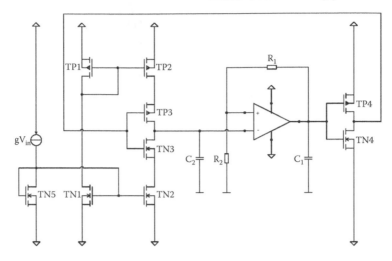

FIGURE 14.23
Voltage-controlled oscillator (VCO) circuit.

Voltage-controlled oscillator. The second circuit we examined is a VCO circuit (Figure 14.23) [19], described by a system of DAEs with 55 continuous variables. In this circuit, the oscillating frequency of the variables v_{C_1} and v_{C_2} is a linear function of the input voltage u_{in}. We study the influence of a time-variant perturbation in C_2, modeled as an input signal, on this frequency. In this example, we show that, in addition to conformance relation, using this framework, we can test a property of the input/output relation. The oscillating period $T \pm \delta$ of v_{C_2} can be expressed using a simple automaton with one clock y in Figure 14.24. The question is to know if given an oscillating trace in $\mathcal{A}$, its corresponding trace in $\mathcal{A}_s$ also oscillates with the same period. This additional automaton can be used to determine test verdicts for the traces in the computed test cases. If an observation sequence corresponds to a path entering the "Error" location, then it causes a "fail" verdict. Since we cannot use finite traces to prove a safety property, the set of observation sequences that cause a "pass" verdict is empty, and therefore the remaining observation sequences (that do not cause a "fail" verdict) cause an "inconclusive" verdict.

We consider a constant input voltage $u_{in} = 1.7$. The coverage measure was defined on the projection of the state space on v_{C_1} and v_{C_2}. The generated test case shows that *after the transient time*, under a time-variant deviation of C_2 which ranges within $\pm 10\%$ of the value of $C_2 = 0.1e - 4$, the variables v_{C_1} and v_{C_2} oscillate with the period $T \in [1.25, 1.258]s$ (with $\varepsilon = 2.8e - 4$). This result is consistent with the result presented in [19]. The number of generated states was 30,000 and the computation time was 14 min. Figure 14.24 shows the explored traces of v_{C_2} over time.

Delta–Sigma circuit. The third example is a third-order Delta–Sigma modulator [6], which is a mixed-signal circuit shown in Figure 14.25. When the input sinusoid is positive and its value is less than 1, the output takes the $+1$ value more often and the quantization error is fed back with negative gain and accumulated in the integrator $\frac{1}{z-1}$. Then, when the accumulated error reaches a certain threshold, the quantizer switches the value of the output to -1 to reduce the mean of the quantization error. A third-order Delta–Sigma modulator is modeled as a hybrid automaton, shown in Figure 14.26. The discrete-time dynamics of the system is as follows: $x(k+1) = Ax(k) + bu(k) - sign(y(k))a$, $y(k) = c_3 x_3(k) + b_4 u(k)$, where $x(k) \in \mathbb{R}^3$ is the integrator states, $u(k) \in \mathbb{R}$ is the input, $y(k) \in \mathbb{R}$ is the input of the

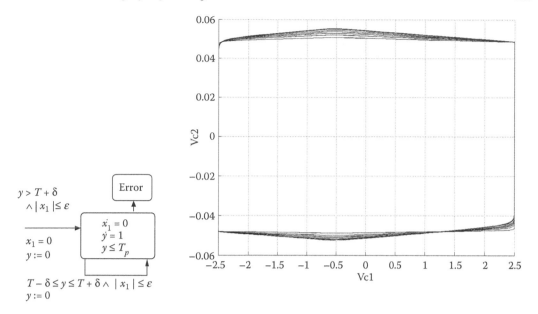

FIGURE 14.24

Left: Automaton for an oscillation specification. Right: Variable v_{C_2} over time. The number of generated states was $30,000$ and the computation time was 14 min.

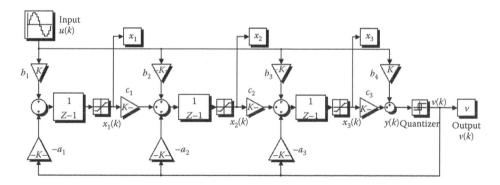

FIGURE 14.25

A third-order modulator: Saturation blocks model saturation of the integrators.

quantizer. Thus, its output is $v(k) = sign(y(k))$, and one can see that whenever v remains constant, the system dynamics is affine continuous. A modulator is stable if under a bounded input, the states of its integrators are bounded.

The test generation algorithm was performed for the initial state $x(0) \in [-0.01, 0.01]^3$ and the input values $u(k) \in [-0.5, 0.5]$. After exploring only 57 states, saturation was already detected. The computation time was less than 1 s. Figure 14.27 shows the values of $(\sup x_1(k))_k$ as a function of the number k of time steps. We can see that the sequence $(\sup x_1(k))_k$ leaves the safe interval $[-x_1^{sat}, x_1^{sat}] = [-0.2, 0.2]$, which indicates the instability of the circuit. This instability for a fixed finite horizon was also detected in [11] using an optimization-based method.

$$c_3 x_3(k) + u(k) \geq 0$$

| $v(k) = -1$ $x(k+1) = Ax(k)+$ $bu(k) + a$ | $v(k) = +1$ $x(k+1) = Ax(k)+$ $bu(k) - a$ |

$$c_3 x_3(k) + u(k) < 0$$

FIGURE 14.26

A hybrid automaton modeling the modulator.

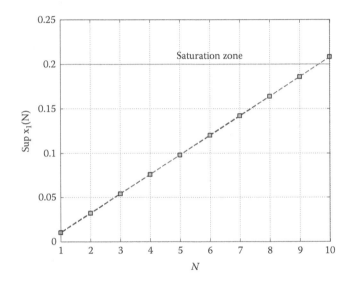

FIGURE 14.27

Test generation result for the Delta–Sigma circuit. The computation time was less than 1 s.

14.14 Conclusion and Related Work

The main results presented in this chapter can be summarized as follows. We proposed a formal framework for conformance testing of hybrid systems. This framework uses the commonly accepted hybrid automaton model. We also proposed a novel coverage measure, which not only is useful as a criterion to evaluate testing quality but also can be used to guide the test generation process. We then developed a number of coverage-guided test generation algorithms for hybrid systems. These algorithms are based on a combination of ideas from robotic path planning, equidistribution theory, algorithmic geometry, and numerical simulation. Based on these results, we implemented a tool for conformance testing of hybrid systems, called HTG, which allowed us to treat a number of case studies from control applications as well as from analog and mixed signal circuits. The experimental results obtained using the tool HTG show its applicability to systems with complex dynamics and its scalability to high-dimensional systems.

The remainder of this section is devoted to related work on hybrid systems testing. Concerning hybrid systems testing, the work [40] developed a theoretical framework for testing hybrid input–output conformance of hybrid transition system models. On an abstract level, this conformance relation is very close to the one we developed in this work. However, no concrete test generation algorithms and implementation were proposed in [40]. Concerning other models, a conformance testing framework was developed for qualitative action systems where the continuous dynamics are described using qualitative differential equations [3].

The paper [36] proposed a framework for generating test cases from simulation of hybrid models specified using the language CHARON [5]. In this work, the test cases are generated by restricting the behaviors of an environment automaton to yield a deterministic testing automaton. A test suite can thus be defined as a finite set of executions of the environment automaton. It is mentioned in [5] that to achieve a desired coverage, nondeterminism in the environment automaton is resolved during the test generation using some randomized algorithm. However, this coverage and the randomized algorithm were not described in detail. Besides testing a real system, another goal of [5] is to apply tests to models, as an alternative validation method. In [24], the testing problem is formulated as to find a piecewise-constant input that steers the system toward some set, which represents a set of bad states.

The RRT algorithm has been used to solve a variety of reachability-related problems such as hybrid systems planning, control, verification, and testing (e.g., see [15, 9, 24, 10, 35] and references therein). Here, we only discuss a comparison of our approach with some existing RRT-based approaches for the validation of continuous and hybrid systems. Concerning the problem of defining a hybrid distance, our hybrid distance is close to that proposed in [24]. The difference is that we use the centroids of the guard sets to define the distance between these sets, while the author of [24] uses the minimal clearance distance between these sets, which is more difficult to compute. To overcome this difficulty, the author proposed to approximate this clearance distance by the diameter of the state space. An advantage of our hybrid distance is that it captures better the average cases, allowing not to always favor the extreme cases. Note also that our hybrid distance d_H does not take into account the system dynamics. It is based on the spatial positions of the states. In [24], the author proposed a time-based metric for two hybrid states, which can be seen as an approximation of the minimal time required to reach one state from another, using the information on the derivatives of the variables. Another distance proposed in [24] is called specification based. This distance is typically defined with respect to some target set specifying some reachability property. It can be, however, observed that for many systems, this "direct" distance may mislead the exploration due to the controllability of the system. In [15, 24] and in our hRRT algorithm, the problem of optimally steering the system toward the goal states was not addressed. In other words, the evolution of the tree is mainly determined by the selection of nearest neighbors. In [9], the problem of computing optimal successors was considered more carefully, and approximate solutions for linear dynamics as well as for some particular cases of nonlinear dynamics were proposed. The authors of [35] proposed a search on a combination of the discrete structure and the coarse-grained decomposition of the continuous state space into regions, in order to determine search directions. This can be thought of as an implicit way of defining a hybrid distance as well as a guiding heuristics.

Concerning test coverage for continuous and hybrid systems, in [15] the authors proposed a coverage measure based on a discretized version of dispersion since the dispersion in general is very expensive to compute. Roughly speaking, the dispersion of a point set with respect to various classes of range spaces, such as balls, is the area of the largest empty range. This measure is defined over a set of grid points with a fixed size δ. For a given test, the spacing s_g of a grid point g is the distance from g to the nearest visited state by the test if it is smaller than δ, and $s_g = \delta$ otherwise. Let S be the sum of the spacings of all the grid points. This means that the value of S is the largest when the set of visited states is empty. Then, the coverage measure is defined in terms of how much the vertices of the tree reduce the value of S. It is important to note that while in our work, the coverage measure is used to guide the simulation, in [15] it is used as a termination criterion. The paper [23] addresses the problem of robust testing by quantifying the robustness of some properties under parameter perturbations. The work in [23] also considers the problem of how to generate test cases with a number of initial state coverage strategies. In addition, the work [18] is similar to ours in the idea of exploiting the existence of metrics on the system state space, which is natural for continuous and hybrid systems. Indeed, by using a concept of approximate bisimulation metrics, one can infer all the possible behaviors of the system in a neighborhood of one trajectory. Hence, by a finite number of simulations, it is possible to decide whether the system is correct under all possible disturbances. However, this approach is applicable only for stable systems and the bisimulation metrics can be effectively computed only for restrictive classes of systems, such as systems with linear continuous dynamics.

Concerning guided exploration, sampling the configuration space has been one of the fundamental issues in probabilistic motion planning. Our idea of guiding the test generation via the sampling process has some similarity with the sampling domain control [42]. As mentioned earlier, the RRT exploration is biased by the Voronoi diagram of the vertices of the tree. If there are obstacles around such vertices, the expansion from them is limited and choosing them frequently can slow down the exploration. In the dynamic-domain RRT algorithm, the domains over which the goal points are sampled must reflect the geometric and differential constraints of the system, and more generally, the controllability of the system. In [31], another method for biasing the exploration was proposed and its main idea is to reduce the dispersion in an incremental manner. This idea is thus very close to the idea of our guiding method in spirit; however, their concrete realizations are different. The method in [31] attempts to lower the dispersion by using K samples in each iteration (instead of a single sample) and then select from them a best sample by taking into account the feasibility of growing the tree toward it. Finally, we mention that a similar idea was used in [15] where the number of successful iterations is used to define an adaptive-biased sampling. To sum up, the novelty in our guiding method is that we use the information about the current coverage of the visited states in order to improve the coverage quality. Additionally, we combine this with controllability information (obtained from the disparity estimation) to obtain a more efficient guiding strategy.

A number of directions for future research can be identified. First, we are interested in defining a measure for trace coverage. Partial observability also must be considered. Convergence rate of the exploration in the test generation algorithm is another interesting theoretical problem to tackle. This problem is particularly difficult especially in the verification context where the system is subject to uncontrollable inputs.

References

1. Acary, V., Bonnefon, O., and Denoyelle, P. (2008). Automatic circuit equation formulation for nonsmooth electrical circuits. Technical report, BIPOP-INRIA, ANR VAL-AMS Report.

2. Acary, V., and Pérignon, F. (2006). SICONOS: A software platform for modeling, simulation, analysis and control of non smooth dynamical system. In *Proceedings of MATH-MOD 2006, 5th Vienna Symposium on Mathematical Modelling*, Vienna. ARGESIM Verlag.

3. Aichernig, B. K., Brandl, H., and Krenn, W. (2009). Qualitative action systems. In *Formal Methods and Software Engineering, 11th International Conference on Formal*

Engineering Methods, ICFEM 2009, Volume 5885 of *Lecture Notes in Computer Science*, Pages: 206–225. Springer, New York.

4. Alur, R., Courcoubetis, C., Halbwachs, N., Henzinger, T. A., Ho, P.-H., Nicollin, X., Olivero, A., Sifakis, J., and Yovine, S. (1995). The algorithmic analysis of hybrid systems. *Theoretical Computer Science*, 138(1):3–34.

5. Alur, R., Dang, T., Esposito, J., Hur, Y., Ivan, F., Kumar, C., Lee, I., Mishra, P., Pappas, G., and Sokolsky, O. (2002). Hierarchical modeling and analysis of embedded systems.

6. Aziz, P. M., Sorensen, H. V., and van der Spiegel, J. (1996). An overview of Sigma-Delta converters. *Signal Processing Magazine, IEEE*, 13(1):61–84.

7. Beck, J. and Chen, W. W. L. (1997). Irregularities of distribution. In *Acta Arithmetica*. Cambridge University Press, UK.

8. Bensalem, S., Bozga, M., Krichen, M., and Tripakis, S. (2005). Testing conformance of real-time applications by automatic generation of observers. *Electr. Notes Theor. Comput. Sci.*, 113:23–43.

9. Bhatia, A. and Frazzoli, E. (2004). Incremental search methods for reachability analysis of continuous and hybrid systems. In *Hybrid Systems: Computation and Control*, Volume 2993 of *Lecture Notes in Computer Science*, Pages: 142–156. Springer, New York.

10. Branicky, M., Curtiss, M., Levine, J., and Morgan S. (2005). Sampling-based reachability algorithms for control and verification of complex systems. In *13th Yale Workshop on Adaptive and Learning Systems*, Yale University, New Haven.

11. Dang, T., Donzé, A., and Maler, O. (2004). Verification of analog and mixed-signal circuits using hybrid systems techniques. In Hu, A.J. and Martin, A.K., editors, *FMCAD'04—Formal Methods for Computer Aided Design*, LNCS 3312, Pages: 21–36. Springer-Verlag, New York.

12. Dang, T. and Nahhal, T. (Nov 2007). Model-based testing of hybrid systems. Technical report, Verimag, IMAG, Grenoble, France.

13. Dang, T. and Nahhal, T. (2009). Coverage-guided test generation for continuous and hybrid systems. *Formal Methods in System Design*, 34(2):183–213.

14. Dobkin, D. and Eppstein, D. (1993). Computing the discrepancy. In *Proceedings of the 9th Annual Symposium on Computational Geometry SCG'93*, Pages: 47–52. ACM Press, New York.

15. Esposito, J., Kim, J. W., and Kumar, V. (July 2004). Adaptive RRTs for validating hybrid robotic control systems. In *Proceedings Workshop on Algorithmic Foundations of Robotics*, Zeist, The Netherlands.

16. Faure, H. (1982). Discrépance de suites associées à un système de numération (en dimension s). *Acta Arithm.*, 41:337–351.

17. Gaudel, M.-C. (1995). Testing can be formal, too. In *6th International Joint Conference CAAP/FASE on Theory and Practice of Software Development*, Pages: 82–96. Springer-Verlag, New York.

18. Girard, A. and Pappas, G. J. (2006). Verification using simulation. In *Hybrid Systems: Computation and Control HSCC'06*, LNCS 3927, Pages: 272–286. Springer, New York.

19. Grabowski, D., Platte, D., Hedrich, L., and Barke, E. (2006). Time constrained verification of analog circuits using model-checking algorithms. *Electr. Notes Theor. Comput. Sci.*, 153(3):37–52.

20. Hall, P. A. V., Zhu, H., and May, J. H. R. (Dec 1997). Software unit test coverage and adequacy. *ACM Computing Surveys (CSUR)*, 29(4):366–427.

21. Hairer, E., Lubich, C., and Roche, M. (1989). The numerical solution of differential-algebraic systems by Runge Kutta methods. In *Lecture Notes in Mathematics 1409*. Springer-Verlag, New York.

22. Henzingerm, T.A. (1995). Hybrid automata with finite bisimulations. In Vaandrager, F. and van Schuppen, J., editors, *Proc. ICALP'95*, LNCS 944, Pages: 324–335. Springer-Verlag, New York.

23. Agung Julius, A., Fainekos, G. E., Anand, M., Lee, I., and Pappas, G. J. (2007). Robust test generation and coverage for hybrid systems. In *Hybrid Systems: Computation and Control*, Volume 4416 of *Lecture Notes in Computer Science*, Pages: 329–342. Springer.

24. Kim, J., Esposito, J., and Kumar, V. (2006). Sampling-based algorithm for testing and validating robot controllers. *Int. J. Rob. Res.*, 25(12):1257–1272.

25. Kuffner, J., and LaValle, S. (April 2000). RRT-connect: an efficient approach to single-query path planning. In *Proc. IEEE Int'l Conf. on Robotics and Automation (ICRA'2000)*, San Francisco, CA.

26. Larsen, K. G., Mikucionis, M., and Nielsen, B. (Sept 2004). Online testing of real-time systems using UPPAAL: status and future work. In Brinksma, E., Grieskamp, W., Tretmans, J., and Weyuker, E., editors, *Perspectives of Model-Based Testing*, Volume 04371 of *Dagstuhl Seminar Proceedings*.

27. LaValle, S. and Kuffner, J. (2000). Rapidly-exploring random trees: progress and prospects. In Workshop on the Algorithmic Foundations of Robotics.

28. LaValle, S. M. and Kuffner, J. J. (May 2001). Randomized kinodynamic planning. *International Journal of Robotics Research*, 20(5):378–400.

29. LaValle, S. M., Branicky, M. S., and Lindemann, S. R. (2004). On the relationship between classical grid search and probabilistic roadmaps. *Intl. Journal of Robotics Research*, 23(7-8):673–692.

30. Lee, D. and Yannakakis, M. (1996). Principles and methods of testing finite state machines—a survey. In *Proceedings of the IEEE*, Volume 84, Pages: 1090–1126.

31. Lindemann, S. R. and LaValle, S. M. (2004). Incrementally reducing dispersion by increasing Voronoi bias in RRTs. In *Proceedings IEEE International Conference on Robotics and Automation*.

32. Mitchell, I. and Tomlin, C. (2000). Level set method for computation in hybrid systems. In Krogh, B. and Lynch, N., editors, *Hybrid Systems: Computation and Control*, LNCS 1790, Pages: 311–323. Springer-Verlag, New York.

33. Nahhal, T. and Dang, T. (2007). Test coverage for continuous and hybrid systems. In *Computer Aided Verification CAV*, Volume 4590 of *Lecture Notes in Computer Science*, Pages: 454–468. Springer, New York.

34. Petrenko, A. and Ulrich, A., editors. (2003). *Test Cases Generation for Nondeterministic Real-Time Systems*, Volume 2931 of *LNCS*. Springer, New York.

35. Plaku, E., Kavraki, L., and Vardi, M. (2007). Hybrid systems: from verification to falsification. In Damm, W. and Hermanns, H., editors, *International Conference on Computer Aided Verification (CAV)*, Volume 4590, Pages: 468–481. Lecture Notes in Computer Science, Springer-Verlag Heidelberg, Berlin, Germany.

36. Tan, L., Kim, J., Sokolsky, O., and Lee. I. (2004). Model-based testing and monitoring for hybrid embedded systems. In *Proceedings of IEEE International Conference on Information Reuse and Integration (IRI'04)*.

37. Thiémard, E. (2001). An algorithm to compute bounds for the star discrepancy. *J. Complexity*, 17(4):850–880.

38. Tretmans, J. (1994). A formal approach to conformance testing. In *Proceedings of the IFIP TC6/WG6.1 Sixth International Workshop on Protocol Test systems VI*, Pages: 257–276. North-Holland Publishing Co., Amsterdam, The Netherlands.

39. Tretmans, J. (1999). Testing concurrent systems: a formal approach. In *CONCUR '99: Proceedings of the 10th International Conference on Concurrency Theory*, Pages: 46–65. Springer-Verlag, London, UK.

40. van Osch, M. (2006). Hybrid input-output conformance and test generation. In Havelund, K., Nunez, M., Rosu, G., and Wolff, B., editors, *Proceedings of FATES/RV 2006*, Lecture Notes in Computer Science 4262, Pages: 70–84. Springer, New York.

41. Wang, X. and Hickernell, F. (2000). Randomized Halton sequences.

42. Yershova, A., Jaillet, L., Simeon, T., and LaValle, S. (2005). Dynamic-domain RRTs: efficient exploration by controlling the sampling domain. In *Proc. of IEEE Int. Conf. Robotics and Automation*.

15

Reactive Testing of Nondeterministic Systems by Test Purpose-Directed Tester

Jüri Vain, Andres Kull, Marko Kääramees, Maili Markvardt, and Kullo Raiend

CONTENTS

15.1   Introduction

15.1.1   On the fly test generation in online testing

Online testing is widely considered to be the most appropriate technique for model-based testing (MBT) of embedded systems where the implementation under test (IUT) is modeled using nondeterministic models [39, 38]. Nondeterminism stems partially from the internal parallel processes of the IUT, timing, and hardware-related asynchronous processes. Other sources of nondeterminism are the higher abstraction level of the model compared to IUT implementation and the ambiguities in the specifications of the IUT. We use the term on the fly in this context to describe the test generation and execution algorithms that compute and send successive stimuli to IUT incrementally at runtime, directed by the test purpose and the observed outputs of the IUT.

The state-space explosion problem experienced by many offline test generation methods is avoided by the on the fly techniques because only a limited part of the state space must be kept track of at any point in time when a test is running. On the other hand, exhaustive planning is difficult on the fly because of the limitations of the available computational resources at the time of test execution.

Thus, developing a planning strategy for industrial strength online testing should address the trade-off between reaction time and online planning depth to arrive at practically feasible test cases.

The simplest approach to on the fly selection of test stimuli in model-based online testing is to apply a so-called random walk strategy where no test sequence has an advantage over the others. The test is performed usually to discover violations of IOCO [35] or TIOCO [6] conformance relations between the IUT model and its implementation. Random exploration of the state space may lead to test cases that are unreasonably long and nevertheless may leave the test purpose unachieved. On the other hand, the advantage of long test cases may be revealed when a single test runs hours or even days. This allows some unexpected and intricate bugs that do not fit under well-defined coverage criteria to be detected.

To overcome the deficiencies of long-lasting testing, usually additional heuristics such as "anti-ant" [25, 29], dynamic approach of DART system [15], inserted assertions [21], path fitness [11], etc. are applied for guiding the exploration of the state space. In the extreme, guiding the selection of test stimuli degenerates to exhaustive planning by solving at each test execution step a full constraint system set by the test purpose and test planning strategy. For instance, the witness trace generated by model checking provides possibly optimal selection of the next test stimulus. The critical issue in the case of explicit state model checking algorithms is the size and complexity of the model leading to the explosion of the state space, especially in cases such as "combination lock" or deep-nested loops in the model [16]. Therefore, model checking-based approaches are used mostly in offline test generation.

In this chapter, we introduce the principle of reactive planning for on the fly selection of test stimuli and the reactive planning tester (RPT) synthesis algorithm for offline construction of those selection rules. As will be shown using experimental results at the end of the chapter, the reactive planning paradigm appears to be a balance between the tradeoffs of using a simple heuristics and the exhaustive planning methods under uncertainty.

15.1.2   Related work

For MBT, a model that represents the IUT specification is required. Finite state machine (FSM) and extended finite state machine (EFSM) are commonly used for the purpose of test case derivation [32]. An FSM can model the control flow of a system. In order to model a system that has both control and data parts (e.g., communication protocols), an extension is necessary. Such systems are represented using an EFSM model [20]. The EFSM model has been widely studied and many methods are available that employ different test data generation methods [22, 24]. Nevertheless, automated test data generation from EFSM model is complicated by the presence of infeasible paths and is an open research problem [31]. In an EFSM model, a given path can be classified as either infeasible or feasible. The existence of some infeasible paths is because of the variable dependencies among the actions and conditions. If a path is infeasible, there is no input test data that can cause this path to be traversed. Thus, if such a path is chosen in order to exercise certain transitions, these transitions are not exercised even if they can be exercised through other feasible paths [20]. While the feasibility of paths is undecidable, there are several techniques that handle them in certain special cases [8, 16, 31].

MBT can be applied for both offline and online generation of test cases. In case of online testing, the test generation procedure derives only one test input at a time from the model and feeds it immediately to the IUT as opposed to deriving a complete test case in advance like in offline testing. In online testing, it is not required to explore the entire state space of the model of the IUT any time the test stimulus is generated. Instead, the decisions about the next actions are made by observing the current output of the IUT [36]. However, online test execution requires more runtime resources for interpreting the model and choosing the stimulus. The online testing methods differ in the way how the test purpose is defined, how the test stimuli are selected on the fly, and the planning effort behind each choice.

The test purpose can be stated in its most abstract form when applying conformance testing. It presumes satisfaction of the IOCO [35] or TIOCO [6] relation [7] and is implemented using a completely random or some heuristic-driven state-space exploration algorithm. A test stimulus to trigger a next transition of the IUT model at a given state is randomly selected from a uniformly distributed set of stimuli. Random choice has been used in the early TorX tool [5], Uppaal–Tron [23, 28], and also in the on the fly testing mode of SpecExplorer [30]. In Reference 13 also, the transition probabilities-directed input selection method is introduced to TorX. More restrictive are the test goal-directed exploration algorithms that reduce the total number of states to be explored in a model. The goal-directed approach is stronger than random exploration in the sense of providing guidance toward a certain set of paths not just an arbitary path in the model. The goal-directed approach was introduced in References 14 and 21, used in testing tool formally elaborated in [10], and later used in TorX [34] and TGV [19].

A further development of the SpecExplorer approach has been introduced in NModel [39, 38], where the IUT model presented as a model program (Model programs in Spec Explorer is written in the high-level specification languages. AsmL or Spec\#) can be composed with scenario models to test certain scenarios that are subsets of all possible behaviors. An "anti-ant" [25]-based algorithm of reinforcement learning [29] is used to cover all specified test paths in the model program.

While distinguishing the on the fly test input selection methods by their planning capability (i.e., the so-called planning horizon), the simplest and fastest method is random choice. The planning horizon of the random choice is zero steps ahead. Nonzero but still short range planning is applied in anti-ant methods that try to avoid already visited paths. For selecting the next move from the possible ones, the anti-ant method takes the least visited one. The planning horizon of the anti-ant algorithm is just 1 step ahead. The anti-ant heuristic-based statespace exploration method is introduced in [23] and used in [25] to cover all transitions of the labeled transition system resulting in the exploration of a model program.

The reactive planning testing approach introduced in this chapter is goal directed like the method used in TorX and TGV. But it can be easily extended to the case where the test purpose is stated in the form of a scenario like in NModel. In addition to the IUT model, such an extension also requires a "test goal" model. Provided the IUT and "test goal" models, both are defined as EFSM; the executable tester automaton is derived as a standard automata product. The test goal model specifies the temporal order that the test items have to be visited in during the test run. The usage of "test goal" model clearly increases the expressiveness of test goals by allowing dynamic resetting test items and repetitive visits to test items when necessitated by the test scenario.

The main difference of RPT [37] compared to other on the fly techniques is that it involves a planner that looks ahead more than one step at a time to reach still unsatisfied parts of the test purpose, whereas the anti-ant approach looks only one step ahead when selecting the least visited outgoing transition from a current state.

Compared with random choice with planning horizon 0 steps ahead and "anti-ant" with 1 step ahead, the planning horizon of RPT can be parametrically tuned. That allows "seeing" the direction of unattended coverage items in the model. Additionally, the RPT is able to guide the model on the fly exploration toward still unexplored areas even in cases when they are "concealed" by parts of the model already traversed. Because of the longer planning horizon, RPT can result in shorter test sequences than random choice and anti-ant methods.

The payback of the shorter test cases is the complexity of the planning constraints to be solved online to guide the selection of test stimuli. Additionally, the advantage of RPT over the pure anti-ant and random choice methods depends on many other factors, such as the model size and structural complexity, the degree of nondeterminism, and the distribution of the coverage items over the model. Model size and complexity affect, in the first place, the size and complexity of the rules the RPT has to execute on the fly. Second, as the IUT specification model may be nondeterministic, it is impossible to predict exactly what paths appear to be more preferable in terms of test length and how long such test should take. Only under the fairness assumption all nondeterministic choices will eventually be possible and the test purpose potentially reachable. Thus, the degree of the nondeterminism of the IUT model is a factor that can render the exhaustive or even deeper planning worthless. For instance, the planning may target reaching a coverage item behind some nondeterministic choice point in the IUT model but because of the nondeterminism, the reachability of the coverage item is not achieved for the run of given length. If there are many nondeterministic choices on the path to the next coverage item, then the nondeterminism can direct the test execution to paths that are more likely but not preferred by the test purpose rules. Therefore, the distribution of coverage items in the model may affect the resulting test sequence length. Depending on the "risk management" strategy, the selection rules in RPT can prefer either potentially shorter paths to reach the test coverage items but including nondeterministic choices, or alternatively, deterministic paths that may be considerably longer but ensure the reachability of test coverage items.

15.2   Preliminaries of Model-Based Reactive Planning Testing

In this section, we shortly introduce the principle of reactive planning introduced in [40] and how the reactive planning can be transferred to model-based online testing.

15.2.1   Reactive planning testing in a nutshell

The concept of *reactive planning* as presented in [40] is motivated by the need for model-based autonomy in applications that must cope with highly dynamic and unpredictable environments. Reactive planning operates in a timely fashion and is applicable in agents operating in these conditions [27]. The tester is reactive in the sense that it reacts to observed outputs of the IUT and to possible changes in the test goals. It attempts to take the system toward the state that satisfies the desired test goals. A model-based executive uses a specification of the system to determine the desired state sequence in three stages - mode identification (MI), mode reconfiguration (MR), and model-based reactive planning (MRP) [40]. MI and MR set the planning problem, identifying initial and target states, while MRP reactively generates a plan solution. MI is a phase where the current state of the system model is identified. In the case of a deterministic transition, MI is trivial, it is just the next state reachable by applying the correct IUT input. In the nondeterministic case, MI can determine the current state by looking at the output of the system provided the output is observable. In the current approach, the MR and the MRP phases are combined into one since both the goal and the next step toward the goal are determined by the same decision tree as will be explained in detail in Section 15.3.4.1. The selection of IUT input to advance toward satisfying the test goal is based on the cost of applying a given input. Further, we characterize this cost using a so-called gain function.

The rationale behind the reactive planning method proposed in this approach lies in combining computationally hard offline planning with time-bounded online planning phases. The offline phase is intended to perform the combinatorially hard planning in advance as much as possible and to record the results of static analysis of a given IUT model and the test goal in the format of compact planning rules that are then applicable online. While the RPT is synthesized, the rules are encoded in the tester model and applied when the test is running. The planning rules synthesized must also ensure proper termination of the test case when a prescribed test purpose is satisfied.

15.2.2 Model-based testing with EFSMs

The key assumption of the RPT synthesis algorithm is that the IUT model is presented as an output observable nondeterministic state machine [26, 33] either in the form of an FSM or an EFSM in which all transition paths are feasible. In [12, 18], it has been shown how to transform an EFSM to one that has no infeasible paths. This has been achieved only for EFSMs in which all variable updates and transition guards are linear. In addition, the problem of determining whether a path in an EFSM is feasible is generally undecidable. Therefore, we limit our approach to EFSMs, which have only linear updates and transition guards.

A test purpose is a specific objective or a property of the IUT that the tester is set out to test. The test purpose is specified in terms of test coverage items. We focus on test purposes that can be defined as a set of "traps" associated with the transitions of the IUT model [16]. The goal of the tester is to generate a test sequence so that all traps are visited at least once during the test.

We synthesize the tester also as an EFSM where the rules for online planning derived during the tester synthesis are encoded into the transition guards of the EFSM as conjuncts called gain guards. The gain guard evaluates to **true** or **false** at the time of the execution of the tester determining if the transition can be taken from the current state or not. The value **true** means that taking the transition is the best possible choice to reach some unvisited traps from the current state. At each step, only the guards associated with the outgoing transitions of the current state of the EFSM are evaluated to select the next transition with the highest gain. Thus, the number of guard conditions that must be evaluated at each step is relatively small.

The gain guard is defined using gain functions and the standard function MAX over gain functions. The gain functions calculate the gain that is a quantitative measure necessary to compare alternative choices of test stimuli on the fly. For each transition that is controllable by the tester, a non-negative gain function is defined that depends on the current bindings of the EFSM context variables. The gain function of a transition computes a value that depends on the distance-weighted reachability of the unvisited traps from the given transition. The gain guard of the tester controllable transition is **true** only when the transition is a prefix of the test sequence with highest gain among those that depart from the current state. If gain functions of several transitions evaluate to the same maximum value, the tester selects one of these transitions either randomly or by the "least visited first" principle. Each transition of the model is considered to have a weight and the cost of testing is proportional to the length of the entire test sequence. Also, the current value (visited or not) of each trap is taken into account.

The resulting tester drives the IUT from one state to the next by generating inputs and by observing the outputs of the IUT. When generating the next input, the tester takes into account which traps have been visited in the model before. The execution of the decision rules at the time of the test execution is significantly faster than finding the efficient test path by state-space exploration algorithms yet still leads to the test sequence that is lengthwise close to optimal.

15.2.3   Extended finite state machine

Synthesis of the RPT-tester is based on a nondeterministic EFSM model of the IUT. The IUT model is restricted to a subclass of EFSMs where all possible sequences of transitions are feasible. In general, an EFSM model can contain infeasible sequences of transitions as the current configuration of context variables at some state may make some of the guards of the outgoing transitions from that state evaluate to false. There are algorithms described in [12, 18] for transforming an EFSM into a form where all paths are feasible.

Definition 1: An EFSM, M is defined as a tuple (S, V, I, O, E), where S is a finite set of states, V is a finite set of variables with finite value domains, I is the finite set of inputs, O is the finite set of outputs, and E is the set of transitions. A configuration of M is a pair (s, σ), where $s \in S$ and $\sigma \in \Sigma$ is a mapping from V to values, and Σ is a finite set of mappings from variable names to their possible values. The initial configuration is (s_0, σ_0), where $s_0 \in S$ is the initial state and $\sigma_0 \in \Sigma$ is the initial assignment. A transition $e \in E$ is a tuple $e = (s, p, a, o, u, q)$, where s is the source state of the transition, q is the target state of the transition $(s, q \in S)$, p is a transition guard that is a logic formula over V, a is the input of M $(a \in I)$, o is the output of M $(o \in O)$, and u is an update function over V.   □

A deterministic EFSM is an EFSM where the output and next state are unambiguously determined by the current state and the input. A nondeterministic EFSM may contain states where the reaction of the EFSM in response to an input is nondeterministic, that is, there are more than one outgoing transitions that are enabled simultaneously.

15.2.4   Modeling the IUT

In this approach, we assume that the model of the IUT is generally an EFSM with all paths to be covered by the test being feasible. We denote the IUT model by M_S. It can be either deterministic or nondeterministic, it can be strongly connected or not. If the model is not strongly connected, then we assume that there exists a reliable reset that allows the IUT to be taken back to the initial state from any state. Since for EFSM models where the variables have finite, countable domains, there exists a transformation from EFSM to FSM [17], and so we present the further details of the RPT synthesis method using simpler FSM model notation. In practice, this leads to a vast state space if we use FSMs even for a small system. Though, as will be demonstrated in Section 15.3.5 by means of RPT and adjustable planning horizon also a FSM-based test synthesis method can scale well to handle industrial size testing problems. Moreover, the transformation from EFSM to FSM is automatic in the method implementation [2] and hidden from the user.

It is essential that the tester can observe the outputs of the IUT for detecting the next state after a nondeterministic transition of the IUT. Therefore, we require that a nondeterministic IUT is output observable [26, 33], which means that even though there may be multiple transitions taken in response to a given input, the output identifies the next state unambiguously. An example of an output observable nondeterministic IUT model is given in Figure 15.1a. The outgoing transitions e_0 and e_1 (e_3 and e_4) of the state s_1 (s_2) have the same input a_0 (a_3), but different outputs o_0 or o_1 (o_3 or o_4).

15.2.5 Modeling the test purpose

A test purpose is a specific objective or a property of the IUT that the tester is set out to test. In general, test purposes are selected based on the correctness criteria stipulated by the specification of the IUT. The goal of specifying test purposes is to establish some degree of confidence that the IUT conforms to the specification. In MBT, the formal model of the IUT is derived from the specification and it is the starting point of the automatic test case generation. Therefore, it should be possible to map the test purposes derived from the specifications of the IUT into test purposes defined in terms of the IUT model. Examples of test purposes are "test a state change from state A to state B in a model," "test whether some selected states of a model are visited," "test whether all transitions are visited at least once in a model," etc. All of the test purposes listed above are specified in terms of the structural elements (coverage items) of the model that should be traversed (covered) during the execution of the test. A tester model is generated from the IUT model attributed with such test purpose. The tester runs until the test purpose is achieved and guides the IUT at runtime toward still unsatisfied parts of the test purpose.

For synthesizing a tester that fulfills a particular test purpose, we extend the original model of the IUT with traps and generate the tester from the extended model of the IUT. The traps are attached to the transitions of the IUT model and they can be used to define which model elements should be visited by the test.

The traps are implemented by trap variables and trap update functions. A trap variable is a Boolean variable initially set to *false*. The trap update functions are attached to the

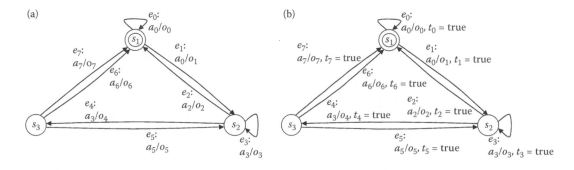

FIGURE 15.1

(a) An output observable nondeterministic IUT model. (b) The above model extended with trap variables.

transitions of the model and they are executed when the transition is visited during the execution of the test. The trap update functions are used to set trap variables to *true*, which denotes visited traps.

The extended model of the IUT M'_S is a tuple $(S_S, V'_S, I_S, O_S, E'_S)$. The extended set of variables V'_S includes variables of the IUT and the trap variables $(V'_S = V_S \cup T)$, where T is a set of trap variables. E'_S is a set of transitions where each element of E'_S is a tuple (s, p', a, o, u', q), where p' is a transition guard that is a logical formula over V'_S, and u' is an update function over V'_S. For the sake of brevity, we further denote the model of the IUT that is extended with trap variables by M_S.

Figure 15.1b presents an example where the IUT model given in Figure 15.1a is extended with trap variables. The example presents a *visit all transitions* test purpose, therefore the traps are attached to all transitions, $T = \{t_0, \ldots, t_7\}$. In this example, $V'_S = T$ and $p_k \equiv true$, $u_k \equiv t_k := true$ for each transition e_k, $k \in \{0, \ldots, 7\}$.

15.2.6   Model of the tester

The tester is synthesized from the extended IUT model M_S. Its structure is derived from the structural elements of M_S – states, transitions, variables, and update functions. We synthesize a tester EFSM M_T as a tuple $(S_T, V_T, I_T, O_T, E_T)$, where S_T is the set of tester states, V_T is the set of tester variables, I_T is the set of tester inputs, O_T is the set of tester outputs, and E_T is the set of tester transitions. Running the test presumes that the tester inputs are connected to the outputs of the IUT and the tester outputs are connected to the inputs of the IUT, that is, $I_T = O_S$ and $O_T = I_S$. Input–output compliance of tester and IUT automata means that the set of context variables of the tester is equal to the set of the context variables of the extended IUT model $(V_T = V'_S)$.

The tester has two types of states—active and passive. The set of active states S^a_T $(S^a_T \subset S_T)$ includes the states where the IUT is idle and the tester controls the test execution. The set of passive states S^p_T $(S^p_T \subset S_T)$ includes the states where the tester is idle and control is on the IUT side. The transitions $e_T \in E_T$ of the tester automaton are defined by a tuple $(s_T, p_T, a_T, o_T, u_T, q_T)$, where p_T is a transition guard that is a logical formula over V_T and u_T is an update function over V_T. We distinguish observable and controllable transitions of the tester. An observable transition e^o is a transition originating from a passive state of the tester. It is defined by a tuple $(s_T, p_T \equiv true, a_T, o_T \equiv nil, u_T, q_T)$, where s_T is a passive state, the transition is always enabled $(p_T \equiv true)$, and it does not expect any output from the tester. A controllable transition e^c is a transition originating from an active state of the tester. It is defined by a tuple $(s_T, p_T, a_T \equiv nil, o_T, u_T \equiv nil, q_T)$, where s_T is an active state, the transition does not contain an input, $p_T \equiv p_S \wedge p_g(V_T)$ is a guard of e^c constructed as a conjunction of the corresponding guard p_S of the extended IUT model M_S and the gain guard $p_g(V_T)$. The purpose of the gain guard is to guide the tester in selecting the next transition from the set of outgoing transitions of the current state to reach the next unvisited trap.

The gain guard must ensure that in the active state of the tester only those outgoing transitions are enabled that have the maximum gain. The enabled transition is the best choice in the sense of the path length from the current state toward fulfilling a still unsatisfied subgoal of the test purpose. We construct the gain guards $p_g(V_T)$ offline using the reachability analysis of the traps from the given transition. The gain guards take into account the number of still unvisited traps and also the gain (the distance weighted reachability) of the respective traps. The tester model can be nondeterministic in the sense that when there are many transitions with equal positive gain, the selection of the transition to be taken next is made randomly from the best choices.

15.3 Synthesis of RPT

This section highlights the basics of RPT synthesis workflow and, thereafter, provides details of each synthesis step.

15.3.1 RPT synthesis in large

The EFSM model of RPT is constructed from an EFSM model of the IUT, from the test purpose that is expressed in terms of trap variable updates attached to the transitions of the IUT model, and from the parameters that define the RPT planning constraints.

The RPT synthesis comprises three basic steps (Figure 15.2): (1) extraction of the RPT control structure, (2) constructing gain guards that includes also construction of gain functions, and (3) reduction of gain guards according to the parameter "planning horizon" that defines the depth of the reachability tree to be pruned.

In the first step, the RPT synthesizer analyzes the structure of the IUT model and generates the RPT control structure that is dual to the IUT control structure. In the second step, the synthesizer finds possibly successful IUT runs regarding the test goal. The tester should make its choice in the current state based on the structure of the tester model and the bindings of the trap variables representing the test goal. The decision rules for on the fly planning are derived by performing reachability analysis from the current state to all trap-equipped transitions by constructing the shortest-path trees. The decision rules are defined for controllable by tester transitions of the model and are encoded in transition guards as conjuncts called gain guards. The gain functions that are terms in the decision rules are derived from the shortest-paths trees in the IUT dual automaton. A shortest-paths tree is constructed for each controllable transition. The root vertex of the tree corresponds to the controllable transition being characterized by a gain function, other vertices present transitions equipped with traps. In case there are branches without traps in the tree that terminate with terminals labeled with traps, the branches are substituted with hyper-edges having weights equal to the length of that branch. By the given construction, the reduced shortest-paths tree (RSPT) represents the shortest paths from the root transition it

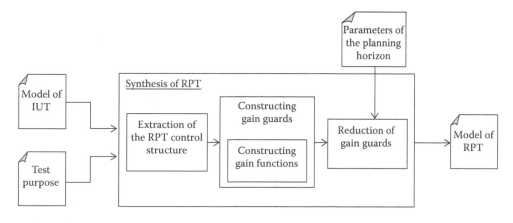

FIGURE 15.2
RPT synthesis workflow.

characterizes to all reachable trap-labeled transitions in the tester model. The gain function also allocates weights to the traps in the tree, and the closer to the root transition, the higher the weight that is given to the trap. Thus, the gain value decreases during the test execution after each trap in the tree is visited.

Since the RSPT in the IUT dual automaton has the longest branch proportional to the length of Euler's path in that automaton graph, the gain function's recurrent structure may be very complex. The final step of the synthesis reduces the gain functions by pruning the RSPT up to some depth that is defined by the "planning horizon" parameter. In Sections 15.3.2–15.3.5, the RPT synthesis steps are described in more detail.

15.3.2   Deriving control structure of the tester

The tester model is constructed as a dual automaton of the IUT model, where the inputs and outputs are reversed. The tester construction algorithm, Algorithm 1, has the following steps. The states of the IUT model are transformed into the active states of the tester model in step 1. For each state s of the IUT, the set of outgoing transitions $E_S^{out}(s)$ is processed in steps 2–5. The transitions of the IUT model are split into two transitions of the tester model—controllable transition $e_T^c \in E_T^c$ and observable transition $e_T^o \in E_T^o$, where E_T^c and E_T^o are the subsets of controllable and observable transitions of the tester. A new intermediate passive state s_p is added between them (steps 6–8).

Let $E_S^{out}(s,a,p)$ denote the subset of the nondeterministic outgoing transitions of the state s, where the IUT input is a and the guard is equivalent to p. The algorithm creates one controllable transition e_T^c for each set $E_S^{out}(s,a,p)$ from state s to the passive state s_p of the tester model (step 7). The controllable transition e_T^c does not have any input and the input of the corresponding transition of the IUT becomes an output of e_T^c.

For each element $e \in E_S^{out}(s,a,p)$, a corresponding observable transition e_T^o is created in steps 8 and 14, where the source state s of e is replaced by s_p, the guard is set to *true*, and the output of the IUT transition becomes the input of the corresponding tester transition.

The processed transition e of the IUT is removed from the set of outgoing transitions $E_S^{out}(s)$ (step 9). From the unprocessed set $E_S^{out}(s)$, the subset $E_S^{out}(s,a,p)$ of remaining nondeterministic transitions with the same input a and a guard equivalent to p is found (step 10). For each $e \in E_S^{out}(s,a,p)$, an observable transition e_T^o is created (steps 12–16).

The gain functions for all controllable transitions of the tester are constructed using the structure of the tester (steps 19–21). Finally, for each controllable transition, a gain guard $p_g(V_T)$ is constructed (step 24) and the conjunction of $p_g(V_T)$ and the guard of e_T^c is set to be the guard of the corresponding transition of the tester (step 25).

The details of the construction of the gain functions and gain guards are discussed in the next subsection.

An example of the tester EFSM created by Algorithm 1 is shown in Figure 15.3. The active states of the tester have the same label as the corresponding states of the IUT and the passive states of the tester are labeled with $s_4, \ldots, s_9$. The controllable (observable) transitions are shown with solid (dashed) lines. For example, the pair of nondeterministic transitions e_0, e_1 of the IUT (see Figure 15.1) produces one controllable transition (s_1, s_4) and two observable transitions from the passive state s_4 of the tester.

For this example, $V_T = T$, where T is the set of trap variables. For example, in Figure 15.3, $p_2^c(T)$ denotes the gain guard of the tester transition e_2^c. Gain guards attached to the controllable transitions of the tester (e.g., $p_2^c(T)$, $p_{34}^c(T)$) guide the tester at runtime to choose the next transition depending on the current trap variable bindings in T.

Algorithm 1 Build control structure of the tester

1: $E_T^c \leftarrow \emptyset$; $E_T^o \leftarrow \emptyset$; $S_T^a \leftarrow S_S$; $S_T^p \leftarrow \emptyset$; $I_T \leftarrow O_S$; $O_T \leftarrow I_S$; $V_T \leftarrow V_S$
2: **for all** $s \in S_S$ **do**
3:   find $E_S^{out}(s)$
4:   **while** $E_S^{out}(s) \neq \emptyset$ **do**
5:     get $e = (s,p,a,o,u,q)$ from $E_S^{out}(s)$
6:     add s_p to S_T^p {passive state}
7:     add $(s,p,\emptyset,a,\emptyset,s_p)$ to E_T^c {controllable transition}
8:     add $(s_p,true,o,\emptyset,u,q)$ to E_T^o {observable trans.}
9:     $E_S^{out}(s) \leftarrow E_S^{out}(s) - \{e\}$
10:     find $E_S^{out}(s,a,p)$ from $E_S^{out}(s)$
11:     $E_S^{out}(s) \leftarrow E_S^{out}(s) - E_S^{out}(s,a,p)$
12:     **while** $E_S^{out}(s,a,p) \neq \emptyset$ **do**
13:       get $e = (s,p,a,o,u,q)$ from $E_S^{out}(s,a,p)$
14:       add $(s_p,true,o,\emptyset,u,q)$ to E_T^o {observable trans.}
15:       $E_S^{out}(s,a,p) \leftarrow E_S^{out}(s,a,p) - \{e\}$
16:     **end while**
17:   **end while**
18: **end for**
19: **for all** $e \in E_T^c$ **do**
20:   construct gain function $g_e(V_T)$
21: **end for**
22: construct dual graph G of the tester model M_T
23: **for all** $e \in E_T^c$ **do**
24:   construct gain guard $p_{g_e}(V_T)$
25:   $p \leftarrow p \wedge p_{g_e}(V_T)$
26: **end for**

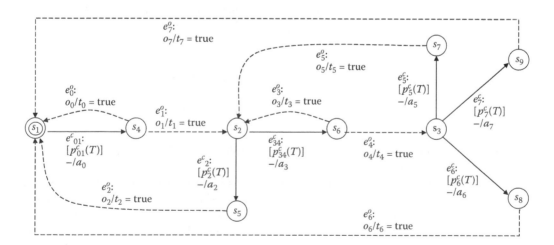

FIGURE 15.3
The EFSM model of the tester for the IUT in Figure 15.1.

15.3.3 Constructing gain guards of transitions

A gain guard $p_g(V_T)$ of a controllable transition of the tester is constructed to meet the following requirements:

- The next move of the tester should be locally optimal with respect to achieving the test purpose from the current state of the tester.

- The tester should terminate after all traps are achieved or all unvisited traps are unreachable from the current state.

The gain guard evaluates to *true* or *false* at the time of the execution of the tester determining if the transition can be taken from the current state or not. The value *true* means that taking the transition is the best possible choice to reach some unvisited traps from the current state. The tester makes its choice in the current state based on the structure of the tester model, the bindings of the trap variables representing the test purpose, and the current bindings of the context variables. We need some quantitative benefit measures to compare different alternative choices. For each controllable transition, $e \in E_T^c$, where E_T^c is the set of all controllable transitions of the tester, we define a non-negative gain function $g_e(V_T)$ that depends on the current bindings of the context variables. The gain function has the following properties:

- $g_e(V_T) = 0$, if taking the transition e from the current state with the current variable bindings does not lead to a state that is closer to any unvisited trap. This condition indicates that it is useless to fire the transition e.

- $g_e(V_T) > 0$, if taking the transition e from the current state with the current variable bindings visits or leads to a state that is closer to at least one unvisited trap. This condition indicates that it is useful to fire the transition e.

- For transitions e_i and e_j with the same source state, $g_{e_i}(V_T) > g_{e_j}(V_T)$ if taking the transition e_i leads to an unvisited trap with smaller cost than taking the transition e_j. This condition indicates that it is cheaper to take the transition e_i rather than e_j to reach the next unvisited trap.

A gain guard for a controllable transition e with the source state s of the tester is defined as

$$p_{g_e}(V_T) \equiv g_e(V_T) = \max_{e_k} g_{e_k}(V_T) \ \wedge \ g_e(V_T) > 0, \tag{15.1}$$

where g_{e_k} is the value of the gain function of the transition $e_k \in E_T^{out}(s)$, in which $E_T^{out}(s) \subset E_T^c$ is the set of outgoing transitions of the state s.

The first predicate in the logical formula (15.1) ensures that the gain guard is *true* only for a transition that leads to some unvisited trap from the current state with the highest gain compared to the gains of the other outgoing transitions of the current state. The second predicate blocks test runs that do not serve the test purpose. The second predicate evaluates to *false* when all unvisited traps from the current state are unreachable or all traps are already visited.

The gain guard of the tester transition enables one or more controllable transitions that should be taken at the subsequent move. If several gain functions evaluate to the same maximum value, the tester selects one of the best transitions at random.

15.3.4 Gain function

In this subsection, we describe how the gain functions are constructed. The required properties of a gain function were specified in the previous subsection. Each transition of the IUT model is considered to have unit weight and the cost of testing is proportional to the length of the test sequence. The gain function of a transition computes a value that depends on the distance-weighted reachability of the unvisited traps from the given transition.

For the sake of efficiency, we implement a heuristic in the gain function that prefers the selection of the path that visits more unvisited traps and is shorter than the alternative ones. Intuitively, in the case of two paths visiting the same number of transitions with unvisited traps and having the same lengths, the path with more traps closer to the beginning of the path is preferred.

In this subsection, $M = (S, V, I, O, E)$ denotes the tester model equipped with trap variables and $e \in E$ is a transition of the tester. We assume that the trap variable $t \in T$ is initialized to *false* and set to *true* by the trap update function u_t associated with the transition e. Therefore, reaching a trap is equivalent to reaching the corresponding transition. A transition e_j is reachable from the transition e_i if there exists a sequence of transitions $\langle e_i, ..., e_j \rangle$, where $e_i, e_j \in E$.

15.3.4.1 Shortest-paths tree

To find the reachable transitions from a given transition, we reduce the reachability problem of the transitions to a single-source shortest-paths problem of a graph [9]. We create a dual graph $G = (V_D, E_D)$ of the tester model as a graph where the vertices V_D correspond to the transitions of the EFSM of the tester, $V_D \cong E$. The edges E_D of the dual graph represent the pairs of subsequent transitions sharing a state in the tester model. If the transition e_i of the tester model is an incoming transition of a state and the transition e_j is an outgoing transition of the same state, there is an edge (e_i, e_j) in the dual graph from vertex e_i to vertex e_j, $(e_i, e_j) \in E_D$.

The analysis of the transition sequences of the tester model M is equivalent to the analysis of the paths of vertices in the dual graph G. In Figure 15.4, there is the dual graph of the tester model in Figure 15.3. For example, after taking the transition e_{01}^c in Figure 15.3, it is possible that either e_0^o or e_1^o follows. In the dual graph in Figure 15.4, this is represented by the existence of the edges to e_0^o and e_1^o from the vertex e_{01}^c.

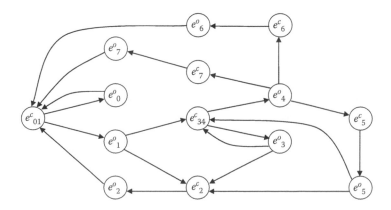

FIGURE 15.4
The dual graph of the tester model in Figure 15.3.

In the dual graph, the shortest-paths tree from e is a tree with the root e that contains the shortest paths to every other vertex that is reachable from e. The shortest-paths tree with the root e derived from the graph G is denoted by $SPT(e, G)$. The shortest-paths tree from a given vertex of the dual graph can be found using known algorithms from graph theory. Running a single-source shortest-paths algorithm for each controllable transition of the tester $|E^c|$ times results in the shortest paths from each controllable transition to every reachable transition.

The dual graph G is an unweighted graph (in this chapter, we do not differentiate between transitions). The breadth-first-search algorithm (e.g., see [9]) is a simple shortest-paths search algorithm that can be applied to unweighted graphs. For a vertex e of the dual graph G, the algorithm produces a tree that is the result of merging the shortest paths from the vertex e to each vertex reachable from it. As we constructed the dual graph such that the vertices of the dual graph correspond to the transitions of the tester model, the shortest path of vertices in the dual graph is the shortest sequence of transitions in the tester model. Each shortest path contains only distinct vertices. Note that the shortest paths and the shortest-paths trees of a graph are not necessarily unique.

The tree $SPT(e, G)$ represents the shortest paths from e to all reachable vertices of G. We assume that the traps of the IUT model are initialized to *false* and a trap variable t is set to *true* by an update function u associated with the transition of the original graph. Therefore, the tree $SPT(e, G)$ represents also the shortest paths starting with the vertex e to all reachable trap assignments. Not all transitions of the tester model contain trap variable update functions. For the evaluation of the reachability of the traps by the paths in the tree $SPT(e, G)$, we can reduce the tree $SPT(e, G)$ to the RSPT denoted by $TR(e, G)$ that includes the root vertex e and only such vertices of $SPT(e, G)$ that contain trap updates. We construct $TR(e, G)$ by replacing those subpaths of $SPT(e, G)$ that do not include trap updates by hyper-edges. A hyper-edge is the shortest subpath containing vertices without trap assignments between the root and a vertex with a trap assignment or two vertexes with trap assignments in the shortest-paths tree.

The RSPT $TR(e, G)$ contains the shortest paths beginning with the transition e to all reachable traps in the dual graph G (and thus also in the tester model). In the reduction of the shortest-paths tree $SPT(e, G)$ to $TR(e, G)$, we label each vertex that contains a trap variable update u_t by the corresponding trap t and replace each subpath containing vertices without trap updates by a hyper-edge (t_i, w, t_j), where t_i is the label of the source vertex, t_j is the label of the destination vertex, and w is the length of the subpath. Also, during the reduction, we remove those subpaths (hyper-edges) that end in the leaf vertices of the tree that do not contain any trap variable updates.

Figure 15.5 (left-hand side) shows the shortest-paths tree $SPT(e_{01}^c, G)$ with the root vertex e_{01}^c for the dual graph in Figure 15.4. For example, the path $\langle e_{01}^c, e_1^o, e_{34}^c, e_4^o, e_6^c, e_6^o \rangle$ from the root vertex e_{01}^c to the vertex e_6^o in the shortest-paths tree in Figure 15.5 is the shortest sequence of transitions beginning with the transition e_{01}^c that reaches e_6^o in the example of the tester model in Figure 15.3.

The RSPT $TR(e_{01}^c, G)$ from the vertex e_{01}^c to the reachable traps for the dual graph in Figure 15.4 is represented in Figure 15.5 (right-hand side). All vertices except the root of the RSPT $TR(e_{01}^c, G)$ are labeled with the trap variables, and the hyper-edges between the vertices are labeled with the number of transitions in the subpath that the hyper-edge represents. The tree $TR(e_{01}^c, G)$ contains the shortest paths beginning with the transition e_{01}^c to all traps in the tester model in the Figure 15.3. For example, the tree $TR(e_{01}^c, G)$ shows that there exists a path beginning with the transition e_{01}^c to the trap t_6, and this path visits traps t_1 and t_4 on the way.

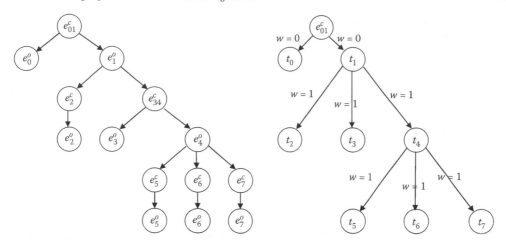

FIGURE 15.5
The shortest-paths tree $SPT(e_{01}^c, G)$ (left-hand side) and the reduced shortest-paths tree $TR(e_{01}^c, G)$ (right-hand side) from the transition e_{01}^c of the graph shown in Figure 15.4.

15.3.4.2 Algorithm of constructing the gain function

The return type of the gain function is non-negative rational $\mathbb{Q}^+$. This follows explicitly from the construction rules of the gain function (see steps below) and from the notion that the set of rational numbers is closed under the addition and *max* operations. The algorithm of construction of the gain function for the transition e of the tester automaton M from the dual graph G is the following:

1. Construct the shortest-paths tree $SPT(e, G)$ for the transition e of the dual graph G.

2. Reduce the shortest-paths tree $SPT(e, G)$ as described in subsection 15.3.4.1. The reduced tree is denoted by $TR(e, G)$. Assign the length of the subpath of each hyper-edge (t_i, w, t_j) of $TR(e, G)$ to w.

3. Represent the reduced tree $TR(e, G)$ as a set of elementary subtrees of height 1 where each elementary subtree is specified by the production rule of the form

$$\nu_i \rightarrow |_{j \in \{1, \ldots, k\}} \nu_j, \tag{15.2}$$

where the nonterminal symbol ν_i denotes the root vertex of the subtree and each ν_j (where $j \in \{1, \ldots, k\}$) denotes a leaf vertex of that subtree, and where k is the branching factor. The vertex ν_0 corresponds to the root vertex e of the reduced tree.

4. Rewrite the right-hand sides of the productions constructed in step 3 as arithmetic terms, thus obtaining the production rule in the form

$$\nu_i \rightarrow (\neg t_i)^\uparrow \cdot \frac{c}{d(\nu_0, \nu_i) + 1} + \max_{j=1, k}(\nu_j), \tag{15.3}$$

where $t_i^\uparrow$ denotes the trap variable t_i lifted to type $\mathbb{N}$, c is a constant for the scaling of the numerical value of the gain function, and $d(\nu_0, \nu_i)$ the distance between vertices ν_0 and ν_i in the labeled tree $TR(e, G)$. The distance is defined by the formula

$$d(\nu_0, \nu_i) = l + \sum_{j=1}^{l} w_j,$$

where l is the number of hyper-edges on the path between ν_0 and ν_i in $TR(e, G)$ and w_j is the value of the label w corresponding to the concrete hyper-edge.

5. For each symbol ν_i denoting a leaf vertex in $TR(e, G)$ define a production rule:

$$\nu_i \rightarrow (\neg t_i)^\uparrow \cdot \frac{c}{d(\nu_0, \nu_i) + 1}. \tag{15.4}$$

6. Apply the production rules (15.3) and (15.4) starting from the root symbol ν_0 of $TR(e, G)$ until all nonterminal symbols ν_i are substituted with the terms that include only terminal symbols $t_i^\uparrow$ and $d(\nu_0, \nu_i)$. Note that $i \in \{0, \dots, n\}$, where n is the number of trap variables in $TR(e, G)$. The root vertex $\nu_0 = e$ of the labeled tree $TR(e, G)$ may not have a trap label. Instead of a trap variable t_i, use a constant $true$ as the label resulting $(\neg true)^\uparrow = 0$ in the rule (15.3).

It must be brought to attention that the gain function characterizes the expected gain only within the planning horizon. The planning horizon is determined by the lengths of the paths in RSPT.

Table 15.1 shows the results of the application of the production rules (15.2), (15.3), and (15.4) to the vertices of the RSPT $TR(e_{01}^c, G)$ in Figure 15.5 (right-hand side). As the root e_{01}^c is not labeled with a trap variable, the transition e_{01}^c does not update any trap, a constant $true$ is used in the production rule (15.3) in the place of the trap variable resulting in $(\neg true)^\uparrow = 0$ in the first row of Table 15.1. Application of the production rules (15.3) and (15.4) to the tree $TR(e_{01}^c, G)$ starting from the root vertex e_{01}^c results in the gain function given in the first row of Table 15.2. Table 15.2 presents the gain functions for the controllable transitions of the tester model (Figure 15.3). The gain guards for all controllable transitions of the tester model are given in Table 15.3. The type lifting functions of the traps have been omitted from the tables for the sake of brevity.

15.3.5 Adjustable planning horizon

Since the gain functions are constructed based on RSPTs, their complexity is in direct correlation with the size of the RSPTs. Consequently, the criterion for covering all transitions sets the number of traps equal to the number of transitions in the IUT model. Considering the fact that the number of transitions in the full-scale IUT model may reach hundreds

TABLE 15.1

Application of the Production Rules to the Elementary Subtrees of Height 1 of the Reduced Shortest Paths tree $TR(e_{01}^c, G)$

Subtree	Production Rule (15.2)	Production Rule (15.3) for Nonleaf or Production Rule (15.4) for Leaf Vertex
e_{01}^c	$e_{01}^c \rightarrow t_0, t_1$	$e_{01}^c \rightarrow 0 \cdot c/1 + max(t_0, t_1)$
t_0	$t_0 \rightarrow$	$t_0 \rightarrow \neg t_0 \cdot c/2$
t_1	$t_1 \rightarrow t_2, t_3, t_4$	$t_1 \rightarrow \neg t_1 \cdot c/2 + max(t_2, t_3, t_4)$
t_2	$t_2 \rightarrow$	$t_2 \rightarrow \neg t_2 \cdot c/4$
t_3	$t_3 \rightarrow$	$t_3 \rightarrow \neg t_3 \cdot c/4$
t_4	$t_4 \rightarrow t_5, t_6, t_7$	$t_4 \rightarrow \neg t_4 \cdot c/4 + max(t_5, t_6, t_7)$
t_5	$t_5 \rightarrow$	$t_5 \rightarrow \neg t_5 \cdot c/6$
t_6	$t_6 \rightarrow$	$t_6 \rightarrow \neg t_6 \cdot c/6$
t_7	$t_7 \rightarrow$	$t_7 \rightarrow \neg t_7 \cdot c/6$

TABLE 15.2
Gain Functions of the Controllable Transitions of the Tester Model

Transition	Gain Function for the Transition
e_{01}^c	$g_{e_{01}^c}(T) \equiv c \cdot max(\neg t_0/2, \neg t_1/2 + max(\neg t_2/4,$ $\neg t_3/4, \neg t_4/4 + max(\neg t_5/6, \neg t_6/6, \neg t_7/6)))$
e_2^c	$g_{e_2^c}(T) \equiv c \cdot (\neg t_2/2 + max(\neg t_0/4,$ $\neg t_1/4 + max(\neg t_3/6, \neg t_4/6 + max(\neg t_5/8, \neg t_6/8, \neg t_7/8))))$
e_{34}^c	$g_{e_{34}^c}(T) \equiv c \cdot max(\neg t_3/2 + \neg t_2/4 + max(\neg t_0/6, \neg t_1/6),$ $\neg t_4/2 + max(\neg t_5/4, \neg t_6/4, \neg t_7/4))$
e_5^c	$g_{e_5^c}(T) \equiv c \cdot (\neg t_5/2 + max(\neg t_2/4 + max(\neg t_0/6, \neg t_1/6),$ $\neg t_3/4, \neg t_4/4 + max(\neg t_6/6, \neg t_7/6)))$
e_6^c	$g_{e_6^c}(T) \equiv c \cdot (\neg t_6/2 + max(\neg t_0/4,$ $\neg t_1/4 + max(\neg t_2/6, \neg t_3/6, \neg t_4/6 + max(\neg t_5/8, \neg t_7/8))))$
e_7^c	$g_{e_7^c}(T) \equiv c \cdot (\neg t_7/2 + max(\neg t_0/4,$ $\neg t_1/4 + max(\neg t_2/6, \neg t_3/6, \neg t_4/6 + max(\neg t_5/8, \neg t_6/8))))$

TABLE 15.3
Gain Guards of the Transitions of the Tester Model

Transition	Gain Guard Formula for the Transition
e_{01}^c	$p_{01}^c(T) \equiv g_{e_{01}^c}(T) = max(g_{e_{01}^c}(T)) \wedge g_{e_{01}^c}(T) > 0$
e_2^c	$p_2^c(T) \equiv g_{e_2^c}(T) = max(g_{e_2^c}(T), g_{e_{34}^c}(T)) \wedge g_{e_2^c}(T) > 0$
e_{34}^c	$p_{34}^c(T) \equiv g_{e_{34}^c}(T) = max(g_{e_2^c}(T), g_{e_{34}^c}(T)) \wedge g_{e_{34}^c}(T) > 0$
e_5^c	$p_5^c(T) \equiv g_{e_5^c}(T) = max(g_{e_5^c}(T), g_{e_6^c}(T), g_{e_7^c}(T)) \wedge g_{e_5^c}(T) > 0$
e_6^c	$p_6^c(T) \equiv g_{e_6^c}(T) = max(g_{e_5^c}(T), g_{e_6^c}(T), g_{e_7^c}(T)) \wedge g_{e_6^c}(T) > 0$
e_7^c	$p_7^c(T) \equiv g_{e_7^c}(T) = max(g_{e_5^c}(T), g_{e_6^c}(T), g_{e_7^c}(T)) \wedge g_{e_7^c}(T) > 0$

or even more, the gain functions generated using RSPTs may grow beyond a size that is feasible to compute during test execution time. To keep the online computation time within acceptable limits, RSPT pruning is added to the RPT synthesis technique. The planning horizon defines the depth of the RSPT to be pruned. Although the pruning of RSPT causes online planning to be incomplete, it renders the RPT method fully scalable regardless of the size of the IUT model and the test goal. Moreover, there is an option to set automatically the planning horizon when specifying at offline computing the upper size limit to the RSPT pruned. Pruning of RSPT reduces the resolution capability of RPT gain functions.

To resolve the priority conflicts between transitions having equal maximum gain values, RPT uses either random or anti-ant choice mechanisms. Both conflict resolution approaches are demonstrated on the City Lighting Controller case study and details discussed in Section 15.5.4.

15.3.6 Complexity of constructing the tester

The complexity of the synthesis of the RPT is determined by the complexity of the construction of the gain functions. For each gain function, the cost of finding the shortest-paths tree for a given transition in the dual graph by breadth-first-search is $O(|V_D| + |E_D|)$ [9], where $|V_D| = |E_T|$ is the number of transitions and $|E_D|$ is the number of transition pairs of the tester model. The number of transition pairs of the tester model is mainly defined by the number of transition pairs of the observable and controllable transitions, which is bounded by $|E_S|^2$. For all controllable transitions of the tester, the upper bound of the complexity of the offline computations of the gain functions is $O(|E_S|^3)$.

At runtime, each choice by the tester takes no more than $O(|E_S|^2)$ arithmetic operations to evaluate the gain functions for the outgoing transitions of the current state.

15.4   Testing Framework for Online Testing with RPT

The testing framework (Figure 15.6) for online testing with RPT consists of the following tools:

- UML CASE tool to build the model of the implementation.

- Test generator tool to generate the abstract RPT from the model and to transform the abstract RPT to TTCN-3 presentation.

- TTCN-3 test tool to execute RPT in TTCN-3 language against the IUT.

In this study, the UML CASE tools Poseidon from Gentleware [4] and Artisan Studio from Artisan Software [3] were used for the modeling. Test generator tool MOTES [2] from Elvior developed in cooperation with TUT was used to generate the abstract RPT and to transform the abstract RPT into TTCN-3 language. In the case studies and performance evaluations of the RPT method, the MessageMagic TTCN-3 test tool [1] from Elvior was used to execute the RPT implementation in TTCN-3 against the IUT.

15.4.1   Creating the IUT model

The model of the IUT for MOTES consists of the EFSM, descriptions of the context variables, and interface specifications.

EFSM is used to model the behavior of the IUT. Because MOTES does not have its own editor for EFSMs, third-party UML CASE tools are used to create them. MOTES is able to import state machines from Poseidon for UML [4] and Artisan Studio [3] CASE tools. EFSMs for MOTES are drawn as flat UML state machine models without parallel

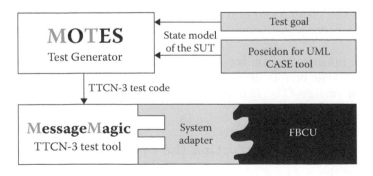

FIGURE 15.6
Testing framework in FBCU case study.

and hierarchical states. UML state machines do not have a formally specified language for the presentation of guard conditions, input events, and actions on transitions. Because it is only possible to generate tests from the formal system model, a formal transition language was developed for MOTES, which is used in transitions of the EFSM.

Context variables are the variables that are used in EFSM to store state information in addition to the control states of the EFSM.

Interface specifications define the input/output ports used in the model. The ports define the IUT interface toward its environment. Port definitions define the data type of the messages that the ports accept. In MOTES, the context variables and ports are defined directly in TTCN-3 files.

15.4.2 Preparing test data

The RPT is automatically generated from the IUT model by MOTES. The test data used by the RPT implementation should be prepared manually. Like context variables and interface definitions, the test data are also defined and instantiated directly in TTCN-3.

15.4.3 Importing the model and test data into MOTES

The test project browser view in MOTES includes a predefined structure of artifact folders for the maintaining of a clear structure for the test generation project.

15.4.4 Defining test purpose

MOTES uses model structural test coverage elements to define test coverage. The following coverage criteria are available:

- Selected elements (states/transitions)

- All transitions

- All n-transition sequences where $n > 2$

Selected elements (selected states and *transitions)* used to define test coverage can be selected from the list of EFSM transitions and states. It is possible to create ordered sets of coverage items and to define how many times each coverage item or a subset of the item should be covered. Complex test scenarios can be created using ordered sets of coverage items. For example, it is possible to define that the test generator should find a test that first covers transition T1, then covers transitions T2 and T3 three times, and finally covers transition T1 once again.

All transitions test coverage defines that the test generator should find a test that covers all of the transitions of the EFSM at least once.

All n-transition sequences is a test coverage criterion that allows long and exhaustive tests that cover all subsequent transition sequences of n transitions to be created. In MOTES, n can be 2 or 3.

15.4.5 Choosing the test generation engine

MOTES includes two test generation engines:

- Model checking engine

- RPT engine

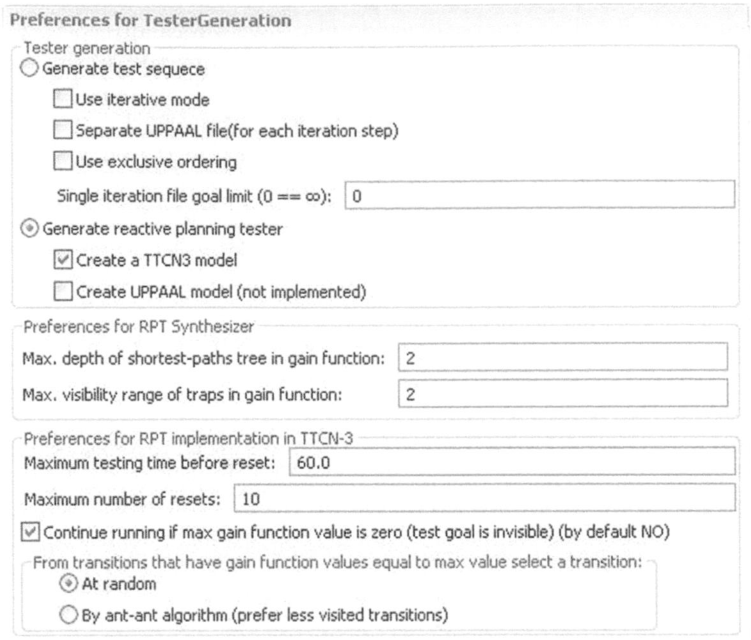

FIGURE 15.7
MOTES test generation preferences.

The test generation preferences view in MOTES is presented in Figure 15.7. The model checking engine can be selected using the radio button **Generate test sequence** and the RPT engine can be selected using the radio button **Generate reactive planning tester**.

The RPT engine implements the test generation method presented in earlier sub-chapters. It does not generate predefined test cases, but rather a RPT which, in the online phase, generates test stimuli on the fly. The RPT can be used to generate tests for deterministic or nondeterministic IUT. The RPT engine uses the reachability analysis in synthesizing the RPT, which is generated in TTCN-3 and can be executed by any TTCN-3 executive. The intelligence of the RPT is encoded in gain functions that characterize the potential choices of test stimuli. In the online testing phase, the RPT must decide which move to make from those that are possible. For each possible move, it calculates the gain function and makes the move that promises maximum gain. The gain function returns a higher value if the next move will lead to a larger amount of unvisited coverage items faster.

The RPT engine can be configured using the following configuration parameters (see also Figure 15.7):

- *Max depth of shortest-paths tree in gain function*. This parameter defines the look-ahead planning horizon. It also defines the maximum path length of the shortest path starting from the root of the shortest-path tree. In other words, it defines the length of the planning horizon in transitions where the RPT looks ahead. The gain functions are calculated taking into account the number of traps and their location in the tree. Planning within the horizon is precise.

- *Max visibility range of traps in gain function*. This visibility parameter defines the visibility horizon of the traps. Within the visibility range, only the reachability of the traps is important, not the exact location, as in the previous parameter. The visibility parameter can also be larger than the planning horizon. In the event that the trap is outside of the planning horizon but within the visibility horizon, the planning algorithm knows that the trap is reachable from the given shortest-path tree but does not know exactly how far away the trap is.

- *Max testing time before reset.* This parameter defines the maximum testing time for a situation where some of the traps have yet to be visited. In nondeterministic models, restarting the IUT may increase the chance of visiting unvisited traps after reset.

- *Max number of resets.* This parameter defines the maximum number of IUT resets that are performed in trying to cover traps that have yet to be covered.

- *Continue running if max gain function value is zero.* This parameter determines whether the tester should continue running when the test goal is invisible or not.

- *From transitions that have gain function values equal to max value select a transition.* This parameter determines the algorithm (random or anti-ant) for selecting the next transition from the alternatives if maximal gain values are equal.

15.4.6 Executing the test generation engine

Once the steps above are completed, the user presses the "G" button and the test generator does the rest. The TTCN-3 test cases or RPT TTCN-3 code are generated. The generated TTCN-3 files can be imported to any TTCN-3 test tool and run against the IUT.

15.4.7 Executing RPT implementation against the IUT

RPT implementation in TTCN-3 can be executed by any TTCN-3 test executive against the IUT. In this study, the MessageMagic TTCN-3 test tool [1] from Elvior was used. MessageMagic supports all TTCN-3 language versions up to the latest (version 4) and TTCN-3 standardized interfaces toward IUT, codecs, and test platforms.

The implementation of RPT includes sufficient information to allow the tester to run the test with the nondeterministic IUT. The tester implementation knows how to stimulate the IUT in each active state of the tester in order to complete the test run successfully. Meeting a test purpose is equivalent to having visited all traps. A test run is started with the goal to visit all specified traps. Execution of the tester starts from the initial state of the tester with the initial values of the context variables. If the current state of the tester is passive, then the tester observes the output of the IUT and takes a transition that matches the output of the IUT. If the current state is active, then the tester evaluates the gain guards of the outgoing transitions of the current state and takes a transition where the gain guard evaluates to *true*. If there is more than one such transition, then one of them is selected randomly. This can happen only if the gain functions of the transitions return an equal gain value meaning that such alternative choices are equally good.

If in the current state of the tester no unvisited traps are reachable, then all the gain functions evaluate to zero, the gain guards become disabled, and the tester run terminates. In the case of a nondeterministic IUT, the coverage of all traps depends on the nondeterministic choices of the IUT. Under the fairness assumption (all nondeterministic transitions are eventually chosen) and when all traps are reachable, the test run will eventually terminate. In practice, the testing time is limited and therefore a test duration limit is specified in addition to the terminationMessageMag condition, where all traps should be visited.

In case the IUT model is not strongly connected, the IUT may arrive at a state from which it is impossible to reach the remainder of the unvisited traps. In this case, the IUT must be reset and the execution of the tester is restarted to visit the remaining unvisited traps. The procedure is repeated till all traps are visited or the limiting test time has passed.

15.5 Performance Evaluation of RPT Using Case Study Experiments

The experiments documented in this section were performed to demonstrate the feasibility of the RPT method and to compare its performance with the random choice and anti-ant methods using an industry scale case study.

15.5.1 The case study

The testing case study developed under the ITEA2 D-MINT project [19] evaluates the MBT technology in the telematics domain. The IUT of the case study is a Feeder Box Control Unit (FBCU) of the street lighting subsystem. The most important functionality of the FBCU is to control street lighting lamps either locally, depending on the local light sensor and calendar, or remotely from the monitoring center. In addition, the controller monitors the feeder box alarms and performs power consumption measurements. The communication between the controller and monitoring center is implemented using GSM communication. The RPT performance evaluation experiments are performed on the power-up procedure of the FBCU.

15.5.2 Model of the IUT

The IUT model implements the power-up scenario of the FBCU. The strongly connected state transition model of the FBCU includes 31 states and 78 transitions. The model is non-deterministic. Pairs of nondeterministic transitions depart from seven states of the model and a triple of nondeterministic transitions departs from one state of the model. The minimum length of the sequences of transitions from the initial state to the farthest transition is 20 transitions, that is, the largest depth of the RSPT for any transition is 20. The model is similar to the model of combination lock with several nested loops. There are a number of possibilities for a scenario that was successful up to a given point to return back to the initial state of the tester model. For example, decreasing the power supply value below the normal level interrupts the power-up procedure.

15.5.3 Planning of experiments

In order to demonstrate the algorithms in different test generation conditions, we varied the test coverage criterion. The tests were generated using two different coverage criteria—all transitions and a single selected transition. The single transition was selected to be the farthest one from the initial state. The location of the single transition was selected on the limit of the maximum planning horizon. Different RPT planning horizons (0–20 steps) were used in the experiments. In case the RPT planning resulted in several equally good subsequent transitions in the experiment with the selected coverage criterion and planning horizon, we used alternatively the anti-ant and random choice methods for choosing the next transition. If the planning horizon is zero, then RPT works like the pure random choice or anti-ant method depending on the option selected in the experiment.

As a characteristic of scalability, we measured the length of test sequences and time spent online on each planning step. The time spent planning is only partially indicative of scalability because it depends on the performance of the RPT executing platform. Still, the measurements provide some insight into the scalability of the method with respect to the planning horizon. In addition to the nondeterministic model, there is always a random component involved in the RPT planning method. Therefore, we performed all experiments in series of 30 measurements and calculated averages and standard deviations over the series.

15.5.4 Results and interpretation of the experiments

The experiments are summarized in Table 15.4. The lengths of the test sequences are given in the form average ± standard deviation of 30 experiments. The results in the first row of Table 15.4 with planning horizon 0 correspond to the results of the pure anti-ant and random choice methods. For estimation of the minimum test sequence length, we modified the examined nondeterministic model to the corresponding deterministic model with the same structure. Eliminating the nondeterminism in the model by introducing mutually exclusive transition guards and using the maximum planning horizon 20, the RPT generated the test sequence with length 207 for all transitions coverage criteria on the modified deterministic model. The minimum length of the test sequence to reach the single selected transition was 20 steps.

The experiment shows that the RPT with maximum planning horizon results on average in a test sequence many times shorter and a considerably lower standard deviation than the anti-ant and random choice tester. For the test goal to cover all transitions of the nondeterministic model, the RPT generated an average test sequence 1.5 times longer than the minimum possible sequence. The difference from the optimum is mainly because of the nondeterminism in the model. Compared to the RPT with the maximum planning horizon,

TABLE 15.4

Average Lengths of the Test Sequences in the Experiments

	All Transitions Test Coverage		Single Transition Test Coverage	
	Planning with Horizon		Planning with Horizon	
	Anti-ant	Random Choice	Anti-ant	Random Choice
0	18345 ± 5311	44595 ± 19550	2199 ± 991	4928 ± 4455
1	18417 ± 4003	19725 ± 7017	2156 ± 1154	6656 ± 5447
2	5120 ± 1678	4935 ± 1875	1276 ± 531	2516 ± 2263
3	4187 ± 978	3610 ± 2538	746 ± 503	1632 ± 1745
4	2504 ± 815	2077 ± 552	821 ± 421	1617 ± 1442
5	2261 ± 612	1276 ± 426	319 ± 233	618 ± 512
6	2288 ± 491	1172 ± 387	182 ± 116	272 ± 188
7	1374 ± 346	762 ± 177	139 ± 74	147 ± 125
8	851 ± 304	548 ± 165	112 ± 75	171 ± 114
9	701 ± 240	395 ± 86	72 ± 25	119 ± 129
10	406 ± 102	329 ± 57	73 ± 29	146 ± 194
11	337 ± 72	311 ± 58	79 ± 30	86 ± 59
12	323 ± 61	284 ± 38	41 ± 15	74 ± 51
13	326 ± 64	298 ± 44	34 ± 8	48 ± 31
14	335 ± 64	295 ± 40	34 ± 9	40 ± 23
15	324 ± 59	295 ± 42	25 ± 4	26 ± 5
16	332 ± 51	291 ± 52	23 ± 2	24 ± 3
17	324 ± 59	284 ± 32	22 ± 2	21 ± 1
18	326 ± 66	307 ± 47	21 ± 1	21 ± 1
19	319 ± 55	287 ± 29	21 ± 1	21 ± 1
20	319 ± 68	305 ± 43	21 ⊥ 1	21 ± 1

the anti-ant and random choice tester generated test sequences were on average 57 and 146 times longer, respectively.

If the test goal is to cover one selected transition, the RPT reached the goal with a length of test sequence that is close to optimal. The anti-ant and random choice tester required on average 104 and 235 times longer test sequences. This experiment shows that the anti-ant tester outperforms the random choice tester by more than twice on average with smaller standard deviation. This confirms the results reported in Reference 25.

The dependency of the test sequence length on the planning horizon is shown in Figure 15.8. Nonsmoothness of the curves is caused by the relatively small number of experiments and large standard deviation of the results. The planning horizon can be reduced to half of the maximum planning horizon without significant loss of average test sequence lengths for all transition coverage criteria in this model. Even if planning few steps ahead significantly shorter test sequences were obtained than in case of the random or anti-ant methods. For instance, when the planning horizon is restricted to two or five steps, the average test sequence length decreases by approximately four or eight times, respectively, compared to the anti-ant and random methods. If the test goal is to cover a single transition, then the test sequence length decreases exponentially to the value of planning horizon.

At planning horizons less than maximum, effects of priority conflict resolution methods do not have clear preference. The anti-ant method performs better for all horizon lengths in case of the single transition coverage criterion (Figure 15.8, right-hand side) and for small values of horizon length in case of all transitions coverage (Figure 15.8, left-hand side). The random choice method performs better on average for horizon lengths from 4 to 10 (Figure 15.8, left-hand side) for this model for the all transitions coverage criterion.

We also measured time spent for planning one step. The average duration of a planning step in milliseconds is shown in Figure 15.9. The computer used for experiments has an Intel Core 2 Duo E6600 processor running at 2.4 GHz. Experiments on the model demonstrate that the growth of planning time with respect to the planning horizon is not more than quadratic. The average time for calculating the gain function values with a maximum planning horizon in one step is less than 9 msec. When the planning horizon is increased to maximum, then the average depth of the shortest-paths trees remains below the maximum horizon and the average planning time stabilizes.

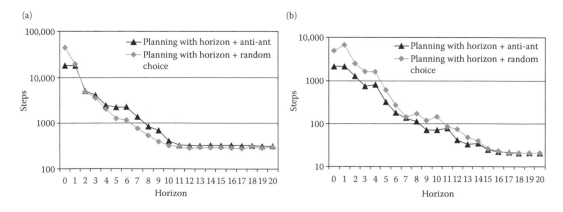

FIGURE 15.8
Average test sequence lengths of the test sequences satisfying the all transitions (left-hand side) and single transition (right-hand side) test goal.

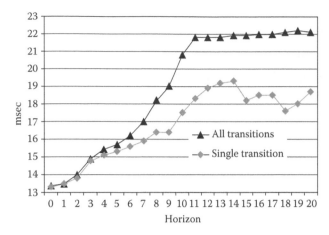

FIGURE 15.9
Average time spent for online planning of the next step.

15.6 Conclusions

In this chapter, we proposed a model-based construction of an online tester for black-box testing of an IUT. The IUT is modeled in terms of an output observable nondeterministic EFSM with the assumption that all transition paths are feasible. The synthesis algorithm constructs a tester that at runtime selects a suboptimal test path by finding the shortest path that covers the test purpose. The principles of reactive planning are implemented in the form of the rules of selecting the shortest paths at runtime. The rules are constructed in advance from the IUT model and the test purpose. The rules are encoded into the guards of the transitions of the tester and are evaluated by the tester in every state when the selection between alternative outgoing transitions should be made. In case multiple guards are enabled, the ambiguities are resolved based on heuristics preselected at RPT generation. As a result, it is not necessary to perform costly model exploration and path finding operations online.

We have studied the performance of the online RPT in comparison with the random choice and anti-ant methods. Based on an industrial scale case study, namely the feeder box controller of a city lighting system, it is demonstrated for the strongly connected state transition model that tuning the planning horizon of RPT allows reaching close to optimal (in terms of test length) behavior that is computationally feasible. The experiments were performed using two different coverage criteria – all transitions and a single selected transition. Different RPT planning horizons (0–20 steps) were used in the experiments. All experiments were performed using both the anti-ant and random choice selection of the subsequent transition in case the RPT planning resulted in several equally good next transitions. Experiments on the case study model show that the RPT significantly outperforms the anti-ant and random choice methods. On average, the RPT with a maximum planning horizon resulted in test sequences length less than a hundredth that of the pure anti-ant and random choice. By increasing the planning horizon, the average length of the test sequences decreases exponentially while the planning time of the next step increases by not more than the length of the planning horizon to the power of 2. The experiments justify applying the reactive planning for industrial scale online testing.

Acknowledgments

This work was partially supported by the ITEA2 D-MINT project, by the Estonian Science Foundation under grant No. 7667, and by the ELIKO Competence Centre project "System Validation and Testing."

References

1. http://messagemagic.elvior.com.

2. http://motes.elvior.com.

3. http://www.artisansoftwaretools.com/products/artisan-studio/.

4. http://www.gentleware.com/products.html.

5. Belinfante, A., Feenstra, J., de Vries, R. G., Tretmans, J., Goga, N., Feijs, L. M. G., Mauw, S., and Heerink, L. (1999). Formal test automation: A simple experiment. In *Proceedings of the IFIP TC6 12th International Workshop on Testing Communicating Systems*, Pages: 179–196. Kluwer, B. V., Deventer, The Netherlands.

6. Brinksma, E. and Tretmans, J. (2001). Testing transition systems: An annotated bibliography. Pages: 187–195.

7. Brandán Briones, L. and Brinksma, E. (2004). A test generation framework for quiescent real-time systems. In *IN FATES201904*, Pages: 64–78. Springer-Verlag GmbH.

8. Chanson, S. T. and Zhu, J. (1993). A unified approach to protocol test sequence generation. *INFOCOM '93. Proceedings of the Twelfth Annual Joint Conference of the IEEE Computer and Communications Societies. Networking: Foundation for the Future.*

9. Cormen, T. H., Stein, C., Rivest, R. L., and Leiserson, C. E. (2001). *Introduction to Algorithms.* McGraw-Hill Higher Education.

10. de Vries, R. G. (2001). Towards formal test purposes. In Tretmans, G. J. and Brinksma, H., editors, *Formal Approaches to Testing of Software 2001 (FATES'01), Aarhus, Denmark*, Volume NS-01-4 of *BRICS Notes Series*, Pages: 61–76, Aarhus, Denkmark.

11. Derderian, K., Hierons, R. M., Harman, M., and Guo, Q. (2010). Estimating the feasibility of transition paths in extended finite state machines. *Automated Software Engg.*, 17(1):33–56.

12. Duale, A. Y. and Ümit Uyar, M. (2004). A method enabling feasible conformance test sequence generation for efsm models. *IEEE Trans. Comput.*, 53(5):614–627.

13. Feijs, L. M. G., Goga, N., and Mauw, S. (2000). Probabilities in the torx test derivation algorithm. In *2000 – 2 nd Workshop on SDL and MSC, pages 173–188. VERIMAG, IRISA, SDL Forum Society*, Pages: 173–188.

14. Ferguson, R. and Korel, B. (1996). Generating test data for distributed software using the chaining approach. *Inf. Softw. Technol.*, 38(5):343–353.

15. Godefroid, P., de Halleux, P., Nori, A. V., Rajamani, S. K., Schulte, W., Tillmann, N., and Levin, M. Y. (2008). Automating software testing using program analysis. *IEEE Software*, 25:30–37.

16. Hamon, G., de Moura, L., and Rushby, J. (2004). Generating efficient test sets with a model checker. In *SEFM '04: Proceedings of the Software Engineering and Formal Methods, Second International Conference*, Pages: 261–270. IEEE Computer Society, Washington, DC.

17. Henniger, O., Ulrich, A., and König, H. (1995). Transformation of estelle modules aiming at test case generation. *Proceedings of the 8th IFIP International Workshop on Protocol Test Systems*.

18. Hierons, R. M., Kim, T. -H., and Ural, H. (2004). On the testability of sdl specifications. *Comput. Netw.*, 44(5):681–700.

19. Jard, C. and Jeron, T. (2005). Tgv: Theory, principles and algorithms: A tool for the automatic synthesis of conformance test cases for nondeterministic reactive systems. *Int. J. Softw. Tools Technol. Transf.*, 7(4):297–315.

20. Kalaji, A., Hierons, R. M., and Swift, S. (2009). A search-based approach for automatic test generation from extended finite state machine (efsm). In *TAIC-PART '09: Proceedings of the 2009 Testing: Academic and Industrial Conference – Practice and Research Techniques*, Pages: 131–132. IEEE Computer Society, Washington, DC.

21. Korel, B. and Al-Yami, A. M. (1996). Assertion-oriented automated test data generation. In *ICSE '96: Proceedings of the 18th International Conference on Software Engineering*, Pages: 71–80. IEEE Computer Society, Washington, DC.

22. Lai, R. (2002). A survey of communication protocol testing. *J. Syst. Softw.*, 62(1):21–46.

23. Larsen, K. G., Mikucionis, M., Nielsen, B., and Skou, A. (2005). Testing real-time embedded software using uppaal-tron: An industrial case study. In *EMSOFT '05: Proceedings of the 5th ACM international conference on Embedded software*, Pages: 299–306. ACM, New York, NY.

24. Lee, D. and Yannakakis, M. (1996). Principles and methods of testing finite state machines – a survey. *Proceedings of the IEEE*, 84(8):1090–1123.

25. Li, H. and Lam, C. (2005). Using anti-ant-like agents to generate test threads from the uml diagrams. In *Proc. Testcom 2005*, LNCS. Springer.

26. Luo, G., von Bochmann, G., and Petrenko, A. (1994). Test selection based on communicating nondeterministic finite-state machines using a generalized wp-method. *IEEE Trans. Softw. Eng.*, 20(2):149–162.

27. Lyons, D. M. and Hendriks, A. J. (1992). Reactive planning. In Shaphiro, S. C., editor, *Encyclopedia of Artificial Intelligence, 2nd edition*, Pages: 1171–1181. John Wiley & Sons.

28. Mikucionis, M., Larsen, K. G., and Nielsen, B. (2004). T-uppaal: Online model-based testing of real-time systems. In *ASE '04: Proceedings of the 19th IEEE International Conference on Automated Software Engineering*, Pages: 396–397. IEEE Computer Society, Washington, DC.

29. Veanes, M., Roy, P., and Campbell, C. (2006). Online testing with reinforcement learning. In *Proceedings of FATES/RV*, Volume 4262 of *LNCS*, Pages: 240–253. Springer.

30. Nachmanson, L., Veanes, M., Schulte, W., Tillmann, N., and Grieskamp, W. (2004). Optimal strategies for testing nondeterministic systems. In *ISSTA04, vol. 29 of Software Engineering Notes*, Pages: 55–64. ACM.

31. Jefferson Offutt, A. and Huffman Hayes, J. (1996). A semantic model of program faults. In *In Proc. of ISSTĂ01996*, Pages: 195–200. ACM Press.

32. Petrenko, A., Boroday, S., and Groz, R. (2004). Confirming configurations in efsm testing. *IEEE Trans. Softw. Eng.*, 30(1):29–42.

33. Starke, P. H. (1972). *Abstract Automata*. Elsevier, North-Holland, Amsterdam.

34. Tretmans, G. J. and Brinksma, H. (2003). Torx: Automated model-based testing. In Hartman, A. and Dussa-Ziegler, K., editors, *First European Conference on Model-Driven Software Engineering*, Pages: 31–43, Nuremberg, Germany.

35. Tretmans, J. (1999). Testing concurrent systems: A formal approach. In *CONCUR '99: Proceedings of the 10th International Conference on Concurrency Theory*, Pages: 46–65. Springer-Verlag, London, UK.

36. Tretmans, J. and Brinksma, E. (2002). Côte de resyste: Automated model based testing. In Schweizer, M., editor, *3rd PROGRESS Workshop on Embedded Systems*, Pages: 246–255. STW Technology Foundation, Utrecht.

37. Vain, J., Raiend, K., Kull, A., and Ernits, J. (2007). Synthesis of test purpose directed reactive planning tester for nondeterministic systems. In *22nd IEEE/ACM International Conference on Automated Software Engineering*, Pages: 363–372. ACM Press.

38. Veanes, M., Campbell, C., Grieskamp, W., Schulte, W., Tillmann, N., and Nachmanson, L. (2008). Model-based testing of object-oriented reactive systems with spec explorer. In *Formal Methods and Testing*, Volume 4949 of *Lecture Notes in Computer Science*, Pages: 39–76. Springer.

39. Veanes, M., Campbell, C., and Schulte, W. (2007). Composition of model programs. In *FORTE 2007, Volume 4574 of LNCS*, Pages: 128–142. Springer.

40. Williams, B. C. and Nayak, P. P. (1997). A reactive planner for a model-based executive. In *Proc. of 15th International Joint Conference on Artificial Intelligence, IJCAI*, Pages: 1178–1185.

16

Model-Based Passive Testing of Safety-Critical Components

Stefan Gruner and Bruce Watson

CONTENTS

16.1 Introduction

In *model-based testing* (MBT, also known as "specification-based" testing, or "model-driven" testing: MDT), the *test cases*, according to which a hardware or software unit (module, component) shall be tested after its development or implementation, are typically not created "ab initio." They are derived by some means of formal reasoning from a formal model, the test model, which is more or less closely related to the specification model of that unit, from which its implementation had been derived (Augusto et al. 2004). This principle is known now since about 20 years; see for example Bochmann et al. (1989) or Richardson et al. (1989). This general idea behind all these approaches is sketched below in Figure 16.1.

Since several years, specialist workshops on MBT are being held, whereby the techniques to support MBT are adopted from diverse areas such as formal verification, model checking, control and data flow analysis, formal grammar analysis, or Markov decision processes. More specifically, those techniques include the following:

- Methods for online and offline test sequence generation

- Methods for test data selection

453

- Runtime verification

- Domain partition analysis

- Combinations of verification and testing

- Models as test oracles

- Scenario-based generation of test cases

- Meta-programming support for testing

- Game-theoretic models of test strategies

- etc.

Two recent textbooks on MTB are Baker et al. (2008), which is suitable also for students and project managers, as well as Broy et al. (2005), which rather addresses topic experts and specialists.

MBT, as a concept or method or technique, is rather general and does not imply anything about the type of system on which it is applied, nor on the specific ways in which the test cases will be eventually applied to the system under test. To this end, we can further distinguish *hardware* systems (modules) versus *software* systems (modules), and *active* testing versus *passive* testing, in relation to the bottom part of Figure 16.1.

The differences between active testing and passive testing can be characterized as follows:

- *Active testing* is the classical, development-related process in which a tester (usually a person) examines the functionality of a unit (system, module) under test *before* that unit is being deployed, whereby that unit does (usually) not have the ability to test itself.

- *Passive testing*, on the contrary, can (and usually does) take place *after* the deployment of that unit and requires either the ability to test itself while in operation, or the existence of a monitor unit that operates simultaneously to the other unit that is subject to passive testing.

Passive testing can also be applied during a longer period of time on a system prototype in a pilot project, before the final release of a system on the basis of such a passively tested prototype. In this way, passive testing could also be useful for improving a system's robustness, that is, the property of "graceful degradation" when the system gets under

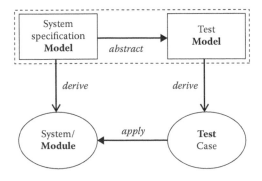

FIGURE 16.1
The principle of model-based testing (MBT).

heavy load or stress, and possibly even for improving a system's resilience, that is, the property of "self-healing" from faults under particular circumstances. The literature on passive testing since Cavalli et al. (2003) is growing steadily, and the reader can easily find further references on the internet.

As it is further discussed below, active testing can take place as black-box testing or white-box testing, whereas passive testing is always a special form of black-box testing— a useful glossary about many further testing-related concepts and notions is provided by *EGEE* and can be found on the internet.* As far as its application type is concerned, passive testing –though in principle applicable anywhere—is even more interesting for the testing of *hardware* components than it is for testing of software modules, and it is also more interesting for *distributed* systems than for classical mono-processor systems. The reason for this special relationship between passive testing on the one hand and hardware systems and distributed systems on the other hand is twofold:

- For the purpose of assuring the quality of *classical software systems*, low-level source code testing—though still necessary– is not the most important method of quality assurance any longer: Software engineers nowadays are (or, at least, should be) guided by the principle of "correctness by construction" (Shukla et al. 2010), which goes hand in hand with valuable and effective development techniques such as source code inspections (Rombach et al. 2008). Those techniques, especially source code inspection, are usually not (or, at least not easily) applicable in the hardware domain.

- *Distributed systems*, especially those based on asynchronous message passing, are notorious for their possibility of *nondeterministic* behavior (at least as far as their micro-activities on the small scale of their communication layer are concerned). This nondeterminism implies that classical tests of type "active testing" applied to such systems are typically *not repeatable*. However, as in the physical sciences, the nonrepeatability of an "experiment" (here a classical active test) diminishes its scientific value and usefulness. In such circumstances (i.e., in which repeatable testing in laboratory conditions is not possible or not feasible), the continuous model-based monitoring of the operations of a running distributed system (i.e., passive testing) can at least *reduce* (or diminish) the risk of applying such kind of systems in safety-critical circumstances and environments.

On these premises, we can now also make plausible why the main intent of this chapter is directed toward model-based, passive testing of *non*software components, though software components will be mentioned as well. In this context,

- Typical *techniques* of passive testing can be regular expressions and finite-state automata (FSA) (Paradkar 2003), grammars and parsing, model checking of process traces specified in terms of the formal language *CSP*, and the like, whereas

- Typical *objects* of passive testing can be "firewalls" (for intrusion detection), memory controllers of processors (with only their "pins" as interfaces), or the emerging technology of *networks on chips* (NOCs), which are also relevant in the context of the abovementioned firewall applications.

All these cases and upcoming application scenarios are especially interesting because the possibilities of their *testability* are *not yet systematically researched* and explored, though

---

*http://egee-jra1-testing.web.cern.ch/egee-jra1-testing/glossary.html.

there are numerous particular contributions and results that point into the direction of a more systematic and comprehensive theory for this domain (Cavalli et al. 2003).

It comes as no surprise that *safety-critical* systems are the typical cases for passive testing. Bruce Watson (coauthor of this chapter), for example, has cooperated with the *CISCO* Systems Corporation on a confidential industry project on the topic of intrusion detection by means of *regular expressions*, which is indeed a special case of passive testing. More recently, another formal-language-based approached to intrusion detection as a kind of passive system testing has been published (Brzezinski 2007).

It should be obvious that passive testing does not come for free—there are computational and economic costs to be invested into the monitor modules that actually implement the passive testing processes—but those additional costs are justified in those safety-critical cases in which other, even larger and higher values (such as human life) are at risk. Moreover, it should be clear—and to this end we issue an explicit warning!—that passive testing *cannot* be used as the proverbial "silver bullet" to the solution of all safety-critical system problems because otherwise a circular meta-problem would immediately arise: Who has meta-tested the MBT devices that are used for the passive testing of the object devices in the correct performance of which we are interested?

The *purpose* of this chapter is mainly *educational*. It is mainly written for university students, and their lecturers, in the faculties of informatics, cybernetics, and/or computer engineering. It is *not* the main purpose of this chapter to publish original results of our own research, as it would be appropriate for a conference or journal paper. Consequently, the largest parts of this chapter consist of literature reviews, whereby we try to keep the presentation "student friendly" (i.e., not too technical nor difficult). Nevertheless, we hope that our chapter will also carry some value as "reference material" for the established researcher in this interesting interdisciplinary field, at the interface between computer software and computer hardware engineering. Industrial technology managers, who are not interested in these "academic details," may speed-read the points of the conclusion section at the end of this chapter as an "executive summary."

16.2   Model-Based Passive Testing of Software Systems

Testing is always based on what is *observable*. Classically, we know two types of software testing, namely *black-box* testing and *white-box* testing (Amman et al. 2008). More recent literature also describes the concepts of *gray-box* testing and *glass-box* testing as hybrid forms or subtle variations of the classical concepts, but those are not important as far as the educational purpose of this chapter is concerned.

- In *white*-box testing of a software module, the tester is allowed to see its program code which is then—at least in modern test suites– also highlighted in different colors, etc., as the run of a program under test is proceeding step by step.

- In *black*-box testing, this kind of observation is either not allowed or not possible. The *observables* in black-box testing are input–output pairs $p \in R$, whereby R is the a priori unknown input–output relation of the module M under test, a relation induced by the *operational semantics* of that module: $R = \{(i, o) \mid i \in I(M), o \in O(M), o = M(i)\}$. Note that the elements $i \in I(M)$ are not required to be finite (or statically limited) objects—they could be data *streams* that are continuously being fed into a running reactive system—which implies that it is also *not* required that $o = M(i)$ represents a terminating computation. In case the computation represented by

$o = M(i)$ is *non*terminating, o is ejected as *side effect* of M (not as its explicit functional return value), which implies that it is also possible for o to be an infinite *stream* (instead of a finite string).

Passive testing is black-box testing. Consequently, the test engineer must carefully prepare a module, which shall be passively tested, in such a way that it will emit a useful string or stream of test data (while running), a string or stream of data which can then be *interpreted* as meaningful *information* about the otherwise invisible operations happening within the confines of the black box. The test engineer should solve this problem *wisely*, such that:

- The test data are *not* emitted *too sparsely* (otherwise too little information can be deduced about the invisible activities in the black box).

- The test data are *not* emitted *too densely* (otherwise the external activity of observation and interpretation becomes too difficult).

- The *auxiliary* additional program code, which is responsible for ejecting the test data for observation during passive testing, should

    - *Not interfere semantically* with the module's main functional code (otherwise the module's specified operational semantics would be violated), and

    - *Not slow down* the execution speed of the module under test to an unacceptably low value (which is especially important in safety-critical real-time applications of such a software module).

The *interpretation* of the emitted test data as "information" about the running module under test is then a matter of the *test model*, which completes this scenario of model-based passive software testing—the same is basically true for model-based passive hardware testing (see subsequent sections of this chapter), with the only difference that the emission of observable test data cannot be implemented by program code in those cases. The *quality* of this test method (and its results) then depends crucially on *two* items, namely,

- The quality (richness, informativeness) of the emitted observable *test data*, and

- The quality (subtlety, distinctiveness) of the *test model* on which the interpretation of the test data depends.

In other words, the richest test data are useless if the test model is too "dumb" to distinguish them, and the finest test model is useless if the emitted test data are too coarse or too sparse for interpretation.* Thus, the construction of a suitable and useful test model as an abstraction of the system's specification model (see Figure 16.1 again) can be regarded as a subproject (of the entire development and testing project) in its own right, which also entails some careful *requirements* elicitation *within* the subproject of test model construction. A recent example of a test requirements elicitation process in support of MBT can be found in Santos-Neto et al. (2007).

---

*There is thus an interesting correspondence between this method of passive testing and the epistemology by the philosopher *Immanuel Kant*, from whom we remember the well-known aphorism: "*Gedanken ohne Inhalt sind leer, Anschauungen ohne Begriffe sind blind*"—thoughts without objects/contents are empty/void, intuitions/observations without theoretical ideas/concepts are blind (Kritik der reinen Vernunft 1781).

Before we get to the discussion of some further literature on model-based passive software testing, let us briefly illustrate these principles, as we have outlined them above, by a very small and simple *example:*

Let $\mathbf{P}$ be a process defined in the well-known notation of *CSP* as follows:

$$\mathbf{P} = (\mathrm{Q} \sqcap \mathrm{R}), \text{whereby}$$
$$\mathrm{Q} = (\boldsymbol{a} \to \mathbf{P}), \text{and}$$
$$\mathrm{R} = (\boldsymbol{b} \to \mathbf{P}).$$

Obviously, this is a nonterminating process in which the "test data" that our "software module" emits are arbitrary long sequences of $\boldsymbol{a}$ and $\boldsymbol{b}$ in any order. As a valid "test model" to this trivial "software module," we could then use the equally trivial regular expression

$$\mathbf{E} = \{\boldsymbol{a}|\boldsymbol{b}\}^*,$$

which is sufficient to "recognize" (identify) all *traces* that can be emitted correctly by process $\mathbf{P}$. Should, for whatever reason, our process emit any different symbol, then $\mathbf{E}$ would not be able to match it any more. Consequently, our "software module" would have *failed* this test.

Alternatively, we could also define an explicit, *positive acceptor* model, for example, by means of the regular expression.

$$\mathbf{A} = \{\boldsymbol{abba}\}^*.$$

What we have called "acceptor" model here is thus also a test model, however, from a different *pragmatic* perspective; depending on whether we want to passive-test "positively" for the presence or "negatively" for the absence of some desired or undesired phenomenon. In this example, we would let our "software module" $\mathbf{P}$ run so long until we observe an emission of a trace t that can be interpreted as a word in the language of $\mathbf{A}$, thus $t \;\epsilon\; L(\mathbf{A})$, in which case we would declare our module as "acceptable" (and possibly stop the process of passive testing at that point). Finally note, at the end of this little example, that $L(\mathbf{A}) \le L(\mathbf{E})$, which is of course a logical matter of *model consistency* under the condition of model refinement.

After we have thus briefly clarified the most basic theoretical principles of model-based passive testing especially in the software domain, we will now discuss somewhat more realistically

- How the technique of *aspect-oriented programming* (AOP) can be used in practice to insert trace-emitting test code into a software module in preparation for passive testing (Subotic et al. 2006), and

- How the *B*-method can be used (as one possibility out of many) for the specification of a test model by means of which a string or stream of test data can be interpreted (Howard et al. 2003).

In the remainder of this section, we will thus discuss several *practical techniques* that can be used and applied in support of model-based passive testing in various scenarios. As it was mentioned already, a more comprehensive systematic taxonomy or classification scheme of methods and techniques for model-based passive testing does, as far as we know, not yet exist and is still a task for "future work" in this field.

16.2.1 Aspect-oriented programming for model-based passive software testing

This subsection, which is by and large a recapitulation of Subotic et al. (2006), outlines how the programming techniques of AOP can be practically used in support of passive testing, especially for the emission of the observable runtime traces on which the technique of passive testing is based. However, this subsection is *not* an introduction into AOP as such. Those readers who are not familiar with AOP are referred to the ample literature on AOP, which has been published since the original papers by Kiczales et al. (1997, 2001).

Aspects are the key concept in AOP. They can be regarded as "pseudo objects" in the sense that they encapsulate internal states and behavior, which are relevant for a software system as a whole, and not only for some of its particular modules. Aspects are thus "woven" into a software system, thereby building a kind of action repository for the so-called *cross-cutting concerns* of the entire system (Subotic et al. 2006). In contrast to the so-called core concerns, which can be sufficiently implemented in functional separation by particular modules according to the software engineering principle of "coupling and cohesion," cross-cutting concerns cannot be elegantly implemented by particular modules without AOP. Examples of those are: logging, tracing, profiling, security policy enforcement, transaction management, and the like (Subotic et al. 2006). Those cross-cutting concerns are typically *dynamic* in their character, which means that they become especially important at system runtime.* The main idea of AOP is then that whenever some particular preconditions are fulfilled *anywhere* in the running software system, an aspect-module related to that precondition will be triggered "on the fly," and some *additional* actions are carried out, which would otherwise not have happened—for example, emitting some trace data which can then be used for the purpose of passive testing. In AOP terminology, the so-called "point-cuts" define the above-mentioned preconditions, whereas the additional (in-woven) actions carried out around those point-cuts are also called the "advice" of the corresponding aspect (Subotic et al. 2006). Figure 16.2 sketches the structure of an aspect definition in the *JAVA*-based AOP language *AspectJ*.

```
public aspect A
{
 pointcut p()

 call ( *Class.do*(...)); // definition of this aspect's "hook"

 before() p()
        {
        //... advice body instructions ...
        }

 after() p()
        {
        //... advice body instructions ...
        }
}
```

FIGURE 16.2
Sketch of an aspect definition in the programming language AspectJ.

---

*There also exist *static* aspects (Subotic et al. 2006), but their discussion is not necessary in the scope of this chapter.

Depending on whether aspects in AOP shall be used for the implementation of functional or nonfunctional system features, we can distinguish *development* aspects from *product* aspects (Subotic et al. 2006). The difference between these types is found in their *pragmatics*, not in their syntax or semantics:

- Development aspects are only "switched on" during the phase of prototyping, typically in support of the programmer's comprehension of the system, especially for the purpose of *active* testing (white-box testing). Development aspects are "switched off" again after the prototype has been found satisfactory and are thus *not* part of the delivered software product.

- Product aspects, on the other hand, are implementations of specified system requirements and can thus *not* be "switched off" when the product system is finally installed and deployed (Subotic et al. 2006).

From this discussion it should be clear that passive testing with AOP can be done either on the basis of development aspects, or on the basis of product aspects. The decision between these two possibilities depends on a system's *requirements* specification:

- If the requirements document stipulates that only a prototype shall be passively tested until satisfaction, then the trace-emitting aspects are clearly classified as development aspects.

- On the other hand, if the requirements document stipulates that also the delivered product shall be subject to continuous passive testing and observation while being in operation, then the trace-emitting aspects are clearly classified as product aspects, which must *not* be switched off when the product system is deployed.

From a theoretical point of view, the execution of aspect code around point-cuts in an aspect-oriented software system must be regarded as a *side effect,* and the usual warnings about the practical risks of side effects are applicable in this context as well. Therefore, we recommend using aspects in passive testing *only* for the purpose of emitting interpretable trace data, but *not* for any other system-relevant computations with effects on that system's internal state. Moreover, it is possible in AOP that two different aspects A and A' can be triggered around the same point-cut P, in which case the *order of execution* of A and A' is generally *not well defined* (unless there are some conflict-resolving meta-rules in place). In the case of AOP-based passive testing, such a point-cut ambiguity could compromise the validity of the emitted test data trace, which could in consequence lead to a false interpretation of the correctness or defectiveness of the system under passive testing. For this reason, we recommend ensuring that the trigger preconditions of any two active aspects are logically disjoint. However, one can also use passive testing to *discover* the appearance of unintended orders of execution of overlapping point-cuts for systems under test, which were programmed by AOP in the first place: if conflicting aspects A and A' at the same point-cut emit observable traces, then a passive testing unit can detect if their order of execution was not as intended. This holds, however, as mentioned before, only if the passive test unit does not itself introduce artifact errors on the basis of this AOP mechanism.

If we carefully navigate around those two pitfalls of AOP—namely, unintended side effects and overlapping trigger preconditions of different active aspects—we can clearly see several *advantages* of using AOP for the purpose of passive software testing:

- The emission of trace data can be elegantly implemented separately in aspect modules, thus keeping the system's core module "clean" from any "nonfunctional litter," and only *one* trace-aspect must be injected into the system instead of inserting trace-generating command lines into the program code of *all* core modules of the system.

- The trace-generating test aspects can be programmed by a group of programmers different from the group of programmers who construct the functional core modules of that system, in accordance with the well-known software engineering principle that development and testing should not be carried out by the same work group (a principle that aims at avoiding the psychological bias of good faith and belief in one's own work).

- The same core system can be differently traced from various viewpoints, simply by replacing one set of trace-aspects A by a different set of trace-aspects A', without any need for cumbersome and error-prone "muddling" with the core modules (Subotic et al. 2006) which are implementing the main functionality of the system under the observation of passive testing.

Having thus described an elegant programming technique for the emission of trace data without any regard to their model-based interpretation, we can now turn our attention to another practical example paper in which the *B*-method was used for the *interpretation* of test traces emitted by a small web-service software system. Once again, we should remark that the underlying principles of this technique have been known for about 20 years now; see, for example, the trace-viewer in Helmbold et al. (1990), or the already mentioned approach of Bochmann et al. (1989).

16.2.2 Model-based trace checking with B and SPIN

The idea and technique of *trace checking* were already known to Jard et al. (1994). Subsequently, this idea became more and further integrated into the emerging "paradigm" of model-based or model-driven software engineering. In a paper from the year 2003, we can find one of the earlier approaches that systematically combined the issues of testing, "tracing," executable formal models, and automated analysis into a methodological framework of MBT, which was then called "model-based trace checking" (Howard et al. 2003), or "trace-driven model checking" in a related publication (Augusto et al. 2004). In this subsection, we shall briefly review the ideas of Howard et al. (2003) as an interesting variant of the general principle that we have already explained above.

As usual in testing in general, the issue also of MBT is the detection of defects, not the proof of their absence. Also, according to the principle of modularity, different test models can be designed for the exploration of different aspects of system behavior, as it was mentioned in the section on AOP of above. The key method of Howard et al. (2003), which was a novel idea at its time of publication, was:

- To implement the system by any suitable method of software development,

- To insert "trace code" (possibly with help of AOP, as explained above) for capturing each significant system operation or event,

- To design one or several formal models of that system in a formal notation for which sufficient tool support exists (a.k.a. "executable formal methods"), and

- To run the generated traces through the available tools to check them against those models (Howard et al. 2003).

This procedure is also interesting in the sense that it inverts the steps of classical MBT: first, the traces are produced from the system under test, and only then are they compared with the model to assess their validity. The web-based case study reported in Howard et al. (2003), which was intended to corroborate the validity of that principle, is interesting for a number of reasons:

- There are several components *distributed* over the internet.

- There are several possible distributed *configurations* of the system, depending on the actual instantiations of the underlying JAVA code classes.

- Each of the participating components has its local version of the working set of data and yet all versions must be kept consistent with each other.

- The system is implemented on top of a basis of complex middleware, including web servers, web-supported data bases, Java server pages, Java servlets, and related techniques (Howard et al. 2003).

Because of the distributed character of the studied system and the complexity of the middleware on which its implementation is based, the testing of such a system is obviously not a trivial task. Indeed such message-passing-based network systems tend to exhibit environmentally induced nondeterminism and, therefore, nonrepeatability in classical (active) tests. For this reason the abovementioned example system can be regarded as a typical case for the application of self-monitoring in the form of trace-based testing. In that case study (Howard et al. 2003), trace-generating program code was not injected by means of AOP, but rather in the form of Java "Beans," because of the web-based character of the distributed system in that case study. At each point where trace information should be captured, two different Java Beans were invoked: one to collect a current state of the system at the point of activation and another to access a web-based data base system (SQL DBMS via JDBC) for storing a "word" of the trace (Howard et al. 2003). The traces in that case study were thus not written into a simple text file, but rather into a purposefully designed *relational database*, for a number of reasons:

- It enables collecting different subsets of trace information according to different needs and interests by different members of the group of test engineers.

- It becomes easier to *parse* and *reformat* the trace entries according to the syntactic input requirements of the checker tools to be used after tracing.

- It becomes possible, thanks to a web-based trace database, to collect trace information from source components that can be located anywhere in the world wide web (Howard et al. 2003).

The problem of *where* in the system's program code to activate trace generation was also discussed in Howard et al. (2003). Obviously, one should record information about a running system whenever there is "news," that is, *whenever the system under test changes its internal state* in an interesting and relevant way. The answer to the question: "what is interesting and relevant?" must obviously be given with reference to the test model on which the entire method is based. As reported in Howard et al. (2003), the test model can—and should—also be used to discover meta-errors, or false-error artifacts, a phenomenon that can occur when the trace-recording itself is not correctly implemented. The discovery of such false-error artifacts can, however, not be based on testing again (Howard et al. 2003). Such discovery would require "rational" techniques, such as code inspection and the careful comparison of the system model with its corresponding implementation, as it was also emphasized again more recently in Rombach et al. (2008).

"B", a model-oriented formal notation (Abrial 1996), was used in Howard et al. (2003) for the definition, construction, and simulation of valid traces. In B, a system is modeled as an abstract state machine with *operations* and *invariants*. B specifications can be written nondeterministically on a high level of abstraction, but can be refined to more specific deterministic specifications. In such a deterministic form, B models can be made executable and thus serve the purpose of rapid prototyping, or—as in the case of Howard et al. (2003)— the generation and analysis of runtime traces.

Indeed, it is interesting to note that testing by trace checking in Howard et al. (2003) was done only in an *indirect*, not in a direct fashion. The checked traces were in fact not the traces of the web-based system that was mentioned above, which we could call in hindsight the "primary" traces. Those primary traces were in fact used to "steer" (control) the executable B-machine, which was designed as a *model of* the web-based software system. That B-machine then generated a trace, which we could now call "secondary" trace, and only that secondary trace was then checked for well-formedness as described in the methodology sections of above. The "oracle machine" for passive testing in this case study was thus a module that implements an executable B-machine (Howard et al. 2003). Needless to say, this *indirect* way of passive testing requires a semantic *correspondence* of the primary software system with its secondary prototype—here the executable B-Machine—which was then passively tested on behalf of the primary software system (Howard et al. 2003). One might perhaps call this technique now, in hindsight, "passive testing of second degree". This structural correspondence between the primary software system and its secondary B-representation was organized in such a way that the corresponding B-machine under test made a state transition whenever one of the trace "beans" in the primary software system was invoked (Howard et al. 2003).

Under these circumstances of indirect passive testing of second degree, whereby a prototype is passively tested on behalf of its corresponding software system of first degree (which is justified by some kind of pre-established structural homomorphism between the primary system and its secondary prototype), the method of Howard et al. (2003) worked as follows:

- Execute a test run of the primary software system to record a primary trace into the trace database.

- Select a (possibly "filtered") trace of "interest" from the database.

- Use this primary trace as input for the B-animation tool and model checker *ProB* (Leuschel et al. 2003), whereby a secondary trace suitable for processing with these tools is generated.

- Let the animator rerun (simulate) the run of the secondary trace in terms of the available state transition operations of the executable B-machine model.

- Regard the MBT as

  - *Passed* if the B-based rerun of the secondary trace was possible.
  - Otherwise regard the MBT as *failed* if no sequence of B-machine operations could be found to reproduce (simulate) the sequence of activities of the trace-writing Java Beans in the primary software system (Howard et al. 2003).

As mentioned above, there must be a *well-defined correspondence* between the primary trace (in the trace database, from the system under test) and the secondary B-machine trace that "emulates" the primary trace and that is checked "on behalf" of the primary trace. Without such a correspondence between the primary and the secondary trace, which resembles the *semantics-preserving* relationship between *concrete* and *abstract* syntax in the domain of compilers, the exercise of checking the secondary (abstract) trace on behalf of the primary (concrete) trace would be futile.

In another branch of the research project in the wider context of Howard et al. (2003), the same trace data from the trace database were extracted (via SQL) to be transformed into a suitable representation for the *Promela*-based model checker *SPIN* (Holzmann 1997). According to the well-known principles of model checking (Clarke et al. 1986, 1999), the *invariants* of a system under test can be expressed in terms of *temporal logic*, which implies that also a temporal logic formula can play the role of a "test model" for such a system. As reported in Howard et al. (2003), it is possible to rewrite a runtime trace of the system under test in terms of the language Promela and use this as an input to SPIN to check the trace against a given temporal logic formula T, to find out whether or not the trace is a legal "inhabitant" of the state-"world" specified by that formula T. In Howard et al. (2003), those formulae were provided manually, "ab initio," and were thus not directly derived from the system's specification. Vice versa, in an inductive way, it was also possible to use those traces, which were already known to be correct, as the concrete basis for the construction of another temporal logic formula T', which could then be used as the abstract "test model" for further experiments in the passive testing of the original system (Howard et al. 2003). Compared with other well-known methods of runtime testing of programs, such as, for example, executable assertions, the indirect method of Howard et al. (2003) has a number of advantages; amongst others,

- It is possible to recheck already existing traces (in the trace database) against other and further system properties, which had not yet been thought of at the time at which those traces were captured.

- One and the same trace (in the trace database) can be reparsed to be checked against a variety of aspects, via a variety of test models, written in a variety of formal notations, supported by a variety of checker tools (Howard et al. 2003).

16.2.3 Passive testing with extended finite-state machines

Finite-state machines (FSM), though not specified in B, are also the basis of the passive testing method by Cavalli et al. (2003). Similar to the approach of (Howard et al. 2003), the approach of Cavalli et al. (2003) also deals with *invariants*, expressed in logic, against which the emitted traces are checked, whereby the possible nondeterminism of the underlying FSM is an additional source of technical difficulty. The key distinction of Cavalli et al. (2003) is the one between the analysis of *control flow* and the analysis of *data flow*, a distinction that is also well known in compiler technology (Aho et al. 2007). Whereas control flow can be passively tested rather easily on the basis of ordinary FSM, data flow is considerably more difficult to test—both in active testing, as well as in passive testing (Cavalli et al. 2003). For this purpose of passively testing data flow properties, the approach of Cavalli et al. (2003) works with *extended* finite-state machines (EFSM) instead of ordinary FSM, whereby the main difference between EFSM and FSM is that the state transitions of an EFSM can have additional *side effects* on the values of a *configuration of variables* (Cavalli et al. 2003) in a similar way in which the state transitions of a stack automaton have side effects on the contents of the stack storage attached to such an automaton (Aho et al. 2007). As in Howard et al. (2003), also Cavalli et al. (2003) considered the possibility of *non*determinism in the operations of the model machines and provide techniques to cope with this difficulty when testing (checking) the EFSM traces against the logic-specified model invariants. A similar approach to model-based system self-checking on the basis of FSA models can be found in Paradkar (2003).

16.3   Network Security by Model-Based Passive Testing

The global computer network is growing day by day, and so does the amount of information and communication that is sent as "traffic" through the net. Accordingly, our vulnerability to malicious misuse of these possibilities has also increased. So-called *firewalls* (also known as "intrusion detection and prevention systems") are devices that are designed to protect individual computers as well as larger subnets from unauthorized external intrusion. They usually consider the stream of network traffic at one of two levels, by

- Examining the *contents* of the network stream, and/or

- Recording (capturing) a stream of metadata about the communication activities at that security point and saving those traces in *log files* in support of an a posteriori analysis; for comparison see Andrews (1998).

These two levels represent the extremes: in the former case, the firewall considers the minute details of the actual network traffic, while in the latter case, much of the detail of the network traffic is projected away, leaving only the "shadow" in the log files. Naturally, solutions in-between are also possible.

Intrusion detection and prevention are conceptually similar to the scenario of passive testing. The difference is mainly a pragmatic one: Whereas in passive testing, we want to find out whether or not a device under test (DUT) is defective, in intrusion detection we want to decide whether or not a communicator "knocking at the door" of a firewall is doing so in a malicious way. We can thus even imagine that a firewall may also be used as a passive testing device for "normal" software, although there is not yet evidence of this application concept in practice. From this description of the scenario, it is obvious that there are two basic possibilities of applying the methods of passive testing to the problem of firewall analysis:

- The firewall could analyze the stream incrementally ("online") as it flows.

- The stream (the actual network contents in a "content-based firewall" or the metadata log files) could be analyzed as one batch ("offline").

Compiler technology (Aho et al. 2007) could then be used for the syntactic analysis (parsing) of the contents of those data "sentences" produced by a firewall. The test model in such a setting is then a formal *grammar* against which those firewall sentences must be *parsed*.* This could be done in a *positive* as well as in a *negative* mode:

- In the positive mode ("green" mode), a formal grammar specifies the syntactic properties of "harmless" sentences, and the test yields the result "acceptable communication" if the sentence can be successfully parsed against that grammar.

- In the negative mode ("red" mode), a formal grammar specifies the syntactic properties that would result from an already well-known pattern X of attempted intrusion. The test would then yield the result "intrusion attempt of type X detected" if a new log can be successfully parsed against that pattern-specific grammar.

The fundamentals of network security are covered in detail in Cheswick et al. (2003), whereas Varghese (2004) provides a detailed treatment of the algorithmics involved.

---

*A student in our research group has recently started his post-graduate project on this topic.

16.3.1 Contents-based analysis

Just as log file analysis suffers from "information loss" as the stream's contents are projected away, *contents-based* analysis can suffer from extreme information overload with several causes:

- Due to the layered approach to networking software, the structured interprocess communication over the network (of the applications using the network) appears as completely "flattened," that is, devoid of internal structure.

- An important strength of modern networking is "packetization," meaning that the network streams (of various applications) are split into small (often fixed sized) chunks that are delivered interspersed, possibly out of order, or even unreliably.

- Modern networks are easily capable of very high throughput up to 10 Gbps (roughly 1GB/second).

Currently, all analysis "models" for intrusion detection are expressed as *regular expressions*. The set of regular expressions (a.k.a. *signatures*) usually number in the thousands, and are crafted for the "red" mode of operation: they characterize unallowable patterns in the network traffic. The first of the overload conditions above means that the regular expressions are also designed to conservatively capture the flattened network stream. The second complication (packetization, etc.) is usually solved with a significant amount of bookkeeping so that the passive testing of the individual network sessions (i.e., between different applications) is dealt with separately; furthermore, such an implementation is capable of reassembling and reordering the packets on the fly. The final complication (of performance) is often solved by brute force: money spent on a full-network-speed firewall comes close to the money spent on the applications it is examining. The alternative (as of this writing ill-explored) is to "sample" the stream and analyze only part of the stream, such as every second byte or only some of the network sessions. As with other imperfect passive testings models, some problems may slip through.

16.3.2 Log file analysis

Research and development in this area are rather new and have not yet progressed very far. Two of the many difficulties that typically arise in this context are the following:

- Classically, parsing requires a *lexically finite* string entity as input, such that the parser can "know" where to begin and where to stop "eating" that input string (Aho et al. 2007). In the potentially endless data streams generated by firewalls, however, those distinguished points are not easy to define and probably require some additional heuristics for their declaration before the parser can start to parse.

- Secondly, the syntactic properties of firewall traces stemming from sophisticated attempts of intrusion might be so complex that they cannot be described by context-free grammars any more. Even if they could still be described in terms of context-*sensitive* grammars, then, however, we would have to face the notorious problem of *general undecidability of context-sensitive parsing* (Salomaa et al. 1997), which we can hope to solve only in particular special cases.

It is needless to emphasize that those two problems could occur in all application scenarios of grammar/parsing-based passive testing, not only the scenario of firewall analysis and network security. Nevertheless research into this direction is on its way, and there are a number of publications, which the reader can refer to as starting points in this context; see for example Brzezinski (2007), Campanile et al. (2007), Gudes et al. (2002), and Memon (2008).

16.4 Model-Based Passive Testing of Hardware

In this section, we consider a number of hardware testing scenarios. In contrast to software, user expectations of hardware are considerably higher, perhaps because of the tangible nature, making MBT crucial. Fortunately, most modern silicon chips are specified, designed, and built with a sequence of formal models, which are then usable in generating (by hand or automatically) the testing model.

16.4.1 Recent trends in hardware development

Thanks to decades of ongoing research advances in chip production, chip densities (the number of transistors packed in a given area) have been doubling every 18–24 months,* in what is well known as *Moore's Law.*[†] This doubling has brought a more than doubling of design complexity, as it was already described in the 1999 edition of the International Technology Roadmap for Semiconductors (Gargini 2000).[‡] A little known fact (outside of the electronics industry) is that such doubling has made modern chips much more sensitive to several effects, most notably the following three.

- *Random quantum effects:* These arise randomly, allowing electrons (e.g.,) to "tunnel" from one place to another, even through insulating material. The probability of tunneling depends on the energy of the tunneler, the material through which it may tunnel, and the distance it may tunnel. This has proven useful in the design of some types of transistors, but has also lead to unpredictable behavior in very small circuits, where the small distances raise the probability significantly.

- *Structural quantum effects:* These are also random effects, whose probabilities rise because of structural properties of a chip design. For example, two parallel wires carrying data create a perfect environment to encourage coupling (electrical or quantum), which allows the data on one wire to crosstalk with the data on the other wire.

- *Timing and race conditions:* These are timing-related problems similar to the ones arising in parallel software systems. Two electrical signals may have nondeterministic arrival times at a gate, potentially yielding undefined or random behavior. In many cases, one of the signals is the clock and the race condition can be partially solved by slowing the circuit's clock rate. Although some race conditions are design problems, more often they arise from manufacturing irregularities or thermal effects in which a wire length or transistor size changes with temperature.

With all of these issues, MBT plays an important role. The interested reader can find much more about functional verification of chips in Wile et al. (2005), while Chiang et al. (2007) gives an excellent treatment of issues in modern nanoscale chips. Last but not least, McFarland (2006) presented an overview of the entire design flow of a microprocessor. In the subsequent subsections, we consider testing and verification of both clocked and unclocked circuits; this is interspersed with a discussion of redundant hardware in highly reliable systems—an area using techniques very similar to model-based verification.

---

*This density doubling is often confused with a doubling of clock speeds, or a halving of prices. While clock speeds did, in fact, double for several years in keeping with Moore's law, recent issues with power consumption, quantum effects in transistors, etc. have made the clock speed plateau. From the economics standpoint, the prices have also not halved—rather, consumers want twice as much "bang for the buck" in an electronics product.

[†]Gordon E. Moore was one of the cofounders of Intel.

[‡]http://www.allbusiness.com/electronics/computer-equipment-computer/6150464-1.html.

16.4.2 Passive testing and verification of clocked circuits

At the time of this writing, the vast majority of digital chips are *clocked* circuits. Such circuits consist of several mutually dependent functional units (e.g., in a microprocessor this includes the register file, the adder, multiplier, memory management, instruction decoder, and so on). Their communication is carefully synchronized (this style of chip is thus called *synchronous*). Ideally, the synchronicity is driven by a chip-wide *clock* that is usually a square-wave electrical signal. Operations as well as data movements through gates only occur at a clock "tick" (at the rising edge or at the falling edge, or on both). As an aside, two practical concessions are made for clocks:

- The clock is never a square wave. Clocks that closely resemble a square wave contain many higher harmonics, which correspondingly require much more power. The power consumption of a chip is presently a leading restriction on chip density,* as higher-powered chips lead to more crosstalk and quantum tunneling, but also thermal failure (melting) of the chip.

- Some parts of the chip may use a local clock that is an integer multiple or divisor of the main clock. In this way, some elements can run faster or slower but still remain essentially synchronous.

The first point is responsible for some of the quantum effects mentioned previously. This second point is taken to extremes in subsequent sections when we consider unclocked chips and NOCs. Careful design and layout are required so that the wire lengths (propagating the clock signal and also data signals) respect the desired clock rate, ensuring that data arrives neither too early or too late. Late or early arrivals lead to race conditions, and borderline race conditions (e.g., as mentioned earlier, depending on the temperature of the chip expanding or contracting the wire lengths) can give nondeterministic behavior, which is inordinately difficult to debug.

Model-based passive testing has long been used in synchronous-chip design, though it is often so ad hoc as to be unrecognizable. Most chips are designed using a version of the following workflow:

i. *Functional specification:* The global functioning of the chip is specified in a high-level language such as VHDL or Verilog.

ii. *High-level simulation:* The specification is directly executed to explore functionality and performance of the chip.

iii. *Compilation:* The VHDL or Verilog is compiled to produce an RTL (register-transfer language) version, at the level of flip-flops, latches, gates, and buses.

iv. *Layout and physical design:* The medium-level RTL specification is further compiled to individual transistors, which are then laid out in detail, respecting wire lengths, individual gate delays, etc., so that proper synchronous behavior is preserved.

v. *Electrical simulation:* To discover design faults before production.

vi. *Production:* That is, the industrial manufacturing of the devices.

vii. *Postproduction testing:* To discover any faults after production of the chip.

---

*The other leading issue is lithographic printing of chip features. At the time of writing, minimum feature sizes are less than one-sixth of the wavelength of laser light used for lithography, leading to resolution and contrast issues, and therefore higher costs in lithography.

MBT is woven into three of the above steps (cf. the numbering above):

- In *step iii:* During compilation, tests on the level of subcomponents are automatically generated as a finite automaton. The automaton makes transitions based on the signals appearing on the component's output pins. The automaton either blocks or transitions to a "dead state" if the outputs are invalid according to the specification. The outputs (for any given clock cycle) of all of the pins are folded into a single automaton alphabet symbol, typically by devoting a single bit to each pin. Clearly, the automaton can be generated to be "liberal"—it accepts more traces than just the allowable ones; this is especially useful when the component may have context-free behavior (e.g., if it has a large internal memory) that the finite automaton must approximate by containment.[*] An important, nonpassive, extension of this consists of using a Mealy machine (an automaton not only with inputs but with outputs on transitions) to *drive* the component as well as monitor it (i.e., recognizer vs. transducer).

- In *step iv:* In practice, using the monitoring automaton requires access to the "pins" of the components. In most cases, the component is *floor-planned* somewhere in the chip's layout such that its inputs and outputs are not close to a physical pin (such pins are usually at the edge of the chip). This gives two possibilities for the passive tester: place it close to the component under test[†] (thereby taking up chip real-estate/area with the passive tester) or give external (off-chip) access to the component's inputs/outputs by routing them to physical pins. The latter option is most often implemented, as it places the monitoring automaton off-chip. Thereby, the infrastructure that is routing signals to physical pins, known as a *scan ring*, allows for the contents of *latches* between components to be read out, usually over a bit-serial link.

- In *step vii:* Testing techniques using the scan ring require some external infrastructure: one standard for such a bit-stream is the JTAG interface, for which there are numerous tools to facilitate demultiplexing and passive testing of the bit-stream by running the monitoring automaton. Such tools are often part of the entire system, allowing a measure of self-testing in the field.

In this last step, we may ask what would happen in the field if a problem is detected while the hardware is running. In the next subsection, we discuss such high-reliability systems that are self-healing. The general design issues for entire "systems-on-chips" is detailed in Jerraya et al. (2004). An excellent overview of modern chip verification is given in Wile et al. (2005).

16.4.3 Passive testing in high-reliability and redundant hardware

As outlined in the previous subsection, lack of robustness plagues systems for a variety of reasons, primarily because of small feature sizes, manufacturing difficulties, etc. Safety- and financial-critical systems have been designed since the 1960s for uptime, robustness, and recovery. As the costs of hardware dropped rapidly, it became feasible to double implement critical hardware: two or more identical units are placed in the system shipped to customers. Such systems bring about some important choices:

- What are the "units" that are duplicated? Depending on the application (of the system), designers" experiments, and cost factors, only some of the components (e.g., memory,

---

[*]It is known from formal language theory, a regular language can be used to approximate a context-free language by containing it, or vice versa.

[†]Also called the *device under test* (DUT) in the literature.

register file, instruction decoder, I/O units, etc.) require duplication. This is referred to as the *sphere of replication*. Intuitively, one would imagine that a larger sphere of replication is necessarily better (providing perhaps total redundancy), however, a small sphere has the advantage of earlier detection of disagreement between the components, and therefore more opportunity for recovery as opposed to simply flagging a failure.

- What does "identical" mean? In the simplest case, "identical" really means two systems designed and constructed identically. This case protects against random errors, but of course not against systematic or design errors. To protect against those possibilities, we require "identical" to mean "functionally or architecturally" identical—but *not* identically implemented. The latter entails completely different costs and is really only used in (some) life-critical systems such as auto-pilots and medical equipment, though even those scenarios are beginning to use identical implementations. See Leveson (1995) for more on such systems.

- How do we know if two units disagree? The outputs of the redundant units are compared to one another in a voting fashion. Obviously, with an even number of units (in the worst case two), we have a limited hope for knowing *which* is malfunctioning; even with an odd number of units, we have the (admittedly very unlikely) scenario where a majority of units fail identically and outvote the good units. The comparators are placed at the border of the sphere of replication. For example, in a redundant register file, the comparator would be at the address lookups and the I/O bus to the register file(s) in which case we may be able to place the comparators off-chip and use the JTAG ports, thereby "replicating" off-the-shelf hardware. In the simplest case, the comparators do bitwise real-time comparisons of the component. Comparators on a large sphere or on diverse underlying implementations are much more complex, making it necessary to take timing differences as well as other issues into account. In such cases, the comparators' complexity approaches that of the system being passively tested. The comparators do *not* form the models in our MBT—it is simply a mechanism; the system is its own testing model.

- What happens when the units disagree? If a unit is malfunctioning, it is taken completely offline. Any hope for recovery and restart depends on there being more than two units duplicated, and also on the size of the sphere, as follows:

  - With a small sphere of replication, the amount of divergence before detection remains rather limited. For example, in a CPU, a redundant instruction decoder will immediately discover disagreement as typically one instruction is decoded every clock cycle. In that case, the good unit(s) will pause and cross-load their internal state into the malfunctioning unit, most often via the JTAG ports or a similar scan ring. From that point, execution continues again.

  - By contrast, a large sphere of replication allows the redundant units to diverge significantly before it is observable to the comparators at the border of the sphere. At that point, it is often too late to recover. Still, high uptime transaction systems (e.g., those running for credit-card processing) do implement such elaborate "state reconstruction," bringing a failed unit back online and synchronized within a few seconds.

Many of these developments were pioneered and perfected in the 1980s by companies such as *IBM, HP,* and *Tandem.* An excellent technical treatment of such systems is given by Mukherjee (2008).

16.4.4   Passive testing of asynchronous (unclocked) circuits

The foregoing sections have shown that clocks (in synchronous chips) are arguably the biggest "evil" of modern chip design and implementation. In the long term, asynchronous circuits (those without a clock) offer the best hope for correctness and high performance. Despite sounding exotic, such circuits were already designed and in use in the 1960s.

Ideally, asynchronous circuits are designed to be robust against arbitrary delays in any part of the circuit. In particular, correctness by construction is a highly desirable property, especially in the face of unknown or unbounded (but finite) delays in wires, gates, etc. The calculi to develop such circuits (known as *delay-insensitive* circuits) is virtually identical to CSP, meaning that the same tools, trace checkers, automated theorem provers, and model-based testers can be used directly. This makes delay-insensitive circuits arguably the most robust type of digital circuit.

The most successful language (derivative of CSP) thus far is Tangram, primarily developed at Philips Research with input from the University of Eindhoven (NL), and now spun-off into *Handshake Solutions Inc.* A delay-insensitive circuit gives the following advantages:

- It can be physically floor-planned without regard to wire lengths (between components), gate delays, etc., and it will still work when manufactured.

- Lithography processes (in chip production) give some natural variation in the geometry (length and width) of individual transistors, which in turn gives variation in the transistors' performance (current used, delay/transition time, etc.). The timing aspects of such variations have no effect on the correctness of delay-insensitive circuits.

MBT of such systems is particularly simple since delay-insensitive circuits are generated from *CSP*-like specifications. A *dual* specification (which is by definition a specification of the *environment*) is generated for each component to subject to MBT. The dual specification can, in turn, be used to generate/compile complementary testing components, which are then integrated into the larger circuit for passive testing. The additional testing components are themselves delay-insensitive and may be floor-planned anywhere, including off-chip for cost savings. Their outputs are most often fed to a *JTAG*-type interface for external monitoring.

Delay-insensitive circuits are but one of the possible forms of asynchronous circuit. Other possible design styles (each with their advantages) are, for example, such as to

- Accept arbitrary wire delays, but finite gate delays.

- Arbitrary gate delays with finite wire delays.

- Handshake circuits, which are partially clocked but exchange additional signals regarding the readiness of signals on the wires.

- A variant of handshake circuits known as bundled asynchronous circuits, in which the timing of data lines is better than the timing of control lines.

Despite their advantages, asynchronous circuits still only occupy a niche. The intrinsic problems of clocked circuits were highlighted already in the 1970s by Charles Molnar, and subsequently explored by Ivan Sutherland and Alain Martin at Caltech, Martin Rem at Eindhoven, Janusz (John) Brzozowski in Waterloo, and Steve Furber at Manchester— see for example, Chaney et al. (1973). Over and above the interaction between research groups, there are a few notable products and spin-off companies: *Sun Microsystems* devoted considerable effort to (partially) asynchronous SPARC processor designs; *Philips* developed a variety of asynchronous chips and later spun-off *Handshake Solutions*, a provider of asynchronous design tools. *ARM* collaborated with the University of Manchester on an asynchronous ARM processor. Much of the asynchronous chips body of knowledge is captured in Brzozowski et al. (1995).

16.4.5 Passive testing of networks on chips

One of the latest alternatives to fully-synchronous chips are *NOCs*. This design paradigm consists of two levels: individual components are designed using one of the well-understood design styles (most likely as synchronous/clocked circuits); the components are connected by a *network*.* This yields some specific advantages:

- The individual components can be quite limited in size, making them easier to design and test individually, including their own clock, clock distribution, and race condition avoidance. Indeed, the components may even be obtained from earlier (fully-clocked) versions of a chip.

- The communication between components is network-based and can therefore "piggyback" on the infrastructure and know-how of distributed systems, especially those with communication between processes via message passing over networks (Tanenbaum et al. 2007).

Very little is known at this time about MBT of such systems, and little automatic support for this is provided by the design-flow tools. There are, however, some promising directions for future research:

- All of the work on network intrusion detection and prevention systems (discussed earlier in this chapter) can be leveraged since the on-chip networks are actually Ethernet-like networks.

- The work on scan rings (e.g., JTAG) can be extended to reading out model-checkable states directly on the on-chip network, providing it to an external network port for analysis.

- Several systems have common underpinnings: NOCs, asynchronous circuits, and *CSP* programs. As a result, a promising research direction is to extend the asynchronous and *CSP* tools (e.g., *Tangram* and *Occam*) with the aim of supporting (and compiling for) NOCs.

- MBT of networks (for collisions, saturation, etc.) is well understood and can be implemented at this level as well. For more on this, see Varghese (2004).

A good introduction and reference to the field of *NOCs* are given in Micheli et al. (2006). Model-based design for *embedded systems* is discussed in Nicolescu et al. (2009), whereby the design models for such systems can be used as the foundations of their test models, as it was sketched above in Figure 16.1.

---

*The alternative connection techniques are most often bus oriented or point to point.

16.5   Extended Example (Exercise for Students)

In this section, we want to study a nontrivial but nevertheless not too difficult example to illustrate some of the features of model-based passive testing, as they have been described and discussed in the previous sections of this chapter.

Let us take a *systolic array* (Kung 1982), which can be used, for example, for (N × N) matrix multiplications, as it is described much in detail in Petkov (1992). Such systolic array examples are relevant also as far as *hardware* components—one of the main themes of this chapter—are concerned, since systolic arrays can be implemented on hardware in the form of *field-programmable gate arrays* (FPGA). In our example, for the purpose of focusing on the key issues, we abstract away from the particular numeric calculations of the matrix multiplication in Petkov (1992), and we regard our systolic array very basically as a simple, rectangular *cellular automaton* (Zhirnov et al. 2008) with the following simple transition rules and properties:

- Cells can be in one of the following states: waiting for work (W), processing (P), terminated (T), or error (E).

- A waiting cell will make the transition W → P if data input is available both at its Northern and Western borders; thereafter, it transmits new data to its Southern and Eastern borders.

- A processing cell will make the transition P → T if it can read a special end-of-input symbol at both its Northern and Western borders, or if its Northern and Western neighbors are both in state T themselves.

For all other cases we declare that something is wrong, which would cause an involved cell to transit into a state of error ($X \rightarrow E$), and we will study whether we will be able to catch those cases by model-based passive testing. However, before we design a suitable recognizer, let us first look at two example runs of such a system, a positive one (correct) and a negative one (faulty), which will give us an intuitive impression of what we actually want to model. In this example, we have N = 4, which means that we have (4 x 4) = 16 computational cells in our systolic array.

Figure 16.3 (consisting of 12 subfigures) depicts a *correct* run of the cellular automaton, whereby a "wave" of computation is "rolling" diagonally through the array from its North-Western to its South-Eastern corner.

Figure 16.4 (consisting of 10 subfigures) depicts—for comparison—a *faulty* run of the automaton, caused by some faulty input (●), which forces the receiving cell into a state of error, which then propagates itself through the array via the cells adjacent to the source of the error.

Figure 16.5 shows a comparison of the correct versus the faulty run, in numbers of active cells (y-axis ↑) at every step of the computation (x-axis →).

In the next step, we have to ask ourselves: *What can be observed during runtime of our system under test?* This is always the guiding question in the methodology of model-based passive testing. In the case of software, we could for example observe the effects of additional "*print* (message)" instructions, which could have been injected into some program code for the purpose of debugging. In our systolic array example (which we want to regard as "hardware" in our scenario), let us assume that we can only watch the following seven external observation "pins" (p1, p2, ..., p7) at the Southern and Eastern border of the of our cellular automaton, as it is depicted in Figure 16.6—in other words: we can only make a *partial* (not a complete) observation of our device.

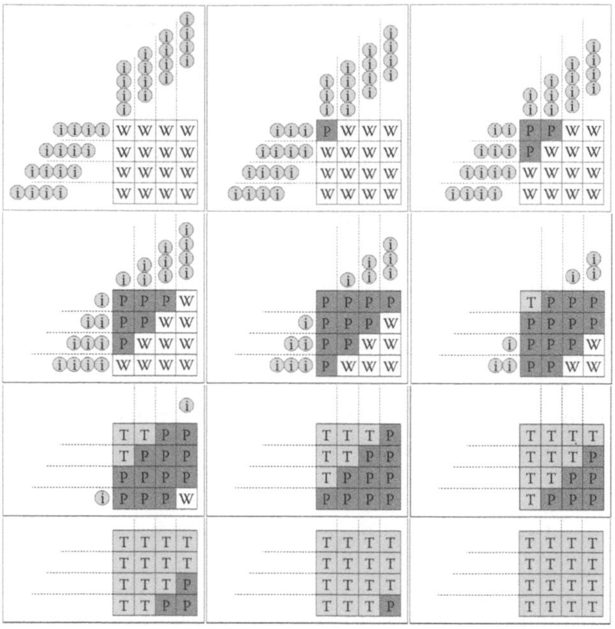

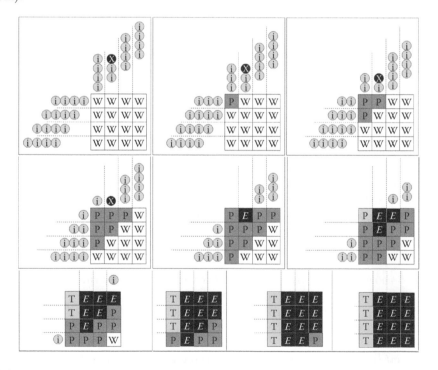

FIGURE 16.3
Correct run of the systolic array in 12 computational steps (from left to right and from top to bottom).

FIGURE 16.4
Faulty run of the systolic array in 10 computational steps (from left to right and from top to bottom).

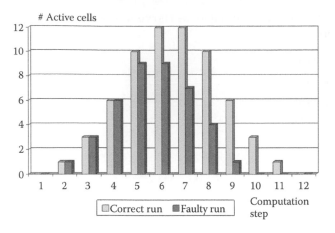

FIGURE 16.5

Cell activities during the correct and during the faulty run of our example.

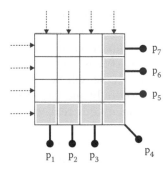

FIGURE 16.6

Seven observation pins, through which the tester observes the behavior of the systolic array at runtime.

During the *correct* run, as shown in Figure 16.3, we can thus observe the following sequence of seven-letter words:

1) WWWWWWW

2) WWWWWWW

3) WWWWWWW

4) WWWWWWW

5) PWWWWWP

6) PPWWWPP

7) PPPWPPP

8) PPPPPPP

9) TPPPPPT

10) TTPPPTT

11) TTTPTTT

12) TTTTTTT

During the *faulty* run, on the other hand, as shown in Figure 16.4, we observe this sequence of seven-letter words:

1) WWWWWWW

2) WWWWWWW

3) WWWWWWW

4) WWWWWWW

5) PWWWWWP

6) PPWWWPP

7) PPPWPP**E**

8) P**E**PPP**EE**

9) **TEEPEEE**

10) **TEEEEEE**

Like in a clinic, where a medical doctor wants to identify diseases by *diagnosis*, each of those seven-letter words is also called a *symptom*, and their union is also called a *syndrome* of the system under observation. This comparison of symptoms and syndromes, as exercised above, can now help us in defining an *alarm system* that monitors the behavior of our cellular automaton. Here, modeling the correct case will be easier than modeling the faulty cases since there is—in our example scenario—only one way of doing a correct computational run of the automaton, whereas there are many different ways and possibilities of running into fault.

As always in scientific modeling, our choice of model also depends on our *purpose*, that is, on what we are actually looking for. The more complex our model, the more details we can "see" and recognize with it—the simpler our model, the less details we can "see" and recognize. Indeed, if we would want to "see everything" in such a test, then our test apparatus could not be less complex than the unit that we actually want to test. However, this would not be feasible in practice. In our example, any correct *individual step* of our 16-cell system is *abstractly represented* by *either* of those seven-letter symptoms, whereas a correct *run* of the system, from start to termination, is *abstractly represented* by the (12×7)-letter syndrome yielded by the correct *concatenation* of those 12 symptoms in the order of the computational steps of our cellular automaton.

There could be other particular application scenarios in which a test engineer could be interested in seeing the occurrence of a specific (sub) *class* of errors (from the many errors that are possible at all). In such a case, our test engineer would have to design a particular recognition model of (or for) that specific class (syndrome) of errors. This is an example of what *philosophers of science* mean when they speak about the "theory-dependence of observations". In our simple example, however, we will not seek to identify particular classes of errors. Here, we are simply interested in the identification of a correct run only, and we will accordingly send a simple, unspecific alarm signal whenever the correct course of events could not be recognized.

Let us thus build a sufficient yet simple formal model of those twelve 7-letter words, each of which represents a correct individual step of the systolic computation of above. It is easy to see that we are here dealing with *palindromes* (because of the symmetric structure of our systolic array), and it is well known that the language of *size-unlimited* palindromes is not a regular language, which means that it cannot be recognized by any memory-less finite automaton (Aho et al. 2007). In our example, however, the *language of all test cases* is a finite set (maximally twelve 7-letter words, one for each step in the trace of the computation),

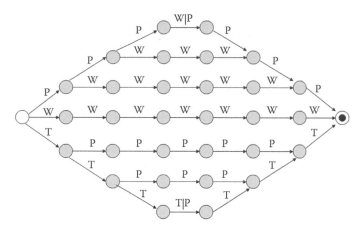

FIGURE 16.7
Acceptor-DFA, for the symptoms of a correct computation in our example.

such that a computationally efficient acceptor module for this language can be specified very easily as deterministic finite automaton (DFA), with 32 states, as it is depicted in Figure 16.7. Not shown in Figure 16.7 is that every nonaccepting state of the DFA also has an "alert exit" for the negative cases, that is, for those characters that would not lead to any of the acceptable symptoms of a correct computational run of our cellular automaton.

It is quite easy to understand now from the constructions of above that our diagnosis model (or test device) has the following property:

> *If a wrong word was detected by the test-DFA,*
> *then the system under observation was faulty.*

However, we may *not* conclude vice versa that the system is OK if no wrong word was detected! The impossibility of such a conclusion is because of the *loss of information* by simplification and abstraction (when we constructed our abstract test model) and is in line with *Dijkstra*'s well-known dictum that testing can only show the presence but cannot prove the absence of errors. For example, our simple test model cannot "see" the actual number-values of the matrix multiplication application within our systolic array; therefore, it can also not "see" if those number-values are correctly multiplied with each other in the inner areas of our systolic array. Our test model in this example can only "see" (via the pins p1, p2, ..., p7) if our seven output cells are in waiting, processing, or terminated states. Moreover, our seven observation pins are also not able to "see" what is going on in the "inner" cells of our DUT. In other words, our test model, though *correct*, is *not* necessarily *complete*—whether or not our model is also *adequate* is then a question of *pragmatics*, that is, our purposes about what we want to do and achieve with it.

16.6 Theoretical Limits of the Passive Testing Method

Every *method* has its limits, and every *methodology* that does not attempt to reflect on the theoretical limits of the methods under its umbrella is not a well-elaborated methodology. Therefore, to "round out" the content of this chapter, we briefly allude to some theoretical limits of passive testing by means of the following small example.

A variant of the well-known *Collatz algorithm* (after the mathematician Lothar Collatz, 1910–1990), can be defined in pseudo code as follows:

INPUT natural number N > 0.

WHILE (N > 1)

    {IF (N even number)

        THEN { N := N/2 }

        ELSE { N := (3 * N) + 1 } }

HALT.

If Collatz's famous conjecture (1937, still unproven to date) is true, then this algorithm will eventually terminate (halt) for *any* (every) given input from the natural numbers.

Let us now try to apply our method of passive testing to Collatz's algorithm and inject trace-emitting instructions at its crucial points, as follows:

INPUT natural number N > 0.

WHILE (N > 1)

    {IF (N even number)

        THEN { N := N/2 ; **PRINT("down")** }

        ELSE { N := (3 * N) + 1 } ; **PRINT("up")** }

HALT.

Under the assumption that Collatz's conjecture is true, it is intuitively evident that this trace-writing variant of the Collatz algorithm implements a function C that maps each input number n to a finite sequence of ups and downs, thus,

$$C : N \rightarrow \{up|down\}^*.$$

We even know with certainty that the last token of every acceptable Collatz-trace must be a "down," namely when the algorithms finally execute the operation 2/2=1 by means of which the loop condition (N > 1) is broken, thus,

$$C : N \rightarrow \{up|down\}^*down.$$

We also know that there exist words in the language L = {up | down}* that do *not* correspond to any possible input number n that could be fed into the Collatz algorithm. Take for example this very short *false* trace "up down." The final action, "down," indicates the operation 2/2 = 1, which means that N = 2 *before* that action. The first action, "up," however, indicates that some action (3*X) + 1 = N must have happened. With N = 2, this implies that X = 1/3, which is not a natural number and thus not allowed in the Collatz domain. In other words, we also know that our characteristic function C cannot be subjective, and every Collatz "module" that would emit the trace "up down" would surely be defective and fail the test.

At this point, the reader should have noticed that in this example, we had indeed intended—however hopelessly—to use the function C as our *test model* by means of which we wanted to passively test and decide whether or not our initial Collatz "module" was defective.

This problem, however, is known to be *computationally irreducible*—in other words: an *accurate* test model to our Collatz "module" *cannot be any simpler or less complex than the module itself*, which we wanted to observe and test in the first place. Thus, we would have to apply the Collatz algorithm itself again, backwards, to find out if we could unwind an observed Collatz-trace step by step until we might arrive at a valid input number, which would then allow us to decide to "accept" the emitted trace. The existence of a *simple and accurate* test model, by means of which we could *correctly and easily* distinguish the "good" from the "bad" up-down traces emitted by our Collatz "module," would imply nothing less than the discovery of an *analytic* correctness proof for Collatz's yet unproven conjecture from the 1930s.

We can thus conclude methodologically that the applicability of model-based passive testing strongly depends on our theoretical ability to discover and construct a suitable test model, and we know at least one counterexample of a computational system—see above—for which such a test model may possibly not even exist.

16.7   Summary

We conclude our discussion of problems and possible solutions in the area of model-based passive testing of safety-critical components with the following points.

- MBT and *passive testing* are closely related to each other, whereby passive testing is usually a special case of MBT. To bring the process of passive testing into operation, formal test models are necessary to specify the required monitor units (see Figure 16.1).

- As far as its *application domain* is concerned, passive testing is better applied to *hardware* components and *distributed systems* rather than software modules for classical mono-processor platforms. Our educative *example*, a *cellular automaton* (systolic array—Section 16.5), was relevant in this regard because systolic arrays are instances of distributed systems, which can also be implemented in hardware as FPGA.

- Nevertheless, model-based passive testing is applicable to *software* modules as well, whereby in this case, we can elegantly combine passive testing with AOP. In this paradigm of programming, the trace-generating auxiliary code can easily be "woven" into the functional modules of a software system under test, as it is, for example, explained in (Subotic et al. 2006) or other literature on AOP.[*]

- In any case, the key idea of passive testing is to let a module (software or hardware) under observation emit a *runtime trace* of *data*, which can be *interpreted* as *information* of the observable operating (running) module about itself. The interpretation of those trace data as test "information" is, of course, based on the definitions of *test model* at the root of this technique. Many different formal techniques and notations can be used to define such test models, such as regular expressions, formal grammars, B, CSP, and

---

[*]Some readers might perhaps have appreciated a more detailed example on trace generation in AOP, but such implementation details would have deviated too far from the purpose and central theme of this chapter, namely the conceptual relationship between abstract trace models and concrete system properties.

the like. The B-notation, for example, was used for the definition of acceptable runtime traces in Howard et al. (2003).

- By means of such kinds of model-based trace interpretations, the actually emitted runtime traces (from the modules under observation in model-based passive testing) are then *classified* as *acceptable* or *nonacceptable*. From a nonacceptable runtime trace, we can conclude that the system under observation did not operate correctly, such that an alarm signal can then be sent. From an acceptable runtime trace, we may, strictly speaking, not conclude anything since testing is—in principle—*not* a suitable method for *proving* the absence of defects with mathematical certainty.

- Model-based passive testing is especially recommendable as an *additional guard* for *safety-critical* systems. Model-based *intrusion detection* by log file analysis of *firewalls* is only one special instance of model-based passive testing in a safety-critical environment.

- A comprehensive theory, or classification, or taxonomy of the various techniques, which can be used to implement model-based passive testing in various software- and/or hardware-based application domains, does—as far as we know—not yet exist and must still be regarded as "future work" for research in this field.

Acknowledgments

Many thanks to the editors of this book for having invited us to write this chapter. Their invitation has provided us with a much-appreciated opportunity to "synthesize" where we often only "analyze"—in other words: the opportunity to bring together under a common conceptual "umbrella" many topics and aspects that are often regarded as rather separate things. Many thanks also to the reviewers from the editorial board of this book: their helpful critique, comments, and remarks about our draft manuscript were instrumental for improving the presentation of our chapter in this publication. Special thanks to Pieter Mosterman for having gone through the tedious task of editorial proofreading, which considerably improved the style of this chapter and "ironed out" a number of "dents." We hope that our readers, especially students and young researchers at the beginning of their careers, will benefit from such a "synthesized" presentation of this complex, rich and future-relevant topic at the interfaces of computer science and computer engineering.

References

Abrial, J. (1996). *The B Book—Assigning Programs to Meanings*. Cambridge University Press.

Aho, A. V., Lam, M. S., Sethi, R., and Ullman, J. D. (2007). *Compilers—Principles, Techniques and Tools*. 2nd ed., Pearson / Addison-Wesley.

Amman, P. and Offut, J. (2008). *Introduction to Software Testing*. Cambridge University Press.

Andrews, J. (1998). *Testing using Log File Analysis—Tools, Methods, and Issues*. Proceedings 13th International Conference on Automated Software Engineering, Pages: 157–167. IEEE Computer Society Press.

Augusto, J. C., Howard, Y., Gravell, A., Ferreira, C., Gruner, S., and Leuschel, M. (2004). *Model-Based Approaches for Validating Business-Critical Systems*. STEP'04 Proceedings of the 11th Annual International Workshop on Software Technology and Engineering Practice. IEEE Computer Society Press.

Baker, P., Dai, Z. R., Grabowski, J., Haugen, Ø., Schieferdecker, I., and Williams, C. (2008). *Model-Driven Testing using the UML Testing Profile*. Springer-Verlag.

Bochmann, G. v., Dssouli, R., and Zhao, J. R. (1989). *Trace Analysis for Conformance and Arbitration Testing*. IEEE Transactions on Software Engineering, Volume 15, Number 11, Pages: 1347–1356.

Broy, M., Jonsson, B., Katoen, J. -P., Leucker, M., and Pretschner, A. (eds.) (2005). *Model-Based Testing of Reactive Systems*. Lecture Notes in Computer Science, Volume 3472 (sub-series: Advanced Lectures), Springer-Verlag.

Brzezinski, K. M. (2007). *Intrusion Detection as Passive Testing—Linguistic Support with TTCN-3*. Lecture Notes in Computer Science, Volume 4579, Pages: 79–88. Springer-Verlag.

Brzozowski, J. A. and Seger, C. -J. (1995). *Asynchronous Circuits*. Monographs in Computer Science, Springer-Verlag.

Campanile, F., Cilardo, A., Coppolino, L., and Romano, L. (2007). *Adaptable Parsing of Real-Time Data Streams*. Proceedings 15th Euromicro Conference on Parallel, Distributed and Network-based Processing, Pages: 412–418.

Cavalli, A., Gervy, C., and Prokopenko, S. (2003). *New Approaches for Passive Testing using and Extended Finite State Machine Specification*. Information and Software Technology, Volume 45, Pages: 837–852. Elsevier.

Chaney, T. J. and Molnar, C. E. (1973). *Anomalous Behavior of Synchronizer and Arbiter Circuits*. IEEE Transactions on Computers, Volume C-22, Number 4, Pages: 421–422.

Cheswick, W. R., Bellovin, S. M., and Rubin, A. D. (2003). *Firewalls and Internet Security—Repelling the Wily Hacker*. 2nd ed., Addison-Wesley.

Chiang, C. and Kawa, J. (2007). *Design for Manufacturability and Yield for Nano-Scale CMOS*. Springer-Verlag.

Clarke, E. M., Emerson, E. A., and Sistla, A. P. (1986). *Automatic Verification of Finite State Concurrent Systems using Temporal Logic Specifications*. ACM Transactions on Programming Languages and Systems, Volume 8, Number 2, Pages: 244–263. ACM Press.

Clarke, E. M., Grumberg, O., and Peled, D. (1999). *Model Checking*. MIT Press.

Dick, J. and Faivre, A. (1993). *Automating the Generation and Sequencing of Test Cases from Model-Based Specifications*. Proceedings FM'93 Conference on Industrial-Strength Formal Methods, Pages: 268–284.

Gargini, P. (2000). *The International Technology for Semiconductors (ITRS)—Past, Present, and Future*. Proceedings GaAs IC: 22nd Annual Gallium Arsenide Integrated Circuit Symposium, Pages: 3–5. IEEE Press.

Gudes, E. and Olivier, M. (2002). *Wrappers—A Mechanism to support State-based Authorization in Web Applications*, in Chen, P.P. (ed.), Data and Knowledge Engineering, Pages: 281–292. Elsevier.

Helmbold, D. P., McDowell, C. E., and Wang, J. Z. (1990). *Trace Viewer—A Graphical Browser for Trace Analysis*. Technical Report UCSC-CRL-90-59, University of California at Santa Cruz, John Baskin School of Engineering.

Holzmann, G. (1997). *The SPIN Model Checker*. IEEE Transactions on Software Engineering, Volume 23, Number 5, Pages: 279–295. IEEE Press.

Howard, Y., Gruner, S., Gravell, A. M., Ferreira, C., and Augusto, J. C. (2003). *Model-Based Trace-Checking*. Proceedings of the SoftTest UK Testing Research II Workshop on Software Testing, Technical Report, University of York.

Jard, C., Jeron, T., Jourdan, G. V., and J. X. Rampon, J. X. (1994). *A General Approach to Trace Checking in Distributed Computing Systems*. Proceedings 14th International Conference on Distributed Systems, Pages: 396–403. IEEE Computer Society Press.

Jerraya, A. and Wolf, W. (2004). *Microprocessor Systems-on-Chips*. Morgan Kaufmann.

Kiczales, G., Lamping, J., Menhdhekar, A., Maeda, C., Lopes, C., Loingtier, J.M., and Irwin, J. (1997). *Aspect-Oriented Programming*. ECOOP'97: Proceedings of the 11th European Conference on Object-Oriented Programming.

Kiczales, G., Hilsdale, E., Hugunin, J., Kersten, M., Palm, J., and Griswold, W. (2001). *Getting started with AspectJ*. Communications of the ACM, Volume 44, Number 10, Pages: 59–65. ACM Press.

Kung, H. T. (1982). *Why Systolic Architectures?* Computer, Volume 15, Number 1, Pages: 37–46. IEEE Computer Society Press.

Leuschel, M. and Butler, M. (2003). *ProB—A Model Checker for B*. Lecture Notes in Computer Science, Volume 2805, Pages: 855–874. Springer-Verlag.

Leveson, N. (1995). *Safeware—System Safety and Computers*. Addison-Wesley.

Memon, A. U. (2008). *Log File Categorization and Anomaly Analysis using Grammar Inference*. M.Sc. Thesis, Queens University of Canada.

McFarland, G. (2006). *Microprocessor Design*. McGraw-Hill Professional.

Micheli, G. de and Benini, L. (2006). *Networks on Chips—Technology and Tools*. Morgan Kaufmann.

Mukherjee, S. (2008). *Architecture Design for Soft Errors*. Morgan Kaufmann.

Nicolescu, G. and Mosterman, P. J. (2009). *Model-Based Design for Embedded Systems*. CRC Press.

Paradkar, A. (2003). *Towards Model-Based Generation of Self-Priming and Self-Checking Conformance Tests for Interactive Systems*. SAC'03 Proceedings of the Annual ACM Symposium on Applied Computing, Page: 1110–1117. ACM Press.

Petkov, N. (1992). *Systolic Parallel Processing*. Advances in Parallel Computing, Volume 5, North Holland Publ.

Richardson, D. J., O'Maley, O., and Tittle, C. (1989). *Approaches to Specification-Based Testing*. TAV3 Proceedings of the 3rd ACM SIGSOFT Symposium on Testing, Analysis and Verification, Pages: 86–96. ACM Press.

Rombach, D. and Seelisch, F. (2008). *Formalisms in Software Engineering—Myths versus Empirical Facts*. Lecture Notes in Computer Science, Volume 5082, Pages: 13–25. Springer-Verlag.

Salomaa, A. and Rozenberg, G. (1997). *Handbook of Formal Languages*. Springer-Verlag.

Santos-Neto, P., Resende, R., and Pádua, C. (2007). *Requirements for Information Systems Model-based Testing*. SAC'07 Proceedings of the Annual ACM Symposium on Applied Computing, Pages: 1409–1415. ACM Press.

Shukla, S. K. and Talpin, J. -P. (2010). *Synthesis of Embedded Software—Frameworks and Methodologies for Correctness by Construction*. Springer-Verlag.

Subotic, S., Bishop, J., and Gruner, S. (2006). *Aspect-Oriented Programming for a Distributed Framework*. South African Computer Journal, Volume 37, Pages: 81–89. Sabinet Publ.

Tanenbaum, A.S. and Steen, M. van (2007). *Distributed Systems—Principles and Paradigms*. 2nd ed., Pearson / Prentice Hall.

Varghese, G. (2004). *Network Algorithmics—An Interdisciplinary Approach to Designing Fast Networked Devices*. Morgan Kaufmann.

Wile, B., Goss, J., and Roessner, W. (2005). *Comprehensive Functional Verification—The Complete Industry Cycle*. Morgan Kaufmann.

Zhirnov, V., Cavin, R., Leeming, G., and Galatsis, K. (2008). *An Assessment of Integrated Digital Cellular Automata Architectures*. Computer, Volume 41, Number 1, Pages: 38–44. IEEE Computer Society Press.

Part V

Testing in Industry

17

Applying Model-Based Testing in the Telecommunication Domain

Fredrik Abbors, Veli-Matti Aho, Jani Koivulainen, Risto Teittinen, and Dragos Truscan

CONTENTS

It is in the public domain that model-based testing (MBT) has been used for years and its benefits have been emphasized by numerous publications.* Despite this, there is still one question pending. If MBT is an excellent way to test, why has MBT not made a major breakthrough in the telecommunication industry? In order to find the answer to this

---

*Results of case studies on industrial context have been explained in numerous publications. Dalal et al. described case studies and experiences in (Dala et al. 1999), and Prenninger, El-Ramly, and Horstmann listed and evaluated selected case studies in Prenninger, El-Ramly, and Horstmann (2005).

question, a MBT methodology was developed for a product testing project at Nokia Siemens Networks (Network 2009).

A systematic methodology was developed during the project. The methodology uses a *system model* instead of test models in contrast with many other MBT systems. In the context of the methodology, the term *system model* refers to models that define the behavior of the system under test (SUT) instead of the behavior of test cases.[*] In addition, during the development, the methodology was constantly evaluated to understand benefits and problematic aspects of the MBT technology.

The methodology includes a process and the supporting tool chain. The methodology was developed for testing functionality of an MSC Server (Mobile Services Switching Centre Server), a network element, using a *functional testing* approach, that is, the product was tested using a *black-box testing* technique.[†] The methodology exploited a so-called *offline testing* approach, where test cases were generated from the models before execution, instead of using an *online testing* approach, where models are interpreted step by step and each step is executed, instantly.[‡]

The content of this chapter will first provide an overview of the project in Section 17.1 that helps understand the following subsections within the chapter. Sections 17.2 through 17.4 focus on process, model development, validation, and transformation aspects. Sections 17.5 through 17.6 describe test generation and test execution aspects. Section 17.7 is devoted to requirement traceability as it spans across the entire process and the tool chain. Conclusions are provided in Section 17.9.

17.1   Overview

Firstly, the overview will provide a high-level view of the process and tools. Next, the SUT and its characteristics are explained. These together provide a detailed context for the work.

17.1.1   Process and tools

The high-level process used in the project is depicted in Figure 17.1. The top level of the process uses the MATERA approach (Abbors 2009a). Four major phases can be identified from the process. First, the requirements are processed and models describing the SUT are created. The models are validated using a set of validation rules in order to improve the quality of the models. Second, the tests are generated from the models. The test generation phase produces executable test scripts. Third, the test scripts are executed with the help of a test execution system. The execution phase produces test logs that are used for further analysis. Fourth, the tests are analyzed in case of failures and requirement coverage tracing is performed. The analysis exploits the test logs and the models. The phases of the process are described in detail in Sections 17.2 through 17.5. In addition, requirement traceability

---

[*]In case the models represent behavior of the tests, the term *test model* is often used. The differences in the two modeling approaches are explained in Chapter 2, Section 2.3 and by Malik et al. (2009).

[†]Beizer defines *functional testing* as an approach that considers implementation as a black box. Hence, the functional testing is often called black-box testing (Beizer 1990).

[‡]Utting, Pretschner, and Legeard describe offline testing in Utting, Pretschner, and Legeard (2006) as a method, where tests are strictly generated before the execution, in contrast with online testing, where a test system reacts on outputs of a SUT dynamically. In the context of model-based verification and MBT the terms *online* and *on the fly* are often used to refer to similar kinds of test execution approaches. Bérard et al. describe the on-the-fly method in Bérnard et al. (2001) as a way to construct parts of a reachability graph on a needs basis rather than constructing a full graph immediately.

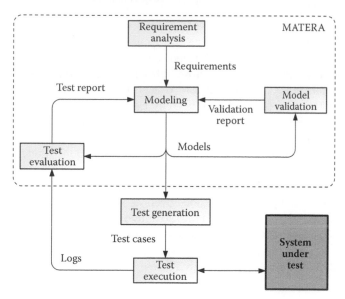

FIGURE 17.1
Process description overview.

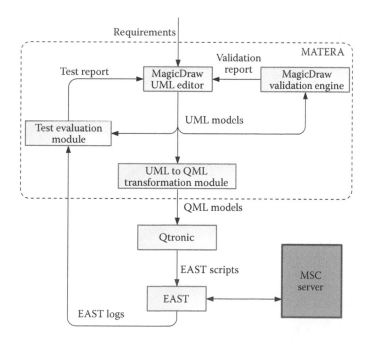

FIGURE 17.2
Tool chain overview.

is discussed in Section 17.6 as it is not restricted to any particular phase but spans across all the phases.

The process is supported by the tool chain depicted in Figure 17.2. The figure resembles Figure 17.1 and therefore indicates the roles of the tools with respect to the described process. The top level is supported by the MATERA framework (Abbors, Bäcklund, and

Truscan 2010), which is developed as a plug-in to the No Magic MagicDraw tool (Magic 2009). In MATERA, the Unified Modeling Language (UML) (Object Management Group models d) is edited, validated, and transformed to Qtronic Modeling Language (QML) (Conformiq 2009b) models and given as an input into the Conformiq Qtronic tool (Conformiq 2009a) for test generation. Qtronic outputs test scripts that are executed with Nethawk EAST (Nethawk 2009). The test logs produced by EAST are analyzed and evaluated against the original models using the MATERA test evaluation function. The relevant details of the tools are provided in Sections 17.2 through 17.5.

The UML models are edited with No Magic MagicDraw tool (Magic 2009). In addition, MagicDraw is used for model validation via custom rules implemented using the Object Constraint Language (OCL) (Object Management Group b). The UML models are transformed with a script into QML models and given as an input into the Conformiq Qtronic tool (Conformiq 2009a) for test generation. Qtronic outputs test scripts are executed with Nethawk EAST (Nethawk 2009). The test logs produced by EAST are analyzed and evaluated against the original models using a test evaluation script. The relevant details of the tools are provided in Sections 17.2 through 17.5.

17.1.2   System under test

The project focused on testing the Mobility Management (MM) feature of a MSC Server, that is, the MSC Server acted as the SUT. It is a key network element of second and third generation mobile telecommunication networks that establishes calls and controls handovers during calls. The MSC Server is capable of handling up to several million users and at the same time provides a near zero downtime. The MSC Server communicates with a number of other network elements as illustrated in Figure 17.3.

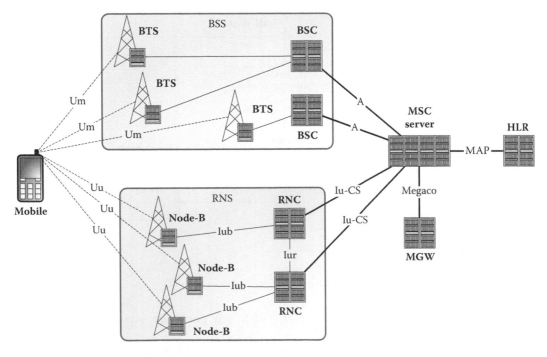

FIGURE 17.3
Project-related mobile telecommunication network elements.

The MSC Server is a typical telecommunication network element from a testing point of view. It has multiple interfaces with other network elements. In addition, the MSC Server communicates with multiple network elements of the same kind, for example, a MSC Server connects to many Radio Network Controllers (RNCs) and Base Station Controllers (BSCs). Communication between the network elements is concurrent and performance scalability aspects are already taken into account in the architecture of the network. The MSC Server also communicates with mobile phones using logical connections, that is, the MSC Server does not have a physical connection with the mobile phones, but the MSC Server uses services provided by other network elements. These elements are part of the radio access networks, that is, Base-Station Subsystem in the second generation network and Radio Network Subsystem (RNS) in the third generation network. The details of the network architecture are specified in the 3GPP Technical Specification 23.002 (The 3rd Generation Partnership Project 2005). Evolution from the second generation GSM systems to the third generation UMTS networks and a detailed description of the latter technology are provided by Kaaranen et al. (2005).

17.2 UML/SysML Modeling Process

The models of the SUT are created following the MATERA approach. The approach starts from the textual requirements of the system, the MSC Server, and incrementally builds a collection of models describing the SUT from different perspectives. In this context, textual requirements refer to the collection of stakeholder requirements, as well as additional documents such as protocol specifications, standards, etc. The modeling phase captures several perspectives (architecture, behavior, data, and test configuration) of the system, at several abstraction levels, by spanning the Functional View and Logical View layers described in Figure 17.4. The functional view defines how the system is expected to behave when it interacts with its users. The logical view describes the logical parts of the system, with behaviors and interactions, without relating to the actual implementation of the system. Each perspective is initially specified on the functional view via the feature, requirements, use case, and sequence diagrams, and it is subsequently refined on the logical view (state machine diagrams, class diagrams, and object diagrams). There are both horizontal (between the perspectives on the same level) and vertical relationships (refinements) among the specification artifacts in this process, as will be illustrated throughout this section.

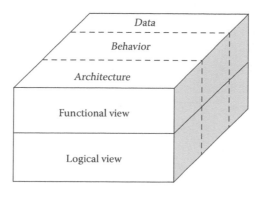

FIGURE 17.4
Modeling perspectives in the NSN case study.

The UML (Object Management Group d) is used as a specification language for system modeling in the project. Additionally, the *requirements diagrams* of the Systems Modeling Language (SysML) (Object Management Group c) are used to capture the requirements of the system in a graphical manner. The No Magic MagicDraw tool was employed to create the models. MagicDraw is a commercial software and system modeling tool, which offers support for both UML and SysML. The tool also offers support for automatic code generation, model validation, model analysis, reporting, etc., and can be extended with the use of various plug-ins.

The system models are created in a systematic manner, based on the MATERA guidelines (Figure 17.5), starting from the textual requirements. The approach consists of five phases. In each phase, a new set of models is created. Each model describes the system from a different perspective. The approach is iterative, so each phase can be visited several times, and the models are constructed incrementally.

The *first phase* deals with the identification of stakeholder requirements, standards, and associated specifications. Since the models are derived from requirements, it is necessary to identify and collect as much relevant information about the system as possible. The system models are later built based on the collected information.

In *phase two*, *feature models* and *requirements models* are created from the information collected in the previous phase. Initially, the features of the SUT are specified using UML Class diagrams (Figure 17.6). The feature models are mainly derived from product requirements and they give a rough outline of which features and functionality the system must be able to perform. Each class describes one feature, whereas *mandatory* and *optional* relationships between features are modeled using aggregation and composition relationships between classes. The feature diagram follows the principles of functional decomposition, where high-level features are decomposed into subfeatures.

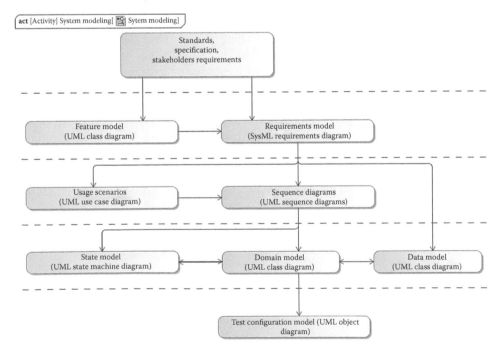

FIGURE 17.5
System modeling process.

A set of requirement models are created using SysML Requirements Diagrams (Figure 17.7). The requirements are specified on several levels of abstraction following principles of functional decomposition. One such model is created for every leaf in the feature diagram. The purpose of these models is to structure the *requirement specifications* corresponding to each feature in a graphical manner. They are structured in requirements diagrams similar to a UML class diagram in which the classes are annotated with the ≪requirement≫ stereotype. A requirement in SysML is specified using different properties including an *id* field, a *textual description*, and the *source* of the requirement. The *textual description* gives a brief explanation of the requirement, while additional details (e.g., technical specifications) are added in the documentation field of each requirement (not visible in the previous figure). The *source* field directs the document to where the requirement has been extracted from.

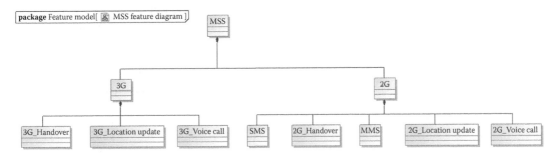

FIGURE 17.6

A feature diagram of the MSC Server.

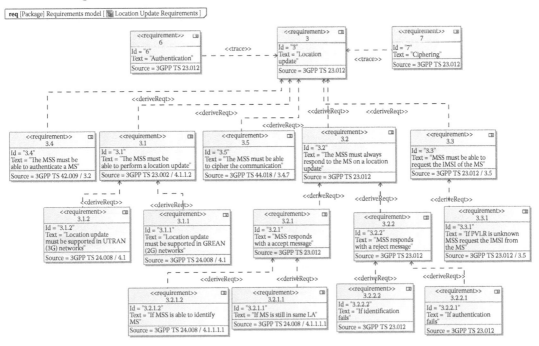

FIGURE 17.7

SysML requirements diagram. (Reproduced from Abbors, F., Backlund, A., and Truscan, D., MATERA—An Integrated Framework for Model-Based Testing, *Proceedings: 2010 17th IEEE Conference and Workshops*, © 2010 IEEE.)

Requirements are traced to other requirements on the same abstraction level or between requirements and other model elements in UML by using relationships such as *DeriveReqt*, *Satisfy*, *Verify*, *Refine*, *Trace*, or *Copy*. If necessary, requirements are also arranged into different categories such as functional, architectural, and communication (data).

In the *third phase*, a *use case model* and a set of *sequence diagrams* are created. The purpose of the use case model (Figure 17.8) is to present a graphical overview of the main functionality of the system. The use case model also identifies the border of the system and the external entities (actors) with which the system must interact. Each use case has a detailed textual description using a tabular format (see Figure 17.9). The description includes fields for precondition, postcondition, actors using the use case, possible sub-use cases, as well as an abstract textual description of the sequence of actions performed by the system when the scenario modeled by the use case is in use. Message sequence charts (MSCs), or sequence diagrams, are illustrated in Figure 17.10 and are primarily used to describe the interactions between different entities in a sequential order, as well as for describing the behavior of a use case, by showing the messages (and their parameters) that are passed between entities for a given use case. Basically the intended usage of MSC is twofold: to discover entities communicating with the system and to identify the message exchange between these entities. The messages exchanged between entities (referred to as *lifelines* in the context of sequence diagrams) are extracted from the protocol specifications referenced by the requirements.

In the *fourth phase*, we define the *domain*, *data*, and *state* models of the SUT. The domain model (Figure 17.11) is represented as a class diagram showing the domain entities as *classes* and their properties (*attributes*). The domain model also describes the *interfaces* that the domain entities use for communicating with one another. Each interface contains a set of messages that can be received by an entity, modeled as class operations. The names of the operations are prefixed with the acronym of the protocol level at which they are used, similar to the approach followed in the sequence diagrams. For instance, the

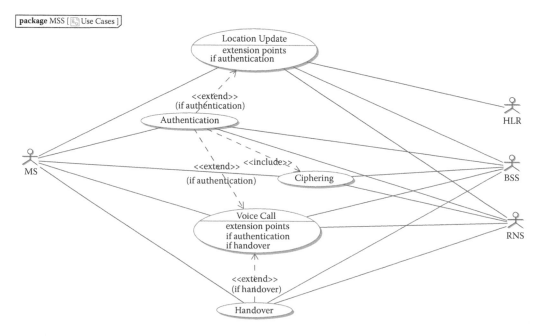

FIGURE 17.8
Use case diagram of the MSC server.

Name	Location Update	
Author	Fredrik Abbors	
Date	1.10.2008	
Actors	MS, HLR, BSS, RNS	
Sub-cases	Authentication, Ciphering	
Description	The MS requests a location update from MSS. The MSS can in some cases initiate authentication and ciphering of the MSS. Finally, the MSS will respond to MS with the location area.	
Pre-conditions	Connection established between MS and MSS.	
Post-conditions	The location area of MS stored/updated in MSS's registers.	
	Actor input	System response
Scenario	1 MS sends requests to MSS to update its location	
	2	MSS responds and accept message containing information about the location area
	3 MS responds with an acknowledge message	

FIGURE 17.9

Tabular description of the location update use case.

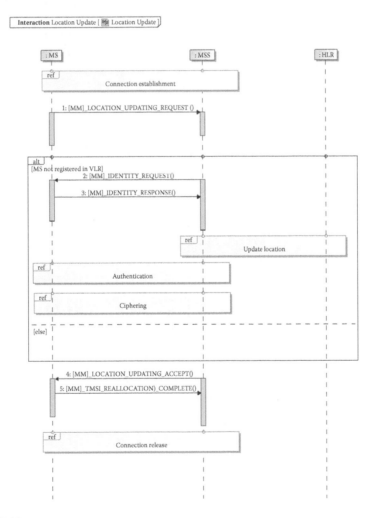

FIGURE 17.10

Sequence diagram describing the location update procedure.

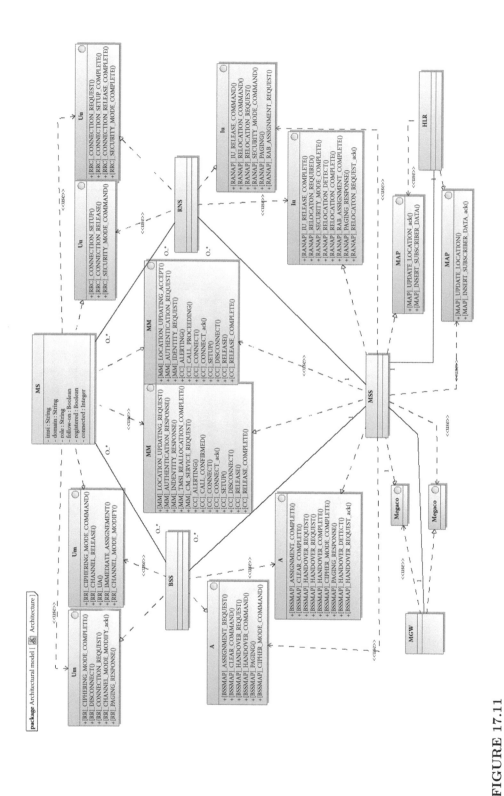

FIGURE 17.11

Class diagram depicting the system domain. (Reproduced from Malik, Q. A., Jääskeläinen, A., Virtanen, H., Katara, M., Abbors, F., Truscan, D., and Lilius, J., Using system models vs. test models in model-based testing. In *Proceedings of "Opiskelijoiden minikonferenssi" at "Tietotekninen tuki ohjelmoinnin opetuksessa,"* © 2009 IEEE.)

[MM]_ prefix depicts a class method used at the MM protocol. The set of messages that can be exchanged between the MSC Server (MSS) and the Mobile Subscriber (MS) at the MM protocol are modeled by two different class interface elements (i.e., MM), one for each direction of communication. The approach allows a clear separation of the communication between different network elements and on different protocol levels. The domain model is built iteratively and is mainly derived from the sequence diagrams and the architectural requirements. Each lifeline in the sequence diagrams generates a new class, whereas the interfaces of the class are obtained from the messages that each lifeline in the sequence diagrams receives.

The *data model* (Figure 17.12) describes the different message types used in the domain model. That is, every message on each interface in the domain model is linked to the corresponding message in the data model. The messages are modeled explicitly via class diagrams. Since, in the telecommunication domain, the main unit of data exchanged between entities is the *message* (or PDU), we focus our attention on how different message types and their structure (*parameters*) can be described. By analyzing the communication requirements and the domain models (where messages have been described as class operation with parameters), we create a data model of the system in which each message type is represented by a *class*, while the parameters of the message are represented as *class attributes*. We structure the message definition based on their corresponding protocols (one diagram per protocol) and use inheritance to model common parameters for a given message. Figure 17.12 shows a class diagram specifying the messages used by the MM protocol. The MM super-class defines the parameters common to all messages, whereas leaf classes define mandatory parameters for each message type. Optional parameters can be added following a similar approach. One important aspect of the data model is that it does not contain all the mandatory fields of a given PDU, but only those necessary to model the SUT at the current abstraction level. The rest of the parameters will be set up during the test generation level when the abstract test cases will be transformed into executable ones (see Section 17.5, Test Concretization).

State models are used to describe behavior of the SUT. The state model of the MSC Server is derived by analyzing sequence diagrams of each use case one by one. That is, for

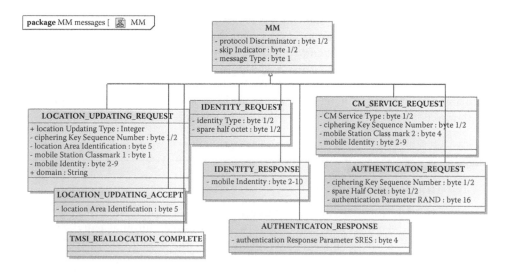

FIGURE 17.12
Data model, defining the structure of MM protocol messages. (Reproduced from Abbors, F., Backlund, A., and Truscan, D., MATERA—An Integrated Framework for Model-Based Testing, *Proceedings: 2010 17th IEEE Conference and Workshops*, © 2010 IEEE.)

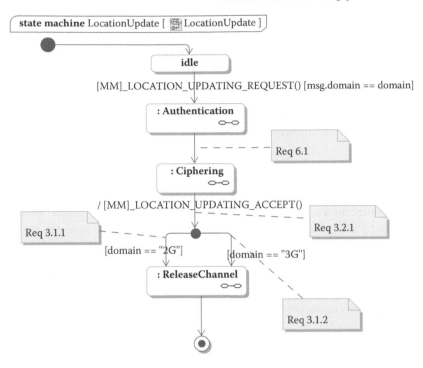

FIGURE 17.13
State machine diagram describing the system behavior.

each sequence diagram, the state of a given object in which each message is received and what messages are sent from that particular state is identified. The former will become *trigger messages*, while the latter will become *actions* on the state machine transitions. By overlapping the states and transitions extracted from each sequence diagram, the full state model of the SUT is obtained. The resulting state model may also contain hierarchical states that will help in reducing the complexity of the model. Figure 17.13 shows a state machine model of the SUT for the location update procedure.

In the *last phase*, a *test configuration model* is created. This model is represented using a UML object diagram and serves to specify the test setup. The elements of this diagram are basically instances of the entities defined in the domain model, showing a particular configuration at a particular point in time. Figure 17.14 shows a test configuration with two mobile phones connected to a 2G network and a 3G network, respectively.

17.3   Model Validation

Humans tend to make mistakes and omit things. Therefore, in modeling, it is necessary for the models to be validated before using them to, for example, automatically generate code or test cases. In our methodology, we take advantage of the model validation functionality of MATERA to check the models for consistency, correctness, and completeness before proceeding to the next step of the process. The idea behind consistency validation is to check for contradictions in the model, for example, a message name in a sequence diagram should match the name of the operation in the class corresponding to that lifeline. Correctness ensures that the models conform to the modeling language (e.g., UML), whereas completeness checks that all necessary information fields have been properly filled out for each

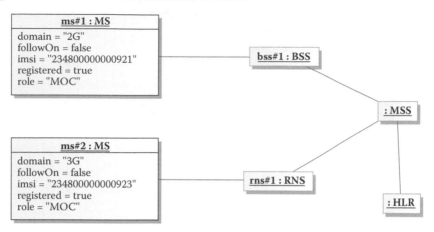

FIGURE 17.14
Test configuration model with two mobile phones connected to the SUT.

element. Model validation can be considered "best practice" in modeling since it increases the quality of the models by ensuring that all relevant information is present or dependencies between elements are correct.

The MATERA framework utilizes the validation engine of MagicDraw for model validation. The engine uses the OCL (Object Management Group b), a formal language for specifying rules that apply to UML models and elements. These rules typically specify invariant conditions that hold true for the system being modeled. Rules written in OCL can be checked against UML models and it can be proved that nothing in the model is violating them. UML, the main modeling language supported by MagicDraw, is accompanied by several predefined suites of validation rules. For example, UML models can be validated for correctness and completeness. There are also additional validation suites for different contexts that can be used. For example, in the SysML context there is a set of validation rules that apply to SysML diagrams. The different validation suites can be checked against either all models or selected models.

Beside predefined validation rules provided by MagicDraw, custom (domain-specific) validation rules can also be created and executed, using the MATERA framework. A complementary set of validation rules was defined (Abbors 2009b) in order to increase the quality of the models with respect to the modeling process. The main purpose of these rules is to ensure a smooth transition to the subsequent steps in the testing process such as generating the input specifications for the test generation tool, or the test generation itself. Next, the validation suites in MATERA and the creation of custom validation rules are described.

An OCL rule normally consist of three parts: (1) a *context* that defines to which language elements the rule applies (e.g., class or state), (2) a *type* that specifies if the rule is a, for instance, an invariant, a precondition, or a postcondition, and (3) the *rule* itself. Optionally, an OCL rule can also contain a *name*. An example of an OCL rule is shown in Listing 17.1.

Listing 17.1
Example of an OCL rule

```
context Region inv initial_and_final_state:
subvertex → exists(v:Vertex|v.oclAsType(Pseudostate).kind =
PseudostateKind::initial) and subvertex → exists(v:Vertex|
v.oclIsTypeOf(FinalState))
```

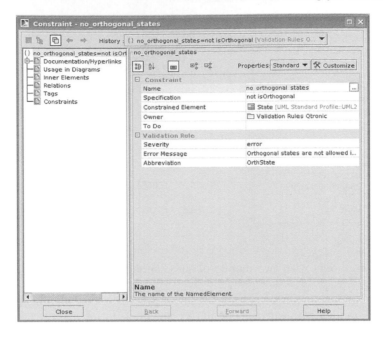

FIGURE 17.15
Screen shot of MagicDraw showing the properties of an OCL rule.

The OCL rule in this example checks that every state machine has an initial and a final state. As one can see, the context of the rule in this example is a `Region` and the type of the rule is *invariant* (`inv`), which means that the expression must always hold true for every instance of type Region. In this example, the OCL rule also has a name, "initial_and_final_state," specified next to the type. By assigning proper names to OCL constraints, it becomes easier to understand the purpose of the constraint and, in addition, it allows the constraint to be referenced by name.

In the validation engine, OCL constraints are treated similarly to model elements. As a result, each constraint has a number of editable properties such as *name*, *specification*, *constrained element*, etc. (see Figure 17.15). The *name* property specifies the name of the constraint. The specification property specifies the rule itself and its type, while the constrained element specifies the context of the constraint.

The validation suites can be invoked at any time during the model creation process. Upon invocation, each rule will be run against the element types or element instances for which it has been defined. If elements violating any rule are found, the user is notified in a Validation Results editor (Figure 17.16). By clicking a failed rule, the elements violating the rules are presented to the user.

17.4 Model Transformation—From UML to QML

The resulting collection of UML models is used for generating the input model for the Conformiq's Qtronic test generation tool, using the MATERA model transformation module. The model in Qtronic is specified using the QML which is discussed more in Section 17.5.1. QML is a mixture of UML state machines and Java, the latter being used as an action

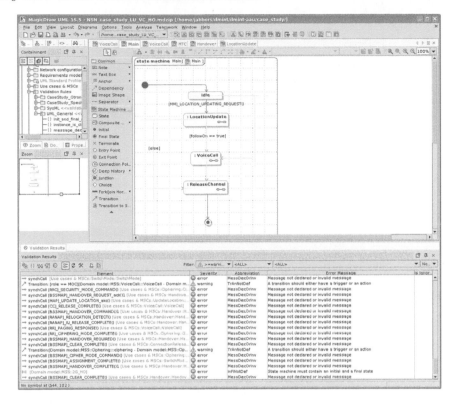

FIGURE 17.16
Screenshot of MagicDraw showing the Validation Results editor. (Reproduced from Abbors, F., Backlund, A., and Truscan, D., MATERA—An Integrated Framework for Model-Based Testing, *Proceedings: 2010 17th IEEE Conference and Workshops*, © 2010 IEEE.)

language. As such, the system models created in MagicDraw are not directly compatible with the Qtronic tool and hence must be transformed into a representation understood by Qtronic, namely QML. The MATERA model transformation module (Abbors et al. 2009) automatically transforms the UML models into the corresponding QML representation. Figure 17.17 shows how different models are mapped onto QML.

17.4.1 Generating the interfaces and ports of the system

In QML, interfaces are specified within a `system`-block. The interfaces describe the ports that can be used to communicate with the environment and which message types can be sent and received on each port. The `Inbound` ports declare messages to be received by the SUT from the environment, whereas the `Outbound` ports declare messages to be sent from the SUT to the outside world.

The ports of the SUT are obtained directly from interface classes in the domain model (see Figure 17.11). `Inbound` messages are taken from the interface realization offered by the SUT, and `Outbound` messages are taken from the interface realization used by the SUT. The name of the ports will be composed of two components, the direction and the interface name. UML operations are listed as messages that are transferred through the ports. The structure of the messages is declared elsewhere as `records`. The partial result of applying the transformation on the MM interfaces in Figure 17.11 is shown in Listing 17.2.

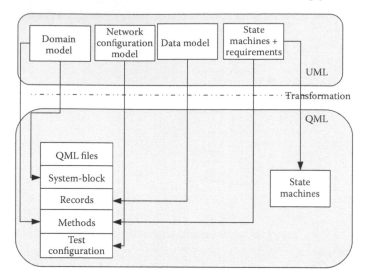

FIGURE 17.17
Mappings from UML to QML.

Listing 17.2
An example of generated Example of QML system-block

```
//System block example
system {
  Inbound MM_in: location_updating_request ,
      authentication_response , identity_response ,
      TMSI_reallocation_complete , CM_service_request , alerting ,
      call_confirmed , connect , connect_ack , setup , disconnect ,
      release , release_complete;
  Outbound MM_out: location_updating_accept ,
      authentication_request , identity_request , alerting ,
      call_proceeding , connect , connect_ack , setup , disconnect ,
      release , release_complete;
}
```


17.4.2   From UML data models to QML message types

In QML, messages are described as records that are used for communicating with the
environment. QML records are user-defined types similar to classes. The fields of a record
can be of type: `byte`, `int`, `boolean`, `long`, `float`, `double`, `char`, `array`, `String`, or of
another record type. In the transformation, records are obtained from classes in the UML
data model. Attributes of the UML classes are transformed into the fields of the record.
Inheritance in UML is reflected in QML using the `extends` relationship. For instance, the
`location_update_request` record in Listing 17.3 is obtained from a class with the same
name in Figure 17.12 following the described approach. The model does not indicate value
ranges of the fields. Instead, the value ranges can be checked by the protocol codecs provided
by the test system.

Listing 17.3

QML record declaration for LOCATION_UPDATING_REQUEST

```
record MM_messages{
  public String protocol_discriminator;
  public String skip_indicator;
  public String message_type;
}
//record inheritance
record location_updating_request extends MM_messages{
  public int location_updating_type;
  public String ciphering_key_sequence_number;
  public String location_area_identification;
  public String mobile_station_classmark_1;
  public String mobile_identity;
  public String domain;
}
```

17.4.3 Mapping the UML state machine to the QML state machine

As mentioned previously, the behavior of the SUT can be specified in Qtronic either textually in QML or graphically using a restricted version of the UML state machines. For simplicity, the latter option was chosen as the target of our transformation. Thus, the transformation is basically a matter of transforming the UML state machine into the corresponding state machine used by the Qtronic tool, which in practice is equivalent with a transformation at the XMI-level. Figure 17.18 shows the same state machine transformed to QML. As one can see, there is a strong similarity between the two models, albeit with small differences. For instance, both state and substate machines are supported and propagated at the Qtronic level. In UML, triggers and actions are declared as methods (selected only from the operations of the interface classes in the domain model).

In QML, triggers are implemented by messages (record instantiations) received on a certain port, whereas actions can be seen as methods of the SUT class definition. This approach allows one to perform further processing of the system data before sending a given message to the output port.* The method generated for the `MM_LOCATION_UPDATING_ACCEPT()` in Figure 17.11 is shown in Listing 17.4.

Listing 17.4

Example of a generated QML method

```
void MM_LOCATION_UPDATING_ACCEPT(){
        MM_LOCATION_UPDATING_ACCEPT location_updating_accept
          ;
        MM_out.send(location_updating_accept);
        return;
    }
```

---

*If necessary, additional QML instructions can be manually inserted into the generated methods, before the .send() statement.

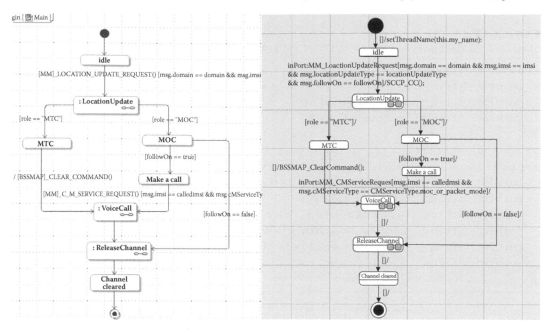

FIGURE 17.18
Example of a UML state machine in MagicDraw (left) and its equivalent in QML (right).

17.4.4   Generating the QML test configuration

As illustrated in Figures 17.11 and 17.14, a :MS is a user of the SUT who communicates with it via the MM interface. The test configuration in our example consists of two MS. In QML, a subscriber is modeled as a record (Listing 17.5) with the same attributes as in the domain model (Figure 17.11). The test configuration is translated into QML in two steps. In the first step, the properties of the test components are extracted from the UML domain model and are declared in the constructor method of the SUT specification class (MSS in our case) as shown in Listing 17.6. In the second step, each test component is instantiated and the properties are initialized with the values taken from the configuration diagram (Listing 17.7). The test components are stored in an array, which is later used for starting concurrent test components of the test system in separate threads.

Listing 17.5
Example of subscriber record

```
record Subscribers {
        public String my_name;
        public boolean followOn;
        public String domain;
        public String role;
        public boolean registered;
        public String imsi;
        }
```

Listing 17.6

Example of subscriber record initialization

```
public InitializeSubscriber(Subscribers sub){
        my_name = sub.my_name;
        followOn = sub.followOn;
        domain = sub.domain;
        role = sub.role;
        registered = sub.registered;
        imsi = sub.imsi;
        }
```

Listing 17.7

main method of the SUT specification class

```
// *** MAIN ***
void main(){
        Subscribers mySubscribers[] = new Subscribers[3];
        mySubscribers[0].my_name = "ms#1";
        mySubscribers[0].followOn = false;
        mySubscribers[0].domain = "2G";
        mySubscribers[0].role = "MOC";
        mySubscribers[0].registered = true;
        mySubscribers[0].imsi = "234800000000921";

        mySubscribers[1].my_name = "ms#2";
        mySubscribers[1].followOn = false;
        mySubscribers[1].domain = "2G";
        mySubscribers[1].role = "MTC";
        mySubscribers[1].registered = true;
        mySubscribers[1].imsi = "234800000000922";

        for(int i = 0;i<=1; i++){
            MSS mss = new MSS(mySubscribers[i]);
            Thread t = new Thread(mss);
            t.start();
        }
}
```

17.4.5 Assigning the state model to the SUT specification

At this point, the only thing that remains to be done is to connect the SUT class specification to the graphical state model. This is done by calling the constructor method for the state machine. Once the state machine has been constructed, the concurrency of the SUT can be tested by starting (via the **Thread.start()** method) separate execution threads for each test component (i.e., subscriber) (Listing 17.8). The approach allows for different MSs to concurrently communicate with the SUT using different configuration parameters. This is necessary as in telecom systems such as the MSC Server, one must test the presence of multiple MSs interacting with the MSC Server (in practice, one MSC Server can serve up to several million users). In addition, we must test calls between pairs of subscribers, where one call requires two subscribers, that is, the caller and the receiver (known as A and B subscribers).

Listing 17.8

State machine instantiation in Qtronic

```
for(int  i  =  0; i <=1;  i++){
            MSS  mss  =  new  MSS( mySubscribers [ i ]) ;
            Thread  t  =  new  Thread (mss) ;
            t . start () ;
        }
```

17.5   Test Generation

The test generation phase follows the modeling phase in the overall process. The test generation exploits the QML models transformed from the UML models. The supporting tool chain implements the test generation phase using the Conformiq Qtronic tool. Qtronic is a tool for Automated Test Design. It derives tests automatically from system models that represent the desired behavior of the SUT. The generated tests are black-box tests and so they evaluate the SUT based on its external behavior, not by monitoring its internal workings.

17.5.1   QML

The systems model given as input into Qtronic is expressed in terms of a language called QML. A model is a collection of the following:

- Textual source files in a Java-compatible but extended notation that describe data types, constants, classes, and their methods.

- UML state-chart diagrams representing the behavioral logic of active classes as an alternative to representing the logic textually.

- Class diagrams as a graphical alternative to declare classes and their relationships.

A QML model is therefore essentially an object-oriented computer program, an abstract implementation of the system to be tested.

The diagrams can be drawn using various tools that Qtronic works with, such as Conformiq Modeler, Enterprise Architect, IBM Rational Software Developer, or IBM Rhapsody. It is also possible to create models completely textually, that is, all the diagram types are optional.

17.5.2   Test generation criteria

Given a system model, Qtronic automatically identifies a number of test cases that together cover the testing goals selected for test generation. Appropriate test input data as well as the correct expected output is automatically calculated and generated by the tool without further input from the user.

For this, Conformiq Qtronic uses semantics-driven methods for generating test suites, which means that test generation is guided by an analysis of the behavior implied by the model, instead of being based on syntactic analysis or simple heuristics. Qtronic uses model-based coverage criteria to select a set of test cases to form a good test suite. The coverage or testing goals are used to guide Qtronic to look for certain behaviors from models or to enable certain behaviors.[*] A test case covers a certain testing goal if execution of the test against the model itself would cause the goal to be exercised. Then, Qtronic uses its capability to simulate the system model to construct test cases, and at the same time, it maps the test cases to the different test goals that result from the coverage settings. It then selects from the constructed test cases a set that covers all of the resulting test goals using a minimal cost test suite. This ensures that the suite is reasonably small and compact, and at the same time, the individual test cases remain relatively short, which eases test execution and debugging. In addition to this, Conformiq Qtronic also prefers covering all test goals as early as possible, that is, after as few messages as possible, providing better separation of concerns between test cases.

A test suite generated by Qtronic has the following characteristics:

- In order to have good error detection capabilities, the generated test suite covers as many testing goals as possible.

- In order to avoid redundant testing, the generated test suite is as compact as possible while individual test cases in the suite are relatively short in order to ease the test execution and debugging.

- In order to provide better separation of concerns between test cases, the test goals are covered as early as possible in test cases.

Qtronic makes the testing goals accessible to the user in the Qtronic user interface (UI). By selecting different testing goals, the user can affect how Qtronic generates test cases. This is the primary vehicle in Qtronic for a user to have a say in how the tests are generated. Figure 17.19 shows testing goals in the Qtronic UI. From this view in the coverage editor the user can see, for example, that the generated test suite covers all requirements (more on requirements below) related to category "1.2" but not all requirements in some other categories. One can also see that 62% of all the states and 60% of the transitions are covered by the test suite.

The selection testing goals that are used depend on how extensive a test suite the user wishes to generate and also on the characteristics of the model and hence the SUT itself. For example, if the model has lots of interesting boundary conditions related to inputs to the system, it is highly recommended for one to enable Boundary Value Analysis as a test generation criterion (see Figure 17.19). With boundary value analysis enabled, Qtronic will attempt to exhaustively generate all possible boundary values based on the various conditions in the model.

Not all testing goals are always reachable. One can, for example, have conflicting if-statements in the model, which cause a certain boundary value case to be statically unreachable. An important aspect of how Qtronic works is that Qtronic will give the user a precise account of which coverage goals were covered by the generated test suite and which ones were left uncovered. Therefore, after test generation, the user knows exactly how well the generated test suite covers the different functionalities of the modeled system (and can react to uncovered areas and change the model if it turns out that there was a defect in the model itself).

---

[*]Both of the terms, *coverage goal* and *testing goal,* are used in this chapter and they mean the same thing.

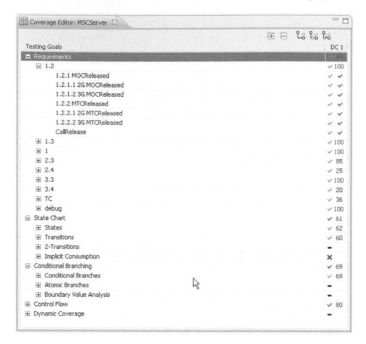

FIGURE 17.19
Testing goals in the Qtronic user interface.

17.5.3 Requirements traceability

Most of the testing goals are related to various properties of the model itself (such as state, transitions, conditional branches, etc.). In addition to these structural coverage criteria, user-defined requirements can be used in models to guide the test generation.

Technically, the requirements are embedded in the model using the requirement keyword, as illustrated in Listing 17.9.

Listing 17.9
An example of a requirement in a model

```
// A requirement embedded in a model
requirement ``This is a requirement'';
```

For the Qtronic algorithm, a requirement is one additional coverage goal in the model to be covered (see discussion on testing goals in Section 17.5). In generating tests, Qtronic will attempt to generate a test suite that covers all requirements embedded in the model at least once, assuming that the user has chosen requirements as a testing goal prior to starting test generation. The notion "covering a requirement" means that there is an execution in the model that passes through the point where the requirement is defined.

At the end of test generation, Qtronic will report how well the requirements in the model were covered by the generated test suite (just like it does for all other testing goals). What is particularly interesting about the requirements coverage is that it provides automatic traceability from the functional requirements of the system all the way through to the generated test cases and test execution. A traceability matrix maps requirements to test cases, allowing a tester to easily pick up test cases to exercise certain functional areas identified by the corresponding functional requirements. An example of a traceability matrix can be seen in Figure 17.20. For example, if the user wishes to execute a test case that

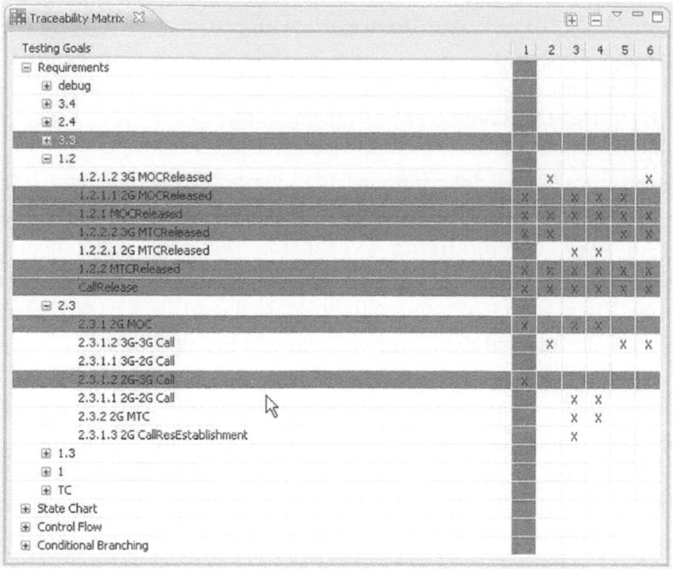

FIGURE 17.20
Traceability matrix in the Qtronic user interface.

exercises a "mobile-originating call from a 2G network to a 3G network," then test case 1 would have to be executed as can be seen from the "Xs" in the highlighted column for test case 1. Furthermore, a failure in test execution can be easily attributed to a functional area by looking at the traceability matrix.

17.5.4 Test concretization

Once the QML system model is created and tests are generated, we have a set of valid test artifacts in terms of test cases, messages, parameters, and testing goals. Test artifacts, from the Qtronic database, must be transformed into an executable format for test execution.

NetHawk's Environment for Automated System Testing (EAST) [3] is used for test execution. EAST embeds the SUT in a virtual network incorporating protocols over interfaces under test. EAST provides a Test Creation Environment (TCE) with graphical programming language for defining tests, test suites, message templates etc., and a Test Execution Environment (TEE) for test execution. In the project described in this chapter, TCE is utilized for creation of the message reference library. TEE is used for executing test cases in Load Testing mode, which provides an execution performance close to real network elements. EAST TEE connects a test execution engine to the desired protocol server required by a protocol under test. A Protocol server is a standalone program simulating a protocol stack below the protocol under test. In this context, *Protocol under test* refers to the level of the modeled behavior of the SUT.

The message reference library is a collection of message descriptions and encoding rules of telecom *protocols under test*. In other words, it is a definition of the run-time behavior of messages. EAST provides most Protocol Data Units (PDU) of telecom protocols in the form of message templates with default content. Each value of the template field in the reference library is accessible through an API. It is up to the user to select the necessary

PDUs from the reference library and create mappings between the parameters from test artifacts and reference library messages. Data models, depicted in Figure 17.12, define the interfaces under test on an abstract level when the message reference library is the actual implementation of PDUs. One could think of the reference library as a refinement model for the data model.

For example, the data model defines a message *MM_LocationUpdateRequest* and Qtronic has generated test artifacts defining parameters and behavior of when the message is sent by the test system. In an executable test case, the event of sending the *MM_LocationUpdateRequest* message will be composed from the implementation of the message in the reference library and of test artifacts calculated by Qtronic. Figure 17.21 shows the relation between reference library and generated test artifacts.

In order to concretize such an executable test case, it is required that there be a tool to combine Qtronic test artifacts and the message reference library refinement data producing EAST executable test cases. For this purpose, a test scripting back-end was implemented. The back-end is connected to Qtronic using an open API. Through the open API, it is possible to create a custom output format (e.g., EAST scripts) from test artifacts and utilize external test libraries such as the EAST reference library. The back-end creates a set of test cases that can be executed on EAST TEE.

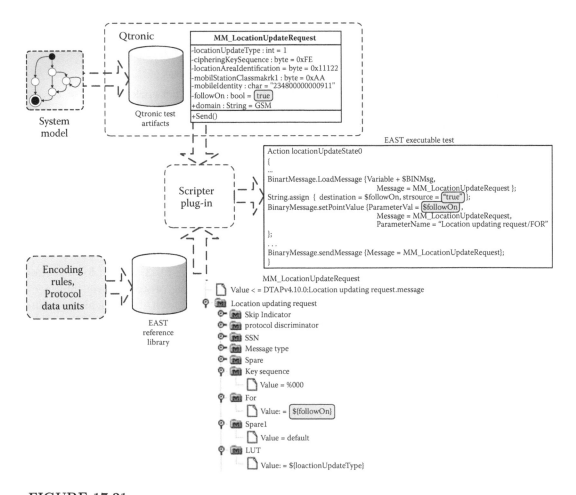

FIGURE 17.21
Test concretization.

The message reference library may have several versions supporting several protocol versions. All versions implement the same API in order to keep maintenance work to a minimum. The API defines the mapping between the reference library messages and the abstract definitions of messages. If the specification version of the protocol under test changes, the reference library version also must be changed. Abstract definitions and the data model should not have to change.

17.6   Test Execution

Test execution is performed with the help of a test system. The test system is illustrated in Figure 17.22. The test system operation is automated with a custom-made Test Automation Script (TAS). The test system is based on the NetHawk EAST Test Execution Environment (TEE) and on *protocol servers* providing connectivity to the SUT. In addition, the message reference library, described in Section 17.4, is providing run-time behavior for PDUs called from EAST test scripts. EAST produces test logs on each test run. The logs are used in the analysis and for coverage tracing purposes later on as will be discussed in Section 17.6. The communication between the MSC Server and the test system is monitored with the Wireshark (Wireshark) protocol analyzer. The message reference library was discussed earlier in Section 17.4 from the test concretization point of view. During the test execution, the message reference library is providing an API for encoding and decoding rules for messages sent and received. For example, if a message *MM_LocationUpdateRequest* is about to be sent by the TEE, the message reference library provides information about online encoding of the message.

The TAS is used for starting related test steps at the same time, for collecting combined coverage results, and for calculating the final verdict of the test case. After the system model is created and test generation performed, the TAS takes care of all steps required in the test execution phase.

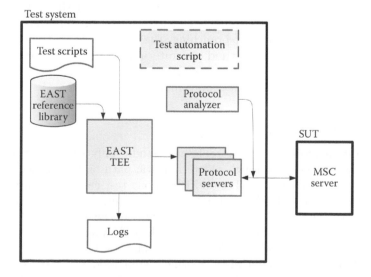

FIGURE 17.22
Architecture of the test system.

17.6.1   Load testing mode

The Load Testing mode has been selected from different execution environments provided by EAST TEE. The Load Testing mode provides an execution engine allowing several simultaneous calls and has a means for identifying different calls by a special Context. The Context in our environment defines a sequence of actions or events having certain unique characteristics. For example, when simulating an originating phone call, the calling party has its own phone number and the called party has its own number. Both, originating and terminating, calls have their own Context that can be addressed by a phone number as phone numbers are unique. Also, while a phone call is ongoing, the Context can be referred to at a protocol level through a specific connection in case of a connection-oriented communication.

17.6.2   Concurrency in model-based testing

Modeling concurrent systems, such as telecom network elements, requires some extra attention in the modeling phase in order to obtain executable tests from the test generation. The SUT is concurrent by nature, which means that several sequences of events are happening at the same time without synchronization between them. This leads to a situation where it is possible to have large amounts of different, but correct, interleaved sequences of elementary procedures. For example, Event A (EA) has 10 elementary procedures that always happen in a known order. Respectively, Event B (EB) has 10 elementary procedures. In a case where EA and EB are happening in parallel, there are $2^{20-1} = 524,288$ possibilities for interleaved elementary procedures of EA and EB. In reality, there are some relationships between EA and EB which narrow down the possibilities. But, when looking from a test generation point of view, it is difficult to create information on models that rule out the possibilities that do not take place in reality. And even after that, it is neither feasible nor wise to generate all possible variations. The most important point is to not generate a test that has one fixed order of parallel events. That is because it is not known beforehand in which order these parallel events happen in the SUT.

For example, originating and terminating calls receive a confirmation when they are connected to each other, but it is not specified which phone receives the confirmation first. Some branching would be necessary in a single test case to cover these types of situations. This will lead to a very complex test case that requires a lot of computational power from the test generation tools. Instead of branching, the solution is to treat originating and terminating calls as separate threads and generate independent test steps for both calls. These steps together form a test case. Both steps are increasing the test coverage (see Section 17.4) of the test case while they are running.

17.6.3   Executable test case

MBT test generation tools such as Qtronic can handle only deterministic behavior. An individual test case should go through a deterministic path on each execution round. But as illustrated in the above example, EA and EB as one test case, there are several options on how a valid test case could behave on interleaved execution leading to nondeterministic behavior. At first, this sounds like a restriction from the test generation tool perspective, but when thinking about this dilemma openly, this is the only way that a test generation tool could produce something reasonable. Concurrency must be handled in the test execution environment and taken into account when creating system models. The system model should define a separate thread for both the originating and terminating phone call. The test generation tool produces one test case where EA and EB are combined in one interleaved sequence of events. In the creation of an executable test in the test concretization phase, described in Section 17.4, executable events of EA and EB are separated on their own test steps. The entire process is depicted in Figure 17.23.

17.6.4 Context

Figure 17.24 depicts the message routing from the SUT to the correct test step in the test system and vice versa. In the picture, the Router is a book keeper of the active Contexts in the test system. It compares the received message type, header, and possible payload to infer the correct Context. The EAST test case from the example above has two test steps with unique phone numbers. There is also a possibility to define a unique protocol transport layer identifier for each test step, for simplicity, is referred to as *connectionID*. Most of the communication between the SUT and the test system is connection oriented and it is easy to determine an identifier, *connectionID*, for each connection during the connection establishment phase. However, because of the nature of the telecom network, there are some connectionless messages exchanged within the network. These messages are mainly used for paging other phones or resources before the connection is known. Here, it is assumed that these broadcast messages carry an address of an entity to which the message belongs. To simplify the example, the address is referred to as *phoneNumber*.

Generated EAST scripts, one script for each test step, are uploaded to the Load Runner. Each test step registers their Context with a run-time database storing the relation between the Context and the test step. In the example Listing 17.10, the Context can be addressed through two context identifiers, *connectionID* and *phoneNumber*.

Listing 17.10
Example of the relation between the Context and the test system

```
String.assign    {    destination = $connectionID , strsource = "1" }
LoadEngine.setContextId   {   Variable = $phoneNumber , type   = "reference" }
String.assign    {    destination = $phoneNumber , strsource = "62030614610001" }
LoadEngine.setContextId   {   Variable = $phoneNumber , type   = "reference" }
```

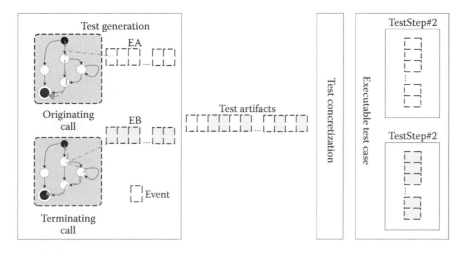

FIGURE 17.23
From the system model to executable test.

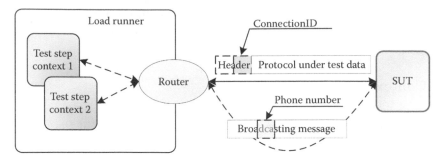

FIGURE 17.24
Context-based message routing.

17.6.5   Run-time behavior

There are two possibilities for the test step to identify its Context. First, it could do a connection establishment after which *connectionID* can be used to address messages through the router. Second, the test step waits for a broadcasting message with *phoneNumber* defining the correct Context after which the connection should be created as in the first case. Broadcast messages are sent to all test steps but discarded if the *phoneNumber* does not match. After the Context is known, the Router depicted in Figure 17.24 knows the destination of each message.

EAST TEE is writing a textual log that contains information of executed events and covered requirements. The log file is an input for the requirement traceability phase described in Section 17.7.

17.7   Requirement Traceability

Conventional test scripting is based on static test cases, that is, a test case has a name and corresponding test code that is not modified significantly between consecutive versions. However, this is not the case when tests are generated from models automatically. In automatic test generation, there are no guarantees that different test generation rounds produce similar test scripts. Thus, as part of the methodology, an approach was developed for tracing the requirements throughout the entire MBT process, from models to test runs and back to models as illustrated in Figure 17.25. This approach provides a consistent way to monitor the modeling, generation, and test execution status, as well as the test completion rates. In addition, tracing of requirements helps understand what kind of impact new or modified requirements have on different artifacts of the testing process. All in all, the requirement tracing process supports short feedback loops that, in turn, support modern product development conventions such as iterative and agile development practices.

The use of the requirements fits well in product testing because the features of the product are defined as a set of requirements. It does not require any additional information on top of the regular product development information. Instead, it uses information that is readily available in the product development. This makes it easier to integrate the MBT tool chain with the current frameworks and with the tools used in product development. In addition, it is also important to be able to evaluate how many tests cover each requirement to support test prioritization and optimization.

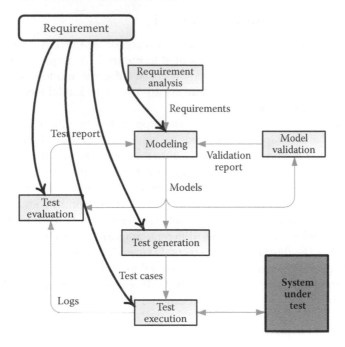

FIGURE 17.25
Tracing requirements throughout the MBT process.

The use of the requirement tracing also helps in the process of test selection and prioritization. Typically in product development, some features (requirements) of the product are more important than others. This affects the testing process in a way that more important features must be tested more extensively compared to others. The prioritization of the test cases can be done in two different phases of the testing process, during test generation or test execution. If the behavioral model of the SUT used for test generation is adorned with priority information, then the test generation tool will be able to generate test cases and calculate their execution order. However, if such a possibility is not supported by the test generation tool, then having the generated test cases tagged with the requirements they are testing will allow the selection of the most important test during execution based on the priority of the requirement. Especially in the telecom domain, where the number of generated test cases tends to be very large, having the possibility to select the most important test cases to be executed based on the requirement to be tested is an important aspect.

At a more detailed level, requirement tracing allows requirements to be propagated to test specifications. Further, functional testing must verify that all requirements have been covered by test. Needless to say, requirements are the keystone in any successful project implementation, and hence, they must be traceable both to models and tests. Tracing requirements during development ensures that all requirements have been implemented and that no functionality has been overlooked. Tracing requirements to tests can even help in identifying missing tests, that is, where critical requirements do not trace to any test. Finally, if a test fails, one can trace the requirement back to the models from where it originated, in order to identify the error. This facilitates the process of identifying which parts of the system model cause a set of test cases to fail. It is described in brief how these aspects are addressed in the following subsections.

17.7.1   Tracing requirements to models

Requirements traceability is one of the key features of MATERA and is based on the approach described in Section 17.2. As presented in Figure 17.7, requirements are structured hierarchically. Top-level requirements are traced to different models (e.g., state machine diagrams), whereas the rest of the requirements are traced to model elements to which they apply (e.g., a transition or a state).

In our modeling process, requirements are propagated to different parts of the models to indicate a relationship between requirements and model elements. For instance, communication-related requirements are traced to data models, architecture-related requirements to architecture models, and functional requirements are initially traced to use case models and then to state models. Figure 17.26 shows an example of how a requirement can be linked to a model element in MagicDraw (e.g., to a transition in a state machine). These links can be useful both for evaluating that all the requirements have been reflected in the models, by showing what elements from different diagrams "implement" a given requirement, and for tracing the requirements to tests. Figure 17.13 presented a partial state model of the SUT in which various transitions have tags depicting requirements that they satisfy.

Once all requirements have been traced to UML elements, validation can be applied to verify that no requirement has been overlooked and that the models are ready for transformation.

17.7.2   Tracing requirements to tests

As mentioned in Section 17.4, Qtronic offers support for tracing requirements during test case generation. MATERA propagates the requirements linked to model elements into QML specifications. During the UML to QML transformation, the requirements are propagated from UML models (namely from state machine models) to QML state machine models. In the current case study, only those requirements that are attached to state transitions are collected from the UML state models. However, nothing prevents the collection of requirements from other UML models.

As such, the requirements are captured from UML transitions and placed on the corresponding transition in QML by using the `requirement`-statement (as explained in Section 17.5). Hierarchy can also be propagated from UML to Qtronic based on the numbering scheme of the requirement.

FIGURE 17.26
Tracing requirement to a model element.

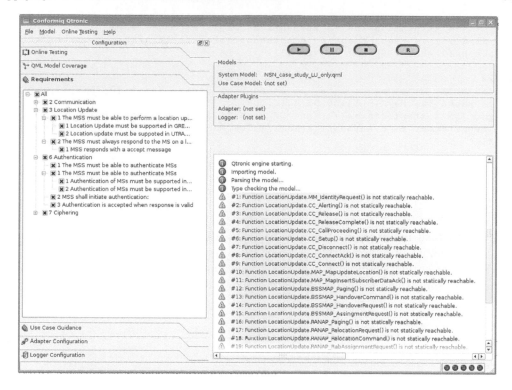

FIGURE 17.27
Example of Qtronic's UI with requirements.

Upon importing a QML model in Qtronic, requirements are displayed hierarchically in Qtronic's UI (Figure 17.27) from where requirements-based test derivation can be pursued. In addition, the user can select which requirements should be covered by Qtronic during test case generation.

During the test case generation, Qtronic propagates the requirements to EAST test scripts via the scripting back-end discussed in Section 17.4. After running the generated test scripts in EAST, the output of the test run is stored in EAST test logs (see Listing 17.11). The test logs contain detailed information on test steps, executed methods, covered requirements, etc.

Listing 17.11
Excerpt from an EAST test log.

```
Log:20
1::1::Symbol: Statement ::Label: Statement :: Script Name: TC_TEST_2G_ms#1_2
    :: When: Tue May 19 14:50:53:451 2009 ::
Assign( ("6 Authentication /1 The MSS must be able to authenticate MSs /1
    Authentication of MSs must be supported in GERAN (2G) networks") ) =
    requirement ( "6 Authentication /1 The MSS must be able to authenticate
    MSs /1 Authentication of MSs must be supported in GERAN (2G) networks" )

Log:21
1::1::Symbol: Statement ::Label: Statement :: Script Name: TC_TEST_2G_ms#1_2
    :: When: Tue May 19 14:50:53:451 2009 ::
```

```
Assign( ("6 Authentication /3 Authentication is accepted when response is
    valid") ) = requirement ( "6 Authentication /3 Authentication is
    accepted when response is valid" )

Log:22
1::1::Symbol: Statement ::Label: Statement :: Script Name: TC_TEST_2G_ms#1_2
    :: When: Tue May 19 14:50:53:451 2009 ::
Assign( ("6 Authentication /1 The MSS must be able to authenticate MSs") ) =
    requirement ( "6 Authentication /1 The MSS must be able to authenticate
    MSs" )
```

17.7.3 Back-tracing of requirements

This approach is the opposite to the ones presented above and it consists of two parts. Firstly, statistical information about the test run regarding the number of passed/failed test cases, number of requirements covered, validated, etc., is collected in a report. The MATERA test evaluation module is used for analyzing the EAST test logs and generating the report (a simplified version of the report is shown in Figure 17.28). During the analysis,

---

Statistics of the test output from EAST[tm]

This HTML log has been automatically generated by a script.

The log contains statistical information about test output from EAST as well as requirements coverage information

Generated on: Fri August 14 14:52:47 2009

Requirements Information

Failed Requirements

3.1.1 Location updated must be supported in GREAN (2G) networks

6.1.1 Authentication of MS's must be supported in GERAN (2G) networks

6.1.2 Authentication of MS's must be supported in UTRAN (3G) networks

7.1 The MSS must be able to cipher the communication with MS's

7.2 The ciphering procedure is initiated by the MSS after a successful authentication procedure

Uncovered Requirements

6.1 The MSS must be able to authenticate MS's

Test Case Information

The testrun generated a total of 10 testcases

Requirement 3.1.1 was present in 2 testcases

Requirement 6.1.1 was present in 1 testcases

Requirement 6.1.2 was present in 1 testcases

Requirement 7.1 was present in 1 testcases

Requirement 7.2 was present in 1 testcases

Model Coverage Information

The testrun covered 4 of 10 requirements in the model

The testrun covered 1 of 3 use cases in the model

The testrun covered 8 of 24 transitions in the model

FIGURE 17.28
Example of statistical information from a test run.

MATERA also collects information from the UML models in order to calculate how different parts of the model have been covered.

Secondly, information concerning failed test cases is collected in order to identify from which parts of the system model a given failed test was generated. In this way, one can see which parts of the system model are not in sync with the real implementation, and therefore with the stake holder requirements. The previously mentioned test evaluation module also retrieves information about requirements from the EAST test logs and generates a set of OCL constraints used by MATERA. The generated constraints have two purposes.

- To locate in the model the uncovered requirements by the test run, in which case their parent requirement will also be located.

- To locate the models and their elements that "implement" a given uncovered or failed requirement.

Listing 17.12 shows an example of generated OCL constraints.

Listing 17.12
Example of generated OCL rules

```
-- Expressions for tracing uncovered or failed requirements in SysML
    requirements diagram:
not(Id = '3') and not(Id = '3.1') and not(Id = '3.1.1') and
not(Id = '6') and not(Id = '6.1') and not(Id = '6.1.1') and
not(Id = '7') and not(Id = '7.1') and not(Id = '7.2')

-- Expressions for finding uncovered or failed requirements placed on
    transitions:
not(clientDependency->exists(a |
(a.supplier.name->includes('3') or a.supplier.name->includes('3.1') or a.
    supplier.name->includes('3.1.1') or a.supplier.name->includes('6') or a.
    supplier.name->includes('6.1') or a.supplier.name->includes('6.1.1') or a
    .supplier.name->includes('7') or a.supplier.name->includes('7.1') or a.
    supplier.name->includes('7.2')))) and clientDependency->notEmpty())
```

The generated constraints are automatically loaded and interpreted in the MATERA framework, which will highlight the targeted elements. Once loaded and executed in MATERA, the first expression will trace and highlight in the diagram editor the set of requirements to which it is attached (see Figure 17.29). Similarly, the second rule will highlight the elements to which the previous requirements were linked. For instance, Figure 17.30 shows a screenshot of MagicDraw displaying the transitions in the state model to which *requirement 3.1.1* is linked.

17.8 Related Work

There is much literature on MBT in general and specific aspects in particular. For instance, Utting, Pretschner, and Legeard (2006) suggests a taxonomy for MBT, while in Hartman (2002) and Utting and Legeard (2006), the authors discuss MBT from a tools perspective, both academic and industrial. MBT techniques and tools have also been researched and developed in the context of the *Automated Generation and Execution of Test Suites for Distributed Component-based Software* (AGEDIS) (http://www.agedis.de) and *Deployment of Model-based Technologies to Industrial Testing* (D-MINT) (http://www.d-mint.org)

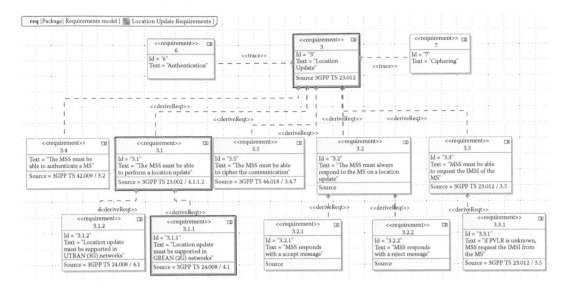

FIGURE 17.29

Back-tracing of requirements to the requirements model. (Reproduced from Abbors, F., Backlund, A., and Truscan, D., MATERA—An Integrated Framework for Model-Based Testing, *Proceedings: 2010 17th IEEE Conference and Workshops,* © 2010 IEEE.)

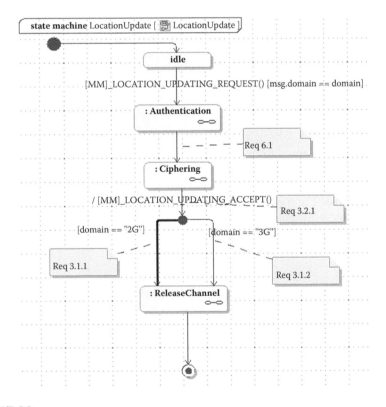

FIGURE 17.30

Back-tracing of requirements to a state machine model.

projects. The projects focused on the applicability of MBT techniques to industrial environments, and numerous publications and reports can be found on the web pages of each project. Both projects were applied to industrial case studies from different application domains, including telecommunications.

For instance, the authors describe in Born et al. (2004) a Model-driven Development and Testing process for Telecommunication Systems. A top-down approach based on CORBA is proposed in which the system is decomposed into increasingly smaller parts that can individually be implemented and tested. They make use of UML and profiles, in particular the Common Object Request Broker Architecture (CORBA) Component Model (CCM) (Object Management Group a), for specifying system models. These models are later on mapped onto software components using code generators, and executed within the CCM platform. Testing is also supported using specific UML profiles, in particular UML2 Testing Profile (U2TP), and by generators. The test design is tightly coupled with the system design so as to be able to reuse information provided in the system design, as soon as it becomes available. Testing is based on contracts built into different components and is derived from the models of the system. Test models are specified using U2TP and are used for the derivation of Testing and Test Control Notation (TTCN-3) tests. The TTCN-3 test are later automatically translated into executable Java byte code.

An approach similar to ours where several telecommunication case studies are evaluated against the Qtronic tool is discussed in Khan and Shang (2009). The main target of the work is to investigate the applicability of MBT in two telecommunication case studies. Another evaluation of MBT via a set of experiments run on Qtronic is described in Nordlund (2010). Both of those investigations draw the conclusion that MBT brings benefits in terms of automation and improved test coverage compared to traditional software testing.

Requirements traceability is a very popular topic in the software engineering and testing communities and has gained momentum in the domain of MBT in the context of automated test generation. However, as requirements change during the development life cycles of software systems, updating and managing traces have become tedious tasks. Researchers have addressed this problem by developing methods for automatic generations of traceability relations Hayes, Dekhtyar, and Osborne (2003), Spanoudakis et al. (2004), Cleland-Huang et al. (2005), Duan and Cleland-Huang (2006) by using information retrieval techniques to link artifacts together based on common keywords that occur in both the requirement description and in a set of searchable documents. Other approaches focus on annotating the model with requirements that are propagated through the test generation process in order to obtain a requirement traceability matrix Bouquet et al. (2005).

Work regarding requirements traceability similar to the MATERA approach is found in Bernard and Legeard (2007). There, the authors suggest an approach in which they annotate the UML system models with ad hoc requirement identifiers and use it to automatically generate a Traceability Matrix at the same time as the generated test cases. Their approach is embedded in the LEIROS Test Designer tool (SmartTesting). However, once the generated test cases are executed, requirements covered by tests are not traced back to the model.

17.9   Conclusions

Our experience indicates that MBT can already be used for productive testing in the telecommunication industry but there are some challenges. The challenges are related more to the tool integration than to MBT as a technology itself. If the challenges will be solved properly, MBT will become even more effective technology within the telecommunication domain.

Initial expectations on MBT were that modeling is difficult and requires a lot of expertise and competence. Our experiences indicate the opposite. Modeling as a competence seems to be fairly easy to learn. At the same time, modeling helps gain and retain an overview of the system behavior much easier than using programming and scripting languages.

Automated test generation is clearly significantly different in comparison to manual test scripting. In test generation, the emphasis is on the design phase instead of the implementation of test cases. Based on our experience, the modifications on the model are propagated into the test cases significantly faster using test generation than manually modifying test scripts. By the same token, automated test generation forces exploiting a different strategy on tracing details, such as requirements, throughout testing compared to manual test scripting. Content of manually written test cases tends to be static and hence the names of the test cases are typically human readable names that testing staff learn fairly quickly. Because of these aspects, mapping of the test scripts and the requirements is fairly static. In the case of automatic test generation, the content of test cases is more dynamic. The content of the test cases may change significantly between test generation rounds and because of that, the names of the test cases do not reflect the content of the tests. Because of this, tracking the details of testing requires a new strategy. In the developed methodology, requirements are used to track test coverage, progress of testing, etc. aspects. Using requirements for tracking is natural because the requirements are present in product development. Hence, use of the requirements for tracking does not necessitate any additional and artificial details for the methodology.

During the development of the methodology, a lot of effort was put into tool integration. Despite the fact that the models were described using standardized modeling languages, such as UML and SysML, it was not possible to create a UML model using one tool and open it with another tool. Compared to text-based programming and testing languages (e.g., C++ and TTCN-3), this is a significant drawback for UML modeling. Model transformations between the tools can be used to circumnavigate this problem. However, seamless use of the tools in, for example, model validation, refactoring, and traceability analysis, requires frequent interaction between the tools. Consequently, transformations will be performed frequently, and hence there would be a tool chain-induced performance penalty. In addition, implementing a number of model transformations increases the costs of the tool chain.

References

Abbors, F. (2009a). An Approach for Tracing Functional Requirements in Model-Based Testing. Master's thesis, Åbo Akademi University.

Abbors, F., Bäcklund, A., and Truscan, D. (2010). MATERA—An integrated framework for model-based testing. In *17th IEEE International Conference and Workshops on Engineering of Computer-Based Systems (ECBS 2010)*, Pages: 321–328. IEEE Computer Society's Conference Publishing Services (CPS), Los Alamitos, CA.

Abbors, F., Pääjärvi, T., Teittinen, R., Truscan, D., and Lilius, J. (2009). Transformational support for model-based testing—From UML to QML. In *Proceedings of 2nd Workshop on Model-based Testing in Practice (MOTIP'09)*, Enschede, the Netherlands.

Abbors, J. (2009b). Increasing the quality of UML Models Used for Automated Test Generation. Master's thesis, Åbo Akademi University.

Beizer, B. (1990). *Software Testing Techniques* (2nd ed.), Van Nostrand Reinhold Co., New York.

Bérnard, B., Bidoit, M., Finkel, A., Laroussinie, F., Petit, A., Petrucci, L., Schnoebelen, P., and McKenzie, P. (2001). *Systems and Software Verification - Model-Checking Techniques and Tools.* Springer-Verlag New York, Inc. New York.

Bernard, E. and Legeard, B. (2007). Requirements traceability in the model-based testing process. *Software Engineering, ser. Lecture Notes in Informatics*, Pages: 45–54.

Born, M., Schieferdecker, I., Gross, H., and Santos, P. (2004). Model-driven development and testing—A case study. In *1st European Workshop on Model Driven Architecture with Emphasis on Industrial Application, number TR-CTIT-04-12 in CTIT Technical Report*, Pages: 97–104. University of Twente, Enschede, the Netherlands.

Bouquet, F., Jaffuel, E., Legeard, B., Peureux, F., and Utting, M. (2005). Requirements traceability in automated test generation: Application to smart card software validation. In *Proceedings of the 1st International Workshop on Advances in Model-Based Testing*, Pages: 1–7. ACM, New York, NY.

Cleland-Huang, J., Settimi, R., Duan, C., and Zou, X. (2005). Utilizing supporting evidence to improve dynamic requirements traceability. In *13th IEEE International Conference on Requirements Engineering, 2005. Proceedings*, Pages: 135–144.

Conformiq (2009a). Conformiq Qtronic. Web page. http://www.conformiq.com/products. php. Accessed on June 13, 2011.

Conformiq (2009b). Conformiq Qtronic User Manual. http://www.conformiq.com/ downloads/Qtronic2xManual.pdf. Accessed on June 13, 2011.

Dalal, S. R., Jain, A., Karunanithi, N., Leaton, J. M., Lott, C. M., Patton, G. C., and Horowitz, B. M. (1999). Model-based testing in practice. In *ICSE '99: Proceedings of the 21st International Conference on Software Engineering*, Pages: 285–294. IEEE Computer Society Press, Los Alamitos, CA.

Duan, C. and Cleland-Huang, J. (2006). Visualization and analysis in automated trace retrieval. In *Requirements Engineering Visualization, 2006. REV'06. First International Workshop on*, Pages: 5–5.

Hartman, A. (2002). Model based test generation tools. *Agedis Consortium, URL: http://www. agedis. de/documents/ModelBasedTestGenerationTools_cs. pdf*. Accessed on June 13, 2011.

Hayes, J., Dekhtyar, A., and Osborne, J. (2003). Improving requirements tracing via information retrieval. In *11th IEEE International Requirements Engineering Conference, 2003. Proceedings*, Pages: 138–147.

Kaaranen, H., Ahtiainen, A., Laitinen, L., Naghian, S., and Niemi, V. (2005). *UMTS Networks, 2nd Edition.* John Wiley & Sons, Ltd, Chichester, UK.

Khan, M. and Shang, S. (2009). Evaluation of Model Based Testing and Conformiq Qtronic. Master's thesis, Linköping University. http://liu.diva-portal.org/smash/ record.jsf?pid=diva2:249286.

No Magic (2009). MagicDraw. Web page. http://www.magicdraw.com/. Accessed on June 13, 2011.

Malik, Q. A., Jääskeläinen, A., Virtanen, H., Katara, M., Abbors, F., Truscan, D., and Lilius, J. (2009). Using system models vs. test models in model-based testing. In *Proceedings of "Opiskelijoiden minikonferenssi" at "Tietotekninen tuki ohjelmoinnin opetuksessa."*

Nethawk (2009). Nethawk EAST simulator. Web page. https://www.nethawk.fi/products/nethawk_simulators/. Accessed on June 13, 2011.

Nokia Siemens Networks (July 2009). Nokia Siemens Networks Company Profile. Web site. http://www.nokiasiemensnetworks.com/about-us/company. Accessed on June 13, 2011.

Nordlund, J. (2010). Model-Based Testing: An Evaluation. Master's thesis, Karlstad University. http://kau.diva-portal.org/smash/record.jsf?pid=diva2:291718. Accessed on June 13, 2011.

Object Management Group. CORBA Component Model. http://www.omg.org/spec/CCM/4.0/. Accessed on June 13, 2011.

Object Management Group. Object Constraint Language (OCL). http://www.omg.org/spec/OCL/2.0/. Accessed on June 13, 2011.

Object Management Group. Systems Modeling Language. http://www.omg.org/spec/SysML/1.1/. Accessed on June 13, 2011.

Object Management Group. Unified Modeling Language (UML). http://www.omg.org/spec/UML/2.1.2/. Accessed on June 13, 2011.

Prenninger, W., El-Ramly, M., and Horstmann, M. (2005). *Model-Based Testing of Reactive Systems*, Chapter Case Studies. Number 3472 in Advance Lectures of Computer Science. Springer, New York.

SmartTesting. CertifyIt, http://smartesting.com/index.php/cms/en/product/certify-it. Accessed on June 13, 2011.

Spanoudakis, G., Zisman, A., Perez-Minana, E., and Krause, P. (2004). Rule-based generation of requirements traceability relations. *The Journal of Systems & Software*, 72(2): 105–127.

The 3rd Generation Partnership Project (2005). 3GPP TS 23.002 Network Architecture, v. 6.10.0. Web site. http://www.3gpp.org/ftp/Specs/html-info/23002.htm. Accessed on June 13, 2011.

Utting, M. and Legeard, B. (2006). *Practical Model-Based Testing: A Tools Approach.* Morgan Kaufmann Publishers Inc., San Francisco, CA.

Utting, M., Pretschner, A., and Legeard, B. (2006). A taxonomy of model-based testing. Working Paper Series, ISSN 1170-478X. http://www.cs.waikato.ac.nz/pubs/wp/2006/uow-cs-wp-2006-04.pdf. Accessed on June 13, 2011.

WireShark. WireShark Network Protocol Analyzer. http://www.wireshark.org/. Accessed on June 13, 2011.

18

Model-Based GUI Testing of Smartphone Applications: Case S60^{TM} and Linux®

Antti Jääskeläinen, Tommi Takala, and Mika Katara

CONTENTS

18.1 Introduction

Smartphones are special types of embedded systems that resemble a desktop environment in the complexity of their applications but embedded systems in their physical size. On the one hand, the size of the keyboard and display makes the graphical user interface (GUI) of the applications somewhat simpler than in their desktop counterparts. Moreover, since smartphones are mass consumer products and the market is highly competitive, the application development is driven by time-to-market deadlines and feature-richness more than zero-defect quality goals. On the other hand, since user experience and perceived quality are extremely important success factors, GUI testing is considered very important. Since hardware must be optimized to provide sufficient performance at lowest possible cost, GUI testing may have to deal with variable response times and similar issues.

Model-based testing can offer many benefits in testing smartphone applications compared to conventional test automation and manual testing, which is still very much in use in this application domain. First, by replacing the task of test design with test modeling, more coverage can be achieved with automatic test generation from the models. Second, test maintenance, one of the main hindrances to automatic GUI testing, becomes easier when changes are restricted to a few component models instead of large test suites. Moreover, the interactions between the applications produce many concurrency-related issues which cannot be adequately tested with conventional methods.

In this chapter, partly based on our previous work (Jääskeläinen et al. 2008a, Jääskeläinen et al. 2008b, Jääskeläinen et al. 2009b), we address these issues from the perspective of two case studies. In the first, we tested applications running on an S60 smartphone. S60 is one of the most widespread smartphone platforms with an open architecture allowing third-party application development. The second case study concerns a mobile Linux platform. Linux is making its way into the smartphone arena in many different variants, which share commonalities based on Linux desktops. In the second case study, we tested a media player application in a Linux environment. For both case studies, we focus on the modeling and adapter development since these appear to be the most critical parts of the model-based testing process in practice. We also discuss the potential problems we see in the industrial deployment of model-based testing technology in the context of GUI testing of smartphone applications. The case studies have been conducted using the TEMA toolset that we have developed. The toolset is available under the MIT open-source license (TEMA Team 2010).

This chapter is structured as follows. First, we begin by introducing the background and theory of our approach. Then, we move on to present the case studies in detail and discuss their results. Towards the end of the chapter, we review related work and conclude with some final remarks.

18.2   Background and Theory

In this section, we start by giving an overview of the TEMA toolset architecture. Then, the basic modeling concepts and the underlying formalism are presented. Next, we discuss the test generation based on models. Finally, the connectivity and adaptation issues pertaining to the automatic execution of the generated tests are addressed.

18.2.1   TEMA toolset architecture

The architecture of the TEMA toolset is presented in Figure 18.1. The toolset contains five different parts: Test Modeling, Test Design, Test Generation, Keyword Execution, and Test Debugging.

The Test Modeling part of the toolset consists of a specific tool designed for creating models in our formalism. The Model Designer application is used to create action machines, refinement machines, and test data, which will be explained in Section 18.2.2. The Test Design part includes a web GUI for creating and starting test runs. For example, the web GUI can be used to design specific use cases that the test run should cover, as shown in Figure 18.2. After a test has been started from the web GUI, the Test Generation part controls the test run by generating test steps from the models according to the chosen rules and generation algorithm and by communicating with the Keyword Execution part. Test generation is presented in more detail in Section 18.2.3. The Keyword Execution part consists of adaptation and test automation tools that execute the test steps on the system under test. There is more discussion about test execution in Section 18.2.4, Connectivity and Adaptation. Finally, the Test Debugging part consists of tools for debugging failed test runs (Heiskanen, Jääskeläinen, and Katara 2010).

18.2.2   Modeling: Action words and keywords

The tests are generated from behavioral models describing the SUT. The semantics of any behavioral model can be represented as a labeled state transition system (LSTS), that is,

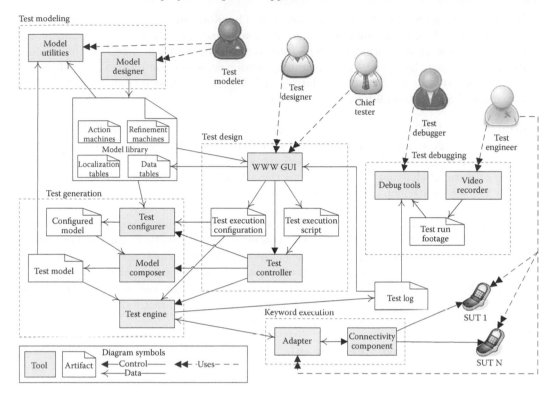

FIGURE 18.1
TEMA toolset architecture.

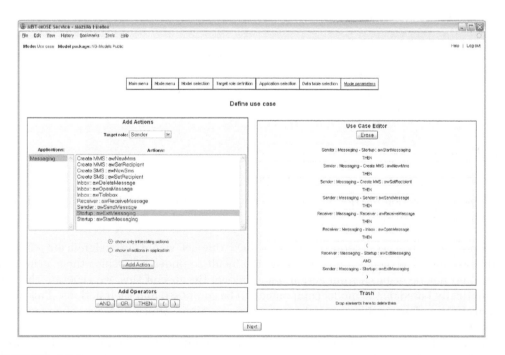

FIGURE 18.2
Use case editor in TEMA web GUI.

a directed graph with labeled edges and vertices. Therefore, by basing the test generation mechanism on LSTSs, the actual models may be created using any behavioral formalism. All that is required is implementing an LSTS interface to give the tools access to the models. State machines are the obvious option and generally sufficient for most SUTs. However, LSTSs are not very good at modeling data, which is generally best included through some other means. LSTS is formally defined as follows:

Definition 1 (LSTS). A *labeled state transition system*, abbreviated to LSTS, is defined as a sextuple $(S, \Sigma, \Delta, \hat{s}, \Pi, val)$, where S is the set of *states*, Σ is the set of *actions* (transition labels), $\Delta \subseteq S \times \Sigma \times S$ is the set of *transitions*, $\hat{s} \in S$ is the *initial state*, Π is the set of *attributes* (state labels), and $val : S \longrightarrow 2^{\Pi}$ is the *attribute evaluation function*, whose value $val(s)$ is the set of attributes in effect in state s. $\qquad\square$

To support modularity, several LSTSs may be combined together in parallel composition, a process that generates a model with the combined functionality of all the component models. In parallel composition, some of the actions of the component models become linked so that they are executed synchronously, while others may be executed independently of other models. Our method of parallel composition is based on a rule set which explicitly defines which actions are synchronized. The parallel composition method was developed by Karsisto (2003), as a generalization of CSP (Roscoe 1998):

Definition 2 (Parallel composition $\|_R$). $\|_R (L_1, \ldots, L_n)$ is the *parallel composition* of LSTSs $L_1, \ldots, L_n$, $L_i = (S_i, \Sigma_i, \Delta_i, \hat{s}_i, \Pi_i, val_i)$, according to *rules* R, with $\forall i, j; 1 \leq i < j \leq n : \Pi_i \cap \Pi_j = \emptyset$. Let Σ_R be a set of resulting actions and $\sqrt{}$ a "pass" symbol such that $\forall i; 1 \leq i \leq n : \sqrt{} \notin \Sigma_i$. The rule set $R \subseteq (\Sigma_1 \cup \{\sqrt{}\}) \times \cdots \times (\Sigma_n \cup \{\sqrt{}\}) \times \Sigma_R$. Now $\|_R (L_1, \ldots, L_n) = repa((S, \Sigma, \Delta, \hat{s}, \Pi, val))$, where

- $S = S_1 \times \cdots \times S_n$

- $\Sigma = \Sigma_R$

- $((s_1, \ldots, s_n), a, (s'_1, \ldots, s'_n)) \in \Delta$ if and only if there is $(a_1, \ldots, a_n, a) \in R$ such that for every i $(1 \leq i \leq n)$ either
    - $(s_i, a_i, s'_i) \in \Delta_i$ or
    - $a_i = \sqrt{}$ and $s_i = s'_i$

- $\hat{s} = (\hat{s}_1, \ldots, \hat{s}_n)$

- $\Pi = \Pi_1 \cup \cdots \cup \Pi_n$

- $val((s_1, \ldots, s_n)) = val_1(s_1) \cup \cdots \cup val_n(s_n)$

- *repa* is a function restricting LSTS to contain only the states that are reachable from the initial state $\hat{s}$. $\qquad\square$

In practice, creating a totally custom-defined rule set would be a significant effort. It would also make the model components more difficult to understand since the synchronizations could not be deduced from action names alone. Because of this, we generate the rules automatically based on the model components to be composed and their actions. This allows us to use only a few specific types of synchronizing actions, which improves the readability of the models.

The models are created according to a two-tier architecture. The models on the upper tier contain an abstract description of the functionality of the SUT. The lower tier defines the implementation of the functionality on the GUI. This separation corresponds to the concepts of action words and keywords (Fewster and Graham 1999, Buwalda 2003), respectively.

Action words describe user actions, such as launching an application or sending a text message. They may be considered small-scale use cases, describing functionality in abstract terms and without reference to GUIs. As such, action words are subject to relatively few changes throughout the product development process.

Keywords describe GUI events, such as key presses and text being displayed on the screen. They are used to define the implementations for the action words. The set of available keywords is platform-specific, as are the implementations of action words. For example, on one device, the sending of a text message could be implemented with a key press to open the menu, more key presses to scroll to the item "Send," and a final key press to accept it. On a different device with a touch screen, the implementation might be to touch the point on the screen with the "Send" button.

The upper tier models, based on action words, are called *action machines*. An action machine may be likened to a process in a single processor multitasking system. One single action machine at a time is active, that is, capable of executing action words, while others remain inactive. This corresponds to the applications on a smartphone, one of which is visible on the screen and the rest hidden in the background. Switching the active model is performed with special synchronization actions, which can also be used for communication between the active and inactive models.

A naive modeling method would be to create an action machine for each application. While this is occasionally viable, in practice, most applications would result in prohibitively huge models with hundreds or even thousands of states. However, synchronizations allow several action machines to be chained into strict interdependence so that they effectively act as a single unit. In this way, various aspects of an application can be separated into different model components.

The lower tier models, called *refinement machines*, contain the keyword implementations for the action words in action machines. Each action machine has its own refinement machine. The keyword implementations are generally designed as loops in a central state, with synchronizations marking the start and end of the refined action word and a sequence of keywords in between. A typical implementation is a linear sequence, but branches and loops are also possible.

The separation into two tiers enhances the maintainability of the models. The functionality of the SUT generally becomes fixed relatively early in the development process, with only minor changes and additions later on. Therefore, once the action machines have been created, they should not require extensive maintenance. If designed carefully, they may even be usable for several different products of the same domain, further reducing the modeling effort. In contrast, different products with different GUIs require separate refinement machines. When the GUI changes, the changes can usually be localized to the refinement machines, which are generally much easier to create and maintain than action machines. A small change in the GUI typically translates to changes in only a few transitions in the models.

To further improve maintainability, GUI text strings are separated from the refinement machines into *localization tables*. A localization table matches a symbolic text string used in the refinement machines into an actual text string according to the specified locale. Localization tables may also be used to store other data of a simple nature.

Data tables provide support for more complicated data requirements. A data table contains a list of structured elements, such as the data of a calendar entry or the contents of a multimedia message. Data tables can be accessed in *data statements*, which are Python code embedded into actions in the models. The data statements provide access to all Python functionality, facilitating the modeling of more data-centric systems with control-based formalisms.

18.2.3   Test generation

There are two basic methods of generating tests from the test model (Hartman et al. 2007). In *offline testing*, the model is first used to create a linear sequence of actions, essentially a test script. This script can then be executed just as a manually designed one. The advantage of offline testing is that it fits easily into existing non-model-based test automation schemes.

The other model-based testing method is *online testing*, in which the execution of the tests is based directly on the model, without separate test scripts. The model is used to generate an action, which is executed immediately before generating the next one. The main advantage is that the result of the execution can be fed back into test generation. This makes online testing well suited for testing nondeterministic systems, such as smartphone applications, where an action may produce different results at different times. For example, a sent multimedia message may fail to leave because of a weak network connection. Online testing is our method of choice, and the rest of the chapter will focus on it.

The actual test generation is based on three things besides the test model: test engine, coverage requirement, and guidance algorithm. The *test engine* acts as an interface for the test model. It keeps track of the current state of execution, provides the list of currently executable actions, and handles the results of the execution of the actions.

The *coverage requirement* is a formal definition of the goal of the test. It serves two purposes: it determines when the test is finished and provides information on what should be done in order to finish. Different types of tests can call for very different coverage requirements. The simplest ones only define a time limit for the test, but no specific tasks that should be accomplished during that time. Others might specify that all actions in the model be executed at least once in order to achieve a basic coverage of the modeled functionality, or require the execution of a sequence of actions corresponding to a use case (as in Figure 18.2). Currently, our tools support coverage requirements based on model actions and execution time, but data coverage has yet to be implemented.

Based on the test model and coverage requirement, it is the task of the *guidance algorithm* to decide which actions to execute. Just as with the coverage requirements, different algorithms are suited for different tests. A simple random heuristic can produce quite useful results and has the advantage of requiring very little time, thus it is good for testing systems with low response times. A use case may be best tested with a graph search algorithm that seeks to fulfill the conditions specified in the coverage requirement as fast as possible. Other guidance algorithms may perform more complicated tasks designed to find specific types of defects, such as running multiple applications side by side in a semirandom fashion in order to locate concurrency-related issues.

18.2.4   Connectivity and adaptation

While model-based testing by definition does not specify how generated tests are executed, it is obvious that their execution should be automatic as well. This is in most cases similar to any other test automation solution, although the online nature of the tests sometimes causes problems with existing offline test automation tools. While our focus has not been in developing industrial-strength test execution tools for smartphones, we have gained some insights during the two case studies. Next, we discuss the connectivity and adaptation issues on a general level. More details about the specific cases are given later when we introduce the case studies.

When the models in online model-based testing are used to generate tests, the produced test steps are presented in an abstract form. In our approach, as discussed above, these test steps are defined with keywords. Keywords are abstract since they do not present detailed information on how the actual action can be executed in a specific device, thus enabling their usage in multiple domains (or at least with multiple versions inside a single domain). To make the keywords executable, an adaptation phase is needed. In this phase, the keywords are turned into actual key presses and other software or hardware functions in the SUT.

The adaptation is a method where the MBT test engine provides the adapter application with a small step (a keyword) to execute and waits for a result. The adapter executes the step in a SUT and then returns a result (a Boolean value) back to the engine. Depending on the result, the generator will decide the next steps that are taken in the model. If the result shows that the SUT has fallen out of synchronization with the model, the execution is usually finished since a defect may have been found. However, it could be possible to continue testing, for example, by reinitializing the SUT.

The keywords used in our system are divided into two different categories: verification keywords and action keywords. The verification keywords are used to examine the state of the device, but do not normally cause any activity in the device (do not change its state). The action keywords are used to control the device, for example, by pressing buttons. The typical verification in our models is the "kw_VerifyText" keyword that returns true if the text given as a parameter is found from the screen. Verification of the text on the screen can be done, for example, by using API access or optical character recognition. Most often the verification is much harder to accomplish than the actions, as is explained in the specific cases later.

A practical problem in the test execution is the response and occasional loading times in the device. If a keyword causes the device to load something and the next keyword is delivered to the device too soon, before the loading has completed, the new keyword may be executed with the device in a wrong state. Consider the following example: A keyword describing a hardware key press is executed in the device, which causes some application to load. The next keyword presumes that the application should be on when it is executed, and tries, for example, to open a specific menu item from the application main menu. Now, if the application is still loading and the menu is not open, the keyword will fail.

The delay issue must be taken into account either in the adapter or in the models. In our system, the delays are managed in both. Common response times that are fairly constant can be handled by defining a specific delay for each keyword type in the adapter. This delay is usually enough if the keyword does not cause any special behavior (like when a key press launches an application). However, if the keyword causes some extra loading, it must be considered in the models, for instance by employing a specific delay keyword.

A better solution to this issue is to always check the state of the device after a keyword that causes any activity to be executed. The state can be checked with the verification keywords described earlier, typically by finding some texts from the screen that identify the expected screen. The verification is continued until it succeeds, but if a specific time limit is exceeded, we can consider that a defect has been found. In some cases, however, finding unique verification texts or other options may be difficult. In the instance of the S60 domain, the two softkey (user-defined shortcut keys) texts combined with the header of the screen usually provide a suitable verification target.

In addition to the delay issues, one must consider whether to distribute the test automation system over a network or not. Our prototype system is distributed over three different machines. The test engine and the test generator are located on a server machine and the client test execution tool resides on a client laptop. The SUT includes an application to connect to the laptop client. The TEMA toolset does not include the client testing tool, so other variations could exist. Communication between the test engine and the client is handled via a simple adapter protocol on top of the TCP/IP, HTTP, or HTTPS protocol. An example usage scenario of the adapter protocol is given in Figure 18.3. Communication between the

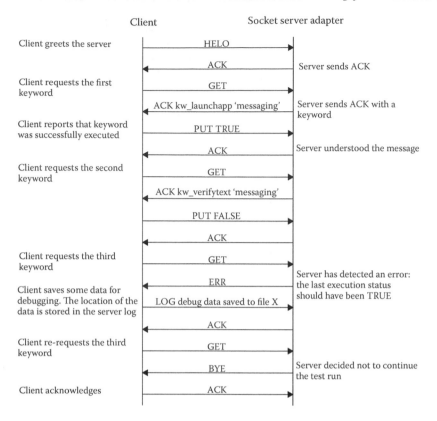

FIGURE 18.3

Adapter protocol communication example. (Adapted from Kervinen, A., Socket server adapter communication protocol, TEMA release package, Tampere University of Technology, 2008.)

client and the SUTs (or any other operation of the adapter) is not specified by our system, but could be achieved with, for example, Bluetooth®, USB, or socket connection.

When creating software for devices that do not yet exist in hardware, testing often needs to be done in an emulator. In our approach, testing focuses on functional testing, and thus an emulator can be a good starting point for testing. However, if the models contain any performance-related assumptions, such as the delays discussed before, these may cause problems when testing on a real hardware (prototype) becomes possible.

18.3   Case Studies

Next, we move on to describe the two case studies. We begin with S60 Third Edition, one of the most widespread smartphone platforms. Then, we describe our experiences with a variant of mobile Linux based on a touch screen. While the newer version of S60 (Fifth Edition), Symbian^3™, and Symbian^4™ use a touch screen, the third is still based on a conventional keyboard.

The mobile Linux variant that we used in this case is a prototype platform by Ixonos which is not, at least not yet, commercially available. However, since many mobile Linux platforms share open-source components inherited from Linux desktops, this platform can be seen as a realistic case study in mobile Linux.

18.3.1   Modeling S60 applications

We have created a relatively comprehensive model library for S60 phones. Included are the applications Calendar, Contacts, File Manager, Gallery, Log, Messaging, Music Player, Notes, RealPlayer®, and Voice Recorder. Later on, we have updated some of these models to support new features, such as tests involving multiple phones. Parallel composition of models of several applications makes it possible to test application interactions and uncovering concurrency-related issues better than using conventional testing methods.

Apart from the manually created models, some models are generated automatically at the beginning of a test run. These include the task switchers, target switcher, and synchronizer. The *task switcher* activates and deactivates the action machines of a single phone as needed and keeps track of the currently active model. Task switchers may be custom-made for platforms that do not support arbitrary switches of applications, but for S60, the automatically generated one works well. The *target switcher* performs a similar task, but for task switchers of different phones. This enables tests involving multiple phones, such as sending a message to another phone. Finally, the synchronizer enables direct connections between models for different phones. That is, when a model sends a text message, the synchronizer is used to transfer control to the models of the intended recipient.

Messaging is a fairly typical smartphone application. Like most applications, its functionality is too diverse to fit practically into a single model, resulting in a number of model components for its different aspects. The components are connected to each other with synchronizations ensuring that they function like a single entity.

The skeleton of the Messaging models comprises the five model components Startup, Main, Inbox, SMS, and MMS. Except for Startup, these directly correspond to different views of the application. Startup (Figures 18.4 and 18.5) is responsible for launching and closing the application. Its functionality could be included in Main, but keeping it in a separate model component makes it easier to adapt the Messaging models for other platforms.

Since the functionality of Inbox is dependent on whether there are any messages within the inbox, we need a method for keeping track of them. For this purpose, we have two models, Messages and Messages Interface. The Messages model (Figure 18.6) is essentially a variable for the existence of messages. The information it holds is abstracted into three possibilities: some messages exist, no messages exist, and unknown. Abstraction is necessary since the number of messages could theoretically be arbitrarily high, but a model can only hold a finite amount of information. It would also be possible to save the information with data statements, or not save it at all and just check it when it is needed. However, encoding it into models makes test generation easier since guidance algorithms can then see it in the structure of the test model.

The sending and receiving of messages are complicated enough to warrant their own models, Sender and Receiver. Their functionality is independent of the message type, so they can be used for both text and multimedia messages. Sender first uses the synchronizer to prime the correct phone to accept connections, and then sends the message and switches control to Receiver on the primed phone. Receiver then opens the message and switches to Inbox.

Messaging is not the only application that can send messages. For example, Gallery can send images in multimedia messages and Contacts can send a message to any of its contacts. It is not sensible to create separate models for all these applications since the functionality is the same in each case. A simple solution would be to connect the SMS and MMS models to all relevant applications simultaneously, something that is easily doable with our synchronization mechanism. However, that would allow only a single application

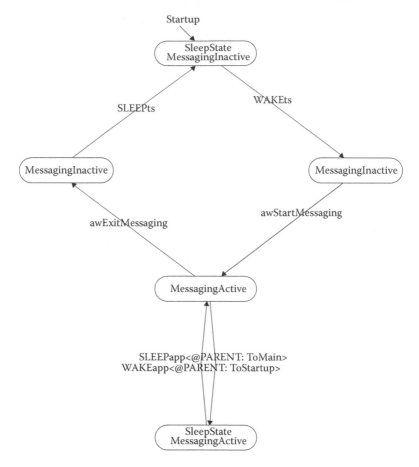

FIGURE 18.4
The Messaging Startup model. The actions with the *aw* prefix are action words and contain the modeled functionality, the *WAKEts* and *SLEEPts* actions connect the model to the task switcher, and the *WAKEapp* and *SLEEPapp* actions connect the Startup model to the Main model.

to write one type of message at a time, reducing opportunities for testing concurrency. A better solution is to make multiple copies of the model so that each application can have its own. Keeping multiple identical copies can cause some problems in model maintenance, but these can be mitigated with proper tool support. Our tools offer a method for linking models together so that changes in one are automatically replicated in others.

18.3.2   Adaptation in S60

In the beginning of our case study, the lack of test automation support in the S60 platform caused a lot of problems in the test execution. Creating a test automation tool from scratch was found too laborious a task, and open-source tools were not available. Thus, a proprietary solution was the only option.

The main problem in this platform is to examine and verify the current state of the device. As there is no API access available, the verification needs to be based on the OCR or bitmap verification. In our S60 models, the verification is based on finding suitable

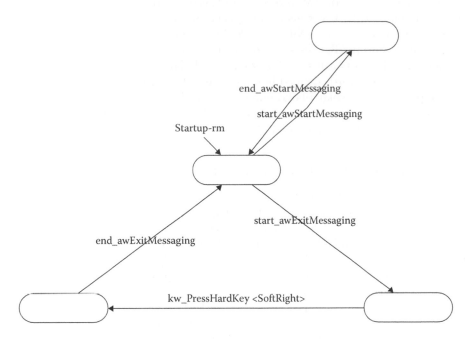

FIGURE 18.5
The refinement machine for the Messaging Startup model. The implementation for the
awStartMessaging action is empty since the generated task switcher launches applications automatically as needed. The action can still be used in coverage requirements and
debugging.

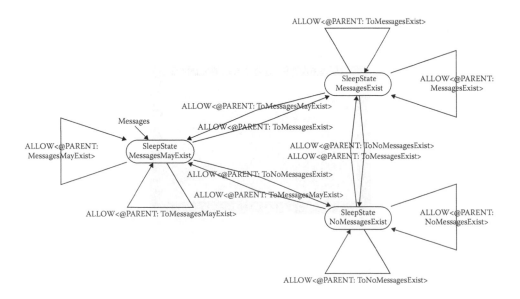

FIGURE 18.6
The Messaging Messages model. Unlike most, the model cannot be activated, but its state
can be changed and queried through the *ALLOW* synchronization actions.

texts on the screen. The tools we have used in the past often had difficulty with the text verification and caused the test runs to fail quickly with false positive results.

The ASTE tool by Nokia® (Nikkanen 2005) has been the most reliable tool we have found so far for executing GUI tests on S60 devices. We have been able to run long period tests without the adapter causing the test runs to fail. The keywords that are used in ASTE are very close to the keywords defined in our models. Therefore, using ASTE required only a simple adaptation. The smartphone under test can be connected using either a USB or Bluetooth connection to the PC running the TEMA adapter. In practice, we found the former connection type to be more reliable in long period tests.

18.3.3 Modeling applications in mobile linux

Creating the action machines for a Linux application did not significantly differ from the earlier modeling efforts on S60. This was to be expected since action machines are designed to be platform-independent. The refinement machines, being platform-specific, have more differences compared to models created earlier. Even they are structurally very similar, though, with the most notable difference being a completely different keyword set. However, the application that was tested, Media Player, has some properties that distinguish it from the majority of smartphone applications.

The core functionality of Media Player (Figure 18.7) is the viewing of video clips, which is a real-time process. Most other smartphone applications have no significant real-time properties: any input would quickly move them into a new stable state. In contrast, when a video playback is begun, Media Player quickly enters a state where it is again responsive to inputs, but alters its state later on when the video clip ends. The end may be expedited by fast forwarding the clip, or delayed by rewinding or pausing it. Since the functionality of the application depends on whether a clip is playing or not, the model needs to either anticipate the ending, or somehow avoid it.

FIGURE 18.7
Media Player in mobile Linux.

We took the latter approach with the few similar applications we had modeled earlier, such as Voice Recorder in an S60 phone. Their models would either forcefully end the playback shortly after it was begun, or do nothing while it played to the end. However, such an approach limits the functionality that can be tested. Since Media Player was now the sole target of the modeling effort, a more flexible approach was called for.

The data statements provided the means to solve the problem. They can be used to execute any Python statements, including the importing of libraries. We therefore imported the datetime library and used it to create timestamps that could be used to measure the progress of the clip. When playback was begun, the time was written into a variable. At any time, we could compare the current time to the starting time, and thereby calculate how far the clip had played. The total lengths of the clips were marked into data tables. When a clip got so close to its end that no new action could be safely initiated, the models would wait for it to end, or, if the clip was paused, fast forward it to the end.

Various actions during the playback affect the timing of the clip. Fast forwarding and rewinding were accounted for by making appropriate adjustments to the starting time; for example, fast forwarding the clip 6 s has the same effect as having started the playback 6 s earlier. Pausing the clip would create a new timestamp for the pause time. When the playback was resumed, the time spent paused was calculated and added to the effective start time. These measures were sufficiently accurate to enable activity during the entire playback, and to even catch one timing-related defect. The action machine side of the measures can be seen in Figure 18.8.

18.3.4   Linux adaptation

The adaptation in the Linux domain proved to be less effort and more efficient than in the S60 case. The main reason for this is that in Linux, we were able to access GUI libraries through an API, and thus error-prone and hard-to-maintain text recognition technologies were avoided. In this case, we tested the software by using an emulator. Another difference with the S60 case was that the GUI was used through a touch screen (with a mouse in the emulator running in a Linux desktop), whereas in S60, different GUI components were only reached by navigating to them using the navigation keys.

Our Linux adaptation is based on accessibility technology. Accessibility technology provides an API access for various helper tools to get information from different GUI widgets. The basis for accessibility has been to support people with disabilities, such as poor eyesight, in using GUIs. However, the same technology also provides a suitable interface for test automation tools. The following overview about accessibility and AT-SPI (Assistive Technology Service Provider Interface) is based on documentation by KDE$^{®}$ Accessibility Project (2010), GNOME^{TM} Accessibility Project (2010), and Linux Foundation (2010).

In the Linux case, the GUI of the tested device was based on the GTK+ GUI toolkit similar to the GNOME desktop environment, which provides support for accessibility. Furthermore, in order to use the accessibility features provided by the GTK+, the AT-SPI technology was used. AT-SPI consists of three parts: the AT-SPI aware applications (application supporting accessibility), accessibility broker (server that stores information about running applications that support accessibility), and assistive technologies (applications that use the accessibility information), as shown in Figure 18.9.

The AT-SPI package delivered with GNOME contains two interfaces for accessible technologies: one written with C (cspi) and one written with Python (pyatspi). Both interfaces provide a means to access different GUI components, for instance to read their text content and create user events. In our adaptation, we used the pyatspi library by creating a keyword interface on top of it. The relations between the adapter and the AT-SPI can be seen in Figure 18.10. Other test automation tools that have been created using cspi and pyatspi are,

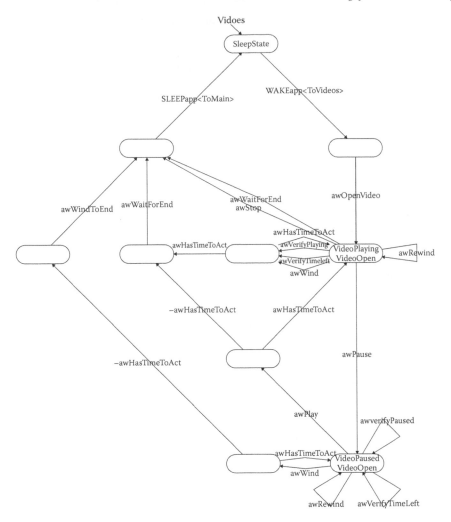

FIGURE 18.8
The Media Player Videos model. This model contains most of the core functionality of Media Player, with the action *awHasTimeToAct* and its negated pair performing the timing checks.

for instance, Dogtail (Fedora® Project 2010) and LDTP (GNU®/Linux Desktop Testing Project 2010).

A big difference when compared to the S60 case study is that the Linux desktop is used with a mouse, or in the case of touch screen, with a stylus. This removes the need to first navigate to the GUI components with the arrow buttons, as was the case with S60. The difference between using touch screen or a regular mouse was found to be small, as both methods include similar pointing and clicking.

The difference in the way of using a GUI also causes differences in the keywords. Most of the created Linux keywords include a reference to the GUI component that is the target for the action. Only when the component has been found from the AT-SPI, can we examine or execute actions on it. In S60, referencing was not needed, as the component was first selected with navigation. Creating a way to refer to a specific GUI widget was perhaps the most challenging part of the adaptation. When referring to components, we used the accessibility names that can be included for the widgets in the Accessibility APIs. This,

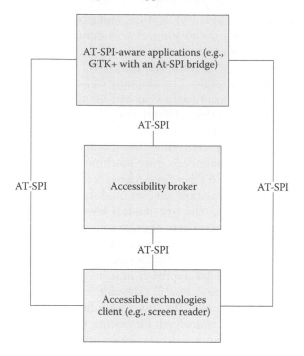

FIGURE 18.9
The structure of AT-SPI.

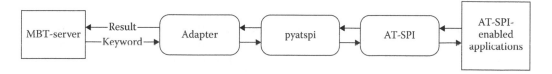

FIGURE 18.10
High-level architecture of the Linux adaptation system.

however, requires input from the programmer, as the names must be defined when the GUI is designed. Some components, such as buttons and labels, obtain accessibility names automatically from their text content, if no explicit name is given. However, relying on these names may cause problems in maintaining test cases or models, as the text contents are likely to change, for example, in different localizations. In addition to referring to components by their accessibility names, we also added a way to refer to components by their text contents (e.g., a button with a specific text). These kinds of references are also locale-specific.

18.4   Related Work

Traditionally, manual testing has been very popular in GUI testing since it can be easily adapted to test new features. Moreover, manual testing can be started earlier in the process, while automatic test execution is not yet applicable because of immaturity of the environment. In GUI testing, especially exploratory testing has proven efficient in finding defects (Kaner, Bach, and Pettichord 2001). In the case of the smartphone domain, the trend is to outsource manual GUI testing to countries with low labor costs.

Considering test automation, so-called capture/replay tools are still used in the desktop GUI testing. However, since they do no support test maintenance well (Kaner, Bach, and Pettichord 2001), they have been largely replaced with keyword-based tools. The problem with the latter type of tools is in choosing the right level of abstraction so that the set of keywords remains as stable as possible. The mapping of keywords to SUT internals is done using a library of keyword implementations that can be updated easily when needed.

Compared to Fewster and Graham (1999) and Buwalda's (2003) approach to using action words and keywords, the main methodological differences in our approach are in using LSTSs and their parallel composition to enable the automatic creation of keyword sequences. For example, Buwalda (2003) recommends state machines and decision tables expressed in spreadsheets for test generation. LSTSs offer a visual formalism that should be quite easy to grasp. Moreover, parallel composition enables modularity and automatic generation of concurrency-related tests, provided the test models have been created with the domain-specific synchronization mechanisms. There are also some minor differences in the terminology compared to Buwalda's approach: in our approach, low-level action words are referred to as keywords, but the most generic keywords can be considered as action words in the sense of functionality. Another similar approach in using keywords and action words is the TestVerb Technology™ in TestQuest Pro™ (Bsquare® Corporation 2010), where action words correspond to TestVerbs forming the test vocabulary used to create high-level tests.

Memon (2001) has proposed a framework for testing GUI applications based on knowledge of GUI components. In this approach, test cases are derived from GUI structure and usage. Test coverage is measured and the correct actions of the GUI are determined by using an oracle based on previously generated test cases and run-time execution information. Belli (2001) and Belli, Budnik, and White (2006) extended state machines to show not only the correct GUI actions but also incorrect transitions. Robinson (1999) popularized the idea of using general purpose GUI test automation tools in conjunction with model-based testing.

Considering the adapter development in conjunction with model-based test automation, Paiva et al. (2005) suggest using a tool for creating a mapping between the model actions and the methods in the SUT. However, this approach is not suitable for us since it works only on programs based on Windows®. Nevertheless, our keywords already implement a layer of abstraction that solves the same problem to a large extent.

From the point of view of the automatic test generation, Bouquet and Legeard (2003) present an approach to generate boundary value tests from formal specifications. As already mentioned, in our approach, the test data are handled in Python statements. While general domain testing algorithms and even constraint solvers could be integrated into our tools in this manner, it makes it difficult to combine test data coverage with state and transition coverage. On the other hand, expressing data as state machines results in huge composed test models, thus we have opted to keep the test data and control separated since it also simplifies the visual appearance of the models, which become easily cluttered.

18.5   Discussion

In this chapter, we have presented our approach for model-based GUI testing of smartphone applications. We have adapted the concept of action words and keywords and combined it with LSTS-based modeling. The main strengths of the method compared to existing techniques are in testing concurrent behaviors of many applications running simultaneously. The evaluation of the results discussed next is qualitative rather than quantitative and partly based on discussions with our industrial partners.

Based on the two case studies, model-based testing is well suited for GUI testing in the smartphone domain. First, it can help to uncover otherwise difficult to find defects. In both case studies, we managed to find defects, even though the applications had been very well tested before our experiments. Some analysis of the defects found from the S60 applications case are reported by Jääskeläinen et al. (2009b). In both cases, a significant portion (roughly two-third) of the issues was found during the test modeling phase. Second, a model-based approach can alleviate the test maintenance effort, which is one of the main hindrances to GUI test automation. However, the case studies did not focus on maintenance. Empirical research is needed to provide quantitative measures on how much improvement can be gained compared to conventional methods.

In the course of the case studies, we identified several issues that must be taken into account when trying to deploy model-based testing in this domain. First, modeling requires expertise that may be difficult to find. Domain-specific modeling languages and model synthesis tools (Jääskeläinen et al. 2009a) can help in model creation, but still an expert test modeler is needed. This problem can be alleviated with proper training. Second, connectivity and adaptation issues require special attention. Some contexts are especially challenging in this respect, while others do not differ from their desktop counterparts. In more detail, the main method for checking the state of the SUT in our approach is to examine whether certain required texts exist on the screen. In this respect, the text recognition technologies and API access work a little differently. The former finds precisely those texts that are found in the current screen capture (presuming that the recognition works properly). The latter is able to find all the texts from all the widgets that currently exist in the application. Although the widgets have accessibility properties that can be used to check whether the widget is currently available, these properties do not always know if the widget is really visible to the user, such as when there is another window partly hiding a widget or the current scroll level of a panel does not reach the widget. In our limited testing scenario, however, we did not encounter these kinds of problems.

Another major issue in this domain is the fast pace of product creation. Since time-to-market is considered of prime importance, testing is often done in an immature test environment. Even though the application to be tested would be of sufficient quality, unfinished hardware and software platform (operating system, GUI libraries, etc.) may prevent effective use of test automation. These kinds of problems are common to all test automation approaches, not only model-based ones. However, it may be difficult to convince the practitioners to switch to model-based automation if it does not seem to provide any added value in these early testing phases. Concerning the actual test execution, it may be wise to save it until the environment is stable enough. Until this stage is reached, scripted and exploratory manual testing may be a better option for GUI testing. Another option is to use an emulator, if one good enough exists, as in our latter case study.

All in all, modeling is known for its ability to reveal defects and it can be started much earlier than the actual test execution. Once the automatic execution of long period tests becomes possible, model-based testing should be able to prove its greater effectiveness over manual testing and conventional test automation approaches.

Acknowledgments

Partial funding from Tekes, Nokia, Ixonos, Symbio, Cybercom Plenware, F-Secure, Qentinel, Prove Expertise, as well as the Academy of Finland (grant number 121012), is gratefully acknowledged. We also appreciate the rest of the TEMA team, as well as Antti Heimola,

Ville Ilvonen, Tuula Pääkkönen, and Matti Salmi for their help in the case studies, and the anonymous reviewers who provided helpful comments.

References

Belli, F. (2001). Finite-state testing of graphical user interfaces. *Proceedings of the 12th International Symposium on Software Reliability Engineering (ISSRE 2001)*, Pages: 34–43. IEEE Computer Society.

Belli, F., Budnik, C. J., and White, L. (2006). Event-based modelling, analysis and testing of user interactions: Approach and case study. *Software Testing, Verification and Reliability* **16**(1), 3–32.

Bouquet, F., and Legeard, B. (2003). Reification of executable test scripts in formal specifation-based test generation: The Java Card transaction mechanism case study. *Proceedings of the 12th International Symposium of Formal Methods Europe (FME 2003)*, Pages: 778–795.

Bsquare Corporation (2011). Automated testing tools, http://www.bsquare.com/auto-mated-testing-tools.aspx (Accessed May 2011).

Buwalda, H. (2003). Action figures. *STQE Magazine*, March/April 2003.

Fedora Project (2011). Dogtail. https://fedorahosted.org/dogtail/ (Accessed May 2011).

Fewster, M., and Graham, D. (1999). *Software Test Automation: Effective Use of Test Execution Tools*. Addison–Wesley.

GNOME Accessibility Project (2010). http://projects.gnome.org/accessibility/ (Accessed Aug. 2010).

GNU/Linux Desktop Testing Project (2011). http://ldtp.freedesktop.org/wiki/ (Accessed May 2011).

Hartman, A., Katara, M., and Olvovsky, S. (2007). Choosing a test modeling language: A survey. *Proceedings of the 2nd International Haifa Verification Conference on Hardware and Software: Verification and Testing (HVC 06)*, number 4383 in *Lecture Notes in Computer Science*, Pages: 204–218, Springer.

Heiskanen, H., Jääskeläinen, A., and Katara, M. (2010). Debug support for model-based GUI testing. *Proceedings of the 3rd International Conference on Software Testing, Verification, and Validation (ICST 2008)*, Pages: 25–34. IEEE Computer Society.

Jääskeläinen, A., Katara, M., Kervinen, A., Heiskanen, H., Maunumaa, M., and Pääkkönen, T. (2008a). Model-based testing service on the web. *Proceedings of the Joint Conference of the 21st IFIP International Conference on Testing of Communicating Systems and the 9th International Workshop on Formal Approaches to Testing of Software (TEST-COM/FATES 2008)*, Volume 5047 of *Lecture Notes in Computer Science*, Pages: 38–53, Springer.

Jääskeläinen, A., Katara, M., Kervinen, A., Maunumaa, M., Pääkkönen, T., Takala, T., and Virtanen, H. (2009b). Automatic GUI test generation for smartphone applications—an evaluation. *Proceedings of the Software Engineering in Practice track of the 31st International Conference on Software Engineering (ICSE 2009)*, Pages: 112–122. IEEE Computer Society (companion volume).

Jääskeläinen, A., Kervinen, A., and Katara, M. (2008b). Creating a test model library for GUI testing of smartphone applications. *Proceedings of the 8th International Conference on Quality Software (QSIC 2008) (short paper)*, Pages: 276–282. IEEE Computer Society.

Jääskeläinen, A., Kervinen, A., Katara, M., Valmari, A., and Virtanen, H. (2009a). Synthesizing test models from test cases. *Proceedings of the 4th International Haifa Verification Conference on Hardware and Software: Verification and Testing (HVC 08)*, Volume 5394 of *Lecture Notes in Computer Science*, Pages: 179–193, Springer.

Kaner, C., Bach, J., and Pettichord, B. (2001). *Lessons Learned in Software Testing: A Context-Driven Approach.* Wiley.

Karsisto, K. (2003). *A New Parallel Composition Operator for Verification Tools.* PhD thesis, Tampere University of Technology.

KDE Accessibility Project (2011). GNOME accessibility architecture (ATK and AT-SPI), http://accessibility.kde.org/developer/atk.php (Accessed May 2011).

Kervinen, A. (2008). Socket server adapter communication protocol, TEMA release package. Tampere University of Technology.

Linux Foundation (2011). ATK/AT-SPI. http://www.linuxfoundation.org/collaborate/workgroups/accessibility/atk/at-spi (Accessed May 2011).

Memon, A. M. (2001). *A Comprehensive Framework for Testing Graphical User Interfaces.* PhD thesis, University of Pittsburgh.

Nikkanen, M. (2005). *Use case based automatic user interface testing in mobile devices.* Master's thesis, Helsinki University of Technology.

Paiva, A., Faria, J. C. P., Tillmann, N., and Vidal, R. F. A. M. (2005). A model-to-implementation mapping tool for automated model-based GUI testing. *Proceedings of the 7th Internatinal Conference on Formal Engineering Methods (ICFEM 2005)*, Pages: 450–464.

Robinson, H. (1999). Finite state model-based testing on a shoestring, Software Testing, Analysis, and Review Conference (STARWEST) 1999.

Roscoe, A. W. (1998). *The Theory and Practice of Concurrency*. Prentice Hall.

TEMA Team (2011). TEMA toolset, http://tema.cs.tut.fi (Accessed May 2011).

19

Model-Based Testing in Embedded Automotive Systems

Paweł Skruch, Mirosław Panek, and Bogdan Kowalczyk

CONTENTS

19.1 Introduction

Increasing complexity and increasing quality and reliability demands, often together with a short time to market, require efficient development methods for automotive software systems. Over the past three decades, model-based development (MBD)* and model-based testing (MBT) both made their way into the automotive industry.

In this chapter, we would like to provide an overview of the state of MBT as implemented in automotive software development based on DELPHI Technical Center Kraków (TCK) experiences. We introduce key concepts specific to MBT in automotive software systems, explain them, and give some examples to illustrate their applicability including some concerns and limitations. It should be noted that the MBT experiences in this chapter are limited to main stream projects (production projects). It should be clear there are other projects conducted by industry (including DELPHI) such as research projects, advanced technology development, and technology trials, which are not the subject of this chapter. Our intention is to provide the reader a clear "as it is" picture of daily engineering experiences related to MBT in automotive software development. That is why we focus only on the main stream projects. Being one of the largest DELPHI software development centers

---

*In this chapter, we use the abbreviation MBD for model-based development. In some publications, the abbreviation MBD is also used for model-based design. We consider MBD a broader term that covers model based requirements engineering, model-based design, and MBT, and we use in this chapter the MBD term in this broader meaning.

worldwide, over the past 10 years we have been involved in a number of projects for different automotive domains and various automotive manufacturers (OEM) in Europe, North America, and Japan, and we think that the presented topics are to a large extent representative for the industry.

Who can benefit from reading this chapter? First, readers not familiar with MBT in automotive who would like to know how it is applied in this industry segment. Second, readers familiar with automotive software development and testing, but not familiar with MBT can obtain a condensed introduction—a good starting point if they are going to investigate this area of testing. Finally, we think this chapter can also be interesting for researchers and MBT tools providers as it gives a perspective on how their ideas and concepts have found their way into the daily engineering work for automotive embedded software testing.

MBD in automotive software development does not have its origins in the CASE tools of 80s or the UML® (OMG UML Spec. 2010) of 90s, but in the world of simulation. This process started in the late 80s in development areas where simulations were used extensively; first in safety and engine control systems, next in body control systems, and later in entertainment/infotainment systems. In the beginning, MBD was limited to simulations and algorithm logic development only. Next, once the tools advanced, portions of the software code started to be automatically generated from the models. Now, models are also used in testing to generate test cases and as test oracles as well. Nowadays, both UML® supported concepts as well as simulation-based modeling are used together. The progress toward MBT in the automotive industry is reflected by the number of publications that have appeared in the past decade (e.g., Rappl et al. (2002), Köhl et al. (2003), Lamberg et al. (2004), Conrad et al. (2005), Gühmann (2005), Mahmud and Alles (2005), Lamberg (2006), Bringmann and Krämer (2008), Navet and Simonot-Lionm (2008), Zander-Nowicka (2008, 2009)). A good portfolio of references related to MBT in the automotive software industry can be found in Zander-Nowicka (2009). Included in this chapter are references not only to primary literature (original research articles authored by the original investigators) but also to publications in which some topics are discussed in greater detail.

Software development processes in the automotive industry are driven by standards like CMMI® (CMMI 2001), SPICE (ISO/IEC 15504 2006), Automotive SPICE® (Automotive SPICE Spec. 2010), Functional Safety (IEC 61508 2005, ISO/DS 26262 2009), etc. In addition, the recently introduced AUTOmotive Open System Architecture (AUTOSAR (AUTOSAR Spec. 2009)) is getting more and more attention in the automotive industry. Those standards require different levels of testing to be applied such as unit testing, integration testing, and system testing. MBT can be applied to all those levels. In this chapter, however, the main focus concentrates on MBT as applied to the system level functional testing (typically conducted through a black-box approach).

After motivations and purpose are presented, the next section of this chapter comprises a brief description of the typical vehicle software system to introduce its key distributed nature and to provide a better understanding of concepts specific to automotive embedded systems testing. In Section 19.3, an overview of how MBD can be incorporated into existing development processes is given. Model-based requirements (MBR) engineering, design, development, and testing scenarios are briefly discussed together because in practice they are often interrelated. In Section 19.4, MBT approaches popular in the automotive industry are further explored, defined, and explained. Section 19.5 categorizes automotive embedded systems behaviors and provides a short presentation of a mathematical formula utilized by various tools as a unified method of simulations and modeling of those basic behaviors. This formula is then used to introduce concepts of abstract and executable test cases and is also referenced by the examples. In Section 19.6, an example Electronic Control Unit (ECU) is presented to illustrate MBT examples provided in Section 19.7. In Section 19.8, a short

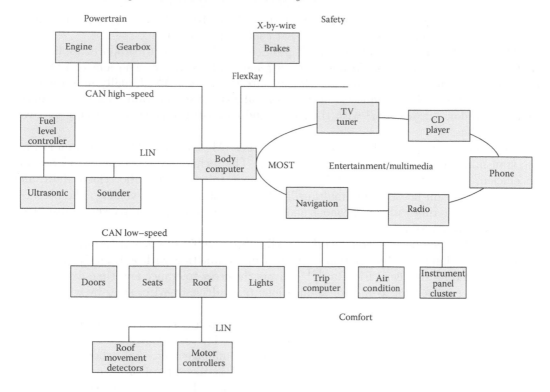

FIGURE 19.1
Typical vehicle software system architecture.

overview of MBT commercial tools popular in the automotive industry is provided. Finally, in Section 19.9, a brief summary and conclusions are presented.

19.2   Typical Vehicle Software System Architecture

A vehicle software system is a distributed, embedded software system where independent microprocessor systems, called ECUs, communicate together using different communication networks. The number of ECUs in vehicles, as well as their complexity, has been increasing from year to year. Nowadays, an average middle class vehicle is equipped with about 30 different co-operating microprocessor systems and the electronics comprises up to 30% of the total vehicle cost. For cars equipped with hybrid powertrains, the electronics cost is even higher, up to 40% (Buchholz 2005, Tung 2007). A typical vehicle software system architecture is shown in Figure 19.1.

Controller Area Network (CAN) is the leading serial bus for embedded control. The protocol was developed specifically for automotive applications and it is now also used in other areas. CAN supports several different physical layers of which the most common and widely used are CAN High Speed (ISO 11898-2 2003) and CAN Low Speed (ISO 11898-3 2006). High-Speed bus (dual wire, up to 1 Mbit/s) is typically used for sharing real-time data such as driver-commanded torque, actual engine torque, steering angle, etc. Low-Speed bus (single wire, up to 125 kbit/s) is typically used for operator-controlled functions, where the system response time requirements are of the order of 100–200 ms such as a roof control module, air conditioning control module, seat control module, etc.

The Local Interconnect Network (LIN) protocol (LIN Spec. 2006) supports applications that are not time critical or do not need extreme fault tolerance. The aim of LIN is to be easy to use and to be a more cost-effective alternative to CAN. Usually in automotive applications, the LIN bus connects smart sensor (ultrasonic sensor, rain sensor) or actuator (wipers, window lift, sounder) with the corresponding control unit that is often a gateway with CAN bus.

Media Oriented System Transport (MOST®) bus (MOST Spec. 2008) is a standard for the transmission of multimedia data in a vehicle. The technology was designed to provide an efficient and cost-effective infotainment backbone to transmit audio, video, data, and control information between any multimedia devices installed in the vehicle. The protocol uses a fiber optic ring with synchronous and asynchronous information packets transmitted at speeds up to 25 Mbit/s.

FlexRay™ communication network (FlexRay Spec. 2005) is a new deterministic, fault-tolerant, and high-speed bus system developed in conjunction with automotive manufacturers and leading suppliers. FlexRay™ delivers the error-tolerance and time-determinism performance required for x-by-wire applications (e.g., drive-by-wire, steer-by-wire, brake-by-wire, etc.).

The central Body Computer Module (BCM) in Figure 19.1 is the primary hub that maintains body functions such as internal and external lighting, security and access control, comfort features for doors and seats, vehicle entry management, and other convenience functions. One of the important tasks of the BCM is the role of the gateway between car communication networks. The gateway not only translates messages between various communication protocols but also supports network management tasks and error detection diagnostics. The role played by the BCM can vary considerably in different automotive system and software architectures. It can be limited to gateway functionality only or it can be responsible for coordination of an entire range of vehicle functions. It should be emphasized that the architecture of vehicle software system is distributed, which implies that typical vehicle functionality is realized by several ECUs communicating with each other.

19.3   Model-Based Development and Testing Process in Automotive

Typically, the automotive software development process (which includes both the code development and the test development processes) is described using a V-model with the automotive SPICE standard referring to it (Hoermann et al. 2008, Automotive SPICE Spec. 2010). In this model, software code development and test development are performed concurrently. Very important for those processes are system requirements, which form an input for both software code development and test development activities. Process standards used in the automotive industry (CMMI 2001, Hoermann et al. 2008, Automotive SPICE Spec. 2010) require bidirectional traceability. This means requirements can be traced to architecture, architecture to design, design to code, and each of these elements of architecture, design, and code to their respective test cases, both forward and backward. This kind of traceability is typically supported by requirements management tools (IBM Rational DOORS Web Page 2010, Geensoft Reqtify Web Page 2010, IBM Rational RequisitePro Web Page 2010).

Figure 19.2 illustrates some possible development scenarios with and without models in use. In the case when models are not used to develop the ECU, software code development

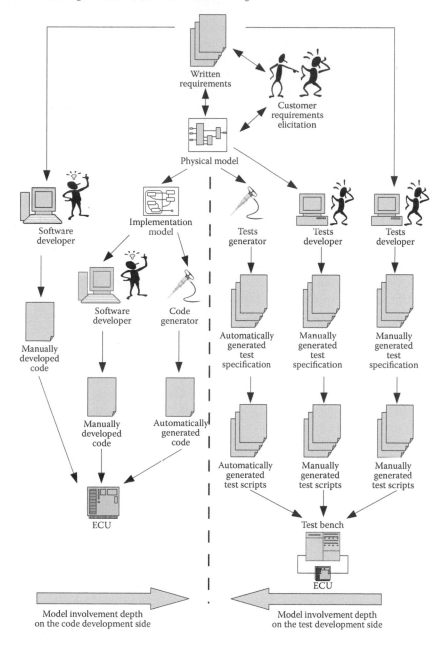

FIGURE 19.2
MBD process.

is based only on written requirements. Software developers manually create code, which is ultimately released and run on the ECU hardware. At the same time, test developers apply various testing methods and techniques to develop system level test cases. The test cases are then run/executed (manually or automatically) against the released software using test benches (also named test harnesses or test beds). Running test cases means applying input signals to the system under test (SUT) and checking the corresponding output signals (responses). The SUT for system level tests is usually the entire ECU.

When models start to be used, various additional variants become possible. First typically a so-called physical model* is created, which captures customer requirements and is used in a very early stage of development to validate system concepts (Lamberg et al. 2004, Tung 2007, Bringmann and Krämer 2008). The term physical model is used here to refer to a functional model, which does not take into account any target implementation constraints. Once a physical model is developed and validated against customer requirements (expectations), it can be used to develop an implementation model (Lamberg et al. 2004, Tung 2007, Bringmann and Krämer 2008).

The term implementation model is used here to refer to a model, which does take into account the various target system implementation constraints (like real-time operating system, scheduling details, signal representations, fixed-point vs. floating-point number representation, etc.). Once the implementation model is ready, it can be either manually transformed into the final code or automatic code generation can be used to generate the final code from the model (Conrad et al. 2005, Tung 2007).

In parallel to the coding process, the physical model can be used by test engineers to develop system level test cases. They are either developed manually or can also be automatically generated from the model using different strategies (Utting et al. 2006, Reid 2006). Next, the test cases are executed (manually or automatically) using the test bench.

The pictograms of an engineer in Figure 19.2 are used to emphasize the labor intensive activities of source code or test case development. Code generation and/or test case generation from the models also require engineers to be involved. The involvement is different however—engineers setup models instead of typing either code or test scripts.

In practice, the entire picture is often far more complicated. It is not possible to develop model(s) that capture all system requirements because of modeling tool limitations, modeled system complexity, time constraints, etc. Models are developed only for some portions of the system and a mixture of approaches shown in Figure 19.2 occurs. Portions of the system are developed using models, while other portions are developed using traditional development methods on both the code and the test development sides. There are often cases when a physical model is developed (e.g., to simulate and validate algorithms and to elicit requirements), however, for various reasons, an implementation model is not developed. In such a case, on the code development side, either written requirements or the physical model is used, but code is developed manually. It is also possible that a model is developed, but there are no system level requirements written. In such a case, the only source of requirements is the model itself. This approach where a model is utilized to elicit requirements and to capture requirements is often referred to as MBR engineering (Nuseibeh and Easterbrook 2000, Berenbach 2004, Soares and Vrancken 2008). Using only a model as a source of requirements is usually impractical and just does not work because in order to understand a complex model and/or complex behavior, engineering teams require some additional description to simplify and accelerate understanding the modeled functionality. Without that deep functional understanding, efficient utilization of models on both software development and test development sides is not possible.

It should also be noted that a real development process being applied strongly depends on the capabilities of the toolset being used. Figure 19.2 shows the case where the modeling tools support both physical and implementation modeling with code and test cases automatically generated. However, there are tools available where the models are more abstract and cannot capture physical and implementation details. Instead, they capture possible system usage or behavior (e.g., Markov chain-based models [Agrawal and Whittaker 1993]

---

*Some publications are using terms like functional abstract model or implementation free model. We decided to use the term physical model, which has its origins in the simulation of the interaction with the physical world and simulation of physical behavior of the SUT itself.

and Use Case-based models). They can be used to automatically generate system level test cases, but it cannot be used at all to automatically generate the code as they are unable to capture the details required. In such situations, the code development side is using different models (or is not using models at all) than those being used on the test development side.

Models used to generate test cases are called test models (Reid 2006). Test models can either model the SUT when they represent SUT intended behavior or can model possible behavior of the environment of the SUT (Utting et al. 2006). Later in this chapter, the main focus will be on the test models that capture the intended behavior of the SUT. This is because the use of MBD in the automotive industry has become very popular and when there is already a model developed to simulate the intended behavior of a device (during requirements elicitation, algorithms development, etc.), a pragmatic approach is to use it also as a test model to generate test cases out of it.

Functional modeling in the automotive industry is dominated by simulation tools. Continuous control systems simulators supported by state charts can efficiently model physical behavior and algorithms of the automotive ECUs. Later in this chapter, more attention is devoted to this modeling methodology. Modeling based on UML® (OMG UML Spec. 2010) and SysML™ (OMG SysML Spec. 2008) is used in the automotive domain to describe and illustrate requirements (SysML™), define system architecture (SysML™) of an ECU, or to design software (UML®). However, UML® and SysML™ cannot be used for efficient simulation, for functional modeling of the control systems, and for the automotive algorithm development. That is why automotive software modeling is dominated by simulation modeling as opposed to other software development domains, which are dominated nowadays by UML®/SysML™ modeling.

Nowadays in the automotive industry, MBT covers: requirements verification and validation (when they are elicited) (Bringmann and Krämer 2008, Zander-Nowicka 2009), automatic test cases generation (Hahn et al. 2003, Lehmann 2003, Tung 2007, Bringmann and Krämer 2008, Zander-Nowicka 2009), and test oracle mechanism development (Binder 1999, El-Far and Whittaker 2001, Reid 2006, Utting et al. 2006, Zander-Nowicka 2009). Those approaches are further explored in later sections of this chapter.

19.4   Model-Based Simulation and Testing Approaches

Modeling and computer simulation today play key roles during development and testing of automotive embedded systems. This trend is stimulated basically by the increasing complexity of the systems and distributed nature of the vehicle software. Moreover, there is a wide range of issues in the system development process that could benefit from an integrated approach using modeling and simulation as opposed to more traditional engineering techniques. Examples are: early fault detection, requirements elicitation, representation of system attributes that cannot be examined in life testing, and fewer physical prototypes. The aim of this section is to describe model-based simulation and testing approaches that have found application in the automotive industry because of their potential to enhance convenience and reduce cost at different stages of the system development life cycle.

The fundamental elements in modeling and simulation of automotive embedded systems are the following:

- The physical system—an ECU that is being developed, clearly defined by its function, parametric characteristic and structure at the proper level.

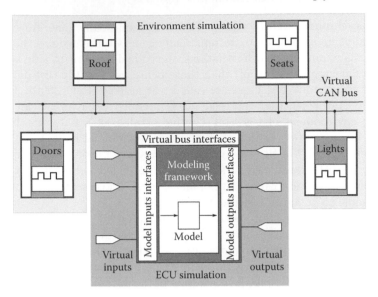

FIGURE 19.3
MiL simulation concept.

- The environment—the physical system's surroundings that may interact with the system by exchanging information or other properties (other ECUs, sensors, actuators, loads, etc.).

- The model—a theoretical construct that contains a detailed and true description of the physical system with mathematical and logical formulas.

- The algorithm—an effective method in the form of a computable set of steps to stimulate the behavior of the system.

- The computer—a programmable machine that implements and executes the algorithm.

Automotive embedded systems are usually tested in the laboratory to assess their functional compliance with specified requirements. The tests consist of exciting the ECU using actuators to simulate its working conditions and measuring the system's response in terms of electrical signals, motion, force, strain, etc. Recently, mathematical models instead of, or in addition to, written requirements have been used to describe the behavior of some parts of the physical system. This new approach has impacted the overall test methodology and tools as well as required new engineering skills. For example, when designing a test bench architecture, it is important to understand all components of the system including its physical and virtual entities and the relationship between them.

The term model-in-the-loop (MiL) (Plummer 2006, Plummer and Dodds 2007, Krupp and Müller 2009) has been used in the automotive industry to describe the concept in which the model of the developed ECU (or part of it) and its environment are simulated without any physical and hardware components. The simulation is based on a functional or implementation model of the ECU and can be performed in an open-loop (without feedback from the environment) setting or a closed-loop (with feedback from the environment) setting. MiL simulation techniques allow testing at early stages of the development cycle to check the model against requirements, obtain reference signals, and verify algorithms. Since the model and the environment usually run on the same machine, a real-time environment is not necessary. Figure 19.3 shows the principle of MiL simulation.

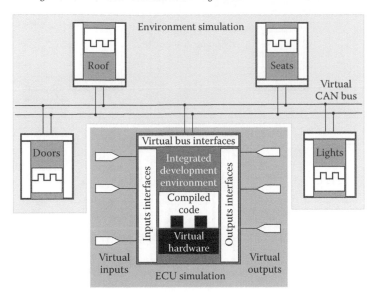

FIGURE 19.4
SiL simulation concept.

In a software-in-the-loop (SiL) concept (Figure 19.4) (Kwon and Choi 1999, Youn et al. 2006, Russo et al. 2007, Chen et al. 2008), the functionality of the developed ECU (or part of it) is already implemented and available as compiled code in the form of a binary. The code can be either handwritten or automatically generated from the model. The ECU hardware, sensors, actuators, loads, and the environment around the ECU are virtual and simulated usually on the same machine where the software runs, therefore a real-time environment is not needed for this concept either. SiL simulation allows analyzing effects because of fixed-point data types and detecting arithmetical problems by comparing the behavior of the ECU code with that of the functional model. In case the SiL simulation also includes a virtual prototype of the ECU hardware (Hellestrand 2005), the hardware design can be, to a large extent, debugged, verified, and validated.

Simulation and testing of an embedded system at a processor-in-the-loop (PiL) level (Liu et al. 2007) means that the embedded software runs in either the target microprocessor or an in-circuit emulator installed in a circuit board to represent at least a portion of the ECU. The virtual environment around the ECU is still simulated usually via real-time computer simulation (Figure 19.5); however, some of the ECU input and output signals are now physical instead of virtual. Tests on the PiL level can verify the system behavior on the target processor, measure code efficiency, and reveal faults that are caused by the target compiler or by the processor architecture. The ability to reuse functional tests from MiL and SiL during PiL testing is important for an efficient development and testing process.

Hardware-in-the-loop (HiL) simulation (Kendall and Jones 1999, Gomez 2001, Li et al. 2002, Bullock et al. 2004, Short and Pont 2008) runs in real time with the final ECU software and hardware. Some parts of the environment around the ECU (that are not yet developed) can still be simulated (Figure 19.6); however, they interact with the ECU via physical digital and analog electrical signals. The main objective of HiL testing is to verify integration of hardware and software components in a real-life environment. It also allows verifying scheduling issues and network aspects in the communication protocol. HiL testing can reveal faults in low-level services of the ECU (such as interfaces with external hardware,

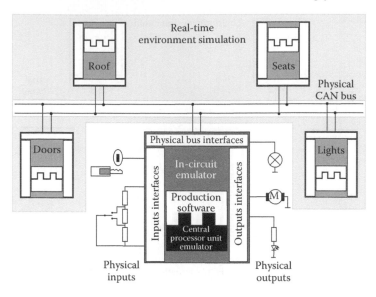

FIGURE 19.5
PiL simulation concept.

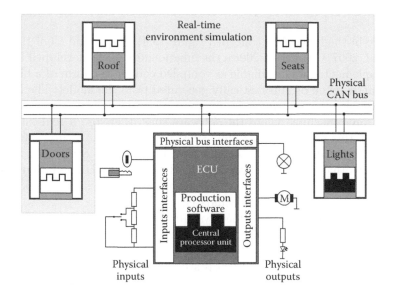

FIGURE 19.6
HiL simulation concept.

sensors, and actuators) which is difficult or impossible to identify at the MiL and SiL testing levels.

The next step is to integrate the ECU with other real ECUs and physical components (electrical, mechanical, or hydraulic). The integration tests are performed in a test rig and focus on communication protocols, performance, and compliance to the vehicle communication standards. Vehicle tests and test drives can be considered as a final verification and integration level.

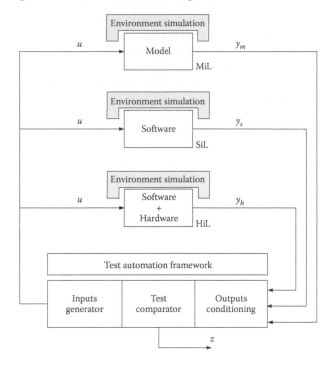

FIGURE 19.7
Integrated MiL-SiL-HiL testing approach.

An integrated MiL-SiL-HiL testing approach is presented in Figure 19.7 (Fey and Strümer 2007, Jaikamal 2009). It shows how different model-based simulation techniques can be used together in MBT of a single ECU. At the beginning of a development project, ECU models shall be validated and then MiL techniques can be considered as a source of test vectors and test stimuli needed in later development phases. Thereafter, the automatic or manual generation of ECU code from the models allows SiL technique to be used. This step brings a higher level of confidence in the implementation of the software without requiring ECU hardware. At this level, functional tests from MiL are reused to verify additional aspects of ECU functionalities. Finally, the software is downloaded to the final hardware and HiL techniques are used to verify that this software behavior is identical to that observed in the MiL and SiL environments.

19.5 Modeling Concept of the System under Test

At the system level, an ECU can be seen as a black box with a set of inputs and outputs that are used to interface with the surrounding environment. Implementation of system requirements by an ECU's software is achieved when responses to the changing inputs seen on the ECU outputs are as prescribed by the requirements.

Assuming the approach described in Section 19.3, application of MBT requires test models to be created that represent the intended behavior of an ECU. A model in the system development life cycle can be considered as a theoretical construction that represents some aspects of the system behavior, with a set of variables and a set of logical and quantitative relationships between them. The model is usually a partial description of the real system and it is built from the requirements of the system. The formal structure of the model is the key element in MBT activities among which the most important are: automatic generation of test cases, test selection, coverage criteria, and test case notation.

An automotive ECU is a control system, typically fairly complex, that is intended to process continuously changing input signals and provide the appropriate output signals based on the inputs. From the control systems point of view, we can categorize ECU behavior into three different types (Douglass 1999, Hahn et al. 2003): discrete combinational (logical operation defined) behavior, discrete sequential (state defined) behavior, and continuous behavior. For the combinational behavior, the outputs of the system depend only on the current state of the discrete inputs and the logic relations/dependencies between the inputs. In the sequential system, the outputs of the system depend not only on the current state of the discrete inputs but also on the history of those inputs (which is preserved and represented by different internal states of the system). Continuous behavior systems are intended to process magnitude-continuous and time-continuous input signals and to produce magnitude- and time-continuous outputs (even when digitally processed). In a continuous system, the response of the system can depend on the history as well. A well-known example of a continuous system is Proportional-Integral-Derivative controller.

Typically, combinational behavior is modeled using Boolean algebra (in practice represented by predicates, logic gates, or graphs such as cause-effect graphs [Binder 1999, Myers 2004]). State behavior is modeled using graphs (usually state charts [Harel 1987, Binder 1999, Douglass 1999]). Continuous behavior is modeled using differential equations (in practice represented by graphs of different Laplace transfer functions [Papoulis 1980]). Automotive ECU behavior is a combination of some or all of the above three basic behaviors as it will be illustrated in the following sections of this chapter. The current state of the art for MBT is that there are efficient algorithms to automatically derive test cases from the combinational models (Binder 1999, Myers 2004) and there are efficient algorithms and methods to automatically derive test cases from state charts models and finite-state machine models (Binder 1999, El-Far and Whittaker 2001), but automatic test case generation for continuous systems is relatively poorly supported by methods and algorithms (Hahn et al. 2003, Bringmann and Krämer 2008, Zander-Nowicka 2009). Testing of those three behavior aspects will be further explored in the next sections.

The modeling concept called state space representation (or input/state/output representation) supports very well the combination of above mentioned basic modeling methods. The state space representation (Ogata 1967, Nise 2007) provides a convenient, compact, and universal way to model and analyze systems with multiple inputs and outputs. The state space model (Figure 19.8) can represent a function, unit, module, system, etc. that is being tested. It is constructed at a certain level of abstraction and describes the functionality of the system at that level (Figure 19.9). The state space model consists of a set of inputs, outputs, and state variables that are expressed as vectors. The approach can be applied to tests at every level (i.e., unit tests, integration tests, and system tests) with formal notation of test design, test specification, and test results. Various commercial-off-the-shelf tools (e.g., MATLAB®/Simulink®, Maple™) as well as proprietary, or open-source tools (e.g., Octave, Scilab, OpenModelica) support the state space notation that is considered a solid foundation to support simulation and mathematical calculation, for one thing because it is amenable to implementing numerical algorithms.

The following equations can express mathematical relationship between inputs, outputs, and states in the system:

$$\frac{dx_c(t)}{dt} = f_c(t, x(t), u(t)), \quad x_c(0) = x_c^0, \tag{19.1}$$

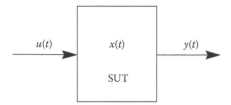

FIGURE 19.8
Model of the SUT.

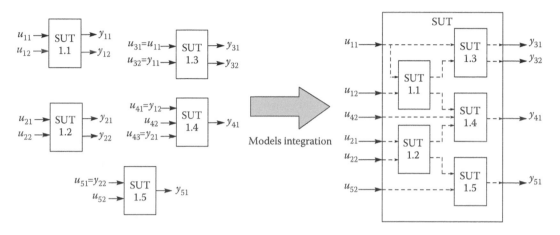

FIGURE 19.9
Models integration concept.

$$x_{\mathrm{d}}[k+1] = f_{\mathrm{d}}(t, x(t), u(t)), \quad x_{\mathrm{d}}[0] = x_{\mathrm{d}}^0, \tag{19.2}$$

$$y(t) = g(t, x(t), u(t)), \tag{19.3}$$

$$x(t) = [x_{\mathrm{c}}(t)\, x_{\mathrm{d}}(t)]^{\mathrm{T}}, \tag{19.4}$$

where $t > 0$ and $k = 0, 1, 2, \ldots$. Here, $x(t) \in \mathbb{R}^n$ is the state vector, $u(t) \in \mathbb{R}^r$ is the control vector, $y(t) \in \mathbb{R}^m$ is the output vector, $\mathbb{R}^n$, $\mathbb{R}^r$, $\mathbb{R}^m$ are real vector spaces of column vectors. The functions $f_{\mathrm{c}}(\cdot)$ and $f_{\mathrm{d}}(\cdot)$ determine transitions between the states for the given initial conditions x_{c}^0 and x_{d}^0, and the function $g(\cdot)$ determines the output. The internal state variables $x(t) = [x_1(t)\, x_2(t) \cdots x_n(t)]^{\mathrm{T}}$ are the smallest possible subset of system variables that represents the entire state of the system at any given time. The first part $x_{\mathrm{c}}(t)$ of the state vector $x(t)$ specifies continuous state variables, the second part $x_{\mathrm{d}}(t)$ specifies discrete state variables, where $x_{\mathrm{d}}(t) = x_{\mathrm{d}}[k] = x_{\mathrm{d}}(kT_{\mathrm{s}})$ for $t \in [kT_{\mathrm{s}}, (k+1)T_{\mathrm{s}})$, T_{s} stands for the sampling interval, $k = 0, 1, 2, \ldots$. In general, both parts may have influence on the calculation of the derivatives of the continuous states (19.1), on the update of the discrete states (19.2), and on the evaluation of the system output (19.3).

The input vector $u(t) = [u_1(t)\, u_2(t) \cdots u_r(t)]^{\mathrm{T}}$ consists of r-bounded elements, where $u_i(t) \in \{u_{i1}, u_{i2}, \ldots, u_{ij}\}$ in case of discrete signals or $u_i(t) \in [u_i^{\min}, u_i^{\max}]$ in case of analog signals, $i = 1, 2, \ldots, r$, $j \geq 2$, $t \geq 0$. The output vector $y(t) = [y_1(t)\, y_2(t) \cdots y_m(t)]^{\mathrm{T}}$ consists of m-bounded elements, where $y_i(t) \in \{y_{i1}, y_{i2}, \ldots, y_{ik}\}$ or $y_i \in [y_i^{\min}, y_i^{\max}]$, $i = 1, 2, \ldots, m$, $k > 2$, $t > 0$. The state variables shall be also considered as bounded variables, that is, $x_i(t) \in \{x_{i1}, x_{i2}, \ldots, x_{il}\}$ or $x_i(t) \in [x_i^{\min}, x_i^{\max}]$, where $i = 1, 2, \ldots, n$, $l \geq 2$, $t \geq 0$.

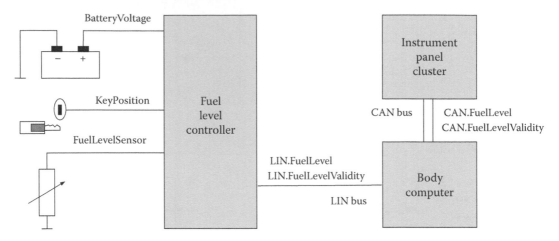

FIGURE 19.10

Functional diagram of the fuel level management system.

It should be noted that the notation 19.1 through 19.4 is consistent with S-function format in the MATLAB®/Simulink® environment. Every model (continuous, discrete, continuous-discrete) created in this environment can be converted to an S-function.

19.6   Fuel Level Controller Example

In this section, the fuel level management system shown in Figure 19.1 is discussed further. The aim of this example is to address complexity issues associated with testing continuous-discrete systems and with the distributed nature of the vehicle electronics architecture. Later sections will use this example to highlight two particular aspects of MBT: (1) the concept of abstract and executable test cases and (2) testing with a model as an oracle.

The fuel level management system is responsible for monitoring fuel consumption and optimizing the refueling of a vehicle. The fuel level in the tank is not easily converted into an accurate indication of the available fuel volume. This relationship is strongly affected by the irregular shape of the fuel tank, dynamic conditions of the vehicle (accelerations, braking, etc.), and oscillations of the indication provided by the sensors. All these factors result in complexity and accuracy problems.

The functional diagram of the fuel level management system is shown in Figure 19.10. The main elements of the system are: a fuel level sensor designed to be mounted on the fuel tank, a Fuel Level Controller (FLC), a BCM, and an Instrument Panel Cluster (IPC). The fuel level sensor is a potentiometer, and it provides a continuous resistance output proportional to the fuel level in the tank. If the tank is full, the sensor shall provide to the control module the resistance of $300\,\Omega$. If the tank is empty, the sensor shall provide the resistance of $50\,\Omega$. The control module shall filter the input signal (*FuelLevelSensor*) and then transmit the filtered value (*LIN.FuelLevel*) and related validity information (*LIN.FuelLevelValidity*) via LIN to BCM. BCM acts as a gateway between two communication buses (LIN and CAN) and transmits the data to IPC which is responsible for fuel level indication to the driver. The fuel level calculation shall be enabled when

$$(KeyPosition > 1) \text{ and } (BatteryVoltage \le 18\,\text{V}) \text{ and } (BatteryVoltage \ge 10\,\text{V}). \quad (19.5)$$

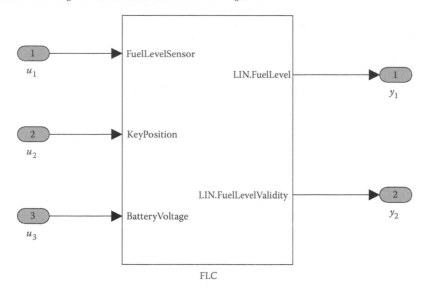

FIGURE 19.11
Input/output representation of the FLC system.

The discrete signal *KeyPosition* is monitored by the system in order to determine the present key position (0 - key is out, 1 - key is in, 2 - ignition on, 3 - engine start). The voltage at the power supply pins is represented by the continuous signal *BatteryVoltage*. If the *BatteryVoltage* > 18 V or *BatteryVoltage* < 10V, then the control module shall switch off the supply voltage for the fuel level sensor and ignore its input signal. If the sensor signal cannot be read (undervoltage, overvoltage) or the signal is invalid (shortcut or invalid range), then the validity information shall reflect this situation, that is, *LIN.FuelLevelValidity* $= 1$, otherwise, *LIN.FuelLevelValidity* $= 0$.

Customer expectation is to have a system that implements a filtering algorithm and provides an accurate indication of the volume of available fuel in a motor vehicle. The role of a systems engineer is to define the method how these requirements are realized in the system. Therefore, a systems engineer in a supplier's organization usually translates high-level requirements into low-level implementation requirements. For example, the filtering algorithm can be modeled by means of algebraic and differential equations. For example, the filtered value can be the result of the following calculation:

$$T\frac{\mathrm{d}x_1(t)}{\mathrm{d}t} + x_1(t) = u_1(t), \quad x_1(0) = x_1^0, \tag{19.6}$$

$$y_1(t) = 0.4x_1(t) - 20, \tag{19.7}$$

where $u_1(t)$ represents the resistance provided by the sensor at time $t > 0$, $y_1(t)$ denotes the filtered value, $x_1(t)$ is the state variable in the system, $T = 600.0$ [s] is the time constant, x_1^0 is the given initial condition.

The functionality of the FLC system can be represented using the state space form as shown in Figure 19.11. Indeed, introducing the notation: $x(t) = [x_1(t)]$, $u(t) = [u_1(t)\,u_2(t)\,u_3(t)]^{\mathrm{T}}$, $y(t) = [y_1(t)\,y_2(t)]^{\mathrm{T}}$, where u_2, u_3, y_2 denote the signals *KeyPosition*, *BatteryVoltage*, and *LIN.FuelLevelValidity*, respectively, yields

$$\frac{\mathrm{d}x_1(t)}{\mathrm{d}t} = -\frac{1}{T}x_1(t) + \frac{1}{T}u_1(t), \tag{19.8}$$

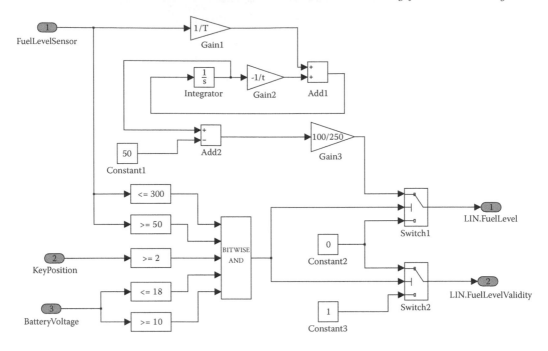

FIGURE 19.12
Physical (functional) model of the FLC system.

$$y_1(t) = \begin{cases} 0.4x_1(t) - 20, & \text{for } u_1(t) \in [50, 300],\, u_2(t) > 1,\, u_3(t) \in [10, 18] \\ 0, & \text{otherwise} \end{cases}, \qquad (19.9)$$

$$y_2(t) = \begin{cases} 0, & \text{for } u_1(t) \in [50, 300],\, u_2(t) > 1,\, u_3(t) \in [10, 18] \\ 1, & \text{otherwise} \end{cases}. \qquad (19.10)$$

It is easy to observe, that the system processes both discrete and continuous signals. The behavior of the system is a combination of two different types: discrete combinational and continuous behaviors. A Simulink® block diagram for this system is shown in Figure 19.12.

19.7   Model-Based Testing Examples

In this chapter, two examples are provided. First, the concept of abstract and executable test cases is illustrated, second, model usage as an oracle is explained.

19.7.1   Testing with abstract and executable test cases

The process of model-based test generation is detailed in Figure 19.13 (Utting and Legeard 2006, Utting et al. 2006). For simplification purposes, model verification aspects are omitted in the picture.

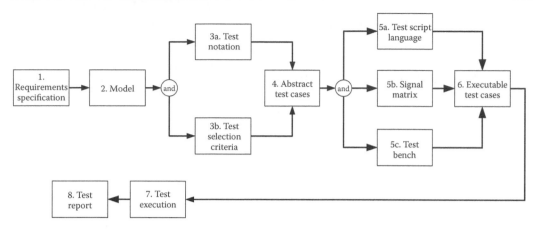

FIGURE 19.13
Overview of MBT activities.

The traditional means of specifying the system behavior is a mixture of textual descriptions and mathematical and logical relationships. This form is often called requirements specification (*Step 1. Requirements Specification*). A model is built based on the requirements and it can reside at different levels of abstraction (*Step 2. Model*). A physical model usually considers the main functional aspects of the system behavior. An implementation model is designed to meet the requirements from a software engineering point of view and it is often used together with a code generator. A test model describes the SUT from a tester point of view. Test cases are derived in whole or in part from the model. The problem how to formally describe test cases is addressed by the test modeling language (*Step 3a. Test Notation*). Test selection criteria (*Step 3b*) refer to the structure of the model and determine the way test cases shall be selected and how test coverage is calculated.

The notation of test cases generated from the models is desired to be independent of both the implementation (test script language) and test platform (test environment). Traditionally, a test sequencer and the test harness tend to cause constraints on the test design process and force a particular way of thinking on a test engineer. This is often because test implementation and execution issues dominate the test design phase. Test engineers usually think how to implement a test case rather than how to design it. Lack of a mathematical mechanism in test formulation complicates the definition of test selection criteria, test coverage calculation, automatic generation of test cases, and proving test generation algorithms. The formal definition of a test modeling language in an abstract mathematical manner should help overcome some of these problems. The output of a test generation tool is called abstract test cases (*Step 4. Abstract Test Cases*) because the test scenarios are described in an abstract manner. The main advantages of abstract test cases are reusability, compactness, and test platform independence.

Using the state space modeling concept of the SUT (Figure 19.8), a single abstract test case can be defined as:

$$(U, Y, T) \quad \text{or} \quad (U, X, Y, T), \tag{19.11}$$

where U stands for a sequence of input functions applied to the system, X describes internal state of the system, Y is a sequence of output functions observed in the system, T describes time base for the input, state, and output functions. For example, in case of discrete systems, the variables in (19.11) have the form:

$$U = \begin{bmatrix} u(t_0) & u(t_1) & \cdots & u(t_k) \end{bmatrix} = \begin{bmatrix} u_1(t_0) & u_1(t_1) & \cdots & u_1(t_k) \\ u_2(t_0) & u_2(t_1) & \cdots & u_2(t_k) \\ \vdots & \vdots & \ddots & \vdots \\ u_r(t_0) & u_r(t_1) & \cdots & u_r(t_k) \end{bmatrix}, \qquad (19.12)$$

$$X = \begin{bmatrix} x(t_0) & x(t_1) & \cdots & x(t_k) \end{bmatrix} = \begin{bmatrix} x_1(t_0) & x_1(t_1) & \cdots & x_1(t_k) \\ x_2(t_0) & x_2(t_1) & \cdots & x_2(t_k) \\ \vdots & \vdots & \ddots & \vdots \\ x_n(t_0) & x_n(t_1) & \cdots & x_n(t_k) \end{bmatrix}, \qquad (19.13)$$

$$Y = \begin{bmatrix} y(t_0) & y(t_1) & \cdots & y(t_k) \end{bmatrix} = \begin{bmatrix} y_1(t_0) & y_1(t_1) & \cdots & y_1(t_k) \\ y_2(t_0) & y_2(t_1) & \cdots & y_2(t_k) \\ \vdots & \vdots & \ddots & \vdots \\ y_m(t_0) & y_m(t_1) & \cdots & y_m(t_k) \end{bmatrix}, \qquad (19.14)$$

$$T = \begin{bmatrix} t_0 & t_1 & \cdots & t_k \end{bmatrix}. \qquad (19.15)$$

The abstract test cases cannot be directly executed against the system because they are at different levels of abstraction. These test cases shall be considered as a formal basis for deriving the executable test cases. For the execution of the abstract test cases (19.11), every stimulus $u(t)$ has to be converted to a physical signal or concrete instance recognized by the system, while the output $y(t)$ has to be matched with the system response. This operation requires a test script language (*Step 5a. Test Script Language*), a signal matrix (*Step 5b. Signal Matrix*), and a test bench (*Step 5c. Test Bench*) to be properly defined. A signal matrix uniquely maps variables of the vectors $u(t)$ and $y(t)$ to physical inputs and outputs. The executable test cases (*Step 6. Executable Test Case*) can then be executed directly against the system (*Step 7. Test Execution*) using the test environment in which the execution takes place. They are usually specified for one particular test platform and it is very difficult to share them with other test platforms. After the execution, a report (*Step 8. Test Report*) is generated that shows the results of the test assessment.

The model represented by the Equations (19.8), (19.9), and (19.10) or the picture of Simulink® blocks (Figure 19.12) can be used to generate abstract test cases. As mentioned in Section 19.5, automated generation of test cases from models processing continuous signals is relatively poorly supported by methods, algorithms, and tools. In fact, using Simulink Design Verifier™ software (R2008b) from MathWorks™ or Reactis Tester (v2006.2) from Reactive Systems® results in the user being informed that the model is incompatible—some of the model items (in this case, Integrator blocks) are not supported. In spite of continuing research on test approaches for continuous and mixed discrete-continuous systems (e.g., Hahn et al. 2003, Zander-Nowicka et al. 2006, Julius et al. 2007, Zander-Nowicka 2009), there is still a lack for patterns, processes, methodologies, and tools that effectively support automatic generation and selection of the correct test cases for such systems. A promising approach uses time- and frequency-domain techniques borrowed from system identification methodology. The key to obtaining a worthwhile result in system identification is choosing recognizable test signals. One of the popular practices is to use a series of single variable step tests. Every input to the system is stepped separately and some step responses are expected on the system output. As an example, the following abstract test case (U,Y,T) has been derived from the model (Figure 19.12):

$$U = \begin{bmatrix} 50 & 50 & 50 & 300 & 300 & 300 & 300 & 300 & 300 & 300 \\ 0 & 1 & 2 & 2 & 2 & 2 & 2 & 2 & 2 & 2 \\ 12 & 12 & 12 & 12 & 12 & 12 & 12 & 12 & 12 & 12 \end{bmatrix}, \qquad (19.16)$$

$$Y = \begin{bmatrix} 0.0 & 0.0 & 0.0 & 0.0 & 1.65 & 8.00 & 15.35 & 48.66 & 71.35 & 91.79 \\ 1 & 1 & 0 & 0 & 0 & 0 & 0 & 0 & 0 & 0 \end{bmatrix}, \tag{19.17}$$

$$T = \begin{bmatrix} 0.0 & 1.0 & 2.0 & 3.0 & 10.0 & 50.0 & 100.0 & 400.0 & 750.0 & 1500.0 \end{bmatrix}. \tag{19.18}$$

Table 19.1 presents the signal matrix that can be used to obtain executable test cases from the abstract test case. The table refers to test bench architecture (*Device No.*, *Slot*, *Line* determine physical inputs and outputs) and network protocol (*ID*, *Length [Bit]*, *Start Bit* determine LIN messages and signals). The signal matrix helps the test sequencer to combine the abstract names of the signals (Table 19.2) with low-level executable commands. The first four steps of the executable test case made from the abstract test case (19.16) through (19.18) are shown in Listing 19.1. The transition from abstract to executable form is usually performed by a special application. In this example, a Perl script has been used to obtain the executable test case and a CANoe environment is used for test execution.

TABLE 19.1

Signal Matrix for the Fuel Level Management System

Variable	ID	Length [Bit]	Start Bit	Device No.	Slot	Line
u_1	-	-	-	4	7	10
y_1	0x24	8	14	-	-	-
y_2	0x24	1	22	-	-	-
u_2	-	-	-	1	1	6
u_3	-	-	-	5	2	1

TABLE 19.2

Signals for the Fuel Level Management System

Signal Name	Variable	Value	Unit
FuelLevelSensor	u_1	$[50, 300]$	$[\Omega]$
LIN.FuelLevelValue	y_1	$[0, 100]$	%
LIN.FuelLevelValidity	y_2	$\{0, 1\}$	-
KeyPosition	u_2	$\{0, 1, 2, 3\}$	-
BatteryVoltage	u_3	$[0.0, 20.0]$	[V]

Listing 19.1

The Executable Test Case Derived from the Abstract Test Case (19.16) through (19.18).

```
/* Part of the test case for testing fuel level management system */
/* STEP 1: */
set_signal(FuelLevelSensor, 50); // The function set_signal() sets the inputs
set_signal(KeyPosition, 0);
set_signal(BatteryVoltage, 12);
check_signal(LIN.FuelLevel, 0.0); // The function check_signal() checks the outputs
check_signal(LIN.FuelLevelValidity, 1);
wait(1.0);

/* STEP 2: */
set_signal(FuelLevelSensor, 50);
set_signal(KeyPosition, 1);
set_signal(BatteryVoltage, 12);
check_signal(LIN.FuelLevel, 0.0);
check_signal(LIN.FuelLevelValidity, 1);
wait(1.0);
```

```
/* STEP 3: */
set_signal(FuelLevelSensor, 50);
set_signal(KeyPosition, 2);
set_signal(BatteryVoltage, 12);
check_signal(LIN.FuelLevel, 0.0);
check_signal(LIN.FuelLevelValidity, 0);
wait(1.0);


/* STEP 4: */
set_signal(FuelLevelSensor, 300);
set_signal(KeyPosition, 2);
set_signal(BatteryVoltage, 12);
check_signal(LIN.FuelLevel, 0.0);
check_signal(LIN.FuelLevelValidity, 0);
wait(7.0);
```

19.7.2   Continuous system testing with a model as an oracle

Testing continuous aspects of automotive control systems creates a number of challenges and still is not sufficiently supported by commercial-off-the-shelf tools and methods (Hahn et al. 2003, Lehmann 2003, Bringmann and Krämer 2008, Zander-Nowicka 2009). One challenge is to judge the results of test cases as pass or fail. According to Binder (1999) and Reid (2006), a test oracle is a reliable source of expected results for a test case. The oracle is then a mechanism used to judge if the results of the executed test case qualify as passed or failed.

For mixed discrete-continuous automotive control systems, various time and magnitude continuous inputs are processed (differentiated, integrated, summed, delayed, combined with discrete inputs, etc.) by an embedded system to produce outputs. Output signals typically cannot be easily predicted by a human being when test cases are developed (a number of differential or difference equations need to be resolved even in a very simple system). In addition, for such mixed discrete-continuous systems, testing methods such as equivalence partitioning or boundary value analysis are typically insufficient. They check the system for some discrete magnitude values and discrete moments of time that are seldom representative. Testing continuous aspects of the automotive system requires test cases that utilize time continuous input signals and time continuous output signals.

The example in this chapter[*] is intended to explain two practical and very typical approaches of MBT. First, this example model is used as an oracle providing the capability to assess the results of test cases in an automatic way. Second, this example also illustrates a very typical case when the test environment of an existing standard automated test bench is extended to accommodate MBT. This integration of MBT with traditional testing approach is required because MBT is usually applied only to a portion of system (the portion that was modeled) with the remaining portion tested without models being used.

The example will use a model of the fuel level controller presented in Section 19.6. The model of the controller is fairly simply and only partially reflects the statements related to difficulties with continuous system testing above. It should be clear that automotive systems are usually far more complicated and this simple model is used only as an illustration of the approach.

As explained in Section 19.6, the fuel level controller processes three inputs to determine the current fuel level in the vehicle tank as well as to calculate the validity signal of this fuel

---

[*]In this chapter, we utilize as an example an approach developed and used by our colleague Wojciech Pluszyński.

level. Both calculated output signals are provided to the rest of the vehicle system through a LIN interface of this ECU. Even for such a simple system, determination of the correct responses for various possible input signal combinations is not easy, but it is a necessity for proving that the ECU correctly implements the algorithm.

In addition to the fuel level calculation, which is the main function of this system, there are other functions that must be tested as well but that have been eliminated in the example since they are not modeled at all (LIN communication protocol, diagnostic protocol, calibration handling, etc.). From a practical point of view, it is recommended to have one test bench architecture to test both the modeled features and the features not modeled.

Figure 19.14 shows a very typical test bench used to test embedded automotive systems. The test sequencer is an application that is reading and executing test cases written in the form of test scripts. Typical sequencer applications are commercial-off-the-shelf tools such as NI^{TM} TestStand^{TM}, IBM$^®$ Rational$^®$ TTCN, Vector Informatik^{TM} CANoe, and dSpace^{TM} AutomationDesk or proprietary solutions typically built around script language interpreters such as Perl, Python, TCL, etc. Test system middleware integrates the sequencer with various plug-ins (drivers) providing connections to the underlying hardware I/O cards utilized to generate the input signals and to capture the output signals. This allows executed test script commands to automatically generate inputs and handle outputs. The signals provided to the ECU being tested (SUT) and those captured from it are stored in the system log file. Executed test sequences (test cases) and their results are stored in test execution log file. With such a test bench, the execution of all or a majority of system level tests can be automated. That is a very typical approach in the automotive industry and is generally the approach used in embedded systems testing. It is easy with this test bench to generate test cases that execute different scenarios for input signals, but, as explained above, it is difficult to judge whether the captured response is the correct one or not (in the case of continuous signals).

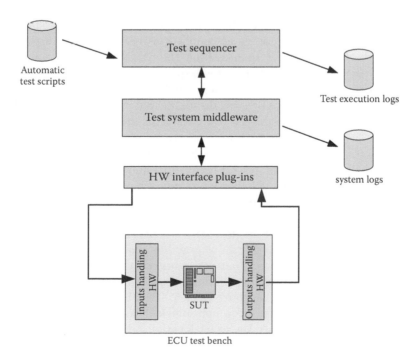

FIGURE 19.14
Test bench for automatic ECU system tests.

There is, however, a model developed for the fuel level controller as described in Section 19.6. Systems engineers using their experiences already proposed a solution, simulated and validated it against customer requirements (using the model-in-the-loop approach), and finally came to agreement with customer that the model represented the product requirements. The algorithm captured in the model fully represents customer requirements so why not use it as an oracle to judge if the algorithm implemented in the ECU being tested correctly implements the customer requirements. This approach of a validated model (a simulator of the implementation [Binder 1999]) being used as an oracle is very popular in the industry and it is often applied. Evaluation of test results with a model used as a reference can be done offline, that is, after test execution, but evaluation made online during test execution is more beneficial. Online test evaluation provides results immediately (once a test case is executed) and without the test engineers spending additional time and effort on the offline analysis of the logs. Additionally, an online evaluation testing strategy can be adapted depending on the results as they occur to make the tests more efficient.

Different solutions are possible to use a model as an oracle in real time, that is, online testing. There are commercial-off-the-shelf tools that allow models to be run in real time. Real-Time Workshop® from MathWorks™ or TargetLink™ from dSpace™, for example, can be used to port a model to a real-time bench. In this chapter, the other possible approach is presented for illustration purposes. In systems where input signal processing is relatively slow, period of 10 ms or slower, and where processing algorithms are not very complicated, it is possible to run a MATLAB®/Simulink® model on a standard personal computer and to utilize its online calculated results as a real-time test oracle.

A MATLAB® interface plug-in can be created using Microsoft COM interface to control the MATLAB®/Simulink® application (de Halleux 2002, Riazi 2003) from within the test sequencer, that is, by automatic test script commands, as shown in Figure 19.15. The original model (e.g., the one from Section 19.6) is instrumented with the model input/output instrumentation blocks inside the MATLAB®/Simulink® environment. Those blocks buffer input and output data and synchronize the model (the oracle) execution (simulation steps) with the SUT. Exactly the same inputs are applied in real time to both the SUT and to the model through the plug-ins. Responses from the SUT and from the model can then be compared by test cases as executed by the sequencer. If the response from the SUT is the same as from the model (according to a defined signal comparison method), the test case is passed. In addition, MATLAB® visualization capabilities can be utilized to generate various kinds of plots based on the logged test data. This greatly simplifies implementation evaluation, failure investigation, and the bug reporting process.

The solution presented in this section can be considered according to Section 19.5 as integration of the HiL tests with MiL tests. Figure 19.16 shows a comparison of the responses obtained from the SUT and the model. The input values have been taken from the test case (19.16)–(19.18) and applied both to the actual ECU, that is, the Fuel Level Controller, and its functional model.

The other aspect of testing with models worth mentioning here is that very often this type of testing enforces what is referred to as gray-box testing. Internal model signals are compared with internal SUT signals. That is because of the system complexity (but can be also because of other reasons, such as an oracle implementation (Zander-Nowicka et al. 2006) or interface-driven testing (Blackburn et al. 2003)). Failure identification based only on external interfaces (black-box) is often not sufficient to identify a bug (it is not sufficient to narrow the scope to identify the exact location and nature of the bug). Gray-box testing requires SUT software to be designed-for-tests (often referred to as instrumented software in the automotive industry) to provide access to some internal signals. In the case of this example, those additional internal signals can be provided through the mentioned LIN interface.

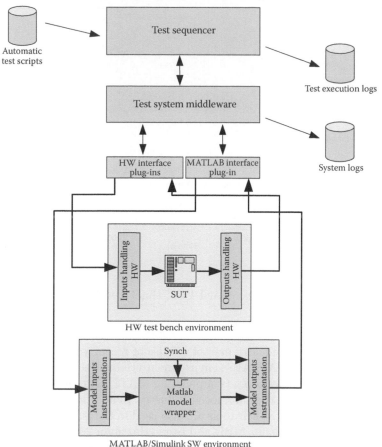

FIGURE 19.15
MATLAB®/Simulink® model integrated into test bench as an oracle.

Model integration into an automatic test bench for real-time model execution also brings additional benefits. The model can be exercised with different inputs and test engineers can learn the desired functionality of the features just by experimenting with model in real-time with physical signals. That supports better requirements understanding and later supports reasoned evaluation of the tests results.

The presented solution (with MATLAB® simulation controlled through the Windows COM interface) is in practice limited to slow systems (such as this fuel level controller), but the concept itself is more general. For time-demanding applications, the models can be run on real-time hardware (instead of slow PC simulation) and again outputs from the tested ECU can be compared with outputs from the model working as an oracle—this means exactly the same pattern can be used, that is, the model is employed as an oracle.

19.8 Tools

MBT would not be possible without supporting tools. In the automotive industry, in addition to various proprietary and academic research solutions (Lehmann 2003, Lamberg et al. 2004, Zander-Nowicka 2009), there are also commercial-off-the-shelf tools used.

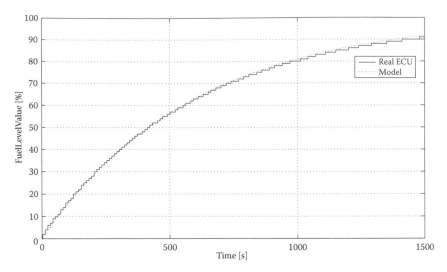

FIGURE 19.16
Comparison of the outputs from the actual ECU (Fuel Level Controller) and its model in
reaction on the same inputs.

According to automotive industry market studies of recent years, the most popular tool
for MBD is MathWorks^{TM} MATLAB$^{®}$ environment (Helmerich et al. 2005). When taking
into account many third-parties toolsets supporting MATLAB$^{®}$, it can be viewed not only
as a tool but also a standard in the automotive industry. In addition to MathWorks^{TM}, there
are other players in the market. Their tools are also used in the automotive industry creating
an even larger portfolio of tools for MBD and tests (Hartman 2002, Zander-Nowicka 2009).
On the other hand, there are also other simulation environments used in the automotive
industry, for example, OpenModelica, SimulationX$^{®}$, MapleSim^{TM}, Dymola$^{®}$ (Dassault
Systemes Web Page 2010, ITI GmbH Web Page 2010, Maplesoft Web Page 2010, Modelica
Spec. 2010), which can be used for modeling various systems. OpenModelica (OpenModelica
Web Page 2010), using the Modelica$^{®}$ language, is a freeware alternative to MATLAB$^{®}$-
based environment tools.

MATLAB$^{®}$ with Simulink$^{®}$ and Stateflow$^{®}$ is a commonly used environment for MBD.
Simulink$^{®}$ supports creating models using control blocks, while for Stateflow$^{®}$ models, a
state chart-based modeling concept is used. Without additional modification or development, Simulink$^{®}$ and Stateflow$^{®}$ are insufficient to generate test cases. There are additional
tools needed. MathWorks^{TM} provides additional functionality in products such as Simulink
Verification and Validation^{TM}, Simulink Design Verifier^{TM}, and SystemTest^{TM}, which are
commonly used in automotive applications and a model-based approach. Using these tools,
automatic test generation and execution can be performed. Models can be evaluated from
a design and functional point of view. Test cases are generated with use of various coverage
criteria; management and analysis of test results are supported as well.

The MaTeLo (Markov Test Logic) tool provided by All4Tec is an example of a tool that
presents an MBT approach based on the functional system usage models represented by
Markov chains. A Markov chain is a state/event diagram where the next step depends on the
current state and on a probabilistic process instead of a deterministic one. A usage model
is created by assigning probabilities for the state transitions based on different possible
system usage profiles. Upon creation of a Markov chain model, a statistical experiment is
run by the tool (using Monte Carlo simulations) to get test cases out of the model as well as
to obtain statistical data from the experiment. A random path through the system states

is generated, which is driven by the transition probabilities and the randomly generated inputs that trigger the transitions. Different algorithms are available to generate test cases.

The IBM® Rational® Statemate® is an example of a graphical design and simulation tool supporting UML® notation for rapid development of embedded systems. Supported with various add-ons, it can be also used for model verification and validation (Statemate ModelChecker) and test vector generation (Statemate ATG).

There are also third-party software providers that develop their MBT tools around popular modeling tools. Reactis® from Reactive Systems® is an example. This tool transforms Simulink®/Stateflow® models into its own internal representation, validates the new model, and then generates tests using various test generation criteria.

Time Partition Testing® (Lehmann 2003) is an example of a tool that supports mixed discrete-continuous systems testing and real-time test execution. It provides one test description for various test approaches (MiL, SiL, HiL). MTest® (Lamberg et al. 2004) uses a classification-tree method for test definition and it is another example of a tool that can be used to test embedded software in each development stage, using MiL, SiL, and PiL test approaches also for mixed discrete-continuous systems.

It should be mentioned that a model-based approach is also supported by various code generation tools such as: TargetLink™ from dSpace™, Real-Time Workshop® from MathWorks™, MATRIXx® from National Instruments®, or Statemate MicroC Code Generator from IBM® Rational®.

19.9 Conclusions

MBD of embedded systems can be considered today as state of the art in many areas of the automotive industry such as safety, powertrain, and body control. MBT is a related feature that supports effective and efficient testing. It should be noted that these new trends in automotive electronics have a big impact on processes, methods, techniques, and tools as well as on the required engineering skills.

A model-based approach will result in the following benefits:

- The time and cost needed to introduce new ECUs can be reduced, and, consequently, the product time-to-market can be shortened (if the model-based methodologies are correctly applied).

- Models can be successfully used when eliciting and clarifying requirements and they can guarantee the correctness of the requirements.

- With the help of modeling and simulation, many system aspects can be verified and validated before real hardware and software are available.

- Physically accurate simulation (virtual prototyping) of electronic, mechanical, and electro-mechanical systems as well as their working environments reduces the number of physical prototypes needed.

- Designers are able to investigate, explore, test, and experience the performance and behavior of an evolving product.

- Code and test cases can be generated from the models, requirements changes can be easily applied to the generated software and tests (especially when they are expressed through executable models), and test reporting with various kinds of metrics can be done automatically as well.

Among the many obstacles to implement a model-based approach on a large scale in production stream projects, the following ones are, in the authors' opinion, especially important in the automotive industry:

- Return on Investment (ROI) for a model-based approach is not guaranteed.

- MBD and testing, especially executable model development, are time-consuming activities. Typically, a model refers only to some aspects of the real system, and two models of the same functionality may be essentially different. The resources (i.e., time and cost) needed for creating a good model for other system aspects may be more than the benefits realized.

- Sometimes the model does not address what is desired to be accomplished. It can be very sensitive to initial conditions or to the values of parameters. Problems occur when the model is too simple or too complex.

- Special training is required to get the project team skilled in both the model-based approach and the tools used.

- The tools are quite expensive in comparison with the traditional software and test development method and MBD, and testing requires that systems, software, and test engineers have the same set of tools.

- Full automation and a push-button approach seem to be very difficult to achieve especially for large projects. Automatically generated test cases are not always obvious and sometimes do not reflect possible system usage.

- Keeping bilateral traceability between test cases and requirements is also difficult, but it is a very basic vehicle manufacturer demand.

- Integration of generated abstract test cases with an existing test automation environment sometimes leads to serious problems and is quite expensive.

- Automated generation of tests from models that process continuous signals is still poorly supported by methods, algorithms, and tools.

- In production projects, MBT still represents a small fraction of overall test laboratory activities.

- Effort estimation methods for MBD have not been well established so far.

To help to overcome the difficulties arising during implementation of an MBT approach, the following suggestions are provided:

- Introduce the MBT approach to only a small portion of a project and over time increase the portion of the project being tested via MBT.

- As a starting point, consider simple functions that have been modeled.

- Start with discrete systems described by state charts and progress to continuous systems working in open-loop and, finally, in closed-loop.

- Build iterative and incremental test models to incorporate the engineers' experiences.

- Do not automatically generate test cases from models that are too complex or difficult to understand initially.

- Monitor the engineering effort from the beginning to have a comparison to the traditional testing method.

- Avoid tools for which the integration with the available test environments is difficult, unclear, and uncertain.

Acknowledgments

We would like to thank our colleague Wojciech Pluszyński from TCK Delphi Poland S.A. for providing the example of an oracle implementation for continuous system testing. Special thanks to Agnieszka Przymusińska, Communication Manager Eastern Europe from Delphi Poland S.A., and David J. Nichols, Technical Manager from Delphi Kokomo Corporate Technology Center, for their roles in the reviewing and editing process.

References

Agrawal, K. and A.J. Whittaker. (1993). Experiences in applying statistical testing to a real-time embedded software system. In *Proceedings of the Pacific Northwest Software Quality Conference*, October, Portland, Pages: 154–170.

Automotive Open System Architecture (AUTOSAR). Specifications, release 4.0. http://www.autosar.org (accessed May 7, 2010).

Automotive SPICE®. (2010). Automotive SPICE® Process Assessment Model (PAM) and Process Reference Model (PRM). http://www.automotivespice.com (accessed May 7, 2010).

Berenbach, B. (2004). Comparison of UML and text based requirements engineering. In *Proceedings of the Conference on Object Oriented Programming Systems Languages and Applications*, Vancouver, Canada, Pages: 247–252.

Binder, R. (1999). *Testing Object-Oriented Systems: Models, Patterns, and Tools*. Addison-Wesley, Boston, MA.

Blackburn, M., Busser, R.D., and A.M. Nauman. (2003). Interface-driven, model-based test automation. *The Journal of Defense Software Engineering* 16(5): 27–30.

Bringmann, E. and A. Krämer. (2008). Model-based testing of automotive systems. In *Proceedings of the 1st International Conference on Software Testing, Verification, and Validation*, April 9–11, Lillehammer, Pages: 485–493.

Buchholz, K. (2005). How demands on low budget. Tech Briefs, Automotive Engineering International. http://www.sae.org/automag/techbriefs/03-2005/1-113-3-40.pdf (accessed May 7, 2010).

Bullock, D., Johnson, B., Wells, R.B., Kyte, M., and Z. Li. (2004). Hardware-in-the-loop simulation. *Transportation Research Part C: Emerging Technologies Journal* 12(1): 73–89.

Chen, X., Salem, M., Das, T., and X. Chen. (2008). Real-time software-in-the-loop simulation for control performance validation. *Simulation* 84(8–9): 457–471.

CMMI® Product Team. (2001). *Capability Maturity Model Integration (CMMI), v1.1, Continuous Representation, CMU/SEI-2002-TR-003*. Pittsburgh: Software Engineering Institute, 2001.

Conrad, M., Fey, I., and S. Sadeghipour. (2005). Systematic model-based testing of embedded automotive software. *Electronic Notes in Theoretical Computer Science* 111: 13–26.

Dassault Systemes. Dymola® Web Page. http://www.3ds.com/products/catia/portfolio/dymola (accessed May 7, 2010).

Douglass, B.P. (1999). UML statecharts. *Embedded Systems Programming* 12(1): 22–42.

El-Far, I.K. and J. Whittaker. (2001). Model-based testing. In *Encyclopedia on Software Engineering*, edited by Marciniak, J.J. John Wiley & Sons, New York, NY.

Fey, I. and I. Strümer. (2007). Quality assurance methods for model-based development: A survey and assessment. SAE Technical Paper Series. Document No. 2007-01-0506.

FlexRay™ Consortium. (2005). FlexRay communications system protocol specification, ver. 2.1, rev. A. http://www.flexray.com (accessed May 7, 2010).

Geensoft. Reqtify® Web Page. http://www.geensys.com/?Outils/Reqtify (accessed May 7, 2010).

Gomez, M. (2001). Hardware-in-the-loop simulation. *Embedded Systems Programming* 14(13).

Gühmann, C. (2005). Model-based testing of automotive electronic control units. In *Proceedings of the 3rd International Conference on Material Testing: Test 2005*, May 10–12, Nürnberg, Germany.

Hahn, G., Philipps, J., Pretschner, A., and T. Stauner. (2003). Tests for mixed discrete-continuous systems. Technical Report TUM-I0301, Institut für Informatik, Technische Universität München.

de Halleux, J. (2002). A class wrapper for Matlab(c) ActiveX Control. http://www.codeproject.com/kb/com/matlabengine.aspx (accessed May 7, 2010).

Harel, D. (1987). Statecharts: A visual formalism for complex systems. *Science of Computer Programming* 8(3): 231–274.

Hartman, A. (2002). Model-based generation tools. AGEDIS Web Page. http://agedis.de (accessed May 7, 2010).

Hellestrand, G. (2005). Model-based development with virtual prototypes. http://www.automotivedesignline.com/howto/showArticle.jhtml?articleID=57701843 (accessed May 7, 2010).

Helmerich, A., Koch, N., and L. Mandel. (2005). Study of worldwide trends and R&D programmes in embedded systems in view of maximising the impact of a technology platform in the area. Final Report for the European Commission, November 18, 2005. Brussels, Belgium.

Hoermann, K., Mueller, M., Dittmann, L., and J. Zimmer. (2008). *Automotive SPICE in Practice: Surviving Interpretation and Assessment*. Rocky Nook, Santa Barbara.

IBM®. Rational DOORS Web Page. http://www-01.ibm.com/software/awdtools/doors/productline (accessed May 7, 2010).

IBM®. Rational RequisitePro Web Page. http://www-01.ibm.com/software/awdtools/reqpro (accessed May 7, 2010).

International Standard IEC 61508. (1998–2005). Functional safety of electrical/electronic/programmable electronic safety-related systems. http://www.iec.ch/zone/fsafety (accessed May 7, 2010).

International Standard ISO 11898-2. (2003). Road vehicles—Controller area network (CAN)—Part 2: High-speed medium access unit. http://www.iso.org (accessed May 7, 2010).

International Standard ISO 11898-3. (2006). Road vehicles—Controller area network (CAN)—Part 3: Low-speed, fault-tolerant, medium-dependent interface. http://www.iso.org (accessed May 7, 2010).

International Standard ISO/DIS 26262. (2009). Road vehicles—Functional safety, part 1 to part 10. http://www.iso.org (accessed May 7, 2010).

International Standard ISO/IEC 15504. (1998–2006). Information Technology—Process assessment, part 1 to part 5. http://www.iso.org (accessed May 7, 2010).

ITI GmbH. SimulationX® Web Page. http://www.iti.de/en/simulationx.html (accessed May 7, 2010).

Jaikamal, V. (2009). Model-based ECU development—An integrated MiL-SiL-HiL approach. SEA Technical Paper Series. Document No. 2009-01-0153.

Julius, A., Fainekos, G.E., Anand, M., Lee, I., and G. Pappas. (2007). Robust test generation and coverage for hybrid systems. In *Proceedings of the 10th International Conference on Hybrid Systems: Computation and Control (HSCC)*, April 2007, Pisa, Italy, Pages: 329–342.

Kendall, I.R. and R.P. Jones. (1999). An investigation into the use of hardware-in-the-loop simulation testing for automotive electronic control systems. *Control Engineering Practice* 7(11): 1343–1356.

Köhl, S., Lemp, D., and M. Plöger. (2003). ECU Network testing by hardware-in-the-loop simulation. *Automotive Electronics (Sonderheft ATZ/MTZ)*.

Krupp A. and W. Müller. (2009). Systematic model-in-the-loop test of embedded control systems. In *Analysis, Architectures and Modelling of Embedded Systems, IFIP Advances in Information and Communication Technology*, 310: 171–184. Springer, Boston, MA.

Kwon, W. and S.G. Choi. (1999). Real-time distributed software-in-the-loop simulation for distributed control systems. In *Proceedings of the IEEE International Symposium of Computer-Aided Control System Design*, August 22–27, Hawaii, Pages: 115–119.

Lamberg, K. (2006). Model-based testing of automotive electronics. In *Proceedings of the Design, Automation and Test in Europe Conference*, March 6–10, Volume 1.

Lamberg, K., Beine, M., Eschmann, M., Otterbach, R., Conrad, M., and I. Fey. (2004). Model-based testing of embedded automotive software using Mtest. SAE Technical Paper Series. Document No. 2004-01-1593.

Lehmann, E. (2003). Time Partition Testing—Systematischer Test des kontinuierlichen Verhaltens eingebetteter Systeme. PhD thesis, Technical University of Berlin.

Li, J., Yu, F., Zhang, J.W., Feng, J.Z., and H.P. Zhao. (2002). The rapid development of a vehicle electronic control system and its application to an antilock braking system based on hardware-in-the-loop simulation. In *Proceedings of the Institution of Mechanical Engineers, Part D, Journal of Automobile Engineering*, 216(2): 95–105.

LIN Consortium. (2006). LIN specification package, rev. 2.1. http://www.lin-subbus.org (accessed May 7, 2010).

Liu, R., Monti, A., Francis, G., Burgos, R., Wang, F., and D. Boroyevich. (2007). Implementing a processor-in-the-loop with a universal controller in the virtual test bed. In *Proceedings of the Power Electronics Specialists Conference*, June 17–21, Orlando, USA, Pages: 945–950.

Mahmud, S.M. and S. Alles. (2005). In-vehicle network architecture for the next-generation vehicles. SAE Technical Paper Series. Document No. 2005-01-1531.

Maplesoft. MapleSim™ Web Page. http://www.maplesoft.com/products/maplesim/index. aspx (accessed May 7, 2010).

Modelica Association. (2010). Modelica®—A unified object-oriented language for physical systems modeling, language specification, version 3.2. http://www.modelica. org/documents/ModelicaSpec32.pdf.

MOST® Cooperation. (2008). MOST dynamic specification, rev. 3.0, 05/2008. http://www. mostcooperation.com (accessed May 7, 2010).

Myers, G. (2004). *The Art of Software Testing. 2nd ed.* John Wiley & Sons, New York, NY.

Navet, N. and F. Simonot-Lionm. (2008). *Automotive Embedded Systems Handbook (Industrial Information Technology)*. CRC Press, Boca Raton.

Nise, N.S. (2007). *Control Systems Engineering. 5th ed.*. John Wiley & Sons, New York, NY.

Nuseibeh, B. and S. Easterbrook. (2000). Requirements engineering: A roadmap. In *Proceedings of the International Conference on Software Engineering*, June 4–11, Limerick, Ireland, Pages: 35–46.

Object Management Group®, Inc. (2008). OMG Systems Modeling Language (OMG SysML™), version 1.1. http://www.omg.org/spec/SysML/1.1 (accessed May 7, 2010).

Object Management Group®, Inc. (2010). OMG Unified Modeling Language™ (OMG UML®), version 2.3. http://www.omg.org/spec/UML/ (accessed August 6, 2010).

Ogata, K. (1967). *State Space Analysis of Control Systems*. Prentice-Hall New York, NY.

OpenModelica. OpenModelica Web Page. http://www.openmodelica.org (accessed May 7, 2010).

Papoulis, A. (1980). *Circuits and Systems: A Modern Approach*. Holt, Rinehart, and Winston, New York, NY.

Plummer, A.R. (2006). Model-in-the-loop testing. In *Proceedings of the Institution of Mechanical Engineering, Part I: Journal of Systems and Control Engineering,* 220(3): 183–199.

Plummer, A.R. and C.J. Dodds. (2007). Model-in-the-loop track simulation. SEA Technical Paper Series. Document No. 2006-01-0732.

Rappl, M., Braun, P., von der Beeck, M., and C. Schroder. (2003). Automotive software development: a model based approach. SAE Technical Paper Series. Document No. 2002-01-0875.

Reid, S. (2006). Model-based testing. In *Proceedings of Test Automation Workshop 2006 (TAW'06)*, August 28–29, Gold Coast, Australia.

Riazi, A. (2003). MATLAB Engine API. http://www.codeproject.com/kb/cpp/matlabeng. aspx (accessed May 7, 2010).

Russo, R., Terzo, M., and F. Timpone. (2007). Software-in-the-loop development and validation of a Cornering Brake Control logic. *Vehicle System Dynamics* 45(2): 149–163.

Short, M. and M.J. Pont. (2008). Assessment of high-integrity embedded automotive control systems using hardware-in-the-loop simulation. *Journal of Systems and Software* 81(7): 1163–1183.

Soares, M. and J. Vrancken. (2008). Model-driven user requirements specification using SysML. *Journal of Software*, 3(6): 57–68.

Tung, J. (2007). Using model-based design to test and verify automotive embedded software. EE Times. http://www.eetimes.com/showArticle.jhtml?articleID=202100792 (accessed May 7, 2010).

Utting, M. and B. Legeard. (2006). *Practical Model-Based Testing: A Tools Approach.* San Francisco: Morgan Kaufmann.

Utting, M., Pretschner, A., and B. Legeard. (2006). A taxonomy of model-based testing. Working paper series, University of Waikato, Department of Computer Science, no. 4, Hamilton, New Zealand. http://www.cs.waikato.ac.nz/pubs/wp/2006/uow-cs-wp-2006-04.pdf (accessed May 7, 2010).

Youn, J., Ma, J., Lee, W., and M. Sunwoo. (2006). Software-in-the-loop simulation environment realization using Matlab/Simulink. SAE Technical Paper Series. Document No. 2006-01-1470.

Zander-Nowicka, J. (2008). Model-based testing of real-time embedded systems for automotive domain. In *Proceedings of the International Symposium on Quality Engineering for Embedded Systems (QEES 2008)*, June 13, Fraunhofer IRB Verlag, Berlin.

Zander-Nowicka, J. (2009). Model-based testing of embedded systems in the automotive domain. PhD thesis, Technical University Berlin.

Zander-Nowicka, J., Schieferdecker, I., and A.M. Perez. (2006). Automotive validation functions for on-line test evaluation of hybrid real-time systems. In *Proceedings of the IEEE 41st Anniversary of the Systems Readiness Technology Conference (AutoTestCon 2006)*, September, Anaheim, USA, Pages: 799-805.

Part VI

Testing at the Lower Levels of Development

20

Testing-Based Translation Validation of Generated Code

Mirko Conrad

CONTENTS

20.1 Introduction

In the following section, an introduction to the embedded software development by application of Model-Based Design is provided. Then, the tool support of this process in the context of the research presented here is discussed. Next, validation and verification of the code generated as a result of the above activities are highlighted and the compliance with functional safety standards is considered. Finally, the contribution of this work is described.

20.1.1 Embedded software development using Model-Based Design

The increasing utilization of *embedded software* in application domains such as automotive, aerospace, and industrial automation has resulted in a staggering complexity that has proven difficult to manage with conventional design approaches. Because of its capability to address the corresponding software complexity and productivity challenges, *Model-Based Design* (Helmerich et al. 2005, Conrad et al. 2005, Conrad and Dörr 2006, MathWorks 2010) is quickly becoming the preferred software engineering paradigm for the development of embedded system software components across application domains. For example, practitioners report productivity gains of 4–10 times when using graphical modeling techniques (Helmerich et al. 2005, Kamga, Herrmann, and Joshi 2007). With the advent of Model-Based Design, an artifact of great value along many axes of use has been introduced into the software life cycle between the specification and the code: the *model*. In a typical Model-Based Design workflow, an initial *executable graphical model* representing the software component under development is refined and augmented until it becomes the blueprint for the final implementation through automatic *code generation*.

20.1.2   Tool support

Simulink$^{®}$* is a widely used Model-Based Design environment. It supports the description of both discrete- and continuous-time models in a graphical block-diagram language. Simulink models can also include state-machine diagrams using Stateflow$^{®}$,[†] or imperative code written in MATLAB^{TM}. An important facility of the Simulink environment is that it provides extensive capabilities for simulation, which attributes an executable semantics to Simulink models.

Furthermore, embedded code generation from Simulink models is supported by Embedded Coder^{TM}[‡] (EC). The code generator is capable of translating Simulink and Stateflow models into production-quality C code that can be deployed onto embedded systems. The generated code is ANSI/ISO C-compliant, enabling it to run on different microprocessors and real-time operating systems.

The Simulink environment is being extensively used in different application domains. According to Helmerich et al. (2005), 50% of the behavioral models for automotive controls are designed using this environment. Because of its popularity, the Simulink environment is used in this chapter to illustrate the proposed testing-based translation validation approach. Further details on modeling with Simulink and Stateflow as well as for code generation with Embedded Coder can be found elsewhere (e.g., Mosterman 2006).

20.1.3   Verification and validation of generated code

Stringent software development workflows are already required to satisfy customer expectations and to ensure the essential quality and reliability of embedded software in many application domains. However, given the high-integrity nature of many state-of-the-art systems, application of verification and validation techniques above and beyond existing software development practices must be considered (Czernye et al. 2004).

For these applications—regardless of the software engineering paradigm utilized—it is crucial that the development of software includes an appropriate combination of *verification and validation* techniques to detect errors and produce confidence in the correct functioning of the software.

In principle, verification and validation of embedded software developed using Model-Based Design and code generation could be carried out in the same way as for manually written code. This, however, would fail to capitalize on the benefits of Model-Based Design to improve process efficiency.

---

*The MathWorks, Inc. Simulink$^{®}$—www.mathworks.com/products/simulink [10/11/20].

†The MathWorks, Inc. Stateflow$^{®}$—www.mathworks.com/products/stateflow [10/11/20].

‡The MathWorks, Inc. Embedded Coder^{TM}—www.mathworks.com/products/embedded-coder [11/05/01].

Introducing executable models into the design process for embedded software provides an opportunity to apply automatic verification and validation techniques earlier in the software life cycle and at a higher level of abstraction. Simulink models are an excellent source of information that can be exploited for various verification and validation techniques, and in terms of process efficiency, engineers should make the most of these opportunities.

Verification and validation activities are used to improve the quality of the embedded software. They cannot be undertaken in isolation but—in order to be effective and efficient—must be integrated into the software development life cycle. Therefore, it is desirable to integrate verification and validation approaches well with the constructive activities of Model-Based Design. In particular, there is a high demand for methods and tools to verify and validate generated code that are easy to incorporate into engineering workflows.

Various authors consider translation validation of generated code as a suitable means to carry out verification and validation of generated code in an engineering context (Edwards 1999, Toeppe et al. 1999, Burnard 2004). In the scope of this chapter, we adopt the notion of *translation validation* introduced by Pnueli, Siegel, and Singerman (1998) to refer to verification/validation approaches where each individual model-to-code translation is followed by a validation phase that verifies that the target code produced by this translation (i.e., code generation/compilation) properly implements the submitted source model. However, we do not imply that formal verification techniques are used to perform this validation. Rather an approach is proposed, where dynamic testing for numerical equivalence between models and generated code constitutes the core part of the translation validation process. An important benefit of this approach in the context of its industrial applicability is that, from the author's experience, a testing-based translation validation approach appears to be a very scalable solution.

20.1.4 Compliance with functional safety standards

The benefits of Model-Based Design apply to a variety of embedded systems, including *high-integrity applications*. In the recent past, engineers have successfully employed graphical modeling with Simulink and code generation to produce software for high-integrity applications (Potter 2004, MathWorks 2005, Arthur 2007, Jablonski et al. 2008, Fey, Müller, and Conrad 2008), but *extra consideration* and *rigor* are needed to address the constraints imposed by functional safety *standards* and to produce the required evidence for compliance demonstration.

A widely adopted safety standard is *IEC 61508*, "Functional Safety of Electrical/Electronic/Programmable Electronic Safety-Related Systems." Originally developed by the process and automation industries, this generic safety standard is now used by many industries. Part three (*IEC 61508-3*) puts forward general objectives and recommendations for software development, verification, and validation. Mapping IEC 61508-3 onto Model-Based Design is not a straightforward task because the standard was assembled in the late 1990s, which was before the broad adoption of Model-Based Design in industry. As a result, the standard does not cover modern software development techniques such as Model-Based Design with code generation.

ISO 26262, "Road Vehicles—Functional Safety," is a more recent, sector-specific functional safety standard for the automotive domain. Part six (*ISO 26262-6*) defines the requirements for product development at the software level for automotive applications. The standard was released in 2011. Applying ISO 26262 to software life cycles using Model-Based Design is more straightforward as the standard also discusses modeling and code generation. The reference software life cycle of the standard still requires tailoring toward Model-Based Design.

The required degree of rigor for software development, verification, and validation varies, depending on how critical the software is. Functional safety standards typically express the degree of rigor in terms of risk levels. IEC 61508 distinguishes four safety integrity levels (SIL), SIL 1 to SIL 4, whereas ISO 26262 knows four automotive safety integrity levels (ASIL), ASIL A to ASIL D.

Neither of the two standards provides comprehensive guidance on how to verify and validate automatically generated code.

20.1.5 Contribution

This chapter proposes a *testing-based translation validation workflow* for generated code in the context of functional safety standards such as IEC 61508 or ISO 26262. This translation validation workflow is intended to comply with applicable requirements of the software safety life cycles defined by IEC 61508-3 (IEC 61508-3 1998) and ISO 26262-6 (ISO 26262-6 2011), as they relate to verification and validation of code generated from executable models. The workflow addresses risk levels that are typical for automotive applications (i.e., SIL 1–SIL 3 according to IEC 61508 and ASIL A–ASIL D according to ISO 26262).

Based on the assumption of a *golden model*, the translation validation process comprises (a) numeric equivalence testing between the generated code and the corresponding model and (b) measures to demonstrate the absence of unintended functionality. The proposed workflow exploits the advantages of Model-Based Design for high-integrity applications. Available tool support is discussed using an example tool chain.

20.2 Model-Based Design Workflow

In the development of embedded application software, graphical *modeling with Simulink and Stateflow* can be used for the conceptual anticipation of the functionality to be implemented. The application software to be developed can be modeled using time-based block diagrams (represented by Simulink blocks) and event-based state machines (represented by Stateflow charts).

Such a graphical model of the application software can be simulated (i.e., executed) within the Simulink Model-Based Design environment. This initial *executable graphical model* representing the software component under development is refined and augmented until it becomes the blueprint for the final implementation through automatic *production code generation* with the Embedded Coder code generator. Throughout this stepwise transformation process (so-called "model elaboration"), the model serves as the primary representation of the embedded software by specifying functionality, capturing design information, and serving as a source for automated code generation.

To elaborate a model in such a stepwise transformation from an early executable specification into a *model suitable for production code generation* and finally its automatic transformation into C code, the model of the application software must be enhanced by adding design information and implementation details. Application software development becomes the successive refinement of models followed by implementation through automatic code generation, as well as compilation and linking as shown in Figure 20.1. Arrows in Figure 20.1 indicate the succession of software life cycle activities.

The modeling notations offered by Simulink and Stateflow are used throughout consecutive modeling stages. This seamless utilization of Simulink and Stateflow models facilitates highly consistent and efficient software development workflows (ISO 26262-6 2011).

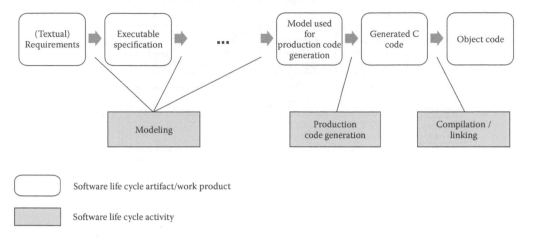

FIGURE 20.1
From requirements to executable code using Model-Based Design.

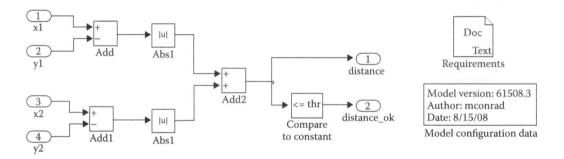

FIGURE 20.2
Example of a Simulink model.

In addition, engineers can carry out different verification and validation activities at the model level, that is, early in the software life cycle before source code and target hardware are available.

Example 1. Figure 20.2 illustrates the modeling notation offered by Simulink. The exemplary model dist_calc calculates the 1-norm distance* between the points (x1, x2) and (y1, y2) and checks if this distance is less than or equal to a given threshold thr. Auxiliary blocks contain the corresponding textual requirements and configuration management data. □

An application software component developed using Model-Based Design with production code generation can be combined with other application software components to form the application software of the embedded system. In combination with the basic software and the hardware of the embedded system, the application software provides the functionality of the embedded system. Figure 20.3 illustrates such a layered architecture by using the example of an electronic control unit (ECU) from the automotive domain.

---

*For two points (x1, x2) and (y1, y2), the 1-norm distance is given as |x1-y1| + |x2-y2|.

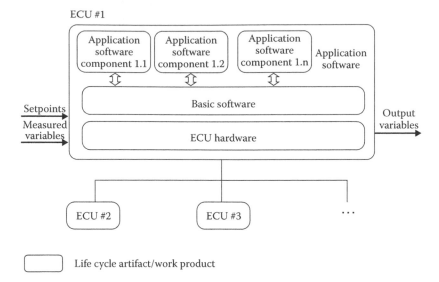

FIGURE 20.3
Overview of an automotive embedded system.

20.3   Verification and Validation of Generated Code

With the advent of code generation, it has become essential to demonstrate that the object code correctly implements the model used for embedded code generation, that is, to show the correctness of the model-to-code transformation process. Two general approaches have been suggested to demonstrate such correctness. It can either be done a priori for all possible translation runs or it can be done after the fact for each individual translation run that is of interest. The former approach is typically referred to as *translator verification*, the latter as *translation validation*.

Originally, it has been suggested to carry out both translator verification as well as translation validation by means of formal proofs. However, the application area of formal proofs is still limited. Modern, industrial strength code generators are of a complexity that is far beyond what can be formally proved, cf. Stürmer et al. (2007). Since code generators in many cases are likely to be more complex than the code generated using them, formal proofs might be better applied to the generated code. Still, even there it turns out that this is only possible for very limited cases.

Recently, systematic (model-based) testing has been proposed as a means for both translator verification and translation validation. Translator verification for code generators through systematic testing (*translator testing*) has been researched in the context of the Automotive Code Validation Suite (AVS) activities, see for instance Stürmer and Conrad (2003), TÜV (2005), Zelenov et al. (2006), Stürmer and Conrad (2005), Pofahl, Sauer, and Busa (2007), Schneider and Slotosch (2007), TÜV (2008), and Schneider, Lovric, and Mai (2009).

This chapter focuses on the alternate approach, the translation validation. In order to easily incorporate processes to demonstrate the correctness of generated code into engineering workflows, the use of systematic (model-based) testing techniques was proposed to perform translation validation (*translation testing*), see for instance Edwards (1999), Burnard (2004), and Conrad (2004).

TABLE 20.1

Methods for Demonstrating the Correctness of the Code Generation Process

Demonstrating Correctness	... For All Translation Runs (i.e., Application independent)	... For a Single Translation Run (i.e., Application specific)
... By using formal proofs	Traditional translator verification	Traditional translation validation
... By using systematic testing	Translator validation through systematic testing (Translator testing)	Translation validation through systematic testing (Translation testing)

Table 20.1 summarizes the different approaches for demonstrating the correctness of the code generation process.

In this chapter, we will adopt the notion of translation validation of generated code as follows (cf. Pnueli, Siegel, and Singerman 1998): Rather than proving in advance that the code generation tool chain, consisting of code generator as well as compiler and linker always produces target code that correctly implements the source model (i.e., translator verification), each individual translation (i.e., a run of the code generation tool chain) is augmented with a validation phase that verifies that the target code properly implements the source model under consideration.

A translation validation approach seems to be better scalable to tool chains currently used in industrial-scale embedded software engineering. Furthermore, it satisfies the requirements of functional safety standards to carry out application-specific verification and validation.

20.4 A Workflow for Application-Specific Verification and Validation of Models and Generated Code

Safety standards such as IEC 61508-3 and ISO 26262-6 call for *application-specific verification and validation* regardless of the tool chain and the development paradigm used.

When using Model-Based Design and production code generation, application-specific verification and validation can be divided into two tasks:

1. To demonstrate that the model used for production code generation is the correct model, that is, that the model correctly implements its (textual) requirements.

2. To demonstrate that the object code to be deployed correctly implements this model.

Different standards provide different levels of detail, to facilitate a practitioner who seeks guidance on how to carry out this activity.

IEC 61508-3 is not very helpful because the software part of the standard dates back to 1998, which is when most software was hand coded. As a result, IEC 61508-3 does not include a notion of model-based testing and code generation, and thus does not provide the required guidance. Rather the objectives and requirements of the standard must be interpreted and mapped onto Model-Based Design (Conrad 2007, Erkkinen and Conrad 2007).

For ISO 26262-6, the situation is more straightforward. This much more recent standard discusses Model-Based Design (referred to as "model-based development" in ISO 26262) and code generation. However, it still assumes a software development process based on manual coding as the default. So when using Model-Based Design, this default process must at least be customized (tailored).

Some companies may use the two standards in parallel. As a consequence, there is a desire to have a single application-specific verification and validation workflow that complies with both standards. Therefore, a reference workflow that describes the verification and validation activities necessary for IEC 61508 and ISO 26262 compliance has been created to facilitate a seamless use of Model-Based Design in the context of these two functional safety standards. Such a reference workflow for application-specific verification and validation of models and generated code may aid users of Simulink and Embedded Coder in defining IEC 61508 and ISO 26262 compliant verification and validation activities as part of an overall software safety life cycle.

Following the above-suggested division of the verification and validation task for applications developed using Model-Based Design and production code generation, the user can subdivide the workflow into the following two steps (cf., Aldrich 2001):

1. *Design verification* that combines suitable verification and validation techniques at the model level to demonstrate that the model is well formed and meets all requirements.

2. *Code verification* that utilizes equivalence testing and other techniques to demonstrate equivalence between the model and the generated code compiled into an executable.

20.4.1   Verification and validation at the model level (design verification)

This chapter focuses on the code verification part (2), while the design verification part (1) is only briefly summarized.

Following the spirit of IEC 61508, design verification is achieved by a combination of reviews, static analyses, and comprehensive functional testing at the model level. Thereby, static analyses mainly address nonfunctional properties and must be augmented by dynamic, functional tests. See Conrad (2008) and Conrad (2009) for more detailed information on the design verification part of the workflow. Design verification results in a *golden model* that is sufficiently validated and verified by the user. In other words, this model is well formed, satisfies its requirements, and does not contain any unintended functionality.

The golden model serves as starting point for the code verification process and therefore is referred to as the *model used for production code generation* (M_{PCG}) in the following.

20.4.2   Verification and validation at the code level (code verification)

The workflow utilizes translation validation through systematic testing (translation testing) to demonstrate that the execution semantics of the model is being preserved during code generation, and compilation and linking. Technically speaking, we are using numerical *equivalence testing* between the model used for production code generation (M_{PCG}) and the executable derived from the generated, cross-compiled C code and additional measures to demonstrate *the absence of unintended functionality*. By comparing behavior and structure

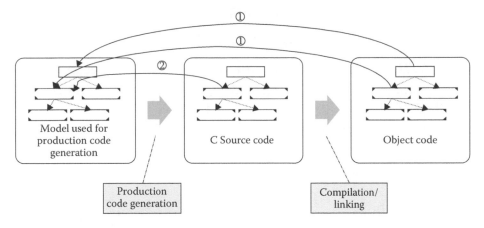

① Equivalence testing
② Prevention of unintended functionality
   a) Model and code coverage comparison
   b) Traceability analysis

FIGURE 20.4
Translation validation process for generated code.

of the model and the generated code, successful code verification indicates proper functioning of the code generator as well as compiler and linker tool chain for the design instance being checked.

Figure 20.4 presents an overview of the proposed translation validation process for generated code. Details are given in Sections 20.4.2.1 and 20.4.2.2.

20.4.2.1 Numerical equivalence testing

Equivalence testing is performed to demonstrate numerical equivalence between the model and the generated code. Equivalence testing between the model used for production code generation and the resulting object code (also known as *comparative testing* or *back-to-back testing*) constitutes a core part of the code verification workflow.

Equivalence testing refers to the execution of the model used for code generation and the object code derived from it via code generation and compilation with the same input stimuli (i.e., test vectors) followed by a numerical comparison of the values at the output (i.e., result vectors).

The validity of the translation process, that is, whether or not the semantics of the model has been preserved during code generation and compilation and linking is determined by comparing the *system reactions* or *result vectors* of the model and the generated code resulting from stimulation with identical timed test vectors $i(t)$. More precisely, the simulation results of the model used for production code generation $o_{M\_PCG}(t)$ are compared with the execution results of the generated and compiled production code $o_{CODE}(t)$.

Figure 20.5 summarizes the equivalence testing approach as described in this section. In-depth discussions of equivalence testing procedures can be found in Stürmer and Conrad (2003), Stürmer et al. (2007), and Conrad (2009). In ISO 26262, equivalence testing (referred to as back-to-back comparison testing between model and code) is listed as a recommended (for ASILs A and B) or highly recommended (for ASILs C and D) method for software unit testing and software integration testing.

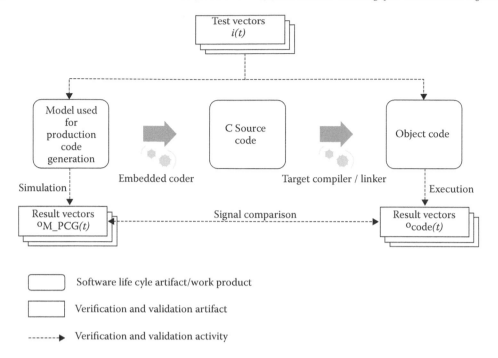

FIGURE 20.5
Numerical equivalence testing approach.

Testing for numerical equivalence is unique because the values of the test vectors expected at the output do not have to be provided (Aldrich 2001). This makes equivalence testing an ideal fit for automation of the test generation.

The following subsections provide detailed information on the equivalence testing procedure.

20.4.2.1.1 *Equivalence test vector generation*

A valid translation requires that the execution of the object code exhibits the same observable effects as the simulation of the model for any given set of test vectors. Since complete testing is impossible because of the prohibitive computational complexity, stimuli (test vectors) shall be sufficient to cover the various structural aspects of the model.

To assess the achieved model coverage at SIL 2 and above (Smith and Simpson 2005), as well as at all ASILs, selected test coverage metrics shall be made available. The extent and scope of structural model coverage must be increased for the higher SILs and ASILs. Table 20.2 lists coverage metrics that can be used to analyze model coverage. It shows, for example, that modified condition/decision coverage at the model level is recommended for SILs 2 and 3 as well as for ASILs A to C (indicated by the "+" entries), whereas it is one of the highly recommended measures for ASIL D (indicated by the "++" entry). For SIL 1, there is no recommendation for or against this measure (indicated by the "o" entry).

Tool Support: Model coverage analysis for Simulink models can be performed by using the Model Coverage Tool in Simulink® Verification and Validation[TM].* Test vectors

---

*The MathWorks, Inc. Simulink® Verification and Validation[TM]—www.mathworks.com/products/simverification [10/11/20].

TABLE 20.2
Model Coverage Analysis

Technique or Measure	IEC 61508			ISO 26262			
	SIL1	SIL2*	SIL3	ASIL A	ASIL B	ASIL C	ASIL D
1a Decision coverage	o	++	++	+	++	++	++
1b Condition coverage	o	+	+	+	+	+	+
1c Modified condition /decision coverage	o	+	+	+	+	+	++
2 Lookup table coverage	o	o	+	o	o	+	+

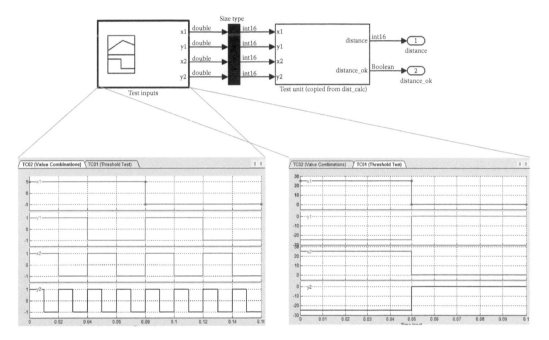

FIGURE 20.6
Example test vectors and test harness model.

resulting from requirements-based testing at the model level can be reused for equivalence testing (cf. Conrad 2004) in this context.

Example 2. Figure 20.6 illustrates the functional tests for a model given in Figure 20.2 that have been created during the design verification phase. The lower part of the figure provides the test vectors $i(t)$ for two functional test cases, the upper part shows a Simulink test harness model that allows for execution of the model with these test cases. The block labeled "Test Inputs" contains the test vectors and allows for their execution. □

The threshold test (TC01) checks whether the distance_ok flag turns on and off properly. The value combinations test (TC02) checks the calculation of the distance by using test vectors utilizing different combinations of positive and input values.

Example 3. Figure 20.7 shows excerpts from the model coverage report that corresponds to the test vectors in Figure 20.6. The report generated by the Model Coverage Tool indicates

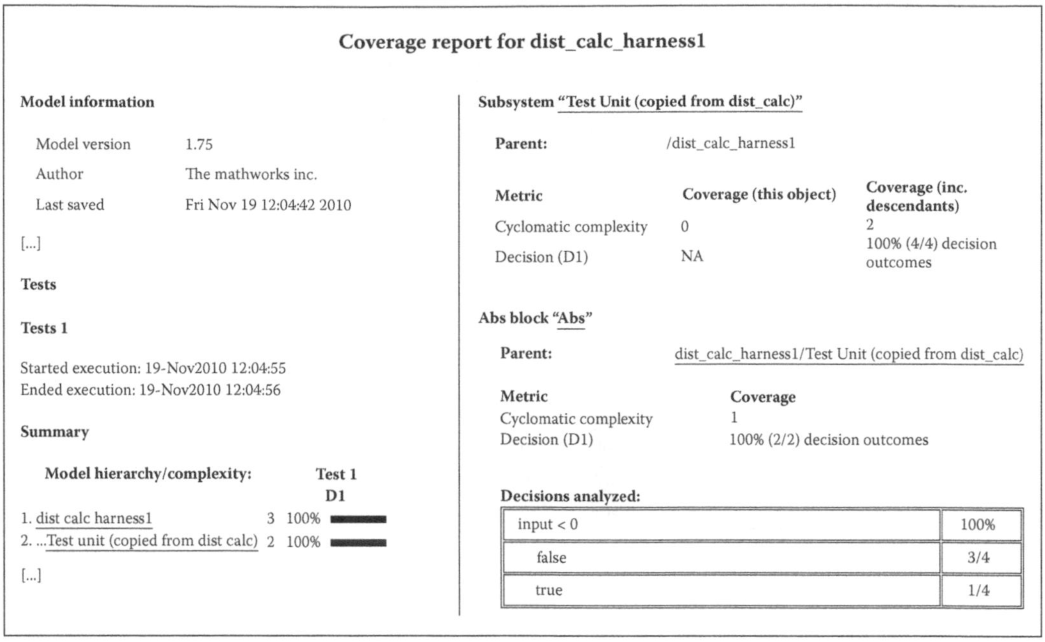

Coverage report for dist_calc_harness1

Model information

Model version	1.75
Author	The mathworks inc.
Last saved	Fri Nov 19 12:04:42 2010

[...]

Tests

Tests 1

Started execution: 19-Nov2010 12:04:55
Ended execution: 19-Nov2010 12:04:56

Summary

Model hierarchy/complexity:		Test 1
		D1
1. dist calc harness1	3	100%
2. ...Test unit (copied from dist calc)	2	100%

[...]

Subsystem "Test Unit (copied from dist_calc)"

Parent:	/dist_calc_harness1

Metric	Coverage (this object)	Coverage (inc. descendants)
Cyclomatic complexity	0	2
Decision (D1)	NA	100% (4/4) decision outcomes

Abs block "Abs"

Parent:	dist_calc_harness1/Test Unit (copied from dist_calc)

Metric	Coverage
Cyclomatic complexity	1
Decision (D1)	100% (2/2) decision outcomes

Decisions analyzed:

input < 0	100%
false	3/4
true	1/4

FIGURE 20.7
Example model coverage report.

that the functional test cases achieve 100% coverage w.r.t. the Decision Coverage metric, the creation of additional test vectors is not necessary for this example.   □

Equivalence testing can be applied to individual as well as to integrated components. Figure 20.8 illustrates equivalence testing at different levels of the model hierarchy. There, three boxes representing the development artifacts present the overall system hierarchy. $i_{COMP}(t)$ and $o_{COMP,*}(t)$ depict test vectors and result vectors for model components; $i_{INT}(t)$ and $o_{INT,*}(t)$ refer to the test vectors and result vectors for the integration stages.

If the coverage achieved with the existing test vectors is insufficient with respect to the selected model coverage metric, additional test vectors that execute the model elements that are not covered shall be created. In practice, the user can iteratively extend the set of test vectors using model coverage analysis until the mandated level of model coverage has been achieved.

If full coverage for the selected metric(s) cannot be achieved, the uncovered parts shall be assessed and justification for uncovered parts shall be provided.

Tool Support: The test generation capability of Simulink® Design Verifier[TM]* can be used to create additional test vectors for equivalence testing.

20.4.2.1.2   Equivalence test execution

The test vectors for equivalence testing shall be used to stimulate both the model used for production code generation and the executable derived from the generated code. The executable shall be tested in an execution environment that corresponds as much as possible to the target environment where the code will be deployed to.

---

*The MathWorks, Inc. Simulink® Design Verifier[TM]—www.mathworks.com/products/sldesignverifier [10/11/20].

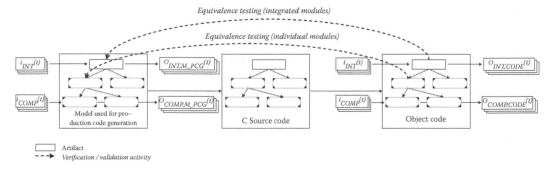

FIGURE 20.8
Equivalence testing of individual and integrated modules.

The executable can be either executed on the target processor or on a target-like processor, for example, by means of Processor-in-the-Loop simulation (i.e., PIL verification), or by means of an instruction set simulator for the target processor (i.e., ISS verification). If feasible, in general, PIL verification is the preferred approach.

If the execution of the resulting object code is not carried out in the target environment, differences between the testing environment and the target environment shall be analyzed in order to make sure that they do not adversely affect the results.

Note that numerical equivalence testing using PIL or ISS verification is not intended to verify real-time aspects of the generated code. Temporal correctness can be addressed after the hardware-software integration has taken place.

Tool Support: The code generation verification capability (Conrad et al. 2010) of Real-Time Workshop Embedded Coder can be used to execute a model and the generated code in a PIL environment.

20.4.2.1.3 Signal comparison

After test execution, the result vectors (simulation results) of the model $o_{M\_PCG}(t)$ shall be compared with the execution results of the generated code $o_{CODE}(t)$.

The simulation results of the model M_{PCG} are used as baseline. They are compared with the result vectors obtained from execution of the object code.

Even in the case of a correct translation of a Simulink/Stateflow model into C code, one cannot always expect identical behavior (equality). Possible reasons include limited precision of floating point numbers, quantization effects when using fixed-point math, and differences between compilers. So, the comparison has to be based on sufficiently similar behavior (sufficient similarity).

A suitable *signal comparison algorithm* shall be selected that is *able to tolerate differences between the result vectors* representing the system responses $o_{M\_PCG}(t)$ and $o_{CODE}(t)$. There is a broad variety of potential comparison algorithms ranging from elementary algorithms such as absolute or relative differencing to more elaborate ones such as the difference matrix method (Conrad, Sadeghipour, and Wiesbrock 2006).

The difference matrix method, for example, allows accounting for bounded differences or partial displacements (shifts) on the time axis. It can be used as a preprocessing algorithm to align $o_{CODE}(t)$ and $o_{M\_PCG}(t)$ by stretching or compressing sections of the y-axis of $o_{CODE}(t)$ to find a best match with $o_{M\_PCG}(t)$. Technically speaking, the algorithm calculates a reparameterization (temporal reordering) of $o_{CODE}(t)$ such that $o_{CODE}(t)$ matches $o_{M\_PCG}(t)$ as well as possible. After using difference matrix preprocessing to better match the time axes of the two signals, elementary algorithms such as absolute or relative differencing can be used to assess the remaining deviations on the value axis (Conrad, Sadeghipour, and Wiesbrock 2006).

Two *result vectors are sufficiently similar* if their difference with respect to a given comparison algorithm is less than or equal to a given threshold. The selection of the comparison algorithm and the definition of the threshold value depend on the application under consideration or even on the characteristics of a given output signal (comparing Boolean outputs, for example, may require different algorithms and thresholds than comparing floating point outputs with potentially denormalized values) and must be documented; see Conrad, Sadeghipour, and Wiesbrock (2006) for an overview of algorithms to consider. Neither IEC 61508-3 nor ISO 26262-6 provide specific guidance here.

In practice, an initial automatic assessment with low thresholds and a subsequent manual review of the discrepancies has been proven successful.

Tool Support: Simulink® Fixed Point™* allows performing bit-true simulations of model portions implemented using fixed-point math to observe the effects of limited range and precision on designs built with Simulink and Stateflow. When used with Embedded Coder, Simulink Fixed Point enables generating pure integer C code from these model portions. The generated code is usually in bit-true agreement with the model used for production code generation, ensuring that the implemented design will perform as it did in simulation. Comparing fixed-point models with fixed-point code simplifies signal comparison activities.

Sufficient similarity serves as a basis for the definition of *functional equivalence* used to determine the numerical correctness of the model-to-code translation. A model and the code generated from this process are functionally equivalent if the simulation of the model and the execution of the executable derived from the generated code lead to sufficiently similar result vectors when both are stimulated with identical test vectors.

Tool Support: Comparison algorithms can be implemented in general-purpose programming or scripting languages such as MATLAB. MEval[†] provides a variety of predefined algorithms for result vector comparison including the difference matrix method.

The Simulation Data Inspector (SDI), a utility integrated into Simulink, supports signal viewing and comparison.

Example 4. Figure 20.9 illustrates the usage of the SDI for signal comparison. In a first step, the two test vectors given in Figure 20.6 were used to stimulate the model in SIM mode and the generated code in PIL mode, respectively. Input and result vectors of these test runs were captured in SDI. In a second step, SDI was used to automatically compare the result vectors $o_{M\_PCG}(t)$ and $o_{CODE}(t)$. The screenshot shows the comparison summary for test case TC01 on the left-hand side and signal and difference plots for one of the result vectors on the right-hand side.   □

20.4.2.2   Prevention of unintended functionality

The second activity within the code verification process is to demonstrate that the generated C code does not perform any *unintended function*. Various techniques are available to achieve this objective. They serve to demonstrate structural equivalence between the model and the source code. By the use of at least one of the following measures, it shall be demonstrated that the generated C code does not perform unintended functions:

---

*The MathWorks, Inc. Simulink® Fixed Point™—www.mathworks.com/products/simfixed [10/11/20].
[†]IT Power Consultants. MEval—www.itpower.de/28-1-MEval.html [10/11/20].

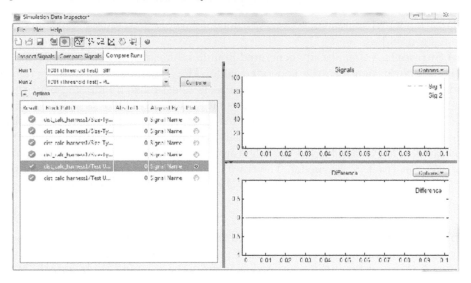

FIGURE 20.9
Example signal comparison.

1. Model and code coverage comparison

2. Traceability review

20.4.2.2a   Model and code coverage comparison

If *model and code coverage comparison* are being used, model and code coverage are measured during equivalence testing and compared against each other. Discrepancies with respect to comparable coverage metrics should be assessed.

To be meaningful, structural coverage metrics comparable with each other should be used on model and code level, respectively. For example, a combination of decision coverage at the model level and branch coverage (C1) at the code level (which is also sometimes termed decision coverage) can be employed (Baresel et al. 2003, Kirner 2009).

Discrepancies between model and code coverage in terms of comparable metrics shall be assessed. If the code coverage achieved is less than the model coverage, unintended functionality may have been introduced.

Tool Support: Code coverage information can be derived from the IDE in use or by applying a stand-alone code coverage tool to the generated source code. The Embedded Coder code generator provides an option to measure code coverage using the BullseyeCoverage™ tool* when running SIL or PIL tests. Model Coverage can be measured using Simulink® Verification and Validation™.

20.4.2.2b   Traceability review

A *traceability analysis* of the generated code can be performed to ensure that all parts of the generated C source code can be traced back to the model used for production code generation. In this case, the generated code is subjected to a *limited review* that exclusively focuses on traceability aspects. During such a traceability review, the traceability matrix

---

*Bullseye Testing Technology™: BullseyeCoverage™—www.bullseye.com [10/11/20].

that relates model to source code elements and *vice versa* is analyzed. *Nontraceable code shall be flagged and assessed.*

Tool Support: Automatically generated Simulink block comments can be used to provide tracing information into the generated code. The Traceability Report section of the Embedded Coder code generation report helps users to establish a mapping between model constructs and the generated code. It lists eliminated and virtual blocks as well as traceable blocks. The "Code-to-block highlighting" option generates hyperlinks within the displayed source code to view the blocks, charts, transitions, and subsystems from which the code was generated. The "Block-to-code highlighting" option allows one to view the generated code for any construct in the model. The traceability matrix generation feature of the IEC Certification Kit* allows exporting complete traceability matrices into Excel where they can be analyzed further.

Example 5. Figure 20.10 shows excerpts from the traceability matrix of the dist_calc model covering requirements, model elements, and generated code. Such a traceability matrix together with the model and the generated code would be examined in a traceability review as described above.   □

Further details of the translation validation workflow are given in Conrad (2009) and (MathWorks 2009a). The detailed workflow documentation is available as part of the IEC Certification Kit. The detailed version of this workflow has been discussed with TÜV[†] SÜD and was a foundation for the in-context certification of Real-Time Workshop Embedded Coder to IEC 61508 and the tool qualification according to ISO 26262 (MathWorks 2008, MathWorks 2009b).

Issues found during the translation validation process could have had a variety of different root causes including configuration management issues, incorrect code generator, compiler and linker settings, or bugs in the tools that have been involved.

Requirements Source	Requirements Location	Model Object Path	Model Object Subsystem	Model Object Type	Model Object Unique ID	Code File Location	Code File Name	Code Function	Code Line Number	Model Optimizatio
dist_calc_req.docx	@OLE_LINK3	dist_calc	dist_calc	Inport	dist_calc:1	c:\...\ dist_calc_ert_rtw	dist_calc.c	dist_calc_step	65	
...	...	...	...	...	...	...	...	...	...	...
dist_calc_req.docx	@OLE_LINK1	dist_calc	dist_calc	Abs	dist_calc:6	c:\...\ dist_calc_ert_rtw	dist_calc.c	dist_calc_step	51	
dist_calc_req.docx	@OLE_LINK1	dist_calc	dist_calc	Abs	dist_calc:6	c:\...\ dist_calc_ert_rtw	dist_calc.c	dist_calc_step	52	
dist_calc_req.docx	@OLE_LINK1	dist_calc	dist_calc	Sum	dist_calc:7	c:\...\ dist_calc_ert_rtw	dist_calc.c	dist_calc_step	67	
dist_calc_req.docx	@OLE_LINK1	dist_calc	dist_calc	Sum	dist_calc:7	c:\...\ dist_calc_ert_rtw	dist_calc.c	dist_calc_step	72	
dist_calc_req.docx	@OLE_LINK1	dist_calc	dist_calc	Sum	dist_calc:8	c:\...\ dist_calc_ert_rtw	dist_calc.c	dist_calc_step	36	
dist_calc_req.docx	@OLE_LINK1	dist_calc	dist_calc	Sum	dist_calc:8	c:\...\ dist_calc_ert_rtw	dist_calc.c	dist_calc_step	40	
dist_calc_req.docx	@OLE_LINK1	dist_calc	dist_calc	Sum	dist_calc:9	c:\...\ dist_calc_ert_rtw	dist_calc.c	dist_calc_step	64	
dist_calc_req.docx	@OLE_LINK1	dist_calc	dist_calc	Sum	dist_calc:9	c:\...\ dist_calc_ert_rtw	dist_calc.c	dist_calc_step	69	
...	...	...	...	...	...	...	...	...	...	...
		dist_calc	dist_calc	Outport	dist_calc:16	c:\...\ dist_calc_ert_rtw	dist_calc.h	Global	47	
...	...	...	...	...	...	...	...	...	...	...

FIGURE 20.10
Example traceability matrix.

---

*The MathWorks, Inc. IEC Certification Kit – www.mathworks.com/products/iec-61508 [10/11/20].

†Technischer Überwachungs Verein, certification authority (notified body) and service provider.

Functional safety standards such as IEC 61508 typically require procedures describing corrective actions on failures of verification or validation activities. Such a procedure would detail the necessary steps to investigate and handle issues that were found during testing-based translation validation. Corrective actions could, for example, include the installation of available bug fixes for the tools involved. If bug fixes are not available, parts of M_{PCG} could be replaced with functional equivalent modeling constructs to work around bugs in the tool chain.

20.5 Related Work

A number of verification and validation approaches to be used in conjunction with Model-Based Design and code generation have been proposed. However, these approaches do not specifically address compliance with IEC 61508-3 or ISO 26262-6.

Earlier work on verification and validation of code generation from models includes Edwards (1999), Burnard (2004), Conrad (2004), and Stürmer, Weinberg, and Conrad (2005). Typically, a combination of translation validation and translator validation approaches has been discussed. However, none of the papers discusses the proposed measures or techniques in the context of IEC 61508 or ISO 26262 requirements.

Translator validation can be considered orthogonal to translation testing discussed in this chapter. A popular means of translator validation is validation testing. Other work (MISRA-C, 2004) suggests documented validation testing as a possible approach to obtain confidence in the tools used. This approach has been applied to Model-Based Design by major automotive companies and the TÜV. A production code generation tool chain including Real-Time Workshop Embedded Coder and a Texas Instruments C compiler successfully passed the resulting AVS, a test suite to validate the quality and robustness of the code generator, and compiler and linker tool chain (TÜV 2005, Pofahl, Sauer, and Busa 2007, TÜV 2008). Underlying concepts on test suites for code generators are laid out for example in Stürmer and Conrad (2003), Stürmer (2006), Zelenov et al. (2006), Stürmer et al. (2007), Schneider and Slotosch (2007), and Schneider, Lovric, and Mai (2009). Test suites may provide additional confidence in the correct functioning of a code generation tool chain. However, due to complexity and other reasons, it is considered infeasible to test a code generation tool chain exhaustively with traditional software testing approaches (Stürmer et al. 2007). According to Burnard (2004), there are more than 2000 ways of implementing a simple arithmetic function such as a := b + c depending on data types, scaling information, and other configurations (e.g., saturation behavior). In practice, this leads often to a combination of a test suite to validate basic aspects and exemplary functionality of the code generation tool chain with application-specific verification and validation techniques.

The interplay between model and code coverage has been examined analytically by Aldrich (2001) and Kirner (2009), and statistically by Baresel et al. (2003).

20.6 Conclusions

The increasing use of Model-Based Design with production code generation calls for new methods and tools to verify and validate generated code in an efficient and effective manner.

Model-Based Design allows for new verification and validation approaches utilizing the existence of executable graphical models as well as multiple executable artifacts that can be checked for equivalence.

For high-integrity application developments, those verification and validation methods/tools must be reconciled with safety standards such as IEC 61508 and ISO 26262. The proposed workflow for verification and validation of models and generated code can be viewed as an instantiation of the verification and validation requirements of IEC 61508-3 and ISO 26262-6, which exploits the opportunities of code generation as an integral part of Model-Based Design.

Users of Simulink and Embedded Coder can use this reference workflow as a pattern to create their own project- and SIL-/ASIL-specific variants of this workflow. More generally, the principles underlying the proposed workflow have a broader scope and can be generalized for other types of executable modeling and code generation.

Note that while adhering to the recommendations in this document will reduce the risk that an error is introduced in the development and not be detected, it is no guarantee that the system being developed will be safe. Conversely, if some of the recommendations in this document are not followed, it does not mean that the system being developed will be unsafe.

References

Arthur, D. L. (2007). Kosten-/Nutzenanalyse der modellbasierten Softwareentwicklung im Automobil.

Aldrich, W. (2002). Using Model Coverage Analysis to Improve the Controls Development Process. Proc. AIAA Modeling and Simulation Technologies Conference, Monterey, CA, USA.

Baresel, A., Conrad, M., Sadeghipour, S., and Wegener, J. (2003). The interplay between model coverage and code coverage. 11. *Europ. Int. Conf. on Software Testing, Analysis and Review* (EuroSTAR '03). Amsterdam, Netherlands.

Burnard, A. (2004). Verifying and validating automatically generated code. *Int. Automotive Conference* (IAC '04), Pages: 71–78. Stuttgart, Germany.

Conrad, M. and Dörr, H. (2006). Model-based development of in-vehicle software. In *Proc. Conf. on Design, Automation and Test in Europe (DATE '06)*, Pages: 89–90. Munich, Germany.

Conrad, M., Fey, I., Grochtmann, M., and Klein, T. (2005). Modellbasierte Entwicklung eingebetteter Fahrzeugsoftware bei DaimlerChrysler. *Inform. Forsch. Entwickl*, 20(1–2):3–10.

Conrad, M. (2004). Modell-basierter Test eingebetteter Software im Automobil: Auswahl und Beschreibung von Testszenarien. PhD Thesis, Deutscher Universitätsverlag, Wiesbaden, Germany.

Conrad, M. (2007). Using Simulink and real-time workshop embedded coder for safety-critical automotive applications. *Proc. Workshop Modellbasierte Entwicklung Eingebetteter Systeme III* (MBEES '07), Pages: 41–50. Schloß Dagstuhl, Germany.

Conrad, M. (2008). Model-based design for IEC 61508: Towards translation validation of generated code. *Proc. Workshop Automotive Software Engineering*: Forschung, Lehre, Industrielle Praxis, co-located with Software Engineering 2008, Munich Germany.

Conrad, M. (2009). Testing-based translation validation of generated code in the context of IEC 61508. *Form. Methods Syst. Des.*, 35(3):389–401.

Conrad, M., Sadeghipour, S., and Wiesbrock, H. (2006). Automatic evaluation of ECU software tests. *SAE 2005 Transactions, Journal of Passenger Cars - Mechanical Systems*, SAE Paper 2005-01-1659, SAE International Warrendale, PA.

Conrad, M., Erkkinen, T., Maier-Komor, T., Sandmann, G., and Pomeroy, M. (May 2010). Code generation verification – Assessing numerical equivalence between Simulink models and generated code. *4th Conf. Simulation and Testing in Algorithm and Software Development for Automobile Electronics*. Berlin, Germany.

Czerny, B. J., D'Ambrosio, J. G., Jacob, P. O., Murray, B. T., and Sundaram, P. (2004). An adaptable software safety process for automotive safety-critical systems. 2004 SAE World Congress, Detroit, MI (SAE Techn. Paper 2004-01-1666).

Erkkinen, T. and Conrad, M. (2007). Safety-critical software development using automatic production code generation. *Proc. SAE World Congress 2007*. Detroit, MI. www.mathworks.com/company/pressroom/articles/article18304.html.

Edwards, P. D. (1999). The use of automatic code generation tools in the development of safety-related embedded systems. *Proc. Vehicle Electronic Systems*, ERA Report 99-0484.

Fey, I., Müller, J., and Conrad, M. (2008). Model-Based Design for safety-related applications. SAE Techn. Paper 2008-21-0033, Convergence 2008, Detroit, MI.

Helmerich, A., Koch, N., Mandel, L. et al. (2005). Study of worldwide trends and R&D programs in embedded systems in view of maximising the impact of a technology platform in the area. Report for the European commission, Brussels, Belgium.

IEC 61508-3 (1998). Int. Standard Functional safety of electrical/electronic/programmable electronic safety-related systems, Part 3: Software requirements. First edition.

ISO 26262-6 (2011). Int. Standard Road vehicles – Functional safety, Part 6: Product development: Software level.

Jablonski, T., Schumann, H., Busse, C., Haussmann, H., Hallmann, U., Dreyer, D., and Schöttler, F. (2008). Die neue elektromechanische Lenkung APA-BS. ATZelektronik 01/2008 Vol. 3 01, Pages: 30–35.

Kamga, J., Herrmann, J., and Joshi, P. (2007). Deployment of model-based technologies to industrial testing. D-MINT automotive case study – Daimler, ITEA 2 Project Deliverable 1.1.

Kirner, R. (2009). Towards preserving model coverage and structural code coverage. *EURASIP Journal on Embedded Systems*.

MISRA-C (2004). Guidelines for the use of the C language in critical systems. MISRA.

Mosterman, P. J. (2006). Automatic code generation: Facilitating new teaching opportunities in engineering education. 36. Annual ASEE/IEEE Frontiers in Education Conf, San Diego, Pages: 1–6.

MathWorks (2005). Alstom generates production code for safety-critical power converter control systems. www.mathworks.com/products/rtwembedded/userstories.html?file= 10591.

MathWorks (2008). The MathWorks Real-Time Workshop Embedded Coder certified by TÜV SÜD Automotive GmbH. Press release.

MathWorks (2009a). Application-specific verification and validation of models and generated code. V1.4.

MathWorks (2009b). The MathWorks Real-Time Workshop Embedded Coder and Polyspace products qualified according to ISO 26262. Press release.

MathWorks (2010). Model-Based Design. www.mathworks.com/model-based-design [10/11/20].

Potter, B. (2004). Use of the MathWorks tool suite to develop DO-178B certified code. ERAU/FAA Software Tools Forum, Daytona Beach.

Pofahl, E., Sauer, T., and Busa, O. (2007). AVS—a test suite for automatically generated code. MathWorks Automotive Conference (MAC '07), Dearborn.

Pnueli, A., Siegel, M., and Singerman, E. (1998). Translation validation. *Proc. 4. Int. Conf. on Tools and Algorithms for the Construction and Analysis of Systems* (TACAS '98), Pages: 151–166. Lisbon, Portugal.

Stürmer, I. and Conrad, M. (2003). Test suite design for code generation tools. 18. *IEEE Int. Conf. on Automated Software Engineering* (ASE '03), Montreal, Canada.

Stürmer, I. and Conrad, M. (2005). Ein Testverfahren für optimierende Codegeneratoren. *Inform. Forsch. Entwickl.* 19(4):213–223.

Stürmer, I., Conrad, M., Dörr, H., and Pepper, P. (2007). Systematic testing of model-based code generators. *IEEE Transactions on Software Engineering*, Pages: 622–634.

Smith, D. J. and Simpson, K. G. L. (2005). Functional safety – a straightforward guide to applying IEC 61508 and related standards. Elsevier Butterworth-Heinemann, Oxford, UK.

Schneider, S. and Slotosch, O. (2007). A validation suite for model-based development tools. 10. *Int. Conf. on Quality Engineering in Software Technology* (CONQUEST 2007), Potsdam, Germany.

Schneider, S., Lovric, T., and Mai, P. S. (2009). The validation suite approach to safety qualification of tools. SAE World Congress 2009, Detroit, MI.

Stürmer, I. (2005). Systematic testing of code generation tools – a test suite-oriented approach for safeguarding automatic code generation. TU Berlin, Germany.

Stürmer, I., Weinberg, D., and Conrad, M. (2005). Overview of existing safeguarding techniques for automatically generated code. 2. *Int. ICSE Workshop on Software Engineering for Automotive Systems* (SEAS '05), St. Louis.

Toeppe, S., Ranville, S., Bostic, D., and Wang, Y. (1999). Practical validation of model based code generation for automotive applications. 18th *AIAA/IEEE/SAE Digital Avionics System Conference* St. Louis, MO.

TÜV (2005). Validation of the MathWorks code generator Real-Time Workshop® Embedded Coder with the Autocode Validation Suite (AVS) v3.0. Report No. 968/EL 211.02/05, TÜV Rheinland.

TÜV (2008). Validation of the MathWorks Real-Time Workshop® Embedded CoderTM product with the Automotive Code Validation Suite (AVS) v4.0. Report No. 968/EL 525.00/08, TÜV Rheinland.

Zelenov, S. V., Petrenko, A. K., Conrad, M., and Fey. (2006). Automatic test generation for model-based code generators. 2. *Int. Symposium on Leveraging Applications of Formal Methods, Verification and Validation* (ISoLA '06). Paphos, Cyprus.

21

Model-Based Testing of Analog Embedded Systems Components

Lee Barford

CONTENTS

Many embedded systems contain analog systems. These analog systems include sensors and actuators, filtering and signal conditioning electronics, and much of the physical layers of both wired and wireless communications interfaces (Marwedel 2006, Ganssle 2008). So, thorough model-based testing (MBT) of an embedded system with analog components involves test of those components.

The topic of analog systems is enormous; after all, every physical system in the universe except for digital electronics is analog. And even digital electronics are analog systems carefully designed to have digital behavior when operated within the correct analog environment (power supply voltages and frequencies, temperature, humidity, etc.). Designers, builders, and testers of embedded systems are primarily concerned with analog systems possessing the following properties:

- The system interfaces both to the world outside the embedded system and to the embedded system through physical interfaces characterized by a finite number of time-varying parameters. For example, an audio speaker has as its interfaces (1) terminals to which a time-varying voltage is applied and (2) the instantaneous sound pressure measured at some point in front of the speaker. Typically, the interface to the embedded system will be a time-varying voltage that is controlled by a digital-to-analog-converter (DAC) or sensed through an analog-to-digital converter (ADC). These interfaces will be referred to as *ports*.

- The system has at least one port. The ports are typically divided into two categories, *input ports* and *output ports*. In a few domains (e.g., microwave devices), all ports have both input-like and output-like behavior and so all ports are just called "ports."

- The system has been designed to have a particular behavior that relates the physical parameters applied to the input ports to those generated at the output ports. That behavior is documented, either in descriptive terms or with a mathematical model (e.g., a circuit diagram including component parameters). The behavior of the system changes extremely slowly compared to the time scales encountered in the normal operation of the embedded system. That is, the behavior can be considered time invariant unless and until the system fails.

The problem discussed in this chapter is that of *fault identification* or *pass-fail test*, or just *test* for short. That is, determining whether an analog system meets its specification as given by some model of correct behavior (the *system model*) or of incorrect behavior (called a *fault model*). Measurements of the behavior of an analog system for the purpose of test are made at the ports. This chapter assumes that analog systems comprising multiple replaceable units with their own ports will be tested unit by unit. Once measurements are made, test is performed through a comparison between the measured behavior of the system and those predicted by the subsystems's model or fault models.

More specifically, analog MBT generally follows a two-phase process. The first, or *pretesting phase*, comprises tasks that must be completed before any test is performed, but are required only once for one analog system design, that is, for one set of analog devices under test (DUTs) that are designed to be identical. The second, or *testing phase*, covers tasks that must be performed for each DUT.

The pretesting phase generally requires the following steps:

System model selection Choose from the appropriate classes of models of system behavior. For example, choices must be made between the following:

- Structural versus behavioral
- Nonparametric versus parametric
- Time domain versus frequency domain
- Linear versus nonlinear models

Fault model selection Choose how faults will be modeled. Faults modeling options include the following:

- Modeling actual physical fault mechanisms or their effects
- Modeling the difference in behavior between the system design and the actual behavior of the DUT

A difference model is a system model, and so choices of difference model come from the same categories as do choices of system models.

Excitation design Choose the physical stimulus that will be applied to the input ports of the DUT during testing.

Simulation from fault models In this step, not present in all MBT approaches, simulate all plausible faults and store the resulting outputs.

The testing phase generally requires these steps, repeated for each DUT:

Measurement Apply the excitation to the DUT's input ports and measure the response at its output ports. If possible, measure the excitation at the input ports as well. Measurement of the excitation allows characterization of any noise or other disturbances at the inputs as well as at the outputs. As noise, disturbances, or measurement uncertainty increase, the input measurement becomes more critical.

System identification Use statistical methods to identify a nonparametric or parametric model for the system or fault model.

Behavioral simulation In this step, not present in all MBT approaches, perform simulations using the system and fault models to predict the behavior of the DUT under excitations not actually applied and measured.

Faulty/not faulty reasoning The system model, fault model, and simulation results are used by an automated reasoner to determine whether or not the DUT is faulty. The automated reasoner may vary considerably in complexity. It could be just conditional (`if-`) statements or a complicated heuristic search algorithm.

Before delving into more detail on these processes, it is necessary to describe the classes of analog system models used in MBT.

21.1 Analog Models

Analog modeling is a vast topic: it could encompass all modeling methods used in physics. Necessarily, then, the presentation here is restricted to models commonly used in MBT. Such models can be categorized as either structural or behavioral models (Graeb 2007). Analog models can also be divided into two categories based on whether or not they contain parameters that must be statistically identified from measurements of the input-output behavior of the system under test.

Parametric models are functions of a number of variables, some of which are parameters fixed for a particular, individual system. That is, a parametric model is a function that can be written $f(a, b, c, \ldots; x_1, x_2, \ldots)$, where the parameters $a, b, c, \ldots$ are fixed for a particular system and $x_1, x_2, \ldots$ vary over time. A well-known parametric model is that of a resistor:

$$i(t) = f(R; v(t)) = Rv(t),$$

where i is current, v is voltage, and the resistance R is the parameter.

Nonparametric models are models built from measured data without explicitly identifying parameters. An example of a nonparametric model would be a table of measured system input and output values or a spline curve fitted to such a table to support interpolation between table values.

21.1.1 Structural models

A structural model describes the behavior of an analog system using the following:

- The components from which it is constructed

- Any necessary physical parameters of the components

- The way components are connected together

Finally, a structural model must support simulation of the system, given a set of parameter values and system inputs.

Structural models often are, or are derived from, computer-aided design (CAD) models. Components may themselves be defined by a structural model. Moreover, a structural modeling scheme must have a set of primitive components whose behavior can be defined mathematically (e.g., as a set of differential equations), or is given by a simulation algorithm.

A common structural model in electronics is a circuit schematic, normally kept in a data structure called a *netlist*. The components are circuit elements such as resistors, capacitors, inductors, transistors, and amplifiers. The physical parameters are the resistances, capacitances, inductances, or other characterization of the voltage-current response of the components.

21.1.2   Behavioral models of linear systems

Behavioral models provide no information about how the analog system operates internally. They describe only the relationships between what happens at the input and output ports as mathematical functions of time.

One of the first questions one must address with in selecting a behavioral model is whether time should be treated as a continuous variable or as a discrete variable. Either choice implies an assumption about the behavior at the system's ports between the times that they are sampled. If time is treated as a discrete variable, then most modeling methods assume that the physical values at the ports remain constant except for the time instants at which they are measured. This is called the *zero-order hold* or ZOH assumption. If time is treated as a continuous variable, the usual implicit assumption is that the behavior at the ports is *band-limited*, that is, that the physical variable at each port can vary only at some maximum rate or frequency. Most analog systems' behavior correspond more closely to the band-limited assumption. So continuous-time models are usually appropriate (Pintelon, Schoukens, and Rolain 2008).

Below, variables $x(t)$ and vectors $\mathbf{x}(t) = (x_1(t), \ldots, x_n(t))$ are inputs. Variables $y(t)$ and vectors $\mathbf{y}(t) = (y_1(t), \ldots, y_n(t))$ are outputs.

Continuous-time analog behavioral models are usually divided into two classes: linear and nonlinear. A system S with inputs $(x_1(t), x_2(t), \ldots, x_n(t))$ and outputs

$$S(x_1, x_2, \ldots, x_n) = (y_1(t), y_2(t), \ldots, y_m(t))$$

is called linear if and only if it has both the *superposition property*, that is,

$$S(x_1(t) + \hat{x}_1(t), x_2 + \hat{x}_2(t), \ldots, x_n(t) + \hat{x}_n(t)) =$$
$$S(x_1(t), x_2(t), \ldots, x_n(t)) + S(\hat{x}_1(t), \hat{x}_2(t), \ldots, \hat{x}_n(t))$$

and is *homogeneous*

$$S(cx_1(t), cx_2(t), \ldots, cx_n(t)) = cS(x_1(t), x_2(t), \ldots, x_n(t))$$

for any constant c. These two properties imply that for single-input, single-output systems S, and any constants $c_1, c_2, \ldots, c_n$,

$$S(c_1 x_1(t) + c_2 x_2(t) + \ldots + c_n x_n(t)) = c_1 S(x_1(t)) + c_2 S(x_2(t)) + \ldots + c_n S(x_n(t)).$$

Our interest is in time-invariant systems, where for any constant t_0,

$$S(x_1(t), x_2(t), \ldots, x_n(t)) = S(x_1(t + t_0), x_2(t + t_0), \ldots, x_n(t + t_0)).$$

Let $\delta(t)$ be the Dirac delta function,

$$\delta(t) = \begin{cases} \infty & \text{if } t = 0, \\ 0 & \text{otherwise} \end{cases}$$

and

$$\int_0^\infty \delta(t)dt = 0.$$

The *impulse response* of the single-input, single-output (SISO) linear, time-invariant (LTI) system S is given by $h(t) = S(\delta(t))$. By time invariance and superposition, the impulse response determines the behavior of S given any input, since

$$x(t) = \int_{-\infty}^\infty x(\tau)\delta(t-\tau)d\tau,$$

so by the superposition property

$$y(t) = \int_{-\infty}^\infty x(\tau)h(t-\tau)d\tau.$$

That is, $y = x * h$, where $*$ represents convolution.

Further, let $\mathcal{F}[\cdot]$ be the Fourier Transform operator (Allen and Mills 2004)

$$\mathcal{F}[f](\omega) = \int_{-\infty}^\infty f(\tau)e^{-2\pi i\omega\tau}\,d\tau,$$

and $\mathcal{F}^{-1}[\cdot]$ its inverse operator

$$\mathcal{F}^{-1}[F](t) = \int_{-\infty}^\infty F(\omega)e^{2\pi i\omega t}\,d\omega.$$

Under nonrestrictive smoothness conditions–that always hold under the band-limited assumption–on functions $f(t)$ and $g(t)$,

$$f * g = \mathcal{F}^{-1}[\mathcal{F}[f]\mathcal{F}[g]],$$

so in particular

$$y = x * h = \mathcal{F}^{-1}[\mathcal{F}[x]\mathcal{F}[h]]. \tag{21.1}$$

The dynamical input-output behavior of a SISO LTI system S is therefore determined by the *frequency response* $H(\omega) = \mathcal{F}[h]$. A large proportion of traditional (nonmodel-based) test procedures for analog LTI systems are based on measuring $H(\omega)$ because that single measurement completely characterizes the dynamical behavior of the system. For the same reason, frequency-domain models are appealing for MBT of LTI systems as well.

An impulse response or a frequency response is a nonparametric model of a SISO LTI system. Parametric models of LTI systems generally are based on systems of ordinary differential equations. In the multiple input, multiple output (MIMO) case, the system of equations is most conveniently kept and manipulated in *state-space* form. Here, let $\mathbf{x}$ and $\mathbf{y}$, vector-valued functions of time, be a time-varying set of inputs and outputs, respectively. Let $\dot{\mathbf{x}}$ be the vector of time derivatives of $\mathbf{x}$. Then, a state-space model for an LTI MIMO system has the form, where $\mathbf{A}$, $\mathbf{B}$, and $\mathbf{C}$ are real matrices,

$$\dot{\mathbf{s}} = \mathbf{As} + \mathbf{Bx}$$
$$\mathbf{y} = \mathbf{Cs},$$

where a dot above a function indicates taking the time derivative. The variables $\mathbf{s}$ are called *state variables* and are not generally measurable from outside the system. The usual general form for an LTI SISO system is as a single differential equation

$$\sum_{i=0}^n a_i \frac{d^i y}{dt^i} = \sum_{i=0}^m b_i \frac{d^i x}{dt^i}. \tag{21.2}$$

Taking the Laplace Transform of this differential equation shows that it has the frequency response

$$H(\omega) = \frac{b_0 + b_1 e^{2\pi i\omega} + \ldots + b_m e^{2\pi m i\omega}}{a_0 + a_1 e^{2\pi i\omega} + \ldots + a_n e^{2\pi n i\omega}}. \tag{21.3}$$

This equation provides the usual general form of a parametric model for a LTI SISO system in the frequency domain. The complex zeros of the polynomial $b(w) = \sum_{i=0}^{m} b_i w^i$ are called the *zeros* of the LTI system (21.2). The complex zeros of the polynomial $a(w) = \sum_{i=0}^{n} a_i w^i$ are called the system's *poles*. The system is called *stable* if the real parts of its poles are all negative. In order that there not be an infinite set of models identical except for the scaling of the parameters, it is common practice to fix arbitrarily the value of one parameter, for example setting $a_0 \equiv 1$.

Equations 21.2 and 21.3 provide the common parametric models for SISO LTI systems. Commercial software is available for system identification of SISO LTI systems in either the time domain form of Equation 21.2 or frequency domain form of Equation 21.3, for example, System Identification Toolbox 2009, Frequency Domain System Identification Toolbox 2009. The excitation design and system identification methods implemented by such software are described in Schoukens and Pintelon (1991) and Ljung (1999).

21.1.3   Behavioral models of nonlinear systems

We will consider behavioral models of both *static* and *dynamic nonlinear systems*. A nonlinear system is static if its output is a function of only its inputs at the current time,

$$\mathbf{y}(t) = \mathbf{f}(\mathbf{x}_1(t), \mathbf{x}_2(t), \ldots, \mathbf{x}_m(t)).$$

A static nonlinear system is also called *weakly nonlinear* (SWNL) if it can adequately be modeled with a low-order polynomial. That is, for the SISO case, it can be modeled as

$$y(t) = \sum_{i=0}^{n} p_i x^i(t) \tag{21.4}$$

for some small n.

The behavior of a dynamic nonlinear system contains unobservable state variables that in effect maintain a memory of past inputs and outputs. That is, it can be represented by a system of nonlinear differential equations

$$\dot{\mathbf{s}} = \mathbf{f}(\mathbf{s}, \mathbf{x})$$

$$\mathbf{y} = \mathbf{s}_0,$$

where $\mathbf{s}$ is the vector of state variables and $\mathbf{f}$ is an unknown, vector-valued function.

System identification software for nonlinear systems is not nearly as readily available as for linear systems. If *a priori* information is available to indicate the dimension of $\mathbf{s}$ and a parametric form for $\mathbf{f}$, then it is preferable to develop a custom procedure for identifying the parameters of $\mathbf{f}$. If such *a priori* information is unavailable, a parametric model can often be obtained using *time-lag embedding*. Time-lag embedding methods are derived from Takens' Theorem (Takens 1981), a theorem showing how to build a model of a nonlinear dynamical system with no inputs and one output. Takens' theorem says that there is an integer $D > 0$, called the *embedding dimension,* so that the following happens: Choose $\tau > 0$. From measurements of y, construct points in $\Re^D$ $(y(t), y(t - \tau), y(t - 2\tau), \ldots, y(t - (D-1)\tau))$ for various values of t. As the number of points increases towards infinity, this set of points in D dimensions will converge toward a manifold that is diffeomorphic to $\mathbf{f}$. Stark et al. (1997) extended Takens' result for systems with input ports. The points to be collected are

$$
\begin{aligned}
\mathbf{p}(t) = (&y(t), y(t - \tau), \ldots, y(t - (D - 1)\tau), \\
&x_1(t), \ldots, x_1(t - (D - 1)\tau), \ldots, \\
&x_m(t), \ldots, x_m(t - (D - 1)\tau)).
\end{aligned}
\tag{21.5}
$$

Once one has collected enough points, some universal approximator is used to fit a function g that predicts the first dimension $y(t)$ of $\mathbf{p}$ from the other $(m + 1)D - 1$ dimensions of $\mathbf{p}$. Common choices for the universal approximator are a least squares polynomial, an artificial neural network, or a support vector machine (Cristianini and Shawe-Taylor 2000). The variable τ is usually set based on the autocorrelation function of $y(t)$. Often τ is taken to be half of the first positive zero of the autocorrelation function of $y(t)$.

The embedding dimension D typically is found by increasing D until no so-called "false nearest neighbors" are found (Kennel et al. 1992). Consider a point p, one point obtained by Equation 21.5. Let $p(t')$ be the nearest neighbor of $p(t)$. Look forward in the sequence of points to the k-th point collected after $p(t)$ and the k-th point collected after $p(t')$, for $k \approx 2$ to 5. If these two points are much further apart than were $p(t)$ and $p(t')$, then $p(t)$ and $p(t')$ are considered false nearest neighbors. The presence of false nearest neighbors means that D is too small. The intuitive argument for the correctness of this approach is that false nearest neighbors occur when state-space trajectories of the system cross, and in deterministic dynamical systems, such trajectories must form a flow that never crosses itself. To eliminate these self-crossings, D must be increased.

Having surveyed continuous-time models used in analog MBT, we are ready to describe MBT methods.

21.2   Structural Model-Based Test

Structural MBT is sometimes used when (1) an accurate structural model of the correct behavior of the DUT is available and (2) a fault model is available that describes all reasonably likely faults. Structural MBT is often encountered in manufacturing test of electronics (Bushnell and Agrawal 2002). An electronic product's component list and netlist give an accurate structural model. Also, in manufacturing test, the DUT is new and is rather unlikely to have faults because of wear and tear, vibration, metal fatigue, corrosion, or other difficult-to-predict mechanisms. An adequate fault model for new electronics comprises:

Component parametric faults A component has a parameter that is out of design tolerance. For example, a 50Ω resistor with a 10% tolerance has an actual resistance of less than 45Ω or more than 55Ω.

Opens Two ports of two components should be connected with a wire or printed circuit line but are not.

Shorts Two nearby wires or printed circuit lines are connected together when they should not be.

Excitation design consists of choosing a small set of excitations from a large dictionary of possible excitations. For example, the dictionary of possible excitations is often sine waves with any of a number of frequencies. Measurement result simulation is performed on the plausible fault scenarios (possibly including multiple faults). A parametric fault is often simulated by setting a component parameter to a value just outside its acceptable design tolerance in both the negative and positive direction. The behavior of the resulting system is then simulated under each of the possibilities. An alternative is using Monte Carlo simulation of the expected variation of all parameters. However, since each draw in the Monte Carlo procedure requires a circuit simulation, the Monte Carlo approach is often rather computationally expensive.

Once the results of these simulations are in hand, an optimization procedure is employed to select from the dictionary of excitations a subset where the most faults result in detectable deviation of the measured system outputs. This optimization procedure is often heuristic.

Consider the simple electronic system illustrated in Figure 21.1. This is arguably the simplest analog system that demonstrates dynamic behavior, that is, where its output depends on history of the input, not just the input's current value. A structural model for the system is the differential equation

$$\frac{dV_{out}}{dt} = \frac{-1}{RC} V_{in}. \tag{21.6}$$

The fault model for this circuit will contain catastrophic faults such as open connections between any two nodes that should be connected or shorts between any two nodes that should not be connected. It will also contain parametric faults, variations of R and C from their design values.

Structural MBT will easily find a catastrophic fault. Given realistic values of R and C, any short or open will result in a large change to the measured output voltages, a change that appears in the simulation results of the catastrophic faults.

However, even for such a simple system, the presence of a parametric fault is rather more difficult to detect. This is because the component parameters R and C appear only as their product RC in the governing equation Equation 21.6. An increase of R from the design value will be masked by a corresponding decrease in C and vice versa. In other words, neither R nor C is observable. More complicated devices usually have a number of nonobservable components.

The usual method for making such components observable during test is to add ports to the analog system (Bushnell and Agrawal 2002). However, it is not possible to make all components observable without modifying the DUT during test. Examination of Figure 21.1 shows that there is no way to add a port at which a voltage measurement will distinguish between R or C being out of tolerance. (The voltage between a and b and between c and d must both be 0 unless there is an open.) Only a current measurement between a and b or between c and d could make both R and C observable. However, making a current measurement between a and c generally involves cutting the connection between them and

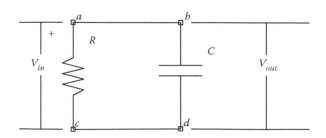

FIGURE 21.1
Simple circuit that illustrates issues with approaches using only structural models.

inserting a current measuring device between them, then remaking the original connection. This disconnecting and reconnecting makes current measurements impractical.

Because of such difficulties and the increasing integration and difficulty of repairing many subsystems especially in electronics, interest is shifting to tests that use behavioral system or fault models and that do not depend on component observability to identify when a DUT is faulty.

21.3   Behavioral Model-Based Test

Behavioral MBT, as its name implies, uses a behavioral model of the system under test rather than performing inference on a structural model. For each DUT, a system model is identified from the measured excitation and response. Various approaches to behavioral MBT differ primarily in how the resulting model is checked.

The pretesting phase of a behavioral MBT process consists of the following steps. First, choose a suitable parametric model for the system model, linear or nonlinear, frequency or time domain. If possible, determine the order of the model during pretesting so that order selection does not have to be performed for every DUT. System model order selection can be based on *a priori* knowledge of the DUT. Otherwise, a number of DUTs can be measured and a model order or small number of possible model orders selected.

Note that the per-DUT phase of test cannot simply be (1) measuring, (2) estimating model parameters, and (3) checking that the model parameters are within some tolerance of nominal parameters. This is because even in the absence of noise and experimental error, the mapping from parameters to input-output behavior is not in general a bijection. Parameters determine the input-output behavior, but more than one set of parameters can produce indistinguishable input-output behaviors. The inverse image of the function from parameter values to input-output behaviors may not even yield connected sets. For example, consider the two polynomials $p1(w) = -0.04w^2 + 1.02w + 0.01$ and $p2(w) = 0.04w^2 + 0.98w - 0.01$. These polynomials are graphed in Figure 21.2. Between 0 and 1, the values of these polynomials are rather close: they differ by at most 0.025 even though their values vary

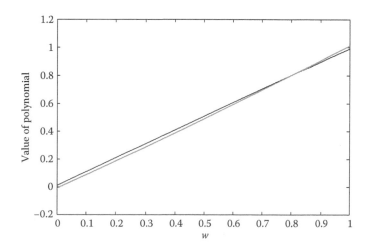

FIGURE 21.2
Two polynomials $p1(w) = -0.04w^2 + 1.02w + 0.01$ and $p2(w) = 0.04w^2 + 0.98w - 0.01$ that have similar values from $w = 0$ to 1 but have rather dissimilar coefficients.

roughly from 0 to 1. But the coefficients of w^2 and of $w^0 = 1$ in the two polynomials differ by twice their magnitude. It is straightforward to construct additional examples where two polynomials are arbitrary close together within some interval, but their parameters differ widely. Many of the system model classes have parameters that are polynomial coefficients: see Equations 21.3 and 21.4. Similarly, in Equation 21.2, the parameters are coefficients of polynomials in the differential operator. If measurement noise has a standard deviation of around 0.01, either $p1$ or $p2$ could be produced by a system identification procedure. But some of their parameters are so different.

How then is faulty/not faulty reasoning to be done? There are a number of rather different strategies for solving that problem in the literature. We will now survey several of the major alternatives. For each alternative, an example MBT system that uses this alternative is presented. Other important factors in the test system design, such as system model selection, and excitation design will also be described.

21.3.1   Faulty/not faulty by model validation

One faulty/not faulty reasoning approach is to avoid creating a system model for each DUT at all. Instead, a system model is identified for a so-called *golden DUT*, a DUT that is believed to be in perfect working order, centered in the acceptable range of all its specifications. The identified model parameters are then used to create an excitation, that, when applied to another DUT during test, ought to produce an output behavior that has certain easy-to-check properties if the DUT is well enough modeled by the system model.

One example of such an easy-to-check property of the output given the chosen excitation would be that the power spectrum of the DUT output is flat–has a certain minimum and maximum power over all measured frequencies. Laquai (2004) describes a test system for phase-locked loops (PLLs) of the sort used in high-speed serial buses, based on this idea. Here, the DUT's input is a two-level digital signal (that is, an analog signal that swings between two voltage levels) generated using an analog parameter ϕ, a phase offset:

$$x(t) = L * (A_1 u(\sin(2\pi f t + \phi)) + A_0),$$

where L is a low-pass filter and

$$u(x) = \begin{cases} 1 & \text{if } x >= 0, \\ 0 & \text{if } x < 0. \end{cases}$$

The output of the DUT is a digital signal that specifies whether a bit error is currently being encountered by a digital radio receiver controlled by the PLL. Essentially, the output signal says whether or not the PLL is locked on to the input signal. The system model of the DUT is a second-order, continuous frequency domain model known from *a priori* knowledge of the design of PLLs to have the form

$$H(\omega) = \frac{\omega_n^2 + c(2\pi i \omega)}{\omega_n^2 + c(2\pi i \omega) + (2\pi i \omega)^2}, \tag{21.7}$$

where ω_n and c are the model parameters. The coefficients of this model are identified by measuring a golden DUT. Figure 21.3 illustrates the shape of a typical power spectrum $|H(\omega)|^2$. A multisine excitation

$$x(t) = \sum_{k=1}^{T} a_k sin(2\pi \omega_k + \phi_k)$$

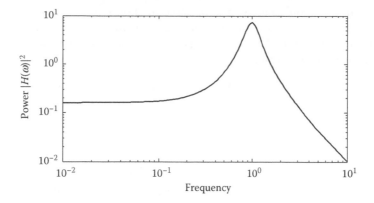

FIGURE 21.3

Shape of power spectrum in model-based exitation design for PLL test.

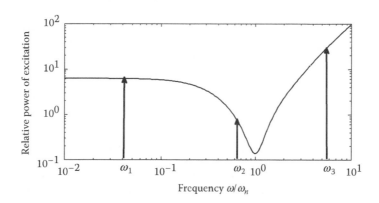

FIGURE 21.4

Selection of power of each tone in multitone excitation for PLL test.

is designed that has power a_k^2 in each tone ω_k that is proportional to $1/|H(\omega_k)|^2$ (Figure 21.4). In order to test a DUT, the excitation $x(t)$ is applied and the power spectrum of the DUT's output is measured at the frequencies $\omega_1, \ldots, \omega_T$. If the model Equation 21.7 holds for the DUT, then power at frequencies ω_k should all be approximately equal. If and only if they are all within a certain bounds of one another, the DUT is nonfaulty. Since Equation 21.7 has two independent parameters, the power must be checked at more than two frequencies. That is, T must be chosen to be greater than two.

The model validation approach is attractive in industrial practice because although the pretest phase requires an iterative numerical algorithm, the test phase does not. The ultimate faulty/not faulty criteria are easily checked. The computation time for each DUT is highly predictable, as is the total time to test each DUT. Such test time predictability is essential, for example, in manufacturing, where test cannot decrease production rates.

Müller et al. (2009) describe another MBT system that uses the model-validation approach to test certain mixed-signal systems (comprising both analog and digital components). The systems are SWNL: they can be represented by nonlinearity that is the composition of an LTI system and weak static nonlinearity Equation 21.4 with $n = 4$. Variation from DUT to DUT is assumed to lie entirely within the nonlinear portion. The LTI portion of each DUT is assumed to be exactly as designed and, therefore, known. For such a low-order polynomial nonlinearity, it is possible to choose a set of tones for the multisine excitation so that the measured complex spectrum output by the DUT $\mathbf{b}$ is related by an affine mapping from the model parameters $\mathbf{p} = (p_0, p_1, p_2, p_3)$. That is, there is a complex matrix $\mathbf{A}_{est}$ and vector $\mathbf{b}^0$ so that

$$\mathbf{b} = \mathbf{A}_{est}\mathbf{p} + \mathbf{b}^0.$$

Similarly, there is a matrix $\mathbf{A}_{pred}$ and vector $\mathbf{y}^0$ so that

$$\hat{\mathbf{y}}(\mathbf{t}) = \mathbf{A}_{pred}\mathbf{p} + \mathbf{y}^0,$$

where $\hat{y}$ is the predicted time-domain response of the DUT given some other excitation that has not been measured. A metric on the difference between this $\hat{y}$ and one measured from a golden DUT is used to determine whether the DUT's behavior is close enough to that of the golden DUT. The matrices and vectors $\mathbf{A}_{est}$, $\mathbf{b}^0$, $\mathbf{A}_{pred}$, and $\mathbf{y}^0$ must be determined during the pretest phase, a process requiring nonlinear optimization methods. However, during the testing phase, only linear equations of a known dimension must be solved. Thus the maximum computation time in the testing phase can be known.

The key issue in designing a test that employs the model validation approach to faulty/nonfaulty reasoning is that it is often challenging to find an easy-to-check property to use given the appropriate system models. In both the test systems described above, deep knowledge of the system behavior was required to identify the easy-to-check property. In the case of the PLL test system, the excitation also had to be designed (the powers of the input tones chosen) to achieve the easy-to-check output property.

21.3.2   Faulty/not faulty reasoning by step-ahead prediction

When the system is modeled solely in the time domain, however, often there is one such easy-to-check property: a prediction of the output from previous inputs and outputs. Prediction a short time into the future is less error-prone than predicting further into the future. So, the prediction is usually of the output next in time, or one step ahead.

Below is a more precise description of faulty/not faulty reasoning by step-ahead prediction for *free-running DUTs*, that is DUTs that have no input port, only output port(s). Because there are no inputs, a model a of free-running system describes the current outputs in terms of past outputs. For example, a time lag embedding model for a free-running system will have the form

$$y(t) = f(y(t - \tau), y(t - 2\tau), y(t - 3\tau), \ldots).$$

During the pretesting phase, a golden DUT is measured and its output recorded for some length of time. During the testing phase, a model is identified for the DUT. The recorded output of the golden DUT is input to the model of the DUT and the predictions checked. That is, if $\hat{y}$ represents the recorded output of the golden DUT and f the model of the DUT, then the differences

$$\hat{y}(t) - f(\hat{y}(t - \tau), \hat{y}(t - 2\tau), \hat{y}(t - 3\tau), \ldots) \tag{21.8}$$

are computed for a number of different times t. If a metric on these differences is sufficiently low, then the DUT passes. Otherwise it fails.

An interesting example of the use of the step-ahead prediction approach is given by Shen et al. (2010). The free-running DUT is a human brain. The measured signal is an electroencephalogram (EEG). The golden DUT is the brain during a normal waking period. Time-lag embedding is used to create a nonlinear model of the DUT from this EEG data.

A support vector machine is used to approximate the nonlinear function f. The test phase comprises continuously checking the same EEG channel to see if the magnitudes of the differences Equation 21.8 get too large. When they do, an abnormal condition such as an epileptic seizure may be occurring.

21.3.3 Faulty/not faulty by checking parameter properties

Sometimes, the system model of a DUT can be re-parameterized so that the new values of the parameters can be checked against straightforward criteria to obtain a faulty/not faulty result. For example, suppose that a test is to determine if an LTI DUT is stable.

Schoukens and Pintelon (1991) and Schoukens, Pintelon, and Rolain (2008) describe doing model-based test for flutter of an aircraft structure. The DUT is a part of an aircraft. It is instrumented and attached to actuators so that an excitation can be applied and the response measured. At a fixed dynamic pressure, that is, fixed true airspeed and air mass density, the part can be considered an LTI system. System identification is done in the frequency domain because the measurements are rather noisy. When identification is complete, what is obtained are the coefficients a_i and b_i in 21.3. For moderate m, all poles are found from the b_i as a solution to a matrix eigenvalue problem or using a routine such the `roots` function in MATLAB® (MATLAB 2009). The real parts of the poles must be sufficiently negative for the test to pass.

21.3.4 Faulty/not faulty by simulating specifications

The correct behavior of some analog systems is given by specifications. A specification is usually a number, a lower or upper bound on a figure of merit obtained from a DUT by applying a particular excitation, measuring the response, and applying some algorithm that yields the same figure of merit. A DUT passes a suite of specification-based tests if each figure of merit is within the specified bounds. Specifications can be derived from knowledge of how the DUT will be used in a larger system. Often, they come from industry practice and government regulations.

Testing a DUT for each of a large number of specifications can be rather time consuming. Yet often the excitation required for each specification is simple. In electronics, the excitation is often a single sine wave.

The idea behind specification-driven MBT is that instead of making many simple measurements of the DUT, one measurement is made using a complicated excitation that exercises the DUT across its entire frequency range of interest. A DUT system model is identified from this single measurement. Then, for each specification, the system model is employed to simulate what the behavior of the DUT would be if it were excited by that specification's excitation. The simulation result is compared against the specification's bound to determine whether the DUT passes or fails that specification. Specification-driven MBT saves time if the total time to perform the measurement needed for system identification, the system identification, and the simulation of each specification test is less than the total time required to perform all of the specification tests one by one.

Barford et al. (2007) describe a specification-driven MBT system for direct-conversion receivers of the radio receivers utilized in mobile phones, wireless computer network interfaces, and the like. A direct-conversion receiver takes a radio-frequency analog signal, usually at several hundred MHz to a few GHz, and reduces its frequency to one that can be sampled by DACs (typically under 100 MHz). This operation is performed by mixing (multiplication by sine waves) and low-pass filtering. An amplifier to strengthen the signal is generally also included. The direct-conversion receivers tested are single-input, two-output systems: the two outputs are taken 90° out of phase from each other, *in-phase (I)* and *quadrature (Q)* channels. A schematic of such receivers is shown in Figure 21.5.

Because of the extreme difference in time scales of the input and the outputs, a *difference modeling* approach is taken. Here, the system model identified from measured data model is the difference between the I and Q signals predicted by simulating a structural model and the measured I and Q signals. The structural model is the CAD model of the receivers. The excitations employed are digital radio signals containing pseudo-random data. For each DUT, time-lag embedding is used to create a nonlinear model of the difference between the DUT's behavior and that of the CAD model (Figure 21.6). During the simulation of the figures of merit, the difference model corrects the output of the CAD model so that the resulting combined model behaves like the DUT (Figure 21.7). The specifications tested are ones traditionally used in radio-engineering to determine that a receiver is sufficiently linear and has little enough distortion to function adequately. One figure of merit, the power

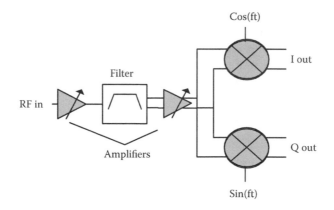

FIGURE 21.5
Schematic of direct conversion receivers tested by specification-driven MBT.

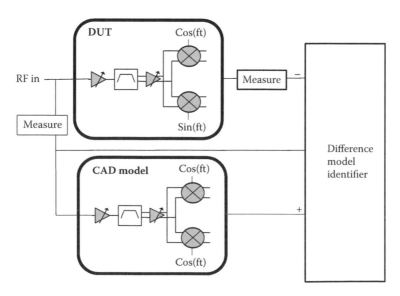

FIGURE 21.6
Identifying the difference model.

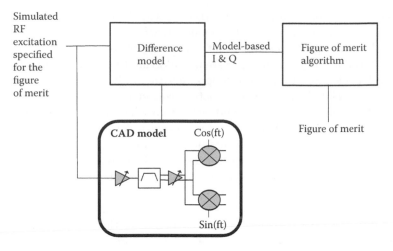

FIGURE 21.7
Simulating a figure of merit using the CAD and difference models.

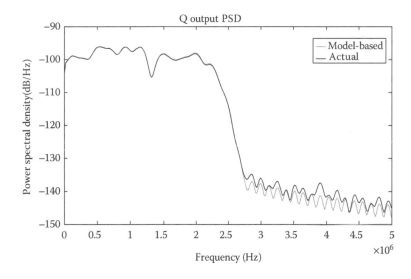

FIGURE 21.8
Actual and model-based Q channel power spectral density.

spectral density (PSD) of the Q channel, simulated from a combined CAD and difference model, is shown in Figure 21.8.

21.4 Conclusions

MBT of analog systems remains an active area of research, driven by diverse needs in fields as disparate as electronics manufacturing, aerospace engineering, and medical research. There is a firm theoretical basis for analog models, drawing on a century's advances in linear and nonlinear dynamics. Thirty years' work in system identification has resulted in methods that are sufficiently automated for use in MBT, provided that DUTs are similar enough to one another to have the same parametric model structure.

A cookbook—a detailed procedure that could be blindly followed—would be desirable for creating a MBT for a given analog system. Unfortunately, at this time, there is no such cookbook. Model selection, excitation design, and the faulty/not faulty reasoning method are all subtly inter-related. In addition, practical considerations such as which ports can be measured at what rate and accuracy and subject to what disturbance further complicate the design of an analog MBT. The test system designer must have a deep knowledge of the DUT and of the relevant modeling and reasoning options.

The rewards, however, for nevertheless using a MBT approach can be substantial. Often, fewer—and less complicated and expensive—measurement instruments may be used compared to nonmodel-based approaches. The total time for test can be reduced because fewer measurements must be made. In the case of embedded systems, use of MBT methods holds out the possibility of greatly enhanced coverage and diagnostic accuracy for built-in self-test.

A number of open questions remain. Two of the more interesting are these: Is there a unifying method for faulty/not faulty determination, or must faulty/not faulty determination inherently be performed in setting-specific ways? And given the progress in MBT of software, included embedded software, is it possible to create unified methods for entire embedded systems including all subsystems of all technologies: analog, mixed-signal, digital, and software?

References

Allen, R. and Mills, D. (2004). *Signal Analysis: Time, Frequency, Scale, and Structure,* Pages: 383–493. IEEE, New York.

Barford, L., Tufillaro, N., Jefferson, S., and Khoche, A. (2007). Model-based test for analog integrated circuits. In *Proc. 2007 IEEE Instrumentation and Measurement Technology Conference,* Pages: 1–6. IEEE, New York.

Bushnell, M. and Agrawal, V. (2002). *Essentials of Electronic Testing for Digital, Memory and Mixed-Signal VLSI Circuits,* Pages: 385–416. Springer, New York.

Cristianini, N. and Shawe-Taylor, J. (2000). *An Introduction To Support Vector Machines (And Other Kernel-Based Learning Methods).* Cambridge University Press, Cambridge.

Frequency Domain System Identification Toolbox. (2009). Version 4.0. Gamax, Inc., Budapest, Hungary.

Ganssle, J. (2008). *Embedded Systems: World Class Designs.* Elsevier, Amsterdam, the Netherlands.

Graeb, H. (2007). *Analog Design Centering and Sizing* 8. Springer, New York.

Kennel, M. et al. (1992). Determining embedding dimension for phase space reconstruction using the method of false nearest neighbors. *Physical Review A* 45(6): 3403-3411. APS, Ridge, New York.

Laquai, B. (2004). A model-based test approach for testing high speed PLLs and phase regulation circuitry in SOC devices. In *Proc. 2004 International Test Conference,* Pages: 764–772. IEEE, New York.

Ljung, L. (1999). *System Identification: Theory for the User,* 2nd ed. Prentice-Hall, Upper Saddle River, New Jersey.

Loukusa, V. et al. (2004). Systems on chips design: System manufacturer point of view. In *Proc. Design, Automation and Test in Europe Conference,* Pages: 3–4. European Design and Automation Association, Brussels, Belgium.

Marwedel, P. (2006). *Embedded System Design.* Springer, New York.

MATLAB®. (2009). Version 2009b, The MathWorks(R), Natick, Massachussetts.

Müller, R. (2009). An approach to linear model-based testing for nonlinear cascaded mixed-signal systems. *Proc. Design, Automation and Test in Europe Conference,* Pages: 1662–1667. European Design and Automation Association, Brussels, Belgium.

Pintelon, R., Schoukens, J., and Rolain, Y. (2008). Frequency domain approach to continuous-time system identification: Some practical aspects. In *Identification of Continuous-Time Models from Sampled Data,* ed. Gamier, H. and Wang, L., Pages: 215–248. Springer, New York.

Schoukens, J. and Pintelon, R. (1991). *Identification of Linear Systems,* Pages: 263–268. Pergamon, Oxford.

Shen, M. et al. (2010). A prediction approach for multichannel EEG signals modeling using local wavelet SVM. *IEEE Trans. Instrumentation and Measurement,* 59(5): 1485–1492. IEEE, New York.

Stark, J. et al. (1997). Takens embedding theorems for forced and stochastic systems. *Nonlinear Analysis: Theory, Methods and Applications* 30(9): 5303–5314. Elsevier, Amsterdam, the Netherlands.

System Identification Toolbox(TM). (2009). Version 7.3.1. The MathWorks(R), Natick, Massachussetts.

Takens, F. (1981). Detecting strange attractors in turbulence. In *Dynamical Systems and Turbulence, Lecture Notes in Mathematics,* ed. Rand, D. and Young, L., 898: 366–381. Springer, New York.

22

Dynamic Verification of SystemC Transactional Models

Laurence Pierre and Luca Ferro

CONTENTS

22.1 Introduction

State-of-the-art Systems on Chip integrate more and more elaborate components such as processors and software, memory, logic, digital signal processors, and sensors. With this increasing complexity, there is a crucial need for new design methodologies. While moderate-size hardware components can still be developed and simulated at the Register Transfer (RT) level using languages, such as VHDL [8] and Verilog [7], the larger scale of more complex hardware/software products requires the Electronic System Level (ESL), see Figure 22.1.

In that context, there is a growing interest in languages like *SystemC* [9, 34] and its transactional level (TLM, Transaction Level Modeling) [46]. SystemC (IEEE standard 1666) is a library of C++ classes for modeling electronic circuits; TLM [29] is an *algorithmic* level that provides communication models for complex data types ("transactions") and communication interfaces. Communications between the components of a SoC are simply implemented as *function (method) calls*. Compared to RT level (RTL), the simulation of TLM models is several orders of magnitude faster. Moreover, the TLM abstraction level enables easier architecture exploration and hardware/software codesign. The resulting early TLM specifications tend to become golden reference models [31], [30], so the need for methodologies for guaranteeing their correctness is compelling. Because of the high level of abstraction of SystemC TLM descriptions, related test and verification activities have similarities with those involved in software development.

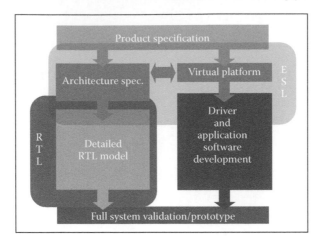

FIGURE 22.1
ESL-based development [13]. John Blyler. The Meaning of ESL Invites Confusion and Competition. Chip Design Magazine, http://chipdesignmag.com/display.php?articleId=67, February/March 2005.

Assertion-Based Verification (ABV) makes use of logic and temporal assertions written in languages such as Property Specification Language (PSL) (IEEE standard 1850) [11] or System Verilog Assertions (SVA) (IEEE standard 1800) [10]. As explained in [26], an assertion is a statement about the design's intended behavior, which must be verified. Verification can be performed using static or dynamic methods: static techniques usually employ model checking [17], whereas dynamic (runtime) approaches are based on simulation. ABV is getting widely adopted in industry since it allows exploiting advanced verification techniques thus providing benefit from a significant reduction in simulation debugging time [27].

ABV is now a well-established technology at the RT level. At this abstraction level, which is close to the hardware implementation, properties of interest target a high level of accuracy, with detailed precisions of the behavior of hardware elements with respect to the synchronization clock. As a simple example, the following PSL assertion (associated with a VHDL RTL design):

$$\text{default clock is (clk'event and clk='1');}$$
$$\text{assert always (req} \rightarrow \text{next[3]ack);}$$

states that, every time a request (*req* signal in the VHDL description) is issued, an acknowledgement (*ack* signal) must occur at the third next clock cycle (referring to the synchronization clock *clk*).

At the transactional (TLM) level, the user addresses more coarse-grained properties, typically related to abstract communication actions. Moreover, there is no synchronization clock, and there is even often no explicit notion of time. An illustrative example of an assertion that can be expressed at this level is: *the intended address is used when a memory transfer occurs.*

Even if various efforts have recently been reported, there still only exist few solutions for ABV at the transactional level. In this chapter, we describe our solution for runtime ABV, which is mainly based on the construction of checkers for temporal assertions, together with an observation mechanism for linking them to the design under verification (DUV). We first give a brief overview of SystemC TLM and we illustrate it with a simple Direct Memory Access (DMA) example. Then, we present the principles of ABV and we briefly describe the

PSL language, before introducing the description of our solution. After mentioning some related work, we conclude with our perspectives.

22.2   Brief Introduction to SystemC

A SystemC model is made of a set of interconnected modules (class *sc_module*). Modules can have input and output ports (class *sc_port*) and can run processes (class *sc_method* or *sc_thread*). Modules are interconnected using *channels* that can be simple instances of *sc_signal* or more elaborate, and possibly user-defined, structures. The use of *sc_export* allows declaring interfaces provided by a module and directly performing function calls on that interface. We use hereinafter the graphical notations of [12], see Figure 22.2.

   Example. We will use as illustrative example a simple DMA controller proposed with the TLM 2.0 draft 1 library (see Figure 22.3). A master programs the DMA through a *router* (this router is a TLM communication channel) to perform transfers between two memories:

- It writes successively into registers of the DMA the source address, the destination address, and the length of the data to be transferred.

- Then, it requests for the transfer by writing into the DMA control register.

Afterwards, it waits for the *dma_irq* signal from the DMA that indicates the completion of the transfer and clears the DMA interrupt by resetting the control register. Once the DMA has received all the necessary data (source address, destination address, etc.), it performs

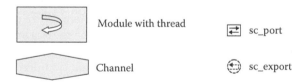

FIGURE 22.2
Graphical notations.

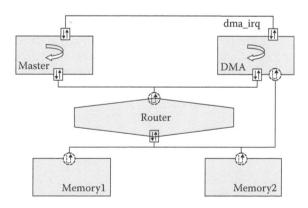

FIGURE 22.3
DMA example.

the transfer between the two memories, through the router, by alternating read and write operations.

The `sc_main` function below is the `main` function of the SystemC program that corresponds to this system. It consists of the declaration of the five modules and of the *sc_signal* `dma_irq`, followed by the statements for the interconnection of these components and by a call to the `sc_start` function that starts the simulation.

```
int sc_main(int argc , char **argv) {
  static const tlm::tlm_endianness SOC_ENDIAN = tlm::TLM_LITTLE_ENDIAN;
  // Component declarations:
  dma_testbench master("testbench",SOC_ENDIAN,0x4000);
  pv_memory mem1("memory1",0x100,SOC_ENDIAN,true);
  pv_memory mem2("memory2",0x100,SOC_ENDIAN);
  pv_dma dma("dma",SOC_ENDIAN);
  pv_router<int,int> router("router");
  sc_signal<bool> dma_irq;

  // Interconnections:
  master.initiator_port(router.target_port);      // master -> router
  master.pv_dma_irq_in(dma_irq);                   // master <-> dma_irq
  dma.initiator_port(router.target_port);          // dma -> router
  dma.pv_dma_irq_out(dma_irq);                     // dma <-> dma_irq
  router.initiator_port(dma.target_port);          // router -> dma
  router.initiator_port(mem1.target_port);         // router -> memory1
  router.initiator_port(mem2.target_port);         // router -> memory2

  sc_start();   // starts simulation
  return 0;
}
```

Simulation depicts the concurrent behavior of the modules of the SystemC description; processes and threads are executed, they can perform computations and communication actions. Here, for example, the master component can mimic the behavior of a processor that uses the DMA component each time memory transfers are required. Testbenches for simulation are therefore programs that the master module executes and that realize memory transfers. Their generation is beyond the scope of this chapter; test programs with a variety of nominal and corner cases can be produced using the solution described in [25].

The initiator ports of the master and of the DMA (of type *pv_initiator_port*) are used to initiate communications; the target ports are instances of *sc_export* that simply convey the communication requests.

The UML diagram of Figure 22.4 gives the class hierarchy for this example. For instance, we can remark that the `pv_memory`, `pv_dma`, `pv_router` and `dma_testbench` (master) components are subclasses of `sc_module`. An important point to note is that the abstract class `pv_if` (an excerpt of its declaration is given below) provides convenience methods that hide the underlying communication protocol through the router. For instance, functions `read` and `write` call the `transport` function of the router (through its target_port) that actually performs the communication. As many SystemC classes, class `pv_if` is a "template" C++ class (its parameters are types for addresses and data).

```
template<typename ADDRESS, typename DATA>
class pv_if {
  public:
    virtual ~pv_if() {}
    virtual tlm::tlm_status read(const ADDRESS& address,       // convenience method "read"
                                 DATA& data,
                                 const unsigned int byte_enable = tlm::NO_BE,
                                 const tlm::tlm_mode mode = tlm::REGULAR,
                                 const unsigned int export_id = 0
                                 ) = 0;
```

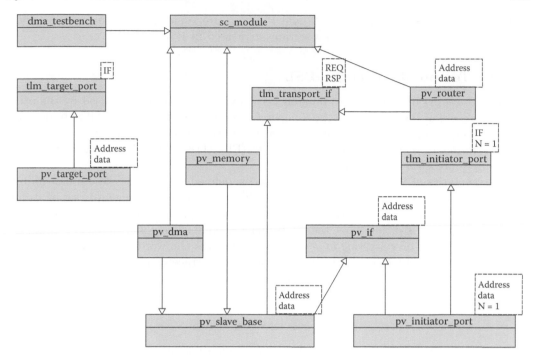

FIGURE 22.4
UML diagram for the DMA example.

```
virtual tlm::tlm_status write(const ADDRESS& address,      // convenience method "write"
                             const DATA& data,
                             const unsigned int byte_enable = tlm::NO_BE,
                             const tlm::tlm_mode mode = tlm::REGULAR,
                             const unsigned int export_id = 0
                            ) = 0;
    ...
};
```

Class `pv_initiator_port` is a concrete subclass of `pv_if`, it implements functions `read` and `write` using the `transport` function of the router. Thus, to initiate a communication, the master uses statements of the form `initiator_port.read(address, data, be, m, i)` and `initiator_port.write(address, data, be, m, i)`.* We will revisit the use of these communication functions in Section 22.4.

22.3   Assertion-Based Verification

ABV aims at guaranteeing that the design obeys properties that are expressed under the form of logic and temporal formulae that capture the design intent [26], [36]. Languages

---

*`Address` is the address of the transfer and `data` are the data to be transferred; the other parameters are not pertinent in the context of this description and are equipped with default values.

such as the IEEE standards SVA [10] and PSL [11] are typically used to formalize these formulae. In this work, we focus on assertions expressed in PSL.

22.3.1   Temporal language—PSL

The *Boolean layer* of PSL enables building basic expressions commonly used by the other layers. The core of the language is the *Temporal layer*, it provides the possibility to describe complex temporal relations, evaluated over a set of evaluation cycles. In the following, we consider formulae of the Foundation Language (FL) class of the Temporal layer, which essentially represents linear temporal logic [22].

The semantics of PSL properties given in [11] is defined with respect to finite or infinite *execution traces*, or words over the alphabet $\Sigma = 2^{\mathbf{P}} \cup \{\top, \bot\}$ ($\mathbf{P}$ is a nonempty set of atomic propositions; $\top$ and $\bot$ are such that, for every Boolean expression b, $\top \vDash b$ and $\bot \nvDash b$). The length of a trace v is denoted $|v|$, the ith evaluation point over v is denoted v^{i-1}, and $v^{i\cdots}$ indicates the suffix of v starting at v^i. Finally, $\bar{v}$ denotes the word obtained by replacing every $\top$ with a $\bot$ and vice versa. A short overview of this trace semantics is presented below (see [11] for more details); φ and ψ are FL formulae.

- Boolean expressions

$$v \models b \iff |v| = 0 \text{ or } v^0 \Vdash b,$$

  where $\Vdash$ is the satisfaction relation for Boolean expressions.

- Logical connectives

$$v \models \neg\varphi \iff \bar{v} \nvDash \varphi$$
$$v \models \varphi \wedge \psi \iff v \models \varphi \text{ and } v \models \psi,$$

- Next operator: the meaning of the next! operator is that the subproperty denoted by its operand should be verified from the next evaluation point of the execution trace (the weak version next is similar, except that the existence of a next evaluation point is not mandatory):

$$v \models next!\,\varphi \iff |v| > 1 \text{ and } v^{1\cdots} \models \varphi$$

- Strong Until, until! or U (classically, the weak until operator until does not impose the occurrence of ψ):

$$v \models [\varphi\,U\,\psi] \iff \exists k < |v| \text{ s.t. } v^{k\cdots} \models \psi \text{ and } \forall j < k, v^{j\cdots} \models \varphi$$

The other operators are built from these primitive ones. As usual, the PSL formula always φ means that φ must be verified on each evaluation point of the trace. The strong next_event! operator requires the satisfaction of φ the next time a Boolean expression b is verified:

$$next\_event!(b)(\varphi) \stackrel{def}{=} [\neg b\,U\,b \wedge \varphi]$$

The semantics of the before operator is also expressed with respect to the until operator:

$$\varphi\,before!\,\psi \stackrel{def}{=} [\neg\psi\,U\,\varphi \wedge \neg\psi]$$

Dynamic verification techniques consider the *"simple subset"* of PSL [11], which conforms to the notion of monotonic advancement of time. This subset restricts the syntax of the assertions that are allowed such that they can be checked during simulation (simulation can "look forward" but never "backward").

22.3.2   RT level tools

Temporal assertions can be checked using formal techniques such as model checking, or using semiformal approaches applied during simulation. Currently available commercial tools can be applied to synchronized VHDL or Verilog descriptions written at the RT level. Among the commercial model checkers, we mention the following:

- *Incisive Formal Verifier* of Cadence [1], which supports the PSL and SVA assertion languages.

- *RuleBase* of IBM [2], model checker for PSL properties.

- *Solidify* of Averant [3], which can also analyze PSL and SVA assertions.

The leading CAD companies have integrated property checking in their RTL simulators, for instance,

- *ModelSim* from Mentor Graphics [4] allows including PSL assertions in VHDL or Verilog descriptions.

- *VCS* from Synopsys [5] supports SVA assertions.

- The *Incisive* platform of Cadence integrates both PSL and SVA.

The tool *FoCs* from IBM [6] generates checkers from PSL assertions in the form of RTL processes in VHDL, Verilog, or SystemC [18]. In the SystemC version, a C++ class is generated for each checker, a method *transition* implements the checker's transition function.

Various academic tools can also produce checkers from temporal assertions for runtime verification (e.g., MBAC [15]). Our team has developed a compositional, formally proven, technique for automatically building synthesizable, hardware assertion checkers from PSL properties [40]. These checkers, produced in VHDL at the RT level, are connected to the DUV and monitor the status of the properties during simulation. The technique, implemented into the *MAAT* module of the *HORUS* environment [43], is based on a VHDL library of primitive PSL operators and on an interconnection method for those operators: checkers are *built compositionally from elementary components* that correspond to the primitive PSL operators. More precisely, a checker for a property $\wp$ is built as a hardware module that takes the following as inputs:

- The *Reset*

- The synchronization clock

- The signals of the DUV that are operands of the PSL operators in $\wp$

Both the library and the interconnection method have been formally proven correct using the PVS theorem prover [47]. This prover has been used to verify that the PSL semantics as given in [11] is respected.

Figure 22.5 illustrates this construction method on the following PSL assertion

$$always\,(A \rightarrow next(B\,before\,C)).$$

The property checker is built as the interconnection of four primitive components: "always," "implies," "next," and "before." The variables A, B, and C of this assertion are linked to signals of the DUV that can be inputs, outputs, or internal signals. Once connected to the DUV, the checker reacts synchronously (using the same clock as the DUV) and monitors the property relative to the values of these signals.

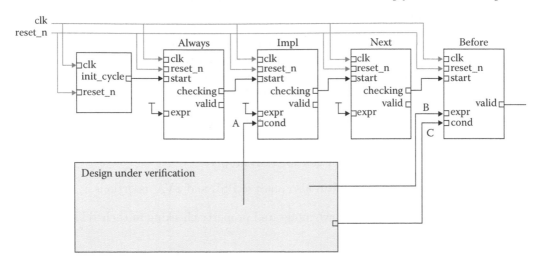

FIGURE 22.5
Example of RTL checker construction.

22.4   Runtime Verification at the Transactional Level

The solution presented in this section to monitor TLM-oriented platforms is inspired by the technology previously described.

22.4.1   Monitor generation

At the synchronous RT level, the observations that constitute the execution traces used by the PSL semantics are made on the clock edges. At the transactional level, we consider assertions that express properties regarding communications, that is, properties associated with transactional events (for instance, *a given transfer eventually completes*, *data are transferred at the correct location*, etc.). Thus, observation points are defined when the assertion needs to be re-evaluated, that is, each time a variable of the assertion has possibly been modified, which means an appropriate event occurs in the corresponding channel. Our traces comprise all the events (communication actions, not SystemC events) that may enable the observation of updated values for the variables involved in the assertion.

Consequently, we use a model that allows observing the transactional events in the system and triggering the checkers at appropriate instants [44]. This model is inspired by the well-known observer pattern: each checker is enclosed in a wrapper that observes the channels and/or signals involved in the property associated with this checker. The channels/signals are thus *observable objects* (or *Subjects*), see Figure 22.6.

To build *composite monitors* for complex properties, we use a variant of the original compositional interconnection method. Compositionality is functional here, instead of structural, but we follow the same general principles [44].

Each *primitive component* has been transformed into a SystemC class (subclass of *Monitor*). To that end, we have translated each VHDL component into a SystemC TLM one, by removing the synchronization clock and by transforming the concurrent set of VHDL instructions into sequential ones (using an interpretation based on the VHDL simulation semantics). This was straightforward, due to the linearity (absence of loops) of all our primitive monitors.

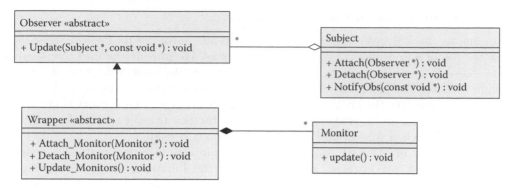

FIGURE 22.6
Model for observing transactions.

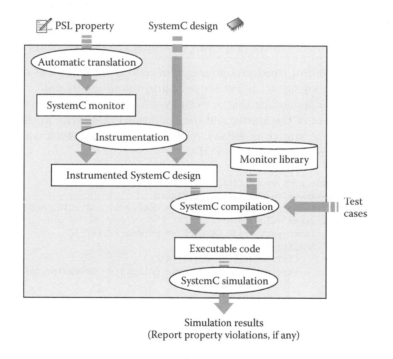

FIGURE 22.7
Overview of the ISIS flow.

Classes *Observer*, *Wrapper*, *Monitor*, and *Subject*, as well as the definitions of the primitive monitors constitute a library `libmon` that is used by our ISIS tool [23], see Figure 22.7. ISIS enables the capture of PSL assertions and performs the automatic construction of the corresponding monitors. Then, the SystemC code of the DUV is instrumented, through a graphical user interface, to relate the variables involved in the properties (formal variables) to the actual variables of the design. This instrumented code is compiled with the library `libmon`, and all that remains is to run the SystemC simulator to get information about the satisfaction or violation of the assertions.

22.4.2   Conditions on the communications

According to the PSL reference manual, the "SystemC flavor" of its *Boolean layer* should allow the use of any SystemC expressions of type *bool*. We provide the user the possibility to use these Boolean expressions to state *conditions on the communications* and we also provide the user the possibility to select the *relevant communications* for each property. Being able to identify the critical communication events in a given context is crucial. For instance, depending on his confidence in a channel, the practitioner may prefer to activate the verifications when this channel is written or when it is read.

From the set of selected communication operations, ISIS automatically generates ad hoc predicates, for instance, write_CALL() for the initiator_port of the example of Section 22.2 to check the occurrence of write actions. Therefore, the Boolean expression initiator_port.write_CALL() can then be used in assertions to express the occurrence of a write action.

More generally, name.fctname_CALL() denotes that the communication function fctname of the element name has just been called. We also use name.fctname_CALL.p# to denote the parameter in position # of function fctname (0 is used for the return value).

Example. Here are two illustrative properties regarding the DMA example of Section 22.2:

- P1: *any time the control register is programmed with a start, an end-of-transfer notification occurs (dma_irq signal) before the next programming attempt.*
  The communication operations that are observed in that case are the write operations in the initiator_port of the master and the operations that write in the dma_irq signal. The property can be stated as follows (0x4000+pv_dma::CTRL_ADDR corresponds to the address of the control register of the DMA):

  ```
  always ((initiator_port.write_CALL()
          && // writing in the control register
             (initiator_port.write_CALL.p1 == 0x4000+pv_dma::CTRL_ADDR)
          && // written value is a "start"
             (initiator_port.write_CALL.p2 == pv_dma::START))
      -> (next! (dma_irq before!
                  (initiator_port.write_CALL()
                  && (initiator_port.write_CALL.p1 == 0x4000+pv_dma::CTRL_ADDR)))))
  ```

- P2: *read and write operations, from and to memory locations, alternate.*
  Here the relevant communication operations are the read and write operations in the memories. The statement of the assertion is:

  ```
  always (next_event(mem1.read_CALL())
                  (next(mem2.write_CALL() before mem1.read_CALL())))
  ```

22.4.3   Modeling layer

The Modeling layer of PSL has not yet been actually considered in ABV approaches. It allows declaring auxiliary signals and variables as well as giving behavior to them. Its SystemC "flavor" consists of those declarations and statements that would be legal in the context of the SystemC module to which the PSL *vunit* is bound [21].

Example. The following assertion *P3* for the DMA example makes use of an auxiliary variable:
P3: *any time a source address is transferred, a read access in the first memory eventually occurs, and the correct address is used.*

The expression of this property requires the memorization of the address transferred to the DMA, to be able to check that this address is actually used when the memory is read. In the property below, `req_src_addr` is the variable that memorizes this value:

```
// Modeling layer:
int req_src_addr;
// Each time an address is transferred into the source register, it is stored
// into req_src_addr:
if (initiator_port.write_CALL() && initiator_port.write_CALL.p1 == 0x4000+pv_dma::SRC_ADDR)
    req_src_addr = initiator_port.write_CALL.p2;
// Assertion:
assert always ((initiator_port.write_CALL() &&
             initiator_port.write_CALL.p1 == 0x4000+pv_dma::SRC_ADDR)
        -> next_event!(mem1.read_CALL())(mem1.read_CALL.p1 == req_src_addr));
```

As illustrated by the previous example, the *Modeling layer* of PSL may be of great interest in the TLM context. However, the reference manual does not provide details about its semantics (it is commonly admitted that the statements of the Modeling layer should be evaluated at each step of the evaluation of the property). To give meaning to an assertion φ augmented with the side effects of a block of statements m in the Modeling layer, we have adopted an operational semantics [24] inspired from the one of [16], instead of the trace semantics given in Section 22.3.1. Like in [16], we consider that the negation operator only appears at the Boolean level, which is consistent with the *simple subset* of PSL.

In the following operational rules, φ, φ', ψ, and ψ' denote FL formulae, b is a Boolean expression, DV denotes the set of all the variables declared in the Modeling layer, m is the block of statements of the Modeling layer, and ρ is the current environment (association of variables of DV with their current values). $[\![\varphi]\!]_\rho^m$ is the interpretation of φ in which the variables of DV are substituted by their values in ρ. Like in [16], the derivation $\overset{\ell}{\to}$ is used: $[\![\varphi]\!]_\rho^m \overset{\ell}{\to} [\![\varphi']\!]_\rho^m$ means that, in order to check if a word starting with the letter ℓ satisfies $[\![\varphi]\!]_\rho^m$, one can check that $[\![\varphi']\!]_\rho^m$ is satisfied by the word without ℓ.

- Logical and (similarly for the logical or):

$$\frac{[\![\varphi]\!]_\rho^m \overset{\ell}{\to} [\![\varphi']\!]_{\rho'}^m \qquad [\![\psi]\!]_\rho^m \overset{\ell}{\to} [\![\psi']\!]_{\rho'}^m}{[\![\varphi \wedge \psi]\!]_\rho^m \overset{\ell}{\to} [\![\varphi' \wedge \psi']\!]_{\rho'}^m}$$

- Next operators:

$$[\![X!\varphi]\!]_\rho^m \overset{\ell}{\to} [\![\varphi]\!]_{\rho'}^m$$
$$[\![X\varphi]\!]_\rho^m \overset{\ell}{\to} [\![\varphi]\!]_{\rho'}^m$$

- Until operators (similar rules for weak and strong versions):

$$\frac{[\![\varphi]\!]_\rho^m \overset{\ell}{\to} [\![\varphi']\!]_{\rho'}^m \qquad [\![\psi]\!]_\rho^m \overset{\ell}{\to} [\![\psi']\!]_{\rho'}^m}{[\![\varphi U \psi]\!]_\rho^m \overset{\ell}{\to} [\![\psi' \vee (\varphi' \wedge (\varphi U \psi))]\!]_{\rho'}^m}$$

where $\rho' = [\![m]\!]_\rho^\ell$.

$[\![m]\!]_\rho^\ell$ denotes the environment obtained after executing the statement block m in the environment ρ and in the context of the atomic propositions of ℓ.

- Boolean expressions: At the transactional level, observation points are related to the communication actions, and letters of the execution traces are more complex than at the

RT level. In particular, these letters can involve the parameters of these communications. To account for this aspect, we slightly modify the notion of letter: we consider that letters are made of pairs $(id, value)$, where id can be the identifier of a variable of the design or an identifier like `name.fctname_CALL.p#` (see section 22.4.2), and $value$ is the corresponding current value. Moreover, such a value may be undefined at a given observation point if the corresponding communication action does not occur at that point. To take this aspect into account, we introduce the notion of *undefined value* (denoted ϑ below). The operational rule for Boolean expressions is thus:

$$\llbracket b \rrbracket_\rho^m \xrightarrow{\ell} \begin{cases} T & \text{if } \ell \Vdash_{\rho'} b \\ F & \text{otherwise} \end{cases}$$

where $\rho' = \llbracket m \rrbracket_\rho^\ell$ and

$$\ell \Vdash_\rho b \Longleftrightarrow \begin{cases} false & \text{if } \exists \text{ a pair } (id, \vartheta) \in \ell|_{I_b} \\ \ell \Vdash b' & \text{otherwise} \end{cases}$$

I_b is the set of identifiers present in the Boolean expression b, and $b' = b_{[DV \leftarrow \rho(DV)][I_b \leftarrow \ell(I_b)]}$ denotes b in which all the identifiers of DV have been substituted by their value in ρ and all the identifiers of I_b have been substituted by their value in ℓ.

With regard to atomic propositions q, the original PSL semantics [11] states that $\ell \Vdash q$ iff $q \in \ell$. It now becomes $\ell \Vdash q$ iff $q_{[DV \leftarrow \rho(DV)][I_q \leftarrow \ell(I_q)]} = true$.

As in [16], iteratively applying these rules on a formula ϕ over the letters of a word w, in an environment ρ, is written $\phi \langle w \rangle_\rho$, where

$$\phi \langle \epsilon \rangle_\rho = \phi \text{ (iteration stops when the word is empty)}$$

$$\phi \langle \ell w \rangle_\rho = \phi' \langle w \rangle_{\rho'}, \text{ where } \llbracket \phi \rrbracket_\rho^m \xrightarrow{\ell} \llbracket \phi' \rrbracket_{\rho'}^m$$

Then, the satisfaction relation, with the initial environment ρ_0, is defined as follows: $w \vdash \phi \Leftrightarrow ok(\phi \langle w \rangle_{\rho_0})$, where ok is a function that calculates whether a given formula has not been contradicted yet w.r.t. the sequence of letters that has already been visited. Since the work of [16] only considers weak operators, this notion of satisfaction is sufficient. We use the same definition for function ok and we extend it for the strong versions of the *next* and *until* operators (the definition is similar to the one of their weak versions):

$$ok(F) = \textbf{false}$$
$$ok(T) = \textbf{true}$$
$$ok(b) = \textbf{true}$$
$$ok(\varphi \wedge \psi) = ok(\varphi) \textbf{ and } ok(\psi)$$
$$ok(\varphi \vee \psi) = ok(\varphi) \textbf{ or } ok(\psi)$$
$$ok(X \varphi) = \textbf{true}$$
$$ok(X! \varphi) = \textbf{true}$$
$$ok(\varphi W \psi) = ok(\varphi) \textbf{ or } ok(\psi)$$
$$ok(\varphi U \psi) = ok(\varphi) \textbf{ or } ok(\psi)$$

Moreover, PSL defines four levels of satisfaction for the dynamic verification context (where traces are finite) including the possibility to support strong operators:

- *Holds* and *Holds strongly* are two satisfaction relations (for simple *Holds*, the property may not hold on any given extension of the trace). Both of them required that "No bad states have been seen" and "All future obligations have been met."

- The *Pending* level is defined as "No bad states have been seen" and "Future obligations have not been met."

- *Fails* means "A bad state has been seen."

The *ok* function of [16] corresponds to the goal "No bad states have been seen." We define a new function *met* that expresses the second goal "All future obligations have been met":

$$met(F) = \textbf{true}$$
$$met(T) = \textbf{true}$$
$$met(b) = \textbf{false}$$
$$met(\varphi \wedge \psi) = met(\varphi) \textbf{ and } met(\psi)$$
$$met(\varphi \vee \psi) = (ok(\varphi) \textbf{ and } met(\varphi)) \textbf{ or } (ok(\psi) \textbf{ and } met(\psi)) \textbf{ or } (met(\varphi) \textbf{ and } met(\psi))$$
$$met(X\,\varphi) = \textbf{true}$$
$$met(X!\,\varphi) = \textbf{false}$$
$$met(\varphi\,W\,\psi) = \textbf{true}$$
$$met(\varphi\,U\,\psi) = met(\varphi) \textbf{ and } ok(\psi) \textbf{ and } met(\psi)$$

Combining the two functions enables to define the *Holds*, *Pending*, and *Fails* levels:

$$Holds_{w,\rho_0}(\phi) = ok(\phi\langle w\rangle_{\rho_0}) \textbf{ and } met(\phi\langle w\rangle_{\rho_0})$$
$$Pending_{w,\rho_0}(\phi) = ok(\phi\langle w\rangle_{\rho_0}) \textbf{ and } \neg met(\phi\langle w\rangle_{\rho_0})$$
$$Fails_{w,\rho_0}(\phi) = \neg ok(\phi\langle w\rangle_{\rho_0})$$

From a practical point of view, the declarations and statements of the Modeling layer are interpreted as associating the variables and their updating statements with the assertion checkers. Since each step of the evaluation of the assertion corresponds to an activation of its checker (monitor), we include the Modeling layer statements in the monitor body (at the beginning of its activation function). Moreover, our monitors for strong operators are equipped with a "pending" output [44]; together with the "valid" output, it allows taking into consideration the levels of satisfaction described above.

22.4.4 Practical results

Considering the DMA example, and its properties proposed in Sections 22.4.2 and 22.4.3, we realized simulations in which 200,000 memory transfers are performed, data of length 64 bytes are transferred (each atomic transfer carries 4 bytes, that is, 16 transfers are needed). Therefore, the monitor for P3, for instance, is evaluated 4.2 million times (21 times for each transfer: 5 writings to the DMA + 16 read operations). The CPU times for simulation* are

- 4.97 s without monitoring.

- 5.18 s for the same simulation with the verification of property P1.

- 6.21 s for the same simulation with the verification of property P2.

- 5.54 s for the same simulation with the verification of property P3.

It is worth mentioning that only 7.35 s are necessary for the simulation with the verification of all the properties together because the observation time is "factorized" when the same communication operations are concerned (these operations are observed only once for all the properties).

---

*On an Intel Core2 Duo (3GHz) under Debian Linux.

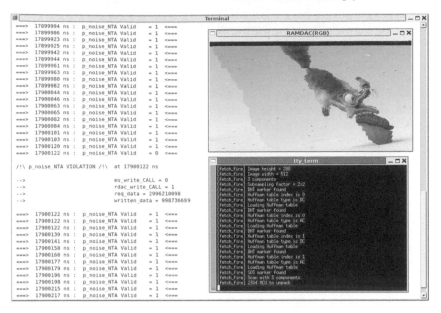

FIGURE 22.8
Instrumented simulation of the MJPEG platform.

Various other examples have been reported in [23], among them a Motion-JPEG decoding platform described at the Transaction Accurate level [28]. It embeds a processor (EU), a global memory, a hardware semaphore RAM, and hardware terminals (TTY). A traffic generator takes its data from a MJPEG file; once decoded, the images are sent to a RAMDAC (viewer). Here is one of the properties that can be verified to check the reliability of the communication channel: *the data that are written on the RAMDAC are exactly the ones that have been transmitted by the EU* (P4). The expression of this property requires the memorization of the data that are transmitted by the EU into `req_data`, to be compared with the data written on the RAMDAC:

```
// Modeling layer:
unsigned int req_data;
if (eu.write_CALL()) req_data = eu.write_CALL.p2;
// Assertion:
assert ALWAYS (eu.write_CALL() -> NEXT_EVENT(rdac.write_CALL())
                                     (req_data == rdac.write_CALL.p2));
```

Figure 22.8 gives a screenshot of a simulation of property P4 with design errors introduced in the communication channel. While the movie on the RAMDAC is not as expected, our monitor reports errors when detected (on the left). Simulations that correspond to 10 s of SystemC time require about 18.67 s of CPU time without monitoring, and 21.64 s with the verification of property P4.

22.5 Related Work

Few results have been proposed regarding the static or dynamic verification of SystemC TLM specifications.

22.5.1 Static analysis

An early formal verification approach for proving linear temporal logic formulae on SystemC designs is proposed in [32], but it is limited to RTL descriptions.

Recent static verification approaches can handle more abstract SystemC descriptions. The model proposed in [37] (Petri-net-based representation) can be used from the initial functional specification to cycle-accurate RTL, including TLM. The SystemC descriptions are translated into this model, then the UPPAAL tool can be used to model check them against a collection of Computation Tree Logic properties. Events, signals, and method calls, as well as the SystemC scheduler itself, are represented as Petri nets: each statement roughly corresponds to one place and one transition. Unfortunately, this solution yields huge Petri nets (e.g., each signal gives rise to a subsystem of about 20 places and 12 transitions), even for rather moderate descriptions.

The method of Habibi and Tahar [35] first captures the design structure and behavior by means of UML diagrams. Then, sequence diagrams are used to represent the expected behavior and to obtain the corresponding PSL assertions (limited to Boolean and SERE expressions). The diagrams are then mapped to Abstract State Machines (ASM) forms. In general, building the entire FSM is not feasible, only portions of the FSM are produced. Optionally, dynamic ABV can be performed on the model by means of property checkers, which are generated in C# from the ASM description. The paper provides only few details on this step. All the reported experiments are restricted to synchronized (clocked) descriptions where communications are operated through signals.

The authors of [41] propose a method where a SystemC TLM design is analyzed and a corresponding structure of communicating parallel machines is built. This intermediate representation, called HPIOM (for "Heterogeneous Parallel Input/Output Machines"), must undergo appropriate abstraction techniques in order to provide tractable models. Nevertheless, there are limitations in the size of the devices that can be handled. Static verification by model checking is possible, but the properties are not expressed using a specification language such as PSL: only a small set of assertions is mentioned, such that "*global deadlock never occurs*" or "*a process never terminates.*"

22.5.2 Runtime verification

A first proposal for constructing checkers for SystemC designs was described in [33], but it was restricted to clocked designs and to assertions that are roughly of the form *a condition holds at a given time point* or *during a given time interval.*

A simulation-based methodology to apply ABV to TL models is proposed in [42]. Using Aspect-Oriented Programming techniques, the simulation is run a first time on the original design and transaction traces are recorded in a value change dump (VCD) file. An important drawback of this approach is that the transactions are simply mapped to Boolean signals (i.e., associating *true* during the transaction and *false* otherwise). It prevents from considering the parameters of the communications, thus considerably restricting the categories of TLM descriptions that can be taken into account. The VCD file is then translated into a Verilog description that is used for a second simulation inside an RTL simulator. SVA assertions can be checked at this level. This leads to a time-consuming solution since the simulation always requires at least two runs.

In [19], the authors explicitly express events through a specialized extension of SVA. A framework to perform runtime verification of transactional SystemC designs is proposed, and a possible implementation using proxies is mentioned in [20]. This approach is appealing, but both the additional semantics constructs and the framework may be unnecessarily complicated. A few experimental results are reported in [20]; even if they cannot be directly compared to those of [23], the simulation time overhead appears to be significant.

In his PhD thesis, Lahbib [39, 38] uses the PSL property checkers generated in C++ by the FoCs tool to monitor PSL assertions in transactional designs. The key feature of this work is the use of specific channel classes, containing instances of the property checkers. During simulation, the original channels can be replaced by the monitoring ones: instead of triggering the checkers on clock ticks, the TLM method *transport* calls on the checker's function *transition* when a transaction is initiated. This approach wraps the checkers in a structure that controls the checker triggering, thus providing a solution similar to our monitoring mechanism. However, the checker is enclosed inside the channel and this induces considerable limitations: the assertions cannot involve several channels (since different ports cannot be monitored simultaneously) and hybrid properties (i.e., including both signals and TLM channels) are not supported.

22.6   Conclusion

We have presented our solution for verifying logic and temporal properties of communications in TLM designs during simulation. The tool is usable with timed and untimed TLM models, and properties can mix various types of channels. The technique is also compatible with globally asynchronous locally synchronous (GALS) and multiclocked systems. Finally, it supports the introduction of auxiliary variables. Storing communication data into auxiliary variables can be crucial for high-level assertions, but it is still insufficient in case of pipelined communications or bursts of successive requests. We have recently proposed a method to deal with re-entrancy in that case [45].

The kind of properties we consider here can be verified in the early design steps. A comprehensive verification flow should facilitate relating such a property to its counterpart at the RT level in order to check the correctness of RTL implementations. This is the purpose of [14] for instance, but the assertions under consideration are restricted to Hoare implications between primary inputs and primary outputs. The refinement of actual temporal properties from TLM to RTL is still an open problem. Our future work includes the exploitation of our various levels of monitors to relate TLM-oriented properties to RTL-oriented ones.

References

1. http://www.cadence.com/products/ld/formal_verifier/pages/default.aspx.

2. http://www.haifa.il.ibm.com/projects/verification/RB_Homepage.

3. http://www.averant.com/products-solidify.html.

4. http://www.model.com/.

5. http://www.synopsys.com/tools/verification/functionalverification/pages/vcs.aspx.

6. http://www.haifa.il.ibm.com/projects/verification/focs.

7. *IEEE Std 1364-1995, IEEE Standard Verilog Language Reference Manual.* (1995). IEEE.

8. *IEEE Std 1076-2000, IEEE Standard VHDL Language Reference Manual.* (2000). IEEE.

9. *IEEE Std 1666-2005, IEEE Standard SystemC Language Reference Manual.* (2005) IEEE.

10. *IEEE Std 1800-2005, IEEE Standard for System Verilog: Unified Hardware Design, Specification and Verification Language.* (2005). IEEE.

11. *IEEE Std 1850-2005, IEEE Standard for Property Specification Language (PSL).* (2005). IEEE.

12. Black, D. and Donovan, J. (2004). *SystemC: From the Ground Up.* Springer.

13. Blyler J. (Feb/Mar 2005). The Meaning of ESL Invites Confusion and Competition. Chip Design Magazine, http://chipdesignmag.com/display.php?articleId=67.

14. Bombieri, N., Fummi, F., Pravadelli, G., and Fedeli, A. (Mar–Apr 2007). Hybrid, incremental assertion-based verification for TLM design flows. *IEEE Design & Test of Computers*, 24(2).

15. Boulé, M. and Zilic, Z. (2008). *Generating Hardware Assertion Checkers: For Hardware Verification, Emulation, Post-Fabrication Debugging and On-Line Monitoring.* Springer.

16. Claessen, K. and Martensson, J. (2004). An operational semantics for weak PSL. In *Proc. FMCAD'04*, Springer, New York, NY.

17. Clarke, E., Grumberg, O., and Peled, D. (2000). *Model Checking.* The MIT Press.

18. Dahan, A., Geist, D., Gluhovsky, L., Pidan, D., Shapir, G., Wolfsthal, Y., Benalycherif, L., Kamdem, R., and Lahbib, Y. (2005). Combining system level modeling with assertion based verification. In *Proc. ISQED'2005*.

19. Ecker, W., Esen, V., and Hull, M. (2006). Specification language for transaction level assertions. In *Proc. HLDVT'06*.

20. Ecker, W., Esen, V., and Hull, M. (2007). Implementation of a transaction level assertion framework in SystemC. In *Proc. DATE'07*.

21. Eisner, C. and Fisman, D. (2006). *A Practical Introduction to PSL.* Springer.

22. Emerson. E. (1991). *Handbook of Theoretical Computer Science (vol. B): Formal Models and Semantics*, chapter "Temporal and modal logic." (1991). The MIT Press.

23. Ferro, L. and Pierre, L. (Sep 2009). ISIS: Runtime verification of TLM platforms. In *Proc. Forum on specification & Design Languages (FDL'09)*, Springer, New York, NY.

24. Ferro, L. and Pierre, L. (Mar 2010). Formal semantics for PSL modeling layer and application to the verification of transactional models. In *Proc. DATE'2010*.

25. Ferro, L., Pierre, L., Ledru, Y., and Du Bousquet, L. (Dec 2008). Generation of test programs for the assertion-based verification of TLM models. In *Proc. IEEE International Design and Test Workshop (IDT'08)*.

26. Foster, H., Krolnik, A., and Lacey, D. (2003). *Assertion-Based Design.* Kluwer Academic Pub.

27. Foster, H. (Jan 2009). Applied assertion-based verification: An industry perspective. *Foundations and Trends in Electronic Design Automation*, 3(1).

28. Gerin, P., Guérin, X., and Pétrot, F. (2008). Efficient implementation of native software simulation for MPSoC. In *Proc. DATE'08*.

29. Ghenassia, F., editor. (2005). *Transaction-Level Modeling with SystemC*. Springer.

30. Glasser, M. (2006). Applying Transaction-Level Models for Design and Testbenches. SOC Central.

31. Goering, R. (2006). Transaction models offer new deal for EDA. EETimes, http://www. eetimes.com/showArticle.jhtml?articleID=181503693.

32. Große, D. and Drechsler, R. (2003). Formal verification of LTL formulas for SystemC designs. In *Proc. ISCAS'2003*.

33. Große, D. and Drechsler, R. (2004). Checkers for SystemC designs. In *Proc. ACM/IEEE International Conference on Formal Methods and Models for Codesign (MEMOCODE'04)*.

34. Grötker, T., Liao, S., Martin, G., and Swan, S. (2002). *System design with SystemC*. Kluwer Academic Pub.

35. Habibi, A. and Tahar, S. (2005). Design for verification of SystemC transaction level models. In *Proc. DATE'2005*.

36. Horgan, J. (2004). Assertion based verification. EDACafe Weekly, http://www10. edacafe.com/nbc/articles/view_weekly.php?articleid=209195.

37. Karlsson, D., Eles, P., and Peng, Z. (2006). Formal verification of SystemC designs using a Petri-net based representation. In *Proc. DATE'2006*.

38. Lahbib, Y., Perrin, A., Maillet-Contoz, L., Clouard, A., Ghenassia, F., and Tourki, R. (2006). Enriching the Boolean and the modeling layers of PSL with SystemC and TLM flavors. In *Proc. FDL'2006*.

39. Lahbib, Y. (2006). *Extension of Assertion-Based Verification Approaches for the Verification of SystemC SoC Models*. PhD thesis, University of Monastir (Tunisia).

40. Morin-Allory, K. and Borrione, D. (2006). Proven correct monitors from PSL specifications. In *Proc. DATE'2006*.

41. Moy, M., Maraninchi, F., and Maillet-Contoz, L. (2006). LusSy: An open tool for the analysis of systems-on-a-chip at the transaction level. *Design Automation for Embedded Systems*.

42. Niemann, B. and Haubelt, C. (2006). Assertion-based verification of transaction level models. In *Proc. ITG/GI/GMM Workshop*.

43. Oddos, Y., Morin-Allory, K., and Borrione, D. (2008). Assertion-based design with Horus. In *Proc. MEMOCODE'2008 (short paper)*.

44. Pierre, L. and Ferro, L. (2008). A tractable and fast method for monitoring SystemC TLM specifications. *IEEE Transactions on Computers*, 57(10).

45. Pierre, L. and Ferro, L. (2010). Enhancing the assertion-based verification of TLM designs with reentrancy. In *Proc. ACM/IEEE International Conference on Formal Methods and Models for Codesign (MEMOCODE'2010)*.

46. Rizzatti, L. (2007). Defining the TLM-to-RTL design flow. EETimes, http://www.eetimes.com/showArticle.jhtml?articleID=196901227.

47. Shankar, N., Owre, S., Rushby, J., and Stringer-Calvert, D. (2001). *PVS Prover Guide*. Computer Science Laboratory, SRI International, Menlo Park.

Index

Note: *Italicized* page references denote figures and tables.